Human Reproductive Biology

THIRD EDITION

Human Reproductive Biology

THIRD EDITION

Richard E. Jones
Professor of Biology Emeritus
University of Colorado
Boulder, Colorado

Kristin H. Lopez
Department of Integrative Physiology
University of Colorado
Boulder, Colorado

AMSTERDAM • BOSTON • HEIDELBERG • LONDON
NEW YORK • OXFORD • PARIS • SAN DIEGO
SAN FRANCISCO • SINGAPORE • SYDNEY • TOKYO
Academic Press is an imprint of Elsevier

Senior Acquisitions Editor: Nancy Maragioglio
Project Manager: Philip Korn
Associate Editor: Kelly Sonnack
Marketing Manager: Trevor Daul
Cover Design: Cate Barr
Composition: Integra Software Pvt. Ltd.

Academic Press is an imprint of Elsevier
30 Corporate Drive, Suite 400, Burlington, MA 01803, USA
525 B Street, Suite 1900, San Diego, California 92101-4495, USA
84 Theobald's Road, London WC1X 8RR, UK

This book is printed on acid-free paper. 

Library of Congress Cataloging-in-Publication Data
Jones, Richard E. (Richard Evan), 1940-
Human reproductive biology / by Richard E. Jones. — 3rd ed.
p. ; cm.
Includes bibliographical references and index.
ISBN-13: 978-0-12-088465-0 ISBN-10: 0-12-088465-8 (hardcover : alk. paper)

1. Human reproduction. 2. Sex. I. Title.
[DNLM: 1. Reproduction—physiology. 2. Sexual Behavior.
WQ 205 J78ha 2006]
QP251.J636 2006
612.6—dc22

2005024432

British Library Cataloguing in Publication Data
A catalogue record for this book is available from the British Library

ISBN-13: 978-0-12-088465-0
ISBN-10: 0-12-088465-8

For all information on all Elsevier Academic Press publications
visit our Web site at www.books.elsevier.com

Printed in China

09 10 9 8 7 6 5

To my wife, Betty, and my four sons
(Evan, Ryan, Peter, and Christopher)
R.E.J.

To Tom and Jessica
K.H.L.

Contents

Preface

A New Contraceptive in the Works! Hormone Replacement Therapy for Menopause Increases the Risk of Heart Disease! Anabolic Steroids Linked to Psychiatric Disorders! Headlines such as these appear in the media almost every day as advances in the field of human reproductive biology are made. Reproductive biology and biomedicine are of primary concern to people of all ages. One goal of this third edition of *Human Reproductive Biology* is to give college students a solid foundation in understanding the human reproductive system so that they can critically evaluate and interpret new findings as they prepare for careers in reproductive biology and medicine, or for their own personal interest.

Scientific research in this area is proceeding with great rapidity. The time between publication of the second edition of *Human Reproductive Biology* and the present (2006) has been loaded with new research findings that have profound influence on our basic understanding in this scientific discipline. In turn, advances in basic biology has a major impact on the practice of reproductive medicine. So the third edition is "pregnant" with new information and has been updated throughout. Our goal is to give you the latest available findings.

In addition to adding new and exciting facts, this edition has been changed in ways that add to the teaching value of the book. Eighteen new Highlight Boxes," presenting especially intriguing topics or areas of special interest, have been added. For the first time, the figures are in full color to help the student interpret and understand the structure and function of the reproductive system. A new chapter on Reproductive Aging has been added. The references at the end of each chapter have been expanded to help the student research a topic for writing term papers or to follow up on a topic of interest in the literature. A dedicated student in a class using this text should know and understand more about human reproduction than about 95% of adults! This book will help prepare students who are considering careers as health care professionals or biomedical researchers. The students will also derive benefit in their personal lives, regardless of career choice. It is our experience that a student would need at least one year of college general biology to gain the greatest benefit from taking a course using *Human Reproductive Biology*.

We have chosen illustrations and tables that offer clear examples of phenomena discussed in the text. Furthermore, we explain each figure fully in the legend, thus saving the reader from searching the text for clarification. A list of illustration and table credits appears at the back of the book. Within the text, key terms are italicized the first time they are used and are also defined in the Glossary.

We are, by profession, reproductive biologists and endocrinologists, and we have used our training and knowledge to the best of our abilities to make this book as scientifically accurate and up-to-date as possible. Although we are not medical doctors, we have attempted to present valid medical information. However, we do not take legal responsibility for any medical information or advice in this book; that is, the readers of this book use medical information and advice contained herein at their own risk, and should always check with their physicians regarding any health or medical problems or treatments.

We wish to note that it is our intention to provide new Highlight Boxes for readers, particularly course instructors and students. Any updates, as warranted by major developments in the field, will be made available on the publisher's Internet site, under the book's dedicated Web pages (http://books.elsevier.com).

We would like to extend our appreciation to the staff at Academic Press/Elsevier for their help and enthusiastic support in the production of this third edition, especially Nancy Maragioglio, Kelly Sonnack, and Philip Korn. A special warm thanks to our colleague and friend Dr. Leif Saul, who used his artistic skill and scientific knowledge to provide us with the many new color figures in this edition.

R.E.J.
K.H.L.

PART ONE

Adult Female and Male Reproductive Systems

CHAPTER ONE

Endocrinology, Brain and Pituitary Gland

Introduction

To understand the biology of reproduction, we must first meet the cast of characters involved in this fascinating process. In Chapters 2 and 4, we will study the anatomical components of the female and male reproductive systems. However, you will discover that, to function properly, these systems require chemical instructions. In fact, nearly every aspect of reproductive biology is regulated by internal molecular messengers called *hormones*. Reproductive hormones signal the reproductive structures to grow and mature. For example, as a boy approaches puberty, circulating levels of hormones called *androgens* rise and cause his reproductive tract to mature. In addition, these hormones induce muscle growth, cause changes in the vocal cords that lower the young man's voice, and initiate adult patterns of body hair growth. Hormones also regulate the timing of reproductive events. In women, the coordinated release of several female hormones orchestrates *ovulation*, the release of an egg from the ovary, approximately every 28 days. This chapter introduces you to the endocrine system, focusing on how the brain and pituitary gland regulate reproductive hormones. Your efforts in studying this material will be repaid as you read further in the text, as an understanding of this topic is essential for you to grasp the information in subsequent chapters.

Endocrine System

There are two kinds of glands in your body. *Exocrine glands* secrete substances into ducts (tiny tubes) that empty into body cavities and onto surfaces. Examples are the sweat and oil glands of the skin, the salivary glands, and the mucous and digestive glands of your stomach and intestines. In contrast, *endocrine glands* do not secrete substances into ducts, which is why they sometimes are called "ductless glands." Instead, endocrine cells secrete products, called hormones, which are released into the adjacent tissue spaces. The hormones then enter the bloodstream and are carried to other regions of the body to exert their effects. Hormones are chemical messengers in that certain tissues in the body are signaled by specific hormones to grow or change their cellular activity.

The science of endocrinology also includes the study of *paracrines*. Paracrines are chemical messengers that are produced by endocrine cells and diffuse to act locally on adjacent target cells with the appropriate receptors. Thus, unlike hormones, paracrines are not carried in the bloodstream (Fig. 1-1).

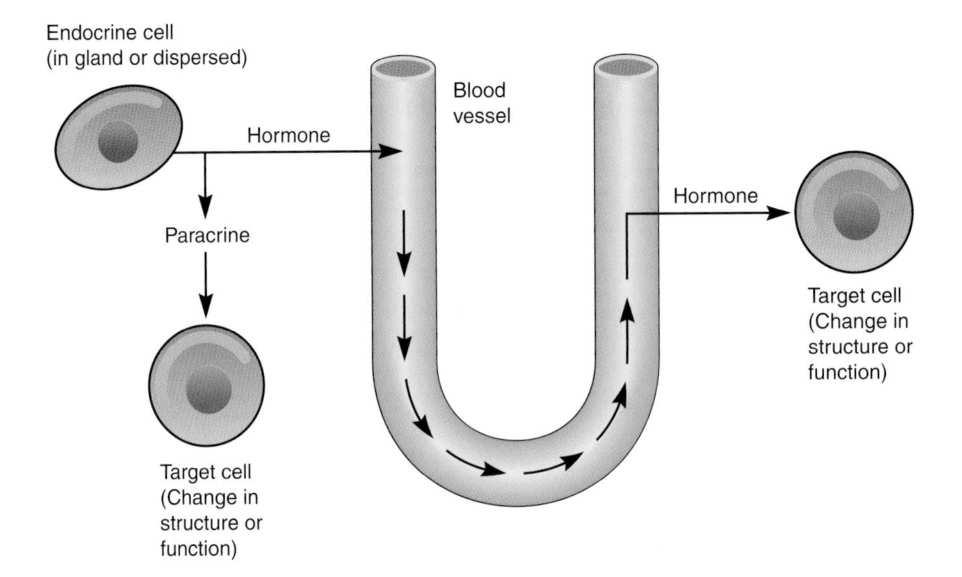

Figure 1-1 Endocrine and paracrine regulation. Target cells for hormones and paracrines must have specific receptors on their cell membrane or in their cytoplasm or nucleus to respond to a particular ligand.

The endocrine system consists of all the endocrine glands and isolated endocrine cells in the body. Included in this system are the pituitary gland (or *hypophysis*), pineal gland, gonads (testes and ovaries), and placenta—all organs of primary importance in human reproduction. In addition, the endocrine system includes the thyroid, parathyroid, and adrenal glands, as well as the hormone-secreting cells of the digestive tract, kidneys, pancreas, and thymus. Figure 1-2 depicts these components of the endocrine system.

Science of Endocrinology

Endocrinology is the study of the endocrine glands and their secretions. Suppose you are an endocrinologist who has an idea that a particular gland has an endocrine function. What would you do to test this hypothesis? A "classical" approach used by early endocrinologists is as follows: (1) remove the gland, (2) observe the effects of gland removal on the body, (3) *replacement therapy*, which involves administration of a preparation of the removed gland, and (4) observe to see if the replacement therapy reverses the effects of gland removal. If the replacement therapy does reverse the effects, what could you conclude?

In the past, the technique of *bioassay* was commonly used to measure indirectly the amount of a given hormone in glandular tissue or blood. With this method, several different amounts of a purified hormone are administered to animals, and the physiological or anatomical changes in target tissues are measured. In this way, a given degree of biological response can be associated with a given amount of hormone administered. Ideally, the response should increase in proportion to the increasing amounts of hormone administered; that is, a *dose–response relationship*, or "standard curve," is obtained. Then, a gland

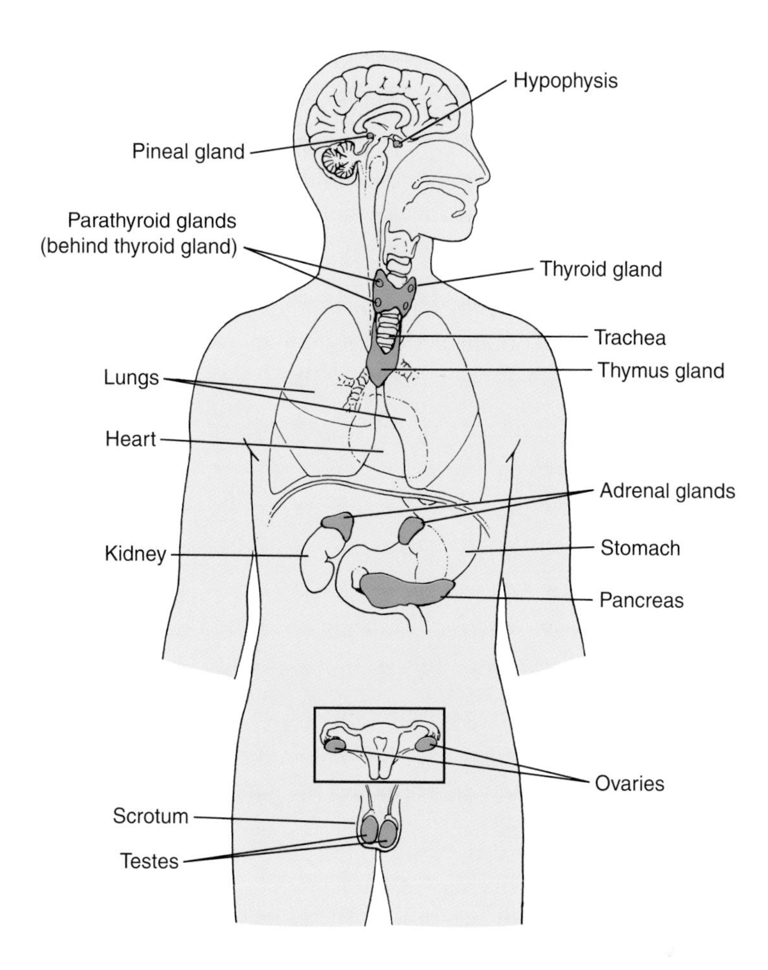

Figure 1-2 Major components of the endocrine system (shown in red). The placenta is not shown.

preparation or blood containing an unknown amount of this hormone is administered to other animals, and the biological response obtained is compared to the dose–response relationship. In this way, the amount of hormone in the gland or blood is determined.

A more direct and accurate way to measure the amount of a hormone in tissues or in blood is *radioimmunoassay*. This assay uses an antibody against the hormone of interest and a radioactive tracer for detection and measurement of the hormone. Radioimmunoassay is used to measure blood levels of hormones in humans and other animals, and it has been a valuable tool in reproductive biology and in tests for disorders of the endocrine system. In 1977, Rosalyn S. Yalow received the Nobel Prize in Physiology and Medicine for her development of this technique.

Radioimmunoassay is now becoming less popular as sensitive nonradioactive methods have been developed. For example, *ELISA* assays are used for measuring a wide variety of hormones and hormone products. *High-performance liquid chromatography* techniques coupled with *spectrophotometric analyses* are used to identify hormones and other regulators in biological fluids. A method used commonly to measure mixed steroids in plasma or urine samples is *gas chromatography* with *mass spectrometry*. Endocrinologists studying the structure and function of endocrine cells use techniques such as *immunocytochemistry* and *immunofluorescent staining* of cells, imaged with the confocal microscope.

Genetic engineering has revolutionized endocrine studies. One major area is the development of probes to measure mRNA production to determine when a gene is activated or shut down by a hormone. Genetically engineered knockout mice and knock-in mice are widely used to investigate problems of sexual differentiation and in behavioral studies of mice and rats. Yeast cells genetically engineered to contain genes for human estrogen receptors, with the help of reporter genes, can be used in assays for measuring estrogenic activity. These new tools have allowed endocrinologists to make advances in our understanding of normal human reproductive physiology, as well as of reproductive disorders such as breast cancer.

Hormones

Hormones have diverse molecular structures. Some hormones are *proteins* or smaller *polypeptides* or *peptides*. These kinds of molecules are made up of chains of amino acids containing oxygen, carbon, hydrogen, and nitrogen. Other hormones are *amines*, which are derivatives of amino acids; these are formed from single amino acids that have been altered chemically. Some hormones are derived from fatty acids. *Steroid* hormones are molecules derived from cholesterol. Male sex hormones (androgens) and female sex hormones (estrogens and progestogens) are examples of steroid hormones. *Androgens* are substances that promote the development and function of the male reproductive structures. *Estrogens* stimulate the maturation and function of the female reproductive structures. *Progestogens* (or *progestins*) are substances that cause the uterus to be secretory.

Receptors

Even though all body tissues may be exposed to hormones, only certain target tissues are responsive to a given hormone. Cells of these target tissues have specific hormone receptors on their surface membrane, or in their cytoplasm or nucleus, that bind to a given hormone. A molecule (such as a hormone) that binds to a receptor is sometimes referred to as its *ligand*. Protein and peptide hormones, including *gonadotropin-releasing hormone* (GnRH), *follicle-stimulating hormone* (FSH), *luteinizing hormone* (LH) and *prolactin* (PRL), bind to receptors embedded in the cell membrane of responsive cells. These receptors are large protein molecules that typically have three major regions, or *domains* (Fig. 1-3). The hormonal signal is received when the ligand attaches to a *ligand-binding site* in the *extracellular domain*, a portion of the receptor that sticks out beyond the cell. A *transmembrane domain* anchors the receptor within the plasma membrane. Finally, the *intracellular domain* is an extension of the receptor protein within the cell cytoplasm. Binding of the ligand to its receptor causes a *conformational* (shape) *change* in the receptor. This triggers a biochemical change in the cytoplasm of the cell, causing the release of a *second messenger*, the first messenger being the hormone itself (Fig. 1-4). Examples of second messengers include cAMP

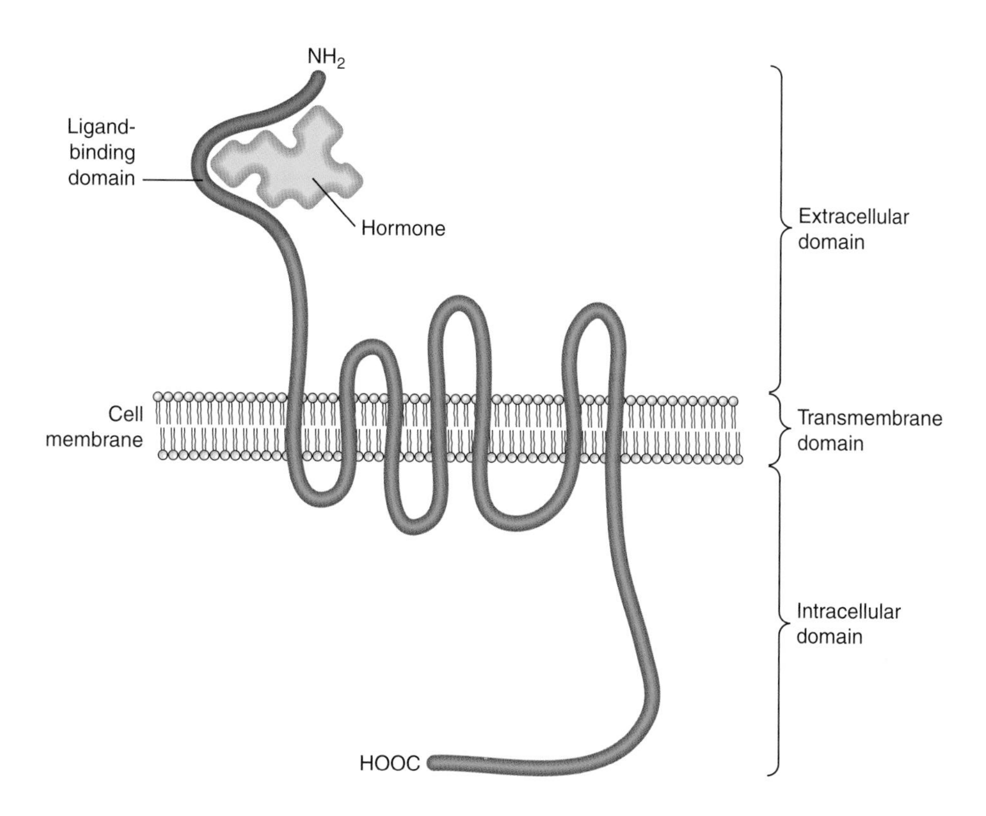

Figure 1-3 Representation of a cell surface receptor molecule showing the *ligand-binding domain* (green) as a portion of the *extracellular domain*. The *transmembrane domain* spans the plasma membrane of the cell, and the *intracellular domain* extends into the cytoplasm.

and Ca^{2+}. This "translation" of the hormonal message to the interior of the cell is called *signal transduction*. Because signal transduction usually involves turning on or off a series of enzymes, a few molecules of a hormone can be *amplified* to alter thousands of molecules inside the cell.

Unlike protein and peptide hormones, which must stay at the cell surface, steroid hormones (such as estrogen, testosterone, and progesterone) are lipid soluble and thus can pass through the phospholipid bilayer of the plasma membrane easily. Steroid receptors are located within the cytoplasm or the nucleus of target cells. When a steroid hormone binds to its receptor (Fig. 1-5), the steroid/receptor complex undergoes a conformational change that exposes a *DNA-binding domain*. This part of the receptor binds to a regulatory region of a steroid-responsive gene, turning on or off transcription of the gene. Because steroids alter gene expression, and transcription and translation of a protein require at least 30 min, the effects of steroids on the body are typically slow (but long-lasting). Whereas the effects of steroids are measured in terms of hours or days, protein and peptide hormones often act within minutes. Some steroids act also through cell-surface receptors.

The biological activity of a given hormone in a particular tissue depends on the local concentration of the hormone. In addition, the activity of the hormone

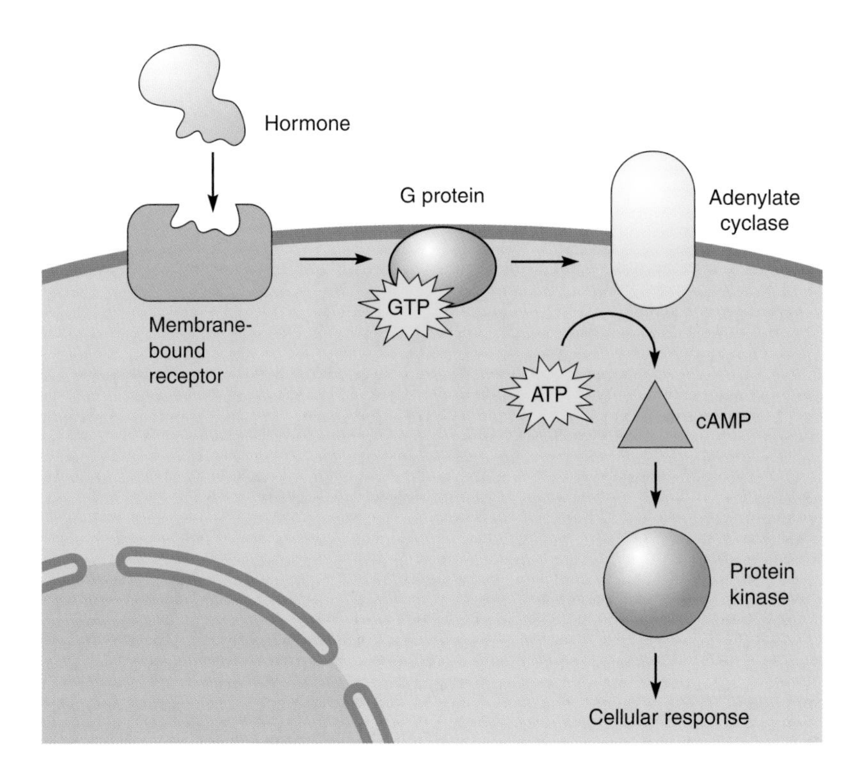

Figure 1-4 Mechanism of action of a peptide hormone. The ligand binds to a cell surface receptor. The signal is transduced to the cell interior, where it modifies the activity of cytoplasmic enzymes.

depends on the number of receptors for that hormone that are present. Cells can gain (*upregulate*) or lose (*downregulate*) receptors, and the timing of some reproductive functions (such as growth of the ovarian follicle) is dependent on the change in number of hormone receptors present.

One might expect that each hormone has its unique receptor, resulting in equal numbers of hormone and receptor types. This is not the case. Many hormone receptors, especially steroid receptors, lack specificity and can accept more than one type of ligand molecule. Theoretically, any molecule that can achieve a three-dimensional fit into the ligand-binding domain of a steroid receptor can affect that receptor. For example, there are three major naturally occurring estrogens, each of which can activate a single estrogen receptor. Because of differences in their chemical structures, however, they bind to the estrogen receptor with different *affinities* (strengths). Estradiol has greatest affinity for the estrogen receptor; for this reason it is considered a potent or "strong" estrogen. Estriol and estrone are "weak" estrogens. Estriol binds to the estrogen receptor about 10% as well as estradiol, and estrone only binds 1% as efficiently.

After being secreted into the circulation, gonadal steroid molecules are quickly attached to *sex steroid-binding globulins* in the blood. Hitching a ride on these circulating proteins allows the lipophilic steroids to circulate more freely

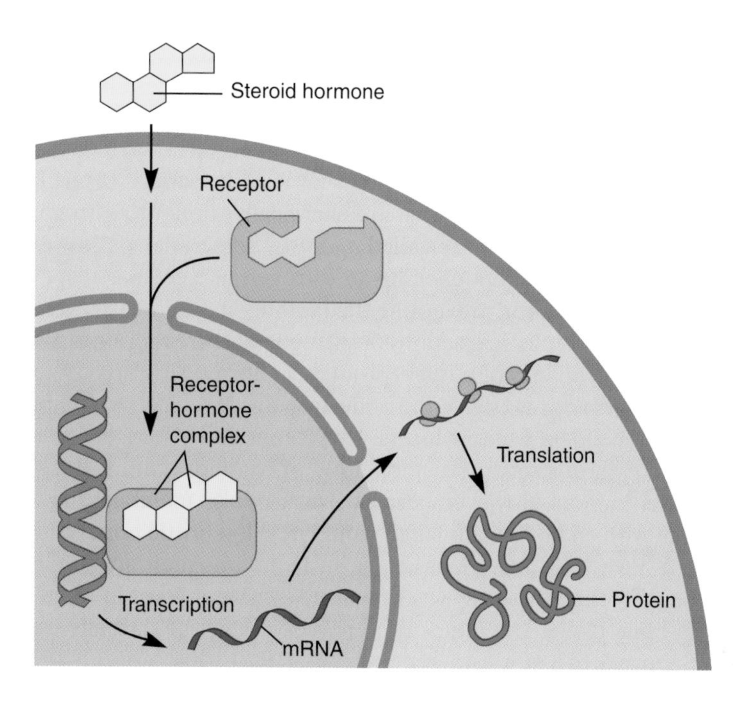

Figure 1-5 Mechanism of action of a steroid hormone. Lipid-soluble steroids enter the cell cytoplasm by diffusion and bind to receptors in the cytoplasm or nucleus. The steroid/receptor complex binds to regulatory regions of DNA, affecting the expression of specific steroid-responsive genes.

in the aquatic bloodstream. The binding proteins release only a certain number of steroid molecules at a given time, freeing them to leave the bloodstream and travel to target cells. Typically only a small proportion of the secreted steroid (2–3% in the case of estradiol) is free in the plasma. The availability of steroid hormones to target cells is regulated by the concentration of carrier proteins, whose levels change under certain conditions such as pregnancy and obesity. The ratio of free to bound steroid is an important factor in the biological activity of a steroid hormone. Thus, to understand the action of a hormone on its target, we must know (1) the local concentration of the hormone, (2) the proportion of hormone available to receptors, and (3) the number of unbound receptors.

Synthetic Hormones

Molecules that activate hormone receptors, thus mimicking the natural hormones produced by endocrine glands, occur in our environment and in the food we eat. For example, in addition to the naturally occurring human estrogens, *exogenous* compounds (from a source outside our bodies) can have estrogenic effects. Weak estrogens synthesized by plants, called *phytoestrogens*, can bind to the estrogen receptor. One of these phytoestrogens is genistein, found in the soybean plant. Scientists have also become aware of numerous man-made chemicals with estrogenic effects. Many of these *xenoestrogens* are pesticides related to DDT, or industrial chemicals such as those used in the synthesis of plastics. An active search is underway to identify these compounds and to determine their possible effects on human health and impacts on wildlife (see HIGHLIGHT box 2-2 "Xenoestrogens and Breast Cancer" in Chapter 2).

The pharmaceutical industry has taken advantage of the lack of specificity of the estrogen receptor to synthesize a wide variety of artificial estrogens. These *analogs* are compounds that are chemically similar to estrogen. Analogs that mimic estrogen-like effects are called *agonists*. Some analogs, however, bind to the estrogen receptor without eliciting an estrogen-like cellular response. These *antagonists* block the receptor, preventing the binding of natural estrogen. Estrogen antagonists oppose the biological actions of estrogen and are therefore also called *antiestrogens*. Synthetic analogs of other reproductive hormones also have been produced.

Tamoxifen is one of the antiestrogens developed to inhibit the growth of estrogen-dependent breast cancers. This compound effectively blocks the ligand-binding site of the estrogen receptor, thus preventing natural estrogen from stimulating breast tumors. However, because estrogen also maintains bone density, it was feared that women taking tamoxifen would develop brittle bones. Surprisingly, the drug did not have this deleterious effect; in fact, tamoxifen actually helped maintain bone density. Thus, the drug acts as an estrogen antagonist in breast tissue but an agonist in bone tissue. This "tamoxifen paradox" may be partially explained by the recent discovery of a new estrogen receptor, ER-β, which appears to have a different tissue distribution than the original receptor, renamed ER-α. (The more abundant ER-α is usually referred to as "the estrogen receptor.") The possibility of developing pharmaceuticals that mimic effects of a hormone in some tissues but block them in others paves the way for so-called "designer drugs."

The Pituitary Gland

The sphenoid bone lies at the base of your skull, and in this bone is a small, cup-shaped depression called the sella turcica ("Turkish saddle"). Lying in this depression is a round ball of tissue, about 1.3 cm (0.5 in.) in diameter, called the *hypophysis* or *pituitary gland* (Fig. 1-6). This gland synthesizes and secretes hormones that travel in the bloodstream and influence many aspects of our body, including the function of other endocrine glands. For example, if the hypophysis is removed (an operation called hypophysectomy), our reproductive system becomes nonfunctional, and even sexual behavior is affected. Therefore, this gland plays a very important role in our reproductive biology. A realization of the importance of the hypophysis led early endocrinologists to call it the "master gland." More recently, however, we have become aware that the activity of this gland, which is connected to the base of the brain by a stalk, is itself influenced greatly by brain messages. Indeed, one might think of the brain as the conductor of a marvelous chemical symphony played by the pituitary orchestra.

Hypothalamo–Neurohypophysial Connection

The hypophysis has two major regions (Fig. 1-7). One is called the *adenohypophysis*, which is discussed later in this chapter. The other is the *neurohypophysis* (*pars nervosa*, or *posterior pituitary gland*). The neurohypophysis is an extension of the brain, and it develops as an outgrowth of the portion of the embryonic

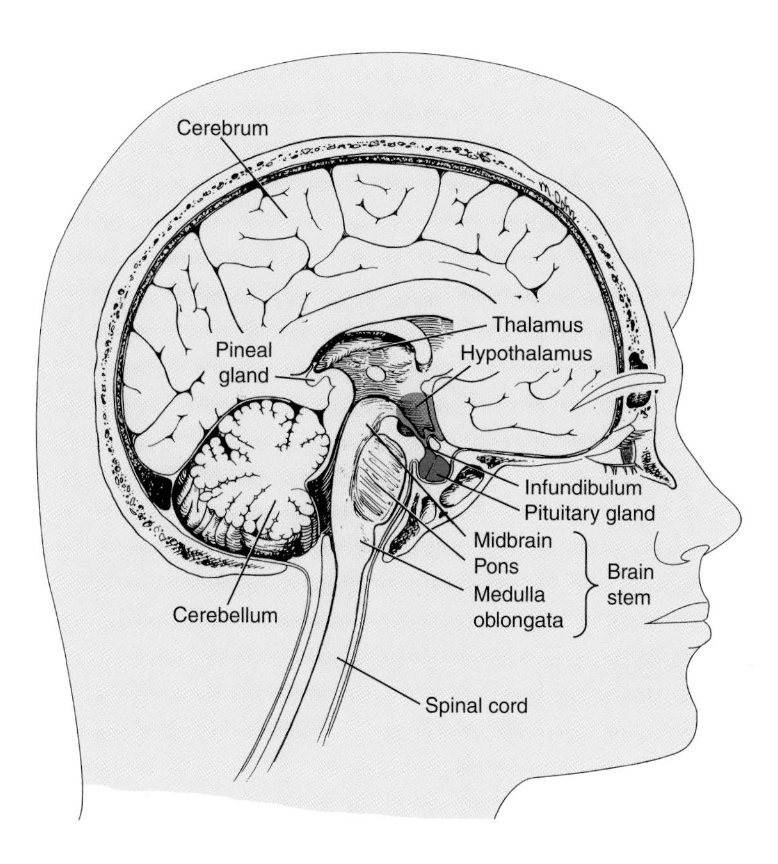

Figure 1-6 Section through the middle of the brain showing the pituitary gland, hypothalamus, and pineal gland. Note that the pituitary gland (hypophysis) rests in a depression in the sphenoid bone and is connected to the hypothalamus by the pituitary stalk.

brain that later becomes the hypothalamus. To understand the function of the neurohypophysis, you first must realize that there are two general types of nerve cells in the body.

Most of the nerve cells, or *neurons*, in our nervous system consist of a cell body (containing the nucleus) along with extensions of the cell called dendrites and axons (Fig. 1-8). *Dendrites* conduct a nerve impulse toward the cell body, which usually is in or near the *central nervous system* (brain and spinal cord). A sensory nerve is really a collection of long dendrites carrying messages to the central nervous system from the periphery. *Axons* carry information away from the cell body. Motor nerves contain axons that stimulate a response in the body, such as muscle contraction or glandular secretion. When one neuron connects with another, information is passed from the first to the second cell; this site of communication is known as a *synapse*. The axonal ending of the first neuron secretes a chemical called a *neurotransmitter*, which travels across the synapse and initiates electrochemical changes leading to nerve impulses in the next neuron.

The other general kind of nerve cell is the *neurosecretory neuron* (Fig. 1-8). A neurosecretory neuron is similar to a regular neuron in that it can conduct a nerve impulse along its axon. The speed of this electrical conduction is, however,

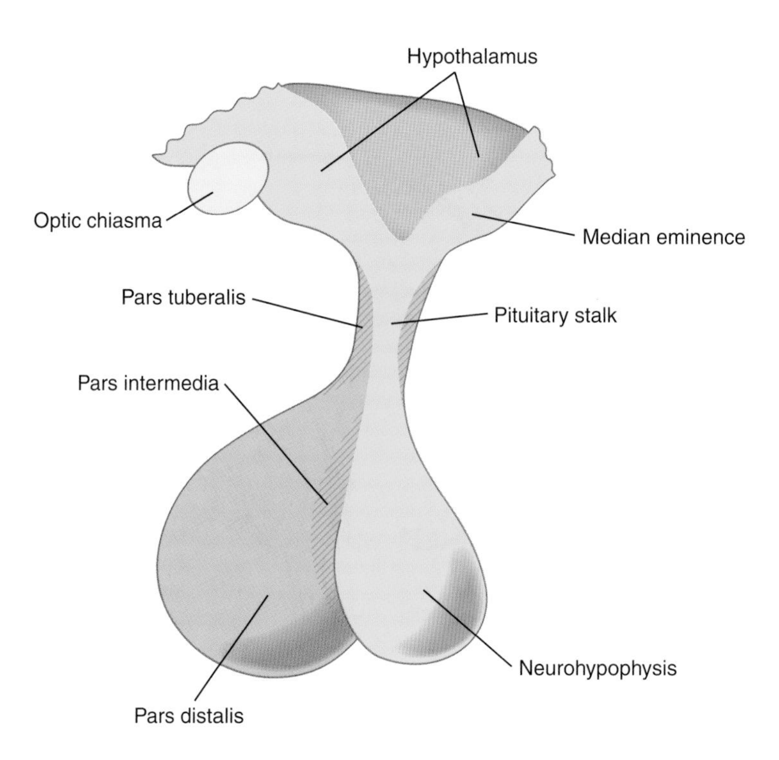

Figure 1-7 Major subdivisions of the human hypophysis (the neurohypophysis and the adenohypophysis) and their relationship to the brain. The pars tuberalis, pars distalis, and pars intermedia all are part of the adenohypophysis. In adult humans, the pars intermedia is often absent. OC, optic chiasma (nerves from the eyes). Anterior is to the left.

much slower than in a regular neuron. Also, neurosecretory neurons are specialized to synthesize large amounts of *neurohormones* in their cell bodies. These neurohormones are then packaged into large granules that travel in the cytoplasm down the axon, and contents of the granules are released into the spaces adjacent to the axon ending. The neurohypophysis contains long axons of neurosecretory neurons surrounded by supporting cells. The cell bodies of these axons lie in the part of the brain called the *hypothalamus*.

The hypothalamus forms the floor and lower walls of the brain (see Fig. 1-6) and contains a fluid-filled cavity, the third ventricle. This ventricle is continuous with the other ventricles in the brain and also with the central canal of the spinal cord. The fluid in the ventricles and central canal is called *cerebrospinal fluid*. The weight of the hypothalamus is only 3/100 that of the whole brain, but it functions in a wide variety of physiological and behavioral activities. For example, there are areas in the hypothalamus that regulate body temperature, thirst, hunger, sleep, response to stress, and aggressive and sexual behaviors.

Of importance to our discussion of the hypophysis is that the cell bodies of the neurosecretory axons in the neurohypophysis lie in paired groups (*neurosecretory nuclei*) in the hypothalamus. More specifically, these are the *supraoptic* and *paraventricular* nuclei. (Note: a neurosecretory nucleus is a group of cell bodies of neurosecretory neurons and should not be confused with "nucleus" as meaning the body within a cell that contains DNA.) The axons of the neurosecretory neurons

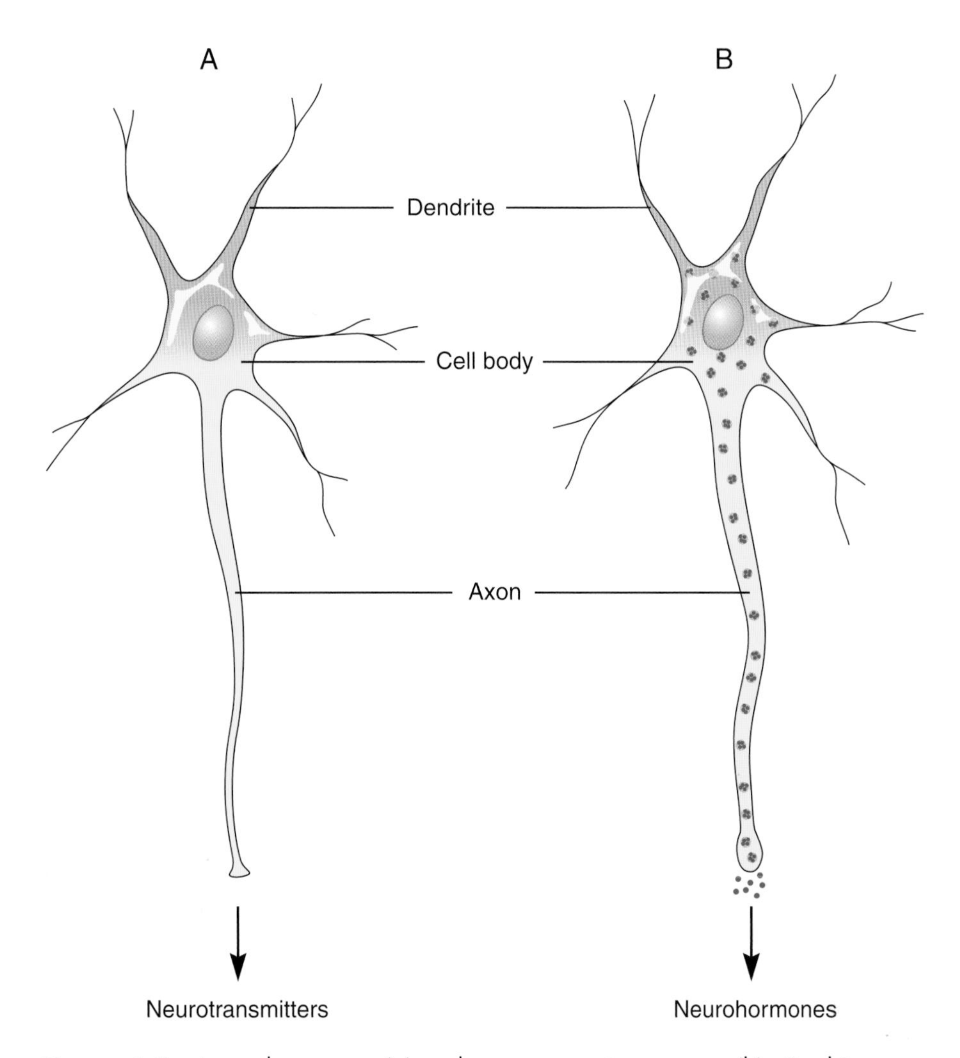

Figure 1-8 A regular neuron (a) and a neurosecretory neuron (b). Dendrites carry nerve impulses toward the cell body, whereas axons carry nerve impulses away from the cell body. Neurotransmitters are secreted by the axon endings of regular neurons, whereas neurohormones (dark dots in b) are released from the axon endings of neurosecretory neurons.

in these nuclei then pass down the *pituitary stalk* (which connects the hypophysis with the brain) and into the neurohypophysis (Fig. 1-9). The granules released by these axons contain two neurohormones—*oxytocin* and *vasopressin* (or *antidiuretic hormone*).

Both oxytocin and vasopressin are polypeptides consisting of nine amino acids. The two neurohormones differ only slightly in the kinds of amino acids in their molecules, but these slight differences result in their having very different effects on our bodies. Oxytocin stimulates contractile cells of the mammary glands so that milk is ejected from the nipples (see Chapter 12). Also, oxytocin causes the smooth muscle of the uterus to contract, thus playing a role in labor and childbirth (see Chapter 11). Vasopressin causes the kidneys to retain water, i.e., the amount of urine formed in the kidneys is reduced and more water remains in the body. Vasopressin also causes blood vessels to constrict and blood pressure to rise. When oxytocin and vasopressin are released

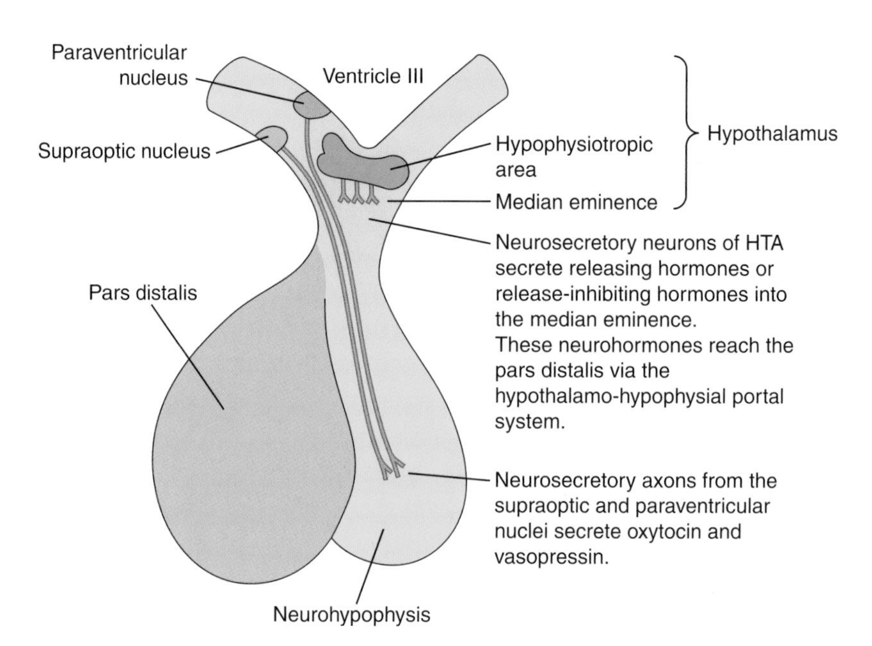

Figure 1-9 Regions of the hypothalamus involved in the function of the hypophysis.

from the axons in the neurohypophysis, these neurohormones enter small blood vessels (capillaries) in the neurohypophysis that drain into larger veins and then enter the general circulation.

Adenohypophysis

The adenohypophysis (*adeno*, meaning "glandular") consists of three regions: the pars distalis, the pars intermedia, and the pars tuberalis (Fig. 1-7). The *pars distalis* (or *anterior pituitary gland*) occupies the major portion (70%) of the adenohypophysis. The *pars intermedia* is a thin band of cells between the pars distalis and the neurohypophysis. In the adult human, the pars intermedia is sparse or absent. The *pars tuberalis* is a group of cells surrounding the pituitary stalk. During embryonic development, the adenohypophysis forms from an invagination (inpocketing) of the cell layer of the embryo that later becomes the roof of the mouth. This invagination of cells then extends toward the neurohypophysis growing from embryonic brain.

The adenohypophysis contains several types of endocrine cells. When we stain the adenohypophysis with laboratory dyes, some cells acquire a pink color. These cells are called *acidophils* (*phil*, meaning "love") because they have an affinity for acid dyes. Some acidophils synthesize and secrete *growth hormone* (GH). This hormone is a large protein that stimulates tissue growth by causing incorporation of amino acids into proteins. The other type of acidophil in the adenohypophysis synthesizes and secretes prolactin, which is also a large protein. As shown in Chapter 12, PRL acts with other hormones to cause the mammary glands in the female to become functional and secrete milk.

Other cells in the adenohypophysis, the *basophils*, stain darkly with basic dyes. These cells synthesize and secrete hormones that are proteins or glycoproteins (large proteins with attached sugar molecules). One of these hormones is the glycoprotein *thyrotropin* (TSH). The abbreviation for thyrotropin comes from the older name, thyroid-stimulating hormone. This hormone causes the thyroid glands to synthesize and secrete thyroid hormones (e.g., *thyroxine*), which in turn control the rate at which our tissues use oxygen. In addition, some of the basophils in the adenohypophysis synthesize and secrete *corticotropin* (ACTH), a polypeptide hormone that travels in the blood to the adrenal glands and causes secretion of adrenal steroid hormones (*corticosteroids*) such as *cortisol.* Cortisol in turn raises blood sugar levels, reduces inflammation, and combats the effects of stress. Other basophils in the adenohypophysis secrete the polypeptide hormones *lipotropin* (LPH) and *melanophore-stimulating hormone* (MSH). Lipotropin breaks down fat to fatty acids and glycerol. MSH causes synthesis of a brown pigment, *melanin*, which is present in cells called *melanophores*. Finally, the basophils of the adenohypophysis secrete two kinds of natural, opioid-like "pain-killers"—the *endorphins* and *enkephalins.*

Of particular interest to our discussion of human reproductive biology are the final two hormones secreted by basophils of the pars distalis. One of these is follicle-stimulating hormone. We shall learn in Chapter 4 that FSH plays a role in sperm production in the testes. In the female, FSH stimulates the ovaries to produce mature germ cells in their enclosed tissue sacs (see Chapter 2). The other hormone is luteinizing hormone. This hormone causes interstitial cells in the testes to synthesize and secrete androgens (see Chapter 4). In the female, LH causes the ovaries to secrete female sex hormones (estrogens and progestogens) and induces the release of an egg from the ovary (see Chapter 2). Most FSH and LH come from cells in the pars distalis, although the pars tuberalis also contains these hormones. Because FSH and LH play vital roles in the function of the gonads, they are grouped under the term *gonadotropic hormones* or "gonadotropins." Figure 1-10 summarizes the hormones secreted by the hypophysis.

Hypothalamo–Adenohypophysial Connection

It has been known for some time that the reproductive cycles of many animals are influenced by environmental factors such as light, behavior, and stress, and the same appears true for humans. For example, it had long been observed that menstrual cycles of women often were altered or even stopped by stressful environmental and psychological stimuli (see Chapter 3). This pointed to an influence of the brain on reproductive physiology. It was not until 1947, however, that J. D. Green and G. W. Harris provided anatomical evidence that neurosecretory neurons in the hypothalamus could influence the function of the adenohypophysis. Recall that some of the neurosecretory neurons in the hypothalamus send their axons down the pituitary stalk and into the neurohypophysis, where they release oxytocin and vasopressin. Other neurosecretory neurons existing in paired nuclei in the hypothalamus, however, do not send their axons down the stalk to the hypophysis. Instead, their axons end in an area in the floor of the hypothalamus near the pituitary

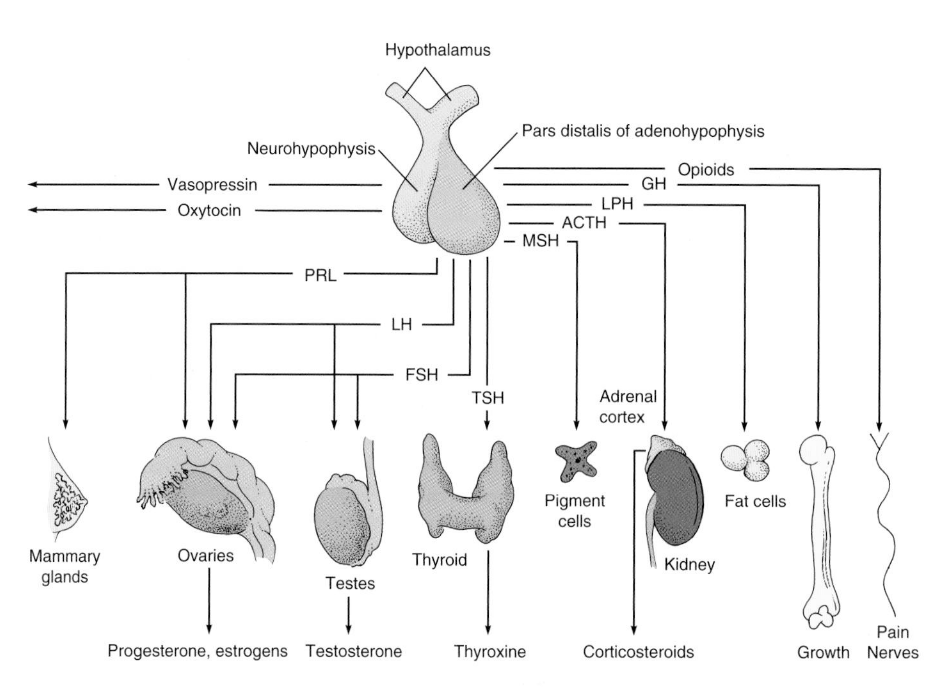

Figure 1-10 The pituitary, connected to the hypothalamus at the base of the brain, has two lobes. The neurohypophysis stores and releases two hormones made in the hypothalamus: oxytocin and vasopressin. Oxytocin causes contraction of smooth muscle in the uterus, breast, and male reproductive tract. Vasopressin acts on the kidneys to cause water retention. The adenohypophysis secretes nine other hormones: growth hormone (GH) promotes growth; corticotropin (ACTH) causes the adrenal cortex to secrete corticosteroid hormones; follicle-stimulating hormone (FSH) and luteinizing hormone (LH) interact to regulate the function of the gonads; prolactin (PRL) causes milk synthesis in the mammary glands; thyrotropic hormone (TSH) stimulates the thyroid gland to secrete thyroxine; lipotropin (LPH) affects fat metabolism; melanophore-stimulating hormone (MSH) stimulates melanin synthesis in pigment cells; and opioids (endorphins and enkephalins) reduce pain.

stalk, called the *median eminence*. (Note: Sometimes the median eminence is considered part of the neurohypophysis.) The cell bodies of these neurons are clustered in several pairs of nuclei in the hypothalamus, and together these nuclei are named the *hypophysiotropic area* (HTA, Fig. 1-9). These nuclei are given this name because the neurosecretory neurons in this region secrete a family of small polypeptides (neurohormones) that either increase or decrease the secretion of hormones secreted by the adenohypophysis.

If the neurohormones controlling adenohypophysial function are released from neurosecretory neurons at the median eminence, how do they reach the adenohypophysis to influence pituitary hormone secretion? In 1930, G. T. Popa and U. Fielding described a specialized system of blood vessels extending from the median eminence to the pars distalis (Fig. 1-9). The superior hypophysial arteries carry blood to the median eminence region. These arteries drain into a cluster of capillaries in the median eminence known as the *primary capillary plexus*. The neurohormones diffuse into these capillaries and into the blood.

Then they are carried down to the pars distalis in small veins. These veins divide into a second capillary bed surrounding the cells of the pars distalis, the *secondary capillary plexus*. The neurohormones then leave the blood through the walls of these capillaries and enter the spaces between the pars distalis cells, where they cause these cells to either increase or decrease hormone synthesis and secretion. This efficient system allows, for example, a very small amount of the neurohormone GnRH, undiluted by the general circulation, to be delivered directly to its target—gonadotropin-secreting cells in the pituitary—and influence their activity precisely.

A *portal system* is a vascular arrangement in which blood flows from one capillary bed to another without going through the heart in its journey. Thus, the vascular system connecting the median eminence with the pars distalis is called the *hypothalamo–hypophysial portal system*, and the small veins connecting the primary and secondary capillary plexi are the *hypophysial portal veins*.

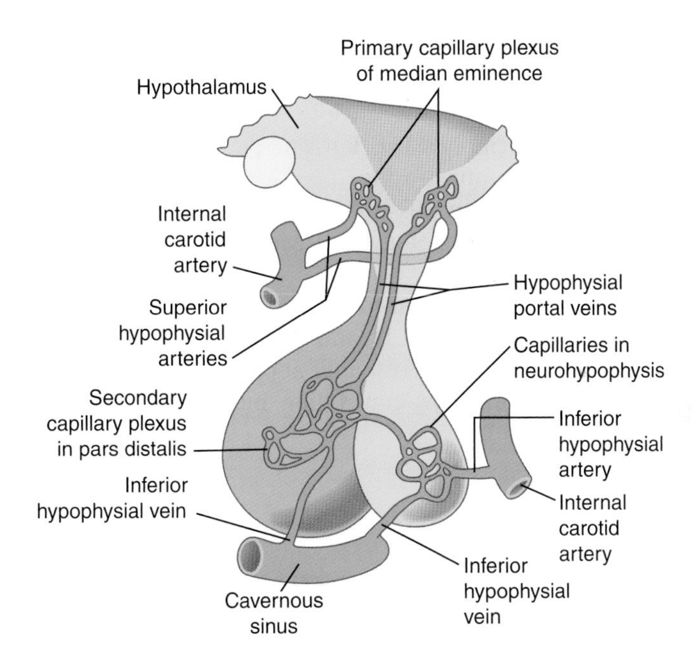

Figure 1-11 The hypothalamo–hypophysial vascular system. Arterial blood enters the median eminence and the neurohypophysis via the superior hypophysial and inferior hypophysial arteries, respectively. Both of these arteries are branches of the internal carotid arteries, major vessels supplying the brain. Neurohormones secreted into the median eminence region enter the blood in the primary capillary plexus. They pass down the hypophysial portal veins to the secondary capillary plexus in the pars distalis. Then they leave the blood and cause the pars distalis cells to secrete or stop secreting hormones. When hormones are secreted by the pars distalis, they leave the hypophysis in the inferior hypophysial veins, which drain into a large vessel, the cavernous sinus. Neurohypophysial hormones enter capillaries in the neurohypophysis, which also drain into the cavernous sinus. Small blood vessels connect the capillaries of the pars distalis and neurohypophysis.

Once the hormones of the adenohypophysis are secreted, they leave the pituitary via the inferior hypophysial vein (Fig. 1-11).

Releasing and Release-Inhibiting Hormones

Neurohormones released by the axons of the hypophysiotropic area can either increase or decrease the synthesis and secretion of hormones of the adenohypophysis. When a neurohormone increases output of a particular adenohypophysial hormone, it is called a *releasing hormone* (RH). For example, the neurohormone that increases the output of thyrotropin is called a *thyrotropin-releasing hormone* (TRH). When a neurohormone lowers the secretion of a particular adenohypophysial hormone, it is termed a *release-inhibiting hormone* (RIH). Thus, the neurohormone that decreases the secretion of prolactin is *prolactin release-inhibiting hormone* (PRIH).

As seen in Table 1-1, each hormone of the adenohypophysis is controlled by a releasing hormone, and some are known to be controlled by both a releasing and a release-inhibiting hormone. Each releasing or release-inhibiting hormone probably is synthesized by a different group of neurosecretory cell bodies in the hypophysiotropic area. The chemical structures of seven of these neurohormones are known. The others are recognized as a result of experiments demonstrating their presence, but their chemical nature is yet to be described. In 1977, Andrew Schally and Roger Guillemin shared the Nobel Prize in Physiology and Medicine for their research on hypothalamic neurohormones.

Of particular interest to us are the neurohormones that control the synthesis and release of FSH, LH, and PRL from the pars distalis. These are discussed in some detail because research about these neurohormones has and will have profound influence in controlling human fertility and treating reproductive disorders.

Table 1-1 Hypothalamic Neurohormones Controlling the Synthesis and Release of Hormones from the Pars Distalis

Neurohormone	Abbreviation
Corticotropin-releasing hormone	CRH
Thyrotropin-releasing hormone	TRH
Gonadotropin-releasing hormone[a]	**GnRH (or LHRH)**[a]
Growth hormone release-inhibiting hormone (or somatostatin)	GHRIH
Growth hormone-releasing hormone	GHRH
Prolactin release-inhibiting hormone	**PRIH**
Prolactin-releasing hormone	**PRH**
Melanophore-stimulating hormone release-inhibiting hormone	MSHRIH
Melanophore-stimulating hormone releasing hormone	MSHRH

[a]GnRH causes release of both FSH and LH and is identical to the LHRH discussed in the scientific literature. Those of particular interest to reproductive biologists are in bold.

Gonadotropin-Releasing Hormone

About 1000 to 3000 neurons in the HTA secrete *luteinizing hormone-releasing hormone* (LHRH). Many of these neurons send their axons to the median eminence to exert control over pituitary LH and FSH secretion. Some LHRH neurons, however, send axons to other brain regions to possibly influence sexual behavior.

Primarily through the efforts of Andrew Schally, the chemical nature of LHRH is known, and it has been synthesized in the laboratory. Many thousands of hypothalami from domestic mammals were obtained to extract and purify this and other neurohormones. LHRH is a polypeptide, consisting of 10 amino acids, and it has been utilized in a wide variety of research and clinical studies. Surprisingly, when LHRH is administered to humans or to laboratory mammals, both LH (Fig. 1-12) and, to a lesser degree, FSH are secreted in increased amounts into the blood. Therefore Schally concluded that there may be only one releasing hormone in humans that controls synthesis and secretion of both LH and FSH. This is despite evidence that the hypothalamus of some mammals contains LHRH as well as a *follicle-stimulating hormone-releasing hormone* (FSHRH) that causes release of mostly FSH and a little LH. Although future research may show that the human hypothalamus secretes both LHRH and FSHRH, let us for now accept that a single releasing hormone increases both LH and FSH secretion from the pituitary; we call this *gonadotropin-releasing hormone*, (GnRH), realizing that this is identical to the LHRH discussed in the scientific literature.

GnRH is actually derived, within the neuron, from a larger protein called *prepro-GnRH* (Fig. 1-13). This protein contains the 10 amino acids of GnRH, plus a "signal sequence" of 23 amino acids (which plays a role in breaking up prepro-GnRH into its component parts), a sequence of three amino acids used for molecular processing, and a 56 amino acid sequence, which is called *GnRH-associated peptide* (GAP). Before secretion of GnRH, the prepro-GnRH molecule is split into GnRH and GAP.

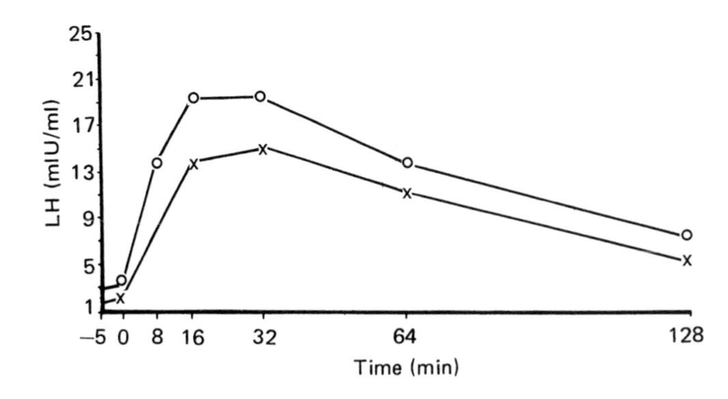

Figure 1-12 When a single injection of GnRH (LHRH) is administered to men (×) and women (O) at time zero, levels of LH rise in the blood, peak after 32 min, and then decline. Levels of FSH also rise after GnRH administration (not shown), but not as high as LH.

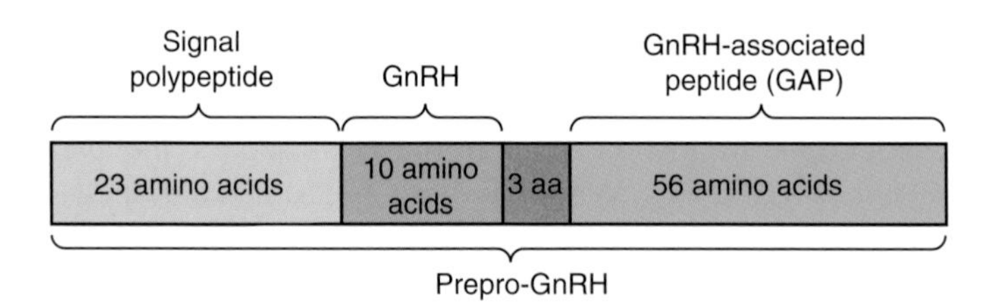

Figure 1-13 Components of prepro-GnRH, the large protein that gives rise to GnRH and GAP.

Chapter 1, Box 1: GnRH Analogs

Once GnRH was discovered, one of the first ideas was to give it to men and women to treat certain kinds of infertility resulting from hypothalamic dysfunction. For example, women who are deficient in GnRH release have gonadotropin levels too low to cause ovulation.

In an attempt to find the most effective type of synthetic GnRH, many molecules similar but not identical to GnRH were manufactured. These are termed *GnRH analogs.* The GnRH analogs that stimulate gonadotropin secretion are called *GnRH agonists* (an agonist is a substance that mimics the action of the naturally occurring hormone). Attempts to treat these patients with injections of GnRH agonists resulted in an initial promising surge in FSH and LH levels. However, to everyone's surprise, most of the GnRH agonists stopped working after about 10 days because they reduced the number of GnRH receptors on FSH- and LH-secreting cells in the pituitary gland. These so-called "GnRH agonists," when given to a person for several days, become *GnRH inhibitory agonists*, a contradiction in terms if there ever was one! It was discovered that GnRH treatments work only when the hormone is administered in pulses about 90 min apart, mimicking the natural secretion pattern of GnRH. Using a small pump placed under the skin of the abdomen or the arm, pulses of synthetic GnRH can stimulate FSH and LH secretion and restore fertility in some cases.

Some GnRH analogs have been found to inhibit gonadotropin secretion by binding to GnRH receptors on FSH- and LH-secreting pituitary cells, but not stimulating gonadotropin secretion. They occupy the receptors and block natural GnRH from binding; these molecules that prevent the action of GnRH are called *GnRH antagonists.* They are like a rusty key stuck in a lock, which will not open the door and will not permit the use of a good key to do so.

The GnRH inhibitory agonists and GnRH antagonists are medically useful for their ability to shut down gonadotropins and, consequently, lower gonadal steroids and inhibit egg and sperm maturation. Typically, these drugs are given as daily or monthly injections, an implant, or a nasal spray. They are employed to treat endometriosis and uterine fibroids by reducing circulating estrogen levels, causing the affected tissues to shrink. They are also being studied as potential contraceptives. In fertility clinics, inhibitory GnRH analogs are used to prevent premature ovulation so its timing can be controlled by the use of fertility drugs in *in vitro* fertilization/gamete intrafallopian transfer procedures (see Chapter 16). Despite their utility, these GnRH agonists often have side effects, including menopausal-like symptoms (hot flashes, insomnia, vaginal dryness), osteoporosis, and headaches, and they can increase the risk of ovarian cysts. The dramatic effects of these molecules in both promoting and suppressing fertility illustrate the central, essential role of GnRH in reproduction.

Continued on next page.

Chapter 1, Box 1 continued.

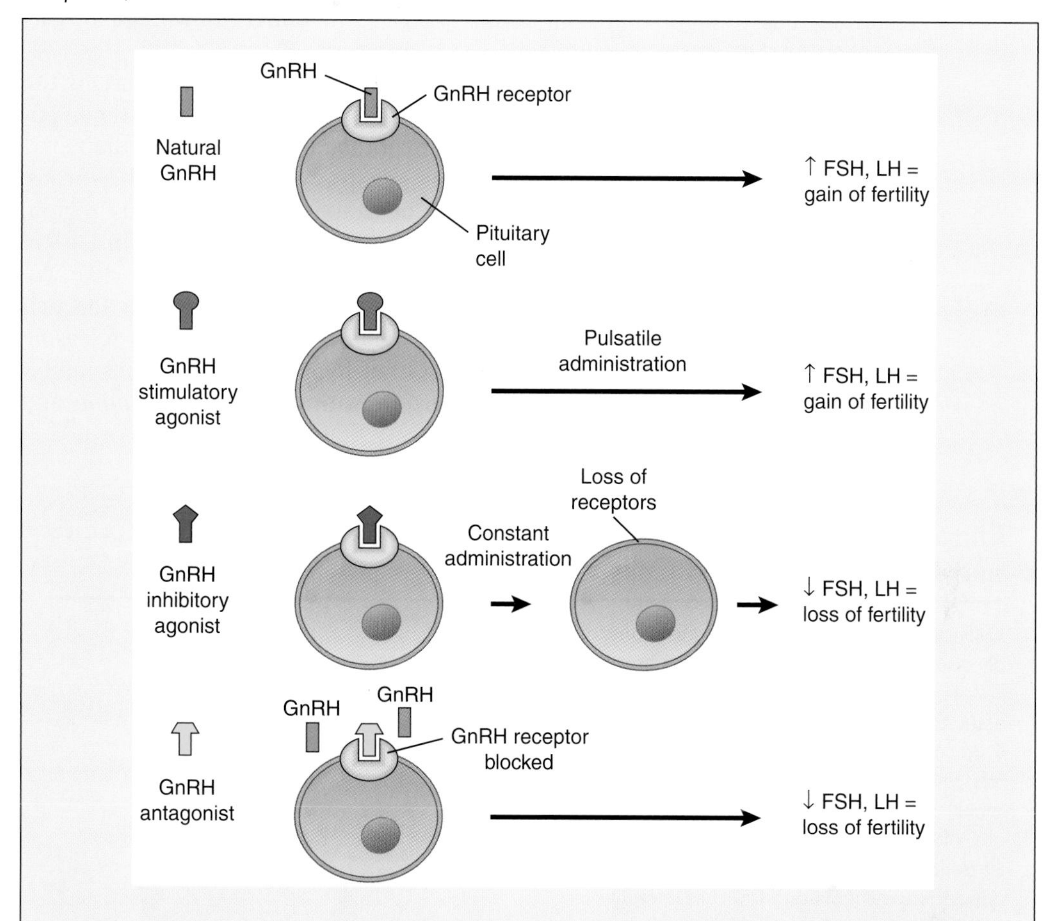

The effects of various GnRH analogs. When GnRH (blue ligand) binds to its cell surface receptors on pituitary cells, the cellular response is increased secretion of gonadotropins. A synthetic GnRH analog binds to the receptor and acts as a stimulatory agonist (green ligand) when administered in pulses. However, when administered continuously, a GnRH agonist (red ligand) can actually act as an *inhibitory agonist,* thus reducing gonadotropin secretion. A GnRH antagonist (yellow ligand) occupies the GnRH receptor without eliciting a cellular response, thus blocking the action of natural GnRH.

The GnRH Pulse Generator and Surge Center

GnRH is not released continuously from the hypothalamus. Instead, it is secreted in pulses every hour or so into the hypophyseal portal system. In response, the pituitary gonadotropes release FSH and LH in a pulsatile fashion. The pulsatile pattern of GnRH secretion is essential for gonadotropin secretion, and thus is central to reproductive function. This is demonstrated in the treatment of men and women whose infertility is caused by insufficient gonadotropin levels. Initially, it was thought that simply giving the patients GnRH agonists would restore fertility. Surprisingly, after initial stimulation of FSH and LH, these agonists stopped working. Only when GnRH agonists are administered in natural pulses by an intravenous pump is normal gonadotropin secretion restored. It is thought that continuous exposure to GnRH downregulates its receptors or the GnRH signaling pathway in pituitary cells.

In humans, the GnRH-secreting cells are mainly located in a part of the HTA called the *arcuate nucleus*, although there are a few elsewhere in the hypothalamus. This nucleus is at the base of the hypothalamus near the median eminence; it also contains regular neurons that synapse with the GnRH neurons. Pulsatile secretion of GnRH is controlled by activity of cells in this region, known as the *GnRH pulse generator*. Whether pulsatility is inherent in GnRH cells alone or is also influenced by synapses with regular neurons is not clear. We do know that neurons in the hypothalamus modify GnRH secretion through several neurotransmitters. For example, *norepinephrine*- and *dopamine*-releasing neuron activity increases GnRH secretion. Other hypothalamic neurons inhibit GnRH secretion through the release of neurotransmitters such as *dopamine* and

Chapter 1, Box 2: Kallmann's Syndrome and the Embryological Origin and Migration of GnRH Cells

Kallmann's syndrome is one of the many possible causes of human infertility (see Chapter 16). People with this syndrome exhibit a curious association of infertility with anosmia (the inability to smell). Examination of their nervous system has revealed an absence of certain olfactory (smell) structures in the brain as well as a lack of GnRH neurons in the hypothalamus; the latter deficiency accounts for their low secretion of gonadotropins (FSH and LH) and infertility.

Kallmann's syndrome is inherited; the gene for this disorder is located on the X chromosome (it is sex linked) and thus it is five to seven times more common in men (see Chapter 5). About 1 in 10,000 men are born with this condition. Fertility of Kallmann's syndrome sufferers can be restored with the administration of GnRH stimulatory agonists, but they still would remain anosmic for life.

What is the explanation for the association of olfactory and reproductive abnormalities in people with this syndrome? During normal embryonic development, the olfactory nerves carry neural information from the olfactory lining in the nasal cavity to the pair of olfactory bulbs at the base of the cerebral hemispheres of the brain. From there, the olfactory sense is carried in nerve fibers of the olfactory tracts to the hypothalamus and other brain areas. Once the olfactory system is formed in

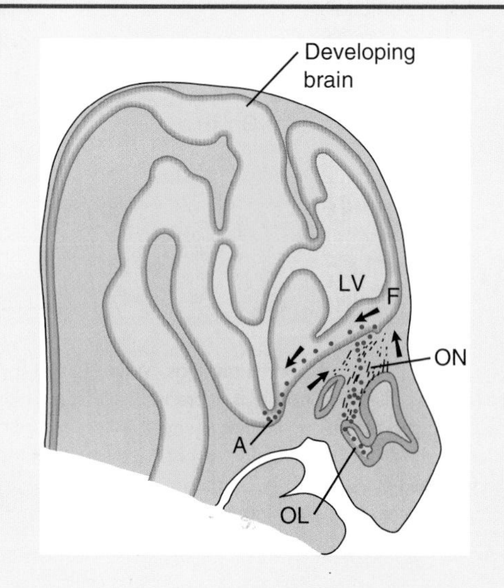

Diagram of a section through the head of a human embryo showing the migration route of GnRH neurons (green dots) from the nasal cavity to the hypothalamus. Failure to migrate, as in Kallmann's syndrome, results in GnRH cells remaining in the nose and in infertility. A, arcuate nucleus of the hypothalamus (GnRH cells migrating to here control gonadotropin secretion in the adult); F, forebrain (route of migration in brain); LV, lateral ventricle; OL, olfactory lining in developing nasal cavity; ON, olfactory nerve fibers (route of migration outside brain); large arrows, migration pathways of GnRH cells.

Continued on next page.

Chapter 1, Box 2 continued.

the embryo (at about day 25 of pregnancy), GnRH cells, which originate in the developing nasal cavity (nose), migrate (move) along the olfactory nerves and tracts to the hypothalamus, where they reside in the arcuate nucleus and prepare to play a role in controlling pituitary FSH and LH secretion and reproduction in the adult. People with Kallmann's syndrome, however, do not develop a normal olfactory system, so the GnRH cells have no "olfactory highway" of nerve fibers to guide them on their journey to the hypothalamus. Thus, the GnRH cells of people with Kallmann's syndrome remain stuck in the nose! Why is development of the GnRH neurons so interwoven with that of the olfactory system? Chapter 8 discusses the important role of social chemical signals ("pheromones") in the reproduction of mammals, including perhaps humans. The intimate association of the development of GnRH cells with the olfactory system may have evolved because of the importance of linking sensory chemical information from the opposite sex to gonadotropin secretion and fertility.

γ-aminobutyric acid (*GABA*). Thus, GnRH secretion can be stimulated or inhibited by a complex pattern of neuronal activity in the brain.

At midcycle in females, GnRH-secreting cells release even more GnRH, resulting in a surge of FSH and LH from the pituitary. The *GnRH surge center* activity controlling this event also resides in the hypothalamus. The importance of the GnRH pulse generator and surge center in the control of the menstrual cycle is discussed in Chapter 3.

Pineal Gland

The *pineal gland*, a single outpocketing from the roof of the brain (see Fig. 1-6), may also influence the release of FSH and LH. In the 17th century, it was believed that this gland was the "seat of the soul." Now we know that the pineal synthesizes and secretes the hormone *melatonin*, which can inhibit the reproductive systems of males and females. Exposure of humans to light suppresses melatonin secretion, whereas exposure to dark increases it. Light does not affect the pineal directly. Instead, light entering the eyes increases the activity of nerves in the accessory optic tracts leading to the brain. These impulses cause the part of the sympathetic nervous system that innervates the pineal gland to decrease release of the neurotransmitter norepinephrine. This then causes a decrease in melatonin synthesis and secretion. Because of this influence of the daily light cycle on melatonin secretion, levels of melatonin in the blood exhibit a daily cycle. In humans, blood melatonin levels are highest during sleep, between 11 PM and 7 AM. When the daily light schedule is shifted by 12 h, it takes 4 or 5 days for the daily rhythm in blood melatonin in humans to shift to the new light cycle. In addition to daily light cycles, sleep and activity patterns can also influence melatonin cycles in humans.

We have much more to learn about the role of the pineal in reproduction. Analogs of melatonin have been made in the laboratory and found to inhibit the human reproductive system when given orally. These analogs could act on the hypothalamus, pituitary gland, or gonads to inhibit reproduction. In fact, melatonin is being used in a birth control pill (see Chapter 14). Melatonin may also play a role in normal puberty (see Chapter 6), as levels of this hormone in the blood of prepubertal children drop markedly just before the onset of puberty.

With all of these important functions, it is surprising that melatonin pills are being sold over the counter in the United States with no control over dosage or consideration of side effects.

Feedback Control of Gonadotropin Secretion

Feedback Systems

In your house or apartment, you probably have a thermostat on your wall that controls the activity of your heating system, and you can set the temperature of your room by manipulating a dial on the thermostat. Now, suppose you set the thermostat at 65 °F. This temperature is called the "set point." The thermostat contains a small strip, made of two metals, that expands or contracts depending on the temperature. If the room temperature drops below 65 °F, the thermostat sends electrical current through the wires leading to the heater, and the heater is activated. When the room temperature reaches 65 °F, the heater shuts off. Thus, the product of the heater (heat) influences the activity of the heater by feeding back on the device in the thermostat that controls the heater activity.

A simple feedback system is depicted in Figure 1-14. A receptor detects changes in the system and translates this information into a message ("input"). In your thermostat, the input travels along wires from the temperature receptor (bimetal strip) and then to a "controller center." The controller center contains the set point and also generates an outgoing message ("output"). Wires leading from the controller center in the thermostat to the heater carry this output

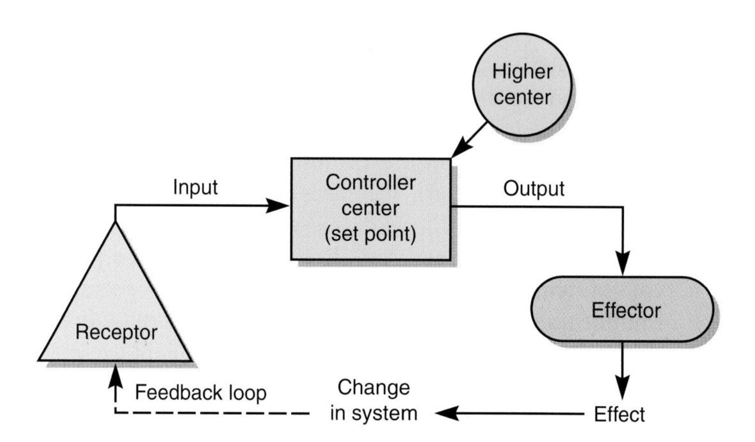

Figure 1-14 A simple feedback system. The receptor detects the level of a particular component of the system and translates it into a message (input) to the controller center. The input is compared by the controller center to its programmed set point, and this center computes whether a regulatory response is required. If necessary, the controller center generates a signal (output) that is transmitted to one or more effectors, which respond by producing some effect. This effect produces a change in the system, which then feeds back as a feedback loop on the controller center after being received by the receptor. Other circuits (higher centers) can modify the activities of the controller center by temporarily altering the set point or by inhibiting the operation of the controller center.

message. These output wires are activated when the room temperature is below the set point. The "effector" (heater) then responds by producing an "effect" (heat). In turn, the effect produces a change in the system (an increase in room temperature) that has a feedback effect on the controller center. This is called a *feedback loop*. In some feedback systems, other circuits ("higher centers" in Fig. 1-14) can modify the activities of the controller center by temporarily altering the set point or by inhibiting the operation of the controller center.

Many aspects of our physiology operate as feedback systems that regulate our internal environment at a steady state; this regulation is called *homeostasis*. Homeostatic control systems in our body operate through *negative feedback*. In the case of pituitary gland function, a negative feedback system is one in which secretion of a pituitary hormone to a level above the set point causes a decrease in secretion of that same pituitary hormone into the blood. In reproductive physiology, however, there are also important occurrences of *positive feedback*, in which the secretion of a pituitary hormone influences the controller center so that secretion of the hormone increases even more. We discuss now in some detail the kinds of positive and negative feedback important in controlling secretion of the gonadotropins from the adenohypophysis.

Regulation of Gonadotropin Secretion by Negative Feedback

As discussed in Chapters 2–4, FSH and LH cause secretion of sex hormones by the *gonads* (testes or ovaries). These steroid hormones (androgens, estrogens, and progestogens) are products (effects) of the action of gonadotropins on the gonads, and it turns out that steroid hormones influence the secretion of LH and FSH by having feedback effects on the systems controlling gonadotropin secretion.

In women, the administration of moderate amounts of estrogen will lower the secretion of FSH and LH into the blood. This negative feedback effect of estrogen (Fig. 1-16) is even more effective when given in combination with high levels of a progestogen. In fact, this is the reason that combination contraceptive pills contain moderate levels of an estrogen and high levels of a progestogen (see Chapter 14). In the normal menstrual cycle, high levels of a progestogen and moderate levels of an estrogen in the blood during the luteal phase of the cycle (the period between ovulation and menstruation) lower gonadotropin secretion by negative feedback, thus preventing ovulation at this time (see Chapter 3).

It may help you to think that the hypothalamus contains a controller center with a *gonadostat* that works like the thermostat in your heating system. This gonadostat contains a set point for levels of estrogen and progestogen reaching it from the blood. When levels of these steroids are higher than the set point, the gonadostat signals the hypophysiotropic area to stop secreting GnRH, and thus secretion of FSH and LH from the adenohypophysis declines. If circulating levels of estrogen and progestogen are below the set point, the gonadostat signals the hypophysiotropic area to release more GnRH and circulating levels of FSH and LH rise. In males, a similar negative feedback of androgens also acts on the hypothalamus.

Because GnRH neurosecretory neurons probably do not have estrogen, progestogen, or androgen receptors, the negative feedback effects of these steroid hormones must act on regular neurons in the brain, which in turn inhibit GnRH cells. Some of the negative feedback effects of the sex steroids can also act directly on the FSH and LH pituitary cells by decreasing their sensitivity to GnRH.

In addition to secreting steroid hormones in response to FSH and LH, the gonads (testes and ovaries) also release glycoprotein hormones that influence gonadotropin secretion. *Inhibin* acts directly on pituitary cells to selectively suppress the secretion of FSH, but not LH. This compound is formed by the dimerization (molecular joining) of α and β subunits. *Activin* is a related molecule, composed of two β subunits of inhibin. It opposes the action of inhibin, stimulating the release of FSH. Activin is also synthesized in the pituitary and brain and may have local (paracrine) effects as well as behaving as a blood-borne hormone. Finally, *follistatin* binds to activin, thus blocking its action.

Positive Feedback

Thus far, we have been talking about negative feedback on pituitary FSH and LH secretion (see Figs. 1-15 and 1-16). There is, however, a stage in the human menstrual cycle when high levels of estrogen in the blood increase the secretion of LH and FSH from the adenohypophysis, resulting in a "surge" of these gonadotropins (primarily LH) in the blood and the subsequent release of an egg

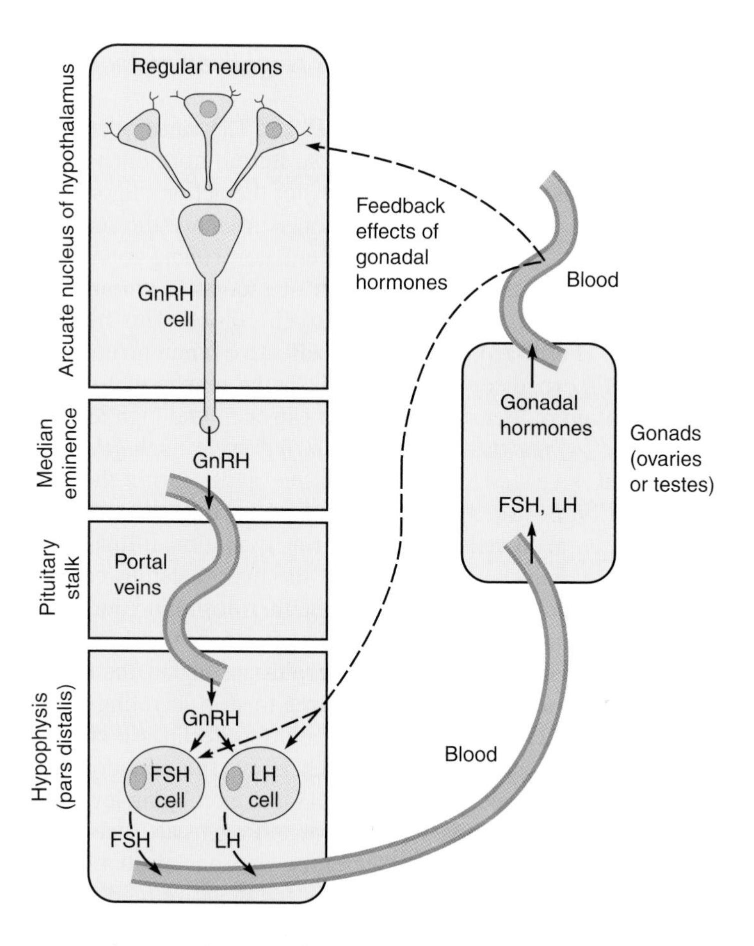

Figure 1-15 Schematic diagram of the control of the reproductive system by the brain and pituitary gland and the sites of feedback by gonadal hormones on this control.

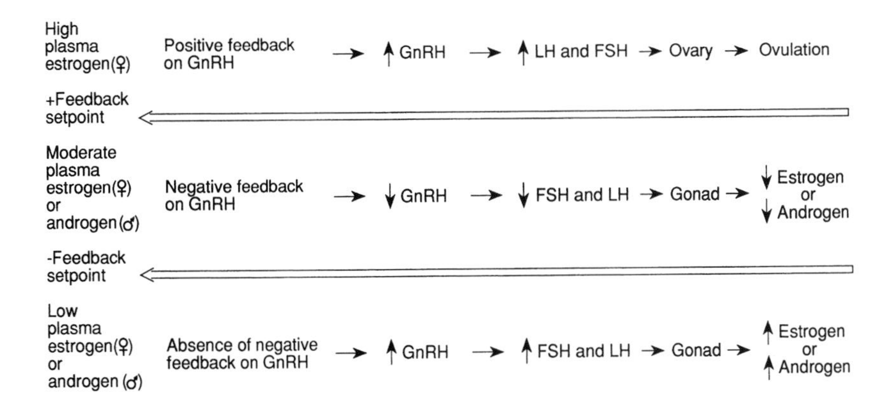

Figure 1-16 The actions of positive and negative feedback on GnRH and, therefore, gonadotropin (FSH and LH) secretion in females and males.

from the ovary (see Chapter 3). It is important to understand that the negative and positive feedback effects of estrogen have separate and different set points. Therefore, high levels of estrogen in the blood (above the positive-feedback set point) have a positive feedback effect on gonadotropin secretion, whereas moderate levels of estrogen (but still above the set point for negative feedback) have a negative effect on gonadotropin secretion (Fig. 1-16). The positive feedback effect of high levels of estrogen operates by stimulating neural activity in the surge center, which then increases GnRH secretion.

The positive and negative feedback effects of gonadal steroids on LH and FSH secretion can operate directly on the pituitary itself as well as on the brain (Fig. 1-15), i.e., the response of FSH and LH secreting cells in the adenohypophysis to GnRH may vary depending on the kinds and amounts of steroid hormones bathing these cells. For example, it has been shown that high estrogen levels increase the sensitivity of the pituitary to GnRH. Progestogens can also increase pituitary sensitivity to GnRH in women. In males, androgens have a negative feedback on gonadotropin secretion not only by influencing the brain, but also by decreasing the response of the adenohypophysis to GnRH.

Control of Prolactin Secretion

The control of PRL secretion by the brain differs in some respects from brain control of LH and FSH secretion. If the hypophysiotropic area of the hypothalamus is destroyed, secretion of PRL from the adenohypophysis increases, whereas secretion of LH and FSH declines. Therefore, there is a *prolactin release-inhibiting hormone* (PRIH) secreted by neurosecretory neurons in the hypophysiotropic area that inhibits prolactin secretion. PRIH may be dopamine or GABA. However, because electrical stimulation of the hypothalamus can increase prolactin secretion, it appears that there is also a PRH. In reality, PRH may be the same neurohormone as TRH, which stimulates thyrotropin secretion, or it may be *vasoactive-intestinal peptide*, a common neurotransmitter in the nervous system. Finally, estrogens increase the response of prolactin-secreting cells in the adenohypophysis to PRH. Knowledge of the hypothalamic control of prolactin secretion is important because of the role of this hormone in milk synthesis by the mammary glands (see Chapter 12) and the association of abnormally high levels of PRL with certain kinds of infertility (see Chapter 16).

Chapter Summary

Exocrine glands secrete their products directly into ducts, whereas endocrine glands release their products (hormones) into the bloodstream. Specific hormones influence the growth and function of certain target tissues. Hormones can be proteins or smaller polypeptides, amines, steroids, or fatty acid derivatives. Methods used in the science of endocrinology include bioassay, radioimmunoassay, nonradioactive methods such as ELISA, and molecular biological techniques. Paracrines are local chemical messengers that are not transported in the blood.

The hypophysis has two major parts: the neurohypophysis and the adenohypophysis. Neurosecretory neurons in the hypothalamus synthesize oxytocin and vasopressin, which travel to the neurohypophysis in neurosecretory cell axons. The adenohypophysis contains three regions: the pars distalis, pars tuberalis, and pars intermedia (reduced or absent in humans). Different cells in the pars distalis secrete the hormones follicle-stimulating hormone, luteinizing hormone, prolactin, corticotropin, growth hormone, thyrotropin, lipotropin, endorphins, and enkephalins. Other pituitary cells secrete melanophore-stimulating hormone, and the pars tuberalis could also secrete FSH and LH.

Neurosecretory neurons in the hypophysiotropic area of the hypothalamus secrete releasing hormones or release-inhibiting hormones into the median eminence region at the base of the hypothalamus. Here, capillaries receive these hormones, which then travel in the blood of the hypothalamo–hypophysial portal system to the endocrine cells of the adenohypophysis. The releasing hormones then increase the secretion of specific adenohypophysial hormones, whereas the release-inhibiting hormones have the opposite effect.

Because luteinizing hormone-releasing hormone increases the secretion of both FSH and LH, it is also called gonadotropin-releasing hormone. Evidence suggests that GnRH plays an important role in human reproduction; this molecule and GnRH analogs are being used to treat infertility and as possible contraceptive agents. The surge center of the hypothalamus causes a surge of LH secretion just before ovulation by increasing GnRH secretion from the HTA. The pineal gland secretes the hormone melatonin, which exerts inhibitory effects on gonadotropin secretion.

Feedback systems control FSH and LH secretion from the adenohypophysis. FSH and LH cause the gonads to secrete gonadal hormones (estrogens, progestogens, androgens, glycoproteins), which can decrease (by negative feedback) further secretion of FSH and LH. Estrogen can also have a positive feedback effect on LH secretion in women. The feedback effects can operate on the surge center, HTA, or adenohypophysis. Prolactin secretion from the pars distalis is controlled by both a prolactin-releasing hormone and a prolactin release-inhibiting hormone from the hypothalamus. Estrogens increase the sensitivity of prolactin-secreting cells in the adenohypophysis to PRH.

Further Reading

Gordon, K., and Hodgen, G. D. (1992). Evolving role of gonadotropin-releasing hormone antagonists. *Trends Endocrinol. Metab.* **3**, 259–263.

Petit, C. (1993). Molecular basis of the X-chromosome-linked Kallmann's syndrome. *Trends Endocrinol. Metab.* **4**, 8–13.

Pollard, J. W. (1999). Modifiers of estrogen action. *Science and Medicine* July/August 1999, 38–47.

Advanced Reading

Adlerkreutz, H., and Mazur, W. (1997). Phyto-estrogens and Western diseases. *Ann. Med.* **29**, 95–120.

Caprio, M., *et al.* (2002). Leptin in reproduction. *Trends Endocrinol. Metabol.* **12**, 65–72.

Gharib, S. D., *et al.* (1990). Molecular biology of the pituitary gonadotropins. *Endocr. Rev.* **11**, 177–199.

Halasz, B., *et al.* (1989). Regulation of the gonadotropin-releasing hormone (GnRH) neuronal system: Morphological aspects. *J. Steroid Biochem.* **33**(4B), 663–668.

Halasz, B., *et al.* (1988). Neural control of ovulation. *Hum. Reprod.* **3**, 33–37.

Hodgen, G. D. (1989). Neuroendocrinology of the normal menstrual cycle. *J. Reprod. Med.* **34**, 68–75.

Kalra, S. P. (1993). Mandatory neuropeptide-steroid signaling for the preovulatory luteinizing hormone-releasing hormone discharge. *Endocr. Rev.* **14**, 507–538.

McDonnell, D. P. (1999). The molecular pharmacology of SERMs. *Trends Endocrinol. Metab.* **10**, 301–311.

Nillsson, S., *et al.* (1998). ERß: A novel estrogen receptor offers the potential for new drug development. *Trends Endocrinol. Metab.* **9**, 387–395.

Ruh, M. F., *et al.* (1997). Failure of cannabinoid compounds to stimulate estrogen receptors. *Biochem. Pharmacol.* **53**, 35–41.

Schwanzel-Fukuda, M., *et al.* (1992). Biology of luteinizing hormone-releasing hormone neurons during and after their migration from the olfactory placode. *Endocr. Rev.* **13**, 623–634.

Stojilkovic, S. S., *et al.* (1994). Gonadotropin-releasing hormone neurons: Intrinsic pulsatility and receptor-mediated regulation. *Trends Endocrinol. Metab.* **5**, 201–209.

CHAPTER **TWO**

The Female Reproductive System

Introduction

The female reproductive system consists of the paired ovaries and oviducts, the uterus, the vagina, the external genitalia, and the mammary glands. All of these structures have evolved for the primary functions of ovulation, fertilization of an ovum by a sperm, and the birth and care of a newborn. The components of this system are integrated structurally and physiologically to these ends.

The anatomical features that distinguish females from males are *female sexual characteristics*. In standard terminology, the female *primary sexual characteristics* are the internal structures of the reproductive system, including the ovaries and the female *sex accessory ducts* (the oviducts, uterus and vagina), as well as the external genitalia (Fig. 2-1). Female *secondary sexual characteristics* include all those external features (except external genitalia) that distinguish an adult female from an adult male. These include enlarged breasts and the characteristic distribution of fat in the torso.

This chapter looks at the anatomy, endocrinology, and disorders of the adult female reproductive system. The menstrual cycle is covered in Chapter 3.

Ovaries

Ovarian Gross Anatomy

The *ovary*, or female gonad, serves two essential functions in female reproduction: development of the female gametes (*oocytes*, or eggs) and the synthesis and release of steroid hormones. The ovaries are paired structures lying on each side of the upper pelvic cavity, against the back of the pelvic wall and near the uterus (Figs. 2-1 and 2-2). These small almond-shaped organs are white or yellowish in color and have a lumpy surface. The mature ovary measures about 2.5 to 5 cm (1–2 in.) in length and 1.5 to 3 cm (0.5–1 in.) wide. The ovaries are innervated by autonomic nerves and receive an especially rich blood supply. They are connected to the uterus and pelvic wall by supportive ligaments.

Ovarian Microanatomy

If a slice is cut out of the ovary and examined under a microscope, it is clear that the ovary is a very complex structure (Fig. 2-3). The external surface of each ovary

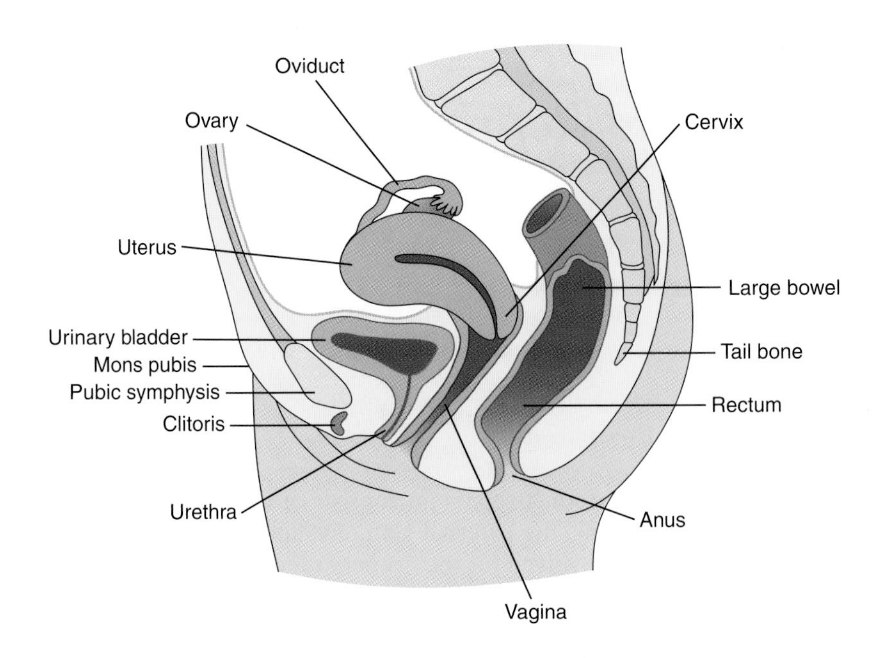

Figure 2-1 Side view of the female pelvic region showing some major components of the reproductive system.

is covered with a thin sheet of tissue, the *surface epithelium*. This epithelium once was called the "germinal epithelium" because it was thought that female germ cells were derived from this tissue. Now it is known that germ cells originate outside of the ovary during embryonic development, as discussed in Chapter 5. Underneath the surface epithelium is a connective tissue framework (the *ovarian stroma*), which is divided into a more dense, outer *ovarian cortex* and a less dense, central *ovarian medulla*. The ovarian medulla contains large, spirally arranged blood vessels, lymphatic vessels, and nerves. The ovarian cortex contains the female germ cells. Steroid-producing *interstitial cells* are embedded in the ovarian stroma.

Each female germ cell, or *oocyte*, is enclosed in a tissue sac, the *ovarian follicle* (*follicle* is Latin for "little bag"). Between the oocyte and the follicular wall is a thin transparent membrane, the *zona pellucida*, which is secreted by the oocyte. We now look at the stages of growth of an individual ovarian follicle.

Stages of Follicular Growth

Most follicles within the adult ovary are very small (50 μm in diameter). These *primordial follicles* consist of an oocyte surrounded by a single layer of approximately 15 flattened *granulosa cells*, the *membrana granulosa*. The primordial follicles lie in the periphery of the ovarian cortex (Fig. 2-3). Relatively few primordial follicles initiate growth at any given time. Those that do grow become *primary follicles*, which are about 100 μm in diameter. This initial growth of the follicle is a consequence of a slight increase in the size of the oocyte, as well as growth of the granulosa layer.

The granulosa is still a single layer of cells in primary follicles, but these cells are now cube shaped instead of flattened (Fig. 2-3). In addition, primary follicles

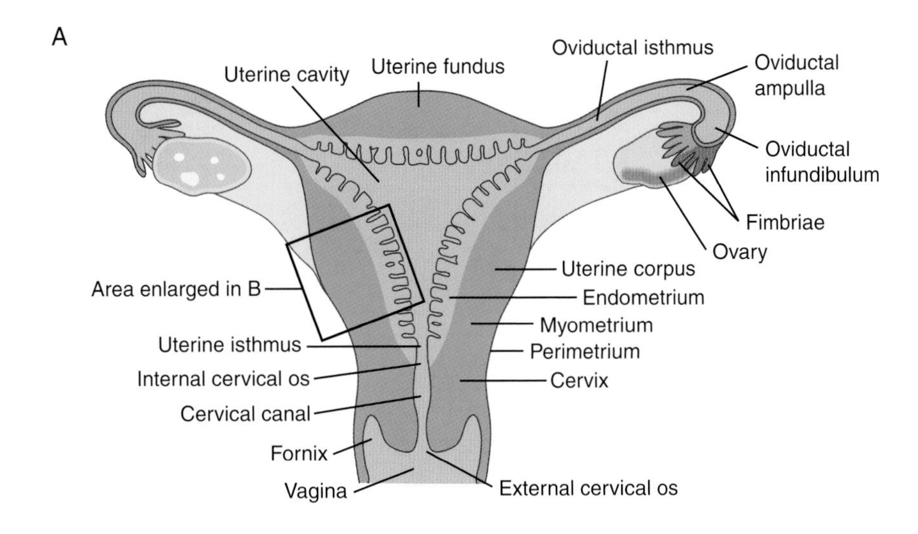

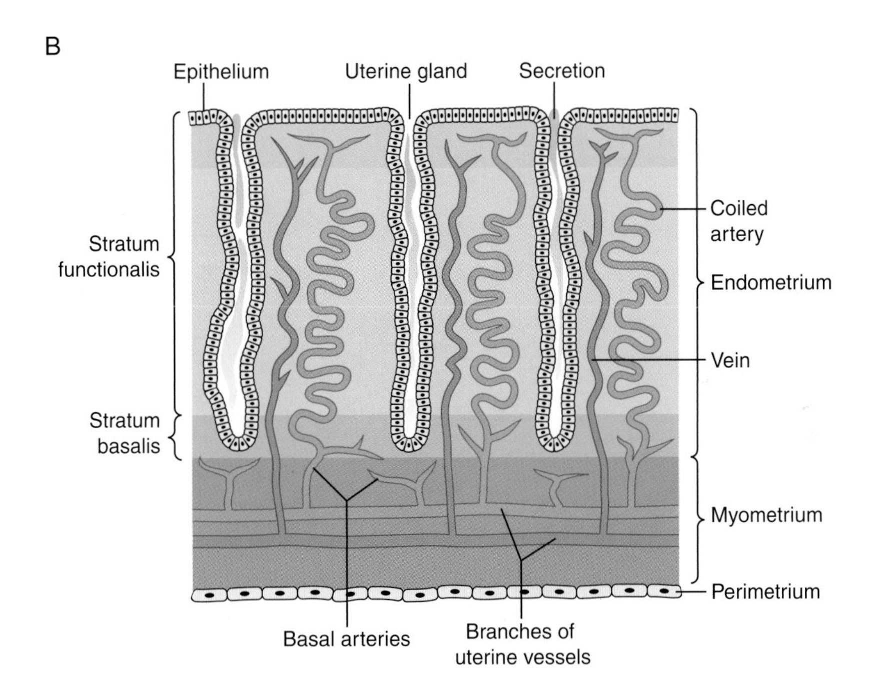

Figure 2-2 (a) A diagrammatic section through the uterus and oviducts. (b) An enlargement of the box in (a) showing the structure of the uterine wall.

acquire a connective tissue covering around the granulosa. This covering, which contains small blood vessels, is the *theca*. As a primary follicle continues to grow, the granulosa cells undergo mitosis and thus *secondary follicles* have a membrana granulosa consisting of two to six cell layers (Fig. 2-3). The theca is still a single layer in secondary follicles. Growth of the follicle to this stage is a relatively slow process, lasting about 4 months.

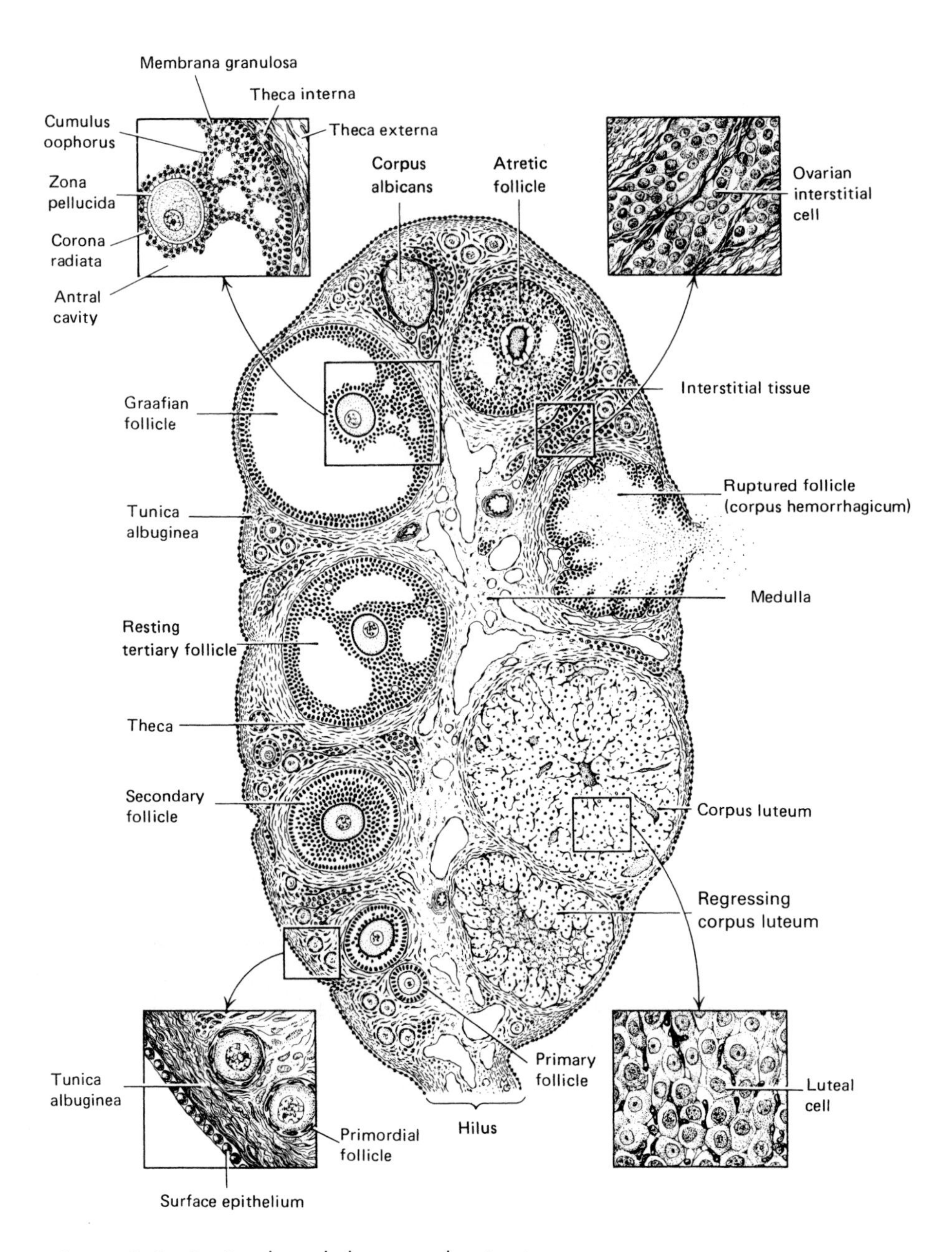

Figure 2-3 Section through the ovary showing its component parts.

A few secondary follicles that avoid becoming atretic grow to a more advanced stage to become *tertiary follicles* (Fig. 2-3). This development takes place over a 2- to 3-month interval. First, the granulosa cells secrete fluid that accumulates between the cells (in follicles about 200 μm in diameter). These fluid spaces then join and a great deal of additional fluid leaves pores in the thecal blood vessels and is added to the fluid space. This fluid-filled space is the *antral cavity*, and the fluid is called *antral fluid* (*follicular fluid*). About 80% of the proteins present in the plasma of the blood in thecal vessels can leave the vessels and diffuse through the follicular wall and into the antral cavity. Antral fluid contains steroid and protein hormones, anticoagulants (substances that prevent blood clotting), enzymes, and electrolytes (ions with positive or negative charges).

Tertiary follicles have a membrana granulosa of several cell layers, and the theca now is differentiated into an inner *theca interna*, containing glandular cells and many small blood vessels, and a *theca externa*, consisting of dense connective tissue and larger blood vessels (Fig. 2-3). The oocyte, accompanied by a mass of granulosa cells called the *cumulus oophorus*, bulges into the antral fluid. A sphere of cumulus cells will remain surrounding the oocyte during ovulation and will accompany the oocyte as it enters the oviduct. Penetration of the cloud of cumulus cells will prove to be one of the obstacles encountered by sperm cells prior to fertilization (Chapter 9). Figure 2-4 shows such an oocyte.

Tertiary follicles can be divided further into categories based on size. Resting tertiary follicles are 1 to 9 mm in diameter, ripe tertiary follicles are about 10 to 14 mm in diameter, and Graafian follicles are about 15 to 25 mm in diameter.

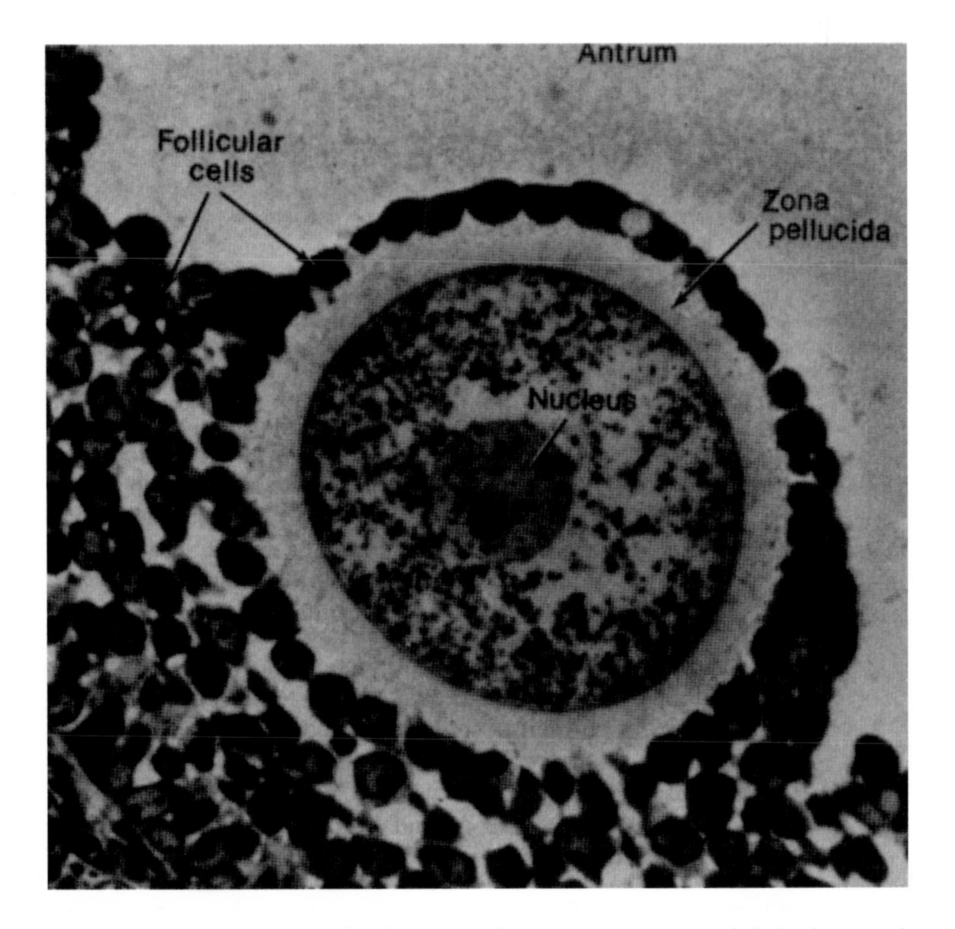

Figure 2-4 Photomicrograph of a region from a human tertiary follicle showing the oocyte surrounded by follicular (granulosa) cells. The lower left corner shows some cells of the theca.

Tertiary follicles continue to grow over a period of approximately 3 weeks and can be divided further into categories based on size and stage of development. *Early antral follicles* have begun to accumulate antral fluid, but the spaces have not yet coalesced into a single antral cavity. *Antral follicles* have a single antrum; the largest and most mature of these follicles are sometimes called *Graafian follicles*. The theca of Graafian follicles is extremely vascular. Only one of these large follicles during each menstrual cycle will be "selected" for final maturation. This *preovulatory follicle* becomes so large that it forms a blister-like bulge on the surface of the ovary before it is ovulated. All other tertiary follicles become atretic and die.

Early in follicular development, granulosa cells extend thin cytoplasmic bridges to the oocyte. The two cell types are connected by *gap junctions*, through which small molecules can pass. It is thought that these may allow chemical communication between cells and allow follicle cells to pass nutrients to the growing oocyte. Similar connections exist between adjacent granulosa cells, and the cells of the follicle remain thus joined throughout its development.

Follicular Atresia

Each ovary contains many follicles, but the number of follicles changes during a female's life. At birth, an infant girl has about two million ovarian follicles in both ovaries. This number is all she will have ever have in her lifetime; no new follicles are formed after birth (although recent evidence questions this long-held notion; see Chapter 7). In fact, over the next 50 or so years the size of her population of follicles will decline steadily. By puberty, only about 200,000 will remain. By age 35 she will have fewer than 100,000 follicles and, by menopause, her follicular supply will be nearly depleted. Typically, a woman ovulates one oocyte per month over an approximately 40-year reproductive life span. This accounts for the loss of 400–500 follicles. What explains the disappearance of the others?

The vast majority of all follicles degenerates and dies in a process known as *atresia*. This is a type of cell death called *apoptosis*, or *programmed cell death*, that plays a role in the selective elimination of many kinds of cells in the body. It is characterized by specific changes in nuclei followed by cell fragmentation. The cellular remains of the granulosa cells and oocyte are removed by leukocytes (white blood cell scavengers), but atretic follicles leave tiny scars in the ovary. The death of a follicle can occur at any stage of follicular growth. Atresia continues throughout a woman's reproductive life, but peaks occur prior to puberty, before menopause, and during certain stages of the menstrual cycle. This vast overproduction of female germ cells followed by a protracted massive die-off is typical of nonhuman mammals as well, although its significance is unclear.

Ovarian Steroid Hormone Synthesis

Ovarian follicles synthesize and secrete steroid hormones that play essential roles in the female. All three major classes of reproductive steroid hormones—estrogens, progestogens, and androgens—are produced by follicles. Before discussing follicular secretion of these hormones, however, we first must see how steroid hormones are synthesized. *Cholesterol* is the precursor for all steroid hormones secreted by the ovary. You have heard about the harmful effects of cholesterol in the body, but this is an example of an important role of this substance. Cholesterol is converted by an enzyme to *pregnenolone*, another precursor steroid. *Steroidogenesis* (biosynthesis of steroids) within the ovary then can follow one of two pathways (Fig. 2-5). In the Δ^5 pathway, which occurs predominantly in large tertiary follicles, pregnenolone is converted to 17-hydroxypregnenolone, then to a weak androgen, *dehydroepiandrosterone* (DHEA), and then to another weak androgen, *androstenedione*. A weak androgen is one that is not very potent in stimulating male tissues. In the second (Δ^4) pathway, pregnenolone is converted to *progesterone*, a progestogen that not only serves as a precursor for other steroids, but enters the female's blood and acts as a hormone on such target tissues as the uterus and mammary glands. Within the ovary, progesterone is converted to 17-hydroxyprogesterone, which then is changed to androstenedione. This Δ^4 pathway is predominant in the corpus luteum, which as we see later is an endocrine gland formed from the follicle after it ovulates.

From here on, both the Δ^5 and the Δ^4 pathways can produce estrogens (Fig. 2-5). Androstenedione is converted to *testosterone* (a potent androgen). Testosterone then can be converted to *estradiol-17β*, the major estrogenic hormone secreted by ovarian follicles and the corpus luteum. (Note: estradiol-17β will simply be called "estradiol" throughout this book.) The enzyme that converts testosterone to estradiol is *aromatase*. Two other estrogens, *estriol* and *estrone*, also can be synthesized in both pathways.

In summary, the ovarian follicles, through the Δ^5 pathway, produce and secrete estradiol and a trace of the other two estrogens. In contrast, the corpus luteum synthesizes and secretes both progesterone and estradiol through the Δ^4 pathway. It is interesting to note that, in either pathway, androgens are precursors for estrogens in the female. In the testis, as shown in Chapter 4, steroidogenesis favors synthesis and secretion of androgens; very little estrogen is produced. Figure 2-6 illustrates the molecular structure of some of these steroid hormones.

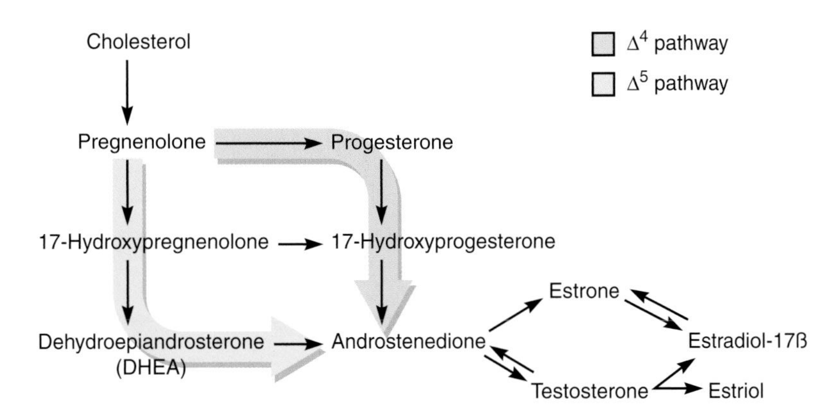

Figure 2-5 Steroidogenesis within the ovary. Conversion of pregnenolone to DHEA, then further synthesis of other androgens and estrogens, is the Δ^5 pathway, predominant in ovarian follicles. Conversion of pregnenolone to progesterone and then to androgens and estrogens is the Δ^4 pathway, predominant in the corpus luteum. Specific enzymes are required for each arrow. Double arrows mean that conversion can go in either direction.

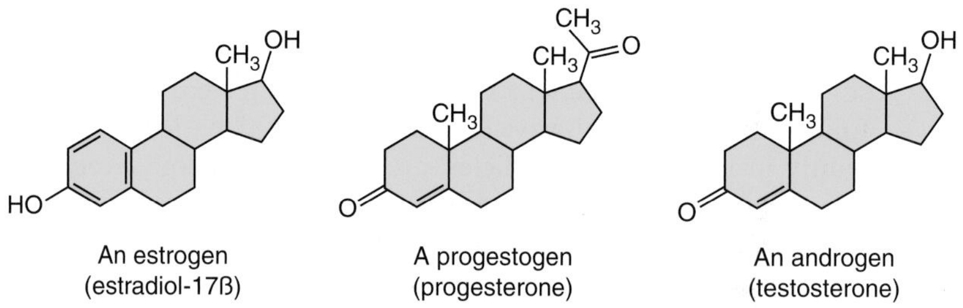

An estrogen (estradiol-17β)

A progestogen (progesterone)

An androgen (testosterone)

Figure 2-6 Chemical structures of three steroid hormones. Note that although these molecules are all derived from cholesterol and differ only slightly in structure, they have quite different actions on tissues.

The theca interna is capable of synthesizing some steroid hormones, but not until the follicle has grown to the late secondary or early tertiary stage. Beginning at this time, the theca interna glandular cells synthesize androstenedione, which then diffuses into the membrana granulosa. The granulosa cells convert this weak androgen to testosterone and then convert the testosterone to estradiol (Fig. 2-7). The estradiol synthesized in granulosa cells diffuses into the

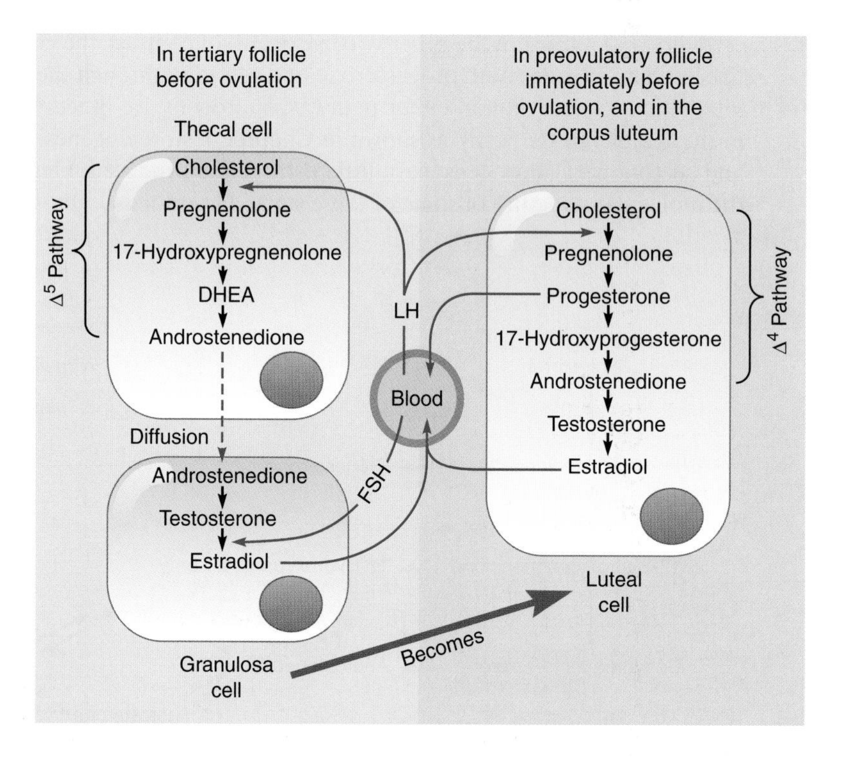

Figure 2-7 Steroidogenesis in the tertiary and preovulatory follicle and in the corpus luteum after ovulation. Note that the tertiary follicle follows the Δ^5 pathway, producing mostly estradiol in the blood. In contrast, the preovulatory follicle (i.e., a Graafian follicle after the LH surge but before ovulation) and the corpus luteum produce a large amount of progesterone and some estradiol, and thus follow the Δ^4 pathway of steroidogenesis.

theca, where it enters blood vessels that transport it to target tissues in other parts of the body. Thus, estrogen synthesis in the follicle depends on the coordinated biochemical activity of two types of cells: thecal cells and granulosa cells. This is known as the *two-cell model* of ovarian steroidogenesis (Fig. 2-7). To summarize, cholesterol precursor molecules are removed from the bloodstream by thecal cells. The metabolic machinery of these cells is specialized to synthesize androgens (primarily androstenedione) by the Δ^5 pathway. Androgens produced by thecal cells diffuse to the granulosa. Subsequently, granulosa cells, which cannot produce their own androgens, take up this androstenedione and use it as a precursor for conversion to estrogens. Each step in the steroidogenic pathway is catalyzed by a specific enzyme. Granulosa and thecal cells differ in their steroidogenic enzyme activity, and their cooperation is necessary for normal steroid production by the ovary.

Not all of the androgens produced by the follicle are converted to estrogens. Low levels of androgens are present in a woman's blood. Weak androgens are predominant; most of these come from the adrenal glands (see Chapter 6). Additional androgens, including testosterone, are secreted from growing follicles and from interstitial cells that originate from the thecal layer of atretic follicles, which does not degenerate with the rest of the follicle. Nonsteroidal hormones produced by the ovary include inhibin, activin, and follistatin.

Hormonal Control of Follicular Growth and Steroidogenesis

Follicular growth and steroid hormone secretion are controlled by two pituitary gonadotropins—follicle-stimulating hormone (FSH) and luteinizing hormone (LH). Remember from Chapter 1 that those cells that are targets for peptide or protein hormones have specific hormone receptor molecules on their cell membranes. The stimulus that induces primordial follicles to initiate growth is not known and is the subject of much study and speculation. The stimulus probably comes from the ovary itself in the form of a local paracrine factor such as a growth factor. Growth of the follicle to the antral stage may be dependent on at least basal levels of gonadotropins; low levels of FSH promote growth of primary and secondary follicles by stimulating the division of granulosa cells. The rapid growth of tertiary follicles is FSH dependent. FSH also stimulates estrogen synthesis by the granulosa cells of tertiary follicles by inducing the production of aromatase, the enzyme that converts testosterone to estrogen.

Thecal cells possess LH receptors, and LH promotes the production of androstenedione in late secondary and tertiary follicles. Thus, in growing follicles, thecal and granulosa cells are dependent on LH and FSH, respectively, for steroid production. As tertiary follicles grow, they release greater quantities of steroid hormones. In fact, the single dominant tertiary follicle produces most of the circulating estrogen. However, just before ovulation, hormone synthesis by the ovary changes. At this time, granulosa cells of the dominant follicle acquire LH receptors, and LH causes these granulosa cells to synthesize a new set of steroidal enzymes. These cells now switch to the Δ^4 pathway so that they begin to secrete progesterone. This transformation is called *luteinization* (Fig. 2-7).

Oocyte Maturation and Ovulation

In the human menstrual cycle, about 20 large tertiary follicles form in both ovaries several days before ovulation. However, only one of these in one ovary actually ovulates. Therefore, about 19 tertiary follicles in both ovaries grow large and then undergo atresia. What causes this atresia is not clear, but one theory argues that the favored follicle is the only one that accumulates high amounts of FSH in its antral fluid. Meanwhile, estrogen secretion increases and inhibits FSH secretion from the pituitary through negative feedback (see Chapter 3). Thus, the other tertiary follicles become deprived of FSH, leading to their degeneration. A surprising new discovery is that ovarian follicles contain gonadotropin-releasing hormone (GnRH) and that this substance inhibits follicular function. Whether or not ovarian GnRH plays a role in follicular atresia, however, is not clear.

Before discussing the process of *ovulation* (extrusion of an oocyte from its follicle), we first must digress for a moment and review a special kind of cell division found in germ cells called *meiosis*. The nucleus of most human cells contains 46 chromosomes, or 23 pairs of homologous chromosomes. This is called the *diploid* condition and is symbolized by "2N." When germ cells first appear in the ovaries of female fetuses (see Chapter 5), they begin the process of meiosis that ultimately produces cells that have only 23 chromosomes, the *haploid* or "N" condition. When a haploid ovum fuses with a haploid sperm, the diploid ("2N") number of chromosomes is restored (see Chapter 9).

The process of *oogenesis* results in the production of mature female *gametes* ("sex cells"). At birth, the ovaries of a female infant contain *primary oocytes* in primordial follicles. These primary oocytes are arrested in the *first meiotic division*, and this *first meiotic arrest* is maintained in the oocytes of follicles of all stages except the most mature. In the menstrual cycle, only one follicle reaches the large Graafian stage every month. Then, as shown in Chapter 3, a surge of LH secretion from the pituitary gland occurs. The LH acts on follicle cells, which trigger the resumption of meiosis (*meiotic maturation*) in the oocyte.

As a first step in the reinitiation of meiosis, the membrane of the oocyte nucleus (*germinal vesicle*) disintegrates, a process called *germinal vesicle breakdown*. Then, the first meiotic division is completed. This is the *reduction division* of meiosis and produces two haploid daughter cells, still within the follicle (Fig. 2-8). Actually, the reduction division is unequal and produces a large haploid *secondary oocyte* (about 190 μm in diameter) and a tiny haploid *first polar body*. This polar body can divide again or remain single; in either case, it degenerates. The secondary oocyte then begins the *second meiotic division*, but this division is again arrested while the oocyte is still within the follicle; this is the *second meiotic arrest*. As discussed in Chapter 9, the second meiotic division is not completed (resulting in a haploid *ootid* and *second polar body*) until after sperm penetration of the ovum, which occurs in the oviduct. Thus, the female germ cell is ovulated as a secondary oocyte (*ovum*, or *egg*) in the second meiotic arrest. The process of oocyte meiosis is summarized in Fig. 2-8.

In humans, germinal vesicle breakdown occurs about 9–12 h after the beginning of the LH surge during the menstrual cycle. Much research has been conducted on humans and other mammals to discover how LH causes resumption of the first meiotic division. Because the oocyte lacks LH receptors, the signal to resume meiosis must begin in the granulosa cells and somehow be passed to the oocyte. What is this signal? One area of investigation focuses on the role of an

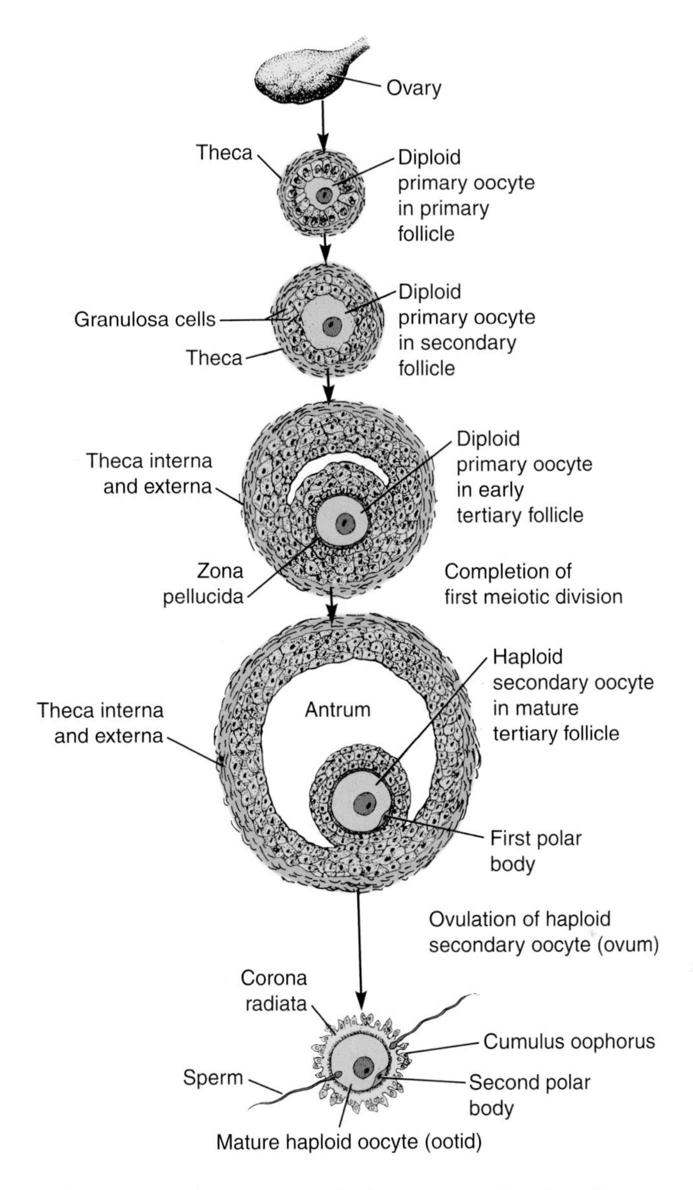

Figure 2-8 The process of oogenesis, which is not completed until a sperm penetrates an ovulated egg.

intracellular messenger molecule, cyclic AMP (cAMP). A certain threshold level of cAMP in the oocyte cytoplasm is needed to maintain meiotic arrest. Thus, cAMP may serve to inhibit the progress of meiosis during the oocyte's long period of stasis in meiosis I. However, the oocyte itself has very limited capability to synthesize cAMP. What is the major source of cAMP within the oocyte? It is thought to be synthesized in granulosa cells and transported to the oocyte through the granulosa/oocyte gap junctions. One of the consequences of the LH surge is the loss of these gap junctions, which would interrupt the flow of cAMP to the oocyte.

Another possible informational molecule transmitting the LH signal from granulosa cells to oocyte is calcium. Evidence from studies of some mammals suggests that LH induces calcium levels in the oocyte to rise, triggering the events of meiotic maturation. Other candidates for signaling molecules have also been proposed. For example, evidence suggests that LH induces meiotic maturation in the oocyte and breaking of the cumulus oophorus from the follicle wall by inducing the granulosa cells in the wall of the follicle to secrete *epidermal growth factors* (EGFs). These growth factors then cause follicle maturation through paracrine action.

Regardless of the molecular signal(s) transmitting the message from LH receptors on granulosa cells to the interior of the oocyte, the actual events of meiotic maturation are controlled by a substance called *maturation-promoting factor* (MPF). Actually, MPF controls both mitosis and meiosis in cells of a wide range of organisms. It is made of two subunits. When the subunits are brought together and the resulting complex activated, MPF causes the germinal vesicle (oocyte nuclear membrane) to break down and the chromosomes to divide.

Now that we have seen how an oocyte begins to mature just before ovulation, how does it escape from the follicle? The general theory is as follows: Just before ovulation, a small, pale (avascular) region, the *stigma*, appears on the follicular surface. In this region, the surface epithelium and thecal layers of the follicle become thinner and dissociated, and the follicular wall exhibits a reduction in tensile ("breaking") strength. Also, the membrana granulosa degenerates in this region. This thinning and weakening of the follicular wall at the stigma appear to be caused by estrogen-induced stimulation of the production of an enzyme (*collagenase*) from the connective tissue cells in this area. Breakdown products of the destroyed connective tissue induce an inflammatory response, with migration of white blood cells and secretion of prostaglandins in this region.

The *prostaglandins* (PG) probably facilitate ovulation by constricting blood vessels and reducing blood supply to the degenerating tissue. Administration of inhibitors of prostaglandin synthesis is known to block ovulation. What are prostaglandins? In 1930, two New York gynecologists discovered that human semen (ejaculate) contains a substance that causes contraction of the human uterus. Shortly thereafter, it was discovered that the active substance in semen was soluble in fat, and it was named *prostaglandin* because it was thought to be secreted by the prostate gland into the semen. We now know that the seminal vesicles (see Chapter 4), not the prostate, are the major source of prostaglandins in semen, and prostaglandins are also produced in almost every tissue of the body.

Prostaglandins are a family of molecules derived from fatty acids. They all contain 20 carbon atoms. The most important kinds of prostaglandins are grouped into three categories—prostaglandins A, F, and E. As discussed later in this book, prostaglandins play several important roles in human reproduction in addition to ovulation. The 1982 Nobel Prize in Physiology and Medicine was awarded to S. K. Bergström and B. I. Samuelesson of Sweden and J. R. Vane of England for their research on prostaglandins.

After the follicular wall has thinned, the pressure within the antral cavity causes the stigma to form a "cone" and then to tear. Contractile, smooth muscle-like cells are found in the follicular wall, but it is not clear what role they play in ovulation. The oocyte, which has become detached from the membrana granulosa and now is floating freely with its cumulus oophorus in the follicular fluid, then oozes out with the escaping fluid through the tear in the follicular wall (Fig. 2-9). Thus, ovulation has occurred.

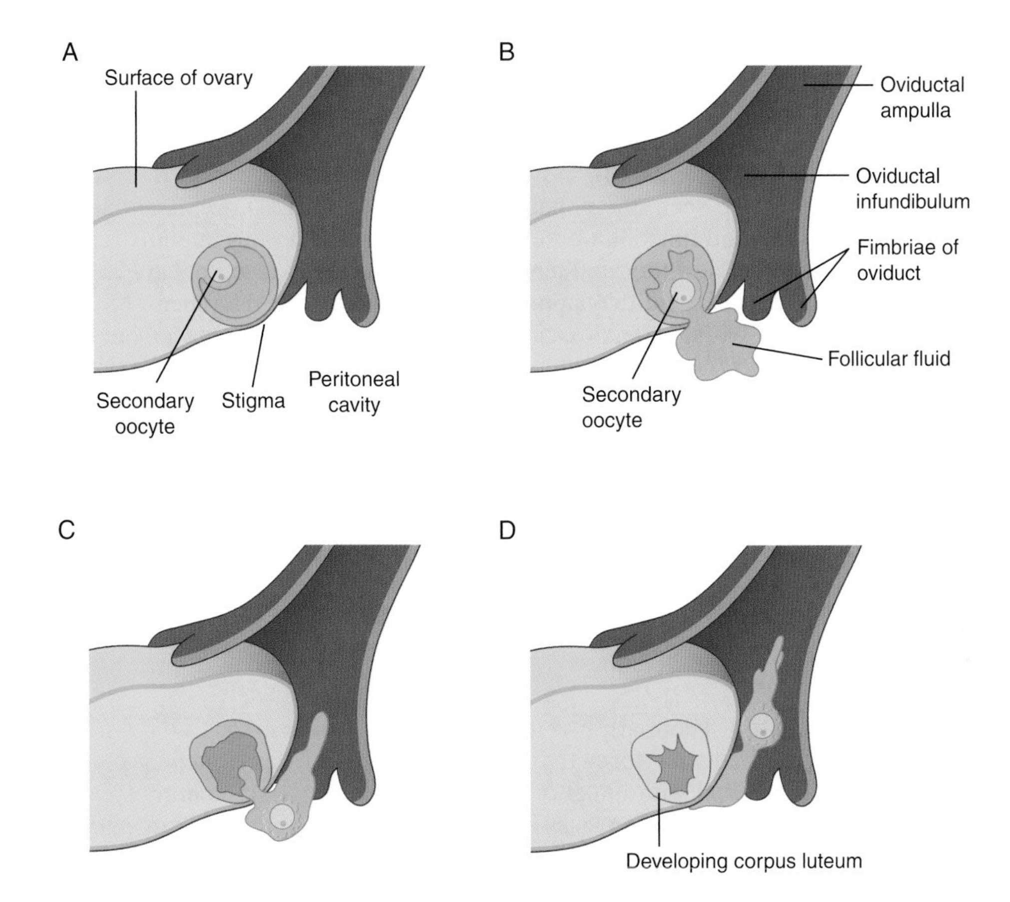

Figure 2-9 Diagrams illustrating ovulation. The stigma (a) ruptures (b) and the ovum is expelled along with follicular (antral) fluid (b and c). The ovulated ovum is arrested in the second meiotic division (d) until it is penetrated by a sperm.

Corpus Luteum

Once ovulation has occurred, the follicular wall remains as a collapsed sac. It then is called the *corpus hemorrhagicum* because a blood clot, derived from the torn blood vessels of the stigma, appears on its surface. The luteinized granulosa cells (*luteal cells*) in this collapsed follicle begin to divide and invade the old antral cavity (Fig. 2-3), thus forming the *corpus luteum* (Latin for "yellow body"). It is yellow because of the presence of pigment in the luteal cells. Then, blood vessels from the thecal layer grow and penetrate the central luteal cell mass.

Cells of the corpus luteum secrete high levels of progesterone and moderate amounts of estradiol; thus, luteal cells follow the Δ^4 steroidogenic pathway. Secretion of both hormones from the corpus luteum requires LH (Fig. 2-7). The pituitary hormone prolactin, along with LH, is essential for corpus luteum function in laboratory mammals, but in women it is only LH that maintains the function of the corpus luteum. As shown in Chapter 3, the corpus luteum forms in the second half of the menstrual cycle, reaches a maximum diameter of 10 to 20 mm, and then degenerates before menstruation. The degenerated corpus luteum fills with connective tissue and then is called a *corpus albicans* (Latin for "white body"). Chapter 10 discusses how estradiol and progesterone prime the uterus for establishing pregnancy if the ovum is fertilized. If pregnancy occurs, the corpus luteum does not die but instead survives to function in the first trimester of pregnancy (see Chapter 10).

As we have seen, the gonadotropins (FSH and LH) are essential for the development and function of the ovarian follicle and corpus luteum. Let us take a moment to review their influences on the ovary. FSH receptors are found only on granulosa cells, and they appear early in follicular development. FSH stimulates growth of the granulosa layer by mitosis and formation of the antrum. Thus, most of follicular growth is FSH dependent. LH receptors are present on early thecal cells and support growth and differentiation of the theca. Shortly before ovulation, LH receptors also appear on granulosa cells; the action of LH on the granulosa leads to luteinization of the granulosa cells, final growth and maturation of the follicle, and ovulation. Finally, the two gonadotropins (FSH and LH) regulate the two types of follicle cells (granulosa and theca) in their cooperative production of steroids (Fig. 2-7). Table 2-1 summarizes the hormonal control of ovarian function.

Ovarian Disorders

Ovarian Cysts There are two major kinds of *ovarian cysts. Cystic follicles* are large fluid-filled sacs formed from unovulated follicles. *Luteinized cysts* are solid masses filled with luteal cells. Both kinds of cysts are common, and often they disappear spontaneously. In some cases, however, they can persist and secrete abnormal amounts of steroid hormones that interfere with fertility. These persistent cysts often must be removed surgically. Use of the combination birth control pill decreases the incidence of benign ovarian cysts (see Chapter 14).

A woman whose ovaries contain many small cysts is said to have *polycystic ovarian syndrome*. Luteinized cysts are common in women with *Stein–Leventhal syndrome*, a form of polycystic ovarian syndrome. In this syndrome, the ability of follicles to convert androgens to estrogens is affected so androgen levels in

Table 2-1 Summary of the Hormonal Control of Ovarian Function

Function	Gonadotropin control
Follicular growth	
Mitosis of granulosa cells	FSH
Formation of theca	LH
Formation of follicular fluid	FSH and LH
Follicular hormone secretion	
Production of androstenedione by theca	LH
Conversion of androstenedione to estradiol by granulosa	FSH
Progesterone production by granulosa (luteinization)	LH
Ovulation and corpus luteum	
Oocyte maturation	LH
Ovulation	LH
Corpus luteum formation and maintenance	LH

these women are high, whereas estrogen levels are low. Because of the lack of negative feedback of estrogen on gonadotropin secretion, high levels of FSH and LH occur in the blood. Associated physical symptoms include lack of ovulation, irregular menstruation, obesity, and growth of male-like body hair (*hirsutism*). Women with polycystic ovarian syndrome are also at high risk for the development of diabetes. The syndrome is fairly common, affecting about 5% of all women. It often persists throughout a woman's reproductive life. Women may be unaware of the problem until they seek medical help for infertility.

A rare ovarian condition involves the formation of *dermoid cysts*, which contain such tissues as hair, nails, and skin glands. One theory is that these cysts result from fertilization of an oocyte while it is still in the follicle; the dermoid cyst represents an embryo that partially develops within the follicle.

Ovarian Cancer Sometimes, body cells can multiply and form a lump or mass called a *tumor*, or *neoplasm*. If the cells in such a mass are normal, it is a *benign tumor*. Examples of benign tumors are the aforementioned luteinized cysts. If, however, the cells lose their ability to control their multiplication, the lump is a *cancer*. These cells usually are abnormal in structure and carry mutations in their genetic material. Some cancerous cells can remain in place and may offer little danger. If, however, cancerous cells break loose from the tumor and travel in the blood or lymphatic system to other parts of the body, a process called *metastasis*, the tumor is *malignant* and new cancers can appear elsewhere in the body. This soon will interfere with an important bodily function and can be fatal.

Ovarian cancer is fairly common and is a dangerous disease among women in the United States. About 25,000 new cases are diagnosed each year, and about 14,000 American women will die each year from this disease. Although any cell type in the ovary could potentially become cancerous, the majority of ovarian cancers (about 90%) result from abnormal epithelial cells on the surface of the ovary. This is thought to be a result of the frequent cell division that this cell layer experiences. When the follicle bursts at ovulation, the ovarian surface in this region ruptures. The tear in the ovarian wall is repaired rapidly by the proliferation of surface cells. This process of rupture and repair occurs every 4 weeks. Because DNA must be replicated before every cell division, the frequent replication increases the likelihood for copying errors (mutations) that can transform a normal cell into a cancer cell. In fact, evidence shows that the more ovulations a woman has experienced in her lifetime, the greater her risk of having ovarian cancer. What can limit the number of ovulations and thus protect against ovarian cancer? Ovulation is suppressed during pregnancy and nursing and while a woman takes oral contraceptives. Thus, women who have had several pregnancies, who have nursed their infants, and who have taken oral contraceptives are at reduced risk for ovarian cancer. Heredity accounts for a small factor (10%) in these cancers. Genetic abnormalities associated with the disease include mutations in the *BRCA1* and *BRCA2* genes (the so-called "breast cancer genes") and overactivity of the *HER-2/neu* gene (also implicated in breast cancer).

Unfortunately, there is no good diagnosis for early stages of ovarian cancer because it is usually asymptomatic. The cancer is often advanced when it is diagnosed (typically about 2 years after its onset) and the epithelial cells may have become detached from the ovary and spread to the surface of other internal organs. Thus, the prognosis for a woman with ovarian cancer is poor. Treatment consists of the surgical removal of all cancerous tissue in combination with chemotherapy. One of the newer chemotherapeutic drugs used to shrink ovarian tumors is *taxol*, a substance first isolated from the Pacific yew tree (*Taxus brevifolia*). It was approved in 1992 and has been used to treat more than 20,000 American women with ovarian cancer, but it took the bark (and the lives) of three 100-year-old trees to treat one patient! The needles of a common related plant, the English yew, were found also to contain taxol, and now taxol and related drugs are synthesized in the laboratory using material from this plant. These drugs are used to treat ovarian, breast, and other cancers but appear to have some adverse side effects on the immune system. An active search for more effective therapies and earlier screening for ovarian cancer is currently underway.

Chapter 2, Box 1: The Estrogen Epidemic

With the advent of modern developed societies, the reproductive biology of women has changed dramatically relative to our hunter-gatherer ancestors, mainly due to cultural effects of the modern world. The age at menarche has decreased markedly (see Chapter 6). This may be a result of the improved nutrition and access to food, along with a decreased necessity for physical activity of children, which allows young girls to grow faster and reach a minimum reproductive weight or fatness level at an earlier age. At the same time, the age at first pregnancy has risen. This may be a reflection of modern expectations regarding age at marriage, broader options for young women as alternatives to early marriage and childbearing, and/or improvements in modern medicine that have decreased infant/child mortality and allowed women to have fewer pregnancies to arrive at a desired family size. Present-day hunter-gatherer women, on average, give birth to more than three times as many children as do women in modern industrialized societies. Because of earlier menarche and later first pregnancy, the interval between menarche and first birth is quadrupled in modern women as compared to hunter-gatherers.

Hunter-gatherer women experience about 160 lifetime ovulation cycles (menstrual cycles) as compared to 450 in women living in developed countries. This is because modern women have a longer reproductive life span (about 38 years) as compared to that of a hunter-gatherer woman (about 31 years) due to earlier menarche and later menopause. In addition, as discussed in Chapter 13, hunter-gatherer women breastfeed their babies 11.6 times longer than does the average breastfeeding woman in developed countries. Frequent breastfeeding suppresses ovulation and thus menstrual cycles, and hunter-gatherer women breastfeed more efficiently for a much longer duration than modern women (2.9 years versus 0.25 years on average). Finally, the fewer children born to modern women means fewer pregnancies in a lifetime; pregnancy suppresses ovulation and menstrual cycles.

Estrogen levels in the blood of women during a menstrual cycle are high, especially in the few days before ovulation (Chapter 3). They are also high in the menarche-to-first pregnancy interval. Therefore, modern women, with longer menarche/first pregnancy intervals and many more menstrual cycles, experience a greatly increased lifetime exposure to estrogen. Hormonal contraceptives, such as the combination pill, do suppress ovulation, but the hormones in these medications are estrogenic. Thus, exposure to estrogens remains high whether or not women are taking hormonal contraceptives. Estrogen-replacement therapy in postmenopausal women also increases the lifetime exposure to estrogens.

Modern women's constant bombardment with estrogen has been accompanied by an ever-increasing rate of the incidence of estrogen-dependent diseases, such as uterine fibroids and endometriosis. The incidence of estrogen-dependent cancers of the breast, endometrium, and ovaries also becomes much more common. In these tissues, estrogen increases cell division and turnover, which

Continued on next page.

Chapter 2, Box 1 continued.

renders cancer more likely to occur. Interestingly, female reproductive cancers are extremely rare in hunter-gatherer women who have not changed their reproductive biology through exposure to the modern agricultural, industrial, and medical world.

As a result, modern women are exposed to much more estrogen in their lifetimes, internally because of the reduction in number of babies born as well as a shorter duration of breastfeeding and externally because of the use of combination pills and hormone replacement therapy. Add to this the lifetime exposure of women to xenoestrogens in pollution (see HIGHLIGHT box in this chapter) and you have an estrogen epidemic, perhaps posing great danger to women in the modern world. Of course, women in developed countries are not about to return to the reproductive lifestyle of the typical hunter-gatherer; this indeed would damage us in another way by increasing the birth rate while decreasing the death rate (see Chapter 13). However, knowledge of the estrogen epidemic perhaps will lead some women, and perhaps Western medicine, toward more healthful practices.

Contrasting Lifetime Reproduction in Hunter-Gatherer Women and U.S. Women

Reproductive variable	Hunter-gatherer women	U.S. women
Age at menarche	16.1	12.5
Age at first birth	19.5	26
Menarche to first pregnancy interval (years)	3.4	13.5
Duration of lactation per birth (years)	2.9	0.25
Lifetime duration of lactation (years)	17.1	0.4
Completed family size	5	1.8
Age at menopause	47	50.5
Total lifetime ovulations	160	450

Oviducts

The *oviducts* are paired tubes extending from near each ovary to the top of the uterus (Fig. 2-2). They also are called *fallopian tubes*, after the 16th century anatomist Fallopius. Each oviduct is about 10 cm (4 in.) long and has a diameter about the size of a drinking straw. An oviduct can be divided into several regions along its length (Fig. 2-2). Nearest the ovary is a funnel-shaped portion, the *oviductal infundibulum*. The opening into the infundibulum, into which the ovulated ovum enters, is the *ostium*. The edges of the infundibulum have finger-like projections, the *fimbriae*. When ovulation occurs, the ovum is captured by the infundibulum and enters the oviduct because of cilia on the oviductal lining that beat in a uterine direction. Proceeding along the tube toward the uterus, there is a wide *oviductal ampulla*, then a narrower *oviductal isthmus*, and finally the part of the oviduct that is embedded within the uterine wall, the *intramural oviduct*. The point at which the oviduct empties into the uterine cavity is termed the *uterotubal junction*.

Each oviduct in cross section has three layers: (1) On the outside is a thin membrane, the *oviductal serosa*. (2) The middle layer consists of smooth muscle (the *oviductal muscularis*). Contraction of this muscle helps transport the ovum and, later, the early embryo toward the uterus. Near the time of ovulation, these contractions are 4 to 8 s apart, whereas they are less frequent during other stages of the menstrual cycle. (3) The third layer, the internal lining of the oviduct, has many folds, with ciliated and nonciliated cells as well as mucous glands on the internal surface. Ciliary beating and mucous secretion play a role in ovum and embryo transportation and in sperm movement up the oviduct (see Chapter 9). Hormones play a role in oviduct function. Estrogens cause secretion of mucus in the oviducts; they also cause the oviductal cilia to beat faster and the smooth muscle to contract more frequently. Progesterone, however, reduces mucous secretion and reduces muscular contraction.

Inflammation of the oviducts, or *salpingitis*, can lead to blockage by scar tissue and subsequent infertility (see Chapter 16). This term comes from the Greek word *salpinx*, which means "tube." Tubal blockage can be repaired in some cases by microsurgery. *Salpingectomy* means surgical removal of the oviducts. *Tubal sterilization* (surgical cutting, tying off, or blocking the oviducts) can be used as a contraceptive measure (see Chapter 14).

Uterus

Uterine Functional Anatomy

The *uterus* (womb) is a single, inverted pear-shaped organ situated in the pelvic cavity above the urinary bladder and in front of the rectum (Figs. 2-1 and 2-2). In *nulliparous* women (those who have never borne a child), the uterus is about 7.5 cm (3 in.) long, 5 cm (2 in.) wide, and 1.75 cm (1 in.) thick. In *multiparous* women (those who have previously borne children), the uterus is larger and its shape is more variable.

The uterus is supported by bands or cords of tissue, the *uterine ligaments*. A pair of *broad ligaments* attach the uterus to the pelvic wall on each side. Paired *uterosacral ligaments* attach the lower end of the uterus to the sacrum ("tail bone"). The *lateral cervical ligaments* connect the cervix and vagina to the pelvic wall. Finally, the paired, cord-like *round ligaments* attach on one end to the uterus near the entrance of the oviducts and on the other end to the lower pelvic wall. These ligaments, in addition to carrying blood vessels and nerves to the uterus, also serve to support the uterus and other pelvic organs in their normal position. We see later in this chapter what happens to the uterus when these ligaments are not doing their job.

If a frontal section is made through the uterus, several distinct regions are revealed (Fig. 2-2). The dome-shaped region above the points of entrance of the oviducts is the *uterine fundus*. The *uterine corpus* or "body" is the tapering central portion, which ends at an external constriction, the *uterine isthmus*. A narrow region, the *uterine cervix*, is adjacent to the vagina.

The wall of the uterine fundus and corpus has three layers of tissue (Fig. 2-2). The external surface of the uterus is covered by a thin membrane, the *perimetrium*. This outer coat, or *serosa*, is continuous with the layer covering the oviducts and other peritoneal organs. Inside the perimetrium is a thick layer of smooth muscle, the *myometrium*, which is thickest in the corpus region. The myometrium is capable of very strong contractions during labor, as discussed in Chapter 11. Its muscle fibers increase in length and number during pregnancy. Internal to the myometrium is the layer of the uterus that lines the uterine cavity.

This layer, the *endometrium*, can be divided further into an internal surface layer, the *stratum functionalis*, and a deeper layer, the *stratum basalis* (Fig. 2-2). The stratum functionalis, consisting of a lining epithelium and *uterine glands*, is shed during menstruation. The underlying stratum basalis is not shed during menstruation but contains blood vessels that produce part of the menstrual flow. After menstruation, the stratum basalis gives rise to a new stratum functionalis. Thus, the endometrium undergoes marked changes in structure and function during the menstrual cycle, and these changes are under hormonal control (see Chapter 3).

The layers of the uterine cervix are similar to those of the uterine fundus and corpus, with a few important exceptions. The cervical myometrium is thinner, and the cervical endometrium is not shed during menstruation. Glands in the lining of the cervix secrete mucus to varying degrees during the menstrual cycle (see Chapter 3). In some stages of the cycle, this mucus forms a plug that retards sperm movement through the cervical canal and prevents infectious microorganisms from entering the uterine cavity. The *cervical canal* is the small channel, within the cervix, that connects the vagina with the uterine cavity. The opening of the cervical canal to the uterine cavity is the *internal cervical os*, whereas the opening of the cervical canal into the vagina is the *external cervical os*. The external os is about the diameter of the head of a kitchen match. The cervix viewed through the vagina appears as a dome, 1.75 to 5.0 cm (1 to 2 in.) in diameter.

Uterine Disorders

Pelvic Infection The uterus is subject to infectious organisms entering through the vagina. Some bacteria, such as those associated with sexually transmitted diseases (see Chapter 18), can cause uterine infection and inflammation that can spread to the oviducts. Such an infection can be dangerous to the fetus during pregnancy and can lead to oviductal scarring and infertility (see Chapters 16 and 18).

Cervical Cancer Cervical cancer is the eighth most common cause of cancer fatalities among U.S. women. There are about 10,370 new cases of cervical cancer diagnosed each year in the United States, and about 3,710 women die each year from this disease. The incidence of cervical cancer is much greater in developing countries, though, where it represents a major but likely preventable cause of death. Before cancer develops, cervical cells exhibit a precancerous condition called *cervical dysplasia*. These cells have an abnormal appearance but are not malignant. These precancerous cells may be removed surgically, although occasionally they can disappear on their own. Pregnant women with cervical dysplasia have a significantly greater likelihood of losing these abnormal cells if they have a vaginal delivery (as opposed to a cesarean delivery). Possibly the stretching of the cervical wall and abrasion during the birth process help detach the cells. True cancer of the cervix can take many years to develop. When these cancerous cells begin to multiply, they cause lesions in the cervical wall accompanied by intermenstrual bleeding and vaginal discharge. If not treated, the cancer can spread to the uterine myometrium and other tissues.

Because of the danger of cervical cancer, cervical cells should be examined using the *Papanicolaou test*, or *Pap smear*, named after Dr. George Papanicolaou, who developed it in 1942. It usually is recommended that a Pap smear be done once a year. Because the development of cervical cancer can take many years, however, it has been argued that a Pap smear needs to be done only every 2 or 3 years. In this test, a few cells from the cervix are removed with a small round brush and fine wooden spatula and examined under a microscope. Detection of cervical precancerous or malignant cells with a Pap smear is not 100% reliable. In fact, about one-third of cases of cervical cancer are missed with one Pap smear. However, new automated methods of examining Pap smears have improved their reliability greatly. A color slide of the cervix ("cervigram") or sampling of cells from the outer wall of the cervix ("cervicography") is often used in conjunction with a Pap smear. Finally a small piece of cervical tissue can also be removed and examined, a procedure called *cervical biopsy*. If cancer is present, the affected area can be removed surgically or a woman can undergo surgical removal of the entire cervix and uterus, an operation called *hysterectomy*. Surgery may be combined with *chemoradiation*, which combines radiation of the pelvic area with chemotherapy of whole body; it has been found that the addition of chemotherapy reduces the death rate of cervical cancer patients.

Most cervical cancers are thought to be caused by *human papillomavirus* (HPV), a family of sexually transmitted viruses. HPV is very common; in some U.S. populations, over half of the women under 35 are infected with the virus. Of course, the vast majority of these women will never develop cervical cancer. Most often, a woman's immune system is able to fight off the infection by age 40. However, women with a persistent, active HPV infection are at higher risk of developing cervical cancer. Because HPV is transmitted sexually, women with a greater number of sexual partners are at increased risk. The incidence of cervical cancer is also greater among women who became sexually active as adolescents. This disease is extremely rare in celibate women. Smoking increases the risk of developing cervical cancer in HPV-infected women.

Of the nearly 100 types of HPV, about 15 are implicated in most cervical cancers. A new DNA test can detect genetic material from these forms of the virus in infected individuals. The test can point to probable cases of cervical cancer, especially in women over 40, although it gives many false positive results (a test result that falsely reports presence of the disease in a healthy individual). Clinical trials of a vaccine against HPV have yielded promising results, and an HPV vaccine likely will be widely available soon.

Cervical Cysts and Polyps Cervical tissue can also form noncancerous growths as the result of past infection or irritation. *Cervical cysts*, for example, are quite common. These cysts look like pea-sized, whitish pimples. They usually are not a problem, but they can be removed by burning (heat cauterization) or by freezing the tissue (cryosurgery). Laser beams also have been used to remove abnormal tissue in the female reproductive tract. *Cervical polyps* sometimes appear as tear-shaped growths extending into the cervical canal. These polyps are benign. If extensive, however, they should be removed because they can interfere with fertility or *vaginal coitus* (sexual intercourse).

Endometrial Cancer Cancer can also develop in the lining of the body of the uterus. Endometrial cancer is more common in older women, in nulliparous women, and in those that are obese or diabetic. There are about 40,880 new cases of uterine cancer each year in the United States, and 7310 women die of this disease. Pap smears detect the presence of endometrial cancer only about half the time. Treatment of menopausal symptoms with estrogens has been shown to increase the incidence of endometrial cancer (see Chapter 3), but use of the combination pill has not been shown to have an influence (see Chapter 14). *Tamoxifen*, a drug used to treat breast cancer (see section on Breast Disorders), can also increase the risk of endometrial cancer. Usually the first symptom of endometrial cancer is irregular menstrual bleeding. An *endometrial biopsy* can be performed to see if cancer is present. If the cancer is in the early stages, the diseased endometrium can be removed using cryosurgery. Uterine cancer also can be treated with radiation therapy, chemotherapy, or hysterectomy.

Endometriosis Sometimes, endometrial tissue can detach from the uterus and lodge in other regions of the body such as the ovaries, oviducts, uterine ligaments, urinary bladder, urethra, and intestines. This is thought to occur during the menstrual flow, when small pieces of endometrial tissue move in a *retrograde* flow up through the oviducts and into the peritoneal cavity rather than out through the vagina. This condition, *endometriosis*, can be very painful because the displaced tissue, which is responsive to steroid hormones, expands and bleeds during each menstrual cycle. This often results in unusually painful menstrual periods (dysmenorrhea), although other symptoms may include chronic pelvic pain, back pain, pain at ovulation or during sexual intercourse, and painful bowel movements.

Endometriosis affects about 10% of American women in their premenopausal years. Unlike many other reproductive disorders, endometriosis often affects young women in their teens, twenties, and thirties. It may go undiagnosed, especially in women who have not begun to receive regular gynecological examinations, as painful menstrual flow is often considered "normal." Endometrial growths can produce inflammation and create scar tissue that can cause infertility. These growths may cover the ovary, preventing ovulation. More commonly, endometrial adhesions can distort or block the oviducts, preventing the normal passage of sperm and/or egg. Infertility affects 30–40% of women with endometriosis.

Endometriosis is diagnosed by *laparoscopy*, a surgical procedure in which a slender, flexible tube carrying a light is inserted into a small slit in the abdomen. Endometrial growths and scar tissue may be removed surgically using laser or small instruments inserted through the laparoscope. Women may also be treated with drugs such as danazol (a synthetic steroid) or a GnRH inhibitory agonist, both of which inhibit FSH and LH secretion and temporarily halt menstrual cycling to reduce the pain and inhibit the growth of displaced endometrial tissue. Pregnancy may cause temporary remission of the symptoms of endometriosis, and the disease may terminate spontaneously, especially after menopause, but no certain cure has been found.

Endometrial Polyps, Hyperplasia, and Fibroids The endometrium can form abnormal growths that are benign (not cancerous). *Endometrial polyps* are mushroom-like growths that extend into the uterine cavity. *Endometrial hyperplasia* is an excessive proliferation of endometrial cells. Both these conditions can be caused by chronic exposure of the endometrium to estrogen without the occurrence of menstruation. *Fibroids* are noncancerous tumors of the smooth muscle layer of the uterus. These benign tumors are common in older women before they reach menopause. All of these types of abnormal growth can cause irregular menstrual bleeding and interfere with fertility. They can be removed by heat cauterization, cryosurgery, or laser beams.

A recent trend is to use *endometrial ablation* instead of hysterectomy for the treatment of uterine bleeding and fibroids in cases where the remainder of the uterus is normal. In this procedure, done under general anesthesia, a small instrument the size of a pen is inserted through the cervical canal and into the uterine cavity. The endometrium is then destroyed by heat cauterization. This procedure is less costly and invasive than a hysterectomy. However, it does cause infertility.

"Tipped" Uterus The position of the uterus can vary in different women. In most cases, the uterus is in an *anteflexed* position, tilted forward at right angles to the vagina (Fig. 2-10). In about one of five women, however, the uterus is in a *retroflexed* position, being tilted backward instead of forward (Fig. 2-10). A retroflexed uterus can cause pain during menstruation or intercourse, although some women experience no adverse symptoms. This condition does not affect a woman's ability to become pregnant and, in some cases, pregnancy can even reposition a retroflexed uterus.

Prolapsed Uterus A prolapsed uterus is one that has slipped down toward the vagina. This often occurs because the uterine ligaments are too relaxed or stretched. Delivery of a large infant, or breech or forceps deliveries (see Chapter 11), sometimes can cause uterine prolapse by stretching or damaging the uterine ligaments. The degree of prolapse can be anywhere from slight to a severe protrusion of the uterus from the vagina. In the latter case, surgery is required to correct the prolapse or remove the uterus.

Flabbiness or overstretching of the *pubococcygeus muscle* can also lead to prolapse of the uterus and other pelvic organs. This muscle is a sling-like band of tissue that connects the pubic bone with the tail bone and forms the floor of the pelvic cavity. Thus, it supports the uterus, part of the vagina, urinary bladder, urethra, and rectum. If this muscle is not taut, the uterus and vagina can sag and there can be uncontrollable leakage of urine from the urethra (*urinary incontinence*). What can a woman with a sagging pelvic organ do about her problem? One answer is the *Kegel exercise*. A physician can show her how to exercise and strengthen the

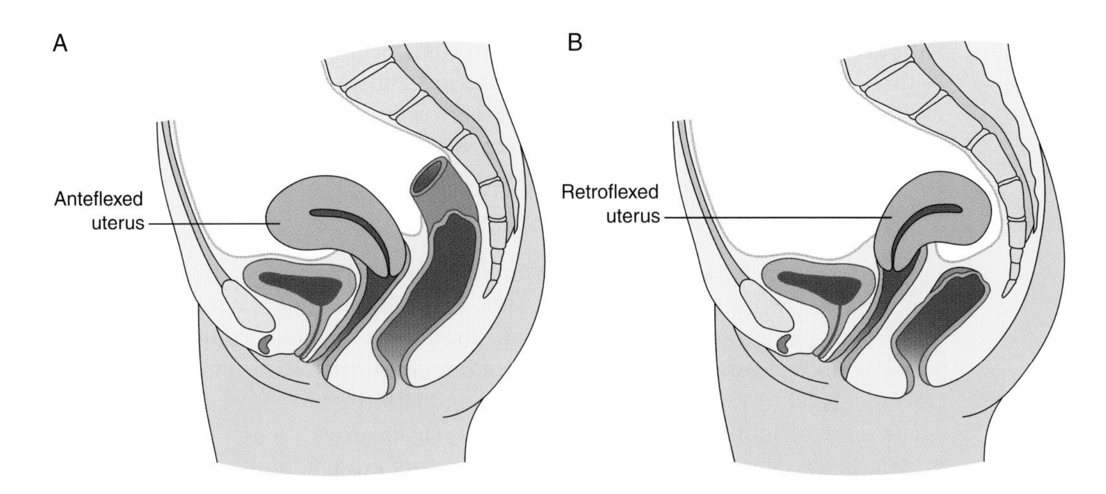

Figure 2-10 Usual and unusual positions of the uterus. (a) The uterus is in the usual anteflexed position, whereas (b) the uterus is tilted backward in a retroflexed position.

pubococcygeus muscle. To do this exercise, a woman first finds this muscle by sitting on a toilet, urinating, and stopping urine flow voluntarily. The muscle that stops urine flow is the pubococcygeus. Then, she should voluntarily contract this muscle 10 times and repeat this six or more times daily. Also, a rubber or plastic device (*vaginal pessary*) can be inserted into the vagina to support this organ as well as the uterus. Vaginal pessaries, however, can irritate the vagina and interfere with coitus, and surgery to repair pelvic support often is the best answer.

Vagina

Structure

The *vagina* is a 10-cm (4-in.)-long tube that lies between the urinary bladder and rectum (Figs. 2-1 and 2-2). It functions as a passageway for the menstrual flow, as a receptacle for the penis during coitus, and as part of the birth canal. The wall of the vagina has folds that allow it to stretch during coitus or childbirth; it is normally collapsed. The vaginal canal leads from the *vulva* (external genitalia) to the *external cervical os*. The opening of the external cervical os into the vagina is circumscribed by a recess called the *fornix*. This recess allows support for a diaphragm contraceptive (see Chapter 14).

A cross section through the vaginal wall shows three layers. The layer next to the *lumen* (interior space) of the vagina is the *tunica mucosa*. This layer includes the epithelial lining of the vagina, which consists of many layers of *squamous* (flattened) cells. This layer expands under the influence of estrogen and undergoes hormone-induced changes in thickness and glycogen content during the menstrual cycle. These changes can be detected by swabbing the lining and looking at the cells under a microscope; this procedure is called a *vaginal smear*. Underlying the epithelium in the tunica mucosa is a layer of dense connective tissue that is rich in elastic fibers. These fibers allow expansion of the vaginal wall. This layer also contains an extensive network of thin-walled veins. It is thought that fluid from these blood vessels seeps through the epithelium into the vaginal canal during sexual arousal, serving to lubricate the vagina.

The middle layer of the vaginal wall is the *tunica muscularis*. As its name indicates, it contains numerous bundles of smooth muscles that are embedded in connective tissue. A sphincter of skeletal muscles at the vaginal opening is under voluntary control. The tunica muscularis is surrounded by a thin outer layer, the *tunica adventitia*. This elastic connective tissue supports nerve bundles, most of which control blood flow and smooth muscle contraction of the vaginal tissue. In addition, some free sensory nerve endings are found deep in the epithelium, mainly near the vaginal opening.

The Vaginal Environment

The vagina is the natural home for several microorganisms (Table 2-2). Some of these bacteria, fungi, and protozoa play important roles in maintaining the vaginal environment. Others are potential disease microorganisms, which normally do not cause problems unless they multiply rapidly. Some of these organisms can be sexually transmitted (see Chapter 18). Normally these microorganisms

Table 2-2 Microorganisms of the Vagina[a]

Lactobacillus acidophilus	Coliform bacilli
Staphylococcus aureus	*Proteus* species
S. epidermidis	*Mima polymorpha*
Fecal streptococci	*Clostridium* species
Streptococcus viridans	*Bacteroides* species
Anaerobic streptococci	*Fusobacterium* species
Neisseria species (other than *N. gonorrhoeae* and *N. meningitidis*)	*Mycoplasma*
Diphtheroids	*Candida* species
Corynebacterium species	*Trichomonas vaginalis*
Hemophilus vaginalis	

[a]Not all these microorganisms are present in a single female at one time.

are controlled by the large population of leukocytes that inhabit and monitor the vaginal environment.

Cells of the vaginal epithelium accumulate large amounts of *glycogen* (a sugar) under the influence of estrogen. As these cells die and are sloughed into the vaginal cavity, they release this glycogen. Certain bacteria present within the vagina and then metabolize the glycogen to lactic acid, rendering the vaginal environment acidic. This acidic condition retards yeast (fungal) infection. If, however, a woman takes certain antibiotics, the bacteria are destroyed. The vaginal environment then becomes more basic, which may lead to yeast infection (see Chapter 18). The vaginal acidity also kills sperm. As discussed in Chapter 9, semen deposition into the vagina during coitus changes the vaginal environment to a more basic condition and allows sperm to survive and move up the female tract.

Female External Genitalia

The female *external genitalia* include the mons pubis, labia majora, labia minora, vaginal introitus, hymen, and clitoris (Fig. 2-11). These organs, collectively called the *vulva*, vary widely in external appearance among different women.

Mons Pubis

The *mons pubis* is a cushion of fatty tissue, covered by skin and pubic hair, that lies over the pubic symphysis. The skin of this area has many touch receptors and only a few pressure receptors. The distribution and the amount of pubic hair vary in different individuals. Usually the pubic hair forms the shape of an inverted pyramid. In about 25% of women, this hair extends in a line up to the navel.

Labia Majora

The *labia majora* ("major lips") are fleshy folds of tissue that extend down from the mons pubis and surround the vaginal and urethral orifices. These folds contain fat, and the pigmented skin has some pubic hair, sweat and oil glands, and fewer touch and pressure receptors than the mons pubis. The labia majora

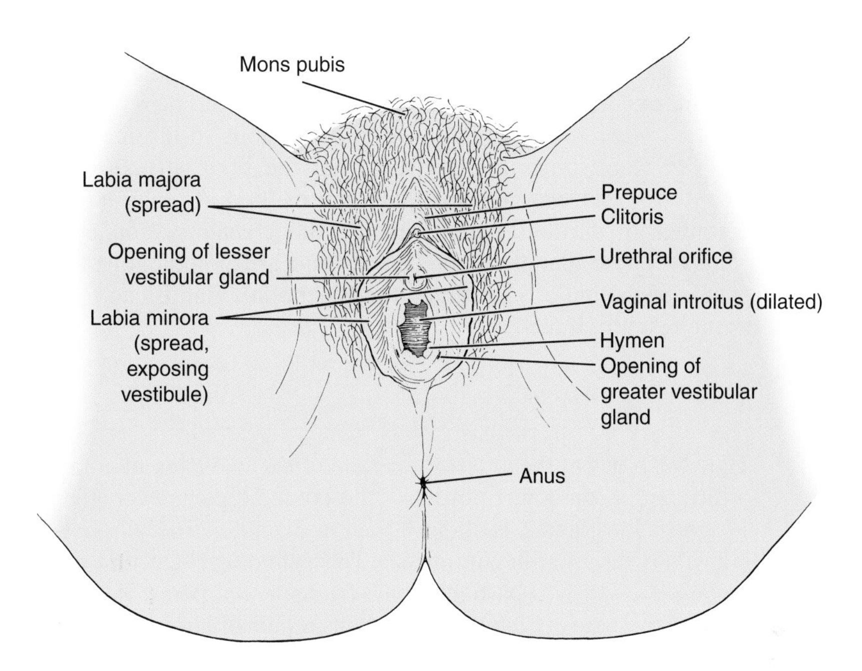

Figure 2-11 The female external genitalia, or vulva.

are homologous to the male scrotum, i.e., they are derived embryologically from the same tissue (see Chapter 5).

Labia Minora

The *labia minora* ("minor lips") are paired folds of smooth tissue underlying the labia majora. They range from light pink to brownish black in color in different individuals. In a sexually unstimulated condition, these tissues cover the vaginal and urethral openings, but upon sexual arousal they become more open (see Chapter 8). The hairless skin of the labia minora has oil glands (but no sweat glands) and a few touch and pressure receptors. In older women or in women who have low estrogen levels, the skin of the labia minora becomes thinner and loses surface moisture.

Vestibule The cavity between the labia minora is the *vestibule*. Most of this cavity is occupied by the opening of the vagina, the *vaginal introitus*. In women who have not previously had coitus, the introitus often is covered partially by a membrane of connective tissue, known as the *hymen*. This tissue often is torn during first coitus, accompanied by minor pain and bleeding. However it also can be broken by a sudden fall or jolt, by insertion of a vaginal tampon, or by active participation in such sports as horseback riding and bicycling. In some women, the hymen can persist even after coitus, especially if the tissue is flexible. Thus, the presence or absence of a hymen is not a reliable indicator of virginity or sexual experience. In rare cases, a wall of tissue completely blocks the introitus, a condition called *imperforate hymen*. The condition is present in about 1 out of 2000 young women. Because an imperforate hymen can block menstrual flow, surgery is required to alleviate the problem.

Urethral Orifice Anterior to the vaginal introitus is the *urethral orifice*. This is where urine passes from the body. Below and to either side of the urethral orifice are openings of two small ducts leading to the paired *lesser vestibular glands* (*Skene's glands*). These glands are homologous to the male prostate gland (i.e., the two gland types are derived from the same structure in the embryo; see Chapter 5) and secrete a small amount of fluid. At each side of the introitus are openings of another pair of glands, the *greater vestibular glands* (*Bartholin's glands*). These glands secrete mucus and are homologous to the bulbourethral glands of the male. Sometimes, the Bartholin's glands can form a cyst or abscess as the result of infection.

Clitoris

The *clitoris* lies at the anterior junction of the two labia minora, above the urethral orifice and at the lower border of the pubic bone. Its average length is about 1 to 1.5 cm (0.5 in.) and it is about 0.5 cm in diameter. There is, however, considerable individual variation in clitoral size. This cylindrical structure has a shaft and glans (enlarged end). It is partially homologous to the penis. The *clitoral shaft*, like the shaft of the penis (Chapter 4), contains a pair of *corpora cavernosa*, spongy cylinders of tissue that fill with blood and cause the clitoris to erect slightly during sexual arousal (see Chapter 8). Another spongy cylinder present in the penis, the *corpus spongiosum*, is not found in the clitoris; this tissue in the female is represented by the labia minora (Chapter 5). The *clitoral glans* is partially covered by the *clitoral prepuce*, which is homologous to a similar structure covering the glans of the penis (see Chapter 4). The clitoris is rich in deep pressure and temperature receptors, but it has only a few touch receptors. Chapter 8 discusses the role of the clitoris and other structures of the female vulva in the female sexual response.

Mammary Glands

The *mammary glands*, also called the breasts or *mammae*, are paired skin glands positioned over ribs two through six on the chest. Their function is to secrete milk during breast-feeding and they also serve as a stimulus for sexual arousal in both males and females. These glands evolved from sweat glands, and so milk is really modified sweat! Embryologically, the mammary glands develop from a *milk line*, which is a chain of potential mammary glands extending from the arm buds to the leg buds of the embryo. In some mammals, several pairs of glands persist in the adult female, depending on the normal litter size. In humans, usually only a single pair persists, but in some individuals more than one pair are present, a condition called *polythelia*. There are records of women having up to eight pairs of functional breasts, and it is estimated that 1 out of 20 males has an extra nipple. Speaking of males, the male mammary glands are usually quiescent, but they are capable of growing and even secreting milk if properly stimulated by certain hormones. Such development of the breasts in males is called *gynecomastia*.

Mammary Gland Functional Anatomy

Each human female breast is covered by skin and contains a variable amount of fat and glandular tissue. The variation in breast size and shape is due to differences in the amount and distribution of fat; breast size does not affect the ability to secrete milk. Each breast contains glandular tissue divided into 15 to 20 lobes separated by fat and ligamentous tissue (Fig. 2-12). The latter tissue (*suspensory ligaments of Cooper*) provides support for the breast, but it tends to be less effective in older women. Each mammary lobe is composed of several *lobules*, which are grape-like clusters of *mammary alveoli* (Figs. 2-12 and 2-13). The *alveolus* is the functional unit of the mammary gland. It is a hollow sphere of milk-secreting cells. Each alveolus receives an extensive blood supply, which provides raw materials for milk synthesis and transports the hormones that control alveolar growth and function. Milk is synthesized and secreted from these alveoli into *secondary mammary tubules*. The secondary tubules from each lobe join to form a *mammary duct*, which empties into a wider *mammary ampulla*, where milk can be stored. The ampullae then empty into a *lactiferous duct*, which opens into the *nipple*. There is one lactiferous duct for each lobe, but some ducts may join before reaching the nipple. Surrounding the nipple is a ring of pigmented skin, the *areola*, which contains oil glands.

Mammary gland development is a long process that extends over many years. Mammary ducts begin to appear during prenatal development. Major growth of the breast begins at puberty, with extensive branching of the glandular tissue and formation of lobules. Maximum development occurs

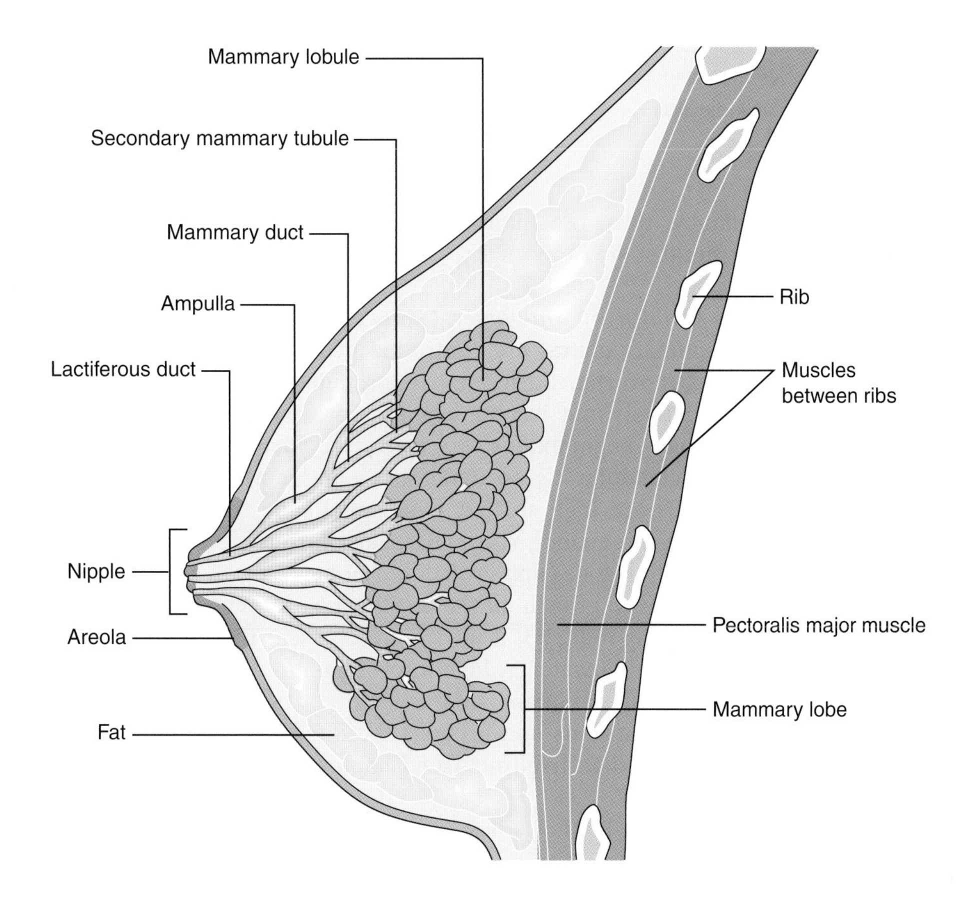

Figure 2-12 Section through the adult female mammary gland.

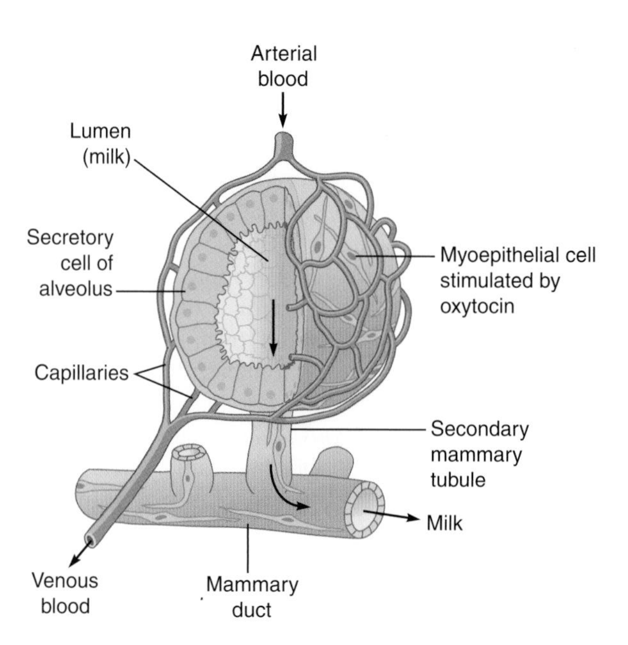

Figure 2-13 Diagrammatic view of a single mammary alveolus. Milk is secreted by epithelial secretory cells into a cavity (lumen) within the alveolus. From here, the milk is carried to the nipple through a series of ducts. The blood supply to the alveolus is important because of its support of alveolus growth and secretion. Also, the blood carries hormones to the alveolus that stimulate its secretion and cause contraction of the myoepithelial cells, leading to milk ejection.

during pregnancy; thus, the mammary glands of nulliparous women never reach this terminal stage of maturity.

Hormonal Control of Mammary Gland Function Growth and development of the mammary glands are controlled by estrogen and progesterone, as well as some extraovarian hormones. In prepubertal females, the mammary gland tissue is relatively inactive, but when the ovaries begin to secrete estrogen the alveoli and ducts begin extensive growth (see Chapter 6). During the menstrual cycle, the glandular tissue undergoes enlargement when estrogen and progesterone levels are high so that the breasts may feel full or tender (see Chapter 3). During pregnancy, rising estrogen and progesterone in the blood, along with adrenal hormones and growth hormone, cause the glandular tissue to enlarge and the ducts to branch. In the last third of pregnancy, the glandular tissue is not yet secreting milk, but it does secrete a clear fluid (*colostrum*), which, in addition to containing water, is high in protein and rich in maternal antibodies. Colostrum nourishes nursing infants before true milk is secreted in the breasts. It provides nutrients and protects the infant against certain infections, especially those of the digestive tract (see Chapter 12).

Suckling Chapter 12 shows how suckling of the nipple maintains prolactin secretion, which causes milk secretion in previously primed mammary tissue. We also see how nursing stimulates the release of oxytocin, which causes ejection of milk from the nipple. The latter is accomplished by the contraction of *myoepithelial cells* that surround each alveolus (Fig. 2-13).

Noncancerous Breast Disorders

The breasts normally have a somewhat lumpy texture under the surface. Most breast lumps come and go, but some persist. Most of these persistent breast lumps are benign fluid-filled cysts. About one in five women between the ages of 25 and 55 will develop these cysts, a condition called *cystic mastitis*. Often a physician will aspirate (remove) the fluid from a cyst with a needle under local anesthetic.

Fibroadenomas are breast lumps that feel like a firm grape; they are benign and are most common in middle-aged women. Often they will be biopsied (a small piece of tissue removed) or removed entirely to be certain that they are only a fibroadenoma and not cancer. There is no correlation of a history of fibroadenoma with later development of breast cancer. Only about 1 in 12 persisting breast lumps are cancerous in women before menopause; after menopause, still less than half are cancerous.

Millions of women have received breast implants (bags of liquid or gel), either for cosmetic reasons or to replace breast tissue lost during mastectomy. The number of women seeking implantation surgery has increased from about 100,000 in 1997 to over 250,000 in 2003. Some 240 different types of implants have been devised over the years, the most popular being the silicone gel implant used since the 1960s. In 1992, however, the U.S. FDA declared a moratorium on their use because studies could not prove that they were safe and effective and because of their suspected links to connective tissue disease, autoimmune disorders (such as lupus and rheumatoid arthritis), and chronic fatigue syndrome. The implants could also cause painful calcium deposits in the breast and could rupture, leaking silicone into the surrounding tissue. Since 1992, the silicone breast implant has been replaced almost completely by the saline implant, which contains harmless salt water. Implants containing soybean oil and plastic gel have come and gone. Now, women are increasingly choosing to have breasts reconstructed from other parts of the body, the so-called "flap reconstruction." However, this is a more difficult operation and not suitable for all women. The complication rate for breast augmentation surgery remains quite high, and lawsuits continue while scientists search for a safe, long-term implant.

Breast Cancer

Risk Breast cancer is the most common form of cancer found in U.S. women and is second only to lung cancer in numbers of cancer deaths. At present, about 212,930 new cases of breast cancer (1690 of these in men) are reported in the United States each year, and about 40,870 women will die each year from this dangerous disease. An estimated 5–10% of breast cancer cases have a pattern of inheritance within a family, whereas the remaining 90–95% of cases have other causes. Considering both inherited and noninherited types, the present risk of an American woman to develop breast cancer in her lifetime is about 1 in 8. Age-specific chances of developing breast cancer are another way of looking at risk. At age 20, a woman in the United States has a 1 in 2044 chance of developing breast cancer in the next 10 years (i.e., by age 30).

A 40 year old has a 1 in 67 chance by age 50, and a 60 year old has a 1 in 28 chance by age 70 (see Table 2-3). Thus, the risk of developing breast cancer increases with age. However, at any given age, these risks never approach 1 in 8. This is because a woman who has reached a certain age cancer free has outlived the odds faced by younger women. Thus, only at birth does a woman have a 1 in 8 chance of having breast cancer sometime in her life. The chance of dying from this disease once one has it gradually increases with advancing age. Early detection and treatment reduce the death risk greatly. Women diagnosed in early stages of breast cancer are much more likely to survive the disease than those whose cancer has spread.

Breast cancer has been increasing over the last two decades in the United States and other developed countries. The number of women in the United States diagnosed with the disease increased by about 4.5% during the 1980s, due at least in part to wider use of screening. The number of diagnosed women continued to increase in the 1990s, but at a slower rate. Death rates, especially among young women, have dropped somewhat in recent years because of earlier detection and improved treatments. Overall breast cancer death rates have dropped 2.3% per year since 1990. Although the incidence of breast cancer is greater in white than in black U.S. women, the mortality rate is higher in black women. A final note: males can also develop breast cancer. There are about 1500 new cases in the United States each year, and about 400 men die each year from this disease. Some male cases also appear to have an inherited basis.

Inherited Breast Cancer As mentioned earlier, about 5–10% of breast cancer cases in the United States exhibit an inherited pattern. Results of a recent study of 117,988 women indicate that a woman's risk of developing breast cancer is 1.8 times that of the general population if her mother developed the disease, 2.3 times if her sister did, and 2.5 times if both her mother and sister did.

Two genes on human chromosomes have been implicated in breast cancer. The *BRCA1* gene, on chromosome 17, codes for a protein involved in repairing errors that are sometimes made when DNA is replicated (copied) during cell division. Expression of this gene normally guards against mutations that may cause breast cancer and ovarian cancer in women. The other gene that relates to breast cancer is *BRCA2* (on chromosome 13). *BRCA2* is associated with inherited breast cancer in women and men, but has less of an influence on ovarian cancer. Inheritance of flawed *BRCA1* and/or *BRCA2* alleles from either one's mother or father leads to a greatly increased (65–85%) risk of breast cancer. Ninety percent of inherited breast cancers are associated with these two genes; other genes linked to breast cancer have also been identified.

Table 2-3 Probability of Breast Cancer in Women

From age	20 to 30	1 out of 2044
	30 to 40	1 out of 257
	40 to 50	1 out of 67
	50 to 60	1 out of 36
	60 to 70	1 out of 28
	70 to 80	1 out of 24
	Ever	1 out of 8

After these breast cancer genes were discovered in 1994 and 1995, it was hoped that one could be tested for the presence of the abnormal proteins coded by these genes to determine if a flawed gene were present. Because each gene has many forms of mutation, it would be difficult to test for all of the abnormal protein forms. However, it has been suggested that one could test for the abnormally short length of these proteins common to people with any of the 38 *BRCA1* mutations. But would a woman want to know that she has abnormal breast cancer genes and therefore live with the fear of developing the disease and face the possible choice of having a *preemptive mastectomy* (removing the breasts before cancer develops)? One advantage of knowing that one has an abnormal *BRCA1* or *BRCA2* gene would be that a woman would know to test for early signs of the disease more frequently (see later).

Noninherited Breast Cancer In addition to age and family history, several factors have been associated with an increased risk of developing breast cancer (Table 2-4). Some are difficult to explain, such as the increased risk for women living in urban as well as more northern regions of the United States, but many appear to be related to total lifetime exposure to estrogens.

During a woman's reproductive years (from menarche to menopause), estrogen levels are elevated each month during the menstrual cycle. Thus, a woman's lifetime exposure to estrogen is related to the number of cycles she experiences during her reproductive years. Entering menarche (the first menstrual cycle) later than the average age of 12 or experiencing menopause earlier than about 50 lowers the overall exposure. Cycles are suspended during pregnancy and frequent breast-feeding, allowing estrogen "breaks" during these times. Therefore, women who have had several children and breast-fed them for a long time period have lower overall exposure (see HIGHLIGHT box 2-1). It has been calculated that a woman's risk of developing breast cancer drops by 7% for each child she has and by 4.3% for each year that she breast-feeds. Pregnancy also activates a cancer-fighting protein called p53, which appears to

Table 2-4 Major Risk Factors for Breast Cancer

Age 50+
Inherited mutations in the *BRCA1* and/or *BRCA2* genes
Family history of breast cancer, especially diagnosed at an early age
Personal history of breast, endometrial, or ovarian cancer
Early menarche (before age 12)
Late menopause (after age 55)
Nulliparity or few pregnancies
First full-term pregnancy after age 30
Little or no breast-feeding
Obesity after menopause
Estrogen replacement therapy after menopause
Oral contraceptive use (?)
Alcohol consumption
High socioeconomic status
Urban residence
Northern U.S. or northern Europe residence

remain in mammary tissue and suppresses breast tumors later in life. The protective effects of bearing children on rates of breast cancer vary with age of first pregnancy. Women who have a child before age 30 have lower breast cancer risk than nulliparous women (those never bearing a child) below age 30. However, those bearing their first child after age 30 have greater risk than do nulliparous women of the same age group (Fig. 2-14).

Chapter 2, Box 2: Xenoestrogens and Breast Cancer

The number of breast cancer cases per year has increased dramatically over the past few decades in industrialized countries. In the United States, for example, 1 in 30 women developed breast cancer in 1940, 1 in 11 in 1980, and 1 in 9 in 1996. A recent discovery that the appearance of feminized male alligators in a lake in Florida correlated with the previous spill of an insecticide that had estrogenic activity has led to an idea that such pollutants, called xenoestrogens, may also be responsible for the increase in breast cancers and other cancers of the female reproductive system in women. It has been known for some time that some chemicals foreign to the body, and often quite different in molecular structure from natural estrogens such as estradiol, can bind to estrogen receptors and either cause abnormal tissue responses or act as antiestrogens (like a rusty key in a lock). Also, some of these, like natural estrogens, can be antiandrogens and disrupt male function (see Chapter 4). Chapter 10 discusses how *diethylstilbesterol* (DES), an artificial estrogen once given to pregnant women to prevent miscarriage, often caused reproductive tract abnormalities in their daughters and sons.

Many environmental pollutants, some now banned from use and some currently in use, have estrogenic activity. Plants also contain xenoestrogens. Although many of these environmental estrogens are not as potent as natural estrogens, they are not broken down as easily in the body and tend to accumulate over time in fatty tissues. Also, they appear to synergize with each other, i.e., their action together is much greater than their action alone if simply added. These environmental estrogens are presented in the following list.

These compounds can be found in soil, water, food, and even breast milk. Many are stored for long periods in the body fat of

Compound	**Use**	**Comment**
Organochlorines		
Atrazine	Weed killer	Used today
Chlordane	Termite killer	Banned in 1988
DDT	Insecticide	Banned in 1972
DDE	Breakdown product of DDT	Some DDEs are estrogens; some are antiandrogens
Endosulfan	Insecticide	Used today
Kepone	Ant and roach poison	Banned in 1977
Methoxychlor	Insecticide	Similar to DDT
Polychlorinated biphenyls (PCBs)	In electrical insulation	Banned, but still in old transformers
Heptachlor	Insecticide	Used today
Dioxins	Used in paper bleaching and defoliants	Produced when garbage is incinerated, can be an antiestrogen in some doses
Dicofol	Insecticide	Used today

Continued on next page.

Chapter 2, Box 2 continued.

Compound	Use	Comment
Plastics		
Bisphenol-A	Breakdown product of polycarbonate; used to line some canned food cans and in dental sealants	Leaches into fluids when hot
Nonylphenol	Plastic softener; used in water jugs and baby bottles	Leaches into fluids at room temperature
Phthalates	Make plastics flexible	In repellent sprays, plastic plumbing, vinyl tile, food wraps, some paper and cardboard
Drugs		
Diethylstilbestrol (DES)	An estrogen used to prevent miscarriage	Banned in 1971
Other synthetic estrogens	Present in waste of people using birth control pills	In effluent from waste-treatment plants
Petroleum products		
Aromatic hydrocarbons	Petroleum; combustion by-products	Can be inhaled from gasoline or car exhausts
Plant estrogens		
Isoflavanes (e.g., coumestrol, daidzein, genistein)	In legumes such as soybeans, peas, beans, clover	No evidence that they increase breast cancer; cause infertility in female sheep
Tetrahydrocannabinol	In the hemp plant	Present in marijuana and hashish

humans and animals. Although some have been banned from use, they still linger in the environment for many years.

What evidence is there that environmental estrogens can cause breast cancer? Let us look at DDT, a weak estrogen. Although many countries have banned the use of DDT (e.g., the United States in 1972), it is still used in some countries and is stored for years in the body fat and lingers in the environment as well. One study of 58 American women with breast cancer compared with 171 cancer-free women matched for menopausal status, age, and education showed that women with 19 parts per billion of DDE (a breakdown product of DDT) in their blood faced four times the cancer risk of women with 2 parts per billion. The DDE concentration was 35% higher in women with breast cancer than in those without. In another study, deaths from breast cancer in Israel among women 44 years or younger dropped 30% after the government banned the use of DDT and related compounds in 1976. Recent research has also shown that DES, genistein (a plant estrogen), and DDT all stimulate cell division of breast cancer cells in the laboratory. It should be mentioned, however, that one other recent study showed no association of blood DDE levels and breast cancer so we have much more to learn about this topic. Chapter 4 discusses the important relationship of environmental estrogens to reproductive abnormalities and infertility in men.

It is thought that increased exposure to estrogens supports the development of breast cancer, other cancers of the female reproductive tract, and even brain cancer; about one-half of breast cancer tumors depend on and require estrogen. Exercise has been shown to reduce the lifetime risk of developing

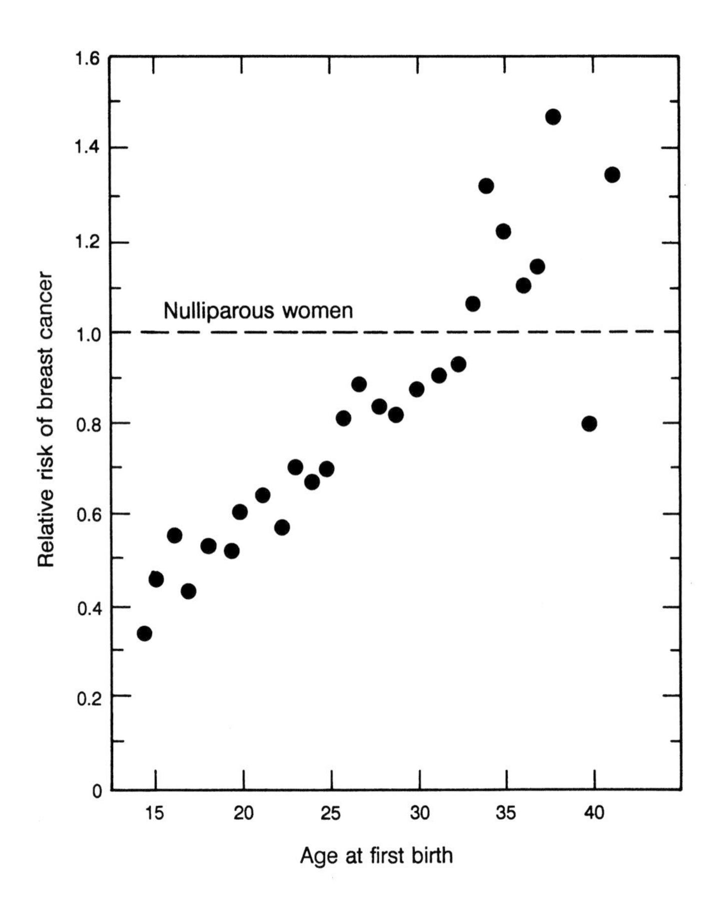

Figure 2-14 Influence of timing of first birth and nulliparity on incidence of breast cancer. Women who first gave birth before age 30 have a lower rate of breast cancer than nulliparous women.

breast and other reproductive cancers. In a recent study, women who exercised 3 or more hours/week in the 10 years after puberty decreased their breast cancer risk 30% by age 40. If these same women exercised until age 40, their risk dropped 60%. The mechanism for the beneficial effect of exercise is not known, but it may involve the fact that exercise tends to lower estrogen production (see Chapter 3). Exercise, however, also tends to reduce body fat, and obesity (especially in postmenopausal women) is associated with higher breast cancer risk. Consumption of alcohol increases breast cancer risk, possibly by affecting circulating estrogens. Similarly, women born weighing more than 8 lbs. have an increased risk of breast cancer as adults, perhaps due to their exposure to estrogen during pregnancy (because many of those babies are born late). Later in this book we will look at the controversy about whether combination birth control pills, which contain estrogen, increase the incidence of breast cancer; most of the evidence suggests that they could in younger women (see Chapter 14). Giving women estrogen after menopause also increases their risk of breast cancer (see Chapter 7). The possible importance of *environmental estrogens* (xenoestrogens) as a cause of the increase in breast cancer (see HIGHLIGHT box 2-2) emphasizes the important role of estrogen in the development of this disease.

Many factors, including genetic, physiological, and environmental, have been shown to correlate with an increased or decreased risk of developing breast cancer. The complexity of sorting out these factors becomes apparent when we examine the incidence of this disease in different races and cultures. People in different areas of the world vary not only genetically but also in a multitude of other ways. Do differences in their breast cancer rates relate to differences in lifestyle, diet, reproductive choices, or exposure to environmental estrogens? Why is the incidence so high in U.S. whites and Hawaiians? Is the relatively low incidence in east Asia due to genes, diet, or environmental factors? Even in one country, it is difficult to explain racial differences. Only further research will dissect the myriad of factors that influence this disease (see Table 2-4).

Detection Although any cell type in the breast could potentially become cancerous, most breast cancers develop in the lobules and milk ducts. Proliferating and abnormal, dead, or dying cells cause the tissue in the cancerous region to alter in various ways. Changes in the breast that persist, such as a lump, swelling, thickening, puckering, skin irritation, inverted nipple, nipple discharge, breast pain, or tenderness, should be thought of as warning signals. One way that a woman can increase the chances of early detection of breast cancer or noncancerous disorders is by a *breast self-examination* once a month after age 20 (see Fig. 2-15). A woman should also ask her physician how often she should have a clinical breast exam, which is an inspection by a physician of the woman's breasts and surrounding areas.

A *mammogram* is an X-ray of the breasts using very low doses of radiation. This procedure can detect breast lumps too small to be felt by a physical exam. There is no clear evidence that the radiation increases the incidence of breast cancer, although a small risk of this is possible with frequent exposure. The American Cancer Society recommends that women aged 20 to 40 should receive a clinical breast exam every 3 years. Women should receive a baseline mammogram at age 35 to 40; those over 40 should have a clinical exam as well as a mammogram annually. Because there is some controversy about the benefit of mammography for women in their forties, a woman of this age should consult her physician about the frequency of mammograms. Mammograms fail to detect about 10% of small breast cancer lumps, but development of new computer-digitizing methods will improve detection in the future. Another method, *thermography*, detects local regions of high temperature in the breast. Benign or malignant tumors produce more heat than surrounding tissues. A sensitive method of ultrasound imaging, called *high-definition imaging* (HDI), is especially good at differentiating benign from cancerous growths in the breast and could, if widely used, reduce the yearly number of breast biopsies in the United States by about 700,000! Because most breast tumors originate in the mammary ducts, methods are being developed to screen duct cells for abnormalities. Finally, it is now possible to study patterns of gene activity in cancerous cells in the hope of predicting which tumors are most likely to spread rapidly.

1 In the bath or shower:

Examine your breasts during bath or shower. With flat fingers move gently over every part of each breast. Check for lumps, hard knots, or thickenings.

2 Before a mirror:

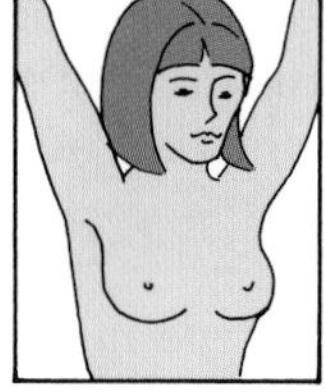

Inspect your breasts with arms at your sides and then with your arms raised overhead. Look for any changes in each breast: a swelling, dimpling of skin or changes in the nipple.

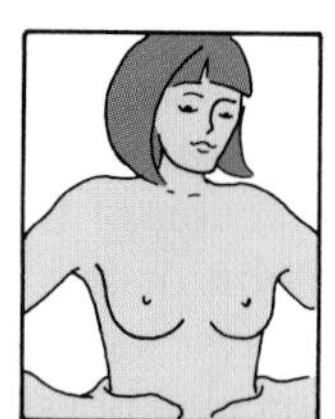

Then, rest palms on hips and press down firmly to flex your chest muscles. Regular inspection shows what is normal for you and will give you confidence in your examination.

3 Lying down:

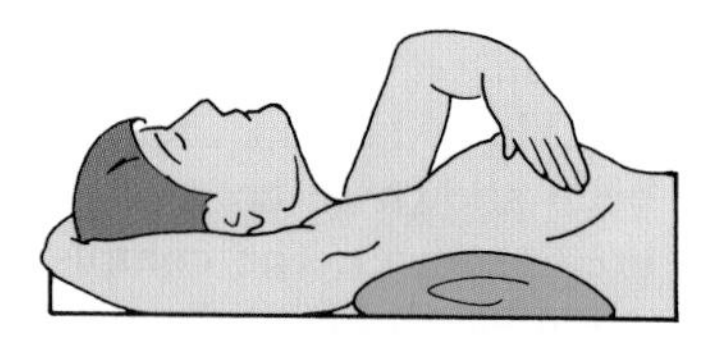

To examine your right breast, put a pillow or folded towel under your right shoulder. Place right hand behind your head - this distributes breast tissue more evenly on the chest. With left hand, fingers flat, press gently in small circular motions around an imaginary clock face. A ridge of firm tissue in the lower curve of each breast is normal. Then move in an inch, toward the nipple, keep circling to examine *every part of your breast*, including nipple. Now slowly repeat procedure on your left breast.

Squeeze the nipple of each breast gently between thumb and index finger. Any discharge, clear or bloody, should be reported to your doctor immediately.

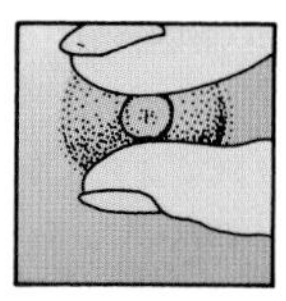

Figure 2-15 How to examine the breasts as a method to detect abnormal lumps and other changes. (Modified with permission of the American Cancer Society.)

Treatment of Breast Cancer

Cancer cells in the mammary ducts that have not spread to surrounding breast tissue are known as *ductal carcinoma in situ* (DCIS). This is the most common form of noninvasive breast cancer and can usually be detected with a mammogram. Although these early cancers may disappear on their own, others develop into invasive cancer. Thus, DCIS is usually treated aggressively; fortunately, most women diagnosed at this early stage of breast cancer can be cured. If the tumor is malignant but has not spread, a *simple mastectomy* (surgical removal of the breast) or *modified radical mastectomy* (removal of the breast and surrounding lymph nodes) often is done. If, however, the tumor is invasive and has spread to surrounding tissues, a *radical mastectomy* involving removal of the breast, underlying pectoral muscle, and axillary lymph nodes can be performed. Cancer specialists have become aware that mastectomies are not always necessary and that a *partial mastectomy* (also known as a *lumpectomy*) may be as effective as more severe operations, especially when combined with radiation or chemotherapy.

A recent study demonstrated that 74% of women who had a partial mastectomy alone (removal of a cancerous tumor plus surrounding tissue and lymph nodes) were free of cancer after 5 years and that this percentage increased to 92% if radiation was given after lumpectomy.

Radiation and chemotherapy have often been used to help prevent breast cancer from reappearing after surgery. Because of the unpleasant side effects of radiation and chemotherapy, however, the drug *tamoxifen* is now used commonly to prevent breast cancer reoccurrence after breast surgery. Women with breast cancer are also given tamoxifen before surgery. This drug, which decreases the spread and reoccurrence of breast cancer in about 40% of those treated, is an antiestrogen, i.e., it blocks estrogen receptors in breast tissue. As mentioned earlier, about 50% of breast cancers require estrogen. Tamoxifen has its effect by shutting down blood vessels that are necessary for tumor growth. Many women are given tamoxifen for 5 to 10 years, and it is now the world's leading cancer medication. Other beneficial effects of tamoxifen are a decrease in *osteoporosis* (brittle bones; see Chapter 7) and a lowering of cholesterol in older women.

Because of minor side effects, including hot flashes and dryness of the vaginal lining, a black cloud is beginning to descend on tamoxifen. More seriously, its use is associated with a 6 to 11 times increase in the incidence of endometrial cancer and also an increase in liver cancer and blood clotting. In fact, in 1996 tamoxifen was officially declared to be a *carcinogen* (cancer-causing agent) by the United Nations. Still, tamoxifen is routinely prescribed for women at high risk for recurring tumors, and there is now some consideration for using tamoxifen to actually prevent breast cancer. Newer antiestrogenic drugs, such as *arimedex* and *raloxifene*, may have fewer side effects. Another drug called *letrozole* may protect women who have survived early stage breast cancer from suffering a relapse of the disease. Results from a large study showed that letrozole was so effective in preventing breast cancer relapse that the study was stopped to make the results known to the public. Unlike estrogenic drugs such as tamoxifen that block the action of natural estrogens, letrozole is an aromatase inhibitor and thus it prevents estrogen synthesis. Taxol, a drug used to treat ovarian cancer, has also been approved to treat some forms of breast cancer (see section on Ovarian Cancer). Scientists are now developing "smart drugs" that target specific proteins associated with especially aggressive tumors; one of these, termed *herceptin*, is being tested on advanced breast cancers.

Chapter Summary

The female reproductive system consists of the ovaries, oviducts, uterus, vagina, external genitalia, and mammary glands. Each ovary consists of an outer ovarian cortex and an inner ovarian medulla. Within the cortex are ovarian follicles in various stages of growth, as well as atretic follicles. Each follicle contains a female germ cell (oocyte) surrounded by a follicular wall consisting of a membrana granulosa and a theca. A membrane, the zona pellucida, separates the granulosa from the oocyte. The number of follicles is fixed by birth, and this number decreases with age because of follicular atresia and ovulation. Follicular growth first involves formation of a primary follicle from a primordial follicle. The primary follicle transforms to a secondary and then a tertiary follicle; the latter contains a fluid-filled antral cavity. The wall of tertiary follicles synthesizes estradiol and progesterone under the influence of FSH and LH. The growth and maturation of follicles involve a complex interaction of FSH, LH, and their receptors on thecal and granulosa cells. The corpus luteum, formed from the ovulated follicle, secretes estradiol and progesterone under the influence of LH.

The diploid primary oocyte contained in most ovarian follicles is arrested in the first meiotic division. A surge of LH at the middle of the menstrual cycle causes completion of the first meiotic division, producing a secondary haploid oocyte and a first polar body. The secondary oocyte begins the second meiotic division while still in the follicle but does not complete this division until after ovulation, when the ovum is penetrated by a sperm. Ovulation is caused by LH and involves a local degradation of the follicular wall. A tear develops at the stigma, and the ovum oozes out with follicular fluid. Abnormal ovarian cysts and ovarian cancer, which can interfere with ovarian function, can be dangerous and even life-threatening.

The oviducts are paired tubes extending from near the ovaries to the uterus. An enlarged oviductal region at the ovarian end, the infundibulum, captures the ovulated egg. Sperm are transported up the oviduct. Fertilization occurs in the oviduct, and the embryo is transported down the oviduct to the uterus. Transport of egg, sperm, and early embryo is aided by contraction of the smooth muscle and the beating of oviductal cilia, both of which are influenced by levels of estradiol and progesterone.

The uterus is a pear-shaped, muscular organ lying in the pelvic cavity. It is divided into three regions—the fundus, corpus, and cervix—and the wall of each region has three layers. The innermost layer, the uterine endometrium, exhibits marked structural changes during the menstrual cycle. Cervical cancer can be detected by a Pap smear, and endometrial cancer by endometrial biopsy. Both forms of uterine cancer can be treated by radiation, chemotherapy, or hysterectomy. Other abnormal uterine growths include endometriosis, cervical cysts and polyps, endometrial polyps and hyperplasia, and uterine fibroids. Disorders of uterine position include retroflexion and prolapse.

The vagina serves as a passageway for menstrual flow, as a receptacle for the penis during vaginal coitus, and as a part of the birth canal. The acidic vaginal environment is maintained by the activity of bacteria in the vaginal lumen.

The female external genitalia (vulva) include the mons pubis, labia majora, labia minora, vaginal introitus, hymen, and clitoris. These structures, and the glands associated with them, are homologous to certain structures in the male. The mammary glands (breasts) consist of glandular tissue and associated ducts embedded in fatty tissue. The growth and function of the glandular tissue are controlled by hormones. Breast cancer is a relatively common disease that might be influenced by lifetime exposure to natural and environmental estrogens as well as inherited genes. Its early detection by breast palpation, X-ray, or thermography can eliminate much of the fatality associated with this condition.

Further Reading

Angell, M. (1994). Do breast implants cause systemic disease? Science in the courtroom. *N. Engl. J. Med.* **330**, 1748–1749.

Boyer, T. G., and Lee, W. (2002). Breast cancer susceptibility genes. *Sci. Med.* **8**, 138–149.

Charlier, C., *et al.* (2003). Breast cancer and serum organochlorine residues. *Occup. Environ. Med.* **60**, 348–351.
Coffey, D. S. (2001). Similarities of prostate and breast cancer: Evolution, diet, and estrogens. *Urology* **57**(Suppl. 1), 31–38.
Conant, E. F, and Maidment, N. D. A. (1996). Breast cancer imaging. *Sci. Am. Sci. Med.* **3**(1), 22–31.
Davis, D.L., and Bradlow, H. L. (1995). Can environmental estrogens cause breast cancer? *Sci. Am.* **173**(4), 166–172.
Ezzell, C. (1994). Breast cancer genes: Cloning BRCA1, mapping BRCA2. *J. NIH Res.* **6**(10), 33–35.
Facklemann, K. A. (1992). The adjuvant advantage: Breast cancer therapies promise a longer life. *Sci. News* **141**, 124–125.
Facklemann, K. A. (1992). Motherhood and cancer: Can hormones protect against breast and other cancers? *Sci. News* **142**, 298–299.
Fackelmann, K. A. (1993). Refiguring the odds: What's a woman's chances of suffering breast cancer? *Sci. News* **144**, 76–77.
Fackelmann, K. A. (1997). The birth of a breast cancer. Do adult diseases start in the womb? *Sci. News* **151**, 108–109.
Guillette, L. J., Jr., and Guillette, E. A. (1996). Environmental contaminants and reproductive abnormalities in wildlife: Implications for public health. *Toxicol. Indust. Health* **12**, 537–550.
Hollander, D. (1995). Risk of ovarian cancer is lessened by childbearing, pill use and hysterectomy. *Family Planning Perspect.* **27**(2), 94–95.
Holloway, M. (1994). An epidemic ignored. *Sci. Am.* (April), 24–26.
Howe, G. R. (1992). High-fat diets and breast cancer risk: The epidemiologic evidence. *J. Am. Med. Assoc.* **268**, 2080–2081.
McLachlan, J. A., and Arnold, S. F. (1996). Environmental estrogens. *Am. Sci.* **84**, 452–461.
MacMahon, B. (1994). Pesticide residues and breast cancer? *J. Natl. Cancer Inst.* **86**, 572–573.
MacMahon, B., *et al.* (1973). Etiology of human breast cancer. A review. *J. Natl. Cancer Inst.* **50**(1), 21–42.
Nicolaou, K. C., *et al.* (1996). Taxoids: New weapons against cancer. *Sci. Am.* **274**(6), 94–98.
Raloff, J. (1993). Ecocancers: Do environmental factors underlie a breast cancer epidemic? *Sci. News* **144**, 10–13.
Raloff, J. (1994). Tamoxifen turmoil. *Sci. News* **146**, 268–269.
Rimer, B. K. (1995). Putting the "informed" in informed consent about mammography. *J. Natl. Cancer Inst.* **87**, 703–704.
Schmidt, K. F. (1993). A better breast test: Bringing digital imaging to mammography. *Sci. News* **143**, 392–393.
Short, R. V. (1994). Human reproduction in an evolutionary context. *In* "Human Reproductive Ecology: Interactions of Environment, Fertility, and Behavior" (K. L. Campbell and J. W. Wood, eds.), pp. 416–424. New York Academy of Sciences, New York.
Shutt, D. A. (1976). The effects of plant oestrogens on animal reproduction. *Endeavour* **35**, 110–113.
Spicer, D. V., and Pike, M. C. (1995). Hormonal manipulation to prevent breast cancer. *Sci. Am.* **2**(4), 58–67.
Stone, R. (1994). Environmental estrogens stir debate. *Science* **265**, 308–310.

Strassmann, B. I. (1999). Menstrual cycling and breast cancer: An evolutionary perspective. *J. Wom. Health* **8**, 193–202.
Taubes, G. (1994). Pesticides and breast cancer: No link? *Science* **264**, 499–500.
Touchette, N. (1994). Trials and errors in breast-cancer research. *J. NIH Res.* **6**(7), 27–29.
Weber, B. L. (1996). Genetic testing for breast cancer. *Sci. Am. Sci. Med.* **3**(1), 12–21.

Advanced Reading

Adashi, E. Y., and Rohan, R. M. (1992). Intraovarian regulation: Peptidergic signaling systems. *Trends Endocrinol. Metab.* **3**, 243–248.
Baker, M. E. (1995). Endocrine activity of plant derived compounds: An evolutionary perspective. *Proc. Soc. Exp. Biol. Med.* **208**, 131–138.
Barbieri, R. (1988). New therapy for endometriosis. *N. Engl. J. Med.* **318**, 512–514.
Boyd, J. (1995). BRCA1: More than a hereditary breast cancer gene. *Nature Genet.* **9**, 335–340.
Brotons, J. A., *et al.* (1995). Xenoestrogens released from lacquer coatings in food cans. *Environ.Health Perspect.* **103**, 608–612.
Brown, M. M., and Lamartiniere, C. A. (1995). Xenoestrogens alter mammary gland differentiation and cell proliferation in the rat. *Environ. Health Perspect.* **103**, 708–713.
Charlier, C., *et al.* (2003). Breast cancer and serum organochlorine residues. *Occup. Environ. Med.* **60**, 348–351.
Clark, C. L., and Sutherland, R. L. (1990). Progestin regulation of cellular proliferation. *Endocr. Rev.* **11**, 266–301.
Cohen, M., *et al.* (1995). Hypotheses: Melatonin/steroid combination contraceptives will prevent breast cancer. *Breast Cancer Res.Treat.* **33**, 257–269.
Collaborative Group on Hormonal Factors in Breast Cancer (2002). Breast cancer and breastfeeding: Collaborative reanalysis of individual data from 47 epidemiological studies in 30 countries, including 50,302 women with breast cancer and 96,973 women without the disease. *Lancet* **360**, 187–195.
Dabre, P. D. (2001). Underarm cosmetics are a cause of breast cancer. *Eur. J. Cancer Prev.* **10**, 389–394.
Demailly, E., *et al.* (1994). High organochlorine body burden in women with estrogen-positive breast cancer. *J. Natl. Cancer Inst.* **86**, 232–234.
Dumeaux, V., *et al.* (2003). Breast cancer and specific types of oral contraceptives: A large Norwegian cohort study. *Intl. J. Cancer* **105**, 844–850.
Eaton, S. B., and Eaton, S. B., III (1999). Breast cancer in an evolutionary context. *In* "Evolutionary Medicine" (W. R. Trevathan *et al.*, eds.), pp. 429–442. Oxford Univ. Press, New York.
Eaton, S. B., *et al.* (1994). Women's reproductive cancers in evolutionary context. *Quart. Rev. Biol.* **69**, 353–367.
Ernster, V. L., *et al.* (1996). Incidence of and treatment for ductal carcinoma in situ of the breast. *J. Am. Med. Assoc.* **275**, 913–918.
Falck, F., *et al.* (1992). Pesticides and polychlorinated biphenyl residues in human breast lipids and their relation to breast cancer. *Arch. Environ. Health* **47**, 143–146.

Gabriel, S. F., *et al.* (1994). Risk of connective-tissue diseases and other disorders after breast implantation. *N. Engl. J. Med.* **330**, 1697–1702.

Guillette, L. J., Jr. (1994). Endocrine-disruptive environmental contaminants and reproduction: Lessons from the study of wildlife. *In* "Women's Health Today: Perspectives on Current Research and Clinical Practice" (D. R. Pakin and L. J. Peddle, eds.), pp. 201–207. Parthenon Publ. Grp., New York.

Haran, E. F., *et al.* (1994). Tamoxifen enhances cell death in implanted MCF7 breast cancer by inhibiting endotheliunr growth. *Cancer Res.* **54**, 5511–5514.

Hirose, K. *et al.* (2003). Dietary factors protective against breast cancer in Japanese premenopausal and postmenopausal women. *Int. J. Cancer* **107**, 276–282.

Hjartaker, A. *et al.* (2001). Childhood and adult milk consumption and risk of premenopausal breast cancer in a cohort of 48,844 women: The Norwegian women and cancer study. *Int. J. Cancer* **93**, 888–893.

Holmes, M. D., *et al.* (1999). Association of dietary intake of fat and fatty acids with risk of breast cancer. *J. Am. Med. Assoc.* **281**, 914–920.

Hunter, D. J., and Kelsey, K. T. (1993). Pesticide residues and breast cancer: The harvest of a silent spring? *J. Natl. Cancer Inst.* **85**, 598–599.

Hunter, D. J., *et al.* (1996). Cohort studies of fat intake and the risk of breast cancer: A pooled analysis. *N. Engl. J. Med.* **334**, 356–361.

Johnson, J., *et al.* (2004). Germline stem cells and follicular renewal in the postnatal mammalian ovary. *Nature* **428**, 145–150.

Krieger, N., *et al.* (1994). Breast cancer and serum organochlorines: A prospective study among white, black, and Asian women. *J. Natl. Cancer Inst.* **86**, 589–599.

Liu, H., *et al.* (1994). Indolo (3, 2-b)carbazole: A dietary derived factor that exhibits both antiestrogenic and estrogenic activity. *J. Natl. Cancer Inst.* **86**, 1758–1765.

Purdom, C. E., *et al.* (1994). Estrogenic effects of effluents from sewage treatment works. *Chemical Ecol.* **8**, 275–285.

Sánchez-Guerrero, J., *et al.* (1995). Silicone breast implants and the risk of connective tissue diseases and symptoms. *N. Engl. J. Med.* **332**, 1666–1670.

Sharpe, R. M., *et al.* (1995). Gestational and lactational exposure of rats to xenoestrogens results in reduced testicular size and sperm production. *Environ. Health Perspect.* **103**, 1136–1143.

Short, R. V. (1994). Human reproduction in an evolutionary context. *In* "Human Reproductive Ecology: Interactions of Environment, Fertility, and Behavior" (K. L. Campbell and J. W. Wood, eds.). *N. Y. Acad. Sci.* **709**, 416–425.

Soto, A. M., *et al.* (1991). *p*-Nonyl-phenol: An estrogenic xenobiotic released from "modified" polystyrene. *Environ. Health Perspect.* **92**, 167–173.

Soto, A. M., *et al.* (1994). The pesticides endosulfan, toxaphene and dieldrin have estrogenic effects on human estrogen-sensitive cells. *Environ. Health Perspect.* **102**, 380–383.

Tawa, K. A., *et al.* (1988). A comparison of the Papanicolaou smear and the cervigram. *Obstet. Gynecol.* **71**, 229–235.

Velicer, C.M., *et al.* (2004). Antibiotic use in relation to the risk of breast cancer. *JAMA* **291**, 827–835.

Wang, Y., *et al.* (1995). Detection of mammary tumor virus *ENV* gene-like sequences in human breast cancer. *Cancer Res.* **55**, 5173–5179.

Whelan, E. A., *et al.* (1994). Menstrual cycle patterns and risk of breast cancer. *Am. J. Epidemiol.* **140**, 1081–1090.
Willett, W. C., *et al.* (1992). Dietary fat and fiber in relation to risk of breast cancer: An 8-year follow-up. *J. Am. Med. Assoc.* **268**, 2037–2044.
Wolff, M. S., *et al.* (1993). Blood levels of organochlorine residues and risk of breast cancer. *J. Natl. Cancer Inst.* **85**, 648–652.

CHAPTER THREE

The Menstrual Cycle

Introduction

Chapters 1 and 2 described how the brain influences secretion of gonadotropic hormones (FSH and LH) from the anterior pituitary gland. These gonadotropins in turn regulate ovarian activity in females by controlling ovarian steroid hormone secretion and by causing maturation of follicles and oocytes and inducing ovulation. Ovarian steroid hormones then regulate the function of the female sex accessory structures and secondary sexual characteristics. However, the whole is greater than the sum of its parts, and the menstrual cycle is the result of finely tuned interactions among the brain, pituitary gland, ovaries, and uterus.

Reproductive Cycles in Mammals

Many mammals exhibit seasonal reproductive cycles, i.e., ovulation occurs only at a certain predictable time of year, and the ovaries are inactive for the remainder of the year. In some of these species, seasonal changes in day length influence reproductive cycles. In human reproduction, there is seasonal variation in the number of births (see Chapter 11).

In most mammals, females are sexually receptive to males only around the time of ovulation, thus ensuring a greater chance for fertilization and pregnancy. Female sexual receptive behavior occurring around the time of ovulation, when estrogen levels are high, is called *estrous behavior* (Greek *oistros*, meaning gadfly or frenzy) or "heat." Animals that exhibit cyclic estrous behavior are said to have *estrous cycles*. (Note: *estrus* is the noun, whereas *estrous* is the adjective.)

In some seasonally breeding mammals, there is only one period of estrus a year (deer, for example); these are called *monestrous species*. During the remainder of the year they are said to be *anestrous* ("without estrus"). Other seasonally breeding animals, such as wild mice, have several estrous cycles within the breeding season, and these are called *polyestrous species*. Some tropical mammals exhibit estrous cycles all year, and humans are similar to these mammals in that they are reproductively active all year. *Menstrual cycles* are found in humans and certain other primates, such as baboons, apes, and monkeys and also in some bats and shrews; uterine bleeding (*menstruation*) occurs during each cycle. There is controversy, however, as to whether human females exhibit an estrous as well as a menstrual cycle, i.e., if women are more receptive to sexual intercourse at the time of ovulation than at other times in the menstrual cycle. In Chapter 8, we look into this controversy.

In some mammals, females are continuously receptive to males but do not ovulate unless stimuli associated with copulation or other male sexual behaviors are present. This phenomenon is called *induced ovulation* in contrast to *spontaneous ovulation*, which is not dependent on male behavior or copulation. In induced ovulators such as rabbits, camels, raccoons, and cats, the act of copulation stimulates sensory nerves in the cervix, the neural message is relayed to the brain, the brain causes luteinizing hormone (LH) secretion from the anterior pituitary gland, and this hormone causes ovulation. In fact, the mere stimulation of the cervix with a glass rod can cause ovulation in these animals. In sexually experienced rabbits, moreover, the presence of a male without copulation is enough to trigger ovulation.

Human females are spontaneous ovulators in that ovulation happens periodically whether *coitus* (copulation; intercourse) occurs or not. However, there is some controversial evidence that coitus can induce ovulation during this spontaneous cycle. For example, more pregnancies result from rape than occur by chance, and it is thought that aggressive coitus may induce ovulation in these cases. Also, the high failure rate of the "rhythm method" of contraception (discussed in Chapter 14) could partly be due to coitus-induced ovulation.

Daily light cycles have a profound influence on the estrous cycle of some mammals. For example, laboratory rats with a 4- or 5-day estrous cycle kept on a 12-h light:12-h dark daily cycle, the lights being turned on at 6:00 AM, ovulate at 2:00 AM on the morning of estrus. Now, if the 12-h light:12-h dark photoperiod is reversed so that the lights turn on at 6:00 PM, the estrous cycles will shift in a few days so that they follow the reversed light cycle. If the rats are kept in constant light, they will be in estrus continuously, and if they are kept in constant darkness, they will be in continuous anestrus. There is no evidence that the human menstrual cycle is influenced by the daily light cycle. Blind women, for example, have normal menstrual cycles.

Is it a coincidence that the average human menstrual cycle length is 29.5 days, the same as the lunar (moon) cycle length? Perhaps women evolved a cycle that, for some reason, was associated with the lunar cycle. A recent study showed that more women than would be predicted by chance ovulate during the dark (new moon) phase of the lunar cycle.

Social interactions also can influence estrous cycles in some mammals. If female rats are kept together without males, their estrous cycles will be irregular and not synchronized. However, if females are grouped together with males, most of the females eventually will reach estrus on the same day, the so-called *Whitten effect*. Remarkably, an analogous situation can occur in humans, and it has been called the "dormitory effect." M. K. McClintock has shown that women grouped together for a period of time in dormitories tend to have synchronized menstrual cycles, i.e., most menstruate at the same time. Also, an increase in dating activities decreases menstrual cycle length. It has been shown that the menstrual cycles of college women who have coitus frequently are less variable in length than those in women who abstain from sex. Some of these social effects on menstrual cycles may be mediated by pheromones (see Chapter 8).

Major Events in the Menstrual Cycle

The menstrual cycle is a highly coordinated series of events that results in ovulation of a single follicle and preparation of the uterus to receive the embryo after fertilization. This cycle is driven by activity of the hypothalamic GnRH pulse generator and surge center. Gonadotropins released from the pituitary induce changes in steroid secretion by the ovaries. In addition to these hormonal fluctuations, structural changes occur in the ovaries and the uterine endometrium. Thus, to understand the menstrual cycle, we must discover how all of these changes in endocrine status, ovarian follicles, and uterine tissues are orchestrated. During these cycles, a woman's reproductive system is in constant flux, and the hormonal changes affect nonreproductive tissues as well.

The word *menstrual* has its root in the Latin word *mensis*, which means "month"; the menstrual cycle averages the length of a lunar month (29.5 days). Only about 10 to 15% of menstrual cycles, however, are exactly 29.5 days. Cycle length can vary greatly in a single woman and even more so among different women. Most cycles are 25 to 30 days long, but some last less than 25 days or longer than 30 days. Younger women tend to have longer cycles than older women. For example, the average cycle length is 35 days in 15 to 19 year olds, 30 days in 30 year olds, and 28 days in 35 year olds. Cycle length is more variable in teenagers and older women than in women in their peak reproductive years. With a cycle length of 28 days, a woman would ovulate about 13 ova per year, or 481 ova in the 37 years from puberty to the end of her reproductive (childbearing) life. Pregnancy and lactation, however, would reduce this total number because ovulation is inhibited during these times (see Chapters 10 and 12). The use of contraceptive methods that stop ovulation (see Chapter 14) would also reduce this number.

Each menstrual cycle can be divided into three main phases: (1) the *menstrual* ("destructive") *phase*, also called *menses*; (2) the *follicular* ("proliferative" or "estrogenic") *phase*; and (3) the *luteal* (or "secretory," "progestational") *phase* (Fig. 3-1). It is customary to designate the first day of the cycle (day 1) as the first day of the menstrual phase. This is the day on which menstruation begins, and it is chosen as a starting point because a woman is more aware of this day than any other in the cycle. The menstrual phase can last from 1 to 6 days, but usually lasts 4 or 5 days. The follicular phase begins at the end of menstruation, or about day 6 if menstruation lasts 5 days. During the follicular phase, the ovaries continue growth that began during the menstrual phase and then secrete estradiol, which in turn causes growth of the uterine endometrium. This phase continues until ovulation, which occurs on about day 14 of a 28-day cycle and on day 16 of a 30-day cycle. Thus, ovulation usually occurs about 14 days before the onset of the next menstruation (day 1 of the next cycle). The luteal phase lasts from ovulation to the beginning of menstruation. It is during this phase that the corpus luteum, which is formed from the wall of the ovulated follicle (see Chapter 2), secretes estradiol and progesterone, which prepare the uterus for the arrival of an embryo. The ovum disintegrates if fertilization does not occur, and its remains are washed out of the body in the menstrual flow. If fertilization does occur, the embryo may implant (embed) in the uterine lining (see Chapter 10), and further menstrual cycles are inhibited during pregnancy. Changes in the oviducts during the menstrual cycle in relation to gamete transport and fertilization are discussed in Chapter 9.

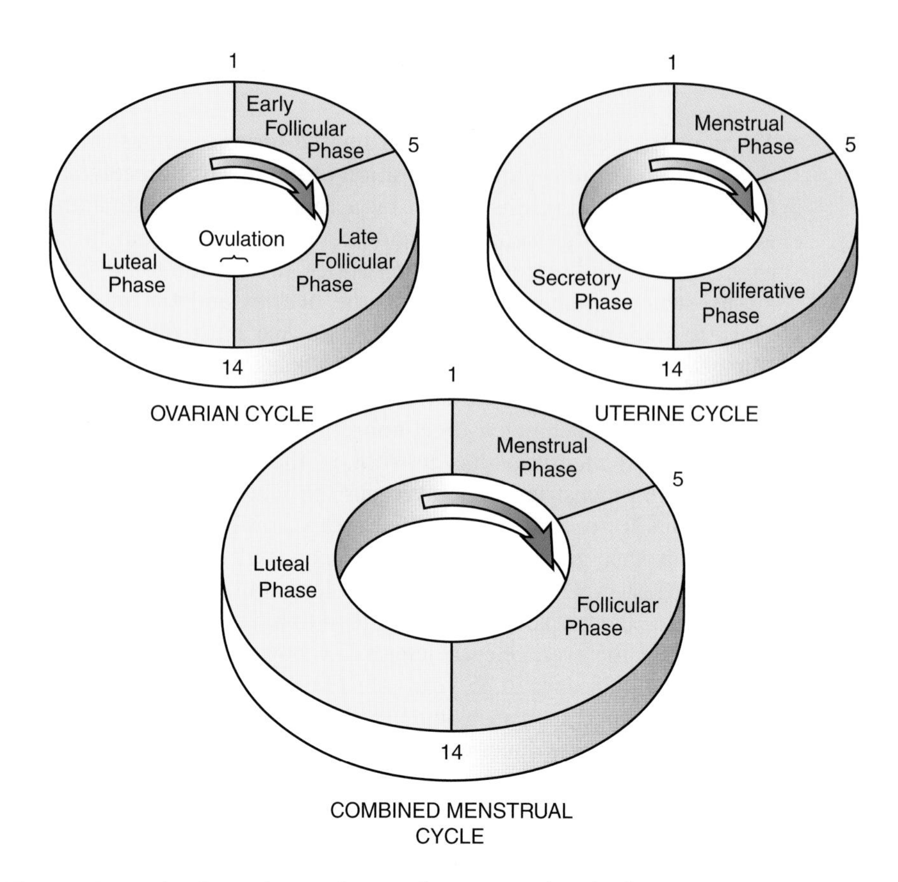

Figure 3-1 The three phases of a 28-day menstrual cycle showing ovarian and uterine events. The first two phases of the *ovarian cycle* (early and late follicular phase) represent growth of selected ovarian follicles. The luteal phase, when the corpus luteum is present, lasts from ovulation to the beginning of the next cycle. The *uterine cycle* describes important uterine events during the 28-day cycle. The menstrual phase, or period of menstrual flow, is followed by the phase of endometrial proliferation. After ovulation, the uterus enters the secretory phase. Both ovarian and uterine events are depicted in the more commonly used *combined menstrual cycle* (lower figure). Here, day 1 is the beginning of the menstrual phase, which lasts about 5 days. The menstrual phase is followed by the follicular phase, which ends at ovulation on about day 14. Next is the luteal phase, ending 14 days later with the beginning of menstruation. The combined menstrual cycle will be used in the text.

The Menstrual Cycle in Detail

Menstrual Phase

During menstruation (Fig. 3-1), part of the lining of the uterus, the *stratum functionalis* of the endometrium, degenerates and is sloughed off. This tissue dies by *ischemia* because the blood vessels that supply it with nutrients and oxygen begin to constrict and dilate spasmodically. These vessels bleed as part of the lining degenerates: 33 to 267 ml (1–8 oz) of blood can be lost during menstruation, but most women lose 33 to 83 ml (1 to 2.5 oz). When a woman suspects that more than 300 ml (about a cup) of blood is lost, she should supplement her diet with iron to prevent possible iron-deficiency anemia (iron is an important component of red blood cells). Women using intrauterine contraceptive devices tend to have a heavier than average menstrual flow, whereas those using the combination pill tend to have a lighter than average flow (see Chapter 14). Excessive menstrual bleeding should be reported to a physician.

Menstrual discharge contains not only blood but also other uterine fluids along with debris from the sloughing endometrium and some cells that have been lost from the lining of the vagina. The debris from the endometrium is thick and often is mistaken for blood clots. Actually, the uterine blood initially clots but then it liquefies and leaves the body in the menstrual flow. Thus, menstrual fluid really is like blood serum. The characteristic odor of menstruation is caused by the action of bacteria on the menstrual flow. A recent theory holds that menstruation evolved to help protect the female against infection by sexually transmitted diseases carried by the male (see HIGHLIGHT box 3-1).

The corpus luteum of the previous cycle has regressed by day 1, and the resultant decrease in blood levels of estradiol and progesterone causes the stratum functionalis of the uterine endometrium to degenerate (Fig. 3-2). The ovaries at day 1 contain only small tertiary follicles, less than 5 mm in diameter, along with several atretic follicles and thousands of smaller follicles (see Chapter 2). By day 3, some of the small tertiary follicles have enlarged to about 10 mm in diameter.

Figure 3-2 summarizes the changes in blood levels of pituitary hormones (FSH and LH) and steroid hormones (estradiol and progesterone) during the menstrual cycle. Study Figure 3-2 closely before you proceed. Note that on day 1 of the cycle, levels of all four hormones are low. By day 3, however, FSH and LH begin to cause the aforementioned increase in size of some follicles at this time. As these follicles grow, they begin to secrete estradiol so that estradiol levels also begin to rise by day 3. Levels of progesterone, however, remain low during the menstrual phase. GnRH pulses, and the resultant LH and FSH pulses in the blood, are of low amplitude and occur about 1 every 100 min (Fig. 3-4).

Follicular Phase

The follicular phase is so named because it is characterized by the rapid growth of ovarian follicles. Under the influence of FSH, some of the tertiary follicles in each ovary increase in size during the follicular phase, and a few reach 14 to 21 mm in diameter by days 10 to 12. Other follicles that have previously undergone rapid growth become atretic (degenerate). By day 13, usually only one large Graafian follicle, 20 to 25 mm in diameter, is present in only one ovary; this large follicle appears like a blister on the surface of one ovary. The remaining large follicles in each ovary become atretic by day 13.

Estradiol is secreted by the larger follicles in each ovary, including those that may later become atretic. The rise in estrogen causes the uterine endometrium to proliferate (thicken) during the follicular phase. Thus, with reference to the uterus, this phase of the menstrual cycle is often called the *proliferative phase*. Under estrogenic influence, uterine glands begin to enlarge (Fig. 3-2). The endometrium becomes more richly supplied with blood vessels, and water accumulates between cells in the tissues, a condition known as *edema*. In addition, the smooth muscle of the myometrium begins to contract mildly in a rhythmic fashion, although a woman usually is not aware of these slight contractions.

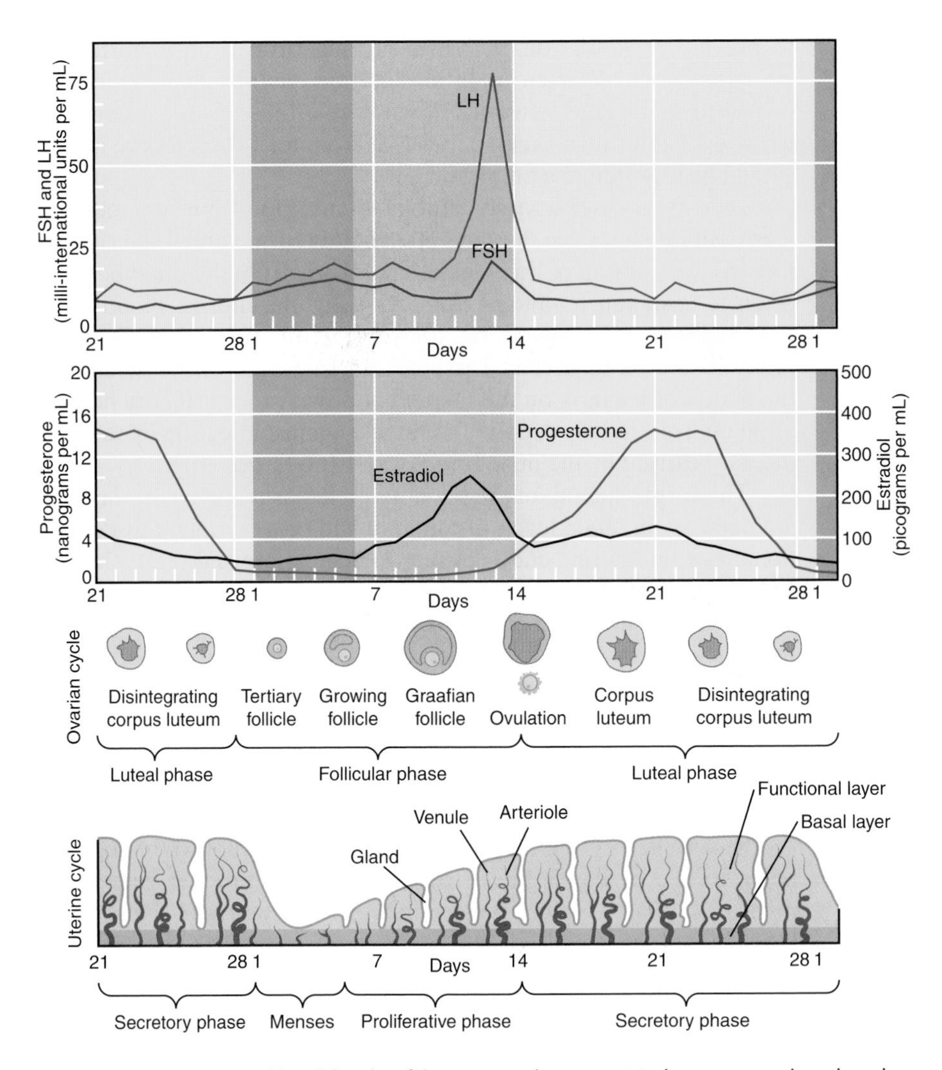

Figure 3-2 Average blood levels of hormones during a 28-day menstrual cycle, along with changes in the ovaries and uterine endometrium. Note that an estradiol peak precedes the LH surge. The ratio of estradiol to progesterone is high during the follicular phase and lower during the luteal phase. The ovaries also secrete small amounts of another estrogen (estrone), androgens, and 17-hydroxyprogesterone (not shown). Levels of progesterone are in nanograms (one thousandth of a milligram) and of estradiol in picograms (one thousandth of a nanogram). Levels of FSH and LH are in milli-international units (one thousandth of an international unit). An international unit is an amount of a hormone that produces a given biological response in a target tissue. Although average FSH and LH blood amounts are shown, the levels of these hormones and GnRH exhibit hourly pulses (Fig. 3-4).

Levels of estradiol in the blood continue to rise throughout most of the follicular phase, reaching a maximum (peak) on day 12 or 13 (Fig. 3-2). About 24 to 48 h after this peak in estradiol, a surge (marked and rapid increase) of LH occurs in the blood (Fig. 3-2). This LH surge initiates the resumption of meiosis in the oocyte within the largest follicle and also causes ovulation of this

Chapter 3, Box 1: Why Women Menstruate

Menstruation is the shedding and bleeding of the uterine endometrium between successive ovulations and loss of this material through the vagina. Menstruation occurs not only in humans but also in many other primates, as well as in several kinds of bats and shrews. Nevertheless, most mammals do not menstruate using the aforementioned definition. Did menstruation evolve with a specific function in humans or is it simply a by-product of some other important process?

Some aspects of menstruation are clearly disadvantageous to women. Valuable iron and other nutrients are lost with the blood flow. The pain and discomfort of menstruation can be distracting or even disabling for some women. In our ancestors, the presence of menstrual blood could have attracted predators. For the menstrual flow to have evolved and been maintained in humans, some scientists argue that a clear adaptive advantage must offset these drawbacks. In recent years, several hypotheses have been offered to propose evolutionary explanations for menstruation. These include the following.

1. Menstruation evolved because it removes sexually transmitted microorganisms that cause infertility and disease. That is, menstruation flushes out bacteria and viruses that are carried into the female reproductive tract by sperm. This theory has several problems. For example, most mammals do not menstruate but still are faced with sexually transmitted diseases. Also, blood is an excellent culture medium for sexually transmitted microorganisms, and some sexually transmitted diseases are acquired more often just after menstruation. Third, women menstruate only infrequently. Hunter-gatherers, the ancestors of modern women 10,000 years ago, rarely had menstrual cycles (see HIGHLIGHT box in Chapter 2); thus this kind of protection from disease would have been highly unreliable. Finally, modern women cycling naturally or undergoing periodic menstruation while on birth control pills still easily acquire sexually transmitted diseases.

2. Menstruation saves the energy it would take to maintain a constantly secretory endometrium. During each cycle, the endometrium thickens and increases in mass and then regresses prior to menstruation. The fully grown endometrium uses about seven times as much oxygen (a measure of energy use) than in the regressed state, and maintenance of the thickened lining would continue to require this additional energy. However, the amount of energy it would take to maintain a secretory endometrium would be much less than it takes to regrow the endometrium during the next follicular phase. There are better reasons why it would not be adaptive to maintain a constant secretory endometrium. There would be no ovulation because the same hormones (estradiol and progesterone) needed to maintain the endometrium also inhibit GnRH and therefore gonadotropin secretion and ovulation. Maintaining a constant secretory endometrium could also cause endometrial hyperplasia (abnormal overgrowth of the endometrium), uterine fibroids, or perhaps endometrial cancer because of the constant bombardment of the uterus by estradiol and progesterone.

3. Menstruation rids a women's body of unimplanted embryos. This would be adaptive because of the potential harm embryonic cells could do to a woman's health in the form of autoimmune diseases (see HIGHLIGHT box in Chapter 10).

4. Menstruation is necessary to allow sperm transport. The thickened, secretory endometrium nearly fills the uterine lumen, and very little uterine fluid is present at this time. Under these conditions, it would be difficult for sperm to move through the uterine lumen. For fertilization to occur effectively, the endometrial lining built up during the previous cycle must first slough off, allowing the uterine space to enlarge and fill with fluid.

5. Menstruation signals to males that a woman is not pregnant. Nonpregnant female mammals typically signal their sexual availability through behavioral estrus. Humans do not exhibit estrus

Continued on next page.

Chapter 3, Box 1 continued.

behavior. Could menstruation be a demonstration to males that a woman has not been impregnated? There are problems with this idea. First, it is most effective to signal sexual preparedness at ovulation, not 2 weeks earlier. Second, the evolution of a new way to show sexual receptivity in humans would be unexpected when estrus is used as an effective signal in other primates.

In wild species that menstruate, most females are pregnant or nursing most of their adult lives. Thus, menstruation is rare and many females pass their genes to the next generation in the absence of any selective effect of menstruation. Perhaps there has been no selection favoring menstruation and that it exists simply as a consequence of other aspects of female reproductive physiology. Humans have a very thick, secretory endometrium due to the invasive nature of implantation in our species (Chapter 11). In Chapter 10 you will learn that the embryo and then fetus develop within the endometrium and not actually in the uterine cavity. This endometrium is large relative to body size in humans and chimps, both of which menstruate. The thickened tissue must diminish and regress before a new cycle occurs. Mammals with no menstruation also build up their endometrium during the luteal phase of the estrous cycle. Endometrial regression is not uncommon, but in most mammals the tissue is simply reabsorbed and the endometrium regresses with little or no bleeding as a new cycle begins. Most of these other mammals, however, do not have a massively thickened endometrium during the luteal phase of their cycle because they have a less invasive implantation and smaller fetal size. An argument against this theory is that although other primates have an invasive implantation and do menstruate, bats and shrews do not have an invasive implantation or a considerably thickened endometrium and also menstruate.

Some think that human menstruation is not an evolved adaptation but simply a way to remove excess blood and regressed uterine tissue. However, the presence of specialized mechanisms of the menstrual process, such as constriction of spiral arteries as well as clotting and then liquefying of the menstrual blood, implies evolved adaptations that are not simply normal consequences of breakdown of the endometrium when a fertilized egg does not implant.

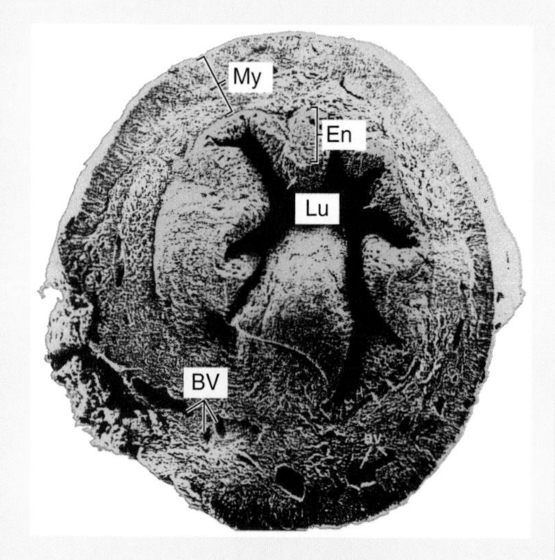

A cross section through the uterus showing the endometrium (En, in red) and the myometrium (My). The stratum functionalis of the endometrium sloughs off during menstruation. Lu, uterine lumen (cavity); BV, blood vessels.

follicle (Chapter 2). The LH surge lasts about 36 h, and ovulation occurs about 9 to 12 h after the peak of the surge. Progesterone and FSH levels remain low in the follicular phase until just before ovulation. At this time, a small FSH surge accompanies the greater LH surge, and progesterone levels rise slightly just before ovulation (Fig. 3-2). This progesterone comes from the wall of the largest follicle under the influence of the LH surge (see Chapter 2) and may (along with estradiol) be important in regulating the LH surge.

Intricate feedback mechanisms operate during the follicular phase. Let us first look at the feedback effects of estradiol on the secretion of gonadotropin-releasing hormone from the hypothalamus. Remember from Chapter 1 that the hypothalamus of the brain may secrete only a single GnRH, which controls secretion of both FSH and LH from the pituitary gland. Now look at Fig. 3-3 and you see that moderate circulating levels of estradiol have a negative feedback effect on GnRH secretion. At the beginning of the follicular phase, when estradiol levels are low, FSH levels rise because of the absence of negative feedback by estradiol on GnRH secretion. After estradiol reaches moderately high levels in the midfollicular phase, it exerts negative feedback on GnRH secretion, and FSH levels begin to decline. Human ovarian follicles contain a substance (inhibin) that inhibits FSH (but not LH) secretion, and this substance may also exert negative feedback effects. A related ovarian peptide called *activin* stimulates FSH secretion. The ovary also produces *follistatin*, a glycoprotein that inhibits FSH secretion by binding to activin and neutralizing its effects. In the last days of the follicular phase, estradiol levels are very high (secreted from several large follicles in each ovary) and now exert a positive feedback on GnRH secretion. This increase in GnRH causes the LH surge on day 14. GnRH (and FSH and LH) pulses increase 10 times in amplitude and occur every 30 min during the LH surge (Fig. 3-4).

Many questions remain about the control of FSH and LH secretion during the follicular phase. For example, why is it that FSH and LH do not cycle in a similar manner (Fig. 3-2) if there is only one GnRH? The answer is not clear, but a possible explanation is that changing estradiol levels not only influence secretion of GnRH from the hypothalamus but also change the sensitivity of the anterior pituitary gland to GnRH. More specifically, it seems that a given level of estradiol may affect the sensitivity of cells in the pituitary that secrete FSH and LH in a

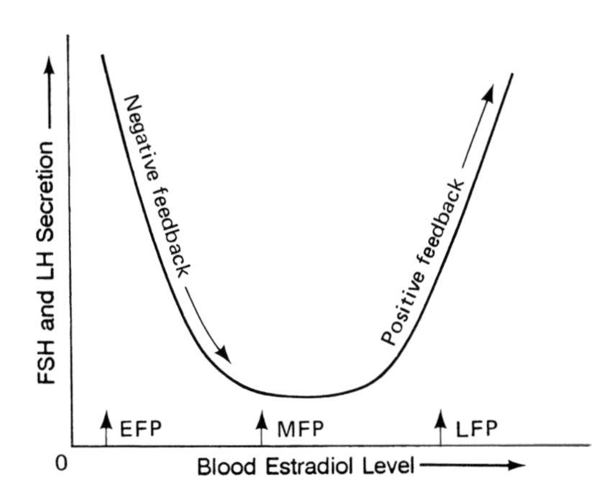

Figure 3-3 The effect of increasing blood levels of estradiol during the follicular phase of the menstrual cycle on secretion of gonadotropin (FSH and LH) from the pituitary. Note that in the early follicular phase (EFP), estradiol levels are low and gonadotropin output is high due to the absence of negative feedback of estradiol on gonadotropin secretion. In the midfollicular phase (MFP), estradiol levels are moderately high and exert a negative feedback on gonadotropin secretion. In the late follicular phase (LFP), the very high estradiol levels now exert a positive feedback on gonadotropin secretion, resulting in the LH surge.

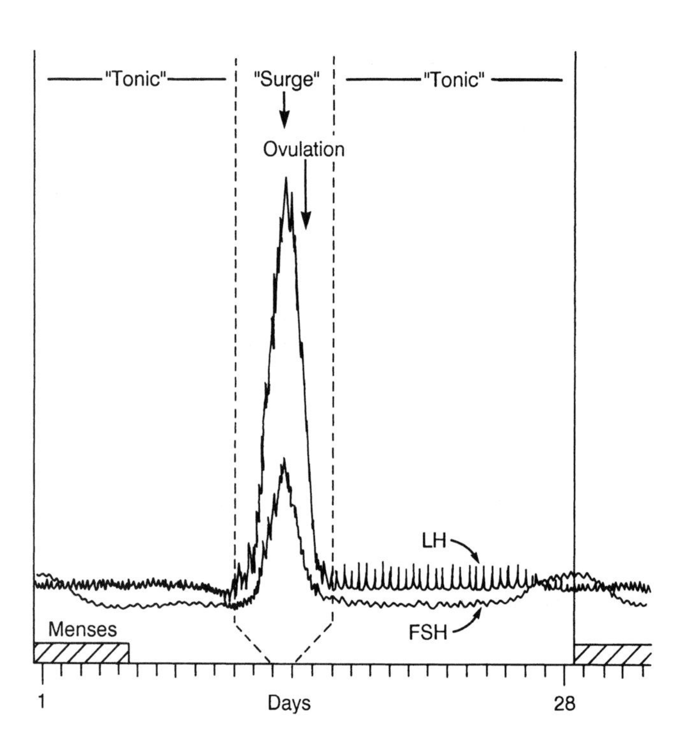

Figure 3-4 The pattern of hourly pulses of LH and FSH in the blood during the menstrual cycle.

different manner. For example, high levels of estradiol in the late follicular phase may have a positive effect on GnRH, but simultaneously decrease the sensitivity of FSH-secreting cells to GnRH. Also, as the pattern of GnRH pulses changes during the cycle (Fig. 3-4), FSH and LH secretion are affected differently. Furthermore, inhibin may selectively inhibit FSH and not LH secretion. Why is there a small FSH surge along with the LH surge on day 14? It may be that the pituitary is unable to release a strong pulse of LH without also releasing some FSH. If an injection of estradiol is given to a woman, an LH surge follows. If, however, this estradiol injection is followed by an injection of progesterone, both FSH and LH surge in the blood. Progesterone alone does not cause either gonadotropin to surge. Finally, estrogen could cause an LH surge by acting directly on the pituitary gland alone instead of on the hypothalamus. Future research on how steroid hormones cause the gonadotropin changes that lead to ovulation is a subject of great interest to clinical scientists because they may be able to use this knowledge to help infertile women ovulate (see Chapter 16).

Look at Fig. 3-2 and notice that estradiol levels drop rapidly just before ovulation. The largest follicle in the ovary reduces its secretion of estradiol and begins to produce progesterone immediately before ovulation. This drop in estradiol can lead to mild uterine bleeding at ovulation, called *spotting* or *breakthrough bleeding*, a somewhat rare phenomenon. Spotting is not menstruation.

There are also many mysteries about the ovarian changes that occur in the follicular phase. For example, although many growing follicles are present in an ovary, we do not understand why eventually only one is selected for ovulation. The other follicles, although growing and initially secreting estrogens, become atretic. We do know that if the large follicle that is going to ovulate is removed early in the follicular phase, another follicle will takes its place. Thus, a follicle that was programmed to die now remains viable and ovulates. We also do not understand the mechanisms controlling alternation of ovulation. If a woman's right ovary ovulates one egg in a particular month, chances are that her left ovary will ovulate the next time around. How this alternation occurs is not clear, but we do know that in women who have had one ovary removed (*ovariectomy*, or *oophorectomy*), the remaining ovary will ovulate every month instead of every 2 months.

Luteal Phase

During the luteal phase, the uterine endometrium becomes thick and spongy (Fig. 3-2) and its glands secrete nutrients that will be used by the embryo—if one is conceived (see Chapter 10). Because the uterus is secretory during the luteal phase, this part of the menstrual cycle is termed the *secretory phase* when referring to endometrial physiology. During this phase, the uterine smooth muscle contracts less frequently than during the follicular phase.

After ovulation, a corpus luteum is formed from the wall of the follicle that ovulated (see Chapter 2). This structure then begins to secrete estradiol and progesterone. The levels of these two hormones rise in the middle of the luteal phase (Fig. 3-2). About 4 days before menstruation begins, the corpus luteum begins to degenerate, and levels of these two steroid hormones decline (Fig. 3-2). It is the combination of high levels of estradiol and progesterone during the luteal phase that maintains the uterus in its secretory condition; when blood levels of these hormones decrease, the endometrium begins degeneration, resulting in the onset of menstruation.

The presence of estradiol and progesterone in the luteal phase results in negative feedback on both FSH and LH secretion. Because of this negative feedback, the levels of FSH and LH are relatively low in the luteal phase (Fig. 3-2). The progesterone:estrogen ratio slows down the GnRH pulse generator to about one pulse every 4 h. This inhibits FSH release and thus restricts follicular development during the luteal phase. You may wonder why a combination of estradiol and progesterone causes release of FSH and LH before ovulation but inhibits secretion of FSH and LH by negative feedback during the luteal phase. Apparently, GnRH is released when progesterone levels are low compared with estradiol levels (day 13), but when progesterone levels are relatively high compared with estradiol in the luteal phase, GnRH secretion is inhibited. Combination pill oral contraceptives contain a little estrogen and a large amount of progestogen because this combination mimics the hormonal condition in the luteal phase and thus prevents ovulation (see Chapter 14). When estradiol and progesterone levels begin to decline at the end of the luteal phase, secretion of GnRH is no longer inhibited and a new cycle begins.

In humans, LH is necessary for formation of the corpus luteum and for its secretory function in the luteal phase. In other words, the low levels of LH in the luteal phase (Fig. 3-2), although not high enough to cause ovulation, are sufficient to maintain the corpus luteum. In some other mammals (such as the rat), LH also causes formation of the corpus luteum, but another pituitary hormone, prolactin

(PRL), is necessary for maintenance of the corpus luteum. There is little evidence, however, that PRL is necessary for function of the human corpus luteum.

What causes the corpus luteum to die at the end of each menstrual cycle? There are several suggested reasons. In sheep, a prostaglandin secreted by the uterus causes the corpus luteum to regress. There is no evidence, however, that uterine prostaglandin is a *luteolytic* ("corpus luteum killing") *factor* in humans, although prostaglandins are present in the human endometrium near the end of the luteal phase, and human corpus luteum cells have prostaglandin receptors. Women who have had their uterus removed for medical reasons have normal hormone cycles, and the life span of their corpus luteum is similar to that in women with a uterus. Therefore, prostaglandins secreted by the uterus probably do not kill the human corpus luteum.

Perhaps the slight drop in LH levels near the end of the luteal phase (Fig. 3-2) causes the corpus luteum to die. However, although injection of LH into a woman can prolong the life of her corpus luteum for a few days, it then dies anyway. The number of LH receptors in the human corpus luteum drops as this structure ages, which may play a role in its death. Changes in the secretion of prolactin could also influence the life of the human corpus luteum, but as was said earlier, we know little about the role of this hormone in the menstrual cycle. Perhaps secretion of an estrogen by the corpus luteum leads to its degeneration because injection of an estrogen directly into the human corpus luteum causes it to die. For some reason, administration of a GnRH inhibitory agonist causes the death of the corpus luteum in women, but the natural role of GnRH in maintaining the corpus luteum is unknown. Recently, it has become known that the human corpus luteum may secrete its own prostaglandin, which kills the luteal cells. How would the prostaglandin do this? Perhaps by stimulating secretion of oxytocin from the luteal cells, which then kills the corpus luteum by reducing blood flow to it. This is the same oxytocin that is also secreted by the neurohypophysis (see Chapter 1). Recent evidence suggests a role of the local messenger molecule nitric oxide (NO) in luteolysis. In the corpus luteum, NO increases the synthesis of prostaglandins and causes apoptosis (programmed cell death) of luteal cells.

About a week before menstruation, the breasts become larger and more sensitive, even painful in some women. This is the result of high levels of estradiol and progesterone in the luteal phase, which cause cell division as well as edema (water retention) in the breast tissue. No milk is secreted, however, probably because not enough prolactin is present (see Chapter 10).

Table 3-1 summarizes the events of the three phases of the menstrual cycle.

Table 3-1 Summary of Events in the Menstrual Cycle, Listed in Order of Occurrence

1. ↑ FSH → follicular growth and estradiol secretion from follicles
2. ↑ Estradiol → proliferative growth of endometrium
   → inhibition of FSH (along with inhibin)
   → stimulation of LH surge
3. LH surge and smaller increase in FSH → ovulation and corpus luteum formation
4. Corpus luteum secretes estradiol and progesterone
5. ↑ Estradiol and progesterone → inhibits FSH and LH
   → secretory phase of uterus
6. Corpus luteum degenerates if fertilization does not take place
7. ↓ Estradiol and progesterone → menstrual flow
8. ↓ Estradiol → ↑ FSH; cycle begins again

Variations in Length of Menstrual Cycle Phases

Figure 3-5 depicts the length of each phase for menstrual cycles that vary in total length (25, 28, 35 days). Note that it is the length of the follicular phase that accounts for most of the variation in total length of the cycle. Although the length of the menstrual phase depicted in Fig. 3-5 is constant, it too can vary among women. In contrast, the length of the luteal phase usually is about 14 days, regardless of the length of the other phases. Thus, ovulation usually occurs 14 days before the onset of menstruation. However, the luteal phase length can also vary in some women.

Recall that progesterone released from the corpus luteum is necessary to maintain the endometrium during the 14 days of a normal luteal phase. If, however, insufficient progesterone is produced or if the endometrium is incapable of responding adequately to the available progesterone, the endometrial lining degenerates prematurely and menstruation begins early (i.e., the luteal phase is shortened). Alternatively, the luteal phase may be of normal length, but progesterone levels are low. Both of these conditions are called *luteal phase deficiency*. Because an inadequate progesterone level prevents sufficient development of the stratum functionalis to support normal implantation, luteal phase deficiency may be a factor in early miscarriage (see Chapter 16). This disorder may be caused by an inadequate LH surge, resulting in failure to ovulate or incomplete luteinization of the follicle.

Methods for Detecting Ovulation

Several methods for detecting ovulation are available for women who wish either to avoid pregnancy or to improve the chances of conception. Although reliable methods are important for those practicing the "rhythm method" of contraception and for determining causes of infertility in women (see Chapters 14 and 16), none so far devised can actually pinpoint the *exact* time of ovulation in advance. First, we look at those methods that a woman can use at home and then at those used more commonly in the laboratory.

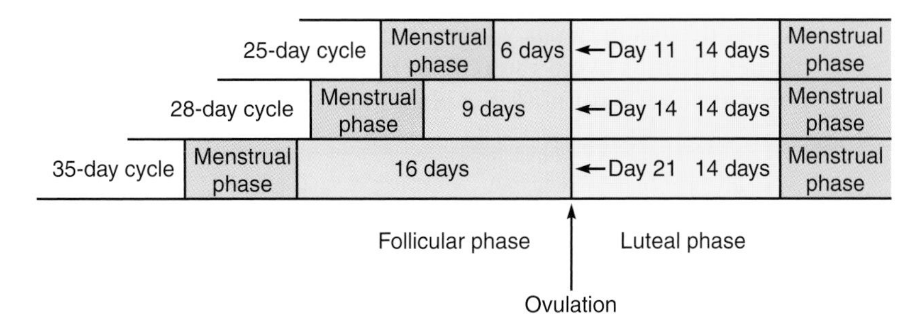

Figure 3-5 Variations in length of menstrual cycle phases. Note that the length of the luteal phase is usually 14 days. Lengths of the follicular phases of these cycles are 6, 9, and 16 days. The length of the menstrual phase, although shown as constant here, can vary.

Home Methods

As mentioned earlier, ovulation usually occurs about 14 days before menstruation begins. If a woman keeps an accurate record of her cycle lengths for at least a year, she may be able to predict the time in her cycle when ovulation most likely occurs. Whether she wants to determine this time to increase her chances of becoming pregnant or as a method of birth control (i.e., to decrease her chances of becoming pregnant), she should take into account the days that sperm would survive and the days that the ovum is fertilizable (see Chapter 7). Details of the calendar method of detecting the time of ovulation are presented in Chapter 14. Given the wide variation in total cycle length and in the length of each cycle phase, however, the calendar method is not very reliable in detecting the time of ovulation.

Some women feel pain in the abdomen around the time of ovulation. This *Mittelschmerz* (German for "midpain") usually is mild and probably is caused by irritation of the abdominal wall by blood and follicular fluid escaping from the ruptured follicle at ovulation. The pain can last for a few minutes to a couple of hours. Because of its mild nature, Mittelschmerz often goes undetected and is not a reliable method for determining the time of ovulation.

Basal (resting) body temperature, after a slight decrease at ovulation, rises 0.3 to 0.5 °C (0.5 to 1.0 °F) during the luteal phase of the cycle. This rise is caused by the increase in progesterone in the blood during this phase. To create a basal body temperature chart, a woman should take her temperature immediately upon awakening every morning, before getting out of bed, using a special basal temperature thermometer. Although this method may show a woman when ovulation has occurred, it is not totally reliable for predicting ovulation, which partly accounts for the high rate of failure of the rhythm method of contraception when body temperature is used as an indicator.

The cervical *mucus*, which is thick and sticky up to 5 to 7 days before and 1 to 3 days after ovulation, becomes more abundant, watery, and stringy near the time of ovulation. Women often notice an increase in mucous flow and a moistening of the vagina near the time of ovulation. Generally, laboratory tests for these changes in mucus are more reliable than personal observation for detecting ovulation time.

Some commercial kits are now available for the detection of ovulation. One type of test kit relies on the detection of LH in the urine. When plasma LH rises before ovulation, some LH appears in the urine. The test kit contains antibodies to LH; when the test substance is mixed with the urine, a color develops if LH is present. These kits are mainly used to determine when ovulation occurs if a woman desires pregnancy (see Chapter 9). If used for contraception (i.e., determining when to avoid intercourse), these tests should be used only in conjunction with other natural family planning methods (see Chapter 14).

Laboratory Methods

Levels of LH or steroid hormones in the blood can be determined in the laboratory. Detection of the LH surge or the increase in blood progesterone levels in the luteal phase can determine if ovulation will occur or has occurred. A breakdown product of progesterone, *pregnanediol*, can also be detected in urine.

As mentioned earlier, tests for the condition of cervical mucus can be performed in the laboratory. Around the time of ovulation, the mucus exhibits a fern-like pattern of crystals of sodium and potassium chloride when dried. This *Fern test* for ovulation is combined with a test for *Spinnbarheit* (threadability) of the mucus and the degree of opening of the cervix to obtain an overall "cervical score." This score is highest when blood estradiol is high and lowest when blood progesterone is high.

The appearance of cells lining the human vagina also changes near the time of ovulation, i.e., cells sloughing from the vagina are more cornified (hardened with the protein keratin) near the time of ovulation when blood estradiol levels are high. Microscopic examination of vaginal smears, therefore, can be an indication of ovulation. Finally, a piece of the endometrium can be removed by a physician. It then can be examined to see if it is in the proliferative or secretory phase. If in the latter, ovulation has occurred. This is called an endometrial biopsy (see Chapter 2).

Premenstrual Syndrome

Symptoms

About 70 to 90% of all women experience some physical and/or emotional difficulties before menstruation begins, although these symptoms are severe in only 5%. These difficulties are called *premenstrual syndrome* (PMS) (a syndrome is a group of symptoms that occur together) and usually begin 3 to 10 days before menstruation. The various physical symptoms include cramps, backache, nausea, dizziness, fatigue, breast tenderness, unpleasant tingling or swelling in the hands and feet, an increase in body weight (due to water retention), acne, and migraine headaches. Emotional components of this syndrome (which are grouped within the term *premenstrual tension*) can include tension, irritability, depression, anxiety, emotional instability, food cravings, and lack of concentration.

Possible Causes and Treatments

The cause of PMS is unknown, and thus treatment typically focuses on the alleviation of major symptoms. Because estradiol and progesterone levels decrease in the late luteal phase, it has been hypothesized that severe premenstrual symptoms are the result of abnormal steroid levels. However, steroid levels of PMS patients are not significantly different from the general female population. Another possibility is that PMS is caused by the precipitous *withdrawal* of steroids, especially of progesterone. Interestingly, women with PMS tend to have a greater fall in blood levels of progesterone than women without the syndrome. Although for many years progesterone was prescribed as a treatment for PMS, studies have shown no effect of progesterone on severe PMS. However, GnRH inhibitory agonists have been found to be effective in treating PMS in some women. This treatment downregulates the pituitary, thus reducing gonadotropin levels and inhibiting steroid production by the ovary. Benefits of this treatment must be weighed against the significant menopausal-like side effects, including hot flashes, headache, vaginal dryness, and loss of bone mineralization.

Whether or not a woman suffers from PMS may be a consequence of individual differences in neurological response to steroid level changes rather than the absolute levels of steroids in circulation. Ovarian steroids are known to affect many types of neurons, including those that release the neurotransmitters serotonin, norepinephrine, GABA, and endorphins. Therefore, treatments that alter brain chemistry have been suggested for severe PMS symptoms. The most effective appears to be *fluoxetine* (Prozac), an antidepressant that increases brain levels of serotonin. Some studies have reported success in treating with an antianxiety medication, *alprazolam* (Xanax).

For milder cases of PMS, lifestyle changes that promote overall good health may be beneficial. Exercise may help, as well as a balanced diet that minimizes salt, sugar, caffeine, and alcohol. Many women find that they can gain some control over their symptoms by keeping a written log of their premenstrual physical and emotional symptoms over a number of months.

Cyclic Changes in Mood and Skill Level

Numerous studies indicate that, on average, women differ in mood and ability depending on the phase of their menstrual cycle. When estrogen levels are high around the time of ovulation, women tend to have more self-esteem, self-confidence, and alertness. In the luteal phase, they tend to become more passive. For the 4 days before menstruation, premenstrual anxiety, depression, irritability, hostility, and feelings of helplessness may occur in women with severe PMS. During menstruation, some aspects of the premenstrual tension disappear, although depression often remains until estrogen begins to rise.

A recent study suggested that when estrogen levels in the blood are low, women have enhanced spatial skills and olfactory sense and poorer complex motor skills and verbal fluency. When estrogen levels are high, the opposite prevails. This information is discussed more fully in Chapter 17. It is important to realize, however, that a woman is not controlled by her physiology because her degree of emotional maturity and life situation can influence and even eliminate these cyclic changes.

Menstrual Difficulties

Dysmenorrhea

Many women (30 to 50%) experience some discomfort during menstruation and a few experience pain that is severe enough to interfere with normal activities. Pain during menstruation is called *dysmenorrhea* or *menstrual cramps,* and it occurs during strong contractions of the uterine smooth muscle. It is common for muscle contraction to be painful and take the form of cramps when blood supply to that muscle is low, such as in the uterus during menstruation. The cause of these cramps is not clear, but prostaglandins present in the menstrual fluid may stimulate the uterine smooth muscles and cause their contraction, thus helping to evacuate the uterus. These prostaglandins, in combination with reduced progesterone levels, may actually initiate menstruation, and it is only when prostaglandin levels are unusually high that menstrual cramping occurs. Evidence that prostaglandins cause menstrual cramps is that aspirin, which inhibits the synthesis of prostaglandins, relieves these cramps and associated back pain in many women. Other antiprostaglandins (ibuprofen, naproxin, mefanamic acid) can be even more effective than aspirin. Exercise can sometimes relieve primary dysmenorrhea by increasing uterine blood flow. Heavy smoking (10 to 30 cigarettes a day) doubles the chance of menstrual cramps.

Dysmenorrhea caused by prostaglandin-induced uterine cramping often is called *primary dysmenorrhea*. In contrast, menstrual pain not caused by prostaglandins, called *secondary dysmenorrhea*, is characterized by lower abdominal pain. It can be caused by an IUD, pelvic infection, uterine tumors, or endometriosis.

Absence of Menstruation

Amenorrhea is the absence of menstruation when it normally should occur. It is called *primary amenorrhea* if a female has not menstruated by age 16. Anorexia nervosa can result in primary amenorrhea. In this condition, a female will eat very little and become emaciated. Inasmuch as young females who are very thin for other reasons can fail to reach puberty at a normal time, the primary amenorrhea of a person with anorexia may be due to the low amount of fat in her body. The role of body fat in the occurrence of primary amenorrhea is discussed in Chapter 6.

Chapter 3, Box 2: Body Fat and Secondary Amenorrhea

A reduction of body fat due to exercise, low food intake, or other factors often leads to temporary infertility (secondary amenorrhea) in adult women. It is estimated that an adult woman's body needs to be about 22% fat (as opposed to protein and water) for menstrual cycles to continue. Even moderate weight loss (10 to 15% of normal weight for height) can stop ovulation and menstrual cycles or cause an intermediate condition, *short luteal phase syndrome*, where ovulation occurs but the corpus luteum dies early. Exercise-induced amenorrhea can cause osteoporosis (see Menopause).

It has been proposed that infertility in lean women was evolutionarily adaptive in prehistoric times, as the 22% fat amount required for menstrual cyclicity represents enough stored energy for pregnancy and nursing with minimal food intake. What about the fact that undernourished women in developing countries can have several children (see Chapter 13)? They, however, only have about 6 or 7 children in their reproductive lifetime, whereas the maximal number of offspring in a well-nourished population that does not use contraception is 10 to 11 children! In addition, the 22% body fat limit stated earlier is based on studies of Caucasian women, and other races may have a different critical body fat content for fertility.

How does the amount of body fat control fertility? Some say that it does not, i.e., something else related to loss of weight, such as a change in exercise, metabolism, or stress, is really the infertility culprit. For example, exercise immediately lowers the blood levels of LH in women. Strong evidence, however, shows that stored body fat directly influences fertility. First, fat (especially in the abdomen, breasts, and bone marrow) can make estrone, an estrogen. In fact, about one-third of a women's blood estrogen comes from fat! It is thought that this fat-derived estrogen (along with ovarian-derived estradiol) is a necessary contributor to the preovulatory estrogen surge that leads to ovulation (see Fig. 3-2). If not enough fat is present, the estrogen surge would be inadequate, neither LH surge nor ovulation would occur, and the

Continued on next page.

Chapter 3, Box 2 continued.

woman would stop cycling. Second, lean women produce more of the estrogen metabolite 2-hydroxyestrone, which does not bind to estrogen receptors, and less of the estrogen metabolite 16-hydroxyestrone, which does bind to estrogen receptors. The lower production of 16-hydroxyestrone, which could also contribute to causing the LH surge, may help explain infertility in lean women. Third, nonlean women have more free estrogen in the blood (not bound to estrogen-binding protein) and so have more estrogen reaching their brain to cause the LH surge. Finally, adipose (fat) cells produce *leptin*, a circulating protein that signals to the hypothalamus information about the amount of stored body fat (see Chapter 6).

How would a low estrogen surge inhibit ovulation and stop menstrual cycles? Most evidence suggests that an inadequate estrogen surge interferes with GnRH pulses necessary for an LH surge to occur. Evidence shows that the low estrogen reaching the brain causes neurons to release the opioid endorphin, which inhibits GnRH secretion. About 70% of exercising women experiencing secondary amenorrhea resume fertility after lessening their exercise. Of the 30% that remain infertile, 65% are restored to fertility by treatment with an endorphin inhibitor (Naloxone).

Leanness and the associated lowering of estrogen exposure can also have benefits. Remember from Chapter 2 that estrogen exposure and obesity can increase the incidence of certain reproductive cancers; exercise has been shown to reduce the incidence of female reproductive cancers. Another point is that too much fat (extreme obesity), as leanness, can lead to infertility.

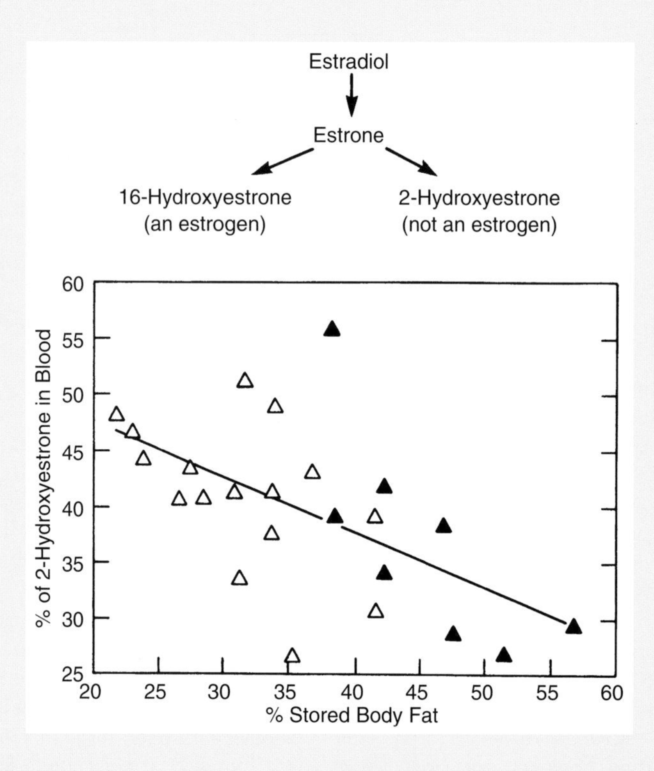

This graph relates the extent of production of the nonestrogenic estradiol metabolite (2-hydroxyestrone) to the amount of stored body fat in female athletes (△) and nonathletes (▲). Note that lean women, athletes or not, produce more of the nonestrogenic metabolite. This can lead to temporary infertility but can also decrease the susceptibility to estrogen-dependent reproductive cancers.

Oligomenorrhea occurs when an adult woman skips one or a few cycles. It is common for women to exhibit oligomenorrhea because of polycystic ovarian syndrome (see Chapter 2), emotional or physical stress, chronic illness, poor nutrition, or intense exercise. Women nearing menopause exhibit irregular menstrual cycles, as do young women who have just begun menstrual cycling. Because oligomenorrhea could be caused by disorders of the reproductive system, a woman should consult a physician if she has a pattern of missed menstrual cycles.

Secondary amenorrhea is when menstruation has failed to occur in an adult woman for at least 6 months. This condition normally occurs, for example, during pregnancy. Another time when secondary amenorrhea occurs is during lactation, because suckling of the breasts tends to inhibit ovulation (see Chapter 12). Secondary amenorrhea, of course, is also a characteristic of a woman after she has reached menopause. Other situations can result in secondary amenorrhea. Women who exercise so much that their body contains a relatively low amount of fat can exhibit secondary amenorrhea (see HIGHLIGHT box 3-2). *Galactorrhea* (abnormal milk secretion from the mammary glands) can be associated with secondary amenorrhea in a few women after they stop taking the combination birth control pill (see Chapter 14). Secondary amenorrhea can also be associated with more serious conditions, such as tumors of the pituitary gland or ovaries, so a woman who has secondary amenorrhea should discuss her condition with a physician.

Menstrual Taboos

For centuries, menstruating women were believed to produce many horrifying effects on the world around them; thus it has been called "the curse." Menstruating women have been considered unclean and unfit for coitus. It was also believed that they spoiled food, drink, and crops. The early Hebrews, for example, punished those who had intercourse with a woman who was menstruating. In medieval times, menstruating women were excluded from going to church and even from entering wine cellars (their presence supposedly spoiled the wine).

In the United States today, only about one-fifth as much coitus occurs during the five days of the menstrual phase as occurs in any other five days of the cycle—even though there is no physical reason for avoiding coitus during menstruation. Advertisements for tampons and sanitary napkins stress that their product hides the menstruating condition, and many women and men are still embarrassed by the subject of menstruation.

Chapter Summary

Many mammals exhibit seasonal reproduction with estrous behavior during the breeding season, and some evidence exists for this phenomenon in human females. Human females ovulate spontaneously, but they also may exhibit induced ovulation in some situations. Daily light cycles have a great influence on estrous cycles of some mammals but not on the menstrual cycle of humans. Social interaction, however, can influence human menstrual cycles.

The human menstrual cycle, which usually is 25 to 35 days long, consists of menstrual, follicular, and luteal phases. Cyclic secretion of hormones from the hypothalamus (gonadotropin-releasing hormone), anterior pituitary gland (FSH, LH), and ovaries (estradiol, progesterone) controls the cycle, and negative and positive feedback plays an important role. The mechanisms causing regression of the corpus luteum at the end of each cycle are not yet understood. Methods used to detect ovulation in the home or laboratory are not reliable in predicting when ovulation will occur.

Many women experience premenstrual syndrome before menstruation; the cause of these difficult biological and neurological symptoms is not clear. Painful menstruation (dysmenorrhea) is another common menstrual disorder and is probably initiated by the effect of prostaglandins on uterine smooth muscle. A missed menstrual period (amenorrhea) can be caused by stress, chronic exercise and loss of body fat, poor nutrition, polycystic ovarian syndrome, or normal reproductive events during pregnancy and lactation. Past taboos about menstruating women still influence present feelings and behavior.

Further Reading

Adami, H.-O., and Persson, I. (1995). Hormone replacement and breast cancer: A remaining controversy? *J. Am. Med. Assoc.* **274**, 178–179.

Anonymous. (1990). New treatment for osteoporosis. *Lancet* **335**, 1065–1066.

Bachmann, G., and Grill, J. (1988). Endocrine and metabolic changes of menopause. *Med. Aspects Hum. Sexual.* **74**, 81.

Bower, B. (1990). Strengthening bone without hormones. *Sci. News* **137**, 334.

Bryner, C. (1989). Recurrent toxic shock syndrome. *Am. Family Physician* **39**, 157–164.

Cutler, W. B. (1980). Lunar and menstrual locking. *Am. J. Obstetr. Gynecol.* **137**, 834–839.

Elias, A. N., and Wilson, A. F. (1995). Exercise and gonadal function. *Hum. Reprod.* **8**, 1747–1761.

Fackelmann, K. A. (1990). Flushing out the mechanism of hot flashes. *Sci. News* **138**, 229.

Fackelmann, K. A. (1990). Menstrual glitches may cause bone loss. *Sci. News* **138**, 279.

Fackelmann, K. (1995). Forever smart: Does estrogen enhance memory? *Sci. News* **147**, 74–75.

Feichtinger, W. (1989). Ultrasound in human conception. *Res. Reprod.* **21**(1), 1–2.

Frisch, R. E. (1988). Fatness and fertility. *Sci. Am.* **258**(3), 88–95.

Frisch, R. E. (1991). Body weight, body fat, and ovulation. *Trends Endocrinol. Metabol.* **2**, 191–197.

Frisch, R. E. (1994). The right weight: Body fat, menarche, and fertility. *Proc. Nutr. Soc.* **53**, 113–129.

Gura, T. (1995). Estrogen: Key player in heart disease among women. *Science* **269**, 771–773.

Matteo, S. (1987). The effect of job stress and job interdependency on menstrual cycle length, regularity, and synchrony. *Psychoneuroendocrinology* **12**, 467–476.

Morris, N. M., *et al.* (1987). Marital sex frequency and midcycle female testosterone. *Arch. Sex. Behav.* **16**, 27–37.
Seachrist, L. (1995). What risk hormones? Conflicting studies reveal problems pinning down breast cancer risks. *Sci. News* **148**, 94–95.
Spicer, D. V., and Pike, M. C. (1995). Hormonal manipulation to prevent breast cancer. *Sci. Am. Sci. Med.* **2**(4), 58–67.
Travis, J. (1997). Why do women menstruate? *Sci. News* **151**, 230–231.
Weiss, R. (1995). Prescription for passion. *Hippocrates* **9**(9), 55–62.
Weller, L., and Weller, A. (1993). Human menstrual synchrony: A critical assessment. *Neurosci. Biobehav. Rev.* **17**, 427–439.

Advanced Reading

Baerwald, A. R., *et al.* (2003). A new model for ovarian follicular development during the human menstrual cycle. *Fertil. Steril.* **80**, 116–122.
Barrett-Conner, E., and Kritz-Silverstein, D. (1993). Estrogen replacement therapy and cognitive function. *J. Am. Med. Assoc.* **269**, 2637–2641.
Brenner, P. (1988). The menopausal syndrome. *Obstet. Gynecol* **72**, 65–115.
Buffet, N., *et al.* (1998). Regulation of the human menstrual cycle. *Front. Neurendocrinol.* **19**, 151–186.
Bush, T. L. (1983). Estrogen use and all-cause mortality. *J. Am. Med. Assoc.* **249**, 903–906.
Clarke, I. J. (1995). The preovulatory LH surge: A case of a neuroendocrine switch. *Trends Endocrinol. Metabol.* **6**, 241–247.
Colditz, G. A., *et al.* (1995). The use of estrogens and progestins and the risk of breast cancer in postmenopausal women. *N. Engl. J. Med.* **332**, 1589–1593.
Dennerstein, L., *et al.* (1984). Mood and the menstrual cycle. *J. Psychiatr. Res.* **18**, 1–12.
De Souza, M.J. (2003). Menstrual disturbances in athletes: A focus on luteal phase defects. *Med. Sci. Sports Exerc.* **35**, 1553–1563.
Downs, L. L. (2002). PMS, psychosis and culpability: Sound or misguided advice. *J. Foren. Sci.* **47**, 1083–1089.
Filicori, M., *et al.* (1993). Role of gonadotrophin releasing hormone secretory dynamics in the control of the human menstrual cycle. *Hum. Reprod.* **8**(Suppl. 2), 62–65.
Finn, C. A. (1998). Menstruation: A nonadaptive consequence of uterine evolution. *Quart. Rev. Biol.* **73**, 163–171.
Florack, E. I. M., *et al.* (1994). The influence of occupational physical activity on the menstrual cycle and fecundability. *Epidemiology* **5**, 14–18.
Freeman, E. W., *et al.* (1995). A double-blind trial of oral progesterone, alprazolam and placebo in treatment of severe premenstrual syndrome. *J. Am. Med. Assoc.* **274**, 51–57.
Frisch, R. E., *et al.* (1993). Magnetic resonance imaging of overall and regional body fat, estrogen metabolism, and ovulation of athletes compared to controls. *J. Clin. Endocrinol. Metabol.* **77**, 471–477.
Hedricks, C. L., *et al.* (1987). Peak coital rate coincides with onset of luteinizing hormone (LH) surge. *Fertil. Steril.* **49**, 235–246.
Hodgen, G. D. (1989). Neuroendocrinology of the normal menstrual cycle. *J. Reprod. Med.* **34**, 68–75.

Howlett, T. A., and Rees, L. H. (1987). Endogenous opioid peptides and human reproduction. *Oxford Rev. Reprod. Biol.* **9**, 260–293.

Kalra, S. P. (1993). Mandatory neuropeptide-steroid signaling for the preovulatory luteinizing hormone-releasing hormone discharge. *Endocr. Rev.* **14**, 507–538.

Kimura, D. (1995). Estrogen replacement therapy may protect against intellectual decline in postmenopausal women. *Horm. Behav.* **29**, 312–321.

Loucks, A. B. (1990). Effects of exercise on the menstrual cycle: Existence and mechanisms. *Med. Sci. Sports Exerc.* **22**, 275–280.

Manning, J. T., *et al.* (1996). Asymmetry and the menstrual cycle. *Ethol. Sociobiol.* **17**, 129–143.

McArthur, J. W. (1985). Endorphins and exercise in females: Possible connection with reproductive dysfunction. *Med. Sci. Sports Exerc.* **17**, 82–88.

Meyer, W. R., *et al.* (1994). The effect of exercise on reproductive function and pregnancy. *Curr. Opin. Obstet. Gynecol.* **6**, 293–299.

Nabulsi, A. A., *et al.* (1993). Association of hormone-replacement therapy with various cardiovascular risk factors in postmenopausal women. *N. Engl. J. Med.* **328**, 1069–1075.

Park, J.-Y., *et al.* (2004). EGF-like growth factors as mediators of LH action in the ovulatory follicle. *Science* **303**, 682–684.

Pause, B. M. (1996). Olfactory information processing during the course of the menstrual cycle. *Biol. Psychol.* **44**, 31–54.

Petitti, D., and Reingold, A. (1988). Tampon characteristics and menstrual toxic shock syndrome. *J. Am. Med. Assoc.* **259**, 678–687.

Postmenopausal Estrogen/Progestin Interventions (PEP) Trial (1995). Effects of estrogen or estrogen/progestin regimens on heart disease risk factors in postmenopausal women. *J. Am. Med. Assoc.* **273**, 199–208.

Profet, M. (1993). Menstruation as a defense against pathogens transported by sperm. *Quart. Rev. Biol.* **68**, 335–386.

Prior, J., and Vigna, Y. (1987). Conditioning exercise and premenstrual symptoms. *J. Reprod. Med.* **32**, 423–428.

Shangold, M. (1985). Causes, evaluation, and management of athletic oligoamenorrhea. *Med. Clin. North Am.* **69**, 83–95.

Siegel, J., *et al.* (1987). Premenstrual tension syndrome symptom clusters: Statistical evaluation of the subsyndromes. *J. Reprod. Med.* **32**, 395–399.

Stanko, S. S., *et al.* (1994). Gonadotropin-releasing hormone neurons: Intrinsic pulsatility and receptor-mediated regulation. *Trends Endocrinol. Metabol.* **5**, 201–208.

Stanford, J. L., *et al.* (1995). Combined estrogen and progestin hormone replacement therapy in relation to risk of breast cancer in middle-aged women. *J. Am. Med. Assoc.* **274**, 137–142.

Steiner, M., *et al.* (1995). Fluoxetine in the treatment of premenstrual dysphoria. *N. Engl. J. Med.* **332**, 1529–1534.

Steiner, R. A., and Cameron, J. L. (1989). Endocrine control of reproduction. *In* "Textbook of Physiology" (H. D. Patton, A. F. Fuchs, B. Hille, A. M. Scher, and R. A. Steiner, eds.), Vol. 2. Saunders, Philadelphia.

Stern, K., and McClintock, M. K. (1998). Regulation of ovulation by human pheromones. *Nature* **392**, 177–179.

Stewart, A. (1987). Clinical and biochemical effects of nutritional supplementation on the premenstrual syndrome. *J. Reprod. Med.* **32**, 435–441.

Tang, M., *et al.* (1996). Effect of oestrogen during menopause on risk and age at onset of Alzheimer's disease. *Lancet* **348**, 429–432.

Trunnell, E., *et al.* (1988). A comparison of the psychological and hormonal factors in women with and without premenstrual syndrome. *J. Abnorm. Psychol.* **97**, 429–436.

Voight, L. F., *et al.* (1991). Progestogen supplementation of exogenous estrogens and risk of endometrial cancer. *Lancet* **338**, 274–277.

Weller, A., and Weller, L. (1997). Menstrual synchrony under optimal conditions: Bedouin families. *J. Comp. Psychol.* **111**, 143–151.

Wise, P. M., *et al.* (1994). Neuroendocrine concomitants of reproductive aging. *Exp. Gerontol.* **29**, 275–283.

CHAPTER FOUR

The Male Reproductive System

Introduction

The male reproductive tract includes a pair of *testes* (*testicles*), which produce the male sex cells (gametes) called *spermatozoa* (sing., *spermatazoon*), or *sperm*. The word *testes* (sing., *testis*) is derived from the Latin word for "witness" or "testify." (In ancient times, men put one hand over their genitals when taking an oath.) The testes, the male sex accessory ducts (which receive, store, and transport sperm), and the sex accessory glands (which add substances to the ducts), as well as the external genitalia, are all male primary sexual characteristics. The general location of these structures is shown in Fig. 4-1. The male secondary sexual characteristics include all those external features (except the external genitalia) that distinguish an adult man from an adult woman. These include larger muscles, greater average height, facial hair, and a lower voice.

Testes

Each testis is oval, with a length of about 4.0 cm (1.5 in.) and a width of 2.5 cm (1.0 in.). On the outside of each testis is a shiny covering, the *tunica vaginalis* (Fig. 4-2). Immediately under the tunica vaginalis is a thin, dense covering of the testis itself, the *tunica albuginea*. Inside each testis are about 250 compartments, *testicular lobules*, which are separated from each other by septa (tissue barriers). Each lobule contains one to three highly coiled *seminiferous tubules* (Fig. 4-2). If a single seminiferous tubule were stretched to its maximal length, it would measure about 30 to 91 cm (1 to 3 ft), and the total length of all the tubules would be longer than a football field! The male germ cells lie next to the inner wall of each tubule, and sperm are produced in the seminiferous tubules. The testes also contain some of the male sex accessory ducts that are discussed later in this chapter.

Each testis is suspended from the body wall by a *spermatic cord* (Fig. 4-2), which penetrates into the pelvic cavity through the *inguinal canal*, the route through which the testes originally descended into the scrotum from the pelvic cavity before birth (see Chapter 5). Each spermatic cord contains one of the accessory ducts (the *vas deferens*), a testicular nerve, and three coiled blood vessels (the

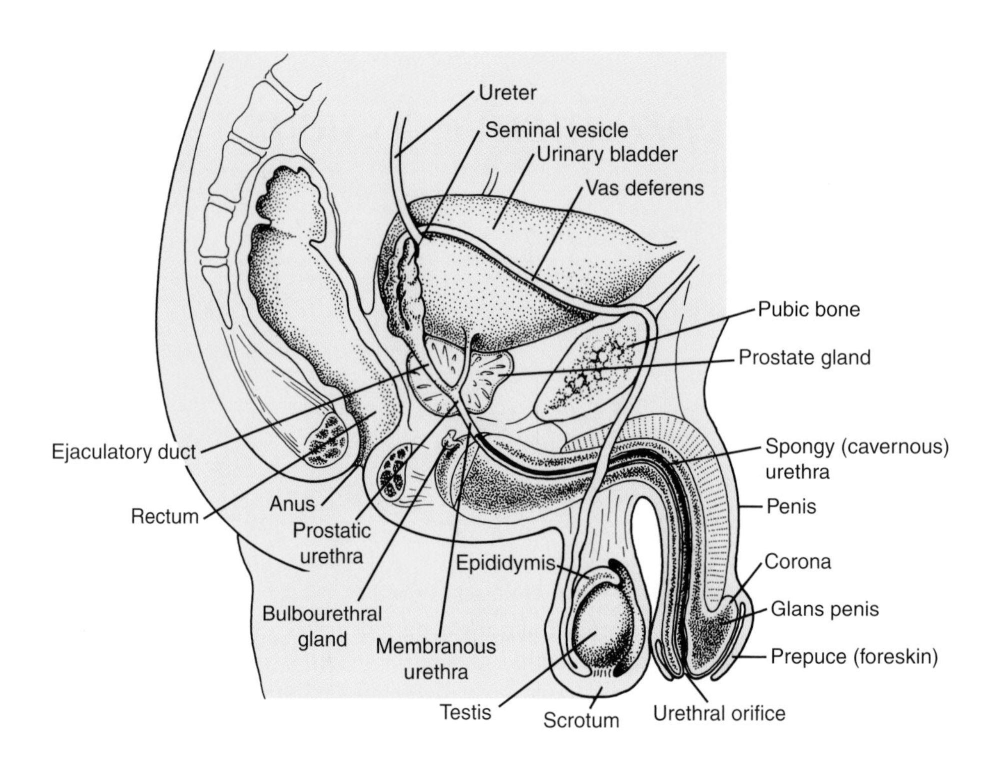

Figure 4-1 A schematic representation of the male pelvic region showing the reproductive organs.

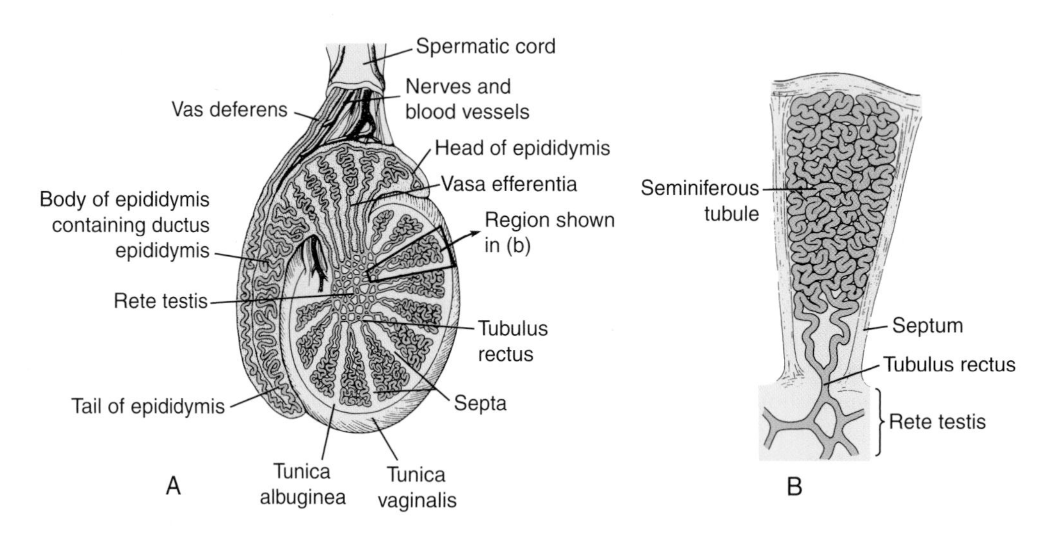

Figure 4-2 (a) Section of the testis, epididymis, vas deferens, and spermatic cord. (b) A seminiferous tubule within a single testicular lobule. Sperm are produced in the seminiferous tubules and then move through the tubuli recti into the rete testis. They are carried through the vasa efferentia into the ductus epididymis and then into the vas deferens. The spermatic cord contains the vas deferens, nerves, and blood vessels.

testicular artery, or spermatic artery, carrying blood to the testis and epididymis, and two *testicular veins*, or spermatic veins, carrying blood away from the testis).

Seminiferous Tubules

Each seminiferous tubule is lined on its inside by the *seminiferous epithelium*, which contains two kinds of cells—male germ cells and Sertoli cells. We will first look at how sperm are produced from the male germ cells and then will discuss the functions of Sertoli cells.

Spermatogenesis and Spermiogenesis The processes by which immature male germ cells produce sperm are called *spermatogenesis* and *spermiogenesis*. First, let us observe spermatogenesis, the process by which a diploid *spermatogonium* (pl., *spermatogonia*) transforms into four haploid *spermatids*. The immature germ cells lying next to the wall (basement membrane; see Fig. 4-3) of each seminiferous tubule are the spermatogonia. These cells multiply by mitosis, a process of cell division by which two diploid daughter cells are derived from a diploid parent cell. Each spermatogonium is diploid and has 46 chromosomes (2 each of 23 different chromosomes). During mitosis, each chromosome duplicates itself, and each member of a duplicated pair passes to a separate progeny cell. Thus, the normal diploid chromosomal makeup is maintained in each new spermatogonium.

As the spermatogonia continue to divide mitotically, some of them change in appearance and begin a different kind of cell division, called meiosis (Figs. 4-3 and 4-4). After a spermatogonium enters meiosis, it becomes a *primary spermatocyte*. A primary spermatocyte then completes the first meiotic division, which is the reduction division because the resultant daughter cells now have only 23 chromosomes (one member of each original pair) instead of 46. After reduction division, the cells are haploid (N) instead of diploid (2N). The haploid cells, called *secondary spermatocytes*, then undergo the second meiotic division to produce four haploid cells, the spermatids. Now the process of spermatogenesis is complete.

The spermatids are subsequently transformed into spermatozoa (sperm cells), a process termed *spermiogenesis*. As a spermatid differentiates into a sperm cell capable of finding and fertilizing an egg (see Chapter 9), an *acrosomal vesicle* is formed by the Golgi apparatus. An elongated *flagellum*, organized by the centriole, grows from the posterior end of the sperm cell. Mitochondria cluster around the flagellum. The nucleus becomes condensed and elongated, and a droplet of excess cytoplasm is extruded from the cell. During spermiogenesis, the sperm heads become embedded in tiny pockets in the Sertoli cell membrane. Later the heads are released from these cells, an event called *spermiation*. Figure 4-4 illustrates these events within the seminiferous tubules.

In humans, the entire process from spermatogonia to spermiation takes 65–75 days. If you look at a given region of one seminiferous tubule, you will see that not all areas of the seminiferous epithelium are in the same stage of spermatogenesis (Fig. 4-3). In fact, the seminiferous tubule is a "production line" of sperm. Earlier stages of spermatogenesis (spermatogonia) are found at the basal side of the tubule; moving toward the tubular lumen, one can observe germ cells in progressively later stages of differentiation. Mature spermatids are nearest the center of the seminiferous tubule, with their tails extending into its lumen. A male is constantly producing new sperm, at a rate of about 1000/second!

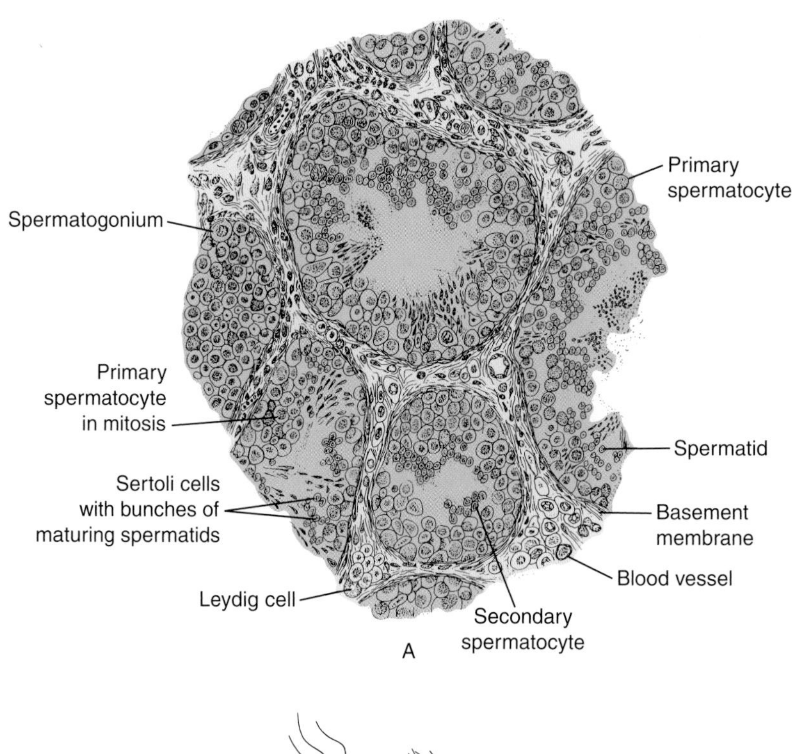

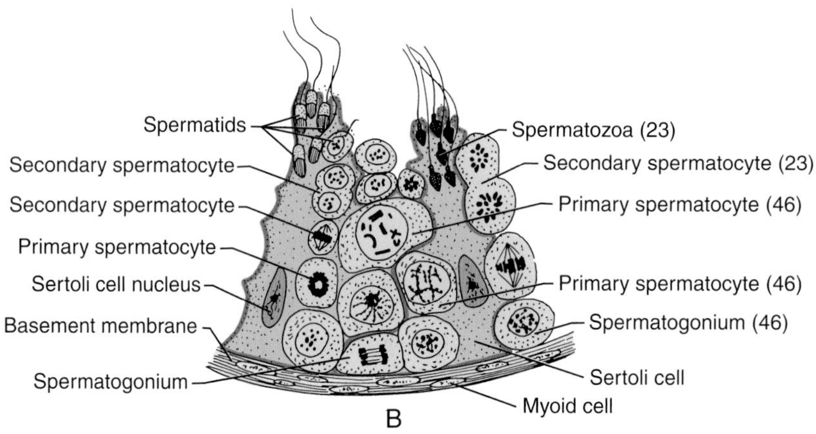

Figure 4-3 (a) Cross section through a region of the human testis showing several seminiferous tubules and interstitial tissue. Note that not all areas in one tubule are in a similar stage of spermatogenesis. Interstitial spaces contain androgen-secreting Leydig (interstitial) cells, connective tissue, and blood vessels. (b) One region of a cross section of a seminiferous tubule showing stages of spermatogenesis. Note spermatogonium dividing by mitosis and the different cell stages occurring during meiosis (number of chromosomes in parentheses). Also note spermatozoa with their heads embedded in Sertoli cells.

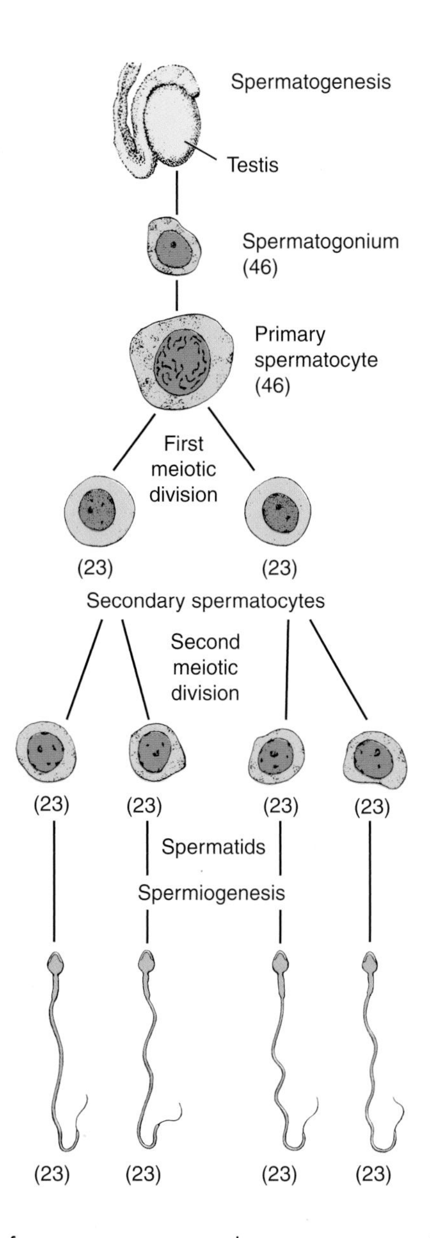

Figure 4-4 Stages of spermatogenesis and spermiogenesis. The number of chromosomes in each cell type is included. Note that the cells transform from a diploid (46) to a haploid (23) chromosomal number during the first meiotic division. Note also that four sperm are produced from one spermatogonium.

Sertoli Cells Sertoli cells, also called *sustentacular cells*, are approximately pyramid-shaped cells lying within the seminiferous epithelium (Fig. 4-3). Their bases lie against the basement membrane of each tubule, and their tips point toward the cavity in the middle of each tubule. These "nurse cells" play an essential role in nurturing and providing structural support for the sperm cells during their development. Spermatogenic cells are embedded in recesses of the Sertoli cells and migrate from the base toward the lumen of the seminiferous tubule as they mature. The plasma membranes of surrounding Sertoli cells adjust to accommodate the gradual movement of the germ cells, and thus the complex shape of the Sertoli cell is constantly in flux. Sperm heads are released from Sertoli cells at spermiation.

Additional functions of Sertoli cells are the secretion of testicular fluid into the tubular cavity and *phagocytosis* (engulfing) of the remains of degenerated germ cells. Tight junctions between the basal portions of adjacent Sertoli cells provide for a *blood–testis barrier*. This prevents the leakage of certain molecules between Sertoli cells and allows these cells to control the chemical composition of testicular fluid within the seminiferous tubules, resulting in a unique microenvironment in the tubules. In addition, the blood–testis barrier protects spermatocytes and spermatids from immune attack, as these haploid cells are recognized as foreign by a man's immune system. Sertoli cells secrete various proteins, including androgen-binding protein, as well as hormones such as inhibin and Müllerian-inhibiting substance (in the fetus; see Chapter 5). These cells produce enzymes that convert testosterone to estrogen and to 5α-dihydrotestosterone (DHT). Sertoli cells also may play an important role in the hormonal control of spermatogenesis, as described later. Because of the dependence of male germ cell development on Sertoli cells, the rate of a man's sperm production is related to the number of Sertoli cells in his testes. Sertoli cell number is determined at puberty; no new Sertoli cells are produced in adult men.

The outer or basal surfaces of the seminiferous tubules are covered with a collagenous *basement membrane* in which scattered *myoid cells* are embedded. It is thought that these muscle-like cells contract weakly to help move sperm cells and fluid through the seminiferous tubules. Like Sertoli cells, myoid cells are influenced by testicular hormones.

Testicular Interstitial Tissue

Because seminiferous tubules are circular in cross section, regions exist outside the tubules. These regions, the interstitial spaces (see Figs. 4-3 and 4-5), contain structures vital to testicular function. Small arteries, capillaries, and small veins are present in the interstitial spaces. This is where blood-borne products (e.g., oxygen, glucose) leave the blood and diffuse into the tubules. These products are important for the germ cells as there are no blood vessels within the tubules themselves. Hormones can also leave the interstitial blood vessels and enter the seminiferous tubules through the basement membrane of the tubules. The basement membrane and the blood–testis barrier, however, act as a barrier to the movement of large molecules from the blood into the tubules. Finally, waste products produced by cells of the tubules pass through the basement membrane and are carried away by the small veins in the interstitial spaces (Figs. 4-3 and 4-5).

The interstitial spaces contain the *Leydig cells*, or testicular *interstitial cells*, which synthesize and secrete androgenic steroid hormones. An androgen is a molecule that will stimulate the growth and the maintenance of male tissues. If levels of steroid hormones in the blood of an adult male are measured, the following androgens are found (listed in micrograms per 100 ml of blood plasma): testosterone (0.7), 5α-dihydrotestosterone (0.05), androstenedione (0.1), and dihydroepiandrosterone (DHEA; 0.5). Testosterone and DHT are potent androgens in that they stimulate androgen-dependent structures in very low dosages. Levels of DHT in the blood, however, are very low so that testosterone

Figure 4-5 A section through a seminiferous tubule taken with a scanning electron microscope (×470). The entire tubule is seen. Note sperm tails in the center of the tubule and Leydig cells between adjacent tubules (lower left).

is the most important androgen in the blood of human males. About 95% of the testosterone comes from the Leydig cells of the testes, with the remainder coming from the adrenal glands. Approximately 98% of the circulating testosterone in men is bound to plasma proteins, especially sex hormone-binding globulin and albumin; the remaining 2% is in free, cell-permeable form. Androstenedione and DHEA are "weak" (not very potent) androgens produced by the adrenal glands, and their role in male reproductive function is not clear. Many changes caused by testosterone in the male, such as the appearance of facial hair and growth of the prostate, penis, and scrotum, are caused by DHT. That is, these target tissues convert testosterone to DHT, and it is DHT that really stimulates the male tissues (see Chapter 5). Muscle is directly affected by testosterone. In most men, testosterone and LH levels fluctuate daily, with higher levels in the morning than in the evening. However, a wide range of variation exists among individuals in circulating levels of testosterone as well as the extent of cyclicity. The diurnal cycle typically diminishes with age (see Chapter 7).

Human males also have estrogens, or "female sex hormones," in their blood. You will recall from Chapter 2 that estrogens are important sex hormones in females. Blood levels of estradiol (an estrogen) in human males are very low (only 0.003 μg/100 ml blood) in relation to androgen levels. The testicular sources of estradiol in males are Sertoli cells. Also, some estrogen in males results from conversion of testosterone to estrogens in tissues other than the testis. For example, many of the effects of testosterone on bone result from the local conversion of testosterone to estradiol. Thus, just as human females produce both "male" and "female" sex hormones, both androgens and estrogens are present in the blood of males. However, the estrogens in human male blood do not "feminize" most men because the androgens are present in a relatively higher proportion (see HIGHLIGHT box).

Chapter 4, Box 1: Estrogens Are Male Sex Hormones

Estrogens (e.g., estradiol) are generally considered to be "female sex hormones," but it turns out that estrogens also play important roles in the reproductive biology and sexual behavior of males as well. For estrogens to act as male hormones, they must be present in male tissues, bind to estrogen receptors (ER-α and ER-β), and induce biological responses in male target cells. The male reproductive tract is loaded with estrogen receptors. They are present in the testis (spermatocytes, spermatids, Sertoli cells, Leydig cells, rete testis), epididymides, prostate, seminal vesicles, and urinary bladder.

The functions of estrogenic activation of these estrogen receptors in the male mostly remain enigmatic. However, it is known that men with a dysfunction of estrogen receptors are infertile. The fluid in the rete testis, vasa efferentia, and epididymis contains a much higher concentration of estradiol than male blood, and estradiol appears to regulate the amount of fluid in these tubules. If estradiol does not do its job, high fluid pressure in the seminiferous tubules can destroy germ cells and cause infertility. Also, estrogens appear to play a causative role in the development of benign prostatic hyperplasia and prostate cancer.

It is well established that many of the effects of testosterone (an androgen) on male function are really effects of estrogen. Testosterone is converted in target cells to estradiol by the enzyme *aromatase*. A good example is in the brain, where most aspects of testosterone-induced male sexual behavior are really caused by a conversion of testosterone to estradiol. That is, estradiol masculinizes the male brain (Chapter 18). Another example is that the stimulatory effects of testosterone on spermatogenesis in the testes are really caused by estrogen; testosterone is converted to estradiol by the Sertoli cells. Thus, it is probably estradiol that stimulates male germ cells and not testosterone, as there are no androgen receptors in the germ cells themselves.

A final way that estrogens may act on male biology is by actually binding to androgen receptors. In androgen target cells, the androgen receptor is bound to a coactivator called ARA70. It appears that an androgen such as testosterone activates the androgen receptor only if ARA70 is attached. To our surprise, however, estradiol can also work through the ARA70–androgen receptor complex to cause an androgenic response in the cell. Estradiol can thus act as an androgen on at least some male tissues. Therefore, estradiol and other estrogens can act as androgens, or be derived from androgens and then have important biological functions in the male. Because the level of estrogens in male blood is very low, most of these effects of estrogens occur after conversion of testosterone to estrogen in target cells.

It should also be mentioned that androgens are important as hormones in women. For example, androgens such as testosterone, DHT, and androstenedione are secreted into female blood, especially at puberty (Chapter 6) and menopause (Chapter 7). These androgens come from the adrenal glands and ovaries. Androgens also rise in a woman's blood near the time of ovulation (Chapter 3). In women, the conversion of androstenedione to estrone (an estrogen) within fat cells plays an important role in female reproductive biology (Chapter 3). The conversion of testosterone to DHT in hair follicles, through the action of the enzyme 5α-reductase, allows the latter to stimulate pubic and axillary hair growth in women as it does in men. So estrogens can be "male sex hormones" and androgens "female sex hormones," and therefore relegation of a sex hormone to one or the other sex is misleading.

Continued on next page.

Chapter 4, Box 1 continued.

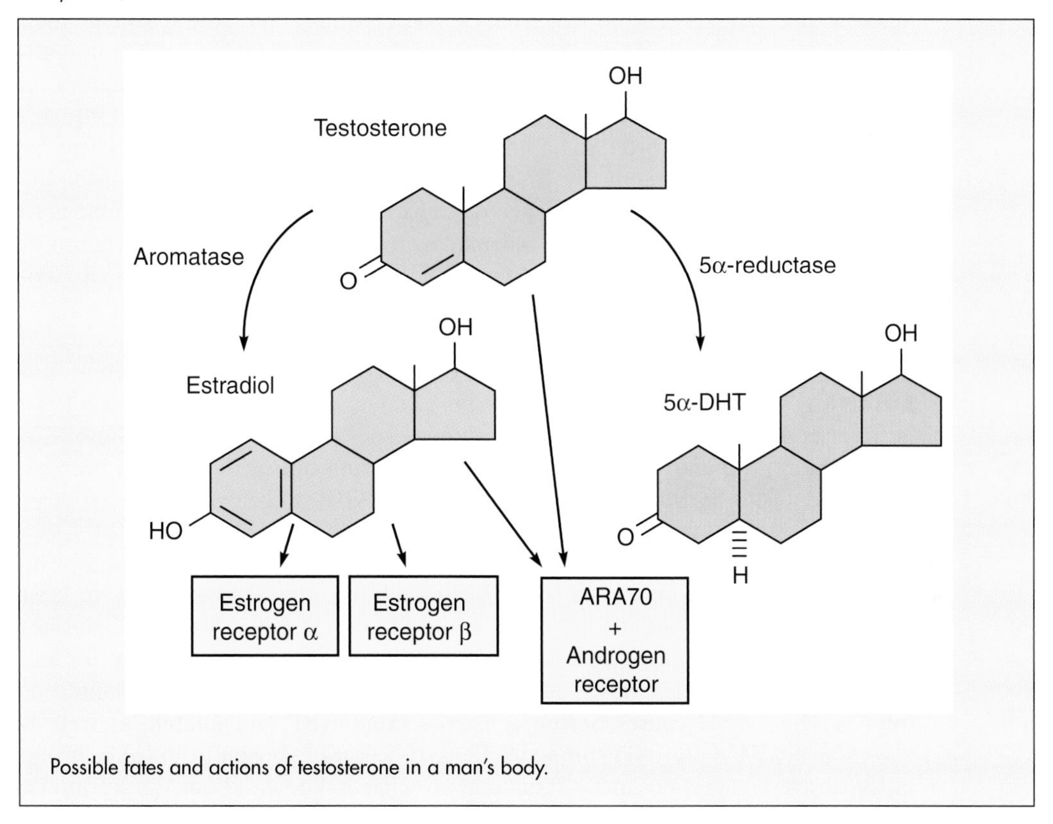

Possible fates and actions of testosterone in a man's body.

Hormonal Control of Testicular Function

The hypothalamus produces pulses of gonadotropin-releasing hormone, which cause the synthesis and secretion of two gonadotropins, follicle-stimulating hormone and luteinizing hormone, from the anterior pituitary gland (see Chapter 1). If the anterior pituitary gland is removed from an adult male (hypophysectomy), or if the anterior pituitary gland is rendered nonfunctional because of tissue damage, the testes will degenerate. No more sperm will be produced, and levels of androgen in the blood will drop considerably because the Leydig cells cease functioning. Similarly, if the testes are removed (*orchidectomy*) or are nonfunctional for some other reason, androgen levels will be very low. (Note: *castration* refers to either orchidectomy of a man or ovariectomy of a woman.) Restoration of full fertility and androgen production in hypophysectomized men requires administration of both FSH and LH; either gonadotropin alone will not work. If both gonadotropins are required for full testicular function, what are the specific roles of each hormone? We do not know much about the answer to this question in humans, so much of what we have to say about this is based on research using laboratory animals such as rats.

In humans and other mammals, LH stimulates the Leydig cells to secrete testosterone. This androgen not only enters the general blood circulation and goes to other areas of the body, but also diffuses from the testicular interstitial spaces through the basement membrane and into the seminiferous tubules, where it enters the Sertoli cells. The Sertoli cells can convert testosterone to DHT. Testosterone and DHT then leave the Sertoli cells and enter the testicular fluid around the germ cells. It is testosterone (and possibly DHT) that stimulates certain phases of spermatogenesis, not LH directly. For example, testosterone in the rat stimulates the first meiotic division (Fig. 4-6), during which diploid primary spermatocytes are converted to haploid secondary spermatocytes. Thus, LH causes this division only indirectly by causing the secretion of testosterone from the Leydig cells. The only direct influence of LH within the seminiferous tubule is to stimulate spermiation (Fig. 4-6). Testicular estrogen, derived by aromatization of testosterone in the Sertoli cells, could also play an important role in spermatogenesis (see HIGHLIGHT box).

Sertoli cells secrete *androgen-binding protein* (ABP). This protein is secreted into the fluid in the tubular lumen, where it combines with testosterone and DHT (Fig. 4-7). The role of ABP may be to concentrate androgens around the germ cells so that these steroids can influence spermatogenesis more effectively. Furthermore, the androgens are kept from diffusing out of the tubules because the blood–testis barrier hinders passage of the ABP–androgen complex out of the tubule.

Why is FSH needed along with LH for complete testicular function? The answer to this question is not known for humans, but there appear to be at least four important functions of FSH in male rats. First, FSH (along with testosterone) may stimulate the mitosis of spermatogonia. Second, FSH is necessary for the transformation of spermatids to spermatozoa (spermiogenesis) in the seminiferous tubules. Third, FSH causes Sertoli cells to secrete ABP and inhibin, as well as testicular fluid. Finally, FSH stimulates the synthesis of estrogen from testosterone in the testis. Figures 4-6 and 4-7 summarize what is known about the hormonal control of spermatogenesis.

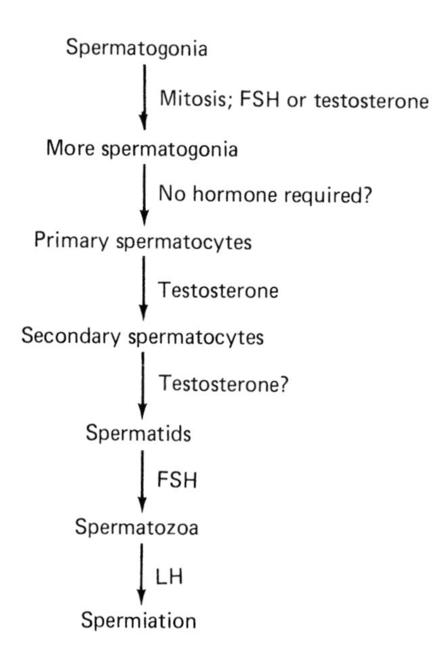

Figure 4-6 Possible hormonal control of spermatogenesis, spermiogenesis, and spermiation. See also the HIGHLIGHT box "Estrogens are Male Sex Hormones."

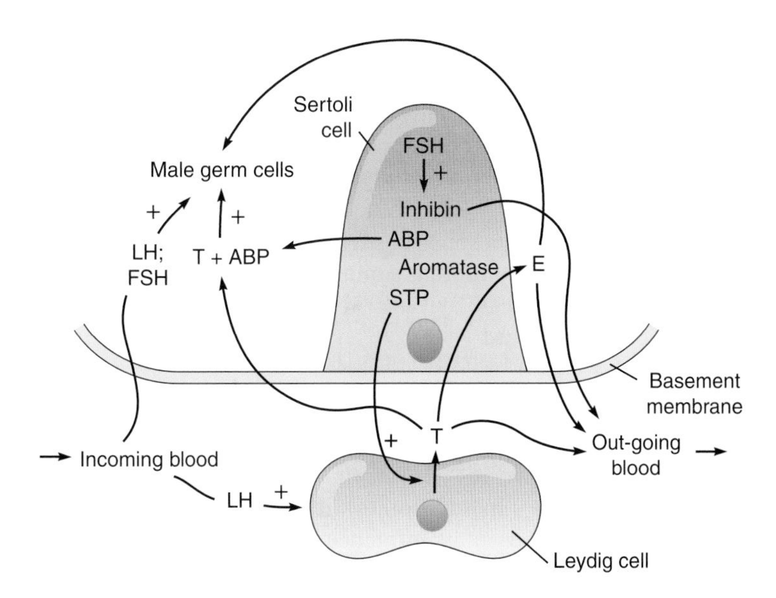

Figure 4-7 Endocrinology of the testes. Leydig cells secrete testosterone (T) in response to LH delivered in the blood flow to the testes. T then diffuses into the seminiferous tubules and enters Sertoli cells to be converted to estradiol (E) by the enzyme aromatase. T is also present in the germ cell region, where some is bound to androgen-binding protein (ABP); some T remains "free" to stimulate steps of spermatogenesis (see Fig. 4-6). In addition, LH and FSH help in sperm production (Fig. 4-6). FSH also binds to Sertoli cells and induces the production of inhibin, ABP, aromatase, and steroidogenesis-stimulating protein (STP). The latter leaves the tubule and helps LH increase T production. Finally, inhibin, E, and T enter the outgoing blood to control target organs in the rest of the body and to exert negative feedback on gonadotropin secretion. Note that E from the Sertoli cell could stimulate male germ cells directly (see HIGHLIGHT box 4-1).

Control of Gonadotropin Secretion in the Male

Blood levels of FSH and LH in men remain relatively constant, although there is some evidence that levels of LH and testosterone exhibit a daily cycle, being 20% higher in the morning. There is a seasonal cycle of blood testosterone in men, with a peak in autumn. Whether this seasonal cycle influences the seasonal variation in frequency of births (see Chapter 11) is not clear.

Feedback effects of steroid hormones regulate LH levels in men. Orchidectomized men exhibit abnormally high levels of FSH and LH in their blood, suggesting that testicular substances have a negative feedback effect on gonadotropin secretion from the adenohypophysis. If testosterone is given to an adult man, LH levels in the blood go down; FSH levels are also reduced to a lesser extent. Thus, testosterone exerts a negative feedback, at least on LH secretion. Under normal conditions, therefore, when testosterone levels drop, LH levels rise and stimulate the Leydig cells to secrete more testosterone; the opposite occurs when testosterone levels rise. When excess androgens are administered to adult men, they can become temporarily infertile but their sex drive is not reduced. Testosterone regulates LH secretion by influencing GnRH release from the hypothalamus and by affecting the sensitivity of the LH-secreting cells in the anterior pituitary to GnRH.

What feedback systems regulate FSH secretion in men? Although the evidence is less clear than for regulation of LH, it appears that FSH secretion is controlled by two negative feedback systems. First, high levels of circulating testosterone exert negative feedback on FSH secretion at the hypothalamic and pituitary levels. In addition, *inhibin*, a glycoprotein secreted by Sertoli cells in response to FSH, plays a role in FSH release. Inhibin suppresses FSH (but not LH) secretion, possibly by decreasing the sensitivity to GnRH of FSH-secreting cells in the anterior pituitary. Finally, there is some evidence that *activin*, a protein closely related to inhibin and produced by both Sertoli and Leydig cells, may have a stimulatory effect on FSH release.

Other Factors That Influence Testicular Function

Other Hormones In addition to the endocrine factors discussed previously, other kinds of hormones may influence the testis. For example, emotional stress is known to inhibit testicular function in men. One way in which this could happen is that adrenal hormones (e.g., cortisol), which are released under stress, may suppress GnRH and gonadotropin secretion. The male anterior pituitary gland releases prolactin (PRL), although the physiological role of this hormone in the testis is not clear. However, men who have abnormally high circulating levels of PRL (e.g., as a result of a pituitary tumor) can have decreased levels of circulating testosterone and lowered fertility. Progesterone also can affect testicular function. Exogenous administration of progesterone has been shown to inhibit spermatogenesis and testosterone secretion by suppressing gonadotropin secretion. In fact, Depo-provera, a progestogen injection used as a female contraceptive (see Chapter 14), has been used to "chemically castrate" male sex offenders. Finally, estradiol may also exert significant negative feedback on the hypothalamus and/or pituitary. The antigonadal effect of estradiol may be especially important in obese men because testosterone can be aromatized to estrogen in adipose (fatty) tissue.

Nutrition and Exercise Although reproduction is an energy-consuming process in both men and women, the energy required for sperm and hormone production in the testis is relatively minor compared to that needed for menstruation, pregnancy, and lactation in women. Most changes in eating patterns (including dieting) do not appear to affect testicular function. However, severe caloric restriction (fasting for 2 or more days) can reduce circulating testosterone and LH levels in men. Despite these hormonal changes, fertility has not been shown to be affected by fasting.

The specific nutrient composition in the diet may also affect testicular function. For example, testosterone levels during the night are slightly lower in vegetarian men than in nonvegetarians. A severe deficiency of certain micronutrients, including vitamins A, D, and E, as well as zinc, has been shown to decrease male fertility.

As discussed in Chapter 3, women who exercise chronically can become temporarily infertile. The effects of exercise on men are less clear. Some studies have shown that endurance runners have lower circulating testosterone and sperm counts, but only the most strenuous exercise over a prolonged time period has any negative effect on male reproductive function. In contrast, short-term strenuous exercise can significantly raise testosterone levels temporarily, after which androgen levels fall back to preexercise levels.

Chemicals, Radiation, and Infections Exposure to toxic substances or certain types of radiation can affect the testes, causing reduced spermatogenesis or even infertility. Certain chemicals, if ingested, can damage the testes. For example, heavy metals such as mercury, lead, manganese, and cadmium can cause complete and irreversible sterility because they block blood flow to the testes. Environmental pollutants, such as organophosphate-based pesticides, can also have adverse effects on testicular function. High levels of radioactivity and chemotherapy can destroy stem cell spermatogonia, thus causing infertility. Exposure to X-rays can destroy some germ cells, sometimes leading to a temporary reduction in fertility.

Certain viruses can infect the testes. For example, mumps is a viral disease that causes painful inflammation and enlargement of the salivary glands. If this virus infects a postpubertal male's testes, orchitis, scarring, and destruction of testicular tissue can result. In severe cases, this may lead to infertility. Certain bacterial infections can affect the testis, especially if the infection is not treated promptly with antibiotics. These include bacteria that cause tuberculosis, enterobacteria (intestinal bacteria), and sexually transmitted bacteria such as *Chlamydia*.

Anabolic Steroids Two of the normal functions of androgens such as testosterone are growth and maintenance of skeletal muscle. Because muscle growth is an *anabolic* process (involving synthesis of macromolecules), these steroids are called *anabolic steroids*. These widely available steroids, such as androstenedione ("andro"), are androgens taken by men and sometimes women, usually in high amounts, to increase muscle size and strength. These drugs do increase muscle mass, but unfortunately they can have serious side effects. In a man, excess androgen in the system can reduce luteinizing hormone secretion from the pituitary gland and can lead to testis shrinkage and infertility. Other side effects include acne, headache, liver tumors, kidney disease, high blood pressure, heart disease, and muscle spasms. Women who take anabolic steroids can also become infertile and suffer all of the aforementioned side effects plus beard growth, voice deepening, breast atrophy, and an increase in clitoris size. Individuals taking large doses of anabolic steroids can develop severe psychiatric complications, including depression, mania, excessive aggression, and psychosis. Fortunately, many of these changes are reversible if steroid usage is stopped.

Testicular Cancer

Although testicular cancer is relatively rare (about 8010 new cases per year in the United States), its incidence has increased by more than 50% in the past 40 years. Testicular cancer is usually diagnosed in men in their twenties and thirties; in fact, it is the leading cause of cancer in males aged 15 to 35. White men, especially northern Europeans, are at particularly high risk. American white males are three times more likely than American blacks to develop this disease. Another risk factor is *cryptorchidism* (undescended testes). Even if the testes eventually descend into the scrotal sac, the delay appears to cause changes in the testicular tissue that predispose it to cancer. Testicular cancer is often detected as a small, hard, painless lump on the testis, or an enlargement of the testis. The majority of these tumors arises from germ cells. Testicular cancer is a potentially dangerous disease but it can be treated successfully, especially in its earlier stages, with radiation, chemotherapy, or surgical removal of one or both testes (*orchidectomy*). With treatment, the survival rate is over 90%. These treatments, as well as the cancerous condition itself, can impair fertility. Sperm counts may be significantly restored after treatment, although this may take months or years. Of course, if both testes are removed, men are infertile. Furthermore, in these men the drop in blood testosterone levels can result in lethargy, hot flashes, impotence, and a reduction in libido. Fortunately, testosterone can be administered (usually in the form of a skin patch) to alleviate these symptoms.

Testosterone and Behavior

A wealth of evidence demonstrates behavioral effects of testosterone on males of many vertebrates, including nonhuman primates. These behaviors include dominance, territoriality, aggression, and mating behavior. Do androgens have similar effects on human behavior? The answer is less clear.

Libido Androgens are associated with *libido* (sex drive) in men. The rise in testosterone in pubertal males is accompanied by an increase in sex drive. Hypogonadal men (with abnormally low testosterone levels) have increased interest in sex after they receive testosterone supplements. However, administration of additional androgens to men within a normal hormonal range does not affect libido. Interestingly, androgens released from the ovaries and adrenal cortex are implicated in libido in women.

Aggression Throughout human societies, men exhibit more aggressive and violent behavior than women. Men are more physically aggressive, commit more murders, and are much more numerous in prison than women. However, these individuals are rare among men, and most men are not violent. Do overtly aggressive men have higher levels of androgens than more pacifistic ones? To help answer this question, several studies have looked for correlations between aggressive behavior and androgen levels. Some of these studies indeed demonstrate a positive, although weak correlation between testosterone and aggressive behavior in men, especially among prison inmates. (In fact, antiandrogens are used to reduce aggressiveness in male sex offenders.) However, other studies show no correlation. Most of these studies examined men whose testosterone measurements fell within the normal range of hormone levels. What happens when men receive higher doses of the hormone? A small but significant proportion of men taking large doses of anabolic steroids experience symptoms of agitation and aggressive urges. That is, only a small subset of men with unusually high androgen levels show aggressive behaviors. One hypothesis to explain this is that testosterone does not cause aggression, but it exaggerates previously existing aggressive tendencies. The extent to which these tendencies are organized in the male brain during embryonic development is not known (see Chapter 17).

Because of conflicting findings, a clear association between androgen levels and behavior has been difficult to demonstrate. In addition, those studies that do show a positive correlation can be interpreted in two ways, relating to cause and effect. Does a rise in testosterone lead to increased aggression or does aggressive behavior elevate testosterone? Studies of competitive interactions among athletes are interesting in this regard. Androgen levels of collegiate wrestlers rise before a wrestling match. After the match, hormone levels remain elevated in the winner but drop in the loser. These studies are used to argue that experience during a competition alters androgen levels. Whatever may be the role of androgens in human aggressive behaviors, it is important to remember that men are not simply the products of their hormone levels! Neurological effects of androgens interact with internal psychological and external social forces to shape behavior.

Sperm Count and Endocrine Disruptors

A 1992 survey of 61 scientific papers around the world, spanning a time from 1938 to 1990 and including information on 15,000 men, revealed an alarming worldwide trend, i.e., various aspects of human male reproductive function and health have been worsening dramatically. For example, the average sperm count in semen was 113 million/ml in 1940, but decreased to 66 million by 1990 (22 million/ml are required to be fertile; Chapter 9). During the same time frame, semen volume per ejaculation decreased from 3.40 to 2.75 ml. Combining these two phenomena, one could conclude that men today produce about one-half the sperm that their grandfathers made! Another study in France reported a decline in the past 20 years in sperm number, motility, and shape normality in semen donated to sperm banks. There has also been a worldwide doubling in the incidence of male developmental abnormalities such as *hypospadias*, when the urine tube (urethra) opens abnormally as a slit on the ventral surface of the penis, and *cryptorchidism* (undescended testes; see Chapter 5). In addition, diseases such as prostate and testicular cancers have more than doubled.

Is there an explanation for these trends? One answer was suggested when biologists began finding male reproductive abnormalities in wildlife. Male alligators in a Florida lake were found to have undersized penises, low testosterone levels in their blood, testicular abnormalities, and many were infertile. There had been a previous spill of pesticides such as DDT, DDE, dicofol, and sulfuric acid into the lake, and it was suggested that some of these chemicals were acting as estrogens or antiandrogens, thus feminizing the male alligators. Other wildlife species may be suffering a similar fate. For example, the number of Florida panthers has dwindled to about 30 animals, and they suffer from a variety of reproductive and endocrine defects, including abnormal sperm, low sperm count, cryptorchidism, and abnormal sex hormone secretion. Evidence shows that the tissues of these panthers contain high levels of environmental pollutants such as DDE, mercury, and polychlorinated biphenyls.

Throughout the world, and especially in industrialized countries, there is widespread exposure to environmental pollutants that are retained in animal tissue and even longer in the environment, even if their use is discontinued. Many of these chemicals mimic or inhibit hormones, and therefore are called *endocrine-disrupting contaminants*, including xenoestrogens. This disruption of developmental, reproductive, and endocrine systems could be responsible for the increase in male reproductive abnormalities in humans as well as in wildlife. Some of these chemicals can be estrogens, antiestrogens, and antiandrogens or they could have effects on other endocrine systems or direct toxic effects on reproductive cells.

Since publication of the 1992 paper showing a worldwide decline in sperm count, these data have been reexamined. New analyses confirm sperm count declines in the United States and Europe but not in non-Western countries, for which data are limited. More importantly, these more recent studies have found that sperm counts vary dramatically among different geographic locations.

Whether these geographical differences relate to the presence of endocrine disruptors in the environment remains to be discovered.

Male Sex Accessory Ducts and Glands

Sex Accessory Ducts

The male sex accessory ducts include the tubuli recti, rete testis, vasa efferentia, ductus epididymis, vas deferens, ejaculatory duct, and urethra (Fig. 4-8). These ducts serve to nurture and transport sperm.

Mature sperm, suspended in testicular fluid, leave the seminiferous tubules and enter the *tubuli recti*. These tubules, in turn, join a network of tubules still within the testis, the *rete testis*. The sperm then enter ducts that eventually leave the testis, the *vasa efferentia* (Fig. 4-2). The vasa efferentia lead to an organ lying outside the testis, the epididymis.

Epididymis The *epididymis* (pl., *epididymides*) is a comma-shaped structure, about 3.8 cm (1.5 in.) long, that lies along the posterior surface of each testis (Figs. 4-2 and 4-8). As the vasa efferentia leave the testis, they enter the larger,

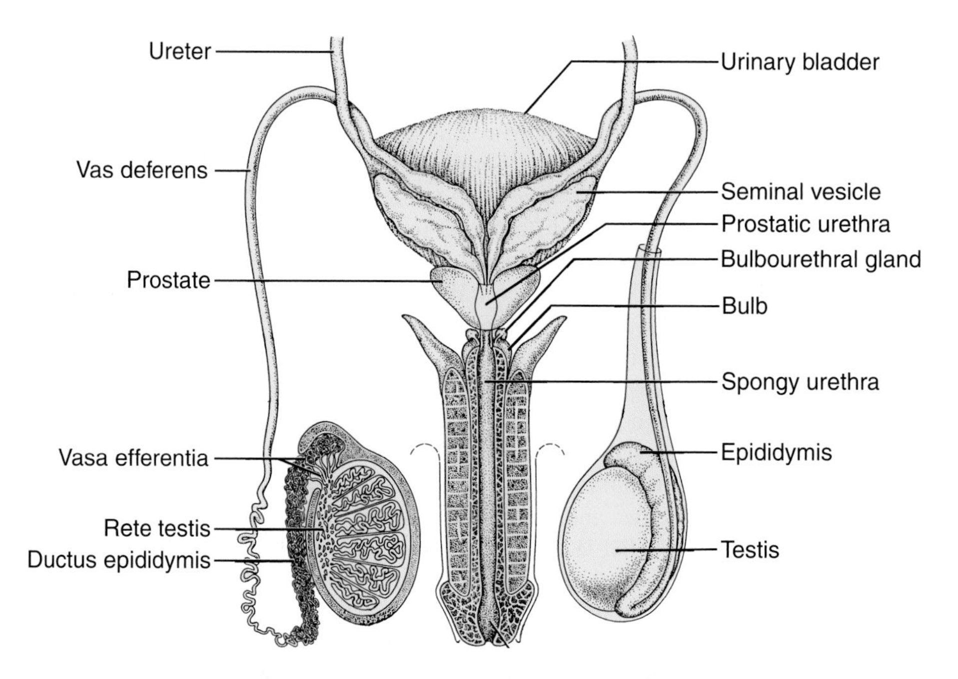

Figure 4-8 The male sex accessory ducts and glands. The structure of the penis is shown in Fig. 4-10.

upper portion ("head") of the epididymis. The vasa efferentia then join to form a single, coiled *ductus epididymis* in the middle region ("body") of the epididymis. This duct then enlarges to form the beginning of the vas deferens in the "tail" region of the epididymis. Tubules within the epididymis secrete important substances that help the sperm survive and mature. While in the body of the epididymis, sperm are nurtured by epididymal secretions and undergo further stages of their maturation. For example, human sperm taken from the head portion of the epididymis can swim, but only in a circle. In contrast, those taken from the body of the epididymis can move forward by swimming in a spiral path.

Vas Deferens The vas deferens (or *ductus deferens*) is a 45-cm (18-in.)-long tube that ascends on the posterior border of each testis, penetrates the body wall through the inguinal canal, and enters the pelvic cavity. Once inside, each vas deferens loops over the urinary bladder and extends down toward the region of the urethra (Figs. 4-1 and 4-8). The end of each vas deferens has an expanded portion, the "ampulla," which serves as a reservoir for sperm. Each vas deferens enters an *ejaculatory duct*, which is 2 cm (1 in.) long. These short ducts then lead into the urethra.

Urethra The urethra is a tube extending from the urinary bladder, through the floor of the pelvic cavity ("urogenital diaphragm"), and then through the length of the penis to its external opening (the *urethral meatus*). Thus, the urethra serves as a passageway for urine. Since, however, the ejaculatory ducts enter the urethra, this tube also transports sperm to the outside. As shown later, the prostate gland surrounds the point where the ejaculatory ducts enter the urethra (Fig. 4-8); that is why this portion is called the *prostatic urethra*. The region of the urethra that passes through the urogenital diaphragm is the *membranous urethra*, and when in the penis this duct is the *spongy* (or *cavernous*) *urethra*.

How do sperm move through these sex accessory ducts? Their journey through the seminiferous tubules, tubuli recti, rete testis, and vasa efferentia is passive, i.e., they do not swim. Instead, the cells lining these ducts have cilia, and the beating of these hair-like processes moves the fluid and its suspended sperm toward the ductus epididymis. Cilia also help the sperm move through the ductus epididymis and vas deferens. The walls of the latter two ducts contain smooth muscle, and this tissue contracts in waves to propel the sperm into the urethra during ejaculation (see Chapter 8). Sperm leave the body through the *urethral orifice*.

Sex Accessory Glands

Male *sex accessory glands* include the seminal vesicles, prostate gland, and bulbourethral glands (Fig. 4-8). These glands secrete substances into ducts that join the sex accessory ducts. Thus, the secretion of these glands, *seminal plasma*, mixes with the sperm to form *semen* or *seminal fluid*.

Seminal Vesicles The *seminal vesicles* are paired pouch-like glands, about 5 cm (2 in.) long, that lie at the base of the urinary bladder. Each seminal vesicle joins the ampulla of the vas deferens to form the ejaculatory duct (Fig. 4-8). These glands secrete an alkaline, viscous fluid rich in the sugar fructose, an important nutrient for sperm (see Chapter 9). A majority of the seminal plasma is secreted by the seminal vesicles.

Prostate Gland The *prostate gland* is a single doughnut-shaped organ about the size of a chestnut. It lies below the urinary bladder and surrounds the prostatic urethra (Fig. 4-8). The alkaline secretion of this gland makes up about 13 to 33% of seminal plasma. The secretion enters the prostatic urethra through many (up to a dozen) tiny ducts.

Bulbourethral Glands The paired *bulbourethral glands* (or *Cowper's glands*), each about the size of a small pea, lie on either side of the membranous urethra (Fig. 4-8). Their ducts empty into the spongy urethra. These glands secrete mucus that lubricates the urethra during ejaculation.

Prostate Disorders The male sex accessory glands, because of their connections to the urethra, can become infected by microorganisms that invade the urinary tract. In fact, bacterial infection of the prostate gland (*bacterial prostatitis*) is common in sexually mature males (see Chapter 18). Once infected, the prostate can enlarge and ejaculation can be painful (see Chapter 8). Other possible symptoms of this condition include fever, chills, rectal discomfort, lower back pain, and an increased urgency of urination (because the enlarged prostate is pressing on the urethra and bladder). This type of prostatitis can often be cured with antibiotics.

The prostate gland, however, can also enlarge without infection. A common cause is *benign prostatic hyperplasia* (BPH). More than one-half of men over 45 have this condition. Although the tissue growth in BPH is not cancerous, men with this condition can experience frequent and difficult night-time urination. Having BPH is not correlated with later development of prostate cancer. A new drug (finasteride) can be used to treat BPH; it lowers production of an androgen (dihydrotestosterone) that causes prostate growth, but it rarely also causes impotence and reduces sex drive. In 1996 the U.S. FDA approved use of an instrument that emits microwaves that can kill excess prostate cells and alleviates difficulty of urination 75% of the time.

Prostate cancer is the second leading cause of cancer-related death (behind lung cancer) in American men. About 1 in 6 men will develop this disorder, and in 2005 232,090 new cases were diagnosed and 30,350 men died from it. Discovery of this cancer usually occurs in men over 65. The incidence of prostate cancer is 30% higher in American blacks than in whites (Fig. 4-9). Of men who die over age 50 from other causes, at autopsy it is discovered that one-third have prostate cancer. The annual incidence of prostate cancer in the United States is increasing (Fig. 4-9).

There is an inherited basis to about 10% of prostate cancer cases. A man's risk is two to four times higher if his father or a brother developed it, and six to eight times higher if both father and brother developed it. One of the genes related to inherited breast cancer (BRCAI; see Chapter 2) also increases the risk of prostate (as well as colon) cancer in men. A gene on chromosome 1 (*hpc1*) and a certain variant of the gene coding for *5α-reductase* also appear to be related to prostate cancer.

Dietary factors appear to influence the risk of prostate cancer. For example, the high levels of an essential fatty acid (a-linolenic acid) found in red meat and some vegetable oils may be a culprit. In general, eating a diet high in saturated fat

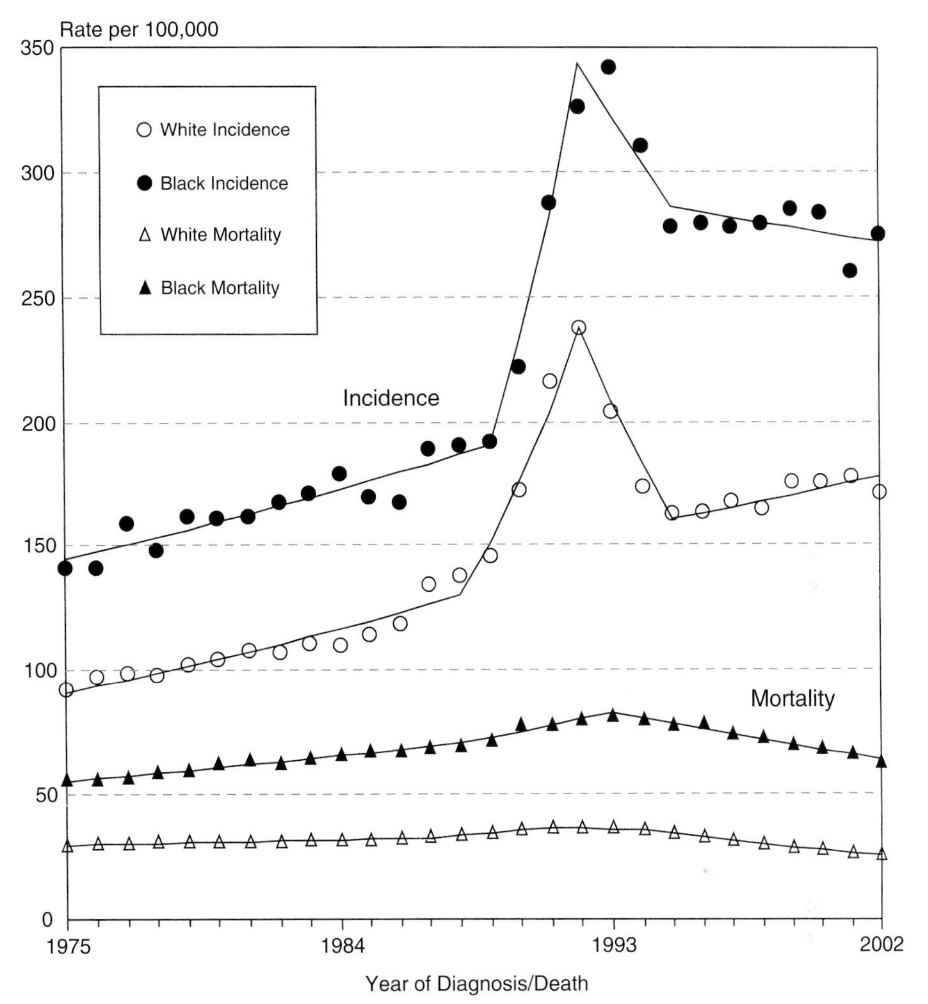

Figure 4-9 A comparison of trends in annual incidence and mortality of prostate cancer in black and white American men. Prostate cancer detection peaked in the early 1990s after the use of prostate-specific antigen screening became widespread following its introduction in 1987. This new method enabled the diagnosis of a large pool of asymptomatic men whose prostate cancer had not been detected previously by a digital rectal examination.

increases the spread of this cancer once it has developed. Dietary antioxidants, found mostly in fruits and vegetables, may be protective.

How is prostate cancer detected? In the past, a physician would insert a gloved finger up the rectum during a physical exam to palpate (feel) for lumps or hardening of the prostate gland. This is called the *digital rectal exam* (DRE). However, DRE alone is not very efficient; 70% of prostate tumors discovered by DRE alone are already in advanced and more dangerous stages. Since 1986, a blood test for a *prostate-specific antigen* (PSA) has been of great help in detecting prostate cancer in early stages. This protein is produced by the normal prostate in low amounts, but for each age there is a PSA level beyond which there is an indication of abnormality. Blood PSA also goes up with BPH, but at a slower rate than with prostate cancer. Combined, DRE and PSA tests can detect 85% of prostate cancers. Ultrasound and a prostatic biopsy can be used to verify the presence of prostate cancer. It should be noted, however, that bacterial prostatitis as well as mechanical pressure on the gland can also increase PSA blood levels and that about 25% of men with abnormal DRE and PSA tests actually do not have prostate cancer. It is now recommended that a man have annual DRE and PSA tests beginning at age 50; men at especially high risk (sub-Saharan African ancestry or a close relative diagnosed with prostate cancer at an early age) should begin having these tests at age 45. Men who have multiple close relatives with prostate cancer should begin screening at age 40. Some physicians feel, however, that men with no symptoms of prostate enlargement, such as urination difficulty, would not benefit from DRE or PSA testing. Because both DRE and PSA testing suffer from high *false positive* results (positive results in the absence of prostate cancer) and *false negatives* (failure to detect cases of prostate cancer), the search is on for new diagnostic tools. One promising blood test that has been shown to detect cancer with high accuracy screens for a pattern of blood proteins instead of a single protein such as PSA. Future tests may also predict which patients have aggressive, fast-growing cases.

Prostate cancer occurs in stages (Table 4-1), with the last stage (metastasis to surrounding lymph nodes and bones) often leading to death in 2 to 5 years. Detected early, however, the survival rate is good. Even without treatment for localized prostate cancer, 87% of men are still alive after 10 years, and with treatment there is a greater survival rate and often the cancer is cured, especially if detected early.

If detected in its early stages, prostate cancer is treated with radiation or radical prostatectomy (surgical removal of the prostate gland; Table 4-1); more advanced stages are treated with radiation and prostatectomy, and some

Table 4-1 Stages and Therapies for Prostate Cancer[a]

	State of disease	Standard therapy
Stage A		
Microscopic cancer within prostate gland	Cancer is confined to one or a few sites within gland	Radiation or radical prostatectomy
Stage B		
Palpable lump(s) within prostate gland	Cancer forms a nodule, or multiple nodules, in gland	Radiation or radical prostatectomy
Stage C		
Large mass involving all or most of prostate gland	Cancer occurs as a continuous mass that may extend beyond the gland	Radiation
Stage D		
Metastatic tumor	Cancer appears in the lymph nodes and/or the pelvis	Orchidectomy; estrogens; antiandrogens; GnRH inhibitory agonists

[a]Modified from Garnick (1994).

men choose orchidectomy to remove the source of androgens that support the tumor cells. Other treatments can be an antiandrogen, estrogens, or a GnRH inhibitory agonist (Table 4-1).

Treating early prostate cancer has been controversial. Unfortunately, it is not possible to predict which cancers will grow rapidly and spread and which will grow slowly. Many men, especially elderly ones, who have been diagnosed with prostate cancer will actually die, not of prostate cancer, but of some other disorder. Some medical professionals recommend close monitoring but withholding aggressive treatment for men with small tumors that have not spread beyond the gland. This strategy is termed "watchful waiting." It avoids potential problems associated with prostate removal, including incontinence and erectile dysfunction (because surgery can disturb the nerves that control erection). Some feel that treating older men with slow-growing prostate cancer should not be done, as the treatment often has negative side effects (e.g., impotence, incontinence, osteoporosis) and men over 70 probably would die of other causes.

Hormonal Control of Sex Accessory Structures

The maintenance and function of male sex accessory ducts and glands are under the control of androgens. Testosterone, for example, stimulates the secretion of sperm nutrients by the epididymis. If a man is orchidectomized, the sex accessory structures will soon atrophy. Some evidence in laboratory mammals also shows that prolactin acts with testosterone to control the function of the sex accessory ducts and glands. Finally, oxytocin and prostaglandins may stimulate contraction of the smooth muscle in these glands and ducts during ejaculation (see Chapter 8).

Penis

The *penis* has an enlarged acorn-shaped end called the *glans penis* and a *penile shaft*, or *body* (Fig. 4-10). The rounded ridge at the back end of the glans penis is called the *corona glandis*. The skin covering the shaft extends in a loose fold over the glans; this is the *penile prepuce*, or *foreskin*. In circumcised men, much of the prepuce has been removed surgically (see Chapter 12). Small glands under the foreskin secrete oil, which is mixed with skin cells to produce *smegma*. Retraction of the prepuce to remove the smegma with soap and water is important, especially in uncircumcised men, because bacteria can thrive in the secretion.

The shaft of the penis contains three cylindrical masses of spongy tissue filled with blood sinuses (Fig. 4-10). Two of these cylinders, one on each side of the top of the shaft, are the corpora cavernosa. The single cylinder at the bottom of the shaft is the corpus spongiosum; the spongy urethra runs through this cylinder. The spongy cylinders fill with blood during an erection (see Chapter 8), and the corpus spongiosum can be felt as a raised ridge on the bottom of the erect penis. The shaft of the penis attaches to the pelvic wall; extensions of the spongy tissue actually are fused to the pelvic bone.

Men and women often are curious or concerned about penis size. The length of the nonerect (flaccid) penis of most men ranges from 8.5 to 10.5 cm (3.3 to 4.1 in.), with an average length of 9.5 cm (3.7 in.). The average length of the erect penis is 16 cm (6.3 in.), with a range of 12.0 to 23.5 cm (4.7 to 9.2 in.). The average circumference of an erect penis at its thickest point is 13.2 cm (5.2 in.). Contrary to

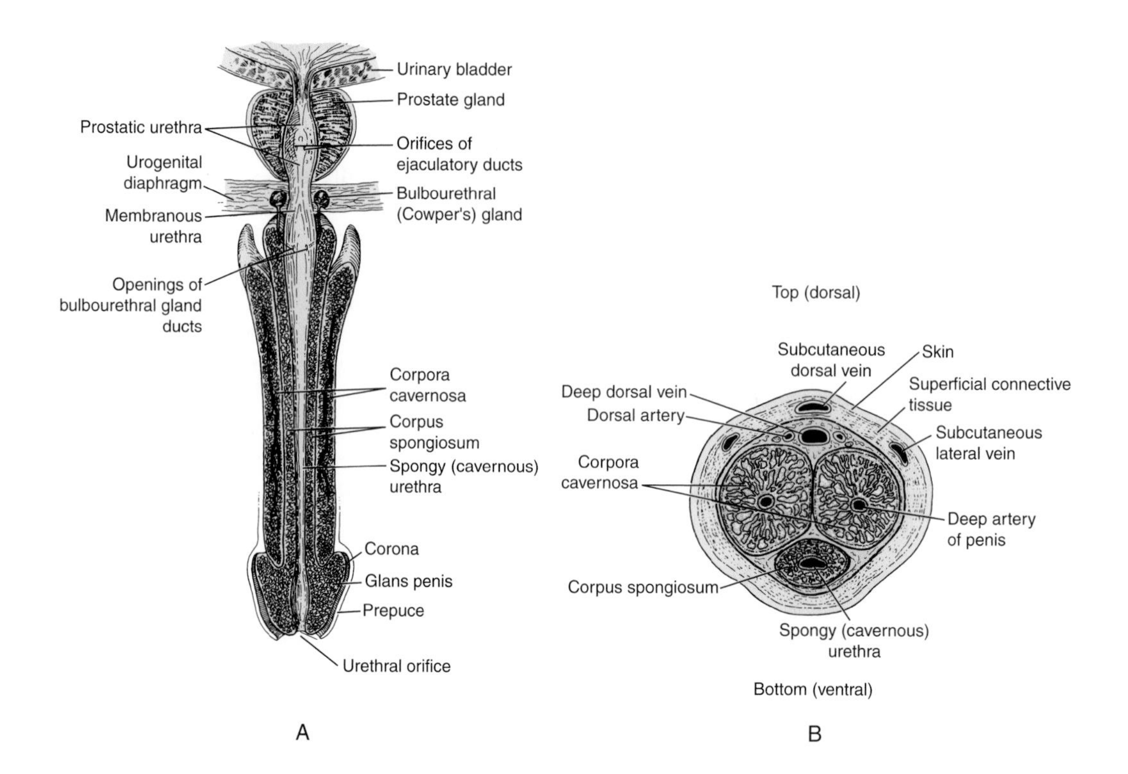

Figure 4-10 The structure of the penis. (a) The sectioned penis is viewed from below. (b) The penis is seen in cross section.

popular belief, there is no correlation between a man's skeletal makeup and his penis size. When a relatively small flaccid penis becomes erect, it enlarges more than a relatively large flaccid penis. Chapter 8 reviews the human sexual response, including what happens to the penis during male sexual arousal.

Scrotum

The *scrotum* is a pouch, suspended from the groin, which contains the testes and some of the male sex accessory ducts. The pouch has two compartments, each housing one testis; the left testis usually hangs lower than the right. The skin of the scrotum is relatively dark and hairless and is separated in the midline by a ridge called the *raphe*. Underneath the skin of the scrotum is a layer of involuntary smooth muscle, the *tunica dartos*. Just under the tunica dartos is another layer of muscle, the *cremaster*, which contracts when the inner thigh is stroked; this "cremasteric reflex" is discussed in Chapter 8. Unlike the tunica dartos, the cremaster is a voluntary striated muscle.

One function of the scrotum is to help maintain the temperature necessary for spermatogenesis by locating the testes outside of the body cavity. The temperature in the human scrotum is 3.1 °C (5.6 °F) lower than the normal temperature within the body (37 °C, or 98.6 °F). Spermatogenesis requires this lower temperature, and an increase in scrotal temperature of 3 °C (5.4 °F) caused by wearing tight underclothing can lead to a decrease in viable sperm and even to an increase in the mutation rate in male germ cells. The testes normally descend from the pelvic cavity into the scrotum before birth (see Chapter 5). If they fail to descend by the time spermatogenesis begins in the child, the sperm will be damaged by the high temperature in the body cavity. Taking a hot bath or sauna can lead to temporary problems with sperm production. Taking a series of hot baths, however, is not a reliable contraceptive measure.

Temperature receptors are located in the scrotum. When the temperature is too low, the tunica dartos muscle contracts, causing the scrotal skin to wrinkle and the testes to ascend. Thus, the scrotum has a smaller surface area for heat loss and the testes now are closer to the warmer groin region. It should be noted that the other muscle in the scrotum, the cremaster, is not involved in temperature regulation. However, this muscle contracts in times of sexual excitement, fear, or anxiety. Numerous sweat glands in the scrotal skin secrete sweat when the scrotal temperature is high, and evaporation of this fluid also helps cool the testes.

Chapter 4, Box 2: Why a Scrotum?

As you will learn in Chapter 5, the testes remain within the abdominal cavity in the fetus until about month 9 of development, at which time they descend through the inguinal canals in the lower body wall to reside in the external scrotum. Mammals are the only vertebrate group that has species with external testes in a scrotum. Although a majority of mammals (2621 species) have this condition, there are many (1582 species) that have internal testes and no scrotum. This latter group includes marine mammals and elephants. There are six positions of testes found in mammals, five of them internal and one external. Why did external scrotal testes evolve in many mammals, including humans? Keeping the testes in an external sac risks exposing them to physical harm, so there must be benefits of having external testes that outweigh this disadvantage. Biologists have come up with many theories, some of which are discussed here.

1. The temperature hypothesis. The temperature within the scrotum is 3.1 °C (5.6 °F) lower than the normal internal body temperature. This hypothesis recognizes that the testes are external because spermatogenesis cannot occur at normal body temperature. However, ancestors of mammals with external testes presumably had internal testes. Therefore, why did scrotal mammals not retain the testicular enzymes that work at the higher temperature and keep the testes inside? All birds and many mammals do have internal testes, so apparently they had no problem evolving these high-temperature testicular enzymes. Also, this theory does not explain the four degrees of testicular descent in mammals with internal testes (see figure).

2. The sperm fitness hypothesis. This hypothesis states that sperm are produced and stored (in the epididymides and vasa deferentia) in cool scrotal temperatures because this environment improves sperm quality. For example, cooler temperatures induce the formation of more mitochondria in each sperm midpiece, with these mitochondria then providing an energy source for more tail beating when the sperm move up the female reproductive tract. Evidence for this is the fact that, in mammals with pelvic internal testes, the epididymides and vasa deferentia, with their stored sperm, project toward the body wall to a scrotum-like skin out-pocketing. However, mammals with abdominal internal testes seem to do fine without this sperm-cooling mechanism.

3. The social signal hypothesis. The scrotum could be used as a social signal during male–male competition or to attract females. This hypothesis is supported by the fact that, in some mammals (e.g., primates),

Continued on next page.

Chapter 4, Box 2 continued.

the scrotum is colorfully pigmented and is used for social display. However, why do many nocturnal and burrowing mammals have external testes? And why is the scrotum of many mammals nonpigmented, covered with fur, and/or otherwise inconspicuous?

4. The clitoral stimulation hypothesis. Female orgasm assists the movement of sperm up the reproductive tract after coitus. Stimulation of the clitoris increases the intensity of female orgasm (Chapter 8). The rear-entry position of mating may allow the scrotum to stimulate the clitoris and, in this way, may produce an orgasm that facilitates sperm movement. Mammals that do not use the rear-entry position (e.g., whales) do not have external testes. Although modern humans utilize many different coital positions (see Chapter 8), rear entry being one of them, many of these may have evolved through cultural rather than natural selection, and it is difficult to know the preferred mounting position of our ancient ancestors. With a few exceptions, mating occurs via rear entry in other primates.

5. The loss of the scrotum hypothesis. Regardless of which of the aforementioned four ideas explains the presence of the scrotum, a new hypothesis suggests that the scrotum and external testes evolved early in the mammalian line. It was only later that descent of testes was selected against when new enzymes that worked at the internal body temperature evolved. Support for this idea is that cetaceans (whales, porpoises) have internal testes but also vestigial inguinal canals, implying that their ancestors had external testes and a scrotum. So perhaps we must explain not only "why a scrotum" but why it was lost and the testes became internal in many mammalian species.

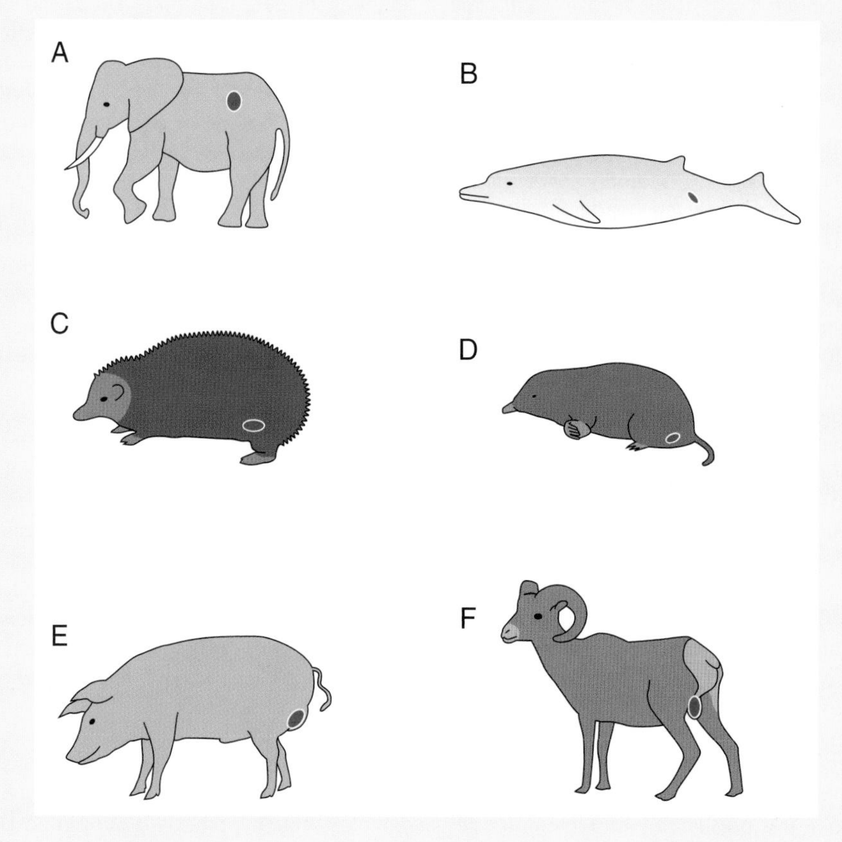

Variation in the anatomical position of mammalian testes. Positions A–D are within the body cavity, whereas E and F are external within a scrotum.

Chapter Summary

The male reproductive tract consists of the testes, the sex accessory ducts and glands, the penis, and the scrotum. Each testis is subdivided into testicular lobules, and each lobule contains seminiferous tubules, which are the site of sperm production (spermatogenesis and spermiogenesis). Sertoli cells are also present within the seminiferous tubules. The interstitial spaces between tubules contain blood vessels and Leydig cells, the source of androgens.

The anterior pituitary gland secretes two gonadotropic hormones, which are both necessary for testicular function. Luteinizing hormone stimulates the secretion of androgens from the Leydig cells, and these androgens control some stages of spermatogenesis. Follicle-stimulating hormone stimulates Sertoli cells in the seminiferous tubules to secrete androgen-binding protein, which concentrates androgens around the germ cells, and estradiol, which may stimulate spermatogenesis. Sertoli cells also support and nourish the male germ cells. Secretion of FSH and LH is controlled by feedback effects of androgens and possibly inhibin produced by the testes. Stress, exercise, nutrition, chemicals, and ionizing radiation can influence testicular function. Testicular and prostate cancer, as well as developmental abnormalities of the reproductive tract, are becoming more common, perhaps due to pollutants that act as endocrine disrupters. Sperm produced in the seminiferous tubules travel through a series of male sex accessory ducts: the tubuli recti, rete testis, vasa efferentia, ductus epididymis, vas deferens, ejaculatory duct, and urethra. Male sex accessory glands (seminal vesicles, prostate, and bulbourethral glands) produce seminal plasma that enters the urethra and mixes with sperm during ejaculation. The male sex accessory structures are under androgenic control.

The penis consists of a glans penis, shaft, and prepuce. The shaft contains spongy tissue that fills with blood during an erection. The scrotum is a pouch of skin and muscle that contains both testes. This sac plays a role in controlling testicular temperature and may play a role in sexual behavior.

Further Reading

Carreau, S., and Levallat, J. (2000). Testicular estrogens and male reproduction. *News Physiol. Sci.* **15**, 195–198.

Culotta, E. (1995). New evidence about feminized alligators. *Science* **267**, 330–331.

Elias, A. N., and Wilson, A. F. (1993). Exercise and gonadal function. *Hum. Reprod.* **8**, 1747–1761.

Facklemann, K. A. (1992). The neglected sex gland. *Sci. News* **142**, 94–95.

Garnick, M. B. (1994). The dilemmas of prostate cancer. *Sci. Am.* **270**(4), 72–81.

Garnick, M. B., and Fair, W. R. (1999). Combating prostate cancer. *Sci. Am.* **10**(2), 100–105.

Guillette, L. J., Jr., and Guillette, E. A. (1996). Environmental contaminants and reproductive abnormalities in wildlife: Implications for public health. *Toxicol. Indust. Health* **12**, 537–550.

Hoberman, J. M., and Yesalis, C. E. (1995). The history of synthetic testosterone. *Sci. Am.* **272**(2), 76–81.

Luoma, J. R. (1995). Havoc in the hormones. *Audubon* July/Aug., 61–67.
Lutz, D. (1996). No conception. *Sciences* **36**(1), 12–15.
Mitwer, M. (1995). Vasectomy appears unlikely to raise men's chances of developing either prostate or testicular cancer. *Family Planning Perspect.* **27**(2), 95–96.
Nicholls, H. (2001). No evidence for "feminizing" effects of soy phytoestrogens. *Trends Endocrinol. Metabol.* **12**, 340.
Pennisi, E. (1996). Homing in on a prostate cancer gene. *Sci. News* **274**, 1301.
Raloff, J. (1994). That feminine touch. *Sci. News* **145**, 56–58.
Raloff, J. (1994). Manhood's cancer. *Sci. News* **145**, 138–140.
Raloff, J. A. (1994). Clues to the rise in testicular cancer. *Sci. News* **145**, 139.
Raloff, J. (1995). Beyond estrogens: Why unmasking hormone-mimicking pollutants proves so challenging. *Sci. News* **148**, 44–46.
Raloff, J. (1997). Radical prostates. Female hormones may play a pivotal role in a distinctly male epidemic. *Sci. News* **151**, 108–109.
Raloff, J. (2000). New concerns about phthalates. *Sci. News* **158**, 152–154.
Rennie, J. (1993). Malignant mimicry. *Sci. Am.* **269**(3), 34–38.
Sharpe, R. M., and Skakkebaek, N. E. (1993). Are estrogens involved in falling sperm counts and disorders of the male reproductive tract? *Lancet* **341**, 1392–1395.
Stone, R. (1994). Environmental estrogens stir debate. *Science* **265**, 308–310.
Stutz, B. D. (1996). Hormonal sabotage. *Nat. History Magazine* **105**(3), 42–49.
Wojciecchowski, A. P. (1992). Evolutionary aspects of mammalian secondary sexual characteristics. *J. Theor. Biol.* **155**, 271–272.

Advanced Reading

Arce, J. C., *et al.* (1993). Subclinical alterations in hormone and semen profile in athletes. *Fertil. Steril.* **59**, 398–404.
Archer, J. (1991). The influence of testosterone on human aggression. *Br. J. Psychol.* **82**, 1–28.
Auger, J. A., *et al.* (1995). Decline in semen quality among fertile men in Paris during the past 50 years. *N. Engl. J. Med.* **332**, 282–285.
Bahrke, M., *et al.* (1996). Psychological and behavioural effects of endogenous testosterone and anabolic-androgenic steroids: An update. *Sports Med.* **6**, 367–390.
Barrett-Conner, E., *et al.* (1999). Bioavailable testosterone and depressed mood in older men: The Ranch Bernardo study. *J. Clin. Endocrinol. Metabol.* **84**, 573–577.
Ben-Jonathaan, N., and Steinmetz, R. (1998). Xenoestrogens: The emerging story of bisphenol A. *Trends Endocrinol. Metabol.* **9**, 124–128.
Boujrad, N., Ogwnegbu, S. O., Garnier, M., Lee, C.-H., Martin, B. M., and Papadopoulos, V. (1995). Identification of a stimulator of steroid hormone synthesis from testis. *Science* **268**, 1609–1612.
Bribiescas, R. G. (2001). Reproductive ecology and life history of the human male. *Ybk. Phys. Anthropol.* **44**, 148–176.
Carey, P., *et al.* (1988). Transdermal testosterone treatment of hypogonadal men. *J. Urol.* **140**, 76–79.
Fleming, C, *et al.* (1993). A decision analysis of alternative treatment strategies for clinically localized prostate cancer. *J. Am. Med. Assoc.* **269**, 2650–2658.

Freeman, S. (1990). The evolution of the scrotum: A new hypothesis. *J. Theoret. Biol.* **145**, 429–445.

Gooren, L.J.G., and Kruijver, F.P.M. (2002). Androgens and male behavior. *Mol. Cell Endocr.* **198**, 31–40.

Guillette, L. J., Jr., *et al.* (1994). Developmental abnormalities of the gonad and abnormal sex hormone concentrations in juvenile alligators from contaminated and control lakes in Florida. *Environ. Health Perspect.* **102**, 680–688.

Hayes, R. B. (1995). Are dietary fat and vasectomy risk factors for prostate cancer? *J. Natl. Cancer Inst.* **87**, 629–631.

Hose J. E., and Guillette, L. J. (1995). Defining the role of pollutants in the disruption of reproduction in wildlife. *Environ. Health Perspect.* **103**(Suppl. 4), 87–91.

Kelce, W. R., *et al.* (1995). Persistent DDT metabolite *p,p'*-DDE is a potent androgen receptor antagonist. *Nature* **375**, 581–585.

Krahn, M. D., *et al.* (1994). Screening for prostate cancer: A decision analytic view. *J. Am. Med. Assoc.* **272**, 773–780.

Kuiper, G. G. J. M., et al. (1998). Estrogen is a male and female hormone. *Sci. Med. July*/Aug., 36–45.

LeBlanc, G. A., *et al.* (1997). Pesticides: Multiple mechanisms of demasculinization. *Mol. Cell. Endocrinol.* **126**, 1–5.

Oesterling, J. E., *et al.* (1993). Serum prostate-specific antigen in a community-based population of healthy men: Establishment of age-specific reference ranges. *J. Am. Med. Assoc.* **270**, 860–864.

Peterson, R. E., *et al.* (1993). Developmental and reproductive toxicity of dioxins and related compounds: Cross-species comparisons. *Crit. Rev. Toxicol.* **23**, 283–335.

Pienta, K. J., and Esper, P. S. (1993). Is dietary fat a risk factor for prostate cancer? *J. Natl. Cancer Inst.* **85**, 1538–1539.

Sharpe, R. M., *et al.* (1995). Gestational and lactational exposure of rats to xenoestrogens results in reduced testicular size and sperm production. *Environ. Health Perspect.* **103**, 1136–1143.

Stauab, N. L., and De Beer, M. (1997). The role of androgens in female vertebrates. *Gen. Comp. Endocrinol.* **108**, 1–24.

Swan, S.H., et al. (1997). Have sperm densities declined? A reanalysis of global trend data. *Environ. Health Persp.* **105**, 1228–1232.

Thibline, I., *et al.* (2000). Cause and manner of death among users of anabolic androgenic steroids. *J. Foren. Sci.* **45**, 16–23.

Werdelin, L., and Nilsonne, A. (1999). The evolution of the scrotum and testicular descent in mammals: A phylogenetic view. *J. Theor. Biol.* **196**, 61–72.

Wheeler, G. D., *et al.* (1991). Endurance training decreases serum testosterone levels in men without change in luteinizing hormone pulsatile release. *J. Clin. Endocrinol. Metabol.* **72**, 422–425.

Yeh, S., *et al.* (1998). From estrogen to androgen receptor: A new pathway for sex hormones in prostate. *Proc. Natl. Acad. Sci. USA* **95**, 5527–5532.

PART TWO

Sexual Differentiation and Development

CHAPTER FIVE

Sexual Differentiation

Introduction

This chapter describes how reproductive structures develop. As described in Chapters 2 and 4, we know that anatomical features of the adult male and female reproductive systems differ greatly. Thus, during embryonic development, males and females must take distinctly different paths in forming their reproductive anatomies. Remarkably, the reproductive systems of genetically male and female embryos are, at an early stage of development, identical. How this identical stage differentiates into the typical (and sometimes atypical) male and female reproductive systems during embryonic and fetal development is the subject of this chapter. Sex differences in brain structure and function, and their development, are discussed in Chapter 17.

Chromosomal Sex

Sex is determined by chromosomes and is established at the time an egg is fertilized by a sperm (see Chapter 9). In females, each diploid cell has 46 chromosomes (23 pairs), including 22 pairs of *autosomes* (nonsex chromosomes) and 1 pair of *sex chromosomes*. The sex chromosomes in females are XX, so the chromosomal type of female cells is designated as 46:XX. This occurs when a haploid (23:X) egg is fertilized by a haploid (23:X) sperm. Male somatic cells are 46:XY (Fig. 5-1), which occurs when a 23:X egg is fertilized by a 23:Y sperm.

X Chromosome

The X chromosome is a fairly large chromosome (Figs. 5-1 and 5-3) and contains many essential genes unrelated to sex determination. Some of these are metabolic enzymes essential for life (so-called "housekeeping genes"). Because females have two X chromosomes and males have only one, it might be assumed that female cells make twice the quantity of proteins encoded in the X chromosome compared with males. This is not the case; in fact, males and females have similar expression of genes on the X chromosome. This is because, early in embryonic development, a female shuts down one X chromosome in each of her cells. In 1949, M. Barr discovered that female cells, but not male cells, contain a small dot of condensed material that represents one of the X chromosomes that has been inactivated (*X-chromosome inactivation*). This inclusion is called the *Barr body* or *sex chromatin* (Fig. 5-2). In 1961, Mary Lyon proposed that one X chromosome is

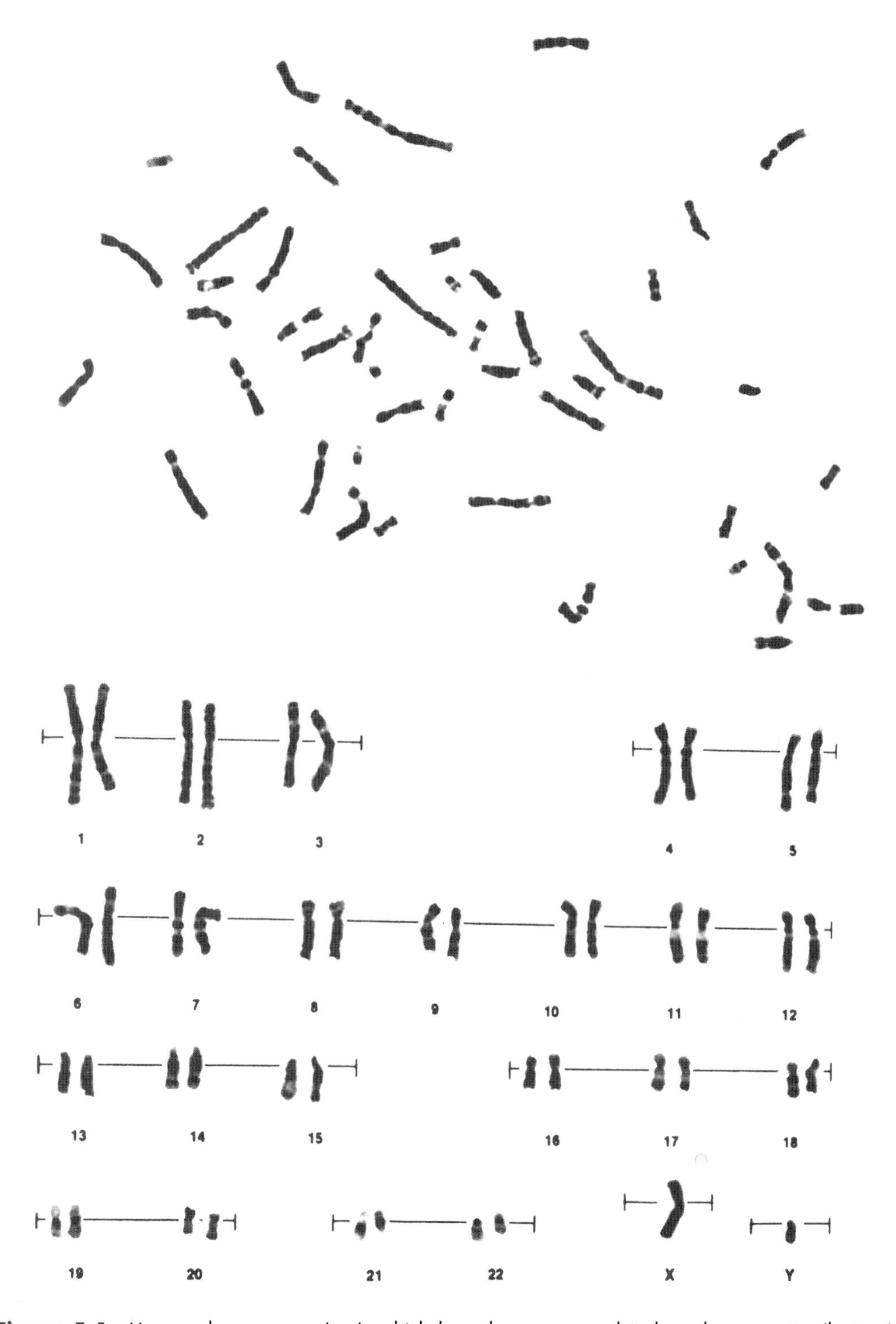

Figure 5-1 Human chromosomes (top), which have been arranged in homologous pairs (bottom). These are from a cell of one of the authors' son (Evan Pryor Jones), which was obtained while he was still a fetus by amniocentesis (see Chapter 10). Note the X and Y chromosomes at lower right.

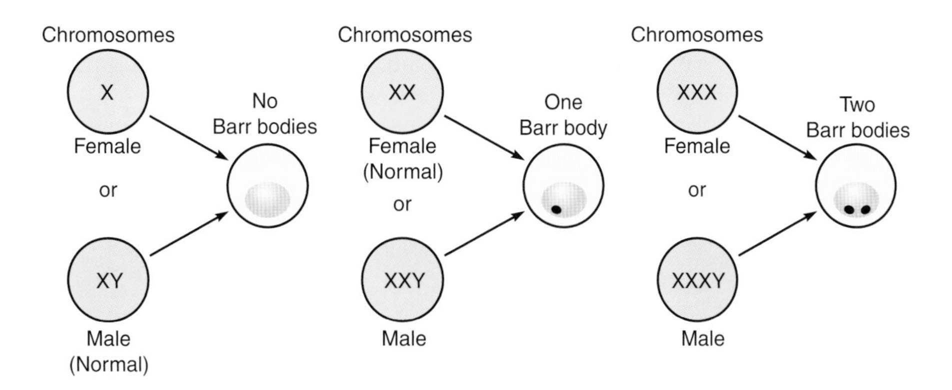

Figure 5-2 Possible variations in number of sex chromosomes in males and females and the resultant number of sex chromatins, or Barr bodies. Note that the number of Barr bodies is one less than the number of X chromosomes.

inactivated randomly in female embryonic cells so that a double gene dosage is avoided. All cells that follow a particular cell line have the same X chromosome inactivated. This means that, in some regions of a woman's body, one of the X chromosomes is active whereas only the other is active in other regions (i.e., women are *mosaics* with respect to X chromosome expression). If more than two X chromosomes are present, as in some conditions produced by errors of fertilization (discussed later in this chapter), all of the X chromosomes are inactivated except one (Fig. 5-2).

The presence of the Barr body is used to determine the sex of fetuses during procedures such as amniocentesis and chorionic villus sampling (see Chapter 10). Also, cells from the lining of the mouth membrane can be checked for the presence of a Barr body (the *buccal smear test*). Fluorescent dyes can be used to identify X and Y chromosomes.

Y Chromosome

The Y chromosome is one of the smallest human chromosomes. Only a minor portion of this chromosome is homologous with the X chromosome, allowing it to pair with the X during meiosis in males (Fig. 5-3). The majority of the Y contains DNA sequences unique to this chromosome. Several genes have been identified on the Y chromosome (see HIGHLIGHT box 5-1). Most of these have essential roles in spermatogenesis; mutations in these genes have been shown to cause infertility in men. The short arm of the Y chromosome contains a *sex-determining region of the Y* (*SRY*) gene that encodes a testis-determining factor (TDF). As its name indicates, the *SRY* gene determines the sex of a human embryo. If *SRY* is present (on a normal Y chromosome), the embryo will form testes and develop as a male. If *SRY* is absent, the embryo will develop as a female. The importance of the *SRY* gene in human sex determination has been revealed by studying a curious situation: the existence of XX males. Approximately 1 out of 20,000 men has an XX genotype. When studied more closely, it was found that a piece of the short arm of the Y had been translocated onto one of the X chromosomes in these men. This piece of Y chromosome contained the *SRY* gene, which caused the development of testes and a male phenotype in these XX individuals. Similarly, cases of XY females have been found. In all of these females, a piece of the short arm of the Y containing the *SRY* gene was missing.

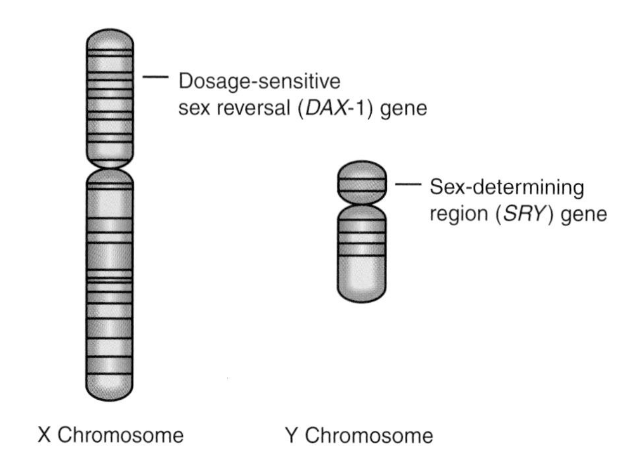

Figure 5-3 Location of genes involved in gonadal sex differentiation. The sex-determining region of the Y (*SRY*) gene codes for the production of testis-determining factor (TDF), which in turn causes testis differentiation. Absence of this gene in an individual lacking the Y chromosome results in the formation of ovaries. The *DAX-1* gene on the X chromosome suppresses *SRY* gene expression, but the normal interaction of *DAX-1* and *SRY* has not been fully discovered.

Chapter 5, Box 1: Why the Y Chromosome?

The development of a testis from an indifferent gonad is triggered by the transient expression of the *SRY* gene on the Y chromosome. It was originally thought that the wimpy Y chromosome was a wasteland, devoid of other genes besides *SRY*. However, it has been shown that the Y chromosome also contains more than 20 genes or gene families that affect spermatogenesis and are also responsible for other aspects of male growth. These 20 genes, as well as the *SRY* gene, are on the part of the Y chromosome that does not trade genes (recombine) with parts of its X chromosome partner. Thus, genes benefiting the male have tended to accumulate on the nonrecombining part of the Y chromosome. About 10 of these Y chromosome genes are expressed only in the testes and here they control spermatogenesis.

Other newly discovered genes on the Y chromosome control developmental and growth processes that give men some of their sexually dimorphic traits. One of these genes increases the rate of fetal growth of male embryos, at least until midpregnancy. Another Y gene has a marked effect on stature; it is located on the long arm of the human Y chromosome and is called "growth control Y, or *GCY*." A third gene on the Y chromosome results in bigger teeth in men by its product acting on dental enamel and tooth thickness. These genes are passed unaltered from father to son, as they do not recombine with genes on the X chromosome. It has been hypothesized that these male growth genes are subject to sexual selection in that they affect body features directly (overall size, body stature, and tooth size) that, in our ancestors

Continued on next page.

Chapter 5, Box 1 continued.

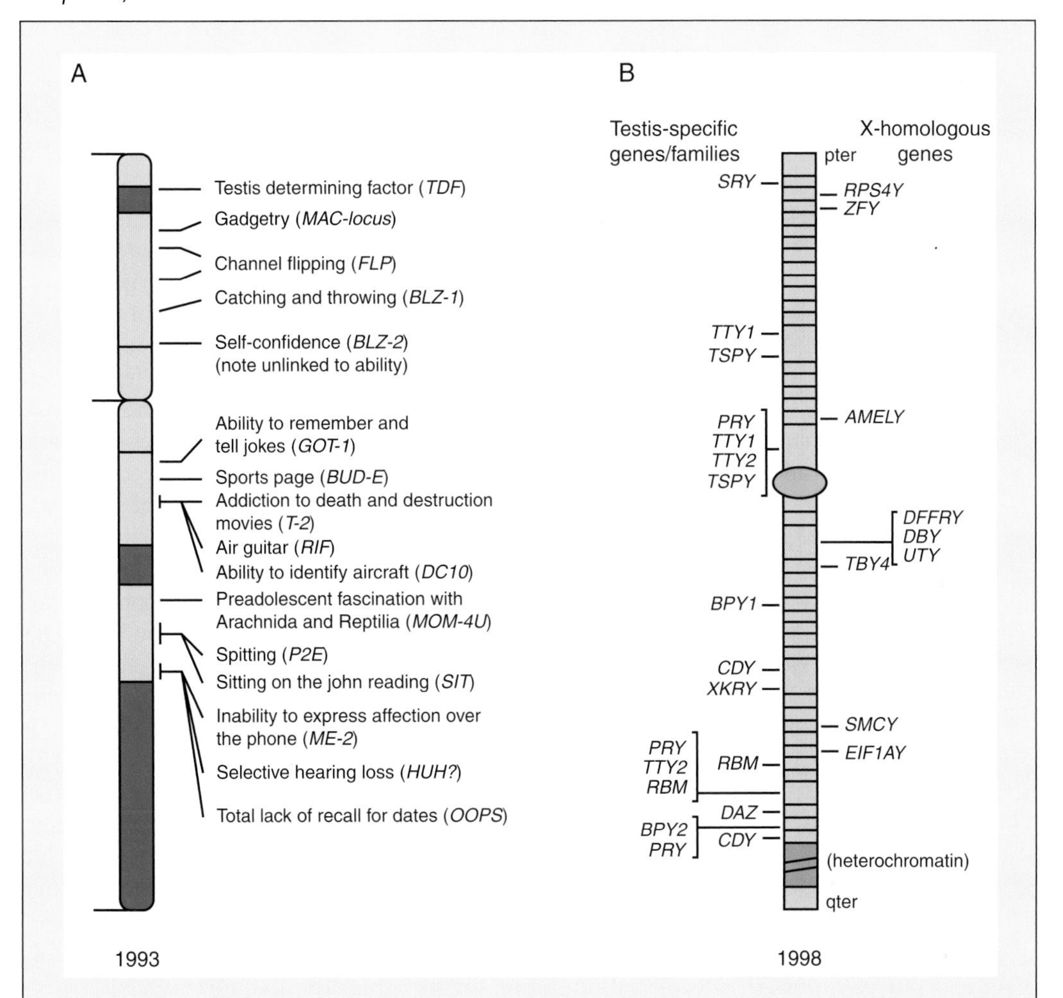

(a) Functions of genes on the human Y chromosome as postulated humorously in 1993. (b) Genes now known to be present on the human Y chromosome, with some expressed only in the testis. Others are homologous to genes on the X chromosome. (Used with permission from Roldan and Gomendio (1999).)

at least, related to male–male competition and female attraction.

As discussed in Chapter 16, one in six married couples in the United States are infertile, and of these, about 20% of the male partners exhibit a failure to produce sperm. It is possible that these infertile men have a deleted or abnormal form of one of the 10 or so genes on the Y chromosome that control various stages of spermatogenesis. One of these 10 genes is called *DAZ*, and *SRY* expression may turn on *DAZ* expression as well as expression of some of the other spermatogenesis genes on the Y chromosomes. *DAZ* is only present on the Y chromosome of primates, including humans, but not in other mammals. This gene may actually be derived from another gene, *DAZL*, which is found on chromosome 3 of both sexes. In primates the *DAZL* gene may have been transposed to the Y chromosome, where it plays a role in spermatogenesis. We do not know the role of *DAZL* in the ovaries.

Sex-Linked Inheritance

As mentioned earlier, the human X chromosome contains numerous genes that have no homologous sequences on the Y. These genes are said to be *sex linked* (or *X linked*), and the phenotypic effects of these genes are called *sex-linked traits*. Females have two alleles for each sex-linked trait (albeit in different cells), whereas males have only one. An example is red–green color blindness (see Fig. 5-4). Normal color vision is determined by the presence of the dominant allele (X^C), whereas the recessive allele (X^c) causes colorblindness. Females who have at least one copy of the dominant allele (X^CX^C homozygous or X^CX^c heterozygous) have normal color vision. They are colorblind only if they inherit two copies of the recessive allele (X^cX^c). The chance of inheriting this recessive allele from both parents is relatively small, and thus colorblindness in females is rare. Although a female heterozygous for this gene has normal vision, she is a *carrier* and passes the recessive allele to half of her children. In contrast, a male has only one X chromosome and thus the allele on this chromosome determines whether he will be colorblind (X^cY) or not (X^CY). Therefore colorblindness, like other sex-linked traits, appears more often in males than in females.

Other sex-linked inherited traits have more serious health implications. For example, 12 protein clotting factors are necessary for normal blood clotting, and people with *classical hemophilia* ("bleeder's disease") lack factor VIII. This form of hemophilia is sex linked, appearing mostly in males. It was present in the royal families of Europe, especially England. Synthetic factor VIII is now available to treat sufferers of this disease. Another example of a sex-linked disorder is *Duchenne's muscular dystrophy*. This is a sex-linked disorder leading to the degeneration of skeletal muscle and death as a teenager. Another sex-linked recessive disorder is *agammaglobulinemia*, in which a person is unable to make antibodies. These males must live their entire lives protected from microbes in a plastic bubble or sterile suit. *Lesch–Nyhan syndrome* is a sex-linked disorder of the nervous system. Lack of an enzyme involved in the synthesis of nucleotides causes abnormal brain function, leading to mental retardation, involuntary writhing, and compulsive self-mutilation (biting of one's own lips and fingertips).

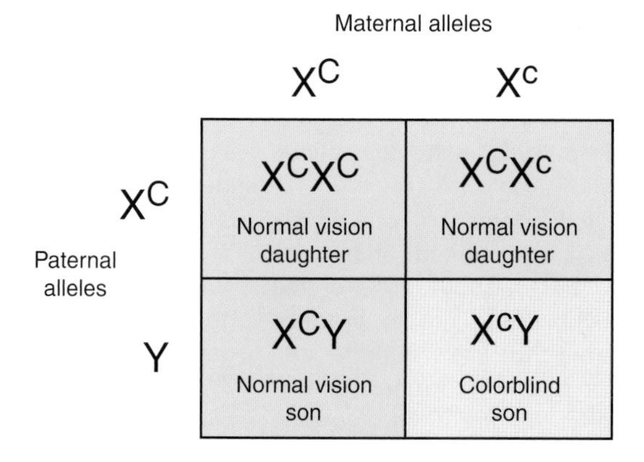

Figure 5-4 Inheritance of a sex-linked trait (red–green color blindness). A normal vision father (X^CY) and a normal vision, carrier mother (X^CX^c) will have normal vision daughters, half of whom will be carriers. Half of their sons will be color blind. X^c represents the color blind allele carried on the X chromosome; X^C represents the normal allele.

Sex-influenced traits occur mostly in one sex. In these conditions, a certain genotype must be present, but the action or expressivity of these genes is controlled by hormones. An example is baldness. Males require only one allele for baldness to be bald because they have high circulating levels of testosterone. Females, however, must be homozygous for bald alleles for baldness to occur because there is less circulating testosterone in their blood.

Development of the Reproductive System

The Sexually Indifferent Stage

The period during embryonic development when the gonads, sex accessory ducts, and external genitalia of both sexes are identical is called the *sexually indifferent stage*. The only way that one can determine the sex of an embryo at this stage is to look for Barr bodies or Y chromosomes in the cells. The gonads develop as paired bulges of tissue near the midline at the back of the abdominal cavity. These are the *genital ridges*, which are similar in both sexes and do not as yet contain germ cells. These ridges then differentiate into *indifferent gonads*, consisting of an outer cortex and an inner medulla (Fig. 5-5). The indifferent gonads appear during the 5th week of development (from fertilization on), when the embryo measures about 6 mm from the top of the head to its rump (the *crown–rump length*, or *CR*). The gonadal indifferent stage lasts until the end of the 7th week of development, when the embryo is 12 mm CR.

During the 6th week of development, the embryo also has indifferent sex accessory ducts, consisting of two pairs of ducts (Fig. 5-6). One pair, the *wolffian ducts* (*mesonephric ducts*), is derived from the ducts of the primitive embryonic kidney, the *mesonephros*. The other pair, the *müllerian ducts* (*paramesonephric ducts*), exists separately from the mesonephros. This indifferent duct stage lasts until the 7th week of development.

Before the 7th week of development, external genitalia are indifferent in both sexes (Fig. 5-7). These indifferent external structures consist of the *genital tubercle* (*phallus*), paired *urogenital folds*, and a *labioscrotal swelling*.

Gonadal Sex Differentiation

During the 4th or 5th week of development, germ cells begin to arrive at the indifferent gonads and to penetrate into these organs. *Primordial germ cells* arise in the region of the developing allantois, an extraembryonic membrane (see Chapter 10), and then migrate by amoeboid movement along the embryonic gut to the indifferent gonads. In humans, there are about 100 of the primordial germ cells that start the journey, but by the time they arrive at the gonads, they number about 1700 because they proliferate en route. Remember that, in genetic males, the primordial germ cells and the tissue of the indifferent gonads are 46:XY, whereas these same structures are 46:XX in genetic females.

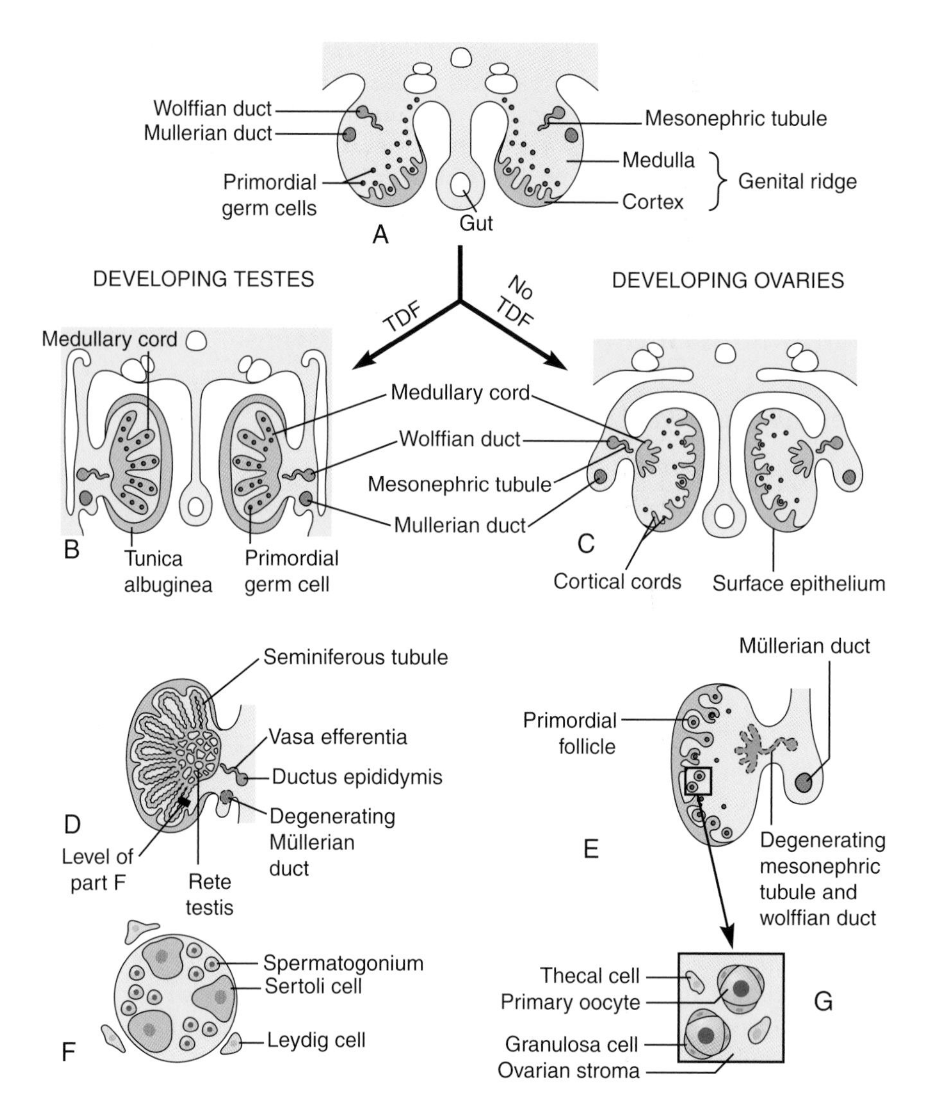

Figure 5-5 Diagram illustrating differentiation of indifferent gonads into testes or ovaries: (a) Indifferent gonads from a 6-week-old embryo; (b) at 7 weeks, showing testes developing under the influence of TDF; (c) at 12 weeks, showing ovaries developing in the absence of TDF; (d) testis at 20 weeks, showing the rete testis and seminiferous tubules derived from medullary cords; (e) ovary at 20 weeks, showing primordial follicles; (f) section of a seminiferous tubule from a 20-week fetus; and (g) section from the ovarian cortex of a 20-week fetus, showing two primordial follicles.

Ovarian Development

If a Y chromosome is not present, the cortex of the indifferent gonad forms ingrowths called *cortical cords*. The primordial germ cells concentrate in the cortical cords, and the medulla degenerates (Fig. 5-5). Differentiation of ovaries occurs a few weeks later than the differentiation of testes. The remnants of the medullary cords persist as *rete ovarii* in the adult ovary. The

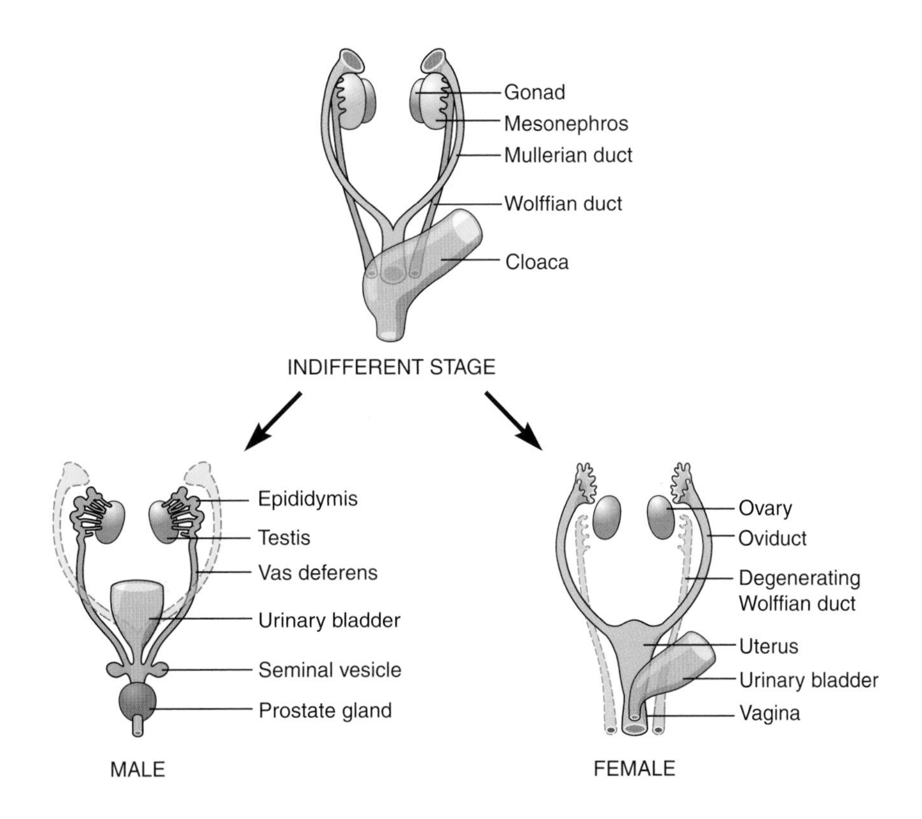

Figure 5-6 Differentiation of sex accessory structures in male and female human embryos.

enlarging cortex becomes an ovary, which contains *oogonia*, cells derived from the primordial germ cells. The oogonia then undergo mitotic proliferation so that there are about 600,000 by 8 weeks of development and 7,000,000 by the 5th month of development. Most of these oogonia degenerate, but some live, enter meiosis, and become oocytes, which are surrounded by a single layer of granulosa cells. Thus, primordial follicles are formed (Fig. 5-5). At birth, all of the female germ cells are oocytes in follicles. Some of these follicles grow to reach the tertiary stage, but then they become atretic. Each ovary contains about 500,000 follicles at birth, and this is all that a female will have for the rest of her life (although a recent finding of oogonia in adult mice raises the possibility that adult women may also retain oogonia potentially capable of giving rise to new germ cells). In contrast, the testes produce many new germ cells (spermatogonia) throughout a male's reproductive life. From the 7th to the 9th month of fetal life, the ovaries descend from an abdominal position to reside in the pelvic cavity.

Testicular Development

If the primordial germ cells and gonadal cells have an XY complement of sex chromosomes, testes develop from the indifferent gonads (Fig. 5-5). This event occurs at about 43 to 50 days of development, when the male embryo is 15 to

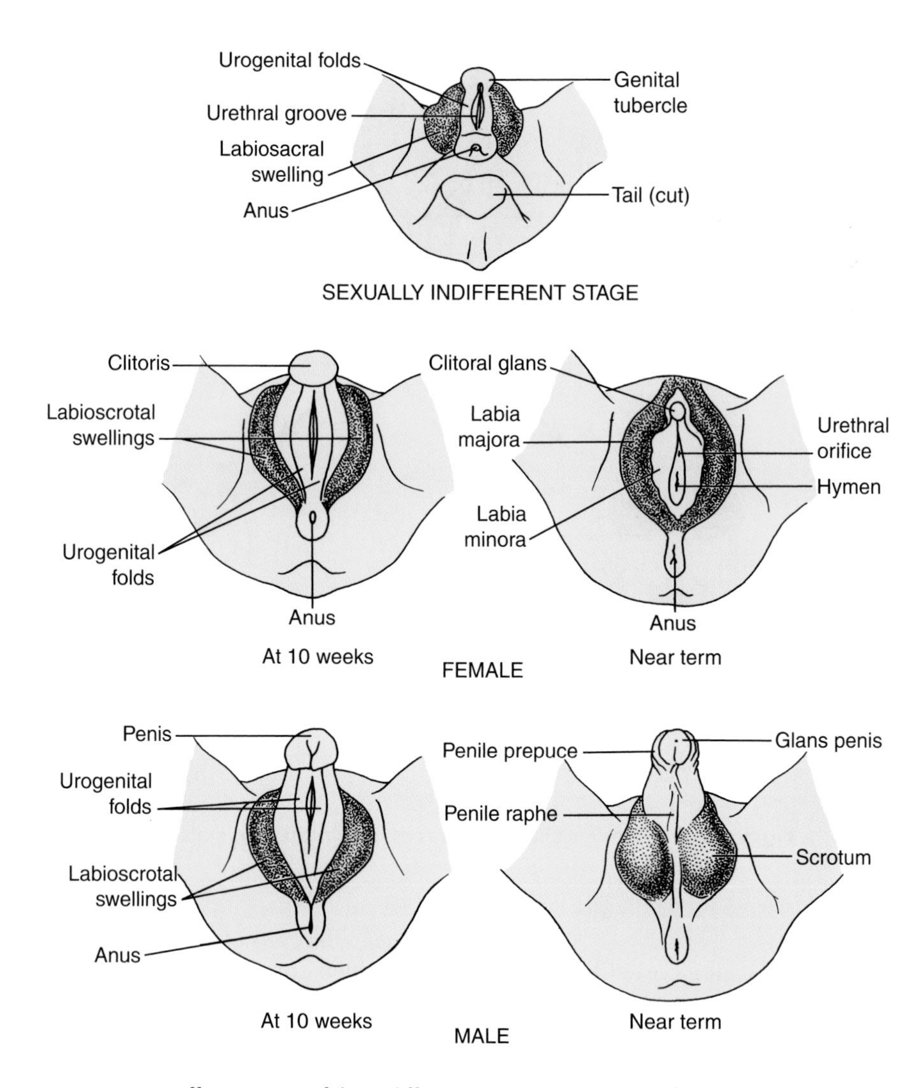

Figure 5-7 Differentiation of the indifferent stage into male and female external genitalia.

17 mm CR (measuring from crown to rump). If a Y chromosome is present, the primordial germ cells concentrate in the medulla of the gonad and the cortex regresses. *Medullary cords* develop within the medulla (Fig. 5-5). These cords contain cells derived from the primordial germ cells, the spermatogonia, and Sertoli cells are also present. Thus, the medullary cords become the seminiferous tubules of the testes (Fig. 5-5). Parts of these cords also form the rete testis. The testosterone-secreting Leydig cells begin to appear between the cords on about day 60 of development (30 mm CR), and these fetal Leydig cells may be stimulated to secrete testosterone by activity of the fetal anterior pituitary gland, which begins secreting at this time. Also, the placenta secretes an LH-like gonadotropin called *human chorionic gonadotropin* or hCG (see Chapter 10), which may pass into the male fetus and cause its Leydig cells to begin secreting testosterone. During the 3rd to the 9th month of fetal life, the testes descend

from an abdominal position into the scrotum. Their journey is stimulated by testosterone. If the testes fail to descend, cryptorchid testes are present (see Chapter 6).

Testis-Determining Factor

How does the presence of a Y chromosome control gonadal differentiation? The *SRY* gene is expressed in the genital ridges just before testicular differentiation in males, but not in females. It is thought to induce the genital ridge epithelium to differentiate into Sertoli cells, perhaps indirectly by causing nearby mesonephric kidney cells to migrate into the genital ridges. The TDF protein appears to act as a transcription factor and thus could potentially regulate the expression of several genes involved in testis formation. These genes, which contain instructions for actually forming the testis, are probably located on the autosomes or even the X chromosome, not on the Y chromosome. The *SRY* gene may serve as a "master switch" that turns on these testis-forming genes. One such gene is called *SOX-9*. It is necessary for testicular development and, if present in XX individuals, can cause testis differentiation even in the absence of *SRY*.

It is important to realize that genes are also actively involved in formation of the ovary and that ovarian differentiation is not simply a "default" state that occurs when the Y chromosome is absent. For example, *DAX1* is located on the small arm of the X chromosome. XY individuals who have a duplication of this portion of the X chromosome develop as females. It is hypothesized that *DAX1* antagonizes *SRY* and causes a syndrome known as *dosage-dependent sex reversal* in these individuals. Thus, *DAX1* may have a role in normal development of the ovary. However, we know very little about the actual function of any of these testis- or ovary-determining genes during embryonic development.

Differentiation of Sex Accessory Ducts and Glands

As mentioned previously, embryos of each sex possess both wolffian and müllerian duct systems until the 7th week of development (Fig. 5-6). If testes develop, they secrete testosterone, which causes the wolffian duct system to develop into the epididymis, vas deferens, seminal vesicles, and ejaculatory duct. The vasa efferentia form from the *mesonephric tubules*, which lead into the mesonephric duct. (The prostate gland and bulbourethral glands develop from part of the urethra, not from the wolffian duct system.) It appears that this effect of testosterone is local and is not accomplished through the general circulation. That is, the right testis secretes testosterone, which diffuses to adjacent wolffian structures and stimulates their development only on that side; the same thing happens on the other side. The embryonic testes also secrete a glycoprotein from the Sertoli cells that causes regression of the müllerian ducts. This chemical is called *müllerian-inhibiting substance* (MIS). A small, pouch-like remnant of the müllerian ducts in males is present within the prostate gland. This remnant is the *prostatic utricle*. Development of the male sex accessory ducts is depicted in Fig. 5-6.

If ovaries develop, no testosterone is secreted and the wolffian ducts regress. Also, because no MIS is secreted, the müllerian ducts develop into a pair of oviducts as well as into the uterus, cervix, and the upper one-third of the vagina (Fig. 5-6). The lower two-thirds of the vagina form from the *urogenital sinus*, which also gives rise to the urinary bladder and urethra. The greater and lesser vestibular glands of the female are derived from urethral tissue.

Given this information, you should be able to explain the results of the following experiments on rats. (1) If a male embryo is orchidectomized (its testes removed), it develops female sex accessory organs derived from the müllerian duct system. (2) If an orchidectomized male embryo is given testosterone, both wolffian and müllerian duct systems develop. (3) If an intact male embryo is given an antiandrogen that blocks the effects of testosterone, neither duct system develops.

Differentiation of External Genitalia

If testes are present in an embryo, the male external genitalia develop during the 8th week of development. That is, the genital tubercle differentiates into most of the penis, the urogenital folds form the ventral aspect of the penis shaft, and the labioscrotal swelling becomes the scrotum (Fig. 5-7). These structures develop because testosterone, secreted by the embryonic testes, is converted to *dihydrotestosterone* (DHT, another androgen) by an enzyme (*5α-reductase*) present in the cells of these tissues. Thus, it is DHT that actually stimulates development of the male external genitalia as well as the bulbourethral and prostate glands. This is in contrast to the wolffian duct derivatives, which contain no 5α-reductase and respond directly to testosterone.

If no testes are present to secrete testosterone (i.e., ovaries are present), the genital tubercle differentiates into the clitoris, the urogenital folds into the labia minora, and the labioscrotal swells into the mons pubis and labia majora (Fig. 5-7). Thus, androgens are required to masculinize the external genitalia; in the absence of androgens, female structures develop. The extent to which the low levels of estrogens released from the developing ovaries play a role in differentiation of the female external genitalia is not known.

Summary of Sexual Determination and Development

Male and female reproductive structures develop from embryonic tissues that are initially sexually "indifferent" (Fig. 5-8). Thus, the reproductive tract of males and females develops from common embryonic tissues; these homologies are listed in Table 5-1. Presence or absence of the Y chromosome determines whether these structures will develop in either a male or a female direction. In XY individuals, expression of the *SRY* gene on the Y chromosome causes the indifferent gonad to develop into a testis. The developing testis then releases hormones that masculinize the internal ducts, glands, and external genitalia. In the absence of *SRY*, the indifferent gonad in XX individuals develops into an ovary. In this case, the reproductive ducts and external

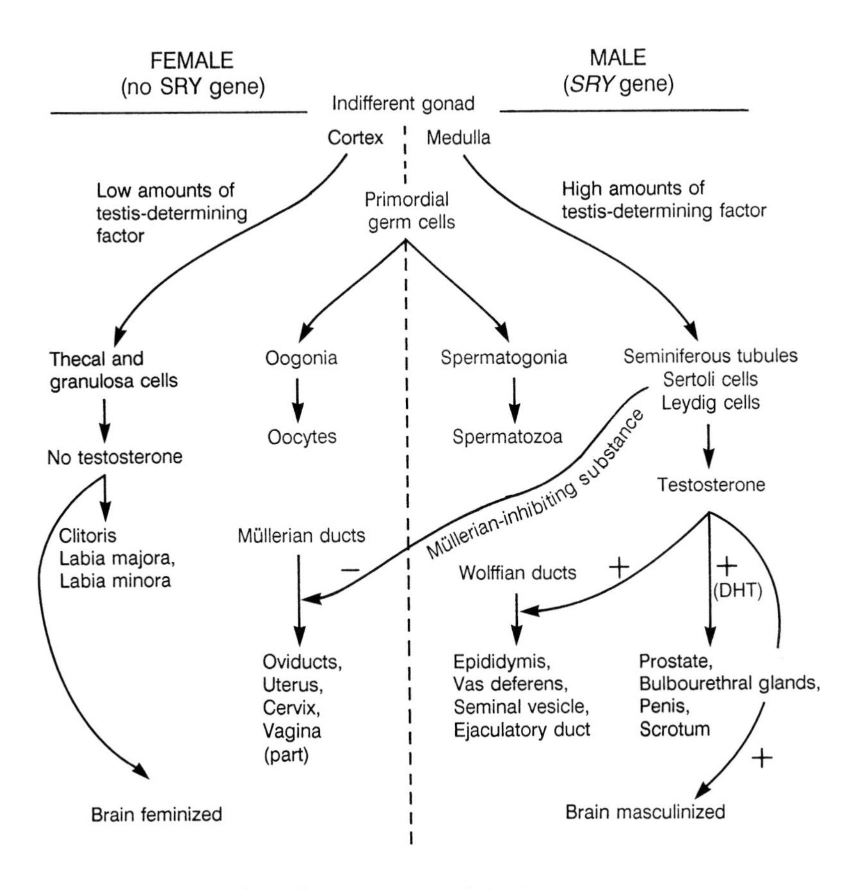

Figure 5-8 Summary of sex determination and development.

Table 5-1 Adult Derivatives of Embryonic Structures Showing Homologies among Portions of the Female and Male Reproductive Systems[a]

Female structure		Indifferent embryonic structure		Male structure
Ovary (from cortex)	←	Indifferent gonad	→	Testis (from medulla)
No functional structures	←	Mesonephric tubules	→	Vasa efferentia
No functional structures	←	Mesonephric ducts (wolffian ducts)	→	Epididymis; vas deferens; ejaculatory duct; seminal vesicles
Oviducts	←	Müllerian ducts	→	No functional structures
Uterus				
Cervix				
Vagina (upper third)[b]				
Clitoris (glans; corpora cavernosa)	←	Genital tubercle (phallus)	→	Penis (glans; corpora cavernosa)
Labia minora	←	Urogenital folds	→	Corpus spongiosum of penis
Labia majora; mons pubis	←	Labioscrotal swellings	→	Scrotum
Lesser and greater vestibular glands	←	Urethral tissue	→	Prostate; bulbourethral glands

[a]Structures derived from the same indifferent embryonic tissue are homologous.

[b]The lower two-thirds of the vagina are derived from the urogenital sinus, which also gives rise to the urinary bladder and urethra.

genitalia develop along a female pathway because they are not exposed to testicular secretions.

Disorders of Sexual Determination and Development

True Hermaphroditism

Sometimes people are born with ambiguous reproductive systems, a phenomenon generally termed *intersexuality*. If a combination of gonadal tissue is present, a person is a *true hermaphrodite*. These individuals often possess an *ovotestis* on one or both sides, which is a gonad that contains a combination of seminiferous tubules and ovarian follicles. This sometimes happens because of an error in fertilization that results in one-half of the individual's cells being 46:XX and the other half being 46:XY. Sometimes true hermaphrodites will have an ovary on one side and a testis on the other side; this person is called a *gynandromorph*. On the side with the ovary, there are müllerian duct derivatives, whereas on the side with the testis, there are wolffian duct derivatives. This is evidence that the effects of the testes on development of the wolffian ducts and regression of the müllerian ducts are local phenomena, with each testis controlling the development of the sex accessory structures on its side only.

Pseudohermaphroditism

A *pseudohermaphrodite* is a person whose gonads agree with the chromosomal sex but who has external genitalia of the opposite sex. Male pseudohermaphrodites have normal testes but incomplete masculinization of the wolffian duct system and external genitalia. One form of this condition is the inherited disorder *testicular feminization syndrome*, also called *androgen insensitivity syndrome* (Fig. 5-9). These individuals have normal testes and male chromosomes (46:XY), and their testes secrete normal amounts of testosterone. However, they have a genetic absence of the receptors for androgens in target tissues. Thus, they develop female-like external genitalia and the testes are not descended. No müllerian duct derivatives are present because their testes secreted MIS during development. Another form of male pseudohermaphroditism has been described in the populations of some small villages in the Dominican Republic (see HIGHLIGHT box 5-2).

Female pseudohermaphroditism occurs when normal ovaries are present but the body is partially masculinized. Examples are individuals with *adrenogenital syndrome*, also known as *congenital adrenal hyperplasia* (Fig. 5-10). This is an inherited disorder that can develop after, but usually occurs before, birth. It accounts for about one-half of all cases of human intersexuality. Genetic females (46:XX) with this disorder have masculinized external genitalia, and pubic and axillary hair develop only a few years after birth. Males can also inherit this disorder. Soon after birth, these males can begin to acquire adult male-like secondary sexual characteristics such as penile enlargement and facial hair.

Adrenogenital syndrome involves an inability of the adrenal glands to secrete steroid hormones such as cortisol. Because cortisol exerts a negative

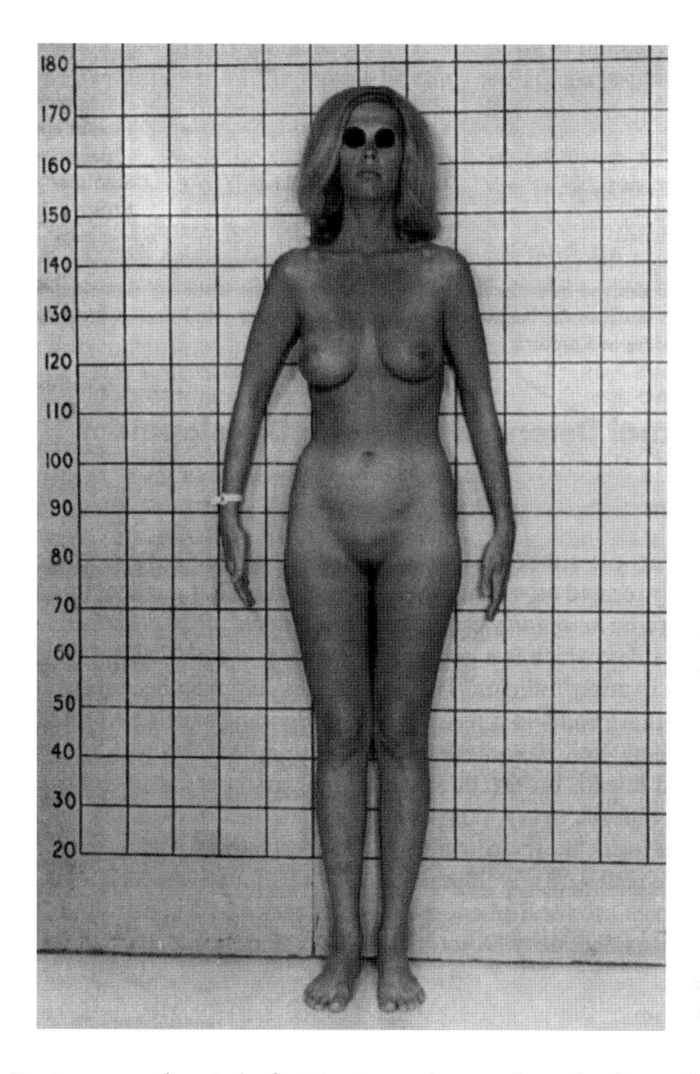

Figure 5-9 A person with testicular feminization syndrome, a form of male pseudohermaphroditism. (Photograph courtesy of Dr. Howard W. Jones.)

Chapter 5, Box 2: Guevedoces

Suppose you were a 12-year-old girl living in an isolated village in the Dominican Republic. You had been worried for awhile that your breasts had not begun to grow and you had not shown any signs of menstruating like some of your friends. Still more frightening were the two lumps appearing under your partially fused labia majora, along with the noticeable increase in the size of your clitoris. Another revelation was that you began to be interested sexually in girls. You are then told that you belong to a group of similar people in your village that have been called *Guevedoces* (penis at 12). The Guevedoces syndrome is relatively common in several isolated, remote villages in the Dominican Republic and a few other places in third world countries. These children, most often raised as girls, are actually male!

The scientific name of the Guevedoces syndrome is *5α-reductase deficiency*. You have

Continued on next page.

Chapter 5, Box 2 continued.

learned in this chapter that 5α-reductase converts testosterone to dihydrotestosterone (DHT) in cells of male external genitalia as well as of the prostate and bulbourethral glands. In contrast, the male sex accessory ducts, as well as seminal vesicles, respond to testosterone directly (see Fig. 5-9). Guevedoces males are genetic males (46:XY) who are born lacking one of the forms of 5α-reductase enzyme. Their testes secrete normal levels of testosterone, but DHT is absent. Because DHT is required for descent of the testes, the testes remain in the abdomen or inguinal canal. As newborns, Guevedoces have normal sex accessory ducts and seminal vesicles and no müllerian duct tissues, but their external genitalia appear female and they are thought to be females. Their genital tubercle is small and clitoris-like. Their labia majora are partially fused and the urogenital sinus ends in a blind vaginal pouch. Their urethra opens to the outside through an opening near the vagina.

After being raised as girls, Guevedoces become masculinized at the time of puberty. Their clitoris grows to penis size and they begin to have erections. Their testes descend into their labia majora, which now fuse to become a scrotum. Spermatogenesis ensues and they can ejaculate because their prostate gland enlarges and sperm can be produced. (However, they are unable to fertilize a woman internally because they ejaculate through their urethral opening, which is not in their penis but is still an opening near their anus.) Also, their muscle size increases, their voice deepens, and they grow male-typical facial and body hair. Chapter 8 discusses the interesting fact that a great majority of Guevedoces exhibit heterosexual male behavior and sexual orientation, even though they were raised as girls.

Why do DHT-dependent tissues of Guevedoces suddenly differentiate and grow at puberty, even though the enzyme to make DHT is genetically absent, even at puberty? The probable answer is that androgen receptors in tissue (which bind both testosterone and DHT) are much more receptive to DHT, i.e., DHT is a more potent androgen than is testosterone. Structures of the fetus nearest the testes (i.e., the epididymides, vasa deferentia, and seminal vesicles) receive high concentrations of testosterone hormone delivered in testicular fluid. Other structures having access only to the lower concentrations of testosterone in the blood (e.g., the genitals and secondary sexual characteristics) amplify their androgen signal by converting testosterone to DHT within the target cells. However, at puberty the blood concentrations of testosterone become so high (see Chapter 6) that these peripheral tissues can now respond to testosterone directly, even without DHT, and the tissues then differentiate and grow.

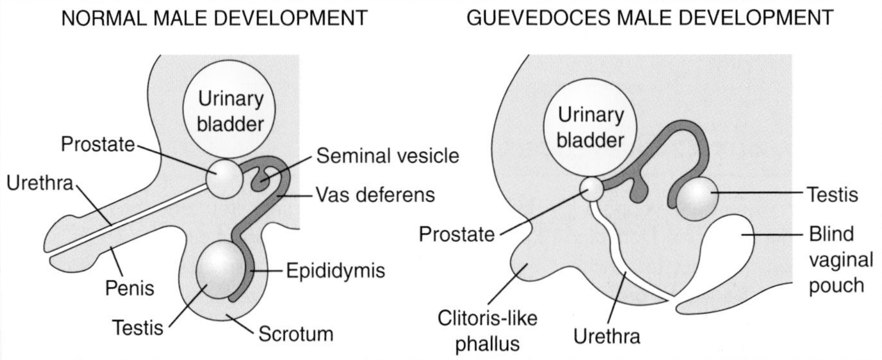

Sexual differentiation and development in a normal male at birth and a Guevedoces male that has 5α-reductase deficiency. Structures dependent on testosterone directly are in black. Note that these are normal in the Guevedoces male, but other structures that require testosterone conversion to DHT (by 5α-reductase), such as the penis, scrotum, and prostate, are feminized and that Guevedoces males are also born with undescended testes and a blind vaginal pouch.

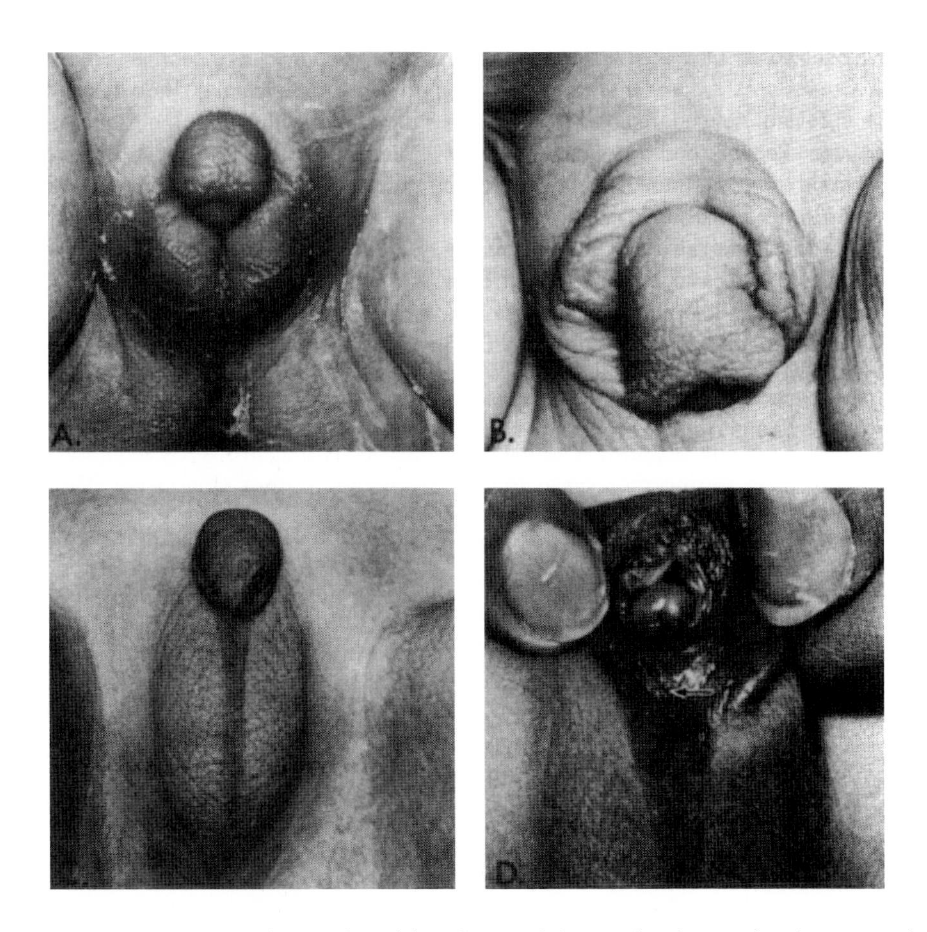

Figure 5-10 External genitalia of female pseudohermaphrodites with adrenogenital syndrome: (a) A newborn female exhibiting enlargement of the clitoris and fusion of the labia majora; (b) a female infant showing considerable enlargement of the clitoris; and (c and d) a 6-year-old girl showing an enlarged clitoris and fusion of the labia majora to form a scrotum. (d) Note the clitoral glans and the opening of the urogenital sinus (arrow).

feedback effect on the secretion of corticotropin (ACTH) from the anterior pituitary gland (see Chapter 1), blood levels of ACTH are very high. However, because the adrenal glands are unable to synthesize and secrete cortisol, they instead secrete large amounts of androgens into the blood, which serve to masculinize the female body and cause precocious sexual development in the male.

If a woman is administered a progestogen during pregnancy, which used to be a fairly common practice to prevent miscarriage, the female fetus can exhibit female pseudohermaphroditism with symptoms similar to those of the adrenogenital syndrome. This is because some synthetic progestogens have androgenic activity. Therefore, combination birth control pills, which contain a synthetic progestogen (see Chapter 14), should be avoided during pregnancy.

Chromosomal Errors and Sex Determination

Other genetic disorders of sex determination can result from errors of fertilization, i.e., the cells of an individual may end up with an abnormal number of sex chromosomes because *nondisjunction* of sex chromosomes can occur during meiosis; i.e., the two sex chromosomes fail to separate during meiosis in the testis or ovary (Fig. 5-11). An example is *Turner's syndrome*, in which the cells are 45:XO (i.e., only one sex chromosome is present) and no Barr body is present. Individuals with Turner's syndrome are sexually infantile with female external genitalia. They have facial abnormalities, a shield-like chest, and a neck that is short, broad, and webbed. They also tend to have cardiovascular and kidney disorders. Their "ovaries" are sterile, consisting mostly of connective tissue; no germ cells are present. A female can have cells that are XXX because of nondisjunction during meiosis in the ovary of her mother (Fig. 5-11). These women are female but sterile.

Nondisjunction of sex chromosomes can produce males with cells of an XXY makeup (47:XXY). In this condition, *Klinefelter's syndrome* (Fig. 5-11), males are sterile and have female-like breasts. About 1 out of 600 males is born with this disorder. These individuals have small external genitalia, cryptorchid testes, and are mentally retarded. Their urethra may also fail to close during development, resulting in an opening on the lower surface of the penis, a condition called hypospadias.

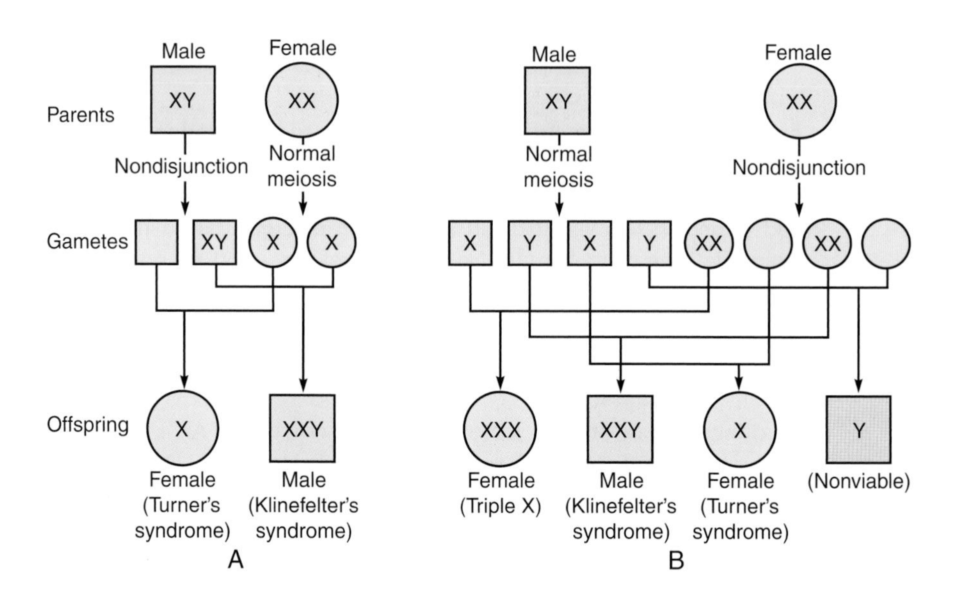

Figure 5-11 Nondisjunction is the failure of a homologous pair of chromosomes to separate during meiosis. (a) Nondisjunction of sex chromosomes during sperm formation. The fertilized ova will have two X chromosomes and one Y chromosome (Klinefelter's syndrome) or only an X chromosome (Turner's syndrome). (b) Nondisjunction of sex chromosomes during meiosis in the ovary results in ova with either two X chromosomes or none. The fertilized ova therefore can exhibit one of four chromosomal aberrations, three of which is viable.

"Supermales" (47:XYY) are caused by fertilization of an X egg by 2 Y sperm (polyspermy; see Chapter 9). These males are very tall and often have acne. They tend to exhibit mental and social adjustment problems at a higher percentage than normal XY males. One in 2000 males has XYY cells. Some statistical evidence shows that the percentage of XYY males (1.8 to 12.0%) in penal institutions is greater than their percentage (0.14 to 0.38%) in the general population. Some controversy, however, surrounds these studies and it is not clear if the greater maladaptive behavior of XYY males is a direct result of their chromosomal abnormality or is due to social problems they had when growing up because of their unusual physical appearance. Apparently the elevated crime rate of XYY men is not related to aggression but may be related to low intelligence.

An inherited disorder of the X chromosome is the second leading cause of mental retardation. This *fragile-X syndrome* occurs in 1 in 2500 births worldwide. In these people, the X chromosome (in either sex) has an abnormally long, fragile arm. In addition to mental retardation, symptoms can include large ears, a long face, connective tissue abnormalities, and behavioral and movement disorders. These symptoms are more severe in males than in females.

Other Problems in Sex Development

Developmental abnormalities of the reproductive tract, other than those mentioned earlier, are rare. One of these, *penile agenesis*, is the absence of a penis due to the fact that a genital tubercle was not formed. In another condition, called *double penis*, there are two penises because of the earlier formation of two genital tubercles or the genital tubercle branches, resulting in a forked, or *bifid penis*. If the pituitary gland of the male fetus is underdeveloped, this can result in a *micropenis*, which often is so small that it is barely noticeable.

Rare abnormalities in the development of the female reproductive tract also can occur. Because the müllerian ducts initially are paired structures, their middle and lower portions must fuse to form the unpaired uterus and vagina. Partial or complete failure of müllerian duct fusion occurs in about 1 in every 1500 newborn females. At one extreme is the *arcuate uterus*, in which almost normal fusion occurs, but a slight nonfusion of the uterus produces a "dent" in the fundus of the uterus. At the other extreme is complete nonfusion of the müllerian ducts, resulting in a *double uterus* as well as two vaginas. Several degrees of nonfusion between these two extremes can occur (Fig. 5-12). For example, the uterus can be forked at the top (*bicornuate uterus*). Many of these conditions do not cause infertility, but are related to an increased risk of miscarriage (see Chapter 10).

Reproductive System in the Newborn

In the way of a summary, let us look at the condition of the reproductive structures in a normal newborn male and female. Despite the complexities of sex determination and development from an indifferent stage, the system works perfectly in most of us.

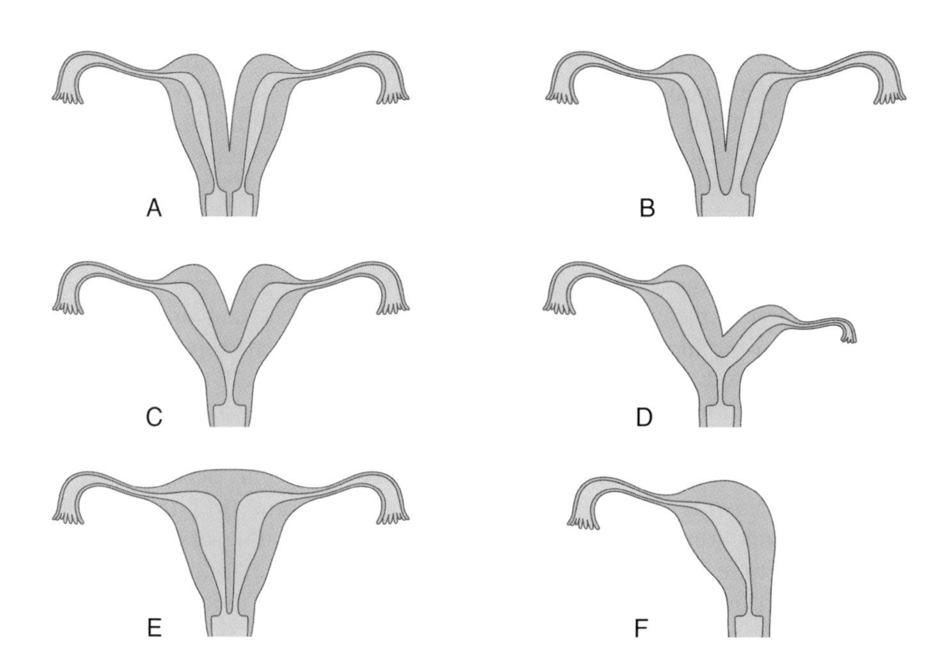

Figure 5-12 Illustrations of various types of congenital uterine abnormalities: (a) double uterus and double vagina, (b) double uterus with single vagina, (c) branched (bicornuate) uterus, (d) bicornuate uterus with a small left branch, (e) septate uterus (divided in the middle), and (f) uterus formed from müllerian duct on one side only.

The newborn male child has a pair of testes that usually have descended into the scrotum. The seminiferous tubules of the testes contain spermatogonia and Sertoli cells but no sperm. The Leydig cells of the testes, after being active in secreting testosterone in the fetal period, are now inactive. The scrotum and the penis are fully formed but small. The male sex accessory ducts and glands are fully formed but inactive. The male secondary sexual characteristics will not appear until puberty, as we shall see in the next chapter.

In newborn females, the ovaries contain a fixed number of oocytes in resting or growing follicles. None of the follicles will ovulate until puberty. The ovaries are now in the pelvic cavity. The female sex accessory ducts are fully formed but inactive. Similarly, the female external genitalia are fully formed but relatively small. The female secondary sexual characteristics will not mature until puberty.

Chapter Summary

The chromosomes of diploid female cells are 46:XX, whereas those of diploid male cells are 46:XY. One X chromosome of the female cells is inactivated, producing a Barr body. Development of the reproductive tract begins with the sexually indifferent stage. The primordial germ cells migrate to the indifferent gonads. If a Y chromosome is present, expression of the *SRY* gene results in synthesis of testis-determining factor, which causes the medulla of the indifferent gonads to form testes. If TDF is lacking, the cortex of the indifferent gonads forms ovaries.

The Y chromosome also has genes that will play a role in spermatogenesis as well as growth and development of some male secondary sexual characteristics.

In male embryos, testosterone secreted by the testes stimulates development of the wolffian ducts into the male sex accessory ducts (epididymis, vas deferens, ejaculatory duct) and the seminal vesicles. Secretion of müllerian-inhibiting substance causes regression of the müllerian ducts. In female embryos, the absence of testosterone and MIS allows the wolffian ducts to regress and the müllerian ducts to develop into the oviducts, uterus, and part of the vagina.

The indifferent external genitalia of the embryo develop into male external genitalia (penis, scrotum) if testosterone is present. This androgen is converted to another androgen, DHT, by tissues of the external genitalia. If testosterone is not present, the female external genitalia develop into a clitoris, labia majora, labia minora, and mons pubis.

Intersexual individuals are born with ambiguous reproductive tracts. True hermaphrodites have ambiguous gonads and external genitalia, whereas pseudohermaphrodites have normal gonads and ambiguous external genitalia. Examples of male pseudohermaphrodites are people with testicular feminization syndrome and the Guevedoces syndrome. Examples of female pseudohermaphrodites are those with adrenogenital syndrome and progestogen-induced masculinization. Other chromosomal abnormalities can also produce developmental abnormalities of the reproductive tract. These include Turner's syndrome (45:XO) and Klinefelter's syndrome (47:XXY). Developmental errors can result in penile agenesis, double penis, bifid penis, micropenis, arcuate uterus, bicornuate uterus, or double uterus and vagina. Finally, sex-linked and sex-influenced genetic traits are related to genes on the X chromosome and are more common in males.

Further Reading

Ezzell, C. (1994). Is there a molecular battle between the sexes? *J. NIH Res.* **6**(1), 21–23.

Imperato-McGenley, J., and Gautier, T. (1986). Inherited 5α-reductase deficiency in man. *Trends Genet.* **2**, 130–133.

Jegalian, K., and Lahn, B. T. (2002). Why is the Y so weird. *Sci. Am.* **284**(2), 56–61.

Marx, J. (1995). Snaring the genes that divide the sexes for mammals. *Science* **269**, 1824–1825.

Advanced Reading

Berta, P., *et al.* (1990). Genetic evidence equating *SRY* and the testis-determining factor. *Nature* **348**, 448–450.

Byskov, A. G. (1986). Differentiation of the mammalian embryonic gonad. *Physiol. Rev.* **66**, 71–117.

Imperato-McGinley, J., *et al.* (1974). Steroid 5α-reductase deficiency in man: An inherited form of male pseudohermaphroditism. *Science* **186**, 1213–1215.

Johnston, J. (1991). Fragile X: Two gene flaws point to mechanism of defect. *J. NIH Res.* **3** (12), 41–42.

Josso, N., and Picard, J. Y. (1986). Anti-müllerian hormone. *Physiol. Rev.* **66**, 1038–1091.

Lyon, M. F. (1993). Gene action in the X-chromosome of the mouse (*Mus musculus* L.). *J. NIH Res.* **5**(2), 72–75.

McElreavey, K., *et al.* (1993). A regulatory cascade hypothesis for mammalian sex determination: SRY represses a negative regulator of male development. *Proc.Nat. Acad. Sci. USA* **90**, 3368.

Nowak, R. (1993). Curious X-inactivation facts about calico cats. *J. NIH Res.* **5**(1), 60–65.

Rabinovici, J., and Jaffe, R. B. (1990). Development and regulation of growth and differentiation of function in human and subhuman primate fetal gonads. *Endocr. Rev.* **11**, 532–557.

Roldan, E. R. S., and Gomendio, M. (1999). The Y chromosome as a battle ground for sexual selection. *Trends Ecol. Evol.* **14**, 58–62.

Sharpe, R. M. (2001). Hormones and testis development and the possible adverse effects of environmental chemicals. *Toxicol. Lett.* **120**, 221–232.

Sinclair, A. H., *et al.* (1990). A gene from the human sex-determining region encodes a protein with homology to a conserved DNA-binding motif. *Nature* **346**, 240–244.

Williams, N. (1995). How males and females achieve X equality. *Science* **269**, 1826–1827.

CHAPTER SIX

Puberty

Introduction

The human life cycle can be divided into five stages: (1) embryonic and fetal existence, (2) infancy and childhood, (3) puberty and adolescence, (4) early and middle adulthood, and (5) late adulthood and old age. This chapter discusses the process of puberty, the internal and environmental factors that influence this process, and the psychological and social adjustments teenagers must make during puberty and adolescence.

Puberty and Its Timing

Puberty (or *sexual maturation*) is the transition period that takes a person from being a sexually immature child to a sexually mature, reproductively fertile adult. The word *puberty* has its root in the Latin word *pubes*, which means "hair." *Pubescence* is the state of the child between the onset of pubertal changes and the completion of sexual maturation. Much is known about timing of the biological events during puberty, yet no "typical" pattern exists. The ages at which the various changes occur vary greatly among individuals. Figure 6-1 illustrates this individual variability in the pubertal sequence. Puberty involves growth and maturation of several body systems and tissues, and the timing of a change in one part of the body (e.g., maturation of the breasts in a female) in relation to a change in another part of the body (e.g., appearance of pubic hair in a female) can vary among individuals.

The Pubertal Process

Table 6-1 summarizes the sequences of pubertal change in a "typical" male and female. The ages given for the various changes are averages, although a wide range of pubertal timing is considered normal. Usually, the earlier puberty is initiated, the earlier sexual maturity is reached. Also, the pubertal sequence begins earlier and ends earlier in females than in males. For example, the body growth spurt begins about 2 years earlier in females, and females become fertile about a year earlier than males (Table 6-1).

Pubertal Changes in Females

The first menstruation (*menarche*) is the most definitive sign of puberty in females. It occurs at an average age of 13 years in the Northern Hemisphere,

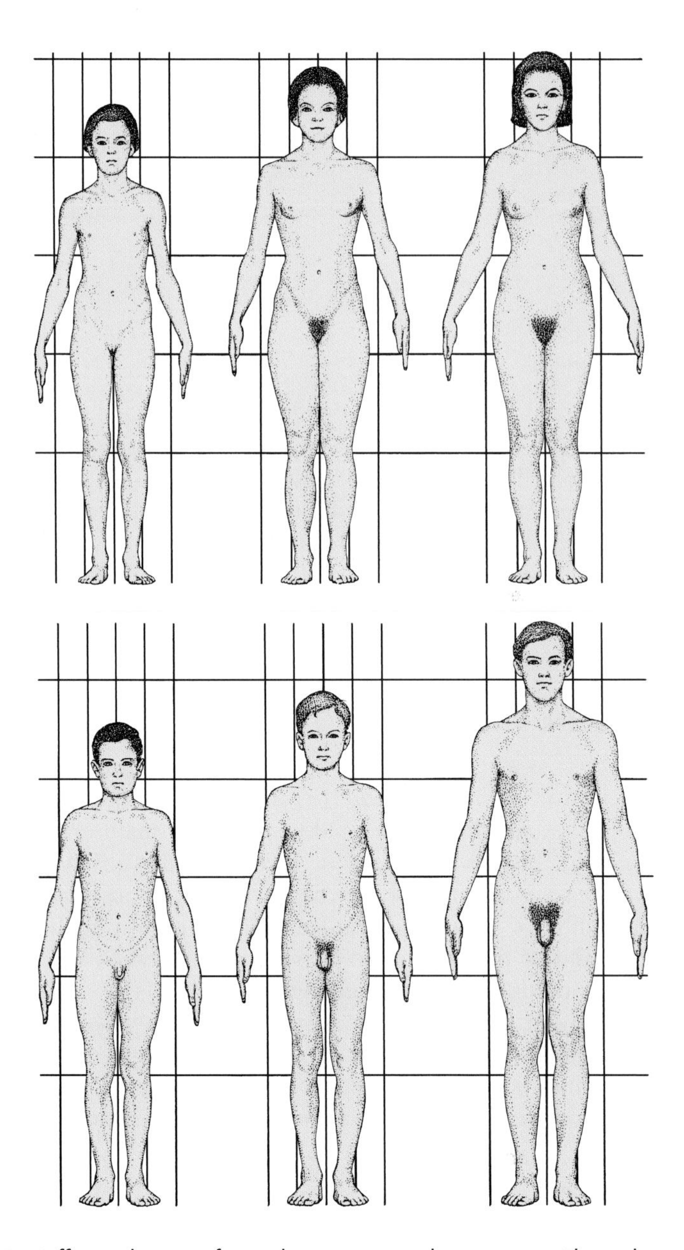

Figure 6-1 Different degrees of sexual maturation at the same age. The males are all 14.75 years old, and the females are all 12.75 years old. This illustrates that the degree of sexual maturity can vary widely in relation to age.

with 95% of females reaching menarche between ages 11 and 15. In the United States, menarche occurs at an average age of 12.3 years. A female's first ovulation can be at the time of menarche, but usually is some months (or as much as 2 years) after menarche. The first menstrual cycles in pubertal females, thus, are often *anovulatory* (infertile). Even though this is the case in most females, there is a slight risk of pregnancy in some who have just experienced menarche, and there are cases of females who have become pregnant before menarche. Also, for several months after menarche, a female may miss some periods.

Table 6-1 "Typical" Sequence of Events during Puberty in a Female and Male[a]

Age	Female	Male	Age
9	Growth spurt begins; initial increase in height and fat deposition	Initial stages of spermatogenesis; Leydig cells appear and begin to secrete androgens	9
10	Initial breast development; pelvis widens; pubic hair begins to appear	Subcutaneous fat deposition increases; testes begin to enlarge	10
11	Maturation and growth of internal reproductive organs (ovaries, oviducts, uterus, vagina); maturation of external genitalia; areola becomes pigmented	Increase in scrotum and penis size; increase in spontaneous erection frequency; first signs of pubic hair; growth of seminal vesicles and prostate gland; skeletal growth spurt begins	11
12	Filling in of breasts; menarche; axillary hair appears	Pubic hair more apparent; nocturnal emissions begin	12
13	First ovulation occurs; skeletal growth rate declines; breast maturation complete; sweat and sebaceous gland development, sometimes with acne	Larynx growth and deepening of voice occur; hair appears in axilla and on upper lip	13
14	Voice deepens slightly	First fertile ejaculation; slight breast enlargement in some individuals	14
15	Adult stature reached	Adult hair pattern, including indentation in front hairline and appearance of chest hair; sweat and sebaceous glands develop, often with acne; loss of body fat occurs	15
		Broadening of shoulders; muscle growth and increased muscle strength	16
		Adult stature reached now or later	17

[a]This table depicts a general sequence, and individuals can vary greatly in the timing of this sequence and still be within the normal range.

Menarche and first ovulation are only steps in a series of changes in pubescent females. The ovaries and female sex accessory structures (oviducts, uterus, vagina) grow and mature (Fig. 6-2). Female secondary sexual characteristics appear, such as growth and widening of the bony pelvis and appearance of pubic and axillary (armpit) hair. Also, the breasts begin to grow and mature. *Sweat* and *sebaceous glands* in the skin become more active, and some females (and males) develop *acne* due to increased activity and inflammation of the sebaceous glands. The *linea nigra* appears as a dark band on the lower abdominal wall; it is especially pronounced in brunettes and dark-skinned females. There is also a slight lowering of the voice in pubescent females but not as much as in males. *Metabolic rate* (the rate at which tissues use oxygen), heart rate, and blood pressure increase in females at puberty.

Many pubescent females deposit fat under the skin. This fat becomes distributed in the hips and breasts as secondary sexual characteristics. Also, a skeletal growth spurt occurs in pubescent females, and both height and weight increase. This bone growth usually ends at about age 15 in females, although some females will continue to grow in height until age 20 or later. High levels of estrogens secreted by the pubescent female eventually halt bone growth.

Pubertal Changes in Males

In males, pubertal changes include growth and maturation of the testes and male sex accessory structures (vas deferens, seminal vesicles, prostate gland; Fig. 6-3). In addition, male secondary sexual characteristics begin to appear. Pubic hair develops, and hair also appears in the axilla, face, chest, and extremities.

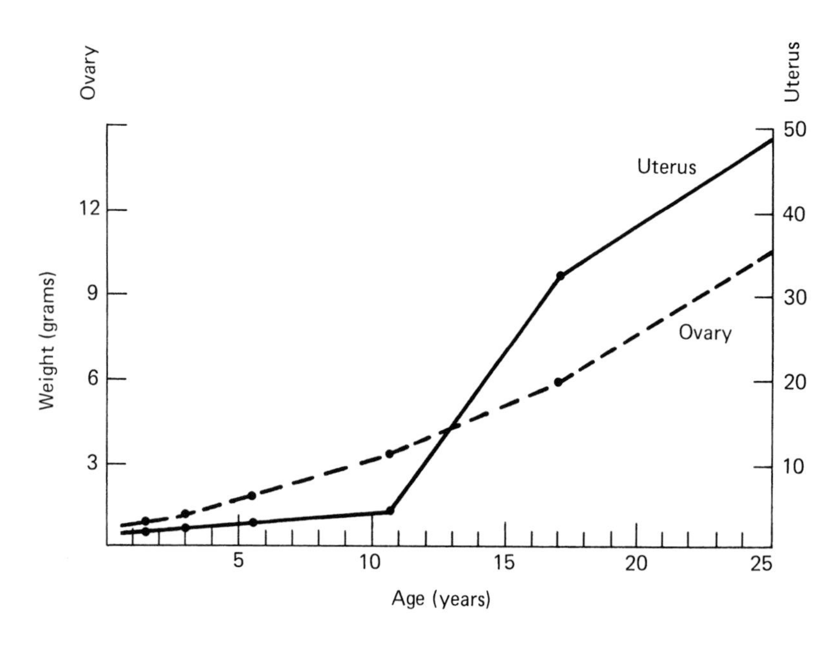

Figure 6-2 Changes in the weight of an ovary and the uterus with age.

Sweat glands develop in the axilla, and they may produce an odor because of bacteria present in the oily secretion. The sebaceous glands become active in the skin of the scrotum, face, back, and chest; acne appears in some individuals. In pubescent males, the nipple becomes pigmented and the areola darkens and widens. The vocal cords in the *larynx* ("voice box") double in length in pubertal males, resulting in a dramatic lowering of the voice. The pitch of the adult male voice is about a full octave below that of adult females.

The scrotum and penis of pubescent males grow markedly. *Spontaneous erections*, which occur in infants and even in male fetuses, increase in frequency during puberty (see Chapter 8). These erections, which can be disturbing to the individual or his parents, occur in response to stressful or emotionally charged stimuli seemingly not related to sexual matters. *Nocturnal emissions* ("wet dreams") begin in pubescent males. These emissions tend to occur during sleep or right after waking up. Initial ejaculations of pubescent males (during nocturnal emissions, masturbation, or coitus) are relatively sperm free, but the possibility always exists that at least some fertile sperm are present. Both spontaneous erections and nocturnal emissions tend to decrease in frequency after puberty.

Pubescent males exhibit a spurt in body growth. Fat is deposited early in the pubertal sequence, without much skeletal or muscle growth. In fact, females aged 10 to 13 usually are taller than males of a similar age. In later stages of the pubertal sequence, however, marked skeletal and muscle growth occurs in males because of the stimulatory effects of testosterone on bone growth. Thus, the average height of adult American males (177.6 cm, or 69.9 in.) is greater than that of females (169.2 cm, or 66.6 in.). At the end of puberty, an average male

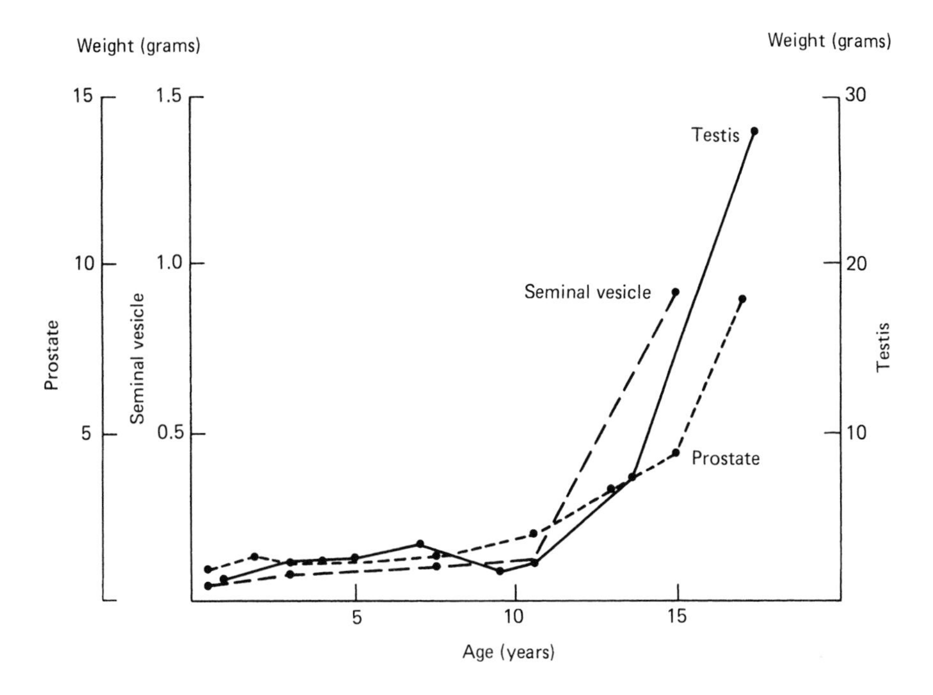

Figure 6-3 Changes in the weight of a single testis, seminal vesicle, and prostate gland with age.

and female of the same height and weight differ in the amount of body fat and water (Table 6-2).

Classification of Pubertal Changes

It often is useful for physicians and others to be able to look at a teenager and determine where he or she is in the pubertal sequence. For this purpose, some anatomical characters that change in a predictable order have been classified into five *Tanner stages*. These characters include the development of pubic hair, penis, and testes in males and pubic hair and breasts in females. Table 6-3 describes these stages, which are shown in Figs. 6-4, 6-5, and 6-6.

Table 6-2 Total Body Water as Percentage of Body Weight, an Index of Fatness: Comparison of an 18-Year-Old Girl and a 15-Year-Old Boy of the Same Height and Weight

Variable	Girl	Boy
Height (cm)	165.0	165.0
Weight (kg)	57.0	57.0
Total body water (liter)	29.5	36.0
Lean body weight (kg)	41.0	50.0
Fat (kg)	16.0	7.0
Fat/body weight (%)	28.0	12.0
Total body water/body weight (%)	51.8	63.0

Table 6-3 Pubertal Development of Certain Anatomical Characters in Females and Males by Tanner Stage

Females		
Stage	**Pubic hair**	**Breasts**
1	Prepubertal	Prepubertal
2	Sparse, lightly pigmented, straight; medial border of labia	Breast and papilla (nipple) elevated as small mound; areolar diameter increased
3	Darker; beginning to curl; increased amount	Breast and areola enlarged; no contour separation
4	Coarse, curly, abundant; but amount less than in adult	Areola and papilla form secondary mound
5	Adult feminine triangle spread to medial surface of thighs	Mature; nipple projects; areola part of general breast contour

Males			
Stage	**Pubic hair**	**Penis**	**Testes**
1	Prepubertal (none)	Prepubertal	Prepubertal
2	Scanty, long, slightly pigmented	Slight enlargement	Enlarged scrotum; texture altered
3	Darker; starts to curl; small amount	Penis longer	Larger
4	Resembles adult type, but less in quantity; coarse, curly	Larger; glans breadth increased	Larger
5	Adult distribution; spread to medial surface of thighs	Adult	Adult

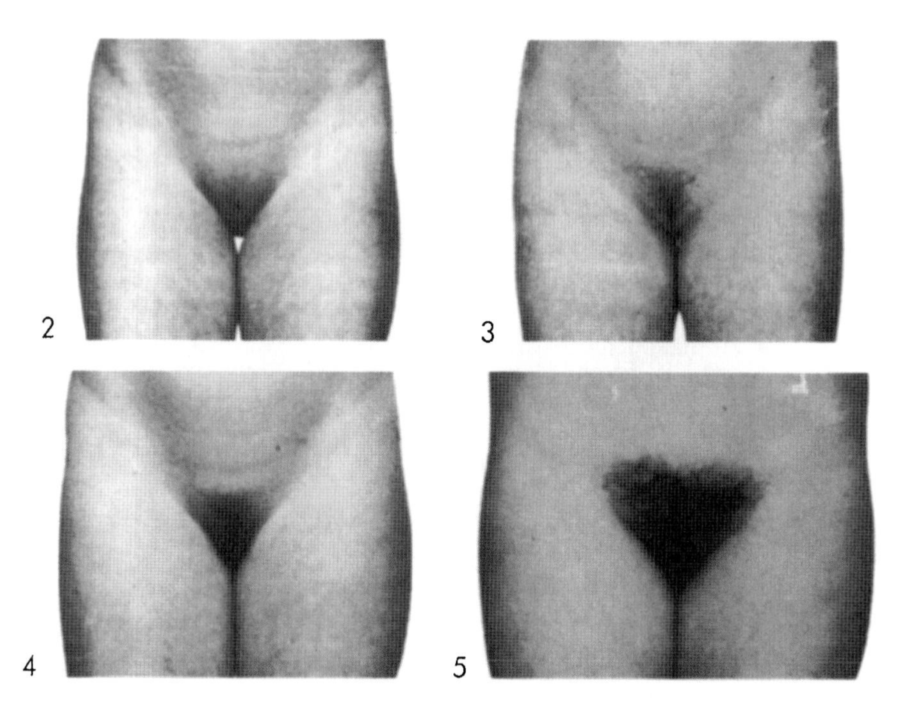

Figure 6-4 Stages of pubic hair development in females. The infantile pattern (stage 1) is not shown. Refer to the description of these stages in Table 6-3.

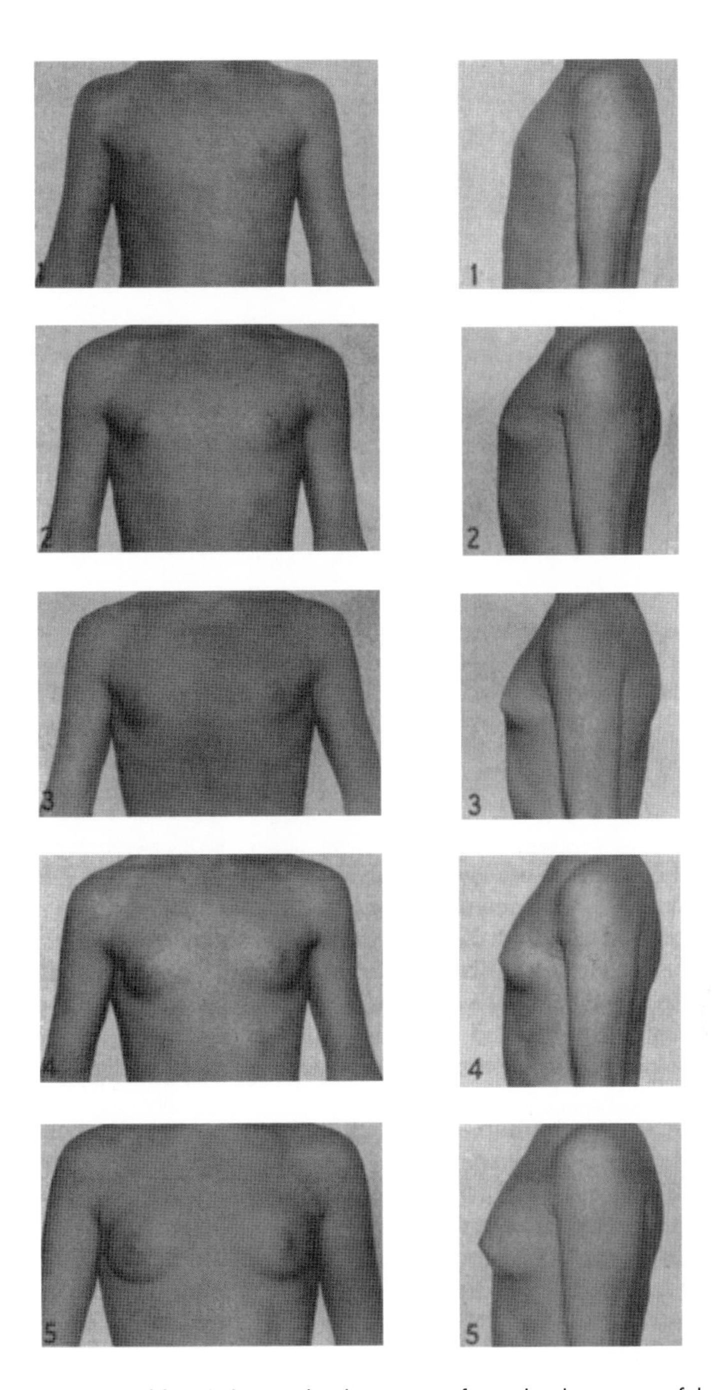

Figure 6-5 Stages of female breast development. Refer to the description of these stages in Table 6-3.

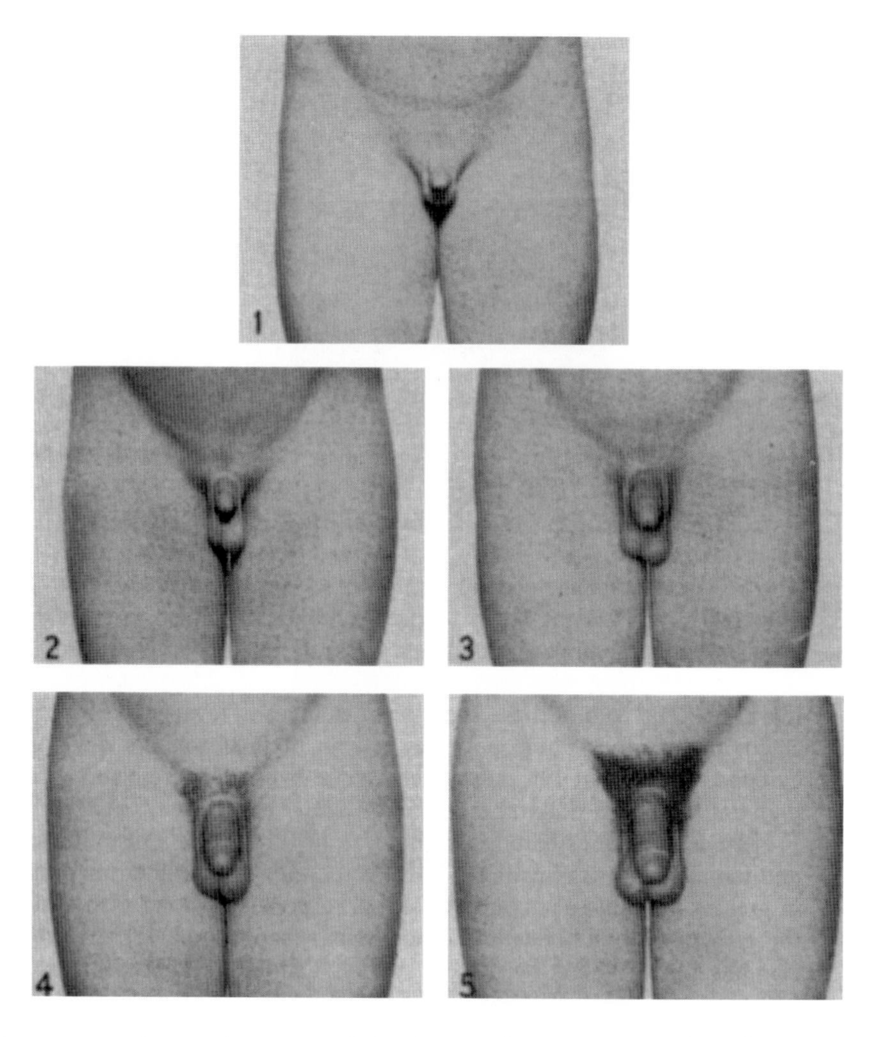

Figure 6-6 Stages of male genital development during puberty. Refer to the description of these stages in Table 6-3.

Gonadal Changes from Birth to Puberty

Ovarian Changes

The ovaries of a newborn female are about 2 cm in length, and each ovary contains about 500,000 ovarian follicles. Most of the oocytes are contained in primordial follicles, but several are in growing follicles; 60% of the ovaries in newborns contain some tertiary follicles. By 1 year of age, the ovaries of all females contain tertiary follicles. No follicles ovulate during childhood, but many growing follicles become cystic or atretic (see Chapter 2). Because of this massive follicular death in the ovaries of female children, the number of follicles remaining in each ovary immediately before puberty is reduced to about 83,000.

The ovaries of pubescent females, however, weigh more than those of a young child (Fig. 6-2) because some remaining healthy follicles have enlarged.

Testicular Changes

Seminiferous tubules in the testes of a newborn male contain only spermatogonia and Sertoli cells. Some Leydig cells are present at birth (these perhaps secrete the surge of testosterone present in the blood of male infants in their 2nd month (see Chapter 5)), but by 6 months of age these cells are almost invisible in the testes. At about 9 years of age, spermatogenesis begins in the tubules and Leydig cells again are visible. Mature spermatozoa are not produced, however, until the 14th or 15th year. At the onset of spermatogenesis, the testes enlarge markedly (Fig. 6-3). Most of this enlargement is due to the increase in diameter of the seminiferous tubules, as the originally solid tubules become hollow and filled with testicular fluid. The volume of a single testis of a 1-year-old boy is about 0.7 ml, that of an 8-year-old boy is about 0.8 ml, and that of a pubescent boy is about 3.0 ml. The testicular volume is 16.5 ml in an adult male, approximately a 24-fold increase!

During the 7th or 8th month of fetal life, the testes usually descend from the abdominal cavity into the scrotum. Testicular descent is under the control of testis secretions, müllerian-inhibiting substance (MIS), and testosterone (converted to DHT; see Chapter 5). They are "pulled" down by a ligament (the *gubernaculum*) that is attached on one end to the testes and at the other end to the scrotum. During their journey, the testes pass through the inguinal canals (two natural openings in the lower abdominal wall), and when they come to rest in the scrotum, the vasa deferentia extend into these canals. The canals then become closed and filled with connective tissue. Incomplete closure of an inguinal canal after testicular descent can lead to an *inguinal hernia*, a condition where a weakness in the abdominal wall remains; a portion of the internal organs may extend through this weakened region into the scrotum. *Hydrocele* occurs when the canal remains slightly open, permitting drainage of abdominal fluid into the scrotal sac. Surgery often is needed to close the canal in such cases.

Cryptorchid Testes

In about 3 to 4% of males, the testes fail to descend by birth (30% in premature newborns). By 1 year of age, about 0.8% of males still have testes within the pelvic cavity, and in 0.3% of cases the testes fail to descend by the time spermatogenesis begins (about age 10). Failure of one testis to descend is about five times as common as failure of both to descend. The process of spermatogenesis requires a temperature 3.1 °C (5.6 °F) lower than that in the abdominal cavity (see Chapter 4). If spermatogenesis begins while the testes remain in the body cavity, temperature damage occurs in these *cryptorchid testes* (*crypt*, meaning "hidden"; *orchid*, meaning "testis"); only Sertoli and Leydig cells remain. Thus, these males are sterile but have fully developed male secondary sex structures because androgen secretion from Leydig cells is normal. Cryptorchid testes often are treated early in childhood because they tend to develop cancerous tissue. Surgery or treatment with gonadotropins or gonadotropin-releasing hormone (GnRH) usually can cause testicular descent before any damage is done; this is usually done early in the second year of life.

Hormone Levels from Birth to Puberty

Chapters 3 and 4 showed that the pituitary gonadotropins—follicle-stimulating hormone (FSH) and luteinizing hormone (LH)—are both necessary for complete function of the ovaries and testes. Gonadotropin levels in the fetus and neonate are quite high, sometimes reaching adult levels. After a year or two, FSH and LH levels decline and remain low during childhood (the so-called "juvenile pause" in hormone secretion). Blood levels of these hormones are not high enough to initiate gonadal function in young children. At 9 to 12 years of age, however, levels of first FSH and then LH begin to rise in the blood.

These increases usually occur 1 or 2 years sooner in females than in males (Figs. 6-7 and 6-8). Not only do LH levels begin to rise in pubescent males and females, but the amplitude of the hour-to-hour changes in secretion of this gonadotropin increases in the day, especially during sleep (Fig. 6-9). In adults, LH levels fluctuate less markedly. The role of these short-term changes in levels of LH during puberty is not understood. They may play a role in final gonadal development, and evidence shows that LH pulses need to be every 1 to 3 h and of a relatively high amplitude for puberty to occur.

Other hormones exhibit an increase in blood levels in prepubertal males and females. Growth hormone (GH) is secreted by the anterior pituitary gland in greater amounts near puberty and is responsible, along with androgens, for the growth of long bones and other tissues (see Chapter 1). Growth hormone also has a major effect on protein synthesis and elevation of blood sugar levels. Children with a GH deficiency exhibit delayed sexual maturation as well as short stature. An increase in secretion of thyroid hormones from the thyroid gland, caused by greater secretion of thyrotropic hormone from the pituitary (see Chapter 1), may account for the rise in metabolic rate in both sexes. Thyroid hormones also are essential for body growth.

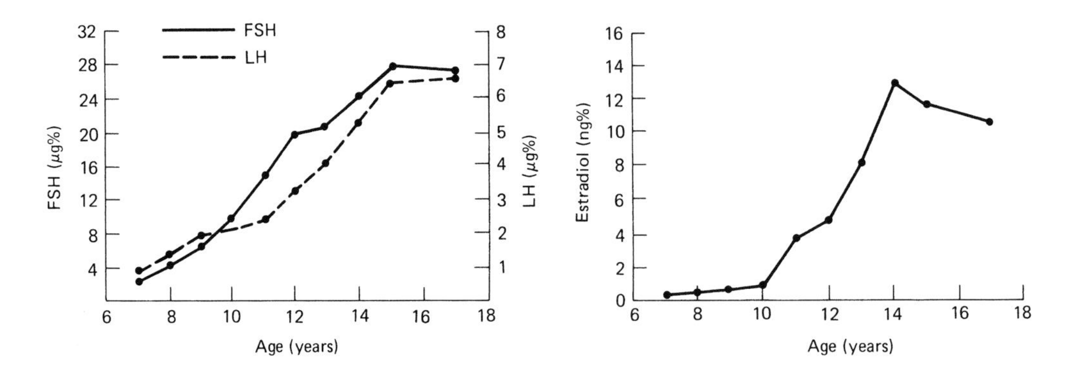

Figure 6-7 Changes in blood levels of FSH, LH, and estradiol in females of different ages. Blood androgen levels are not depicted, but these also rise as puberty nears. Note that the values for hormone levels are in nanograms percent (ng %) or micrograms percent (μg%). A nanogram is $\frac{1}{1000}$ of a microgram, and 1 ng% is equivalent to 1.0 ng in 100 ml of blood plasma. Thus not many hormone molecules are needed to change biological function.

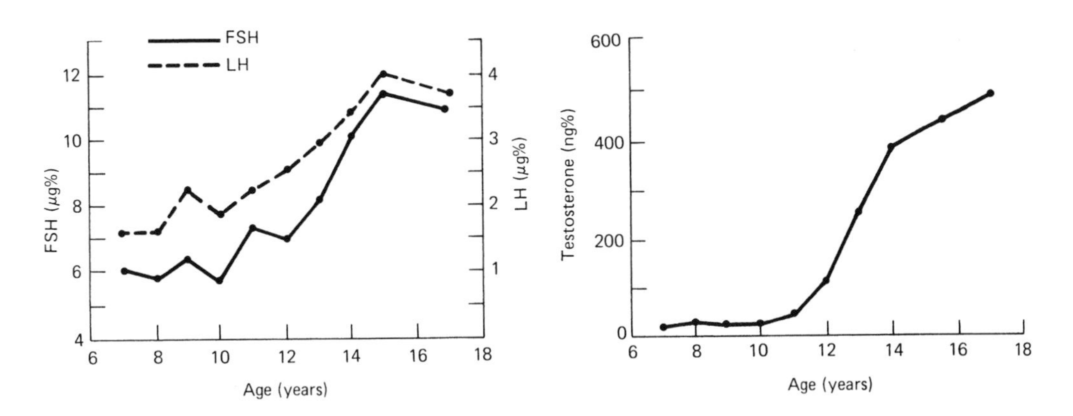

Figure 6-8 Changes in blood levels of FSH, LH, and testosterone in males of different ages. Blood estrogen levels are not depicted, but these also rise slightly as puberty nears. For an explanation of ng% and μg%, see the legend to Fig. 6-7.

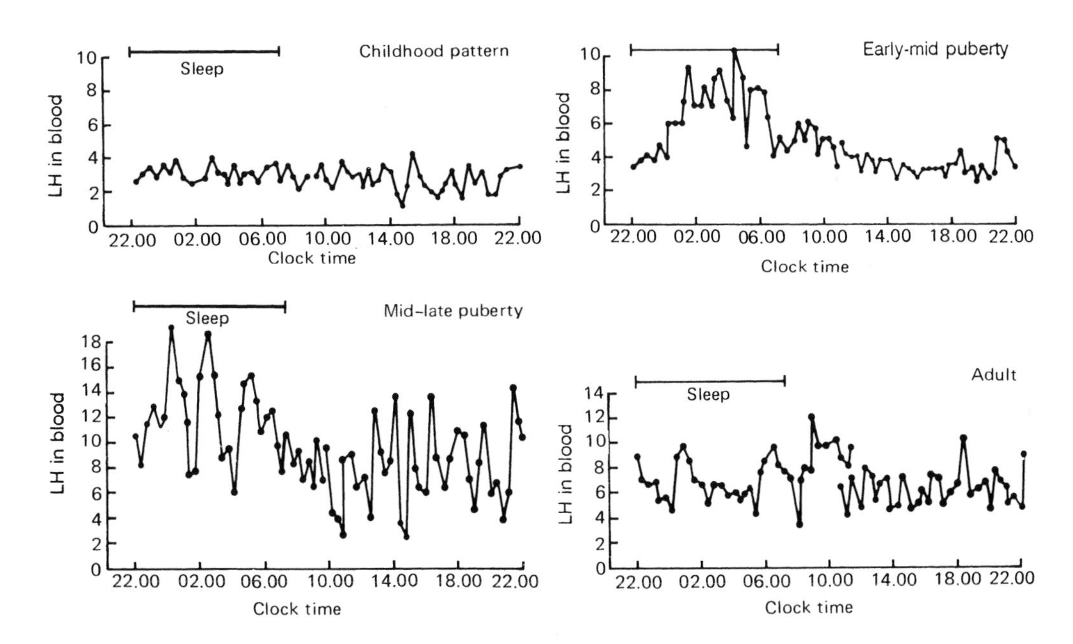

Figure 6-9 Daily patterns of LH levels in blood of females at various stages of sexual maturation. Note that during puberty, LH levels rise during sleep and exhibit marked hour-to-hour fluctuations. Average levels and fluctuations in both day and night then increase during late puberty, before decreasing somewhat in the adult. LH concentrations are in milli-IUs per milliliter of blood (see Fig. 3-2.)

Hormones in Females

Menarche and First Ovulation Near the time of the first ovulation in females, there is a sudden surge of LH and, to a lesser extent, of FSH. This first preovulatory surge of gonadotropin secretion may not be sufficient to cause ovulation, but it does cause enough cyclic variation in ovarian estrogen production so that menstruation occurs (see Chapter 3). Thus, as mentioned earlier, the initial menstrual cycle of pubescent females often is anovulatory. Eventually, the endocrine system and brain mature so that the LH surge is sufficient to cause the first ovulation.

Estrogens Levels of estrogens (estradiol and estrone) begin to rise in pubescent females, causing the onset of breast maturation (Table 6-1), and they continue to increase as the female approaches menarche (Fig. 6-7). The major source of estradiol is the large, growing follicles in the ovary, maturing under the influence of FSH and LH. Estrone is produced by body fat. Estrogens initiate development of the mammary glands, cause growth of the bony pelvis, and help deposit subcutaneous fat. These steroid hormones also cause growth of the external genitalia and of the vagina, oviduct, and uterus.

Androgens Levels of the androgens androstenedione and dehydroepiandrosterone also increase in the blood of pubescent females, but the ovary is not the main source of these so-called "weak" androgens. The adrenal glands secrete large quantities of these hormones during puberty. In both sexes, the release of weak androgens (especially DHEA) from the adrenal glands is the first endocrine change during puberty and is termed *adrenarche*. Adrenal androgens are responsible for some of the changes in the female during sexual maturation, including the growth of pubic and axillary hair, slight lowering of the voice, development of sebaceous glands and acne, and growth of long bones. In addition, these androgens may increase sex drive in pubescent girls (see Chapter 8). When these weak androgens arrive at some target tissues of the female, they are converted by enzymes to testosterone (see Chapter 2), which then makes the tissues grow and mature. In this way, masculinization of nontarget tissues by high blood levels of the potent androgen testosterone is avoided.

Hormone Levels in Males

Gonadotropins and Androgens The increase in secretion of FSH and LH in males at about 10 years of age is responsible for the onset of spermatogenesis and the rise in androgen secretion (mainly testosterone) from the testes (Fig. 6-8). Testicular androgens, along with adrenal androgens, cause growth of the sex accessory structures (seminal vesicles, prostate gland, penis, scrotum) and secondary sexual characteristics (pubic, axillary, and facial hair; larynx growth). Also, testosterone causes retention of nitrogen, calcium, and phosphorus in the body, thus supporting the bone and muscle growth seen in sexually maturing males. Spontaneous penile erections become more frequent as puberty approaches due to the effects of androgens on growth and touch sensitivity of the penis and on the male sex drive (see Chapter 8).

Estrogens Blood estrogen levels also rise slightly in males before and during puberty. The source of these estrogens is probably the Sertoli cells of the testes (see Chapter 4). The presence of estrogens in males near puberty explains why some pubertal males exhibit slight growth of the mammary glands (*gynecomastia*). In this condition, a small (1 to 2 cm in diameter) lump appears behind the nipple, but it usually goes away in about 2 years. Thus, levels of both male and female sex hormones rise in both sexes, and the relative levels of the two hormones determine if secondary sexual characteristics develop in a male or female direction. The interaction of androgens and estrogens in sexual maturation remains an exciting area of research.

Androgens and Acne

Rising androgen levels during puberty in both males and females increase the secretion of the skin's sebaceous glands. These glands, which normally secrete an oily substance termed *sebum*, can become clogged and infected with bacteria, producing *acne*—pimples, blackheads and cysts on the face, chest, or back. Most adolescents have some acne. Usually it clears up in a few years, but even adults in their twenties, thirties, forties, and beyond can have acne. Young people experiencing this condition may suffer from embarrassment and loss of self-esteem. Contrary to popular belief, acne is not caused by improper hygiene, long hair, or ordinary stress, nor is there any scientific evidence that any specific food causes acne. Some treatments used to alleviate acne include antibiotics and medications containing benzoyl peroxide, salicylic acid, or sulfur. Cases of severe acne may be treated with female hormones or medications that decrease the effects of androgens. An oral medication called *isotretinoin* (Accutane) may be used to treat acne that has not responded to other medications. Among the side effects are severe birth defects, so prevention of pregnancy during isotretinoin treatment is essential.

What Mechanisms Cause Puberty?

The pubertal changes in secondary sex characteristics (such as breast development, fat deposition, development of the genitalia, changes in the larynx, and body hair growth) result from increasing levels of gonadal steroids. Sex steroid levels, in turn, are influenced by rising gonadotropins. As discussed earlier, both FSH and LH increase during puberty. In early puberty, the rise in gonadotropin secretion is observable only at night. Nocturnal pulses of LH, and to a lesser extent of FSH, exhibit greatly increased amplitude and probably also increased frequency. The significance of this diurnal pattern of gonadotropin release in early puberty (high levels at night, with lower levels during the daytime) is not understood. In later stages of puberty, FSH and LH levels increase during the day also, approaching the adult pattern of relatively high levels of pulsatile gonadotropin secretion lacking a diurnal rhythm. As discussed in Chapter 1, FSH and LH are released from the pituitary in response to the pulsatile secretion of GnRH from the hypothalamus. Thus, puberty involves increased release of GnRH, gonadotropins, and sex steroids from the hypothalamus, pituitary, and gonads, respectively.

What changes in the hypothalamo/pituitary/gonadal axis are responsible for initiating puberty? The gonads gain the ability to respond to gonadotropins very early in life. In response to circulating gonadotropins, follicular growth and atresia can occur even in the fetal and neonatal ovary, and testosterone and estrogen secretion can reach adult levels in the neonate. Similarly, the fetal pituitary is capable of secreting gonadotropic hormones in response to GnRH stimulation (although the pituitary becomes more sensitive to GnRH in adulthood). Thus, both the pituitary and the gonads are capable of responding to endocrine signals well before the onset of puberty. Pulsatile secretion of GnRH by the hypothalamus begins during the second trimester of fetal development. Because the hypothalamus, pituitary, and gonads are capable of endocrine function from birth, no obvious "maturation" of these endocrine organs is necessary for puberty to begin.

Currently, evidence suggests that, during childhood, sexual maturation may be suppressed by two mechanisms. The first involves a change in the brain's "gonadostat." As described in Chapter 1, estrogens and androgens exert negative feedback on the production of GnRH from the hypothalamus. Steroid hormones circulating in the blood of children exercise this negative feedback, resulting in very low levels of GnRH and, consequently, low FSH and LH. It is thought that the hypothalamus is very sensitive to circulating steroids during childhood. Therefore, even the low levels of androgens or estrogens in young children are enough to effectively inhibit gonadotropin secretion.

The negative feedback effects of gonadal steroids operate in relation to a set point in the gonadostat of the brain (hypothalamus). One hypothesis explaining the prepubertal increase in gonadotropin secretion in both sexes is that the sensitivity of the hypothalamus to steroidal negative feedback decreases as puberty approaches. That is, the set point for negative feedback increases so that higher concentrations of steroid hormones in the blood are required to decrease gonadotropin secretion from the anterior pituitary gland. (Similarly, as you increase the set point of the thermostat in your house, it takes a higher room temperature to shut off the heater.) Thus, even though circulating levels of steroids rise because of the increase in gonadotropin secretion, the levels of these steroids are not high enough to block a further rise in gonadotropin secretion. It has been hypothesized that the hypothalamus of children is 6 to 15 times more sensitive to steroidal negative feedback than that of an adult. How is this known? One piece of evidence is that it takes a much smaller amount of injected estrogen to lower blood gonadotropin levels in girls than in adult women.

Although a change in sensitivity of the gonadostat does appear to have an important role in the onset of puberty, evidence also shows that the GnRH pulse generator may be released from central inhibition at this time. According to this hypothesis, the GnRH pulse generator is actively inhibited by other brain areas during childhood. This inhibition is lifted at puberty, allowing the hypothalamus to generate GnRH in an adult pattern of specific amplitude and frequency. For example, lesions or tumors of specific brain regions in female rats cause early puberty, which suggests that these particular brain areas normally inhibit the secretion of GnRH until puberty. Children born without gonads (lacking the steroids for negative feedback) nevertheless have very low gonadotropin levels during childhood, and GnRH secretion occurs at the normal age. Thus, changes in the brain may release inhibition of the GnRH pulse generator at the onset of puberty. The timing of these changes may be influenced by the environment (see next section).

At puberty, another change occurs in girls (but not in boys) such that estrogens can also have a *positive* feedback effect on gonadotropin secretion. High blood estrogen levels cause a surge of LH and FSH secretion that induces ovulation in adult women. This positive feedback effect of estrogen is not present in younger females. If estrogens are given to young girls, no gonadotropin surge follows, but if given to girls in late puberty, an FSH and LH surge results immediately. Therefore, the positive feedback effect of estrogen begins to appear just before menarche and is fully operable late in the pubertal sequence; it is responsible for the first ovulation. The appearance of positive feedback may be a consequence of the maturation of the hypothalamic GnRH surge center or the ability of the pituitary to now synthesize and store adequate amounts of gonadotropins. The regulation of puberty is better understood in females than in males. However, it appears to be similar in both sexes, except that no gonadotropin surge occurs in males (see Chapter 17).

Environmental Factors and Puberty

Environmental factors that influence the physiology of the child can also affect the sequence of sexual maturation and puberty. Of the many possible factors, nutrition, day length, stress, climate, and social interaction are examined. You should be aware, however, that it is very difficult to separate the effects of one environmental factor from those of other factors that vary along with it.

The age of menarche in females generally has shown a steady decline worldwide, at a rate of about 3 months per decade over the last century (Fig. 6-10), although this rate of decline may be an overestimate. In the United States, the average age of menarche was 14.2 years in 1900, but currently it is estimated to

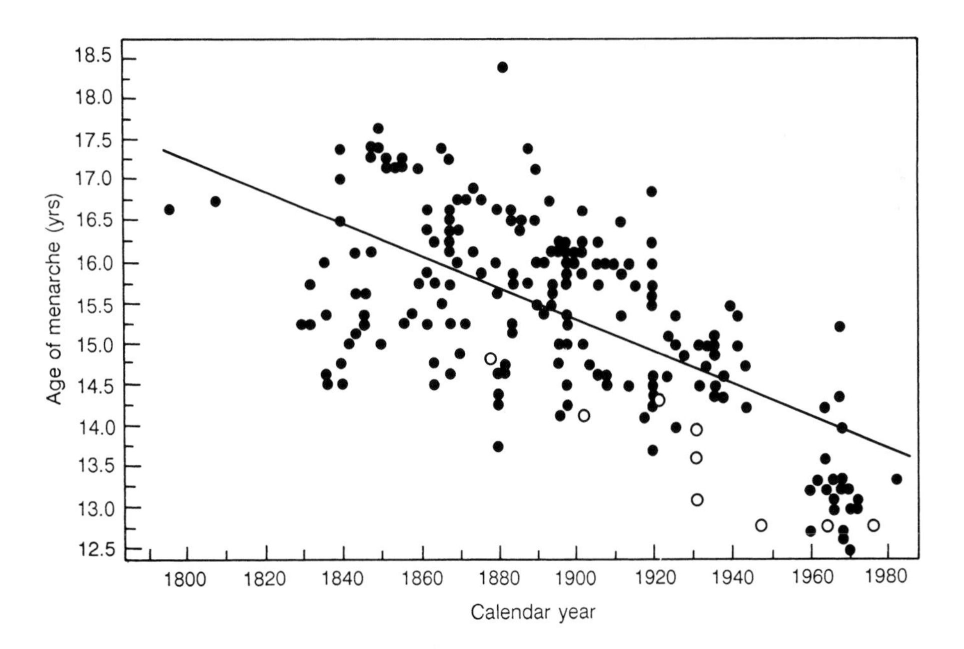

Figure 6-10 Average age of menarche, 1790 to 1980. ○, United States; ●, various European countries. The regression line represents the average rate of decrease, which has been 3 to 4 months per decade. The average age of menarche in the United States in the late 1990s was 12.3 years, which appears to be stable at present. Age of menarche has not been decreasing in underdeveloped countries, probably because of malnutrition.

be 12.7 for Caucasian girls (age 12.2 for African-American girls). Menarche now commonly occurs as early as age 8. The average age of pubertal development in white girls in the United States begins just before age 10 and in black girls just before age 9. Evidence also suggests that puberty in males now occurs at an earlier age. In all probability, there is a biological limit to this decrease, and this limit may have already been reached in some countries (e.g., the United States). The decline in age at puberty has had profound implications for human reproductive biology and sexuality in that the period of adolescent sexuality is lengthened. In addition, there are important health implications. For example, earlier timing of puberty is associated with an increased risk of breast cancer later in life.

Nutrition

What is responsible for the decline in the age of puberty? Some evidence shows that improved nutrition and health in childhood are factors. Females in countries with poorer nutrition tend to exhibit a relatively delayed puberty. In fact, acute starvation prevents puberty. Also, poorer nutrition in rural compared with urban regions, in lower social classes, and in larger families may account for the later age at puberty seen in these groups. It is well known that teenage girls with *anorexia nervosa*, a disorder in which food intake is reduced greatly, reach puberty at a relatively late age, as do lean female athletes and ballet dancers.

How does nutrition affect the mechanisms for regulating the onset of sexual maturation? One suggestion is that a certain amount of fat is needed in the body of females before menarche can occur. Obese girls tend to reach puberty at an earlier age than other girls. There even is some evidence that a girl must be at a critical weight of about 45 kg (104 lb), and 11 kg (24 lb) of this must be fat (17%) before sexual maturation can begin. Having too little fat can delay menarche to age 19 or 20, or rarely even age 30. It may be that, through natural selection, females have evolved a mechanism ensuring that they have enough energy stored to give birth successfully. Much controversy exists about this *critical body fat hypothesis*, and some think that a critical metabolic rate, rather than a critical body weight or composition, permits puberty. If a critical body weight or critical fat mass is necessary to enter puberty, how does the metabolic system signal the reproductive system to proceed with pubertal development? One proposed signal is a protein called *leptin*, which is produced by adipose (fat) cells. Increased leptin levels suppress appetite through negative feedback to the hypothalamus. Mice lacking the gene for leptin eat voraciously and are obese. These mice are also infertile, but if leptin is administered, their fertility is restored.

Humans lacking the ability to produce leptin or its receptor can fail to enter puberty. In contrast, overweight children, who have higher leptin levels than children of average weight, tend to enter puberty earlier than their lean peers. This is especially true for overweight girls. Thus, a threshold level of leptin may be necessary for the initiation of puberty, i.e., leptin may have a permissive effect on puberty. This may help explain why severely malnourished girls such as those with anorexia nervosa, who have very low leptin levels, often have delayed puberty. In both boys and girls, circulating leptin levels increase before the onset of puberty. Because leptin-binding proteins in the blood decrease during late childhood, bioavailable leptin may increase dramatically at this time. Leptin receptors are found in hypothalamus and pititary as well as gonadal cells, and studies show that leptin increases the rate of GnRH pulses. In adults, leptin levels are higher in females than in males, and these levels remain into old age.

The hope that leptin could be used as an appetite-suppressing therapy for obesity has not been realized. Overweight individuals have unusually high circulating leptin levels, but this increase in leptin does not result in reduced caloric intake or increased metabolic rate. It is possible that obese individuals are leptin resistant. More work will be needed to discover how this molecule influences the interactions among weight, body fat, and reproductive status.

Day Length and Season

Recall that darkness may increase the synthesis and secretion of melatonin and other pineal gland secretions and that these chemicals may then inhibit reproductive function (see Chapter 1). Light has the opposite effect. Therefore, do light–dark cycles, through their influence on pineal gland function, play a role in the onset of puberty? Has the artificial increase in day length in recent years, as a result of the increased use of artificial lighting, been at least partially responsible for the recent decline in the age of puberty? It appears that children who have lesions of the pineal gland, rendering it nonfunctional, undergo precocious puberty. The absence of circulating pineal secretions in these children may explain their early puberty. In contrast, about one-third of children with secretory pineal tumors exhibit delayed puberty; perhaps their relatively high levels of pineal secretions inhibit sexual maturation. Paradoxically, however, blind children with limited light perception reach puberty at an earlier age than do children with normal vision, and females with no light perception exhibit the earliest menarche. Theoretically, at least, because darkness causes melatonin secretion, these blind children should exhibit delayed maturation, but this is not the case. Deaf children also tend to reach puberty early; thus, sensory deprivation in general may accelerate sexual maturity. Obviously, we have much to learn about the possible role of day length and other environmental factors in regulating puberty in humans. Some evidence suggests that more females reach menarche at a certain time of year; this time varies among geographic regions. Whether or not day length influences this seasonality of puberty is not known. Overall, children undergoing puberty should think about the effects of "over-the-counter" melatonin before they use it.

Stressors

In laboratory animals, various kinds of environmental stressors (emotional or physical) can delay sexual maturity. A *stressor* is any set of circumstances that disturbs the normal homeostasis in the body. *Stress*, however, is the physiological effect of a stressor. Any kind of long-term stressor produces a set of physiological stress responses that Hans Selye has termed the *general adaptation syndrome*. Although this syndrome encompasses a variety of responses, a major component is enlargement of the adrenal glands and increased secretion of adrenal steroid hormones such as cortisol. Some believe that this increased adrenal function may inhibit the reproductive system. The evidence for this is controversial. There is some information, however, that stressors (emotional or physical) can delay puberty in humans. Evidence also exists that stress during infancy and childhood can accelerate puberty. Finally, studies have implicated family stress in accelerated puberty in girls. Girls who grow up in the absence of a biological father, especially with a stepfather present, tend to enter puberty early (see HIGHLIGHT box 6-1).

Chapter 6, Box 1: Early Childhood Stress and Sexual Maturation

There are many examples of animals responding to environmental situations by changing their reproductive physiology and behavior. The evolutionary explanation for these changes in reproductive strategy is that individuals increase their chance of passing on their genes (through offspring) to the next generation. One example of such a response is that if life is stressful as a juvenile, it could be adaptive to exhibit a more rapid sexual maturation and to reproduce before one is affected adversely by the stressful environment. An experimental example occurs when baby rat pups are stressed by removing them from their mother and placing them in a new cage for 15 min a day for the first 3 weeks of life. After this "early handling," the pups open their eyes sooner, reach sexual maturity ("puberty") at an earlier age, are more active, and respond less to stressors as adults. This early handling effect does not occur if the pups are not handled until after 3 weeks of age. Therefore, there is a critical period in early infancy when stress has a beneficial effect. Similarly, early handled kittens exhibit accelerated maturation: they open their eyes earlier, emerge from the nest box earlier, and reach sexual maturity at an earlier age.

Evidence also exists that stress during human infancy can accelerate puberty. For example, J. W. M. Whiting studied 50 societies to determine whether infant stress was present. The kinds of stressors included mother–infant separation, pain due to ornamentation or scraping with objects, and exposure to extreme temperatures or loud noises. After eliminating other factors such as climate and nutrition, he found that the age of menarche in societies that stressed their infants was 12.75 years, whereas that of those unstressed was 14.0 years. Such early stressors must not, however, be extremely severe, as it is known that severe stress during infancy can cause physical deterioration, slowing of growth, and emotional disorders.

Evidence suggests that psychological stress during the first 5 to 7 years of life accelerates sexual maturation in girls (boys are more difficult to study because they have no easily quantifiable puberty sign, such as menarche, to record). These studies indicated that young girls with excess parental conflict, parent–child conflict, or absence of the father (but not the mother) and presence of a stepfather exhibited the following traits compared to girls being in relatively conflict-free, two-parent households.

1. They passed through puberty stages more rapidly.
2. They reached menarche about 5 months earlier, on average.
3. They had their first sexual intercourse at an earlier age.
4. They had children at an earlier age.
5. They spent less effort and time with their children.
6. They formed less stable bonds with their partners.

Also, boys with absent fathers tended to exhibit exaggerated and stereotypical masculine behavior during adolescence. The idea is that stressed children perceive others as untrustworthy, relationships opportunistic and selfish, and resources scarce and unpredictable. The biologically adaptive response then is to mature earlier sexually and orient toward short-term relationships, i.e., to allocate energy to growth and development and not to parenting.

Could this be an example of evolutionary strategy that is, or at least was before humans became "domesticated," an adaptive response?

Continued on next page.

Chapter 6, Box 1 continued.

Is there a critical period to this effect, after which stress delays sexual maturation? How does early stress accelerate the pubertal process in the brain? How does childhood stress interact with other factors (e.g., inheritance, exercise, nutrition) to influence sexual maturation? Does this early stress phenomenon have something to do with the recent decrease in the age of puberty and the increase in teenage pregnancies? These questions and others remain.

Theory of How Early Childhood Family Stressors Accelerate Sexual Maturation and Influence Sexual Activity[a]

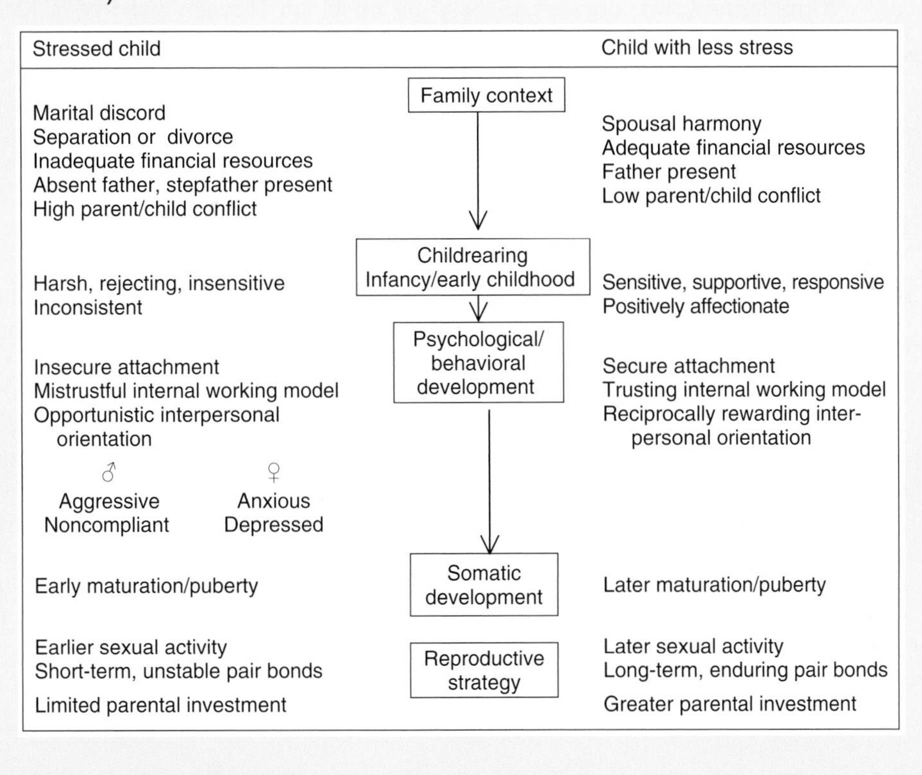

[a]Adapted from Belsky *et al.* (1991).

Environmental Pollutants

Some of the chemicals used in the synthesis of plastics and in other industrial processes act as weak estrogens. Exposure of children to these widespread "xenoestrogens" may accelerate puberty, although this idea is controversial. In overweight children, the weak estrogens synthesized by adipose tissue may add to the estrogenic load. It is unclear whether the proposed puberty-accelerating effects of exposure to these estrogens represent "real" puberty (early ovulation) or "pseudopuberty" (precocious breast development and uterine bleeding). If environmental estrogens accelerate sexual development significantly, one might expect to see an epidemic of gynecomastia (breast enlargement) also in boys.

However, environmental pollutants (e.g., dioxins, polychlorinated biphenyls, or PCBs) have been implicated in delaying male puberty. More work is needed to decipher the effects of endocrine-disrupting environmental contaminants on pubertal development.

Climate and Altitude

It used to be thought that children in warmer regions reached puberty at an earlier age than those in colder areas. However, careful statistical comparisons have shown that these differences are due to other factors besides temperature. Altitude, however, appears to have an effect on the age puberty is reached. Indeed, for each 100-m increase in altitude, puberty is delayed by about 3 months. The mechanisms by which high altitude suppresses the sexual maturation process are not clearly known and may be related to poorer nutrition or greater energy expenditure at higher elevations.

Inheritance and Age of Puberty

Undoubtedly, inheritance interacts with environmental factors to determine the age of sexual maturation. In fact, the differences in age of menarche among different countries (Fig. 6-10) may be related to genetic differences in populations as well as to differences in childhood health and nutrition. It has been estimated that 15% of the variance in age at puberty is due to genetic differences (see HIGHLIGHT box 6-2). For example, age at menarche tends to be similar in mothers and daughters. Moreover, in identical twin females (identical genotypes), the age of menarche usually differs by a maximum of 2 months. Any difference here is probably a measure of slight differences in birth weight or in exposure to slightly different social or physical environments during childhood. In contrast, the difference between age at menarche in fraternal twin girls (with different genetic makeups) can be up to 8 months. In the United States, black girls reach puberty approximately 1 year earlier than white girls.

Chapter 6, Box 2: Puberty Genes

Among girls, there is a wide age variation in the onset of the pubertal process. Breast enlargement begins about a year before menarche. Some girls begin this process at an early age, whereas others begin breast enlargement later but still within the normal range. Many feel that environmental factors (diet, lighting, stress) influence the timing of puberty. It is likely that childhood obesity, caused by overeating and lack of exercise, is a key player in the early onset of menarche seen in developed countries. The extra fat, as discussed in this chapter, would synthesize more estrogen and thus initiate puberty at an earlier time.

Genes also appear to play a role in the onset of puberty. A recent study of children with precocious puberty examined the timing of puberty of their family members. For nearly a third of the children, their early puberty had a familial pattern. Results of the study suggest an autosomal dominant inheritance of a gene or genes related to precocious puberty. There is greater penetrance (phenotypic expression)

Continued on next page.

Chapter 6, Box 2 continued.

of the gene(s) in girls than in boys. Perhaps this helps explain why precocious puberty is found so much more frequently in girls as in boys—10 times as often in girls.

In females, one would guess that genes that influence estrogen secretion or breakdown (metabolism) of estrogens would have an important role in the timing of puberty. Surprisingly, scientists have discovered an unlikely culprit that affects the age of puberty, a gene that accelerates the breakdown of testosterone (an androgen) in a girl's body.

A variant of the gene *CYP3A4* influences the production of a liver enzyme that breaks down testosterone. This gene has two alleles: a high activity and a low activity form. In a study of 192 girls aged 9–10, 90% of girls with two copies of the high activity form of this gene begin breast development by age 9.5. In contrast, 56% of those with only one copy of the high activity allele begin puberty by age 9.5, and only 40% begin breast development by age 9.5 if they have no copies of the high activity allele. Thus, presence of the high activity allele is correlated with early pubertal development. Scientists found a greater proportion of the high activity form of this gene in black and Hispanic girls than in Caucasians, with the latter tending to reach puberty later than the former two populations.

How does relatively lower testosterone (more copies of the high activity form of the gene) cause early puberty in girls? It has long been known that estrogen levels rise and circulating androgens fall during puberty in girls. Reduction of the testosterone levels in the blood of young girls would lead to a higher ratio of estrogen to testosterone levels in their blood. This, in turn, may then lead to an earlier initiation of breast development and the remainder of the pubertal process. Surprisingly, no genes yet studied control levels of estrogen in a young girl's blood so it looks like genes controlling testosterone levels play the major role.

Hypothetically, the reverse story could be true in boys. That is, boys who begin puberty at an earlier age could have one or more copies of the low activity allele of the *CYP3A4* gene. They would have less of the liver enzyme that metabolizes testosterone and thus would have high blood testosterone levels and enter puberty earlier. Studies on the *CYP3A4* gene variants in boys, however, have not yet been done. Also needed are studies of the possible interaction of environmental factors and puberty genes.

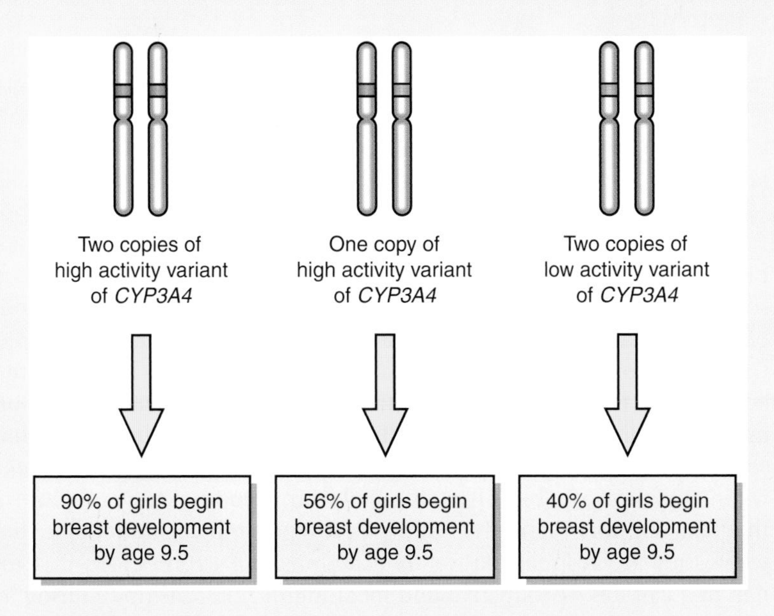

Effects of inheritance of *CYP3A4* gene variants on timing of puberty in girls.

Puberty and Psychosocial Adjustment

Young people must adjust to remarkable physiological, anatomical, and psychological transformations during the process of puberty. Their bodies change rapidly, and thus their *body image* (how they perceive their bodies, especially in relation to the bodies of their peers) also changes. Any deviations of their bodies from what they or their peers consider "normal" can lead to low self-esteem. An individual who reaches puberty later or earlier than his or her peers can suffer psychological pain. For example, females will worry about the size of their breasts and other aspects of their figure and will not feel feminine if these are not in line with the norm of their peers. Similarly, a male may worry about the size of his penis or his physique; if these are not within the norm, he will feel that something is wrong with him. Young people should realize that there is considerable variation in the timing of the stages of puberty among different individuals and that most people develop into "normal" adults in the course of their sexual and physical maturation.

Medically speaking, puberty that occurs abnormally early is considered *precocious*. Precocious puberty used to be defined as the onset of breast development before age 8. Now, a diagnosis of precocious puberty in girls is given if either breast development or pubic hair growth occurs before age 7 in white girls or age 6 in black girls. For boys, precocious puberty is still defined as sexual development before age 9. Precocious puberty is reported in girls 10 times as often as in boys. However, puberty is considered *delayed* in males if no testicular growth has occurred by age 14 or if no skeletal growth spurt has happened by age 18. In females, puberty is delayed if no breast growth occurs by age 14 or if no skeletal growth spurt appears by age 15. A person experiencing precocious or delayed puberty should be examined by a physician to ascertain the cause and possible treatment (e.g., with GnRH stimulatory or inhibitory agonists; see Chapter 1).

In addition to being a time of dramatic reproductive and somatic development, key brain growth occurs in teenagers. Gray matter increases in some brain areas, followed by neuronal cell death. This remarkable remodeling of the brain is especially apparent in the frontal lobe, a region of the brain involved in regulating emotions and planning/organizing actions, as well as inhibiting impulsive behaviors.

Adolescence is the period between puberty and adulthood when a good deal of social learning takes place. The length of this period of youth is determined socially as well as biologically. In the United States, it is being extended because of the decrease in age of biological puberty and the lengthening in educational training before adulthood. In some primitive societies, however, the period of adolescence is brief or nonexistent; teenagers are expected to be adults in sexual matters and in other areas of life. In the United States, being an adult can mean several things. Biologically, teenagers are adult after they have reached puberty, when they are capable of having children. Economically, they are adult when they can support themselves and possibly a family. Emotionally, they are adult when they are responsible for their actions, can express love in a mature manner, and can have productive and meaningful relationships. During adolescence, teenagers must achieve economic and emotional adulthood, deal with separation from family, develop realistic vocational goals, and come to terms with their emerging sexuality. In respect to sexuality, they must accept their sexual fantasies, ambivalence about intercourse, concerns about sexual adequacy, confusion about love, and fears about sexually transmitted diseases. In addition, they must achieve these adjustments in a society that expects adult behavior from them but often does not treat them as adults. Thus, it is an attestation to human understanding and flexibility that most adolescents become mature, capable members of society.

Chapter Summary

The human life cycle can be divided into five stages: (1) embryonic and fetal existence, (2) infancy and childhood, (3) puberty and adolescence, (4) early and middle adulthood, and (5) late adulthood and old age. Puberty is the transition a young person makes from a child to a reproductively mature adult. This process involves the growth and development of the male or female gonads, sex accessory structures, and secondary sexual characteristics. The timing of puberty varies greatly among different individuals.

In both sexes, the blood levels of steroid hormones (androgens and estrogens) and of pituitary gonadotropins (FSH and LH) rise before puberty. The rising levels of steroid hormones cause the physiological and anatomical changes of puberty. Variations in the negative and the positive feedback of steroid hormones on pituitary secretion of gonadotropins control the onset of puberty. The brain also influences the onset of puberty through genetically programmed development interacting with environmental factors such as climate, nutrition, day length, stressors, health, and social interaction. The decline in age of puberty in developed countries is thought to be caused in part by improved nutrition and health care.

A teenager's self-esteem and adjustment are influenced by his or her body image. During the period of adolescence, a teenager must develop economic and emotional adulthood, deal with separation from family, plan a vocation, and come to terms with his or her sexuality. Adolescents must also adjust to conflicting standards of society about how they should behave and what is considered to be adult behavior.

Further Reading

Frisch, R. E. (1988). Fatness and fertility. *Sci. Am.* **258**(3), 88–95.

Frisch, R. E. (1991). Body weight, body fat, and ovulation. *Trends. Endocrinol. Metab.* **2**, 191–197.

Hooper, C. (1991). The birds, the bees, and human sexual strategies. *J. NIH Res.* **3**(10), 54–60.

Landauer, T. K., and Whiting, J. W. (1964). Infantile stimulation and adult stature of human males. *Am. Anthropol.* **66**, 1007–1028.

Meany, M. J., *et al.* (1991). The effects of neonatal handling on the development of the adrenocortical response to stress: Implications for neuropathology and cognitive defects later in life. *Psychoneuroendocrinology* **16**, 85–103.

Raloff, J. (2000). New concerns about phthalates. Ingredients of common plastics may harm boys as they develop. *Sci. News* **158**, 152–154

Saenger, P., and Reiter, E. D. (1992). Management of cryptorchidism. *Trends Endocrinol. Metab.* **3**, 249–253.

Whiting, J. W. M. (1965). Menarcheal age and infant stress in humans. *In* "Sex and Behavior" (F. A. Beach, ed.). Wiley, New York.

Advanced Reading

Belsky, J., *et al.* (1991). Childhood experience, interpersonal development, and reproductive strategy: An evolutionary theory of socialization. *Child Dev.* **62**, 647–670.

Bolton, F., and MacEachron, A. (1988). Adolescent male sexuality: A developmental perspective. *J. Adolesc. Res.* **3**, 259–273.

Bourguignon, J.-P. (1995). The neuroendocrinology of puberty. *Growth Genet. Horm.* **11**(3), 1–6.

Brown, D. E., *et al.* (1996). Menarche age, fatness, and fat distribution in Hawaiian adolescents. *Am. J. Phys. Anthrol.* **99**, 239–247.

Bullough, V. L. (1981). Age at menarche: A misunderstanding. *Science* **213**, 365–366.

Ellis, B. J., and Garber, J. (2000). Psychosocial antecedents of variation in girls' pubertal timing: Maternal depression, stepfather presence, and marital and family stress. *Child Dev.* **71**, 485–501.

Fishman, J. (1989). Fatness, puberty, and ovulation. *N. Engl. J. Med.* **303**, 42–42.

Frisch, R. E. (1994). The right weight: Body fat, menarche, and fertility. *Proc. Nutr. Soc.* **53**, 113–129.

Gale, C.R. (2004). Critical periods of brain growth and cognitive function in children. *Brain* **127**, 321–329.

Herman-Giddens, M. E., *et al.* (1997). Secondary sexual characteristics and menses in young girls seen in office practice: A study from the Pediatric Research in Office Settings network. *Pediatrics* **99**, 505–512.

Herman-Giddens, M. E., *et al.* (2001). Secondary sexual characteristics in boys: Estimates from the national health and nutrition survey III, 1988–1994. *Arch. Pediatr. Adolesc. Med.* **155**, 1022–1028.

Howdeshell, K. L., et al. (1999). Exposure to bisphenol A advances puberty. *Nature* **401**, 763–764.

Kaplowitz, P. B., *et al.* (1999). Reexamination of the age limit for defining when puberty is precocious in girls in the United States: Implications for evaluation and treatment. *Pediatrics* **104**, 936–941.

Meier, G. (1961). Infantile handling and development of Siamese kittens. *J. Comp. Physiol. Psychol.* **54**, 283–286.

Merzenich, H., *et al.* (1993). Dietary fat and sports as determinants for age at menarche. *Am. J. Epidemiol.* **138**, 217–224.

Ojeda, S. R. (1991). The mystery of mammalian puberty: How much more do we know? *Perspect. Biol. Med.* **34**, 365–383.

Stanhope, R., and Brook, C. (1988). An evaluation of hormonal changes at puberty in man. *J. Endocrinol.* **116**, 301–305.

Steinberg, L. (1988). Reciprocal relation between parent-child distance and pubertal maturation. *Dev. Psychobiol.* **24**, 122.

Tanner, J.M., and Davies, P.S. (1985). Clinical longitudinal standards for height and height velocity for North American children. *J. Pediatr.* **107**(3), 317–329.

Thornburg, H., and Aras, Z. (1986). Physical characteristics of developing adolescents. *J. Adolesc. Res.* **1**, 47–78.
Udry, J. R., *et al.* (1985). Serum androgenic hormones motivate sexual behavior in adolescent boys. *Fertil. Steril.* **43**, 90–94.
Wyshak, G., and Frisch, R. E. (1982). Evidence for a secular trend in age of menarche. *N. Engl J. Med.* **303**, 1033–1035.

CHAPTER SEVEN

Reproductive Aging

Introduction

The reproductive capacity of a woman varies greatly over her lifetime, diminishing over several years and finally coming to an end at menopause. Over a million U.S. women enter menopause each year. Because the average life span of humans has increased, especially in the past 100 years, more women will live a larger proportion of their lives postreproductively. Male fertility often continues through adult life, although it also diminishes with age. This chapter examines the aging of the reproductive system, comparing the two sexes in this regard. We discuss the patterns and consequences of endocrine changes at menopause and andropause and look at the pros and cons of therapeutically replacing the lost reproductive hormones in older adults.

Menopause

Most women who survive to the age of 50 can expect to live 20, 30, or even 40 more years in reasonably good health. A typical 50-year-old woman's heart vigorously continues to pump blood, other internal organs such as her kidneys and liver are properly functioning, and her brain is active. In contrast, her ovaries undergo an abrupt decline, resulting in their degeneration and complete loss of function. Her follicles deteriorate and her egg supply is lost, rendering her infertile. Levels of ovarian hormones drop, affecting steroid-responsive tissues, especially the uterus, vagina, and breast tissue. Loss of ovarian steroids causes menstrual cycles to cease. Thus, a woman's reproductive system ages prematurely in comparison with the rest of her body. This period of reproductive decline is called *menopause*.

Menopause is an unusual feature of human life history; it is not typical of other animals. Ectothermic vertebrate females (fishes, amphibians, and reptiles) generally have the ability to form new oocytes throughout their lifetime and their reproductive life span is not limited by availability of oocytes. Although other mammals have a limited oocyte supply at birth, they typically remain fertile throughout their adult lives, although sometimes at a gradually diminishing rate. Thus, most other animals continue to reproduce as long as they live. Only human females (and perhaps a couple of whale species) live a significant portion of their lives postreproductively. Not only do women differ in this way from other animals, their reproductive aging contrasts with that of human males. There is some overall decline in reproductive function in certain older men (see discussion later), but men have been known to father children well into their nineties!

Timing of Menopause

Menopause is the permanent end of menstrual cycling associated with the loss of ovarian follicular activity. A woman is diagnosed in retrospect as having entered menopause after she has not had a menstrual cycle for a full year. In the United States, the average age of menopause is 52, with a normal range of 45–55. Women who smoke cigarettes enter menopause almost 2 years earlier than nonsmokers. Evidence suggests that the number of ovulatory cycles experienced by a woman over her reproductive lifetime influences the timing of menopause. Women with a greater number of total ovulations may enter menopause earlier. For example, individuals with short (and therefore more frequent) cycles enter menopause earlier than those with long cycles. Nulliparous women and those with few children tend to have earlier menopause than women who have had several full-term pregnancies. The use of oral contraceptives (which inhibit ovulation) tends to delay menopause. Thus, there seems to be an effect of the number of lifetime ovulations on the timing of menopause. Does each woman have a limited number of ovulations after which her ovaries are exhausted? If so, one would predict that girls who entered menarch early would also experience a relatively early menopause, but studies have not supported this hypothesis. Age at menopause is also unrelated to birth weight.

Female Reproductive Age

Because a woman's fertility changes throughout her lifetime, it is helpful for women to have information about their reproductive status to make informed choices regarding the timing of pregnancy and/or contraceptive use. To help provide this information to individual women, researchers have developed a staging method similar to the method used to determine the stages of puberty (Chapter 6). Two factors are taken into account when determining "reproductive age." One is the length and regularity of menstrual cycles. The second is the measurement of circulating levels of FSH. These stages are shown in Table 7-1.

Table 7-1 Stages of a Woman's Reproductive Life

Stage
O. Prereproductive
I. Early reproductive Menstrual cycles are variable to regular, FSH levels are normal
II. Peak reproductive Menstrual cycles are regular; FSH levels are normal
III. Late reproductive Menstrual cycles are regular; FSH levels are elevated
IV. Early menopausal transition Length of cycles varies by 7 days or more; FSH levels are elevated
V. Late menopausal transition Two or more cycles missed; FSH levels are elevated
VI. Early postmenopause Lasts for 4 years after menopause begins; FSH levels are elevated
VII. Late postmenopause Lasts until death; FSH levels remain elevated

There are some drawbacks to the use of this staging method. Note that there is no age range for a given reproductive stage. This is because there is a great deal of individual variation in the age at which women enter reproductive stages and also in the length of each stage. For example, if a woman in the late reproductive stage decides to have a baby, it is not possible to predict whether she will have 5 months or 5 years to become pregnant. Furthermore, women who have conditions that may alter their endocrine state (e.g., those who are extremely thin or obese, extremely athletic, have irregular cycles, have endometriosis or uterine fibroids, are heavy smokers, or who have had a hysterectomy) may be difficult to stage. Nevertheless, an awareness of the natural progression of the female reproductive stages may help women move more comfortably through their reproductive life span and beyond.

Chapter 7, Box 1: Egg Aging

In women, the chances of infertility increase with age. For example, infertility rates in American women are 4.1% at ages 20 to 24, 5.5% at ages 25 to 29, 9.4% at ages 30 to 34, and 19.7% at ages 35 to 39. One major theory about the cause of this increase in female infertility with age is that there is an increase in the incidence of chromosomal abnormalities in ovulated eggs as they get older. Remember from Chapter 5 that each ovary has about 500,000 oocytes in follicles at birth and that these primary oocytes are arrested in prophase of the first meiotic division for up to 50 years unless they die by atresia or are ovulated. As a result, most of the oocytes in the ovary age for many years or even decades, and apparently this aging can cause problems in the selected few that finally reinitiate meiosis and ovulate.

Problems in meiosis can produce embryos that have too many sets of chromosomes, having cells that are 3N or even 4N instead of 2N. This is called *polyploidy*, and all of these polyploid embryos are lost through spontaneous abortion. A more common chromosomal error is *aneuploidy*, in which both members of a homologous pair of chromosomes end up in one cell during meiosis (resulting in embryonic *trisomy*), with the other cell of the abnormal division having no chromosomes of that pair (resulting in *monosomy*). Perhaps the homologous sets of chromosomes are unable to separate as well during meiosis in older oocytes, thus leading to aneuploidy. Although most trisomic embryos are aborted spontaneously, trisomies of chromosomes 13, 18, and 21 are often born, but with severe developmental and mental disorders. Trisomies of sex chromosomes (XXY, XYY, XXX) are discussed in Chapter 5. Monosomic newborns with Turner's syndrome (XO) develop mental and developmental abnormalities (Chapter 5).

Because the incidence of chromosomal abnormalities in newborns increases with the mother's age, one theory has been that there are more oocytes ovulated with chromosomal abnormalities in older women, which lowers female fertility. Central to this hypothesis is that certain preembryos are rejected by the uterus before or during implantation through some as yet unknown mechanism. Even if these abnormal embryos implant and a pregnancy is established, many soon miscarry early in the first trimester. Most estimates suggest that chromosomal abnormalities account for about one-half of early miscarriages after a recognized pregnancy, and aneuploidy explains about 60% of these. The number of preembryos that have chromosomal abnormalities and fail to implant is high (by one estimate, about 31% of all conceptions).

Actually, a woman's fertility depends not on how old a woman is, but how close in years she is to menopause. Up to age 40, the main cause of infertility in women is not the failure to ovulate or the inability of the uterus to implant an egg, but instead the increasing loss of preembryos or early implanted

Continued on next page.

Chapter 7, Box 1 continued.

embryos, many of which have chromosomal abnormalities. Only after age 40 does female infertility increase rapidly due to a failing of ovarian function and of the ability to conceive. Thus, the decline in ovulated egg quality precedes the decline in ovulated egg quantity.

There is, however, another theory to explain the increase in aneuploid births in older women. This theory says that the higher incidence of spontaneous abortion and birth defects in older women is due to a decrease in the efficiency of the mechanism that causes spontaneous abortion of abnormal embryos. This theory assumes that a mother has a screening system, probably of uterine origin, that regulates offspring quality, and that this system falters with age. This relaxed screening of abnormal embryos then causes the observed increase in defective embryos in miscarriages and the increase in genetically abnormal fetuses carried to term in older mothers. The evolutionary explanation for this relaxation could be that, since an older woman is nearing menopause, it benefits her to relax the screen and to allow more successful implantation, taking the good embryos with the bad, since this is her final chance to pass on her genes. This theory is in contrast to the first theory because it predicts that older and younger women ovulate the same percentage of embryos with chromosomal abnormalities.

What about the sperm of older men? Aneuploidy and other chromosomal abnormalities do occur during meiosis in the testes, with increased frequency in older men. It is likely that a great majority of these abnormal sperm do not fertilize an egg. Some do, however, as evidenced by the increase in miscarriages associated with fertilization by sperm from older men. Male fertility decreases with age. At age 24, the probability of a man to produce a pregnancy within a year of frequent, unprotected coitus is about 92%. From then on, this probability decreases about 3% for every subsequent year, regardless of the age of the woman.

Maternal age (years)	**Down syndrome**	**Any**
15 to 24	1/1300	1/500
25 to 29	1/1100	1/385
35	1/350	1/178
40	1/100	1/63
45	1/25	1/18

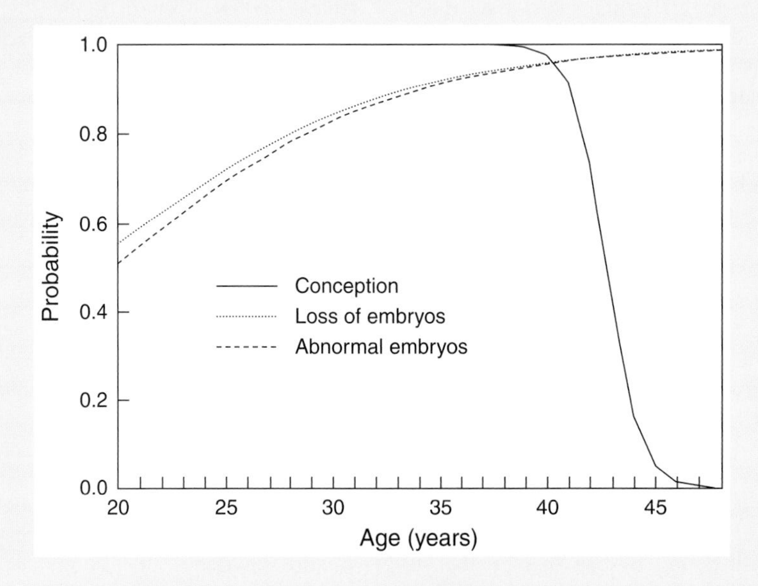

Relative probabilities of conception, loss of embryos (and preembryos), and abnormal embryos (and preembryos) at different female ages. Note that difficulties in conception do not explain the decline in fertility with age until after age 40.

Perimenopause

Although the average age of menopause for American women is 52, a woman can begin to experience premenopausal symptoms in her forties (average age, 47.5) or even late thirties. Symptoms can include oligomenorrhea, breast pain, hot flashes, and vaginal dryness. This stage preceding true menopause is called *perimenopause*; it lasts an average of 4 years. Women who are smokers tend to have an earlier and shorter perimenopause. Although a woman's periods can be irregular during perimenopause and there may be shorter cycles, she still is ovulating and should use contraceptives to avoid pregnancy during this time. In fact, women using the combination birth control pill (Chapter 14) likely will not experience as many perimenopausal symptoms.

Premature Menopause

About 1 in 100 women enter menopause unusually early, before age 40. This condition, called *premature ovarian failure*, most often has a genetic cause. Premature menopause can also be caused by the chemotherapy and radiation used to treat cancers or by autoimmune disorders in which the body destroys its own tissues. A woman who experiences premature ovarian failure will have the same menopausal symptoms as those who enter menopause at a later time.

Symptoms of Menopause

Before and after reaching menopause, at least 85% of women have symptoms, the occurrence and intensity of which vary among individuals. These are caused directly or indirectly by changes in circulating hormones, especially the drop in circulating estrogen. The symptoms may end within a year or they may last for several years. The most common symptom of menopause, experienced by over 75% of menopausal women in the United States, is the occurrence of *hot flashes* (or *hot flushes*). These are sudden, intense feelings of heat experienced in the chest and face, sometimes spreading over the body. They may be accompanied by profuse sweating. Hot flashes may occur at any time of the day or night, as often as every 10 min or less frequently, and they may last for 30 s to several minutes. For most women, hot flash symptoms last for more than a year, and many postmenopausal women continue to suffer hot flashes for 5 years or more. Some women experience hot flashes as only a minor nuisance; for other women, they cause major disturbances in sleep patterns and interfere with normal daily activities.

Other physical symptoms of menopause include slight shrinking of the external genitalia, breasts, and uterus. The vagina can decrease in size, and its internal lining becomes thinner and drier. For some women, loss of vaginal lubrication can lead to pain during sexual intercourse. An increased pH of vaginal fluids can make the vagina more prone to infection (*vaginitis*; see Chapter 18). Lower urinary tract disorders such as urinary tract infections and incontinence are common. Other physical changes, such as abdominal weight gain, deepening of the voice, and the appearance of hair on the chin and upper lip, may occur during and after menopause.

The phrase *female climacteric* is used to refer to all of the changes (psychological and emotional as well as hormonal and physical) of menopause. Menopause can be a period of psychological adjustment for some women. Emotional and behavioral symptoms that can occur include irritability, insomnia, and fatigue. Whether these symptoms are a direct result of hormonal changes is not clear. Many of the psychological responses to menopause may result from other changes at menopause. Hot flashes that can waken a woman repeatedly during the night may be a major cause of insomnia, fatigue, and mood swings. The psychological effects of a woman's loss of fertility, which may represent an important aspect of her feminine identity, can affect her emotional state negatively at this time. One positive note for most women is the end of the monthly menstrual flow and, for some, the PMS associated with menstrual cycling (see Chapter 3). A study of 2500 women reported that 70% had positive or at least neutral feelings about end of their periods. It must be emphasized that menopause is a normal phase of the human life cycle, and most women adjust to this change in a healthy and comfortable way. Later in this chapter we learn that older men can also suffer from a more gradual decline in sex hormones, with associated symptoms (andropause).

Endocrine Changes during Menopause

During perimenopause, the first change seen in a woman's endocrine status is a decline in the release of inhibin from her ovaries. Inhibin is a glycoprotein that selectively suppresses the release of FSH from the anterior pituitary. When the pituitary is released from this inhibition, the secretion of FSH rises. Higher levels of FSH can be detected in the blood of women in their forties who are still menstruating regularly and showing no overt signs of menopause (Fig. 7-1). During perimenopause, a woman usually continues to menstruate, although her cycles may become irregular. This may be because the higher levels of FSH increase the rate of follicular maturation, thus shortening the follicular phase and speeding up cycles. The cycle length, however, can be quite variable, with some cycles longer and others shorter than normal.

During perimenopause, a woman has an increasing number of cycles that are anovulatory. In fact, female fertility declines beginning at age 27 and continues to decline during perimenopause. In addition to having fewer eggs ovulating, the quality of a woman's eggs declines. By age 40, half of a woman's eggs are chromosomally abnormal, and 2 years later that figure rises to 90%. During this period of decreased fertility, it is common to have missed periods. However, ovulation could occur unpredictably, even after the last menstrual period.

After a woman's cycles have ended and she is in true menopause, the ovaries cease to respond to FSH and LH, possibly partly because their blood supply decreases. In addition, most of a woman's stock of ovarian germ cells is gone by this stage. The depletion of ovarian follicles results in a decrease in the production of estradiol to very low levels (Fig. 7-1). Because estradiol levels are so low, there is little if any negative feedback on gonadotropin secretion. Thus, in the absence of estrogen feedback, levels of FSH continue to rise. Secretion of LH also rises during menopause. In fact, the pituitary gland secretes 10 times as much FSH and 4 times as much LH as in younger women. When these

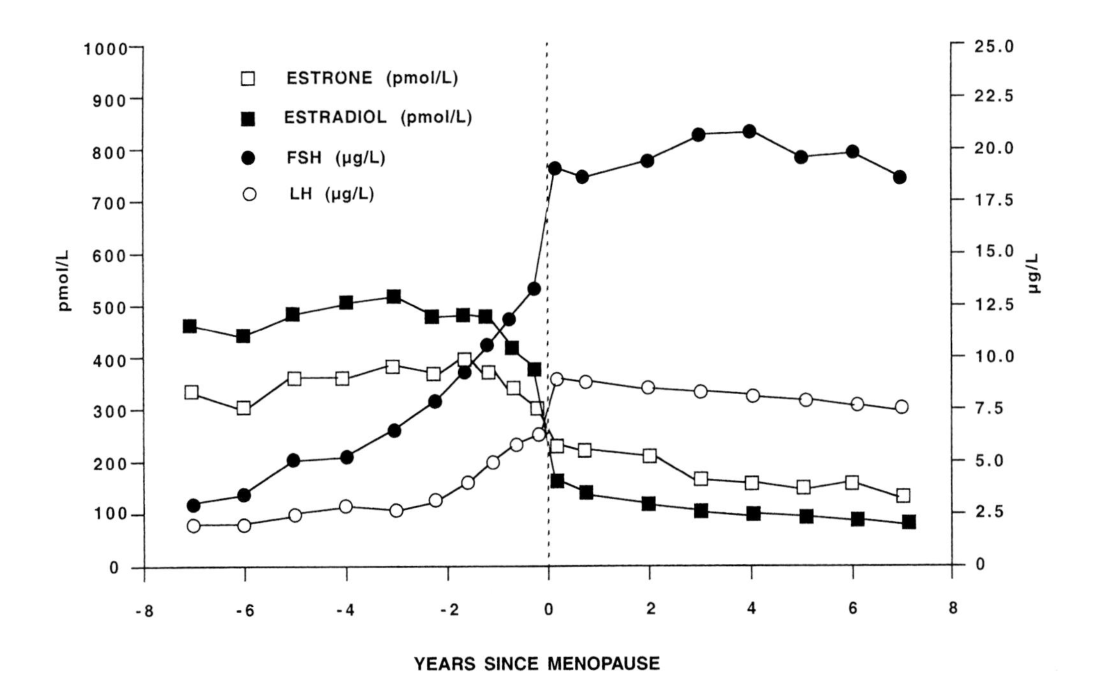

Figure 7-1 Circulating levels of FSH, LH, estradiol, and estrone in women before, during, and after menopause. Note the steep decline in estrogen. The increase in circulating gonadotropins, especially FSH, seen during perimenopause remains elevated for years after menopause.

gonadotropins are extracted from the urine of postmenopausal women, the extract is called *human menopausal gonadotropin*. GnRH levels in the hypothalamus are low, and some studies have demonstrated a decrease in GnRH pulse frequency.

Androgen levels fall gradually as women age, but this decrease is thought to be unrelated to menopause. However, the adrenal glands and ovaries of postmenopausal women continue to secrete androgens. These hormones, in the absence of estrogens, cause some menopausal symptoms such as voice deepening, enlargement of the clitoris, and appearance of facial hair.

Because of the dramatic drop in estrogen secretion during menopause, an average woman in the United States spends about one-third of her life without estrogens, which influence over 300 bodily functions! It had long been thought that hot flashes were caused directly by the abrupt lowering of estradiol levels, but now we know that a woman's sympathetic nervous system is more active after menopause because of low estrogen. This causes the dilation (widening) of blood vessels in the skin, which results in warm blood being carried to these regions. Sympathetic activation also causes sweating, as well as a 1 °C rise in body temperature and an increase in heart rate during the flashes. Hour-to-hour changes in the secretion of LH from the pituitary gland of postmenopausal women have also been associated with hot flashes.

We have described many changes in a woman's body that occur during menopause, but what actually causes menopause? Surprisingly, we are not sure. One theory suggests that a reduction in number of oocytes drives menopause.

Clearly, the depletion of ovarian follicles below some critical number decreases fertility, and the drop in estrogen levels resulting from follicle depletion changes the endocrine feedback on the hypothalamus and pituitary. Recently, however, scientists have focused on the neural control of menopause, suggesting that age-related changes in the central nervous system may secondarily affect the hypothalamus, pituitary, and ovaries.

Chance of Pregnancy

When menstrual cycles begin to decline as menopause approaches, and even after menstruation has ceased, a woman can still ovulate. Therefore, some method of birth control should be used for at least a year after menopause has occurred if possible pregnancy is to be avoided. Not only does a woman of this age usually not want a baby, but older women tend to have more miscarriages and stillbirths, and babies born to older women have an increased tendency to have birth defects.

However, women today live longer and in better health than in previous generations. Many women now delay childbearing because of their careers, education, and desire to pursue their interests before starting a family. Increasingly, women are becoming ready for childbearing at an older age, only to find that they are in perimenopause and that their fertility is decreasing rapidly. For example, after the age of 42, a woman has less than a 10% chance of becoming pregnant, according to the Centers for Disease Control. Thus, the timing of menopause is becoming a greater constraint on women. In the United States, one in five women between the ages of 40 and 44 is childless, and the rate of childlessness has doubled in the past 20 years. Of course, not all of these childless women desire to have a family. However, an understanding of the timing of the decline in female fertility (see also Chapter 16) will help women and their partners plan childbearing.

Is it possible to push the frontiers of female fertility into a woman's forties, fifties, and beyond? In 1994, a 63-year-old Italian woman gave birth to a baby using an egg donated by a younger woman (see Chapter 16). Since then, many other older women have used this method to become pregnant. In the future, it may be possible for women to prolong their own fertility after natural menopause. Studies show that, contrary to long-standing views, some female mammals appear to retain ovarian stem cells well into adulthood. If this holds true for humans, it may be possible to develop therapies that prolong fertility through the proliferation and maintenance of these stem cells. This approach may also lessen or delay menopausal symptoms.

Some have questioned whether an older woman has a "right" to raise a child. Others feel that this is an outdated argument. Indeed, many grandparents are now raising their grandchildren successfully. It has been shown that women in their fifties, after undergoing IVF treatment with donor eggs (see Chapter 16), are able to carry a pregnancy as successfully as younger women. Older parents tend to have the financial and emotional stability necessary for child-rearing, and women (and men) who have the advantage of good health care and living conditions can now expect to have more years of excellent health than past generations. However, as science and medicine stretch the boundaries of menopause, these ethical and practical arguments will continue.

Chapter 7, Box 2: The Evolution of Menopause in Humans

Human females are unique among mammals in that the ovaries stop functioning after about 50 years (menopause), which is often followed by a long postreproductive life. Only a very few other mammals undergo menopause (e.g., the short-finned pilot whale, the African elephant, and African chimpanzee). There are two general ideas about the presence of menopause in humans: menopause (1) is a deleterious effect of civilization or (2) is an adaptive trait in our species that evolved under the forces of natural selection.

The idea that menopause is a deleterious disease is inspired by the fact that the risk of some life-threatening disorders such as breast cancer, osteoporosis, and cardiovascular disease increases markedly after menopause, when the ovaries stop producing estrogen. This theory says that menopause is a disease of civilization resulting from the increase in life expectancy, which in turn is caused by better sanitation, nutrition, and health care in the modern world. Implied is that our hunter-gatherer female ancestors died on average much earlier than the 50 years it would have taken for their ovaries to run out of follicles. Studies of the bones of our hunter-gatherer ancestors have suggested that they did not live past their thirties, which supports the aforementioned theory. However, older skeletons do not preserve as well, and it is now believed from studies of present hunter-gatherers that a short *average* life span does not mean that no postmenopausal women were present. In fact, in present-day hunter-gatherers the life expectancy from birth averages in the thirties, but the average life span for women who have already reached age 45 is 50 to 60. And this is an average, so there are many women who live past their seventies. The lower average life expectancy from birth is because many children die before age 15 in present hunter-gatherer cultures so there probably were many women in our hunter-gatherer ancestors who were postmenopausal, and menopause is not simply an artifact of prolonging life span in modern societies.

Instead, menopause may have evolved as an adaptive trait in our species. What might be the adaptive advantage of menopause? First, it limits the ovulation of oocytes with chromosomal abnormalities, which increase with age in humans. This argument is challenged by the fact that most older mothers have genetically normal children despite the increased risk of chromosomal errors. Second, menopause ensures that mothers are young enough to withstand the stresses of pregnancy, birth, and childcare. One might argue that any chance to have additional offspring would increase a woman's evolutionary success. However, risking death during childbirth might jeopardize the well-being of her existing older children by abruptly ending her continuing care of them. The increasing possibility of unsuccessful pregnancies or nursing difficulties in older mothers would also drain important maternal resources. The trade-off may favor nurturing her existing children and foregoing additional pregnancies. A third adaptive advantage of menopause is related to this concept. The idea is that older females could contribute more of their genes to the population by investing their behavioral energy in kin, especially their grandchildren. This would have been their "best" evolutionary choice at an older age. This is called the *grandmother hypothesis* and is currently the most prevalent hypothesis about the evolution of human menopause. Older women who help their children care for their own offspring would allow their daughters (and perhaps daughters-in-law) to have more children. This would increase a grandmother's own evolutionary success. Grandmothers even today often care for their grandchildren and perform duties perhaps similar in general to our hunter-gatherer ancestors, e.g., gathering food, patrolling against danger, assigning tasks, nurturing (holding and rocking), assessing, coaching, care-taking, and managing. Our ancestral grandmothers may have helped their children, and therefore their grandchildren, by participating in foraging and hunting activities. In today's modern world, they can watch the children while their parents work. In this and other ways, grandmothers can assist in passing on their genes to future generations, the measure of evolutionary success.

Continued on next page.

Chapter 7, Box 2 continued.

What about older men? As you know, the reproductive system of older men does not suddenly cease to function as it does in women. Instead, andropause in men involves a slow, highly variable decline in sperm quality and quantity as well as plasma testosterone levels. Nevertheless, men of our hunter-gatherer ancestors lived as long as women, and there must have been an age period when they could "help their genes" by assisting in the care of their grandchildren through some of the same activities as older women instead of going on relatively unsuccessful hunts. In fact, this is so in present-day hunter-gatherer societies. Perhaps an additional important role of elders of both sexes was as a repository of information critical to the survival of the group, such as how to survive a sporadic climatic event that may occur only once a lifetime. So perhaps there should also be a "grandfather hypothesis."

Osteoporosis and Other Postmenopausal Disorders

During menopause and the postmenopausal state, calcium and phosphorus are lost from women's bones at a greater rate, and a condition called *osteoporosis* can develop. In this condition, the bones become weak and prone to fracture, and a woman may lose height because her vertebrae compress.

In the United States, about one in four women over age 50 develops osteoporosis. This means that today about 25 million American women have this disease. Osteoporosis results in about 1.3 million bone fractures, leading to 30,000 deaths each year. Blacks, Hispanics, and Asians tend to suffer less from osteoporosis. The primary cause of osteoporosis is a lowering of estrogen levels in a woman's blood. This can occur after natural menopause, after "induced" menopause (i.e., surgical removal of the ovaries for some other reason), or during amenorrhea. Factors that increase the risk of osteoporosis include low physical activity or chronic intense exercise, low calcium intake, smoking, the use of alcohol or caffeine, a fair complexion, being thin with small bones, never having been pregnant, early menopause (natural or induced), and a family history of osteoporosis. Obese women tend to have fewer problems with osteoporosis, perhaps because they have higher estrogen production from fat cells.

Older women are at increased risk for a variety of other diseases, which may be caused by hormonal changes at menopause or may simply be associated with the aging process. For example, women seldom have cardiovascular disease before menopause, but the risk increases greatly in postmenopausal women (possibly related to postmenopausal abdominal fat deposition), making heart disease the leading cause of death in U.S. women. The incidence of cancer, including cancers of the reproductive organs, increases after menopause. Another group of diseases found predominantly in postmenopausal women are various types of dementia, including Alzheimer's. Because of the association between the natural loss of hormones at menopause and the onset of these diseases, it was hoped that replacing some of a woman's lost hormones artificially would prevent the diseases. Indeed, early studies, many of which were observational, demonstrated a widespread benefit of hormone replacement. However, as we shall see, the earlier promise of hormones as a "fountain of youth" has been sobered by the results of more recent studies.

Treatments for Menopause: Benefits and Risks

Some women have mild menopausal symptoms and may not seek medical treatment. About half of women who do have moderate to severe symptoms before, during, or after menopause can take *hormone replacement therapy* (HRT) to alleviate the symptoms. An estrogen or a combination of estrogen plus progestogen is taken to "replace" the hormones released previously by the ovaries. Estrogens are delivered by pill form or through a skin patch. The estrogen may be synthesized from a plant source or obtained from the urine of pregnant horses. The latter, such as Premarin (a mixture of 6 estrogens derived from *pre*gnant *mar*e's ur*ine*) is administered most commonly as a pill. Women who are still cycling often take an estrogen pill daily, adding progestogen for 12–14 days a month, followed by progestogen withdrawal to allow menstruation to occur. A daily estrogen plus progestogen pill is given most commonly to women who have stopped cycling. Estrogen-containing vaginal creams are also used to decrease vaginal dryness and lessen painful coitus.

Hormone replacement therapy is very effective in alleviating the hot flashes, night sweats, and sleep disturbances of menopause. These symptoms are the main reason for prescribing HRT. In addition, the hormone(s) relieves urogenital discomfort (vaginal dryness and pain and frequent urination). Early HRT, especially if taken starting in the first 3 years after menopause, reduces the spine and hip fractures due to osteoporosis. However, one must continue to take estrogen for the benefits to persist; the bones begin to lose density when HRT is discontinued. Other nonhormonal treatments for osteoporosis are available. Combined estrogen plus progestogen (but not estrogen alone) appears to reduce the risk of colorectal cancer. It is not known whether the protective benefits of HRT on this type of cancer continue after a woman stops taking the hormones.

Earlier studies had indicated that estrogen lowered the incidence and mortality rate of cardiovascular disease in postmenopausal women. Based on this information, many women used HRT to protect against heart attacks and stroke, as well as osteoporosis, even if their menopausal symptoms were not distressing. However, in 2002, a large federally funded study examining the health effects of HRT on 16,000 postmenopausal women was halted abruptly when it found that women taking estrogen and progestogen were actually at higher risk of heart disease, strokes, and blood clots. Combined hormonal treatment increased the risk of heart disease by 29% and the risk of stroke by 41%. Thus, HRT should not be used to protect against cardiovascular disease. After the results of this study were made public, many women discontinued their use of HRT.

Earlier studies also reported that HRT increased short-term memory and the ability to learn new tasks and decreased the risk of developing Alzheimer's disease by 40%. However, a large controlled study of nearly 3000 women recently found that *estrogen replacement therapy* (ERT) in postmenopausal women who have had a hysterectomy not only failed to prevent memory loss, but actually increased the risk of dementia.

Taking HRT appears to increase the risk of developing breast cancer, increasing the risk about 2.3% for every year of use. A study of 122,000 American nurses indicated that those aged 55 to 59 who were taking ERT for at least 5 years had a 54% increased risk of breast cancer, and ERT at age 60 to 64 years increased the risk to 71%. In this study, there was no ERT effect on breast cancer if used less than 5 years. Another recent study also showed no effect of ERT on breast cancer.

Other studies have concluded that the risk of breast cancer from combined estrogen plus progestogen use is greater than that from the use of estrogen only. The good news is that, 5 years after having stopped HRT treatment, a woman's risk of breast cancer is no greater than that of a woman who never used HRT.

A strong correlation exists between ERT and the increased incidence of endometrial cancer. Estrogen alone for at least 1 year increases risk of this cancer 6 to 14 times, depending on estrogen dose and duration of treatment. It is not clear if this increased risk goes away after ERT is discontinued. Unlike for breast cancer, however, adding a progestogen to the estrogen decreases the estrogenic effect on endometrial cancer significantly. Therefore, it is recommended that women who have their uterus (i.e., those who have not had a hysterectomy) should use the combined HRT instead of estrogen alone (ERT).

One study of 240,000 American women showed that ERT given for at least 6 years increases the risk of fatal ovarian cancer by 40%, and this risk increased to 70% if ERT continued for 11 years. This was confirmed by another study that found a 60% greater risk of developing ovarian cancer in women taking estrogen-only hormone replacement therapy for 10 years or more. The effect of combined estrogen plus progestogen HRT on ovarian cancer is inconclusive.

Finally, ERT and combined HRT increase the risk of gallbladder disease, and ERT appears to cause the growth of benign uterine fibroids.

In summary, HRT has benefits for some women but can have dangerous side effects in others (Table 7-2). Certainly, a woman's individual and family history of cancer of the reproductive tract should be considered before she chooses HRT. One must also consider relative risks. HRT raises the risk of heart disease, which kills 250,000 American women each year, and also increases the risk of breast cancer, which kills 46,000 women each year. Thus, women who are at high risk for heart disease or who have a family history of breast cancer are usually discouraged from using HRT. However, estrogen maintains bone density, protecting against osteoporosis and the resulting hip and spinal fractures that can be debilitating for older women and kill 30,000 women per year. Some women may choose to take HRT for 1 or more years to alleviate menopausal symptoms and then discontinue its use because of associated health risks. Those entering menopause prematurely probably can safely take HRT longer. Whether to add a progestogen to the ERT is also a choice. Both estrogen and progestogen cause mitosis of mammary gland cells. In contrast, estrogen increases, but progestogen decreases, mitosis of endometrial cells. Because increased mitosis increases the chance of genetic error and cancer, progestogen supports the increased risk of breast cancer after ERT but reduces the risk of endometrial cancer after ERT. Progestogen may even enhance the

Table 7-2 Possible Benefits and Risks of Estrogen Replacement Therapy for Menopause

Benefits	Risks
Relieves hot flashes, night sweats, insomnia	Increased risk of endometrial cancer (unless a progestogen is used with estrogen)
Relieves vaginal and skin dryness, genital shrinking	Increased risk of breast cancer
Reduced osteoporosis (especially if a progestogen is also used)	Increased risk of heart disease
Reduced risk of colorectal cancer	Increased risk of gallbladder disease
	Increased risk of fatal ovarian cancer
	Increased risk of benign uterine fibroids

beneficial effects of ERT in relation to osteoporosis. However, the addition of progestogen increases the incidence of premenstrual syndrome, menstrual cramps, and unpredictable menstrual bleeding. Most women today take a progestogen as part of their HRT. Some recommend "natural" progesterone derived from a progesterone building block in plants such as the wild yam.

Often women choose other remedies besides ERT for menopausal symptoms. Factors that reduce the risk of osteoporosis include weight-bearing exercise as well as prescription drugs that inhibit bone breakdown, such as calcitonin, etidronate sodium, and alendronate. Calcium supplements appear to help, especially later on in postmenopausal years and when accompanied by weight-bearing exercise. The National Institutes of Health recommend 1000 to 1500 mg per day of calcium for postmenopausal women and 1000 mg per day for women aged 25 to 50. These calcium requirements can be obtained by diet (about 400 mg per day, excluding dairy products) or calcium supplements, but a caution is that greater than 2000 mg per day can cause kidney damage. Potassium biocarbonate taken orally has been shown to decrease bone loss. Natural foods also contain hormones that may alleviate menopausal symptoms. Many of these contain small amounts of phytoestrogens (Chapter 2); examples include soy products. Wild yam root also contains a precursor of estrogens (see Chapter 2) that has been proposed as an antiaging substance.

In summary, the Food and Drug Administration now recommends that women with moderate to severe menopausal symptoms use HRT at the lowest effective dose and for the shortest possible time. Results from the Women's Health Initiative dispelled the notion that HRT should be used broadly as an antiaging drug. HRT may be prescribed for women at high risk of osteoporosis, but hormone therapy should not be used to prevent heart disease or dementia.

Andropause

Testicular Function in Old Age

Some men undergo a phenomenon that is similar to menopause in women; it is called *andropause*. In women, menopause usually occurs over a period of a few months or years, but andropause usually occurs over decades, usually between the ages of 48 and 70. It may begin in a man's forties or later. One sign of andropause is that erections may take longer to achieve and it may require more active stimulation of foreplay to enable a couple to have sexual intercourse (see Chapter 8). These and other symptoms are at least partly a result of a decline in testosterone levels in some older men. Male sexual characteristics also may change: the voice may rise in pitch, facial hair growth may decrease, and the scrotum and penis may shrink. Sex accessory structures, such as the seminal vesicles and other glands, may become reduced in size. There is also a decline in muscle mass and strength and osteoporosis can begin. Because of these changes, some older men become depressed and irritable. Men over 70 who are sexually active have higher blood testosterone levels than those abstaining from sex, but it is not clear if the high levels in these men increased their sexual motivation or if their sexual activity increased their testosterone levels. An older man's sex life should not suffer if he accepts andropause in the same positive way that a woman is encouraged to accept menopause.

The reduction in sex hormones in men is gradual. It is estimated that male testosterone levels typically drop 1% per year after age 40. However, this relatively small decrease may underestimate the physiological effects of testosterone reduction during aging. Because lower levels of steroid-binding proteins also occur in the blood of older men, bioavailable testosterone might drop by more than 1% per year. One study found that 30% of men age 60–70 and 70% of men age 70–80 have low bioavailable testosterone. A primary cause of lowered testosterone appears to be a decrease in the ability of the testes to respond to pituitary gonadotropins. There is also a reduction in the number of Leydig cells in the testes that produce testosterone. In some older men, lower blood levels of testosterone cause a decrease in negative feedback of testosterone on LH secretion, resulting in higher LH levels. Blood levels of estrogens also rise in older men. The testes may be less functional because their blood supply is, for some unknown reason, reduced. The seminiferous tubules show damage in some older men, and semen volume, as well as sperm motility, decreases with age. However, some men continue to be fertile into very old age.

Treatment of older men with an androgen (*androgen replacement therapy*; ART), along with short-term psychological counseling, may reverse some of the symptoms associated with the reduction in testosterone secretion. Testosterone can be given as an injection, but the rapid increase in circulating levels of the hormone could have adverse side effects. Thus, testosterone is usually is given as a skin patch or a rub-on gel. Another androgen, DHEA (see Chapter 4), is also popular as a nonprescription remedy for andropause.

Studies indicate that ART can protect against osteoporosis in older men. One study showed that testosterone increases bone density in the lower spine by about 10% and in the hip by at least 2%. As an additional protection against osteoporosis in men, the National Institutes of Health recommend a calcium intake (through diet or calcium supplements) of 1000 mg daily for men aged 25 to 65 and 1000 to 1500 mg daily for men over 65, not to exceed 2000 mg/day. Weight-bearing exercise can assist these other therapies.

The use of ART in older men has been shown to increase muscle mass and to decrease body fat. In younger men with decreased testosterone due to testicular cancer or injury, ART can cause the secondary sex characteristics and sex accessory structures to return to their normal condition. In addition, sex drive and the ability to have an erection improve in these men. It is not yet clear if ART can significantly improve energy, sex drive, sexual function, or depression of older men. More research is needed to examine the effects of hormone replacement in older men; currently neither the potential benefits nor the potential risks are well understood. One of the concerns to be addressed is the possibility that ART could increase the risk of prostate cancer or exacerbate existing (possibly undiagnosed) prostate cancer.

Chapter Summary

Menopause occurs as menstrual cycles cease when a woman reaches 45 to 55 years of age. At this time, ovarian function declines, resulting in a syndrome of physical and possibly neurological changes. Menopause is preceded by perimenopause, a stage of gradual loss of ovarian function, which lasts about 4 years. Symptoms of menopause include hot flashes, sleep disturbances, weight gain, and loss of bone mineralization. Other changes in female physiology during this time may be caused by menopause or by the aging process. Endocrine changes during menopause include an early drop in inhibin, followed by increases in FSH and LH and a steep decline in estrogen; the loss of estrogen probably causes most of the menopausal symptoms. Estrogen replacement therapy relieves the symptoms of menopause but may increase the risk of breast, endometrial, and ovarian cancer, as well as cardiovascular disease. This treatment, however, lowers the risk of bone fractures in postmenopausal women. In early menopause, women have irregular cycles but the possibility of pregnancy exists.

Women differ from most other vertebrates in that they live for a significant portion of their lives postreproductively. The evolution of menopause in humans may be related to the role of grandmothers in human societies.

The testes of older men may decline in function, and the resultant hormonal changes can influence some men physiologically and psychologically, a process called andropause. Treatment with testosterone can reverse some of these symptoms, but more research must be done to understand the potential benefits and risks of androgen-replacement therapy.

Further Reading

Cohen, J. (2002). Sorting out chromosome errors. *Science* **296**, 2164–2166.

Enserink, M. (2002). The vanishing promises of hormone replacement. *Science* **297**, 325–326.

Forbes, L. S. (1997). The evolutionary biology of spontaneous abortion in humans. *Trends Ecol. Evol.* **12**, 446–450.

Hawkes, K. (2003). Grandmothers and the evolution of human longevity. *Am. J. Hum. Biol.* **15**, 380–400.

Johnson, J., *et al.* (2004). Germline stem cell and follicular renewal in the postnatal mammalian ovary. *Nature* **428**, 145–150.

Jordan, V. C. (1998). Designer estrogens. *Sci. Am.* **279**(4), 60–69.

Marlowe, F. (2000). The patriarch hypothesis: An alternative explanation of menopause. *Hum. Nature* **11**, 27–42.

Perls, T. T., et al. (1997). Middle-aged mothers live longer. *Nature* **389**, 133.

Simon, H. B. (1999). Longevity: The ultimate gender gap. *Sci. Am.* **10**(2), 106–112.

Tenover, J. S. (2003). Declining testicular function in aging men. *Int. J. Impot. Res.* (Suppl. 4), S3–S8.

Travis, J. (2000). Boning up: Turning on cells that build bone and turning off cells that destroy it. *Sci. News* **157**, 41–43.

Advanced Reading

Blurton Jones, N. G., et al. (2002). Antiguity of postreproductive life: Are there modern impacts on hunter-gatherer postreproductive life spans? *Am. J. Hum. Biol.* **14**, 184–205.

Caspari, R., and Lee, S. (2004). Older age becomes common late in evolution. *Proc. Natl. Acad. Sci. USA* **101**, 10895–10900.

De la Rochebrochard, E., and Thonneau, P. (2002). Paternal age and maternal age are risk factors for miscarriage: Results of a multicentre European study. *Hum Reprod.* **17**, 1649–1656.

Holman, D. J., and Wood, J. W. (2001). Pregnancy loss and fecundability in women. *In* "Reproductive Ecology and Human Evolution" (P. Ellison, ed.). Aldine de Gruyter, New York.

Lacey, J. V., *et al.* (2002). Menopausal hormone replacement therapy and risk of ovarian cancer. *J. Am. Med. Assoc.* **288**, 334–341.

Leidy, L. E. (1999). Menopause in evolutionary perspective. *In* "Evolutionary Medicine" (W. R. Trevathan *et al.*, eds.), pp. 407–428. Oxford Univ. Press, New York.

Orwell, E. S., and Klein, R. F. (1995). Osteoporosis in men. *Endocr. Rev.* **16**, 87–116.

Peccei, J. S. (2000). Genetic correlation between ages of menarche and menopause. *Hum. Nature* **11**, 43–63.

Rossouw, M. B. ChB., *et al.* (2002). Risks and benefits of estrogen plus progestin in healthy postmenopausal women: Principal results from the women's health initiative randomized controlled trial. *J. Am. Med. Assoc.* **288**, 321–333.

Sievert, L. L. (2001). Aging and reproductive senescence. *In* "Reproductive Ecology and Human Evolution" (Ellison, P. T., ed.), pp. 267–292. Aldine de Gruuyter, New York.

Sievert, L. L., *et al.* (2001). Marital status and age at natural menopause: Considering pheromonal influence. *Am J. Hum. Biol.* **13**, 479–485.

Swerdloff, R. S., *et al.* (1992). Effect of androgens on the brain and other organs during development and aging. *Psychoneuroendocrinology* **17**, 375–383.

PART THREE

Procreation

CHAPTER **EIGHT**

The Human Sexual Response

Introduction

Previous chapters described how the human egg and sperm are formed and become mature. These gametes are then brought together during the sexual response cycle, which involves not only the mechanics of coitus (sexual intercourse), but also sexual arousal and other behaviors surrounding sexuality. This chapter discusses the response cycle in men and women and shows how malfunction of the cycle can occur. We also discuss how hormones, pheromones, and therapeutic and nontherapeutic drugs can influence our sex drive and the sexual response cycle.

Sex Roles

Factors Influencing Sex Roles

Sex roles are a product of our biological nature, how we perceive this nature, and how we present our sexuality to others. First, our biological sex determines our anatomical and physiological *femaleness* or *maleness*. If our cells have two X chromosomes we develop ovaries, female sex accessory structures, and female secondary sexual characteristics (see Chapter 5). If our cells have one X and one Y chromosome, we develop testes, male sex accessory structures, and male secondary sexual characteristics. In addition to these anatomical and physiological sex differences, our biological maleness and femaleness also include sex differences in brain function that could influence our sexual behavior (see Chapter 17).

Gender Identity One's *gender identity* is the psychological belief or awareness that one is biologically female or male. It is the private experience of one's biological sex. *Sex role* (or *gender role*) is the outward expression of gender identity. It is the way we present our gender identity publicly through our behavior (sexual or otherwise), speech, dress, and so on. We are deemed *feminine* if our sex role fits what society defines as the female role and *masculine* if the form of our sex role fits society's definition of what is male. The term *androgyny* refers to one having both masculine and feminine characteristics. Our sex role is influenced greatly by our culture, economic status, social life, the nature of our home and family, and religious beliefs as well as biology.

Nature or Nurture Controversy has existed over whether our sex role is mainly a product of our biology (our "nature") or of our learning and experience (our "nurture"). Are we feminine or masculine because of our biological makeup or because of our rearing and present environment? In the past, most favored the theory that nurture plays a more important role. Anthropological studies had shown that sex role is highly influenced by social training and culture.

A modern theory of gender identity and sex role describes a continuous interaction of biological and social influences from conception. This theory holds that chromosomal and hormonal makeup determines whether the genitals (and perhaps the brain) develop in a male or female direction. The "genital dimorphism stage," a biologically determined event (see Chapter 5), is followed by the increasing influence of psychological and social input. How individuals perceive their bodies (i.e., whether they have male or female sex structures) leads to a body image of maleness or femaleness (gender identity). Generally, this awareness of one's maleness or femaleness begins around 18 months of age. The expression of this gender identity through sex roles develops later and is influenced greatly by familial and social factors. That is, one's sexual behavior, appearance, and even the way one thinks are a product of training.

Children tend to imitate the parent of the same sex, and parents and peers behave differently toward boys and girls, reinforcing or punishing certain behaviors. Although there are sex differences in behavior at birth (Chapter 17), any role of these differences in the development of sex role is conjectural at this time. Thus, sex role is the result of some undefined interaction between biological sex, environmental input, and a person's awareness of self and his or her relationship to others. This theory is still undergoing revision, and much remains to be learned about this fascinating and important subject. For example, *transsexuals* (people with a gender identity opposite from their biological sex) have recently been shown to have a part of the brain typical of the opposite sex (see Chapter 17).

Sexual Arousal

One is sexually aroused when environmental factors or internal thoughts initiate a sexual response. *Erotic stimuli* are those factors in the environment that are sexually arousing to an individual. A detailed survey of the kinds of erotic stimuli in our species would be impossible because what is erotic differs greatly among individuals and is highly subject to past experience. What excites one person may not be sexually arousing to another.

Cultural Influence and Individual Variation

There is certainly a cultural influence on what is perceived as erotic. In the United States, the stereotypes of the sexually attractive woman and man are perpetuated by the media, advertisements, and movies. But what may be true in our culture may not be true in other cultures. For example, many American men are sexually aroused by the sight of female breasts, but in cultures where the women do not commonly wear clothing on their upper bodies, this may not be so. Also, in our culture, being thin is considered sexier than being fat, but this is not true in some other cultures. Nevertheless, there also appear to be universal (cross-cultural) common patterns of what is sexually attractive in each sex. Within a given culture, there can be great individual variation in what is perceived as erotic. One woman may be aroused by a man's open shirt and hairy chest, whereas another may especially like men's forearms. One man may be excited by the sight of a woman's ankles, whereas another is stimulated by long hair.

Erotic Stimuli

Sound can be erotic; soft music can set the scene for sexual interaction, as can the rhythmic beat of hard rock. The taste of certain food or drink can be associated with past sexual encounters and can be sexually arousing. Visual images are erotic to many. Although humans are not considered to rely on smell as much as other mammals, smells can be associated with past sexual encounters and can be arousing, as evidenced by the commercial sales of scents, perfumes, and colognes. Also, certain chemical signals exuded by our bodies may play a role in our sexual biology (see HIGHLIGHT box 8-1).

Erogenous Zones

Erotic stimuli can be perceived by all of our senses: vision, hearing, smell, touch, and even taste. Touch (or tactile) stimuli are important for sexual arousal in both sexes. The body is particularly sensitive sexually in certain regions; these are known as the *erogenous zones*. In males, examples are the glans, corona, and lower side of the penis. In females, examples include the clitoris, mons, labia minora, and lower third of the vagina. The upper two thirds of the vaginal wall are relatively insensitive to touch. Erogenous zones in both sexes include the nipples, lips, tongue, ear lobes, anus, buttocks, inner thighs, and even the back of the knees, soles of the feet and center of the back. There is, of course, individual variation in the sensitivity of these areas.

Proceptive Behavior

Proceptive behavior is the scientific term describing courtship, flirting, seduction, and even foreplay in humans. Although human proceptive behavior is influenced by culture and tradition, it also has some features that appear to be universal in all human cultures and therefore could have evolved as patterns in our ancestors (just as may also be true for human sexual attraction and mating strategies). These cross-cultural features include eye contact, a slight smile, talking about superficial things with a high degree of animation, rotating the face toward each other, moving closer together, moistening the lips, and perhaps revealing or emphasizing parts of the body. In addition, a courting couple can "accidentally" brush together or touch, as well as mirror and synchronize with the other's postures and movements.

Kissing the lips can, of course, merely be a sign of affection and not have sexual overtones. Kissing, however, is a common form of sexual arousal in the United States and other countries. This can be done simply by a couple pressing their lips together with gentle movement. In "French kissing," the tongue of one or both partners is inserted and moved within the other's oral cavity. In other cultures, kissing is not a usual form of sexual arousal. People in Japan, China, and Polynesia, for example, did not kiss until they had contact with Westerners. In some cultures, such as the Thonga of Africa, kissing the lips is looked upon as repulsive.

Mild pain also can be an erotic stimulus. Gentle nibbling, biting, pinching, and scratching can be sexually arousing for some couples. In some cultures, minor pain often is associated with arousing interactions. For example, the Apinaye females of South America bite the partner's eyebrow as part of the sexual encounter, and the Trukese women of the South Pacific poke a finger in the ear of their partner. In our culture, forms of sexual interaction that cause intense pain are considered deviant.

The partners in a sexual interplay are responsible for giving effective erotic stimuli to each other in the process of *foreplay*, or "petting." To be effective, foreplay requires open expression of emotions, communication, and consideration of place, pace, and style. During foreplay, several means of arousing one's partner can be utilized, some more common than others. For many young people, especially those wishing to avoid coitus, foreplay is the "only play." For many other couples, however, foreplay is a prelude to coitus. In one study, foreplay of married couples for 1 to 10 min led to 40% of the wives having orgasm during coitus. This percentage rose to 50% with 12 to 20 min of foreplay and to 60% with greater than 20 min of foreplay. It is, of course, not only the time spent in foreplay that is important. It is the quality of the stimuli, including a good dose of love and affection, that is especially satisfying and arousing.

The Sexual Response Cycle

In the not too distant past, the human sexual response was little understood and was not the subject of much scientific observation. Alfred Kinsey had opened the door to the scientific study of human sexuality, but it was not until the studies of William Masters and Virginia Johnson that we began to know more about the physiology of the human sexual response. Although their methods have been criticized by some, their research has had a profound effect on the understanding of ourselves, on research in human sexuality, and on the treatment of sexual dysfunction. Most of what is said here about the sexual response cycle will, therefore, be based on the studies of Masters and Johnson.

In both sexes, the *sexual response cycle* can be divided into four phases: *excitement, plateau, orgasmic*, and *resolution*. Figure 8-1 depicts these phases in the female and male. Although these are described as distinct phases, you should realize that they flow into one another as a continuous cycle if effective erotic stimuli are present. If these stimuli are not adequate, however, the initial phases are not followed by the final phases. Our discussion focuses on the sexual response cycle during heterosexual coitus; however, the full cycle in either sex can occur during masturbation or homosexual sex as well.

The Female Sexual Response Cycle

The female sexual response cycle is similar to that of the male in that it is divided into four phases. As we see, however, there are some important differences between male and female cycles.

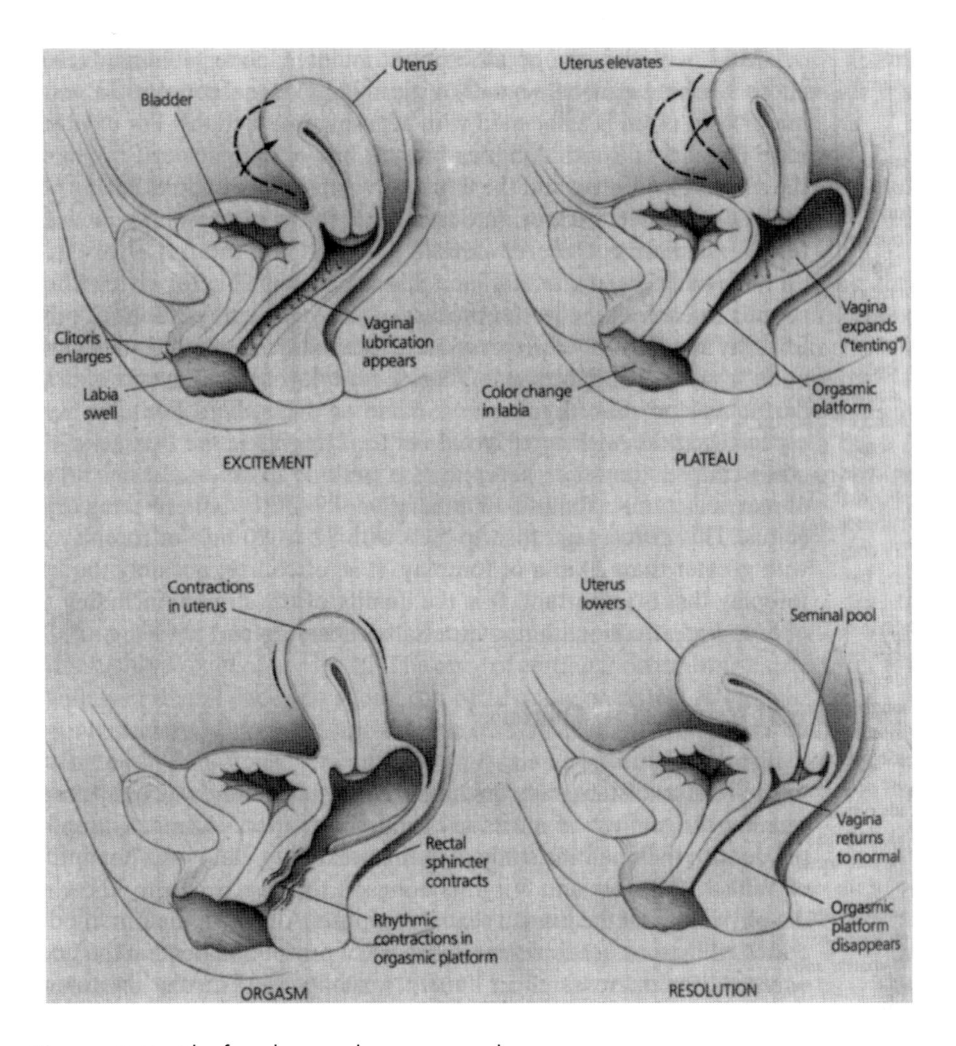

Figure 8-1 The female sexual response cycle.

Excitement Phase The female excitement phase (Fig. 8-1) is initiated by the presence of effective erotic stimuli. The first change, usually occurring within 10 to 30 s, is *vaginal lubrication*, i.e., the membrane lining the vagina becomes more moist. It used to be thought that this was caused by secretions from the Bartholin's glands, but the work of Masters and Johnson showed this not to be the case. Instead, the fluid leaks out of blood vessels present in the vaginal wall. Other responses occurring during the female excitement phase include the following.

1. The inner two-thirds of the vaginal barrel begin to increase in length and width. Thus, the vaginal cavity, which is closed at rest, begins to widen.
2. The body of the uterus ascends (the *tenting effect*), pulling the cervix away from the vagina and thus further increasing vaginal length. There can also be rapid, irregular uterine contractions (fibrillation). These uterine contractions are not painful. The size of the uterus also increases due to *vasocongestion* (pooling of blood in blood vessels).

3. The walls of the vagina become engorged with blood and become darker in color.
4. The shaft of the clitoris increases in diameter (but rarely in length) and there may be a slight *tumescence* (swelling) of the clitoral glans due to vasocongestion.
5. The labia minora become engorged with blood and their size increases considerably.
6. The labia majora, which at rest lie over the vestibule, flatten out and retract from the midline.
7. The nipples become erect, the areola becomes wide and darker, and the size of the breasts increases about 25% due to fluid accumulation.
8. A *sex flush* begins to appear in about 74% of women, i.e., areas of skin become reddened due to dilation of blood vessels. It usually begins on the abdomen and throat and then spreads to the chest, face, and even the shoulders, arms, and thighs.
9. There is an overall increase in tension in voluntary and involuntary muscles (*myotonia*).

Plateau Phase During the female plateau phase (Fig. 8-1), the following changes occur if effective erotic stimuli are present.

1. The wall of the outer one-third of the vagina becomes greatly engorged with blood so that the vaginal cavity is reduced from that in the excitement phase. Also, the labia minora become more engorged with blood and thus become redder and larger. These changes in the outer third of the vagina and the labia minora are called the *orgasmic platform* because they indicate that orgasm is imminent.

2. The clitoris retracts to be completely covered by the clitoral hood, and its length decreases by about 50%. Thus, from this stage on, the clitoris can be stimulated directly only through the hood or mons and stimulated indirectly by tension applied to the labia minora.

3. Uterine fibrillation continues and may increase in intensity. The uterus also elevates further.

4. The nipples become more erect and the areola darker; the breasts reach their maximal size.

5. The sex flush, if present, spreads and becomes more intense.

6. Heart rate, blood pressure, and the depth and rate of breathing increase.

7. There is a further increase in muscular tension.

Orgasmic Phase As discussed later in this chapter, not all women have an orgasm all the time. The female orgasmic phase, if stimulated by coitus, usually occurs 10 to 20 min after *intromission* (penetration of the penis into the vagina). The word *orgasm* ("climax") comes from the Greek word *orgasmos*, which means "to swell" or "be lustful." An orgasm in either sex can be one of the most intense and pleasurable of human experiences. We talk more about the experience of orgasm later, but now let us look at some physiological changes that occur (Fig. 8-1).

1. Strong muscular contractions occur in the outer one-third of the vaginal wall. The first contraction lasts about 2 to 4 s and is followed by rhythmic contractions at intervals of 0.8 s, the same frequency as the muscular contractions during male ejaculation. There can be 3 to 15 of these contractions, and the intensity of the initial ones is greater than that of later ones. The rectal sphincter can also exhibit rhythmic contractions at 0.8-s intervals.

2. The inner two-thirds of the vagina often expand, which facilitates movement of the penis within it.

3. Rhythmic contractions of the uterus occur, probably brought about by release of the hormone oxytocin.

4. The sex flush, if present, peaks in intensity and distribution.

5. The heart rate, blood pressure, and depth and rate of breathing peak at rates similar to those during male orgasm (see later).

6. There may be strong involuntary muscle contractions and clutching or clawing motions of the hands and feet.

There also is a release of neuromuscular tension. The conditions of the labia minora, labia majora, clitoris, and breasts remain similar to those in the plateau phase.

One major difference between the female and the male sexual response cycle (see Fig. 8-2) is that the female does not have a refractory period right after orgasm, which the male has, as discussed later. Kinsey first reported that only about 14% of women in his study had multiple orgasms if effective stimuli were present, but probably more women are physically capable of this experience. Such women report

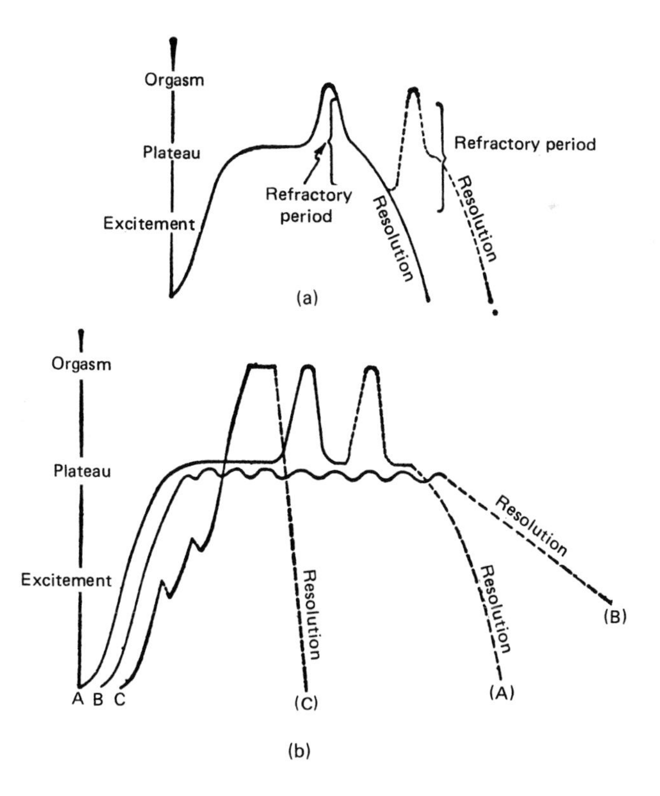

Figure 8-2 The sexual response cycle in men (a) and women (b). Curves represent degrees of sexual arousal. The three different female cycle patterns in (b) are discussed in the text.

that later orgasms in the sequence are more intense than the initial one. A few women can have *status orgasmus*, which is a sustained orgasm lasting up to a minute.

The experience of orgasm can vary in one woman and among different women. S. Hite reported that this experience often occurs in three stages. First, women experience a sensation of "suspension," lasting only an instant, followed by a feeling of intense sensual awareness, oriented at the clitoris and radiating upward into the pelvis. In the second stage, there is a sensation of warmth, beginning in the pelvis and spreading to other parts of the body. Finally, there is pelvic throbbing, focusing in the vagina and lower pelvis. Other experiences, varying from one woman to another, include mild twitching of the extremities, body rigidity, facial grimacing, and uttering of groans, screams, laughter, or crying. Some women even report that they lose consciousness briefly. Surveys of the experiences of orgasm in women and men suggest that the pleasurable feelings of both sexes during orgasm are similar. In fact, one study of positron emission tomography scans of the brains of men and women during orgasm showed nearly identical activity in brain areas associated with pleasure, which was similar to that of a person under the influence of heroin! Orgasm in both sexes is stimulated by stretching of the pelvic muscles (due to vasocongestion) and by stimulation of the clitoris and vagina or penis.

There may be several kinds of orgasm in women. One kind, the *clitoral orgasm*, was described by Masters and Johnson. This type results from stimulation of the clitoris during coitus or masturbation. *Vaginal orgasms*, however, are thought to be the result of direct stimulation of the vaginal wall. There may be a small region in the front wall of the vagina that, when stimulated, can produce sexual arousal and orgasm. This region is termed the *Grafenberg spot*. The orgasm that results from stimulation of this region involves intense contraction of the uterus and pubococcygeus muscle and has been called an *A-frame orgasm* or *uterine orgasm*. In reality, most orgasms probably involve a blend of the aforementioned kinds. A woman certainly should not suffer from "performance anxiety" if she does or does not have a particular kind of orgasm.

Does *female ejaculation* occur during orgasm? Recent studies indicate that about 10% of women expel a small amount of fluid into the vestibule during orgasm. This fluid actually comes from the lesser vestibular (Skene's) glands near the urethral opening. These glands are homologous to the prostate gland of the male (see Chapter 5).

It is not true that the size of a man's penis bears a direct relationship to sexual satisfaction in the female. This is because the vagina adapts in size. However, extremely small penises may not provide sufficient stimulation, whereas extremely large penises may cause discomfort. It should also be mentioned that there is no benefit to simultaneous orgasm in a man and woman unless this is an achievable and pleasurable goal of a couple's sex life. In fact, if a woman enjoys multiple orgasms, it may be necessary that the man delay his orgasm.

It is sometimes the case that the vagina becomes so relaxed that it leads to less sexual stimulation during coitus. This is a common complaint of women who have had several children. In this case, the couple can try new coital positions, and the woman can exercise the pubococcygeus muscle (see Chapter 2) to strengthen the vaginal wall.

Resolution Phase After orgasm, and if there are no effective erotic stimuli present, the woman's system returns to normal during the resolution phase

(Fig. 8-1). Some symptoms return to normal rapidly. In less than 10 s after orgasm, vaginal contractions cease and the clitoris leaves its retracted position. The heart rate, blood pressure, and respiration decline quickly to resting levels. Also, the labia minora return to their original color, usually within 2 min. The internal cervical os dilates immediately after orgasm, perhaps to allow sperm to move into the uterus (see Chapter 9). Muscle tension decreases in about 5 min, and the breasts decrease in size in 5 to 10 min. Vasocongestion in the clitoris, vagina, and labia minora ebbs in 5 to 10 min, and the uterus usually returns to its normal size and position by this time. The labia majora return to their resting condition in about an hour. About one-third of women sweat profusely after orgasm, and many have an intense desire to sleep.

Individual Variation The experience of orgasm differs among women and also at different times in one woman. Figure 8-2 shows three variations in the female sexual response cycle. In pattern A, a woman goes through a complete cycle, including multiple orgasm. In pattern B, a woman reaches a plateau and then goes into resolution without reaching orgasm. This pattern often occurs in inexperienced women or if inadequate stimuli are present. In pattern C, stimuli produce an early, intense orgasm.

There is considerable variation in the ability to have an orgasm. About one in three women seldom or never reach an orgasm during coitus. Although many of these women are able to experience orgasm using other kinds of sexual stimulation, one in five women report rarely or never achieving climax during masturbation. Recent twin studies show that the ability of women to achieve orgasm is partly determined genetically.

The Male Sexual Response Cycle

The sexual response cycle is quite similar in all males, with individual men differing in the duration more than the intensity of each phase. Also, the physiological changes in the different phases of the cycle are similar, regardless of the nature of the stimuli present and regardless of whether the cycle is initiated by masturbation or by heterosexual or homosexual behavior.

Excitement Phase The excitement phase of the male sexual response cycle (Fig. 8-3) can be initiated by any effective erotic stimulus. The first thing that happens is that an *erection* begins. The penis stiffens and increases in length and diameter. Thus, the penis is said to become tumescent. It should be noted that an erection can also occur without erotic stimuli being present. For example, it is very common for men to gain an erection about every 30 to 90 min at night, when rapid eye movement sleep occurs. Also, many times a man can wake up in the morning with an erection. The reason for this "morning erection" is not known, but is probably not caused by a full bladder. Spontaneous nonsexual erections can also occur if the urinary bladder or prostate gland is infected or inflamed, or they can occur in pubescent males (see Chapter 6).

An erection involves a basic biological phenomenon that, as mentioned earlier, also occurs in the female sexual response cycle. This phenomenon, vasocongestion, occurs when the flow of blood into a tissue in the arterial vessels is greater than the amount of blood that leaves the tissue in the venous drainage.

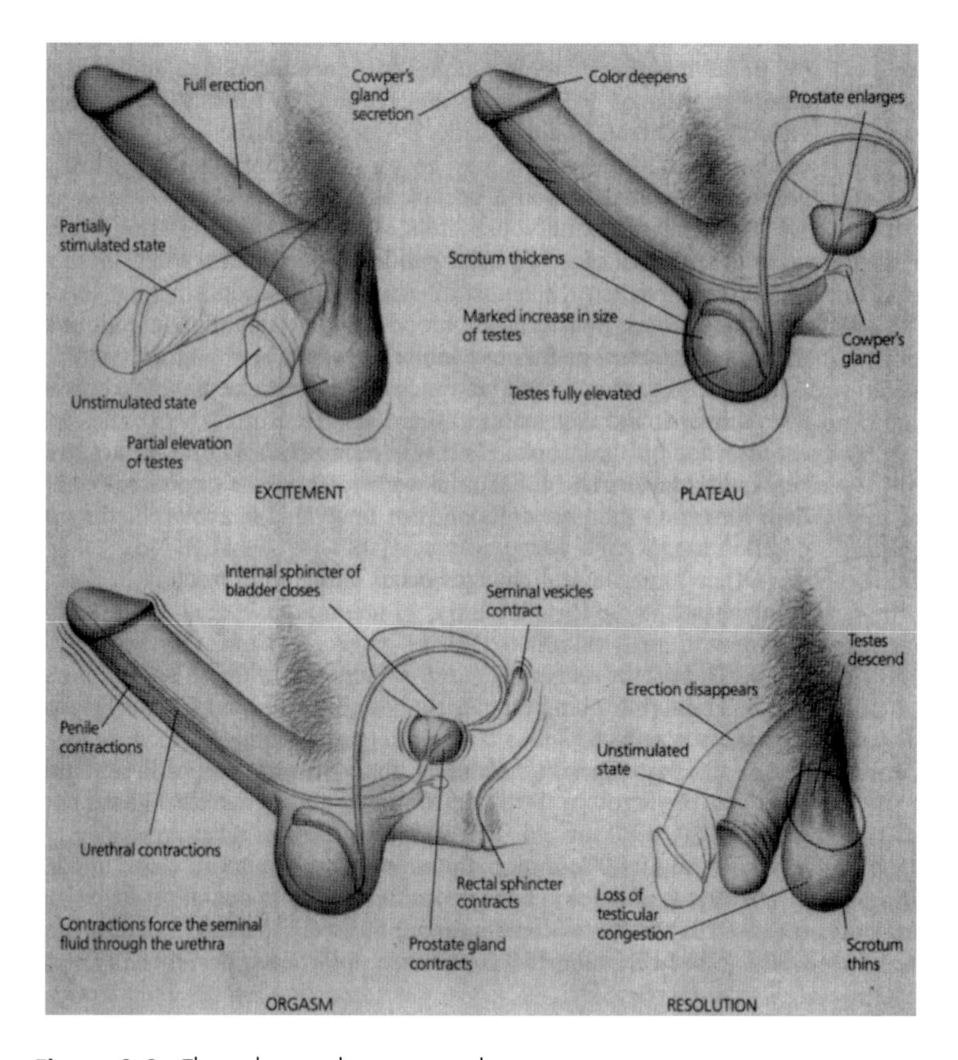

Figure 8-3 The male sexual response cycle.

This results in pooling and engorgement of blood in the tissue. Erotic stimuli initiate nerve impulses that travel directly to the spinal cord or to the brain and then to the spinal cord. This initiates an *erection reflex* by activating an *erection center* (in the lower end of the spinal cord) that contains neurons that control erection. These neurons send their axons to the blood vessels (arterioles) that supply the erectile tissue in the penis. Erotic stimuli cause the parasympathetic nerves of the erection center to dominate, and these neurons release acetylcholine that causes the arterioles to dilate. This results in vasocongestion in the blood vessels contained in the corpora cavernosa and corpus spongiosum of the penis, and the engorgement of blood in these spongy tissues causes penile tumescence. Recent studies show that the neurotransmitter vasoactive intestinal peptide (VIP) is released by parasympathetic nerves along with acetylcholine and that intravenous administration of VIP induces an erection. In turn, VIP may have its effect via other molecules such as cyclic GMP and nitric oxide.

Depending on the intensity and effectiveness of the stimuli, an erection may be gained partially and then lost repeatedly before a maximal response occurs.

The sympathetic nervous system, which is dominant during stress, also innervates the smooth muscle of the penile blood vessels. When these neurons are active, they release the neurotransmitter norepinephrine, which contracts the penile arterioles, thus reducing blood flow and inhibiting erection. This may be the way that stress or fear can inhibit erection.

The ability to maintain an erection without ejaculating seems to vary with age. Kinsey found that males in their late teens or early twenties could hold an erection for up to an hour. This was reduced to 30 min in men from 45 to 50 years old. Masters and Johnson, however, found the opposite, i.e., older men take longer to gain an erection, but once this is achieved, they maintain an erection longer than younger men.

Other physiological changes occur along with erection in the male excitement phase.

1. The urethral opening (urethral orifice) widens.
2. The scrotal skin becomes congested and thickened, and thus the scrotal diameter is reduced.
3. The testes become elevated due to contraction of the cremaster muscle in the scrotum. Stroking the inner thighs can also cause contraction of this scrotal muscle. This is the *cremasteric reflex*.
4. In about 60% of men, the nipples become more erect.
5. Areas of the skin become reddened due to dilatation of the blood vessels. This sex flush occurs in about 50 to 60% of men.
6. The heart rate, blood pressure, and depth and rate of breathing begin to increase.
7. There is an increase in tension of voluntary and involuntary muscles (myotonia).

Plateau Phase The next phase of the male sexual response cycle, given the continued presence of erotic stimuli, is the plateau phase (Fig. 8-3). In this phase, an erection continues and the following changes occur.

1. There is a slight increase in the size of the glans of the penis, and its color deepens. The coronal ridge (corona glandis) also tends to swell.
2. The *urethral bulb* (enlarged end of the urethra in males) enlarges to three times its normal size.
3. There may be preorgasmic emission, from the Cowper's glands, of a few drops of semen. Although slight in volume, this first stage of ejaculation could contain some sperm (see Chapter 9).
4. The testes become more elevated, rotate slightly, and come to lie closer to the groin. Also, the volume of the testes increases by about 50% due to accumulation of fluid.
5. The prostate gland enlarges.
6. The sex flush, if present, spreads and increases in intensity.
7. There is a further increase in heart rate, blood pressure, and the depth and rate of breathing.
8. There is more tension of voluntary and involuntary muscles.

Orgasmic Phase The male now enters the orgasmic phase (Fig. 8-3).

1. There is a loss of voluntary control of muscles and a release of neuromuscular tension. There may be clutching or clawing motions of the hands and feet.

2. *Ejaculation* is the expulsion of semen and is controlled by an *ejaculation reflex*. There is an *ejaculatory center* (or *spinal nucleus of the bulbocavernosus*) in the spinal cord, located higher up than the erection center. When activated, this center sends sympathetic neural stimulation to the bulbocavernosus muscle at the base of the penis. Ejaculation occurs in two phases. First there is a specific sequence of contraction of smooth muscle in the walls of the testes, epididymides, vas deferens, ejaculatory duct, seminal vesicles, prostate gland, bulbourethral glands, and urethra. These contractions expel semen into the urethral bulb. Simultaneously, a muscular sphincter that guards the opening of the urethra into the urinary bladder contracts, thus preventing urine from entering the urethra and semen from entering the bladder. This series of events constitutes the *emission stage of ejaculation* and is experienced by a male as a sensation of imminent ejaculation. These contractions may be influenced by the hormone oxytocin and by the presence of prostaglandins in the seminal fluid. The *expulsion stage of ejaculation* begins next, with rhythmic contractions of the penis and *bulbocavernosus muscle*, which lies at the base of the penis. The first three or four of these contractions are intense and result in a forceful expulsion of the majority of the semen from the urethra. The contractions that follow are less intense and produce smaller spurts of semen. These expulsion contractions are approximately 0.8 s apart.

3. The testes are at their maximal elevation.

4. The heart rate peaks as high as 180 beats per minute (from a resting rate of about 70). The blood pressure peaks at about 200 over 110 (from a resting pressure of about 130 over 70). The respiratory rate peaks at about 41 breaths a minute (from a resting rate of about 12 per minute).

5. The sex flush, if present, peaks in intensity and distribution.

For most men, ejaculation is an essential component of the experience of orgasm. Orgasm, however, can occur without ejaculation. For example, during retrograde ejaculation (see Chapter 14), the emission stage occurs but not the expulsion stage, resulting in a "dry orgasm" when the semen enters the urinary bladder. This can be due to physical damage to the urethra or to a relaxed urinary sphincter muscle. Also, coitus reservatus (see Chapter 14) has been practiced as a birth control method by some people, e.g., in India. In this method, men learn to approach ejaculation repeatedly with no expulsion.

Immediately after ejaculation, the male (unlike the female) enters a *refractory period* (Fig. 8-2). During this period, potentially erotic stimuli are not effective in causing or maintaining an erection until sexual tension decreases to near resting levels. This refractory period may last only a few minutes in a young man but may take more than an hour in an older man. Thus, a younger man may be able to have several orgasms, each separated by a few minutes. The amount of semen, however, is less in later ejaculations. The duration of the refractory period in young men is about 10 min but can be influenced by fatigue and amount of sexual stimulation. Kinsey found that 6 to 8% of the men he studied had more than one orgasm during one sexual encounter, and these men reported that the initial orgasm was the most pleasurable.

There is a misconception that many older men die of a heart attack during coitus. Actually, the risk of having a heart attack in a man within any 2 h is 10 in 1 million, and this risk is 20 in 1 million within 2 h of beginning sex.

Resolution Phase During and after the refractory period (and if no effective erotic stimuli are present), the male goes through the resolution phase, in which the arousal mechanisms return to a resting state (Fig. 8-3). In this phase, the erection is lost because the erection center is now dominated by the activity of sympathetic neurons. This causes the arterioles supplying the penile spongy tissue to constrict, thus reducing vasocongestion. About 50% of penis size is lost rapidly. Other responses that occur rapidly include disappearance of the muscle tension and sex flush and a lowering of heart rate, blood pressure, and respiratory rate (all usually in about 5 min). Other changes taking a longer time include final reduction in penis size, relaxation of the scrotum, descent of the testes, and loss of nipple erection. About one-third of men sweat over their body, and many experience an intense desire to sleep. The entire resolution phase can take up to 2 h. Close physical contact with the partner, such as keeping the penis within the vagina, touching, and caressing, can delay male resolution. A desire or attempt to urinate can speed up resolution.

Why Did Orgasm Evolve?

To humans, orgasm is an intensely pleasurable experience, but is it directly necessary for reproduction? The answer is no. As discussed in Chapter 9, female orgasm is not necessary for fertilization to occur, and some men can ejaculate (fertilize) without having an orgasm.

One theory about the evolution and adaptive value of orgasm is as follows. Most men experience orgasm when they ejaculate, whereas fewer than half of American women experience orgasm each time they have sex. A vast majority of women do not have orgasm unless they receive effective clitoral stimulation, and one idea is that only a man who is caring, knowledgeable, and sensitive can assist his partner in orgasm. The orgasmic response in the woman would then be a reward to the man; i.e., it would make sex more pleasurable for him. Thus, a pair bond based on caring, sensitivity, and pleasure is mediated at least partially by female orgasm. Female orgasm then may have evolved as a mechanism of mate choice, ensuring that a woman's long-term partner is sensitive to her needs (sexual and otherwise) and thus more likely to be a good provider for their offspring. Other theories of female orgasm include the idea that muscular contractions during orgasm may help draw sperm into the uterus. Finally, the sense of relaxation and sleepiness often produced by orgasm may promote sperm retention by the female reproductive tract.

Coitus (Sexual Intercourse)

Coitus (Latin *coitio*, meaning "a coming together") is, for many of us, a vehicle for the expression of emotion and intimacy. Strictly speaking, coitus (or sexual intercourse) is the penetration of the vagina by the penis, which can be called *vaginal coitus* (Fig. 8-4). However, the term *coitus* is also used for other forms

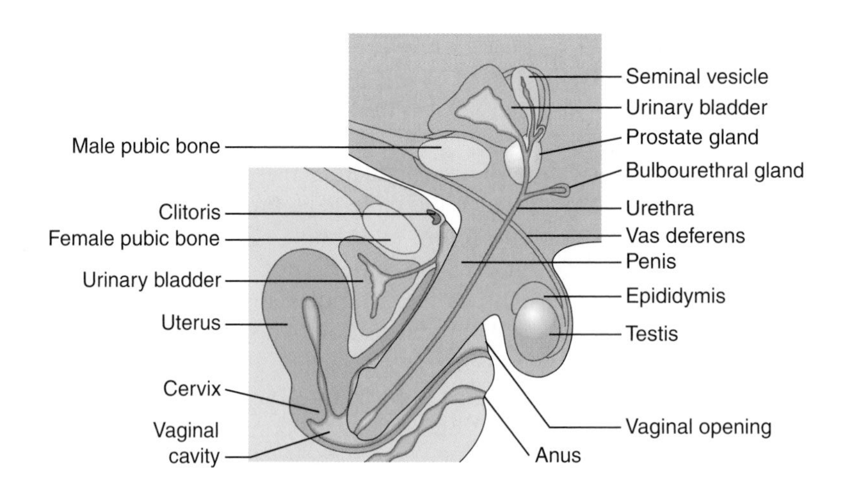

Figure 8-4 Representation of the erect penis inserted into the vagina during vaginal coitus.

of sexual contact, including *oral coitus* (oral–genital contact), *femoral coitus* (when the penis is inserted between the thighs), *mammary coitus* (when the penis is inserted between the breasts), and *anal coitus* (insertion of the penis into the rectum). There are many common slang phrases for coitus, such as "making love," "going to bed," and other more descriptive phrases. Legally, *fornication* is the voluntary coitus between an adult man and woman who are unmarried. *Adultery* is voluntary coitus between two people, at least one of whom is married to someone else. *Sodomy* means different things in different states; it usually refers to anal or oral coitus, but also can mean "acts against nature" such as coitus with an animal. Finally, *masturbation*, which is not a form of coitus, is the act of deriving sexual pleasure from self-stimulation of the genitals.

In anal coitus, the penis penetrates the anus and is moved within the rectum. This method of coitus is common in male homosexuals and in some heterosexual couples. A heterosexual couple should use a condom and never switch from anal to vaginal coitus before washing the penis, as the rectum contains microorganisms that could infect the female reproductive tract (see Chapter 18). The walls of the rectum are not as well lubricated as are those of the vagina, and the anal sphincter is constricted. Therefore, lubrication of the anus and penis with saliva or a sterile lubricant is common.

Oral coitus is contact of the mouth with the genital organs. When the mouth of the partner touches the genitals of a female, it is called *cunnilingus* (Latin *cunnus*, meaning "vulva"; *lingere*, meaning "to lick"). Cunnilingus is practiced in several cultures. One danger of this form of oral coitus is the possibility of air being blown into the vagina, as air bubbles could enter the bloodstream and could be dangerous. Therefore, air should not be blown into the vagina.

Fellatio (Latin *fellare*, meaning "to suck") is the oral manipulation of the penis or scrotum by a sexual partner. Some worry about the adverse effects of swallowing the semen, as it can contain microorganisms such as HIV (see Chapter 18). Obviously, a woman cannot get pregnant from this form of coitus.

Hormones and Sexual Behavior

In many animals, certain hormones need to be present for sexual behavior to be expressed. There are essentially two ways that a hormone can stimulate (or inhibit) sexual behavior. First, the hormone can act directly on the brain to affect sensitivity or activity of neurons that influence *sex drive*, or *libido*. This is called a *central effect of a hormone*. Second, a hormone can affect the sensitivity or growth of peripheral tissues, such as skin or muscle, that are involved in sexual behavior. This is called a *peripheral effect of a hormone*. In both instances, however, proper erotic stimuli are necessary for the behavior to be expressed, with the hormones simply increasing or decreasing sensitivity and/or response to these stimuli. The region of the human brain that influences sexual behavior is in the *limbic system*, which includes the thalamus, amygdala, hippocampus, and part of the hypothalamus and cerebral cortex (see Chapter 17). In fact, when electrical stimulation is applied to the human limbic system during brain surgery, a sexual response can be elicited. We now discuss the role of hormones in the human sex drive and the sexual response.

Hormones and Male Sexual Behavior

In other mammals, the androgen testosterone stimulates the male sex drive. If adults of these animals are orchidectomized (the testes are removed), the sex drive disappears. That is, the males show no behavioral response to stimuli from a female. Administration of testosterone can restore sex drive in these orchidectomized males. This androgen has a central effect on the brain and also increases the sensitivity of the penis to tactile stimuli. Recently, it has been discovered that testosterone does not directly affect male sex drive in the brain of laboratory mammals. Instead, testosterone is converted enzymatically to estradiol by cells in part of the limbic system, and it is this estrogen that actually increases male sex drive. Whether or not this is true in humans is not clearly known (Chapter 17).

Testosterone also influences the sex drive in human males, but this influence is affected markedly by learning. Orchidectomy of human males before puberty generally decreases the development of male sex drive at puberty. In ancient China and Egypt, some young males were orchidectomized. They not only failed to develop male secondary sexual characteristics, but also were reported to have little or no sex drive. These men, called *eunuchs* (Greek for "guardian of the bed"), then were trusted to guard females in royal harems.

If a man is orchidectomized after he has reached puberty, the usual results are a slowly declining sex drive and a gradual loss of the ability to have an erection. The extent of the loss of libido varies from one man to the next, emphasizing the role of learning in male sexual motivation and that androgens may not be as necessary after the original behavior pattern develops. Usually, penis size and the extent of facial and body hair are not influenced by orchidectomy. Administration of an androgen to an orchidectomized man who suffers a decrease in sex drive can restore his libido and the ability to have and maintain an erection. In normal men, the higher the testosterone levels in the blood, the less time it takes to achieve maximal penile erection. If a man is given excess androgen, it is not known to produce an excessive sex drive.

Chapter 8, Box 1: Human Pheromones and the Vomeronasal Organ

The predominant human sexual senses are often considered to be vision, touch, and hearing. To other mammals, however, the sense of smell plays a critical role in sexual interactions. When a member of a species emits a chemical that changes the physiology or behavior of another member of the same species, this chemical is called a *pheromone*. For example, there is a substance in the urine of adult male mice that accelerates the onset of puberty in female mice and also synchronizes the estrous cycles of adult females. A male rat can tell when a female is in estrus by a pheromone in her urine. Similarly, a female Rhesus monkey that is ovulating exudes pheromones ("copulins") that increase the sex drive in male monkeys. Vaginal secretions of female hamsters contain "aphrodisin," a chemical that attracts males. The saliva of male pigs contains an androgen that makes sows sexually receptive.

We are just beginning to investigate the idea that pheromones influence human behavior, a possibility that the manufacturers of perfumes and colognes have assumed for years. Some artificial, musk-like odors can be smelled only by adult women and not by children or males. What is more, adult women can smell them only near the time of ovulation. Is it possible that men produce a musky pheromone that increases female sex drive and that women differ in their ability to sense this smell in accordance with the stages of the menstrual cycle? There are glands on the male skin, axilla, penis, and scrotum that could produce pheromones.

Women may also produce pheromones. The vaginal secretion of women, especially when they are near the time of ovulation, contains several chemicals that are known to be copulins in monkeys. We do not know of any behavioral effects of these potential pheromones on men. However, evidence indicates that vaginal secretions applied to the bodies of women increase their sexual attractiveness. Is it possible that feminine hygiene sprays are covering up something biologically meaningful? Another study showed that female body odors smell more pleasant to males when they are from women in the follicular phase of the menstrual cycle (i.e., before ovulation) than in the luteal phase, suggesting that men may be able to sense when a woman is fertile by her pheromones.

We already discussed (Chapter 3) how women housed together tend to have menstrual cycles that are synchronized, and a recent study suggests that an odor produced by the armpits (axilla) of women may cause this synchrony. Other studies indicate that the frequency of exposure to men can influence menstrual regularity and length, and some evidence suggests that this male effect is caused by an axillary pheromone.

In most mammals, pheromones are sensed by special sensory organs in the floor of the right and left internal nasal cavities, separate from the olfactory cells on the roof of the nasal cavity that we use to smell nonsexual odors. These special pheromone sense organs, the *vomeronasal organs* (VNO), were until recently thought only to be present in the human fetus during the first trimester, after which they presumably degenerated. Recent research, however, has revealed that adult humans have a small VNO on the floor of the nasal cavity on each side of the nasal septum (the wall dividing right and left internal nasal cavities). Each VNO opens into the nasal cavity via a tiny pit. Furthermore, the special cells of the adult human VNO have all the correct nerve sensory cells and tracts leading to the hypothalamus and other areas of our brain involved with sex (see Chapter 17). In fact, the GnRH cells that eventually control reproduction in the adult migrate along these tracts to reach the hypothalamus (see Chapter 1), thus showing an evolutionary link between pheromones and human reproduction. If one records electrical activity in the human VNO, there is an increase when the VNO is exposed to human skin extracts or axillary sweat. These extracts increase heart and respiratory rates at the same time and even influence LH secretion. Furthermore, skin extracts containing

Continued on next page.

Chapter 8, Box 1 continued.

androstenes (steroids related to androgens) cause the female VNO to respond more than the male VNO, whereas *estrenes* (steroids related to estrogens) cause a greater VNO response in males. Whether these or other pheromones play a role in human sexual behavior, however, remains to be determined.

A final note about human pheromones: Recent research has shown that women are attracted to the scents of men who are most unlike themselves in *major histocompatibility complex genes* (MHCs). These genes encode cell surface proteins that offer a "display" of proteins being made inside the cells. Thus the immune system can detect foreign proteins (such as those from a virus or toxin) in a cell and destroy the cell accordingly. The more dissimilar the mate's MHCs, the better their offsprings immune system will be at detecting foreign proteins. In fact, human mates with similar MHCs tend to be less fertile and have more miscarriages. Therefore, it could be an evolved, adaptive trait that humans can "smell" different MHCs (or their associated chemical cues) in a potential mate and avoid similarity. It is not known if this MHC detection involves use of the vomeronasal organ.

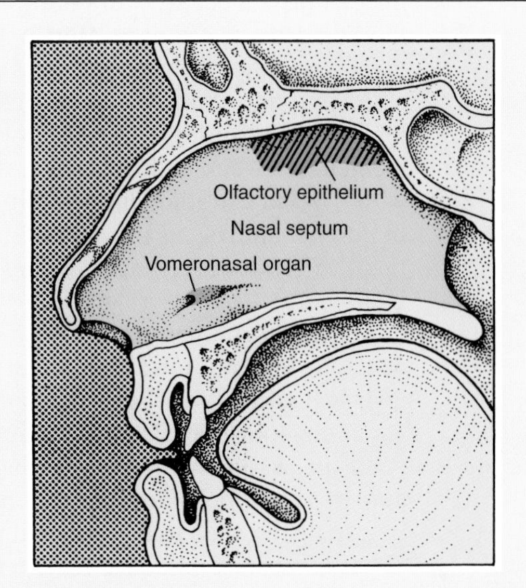

The anatomical location of the human vomeronasal organ (VNO). Note that the paired VNOs are located about a half inch into the nasal cavity, on its floor on each side of the nasal septum (the midline partition that divides the nasal cavities into right and left). Each VNO is a sac with an opening about 1 mm in diameter. Note also that the olfactory epithelium is on the roof, not the floor, of the nasal cavities on each side.

Many factors can influence the levels of testosterone in a man's blood and, therefore, potentially his sex drive. First, there is a daily rhythm of testosterone secretion, peaking in the morning (see Chapter 4). Also, anticipation or performance of sex may increase testosterone secretion. Erotic arousal stimulates LH secretion in men, which may raise testosterone levels. In Rhesus monkeys, dominant males have relatively high testosterone levels, which decrease when they become subordinate. This also may be true, to a modified extent, in humans. Finally, there is a tendency for blood testosterone levels to decrease with age. Sexually active 70-year-old men have higher testosterone levels than sexually nonactive men of the same age. We discussed the subject of "andropause" in Chapter 7.

Hormones and Female Sexual Behavior

In females of many animal species, it is an estrogen (sometimes acting with progesterone) that increases the sex drive. In human females, however, the story appears to be more complicated. If an adult woman has her ovaries removed (ovariectomy), it generally has little or no influence on her libido, although her breasts, vagina, and uterus will shrink because of the absence of estrogen. If, however, her adrenal glands are removed along with her ovaries, her sex drive decreases dramatically. Then, if this woman is given an androgen, her sex drive is restored. Remember from previous chapters that the adrenal glands secrete "weak androgens" [predominantly dehydroepiandrosterone (DHEA)] and that the ovaries also secrete testosterone, with a peak in blood levels of this androgen around ovulation. It is now generally thought that, in women, the weak androgens secreted by the adrenal glands, perhaps in concert with ovarian estradiol and ovarian testosterone, control the sex drive. Secretion of adrenal androgens in women is greatest from puberty through the late twenties, declines between ages 30 and 50, and then remains at steady low levels in later years.

Other hormones also seem to affect the female sex drive. Progesterone, for example, tends to lower female libido, and combination birth control pills with high amounts of a progesterone may also have this effect. In addition, gonadotro-pin-releasing hormone (GnRH) directly stimulates female sexual behavior in some mammals. We have much to learn about the role of GnRH in human sexual behavior, but it is possible that GnRH stimulatory agonists can eventually be used to increase human sex drive in patients with low libido.

If hormones influence female libido, there could be a variation in sex drive during the menstrual cycle. The evidence for this possibility, however, is controversial. For example, some studies have used frequency of coitus during the menstrual cycle as an indicator of sex drive. It is also clear that coitus is influenced by many psychological factors, such as opportunity, fear of pregnancy, illness, and fatigue. Kinsey as well as Masters and Johnson found an increase in frequency of coitus during the 3 or 4 days before menstruation begins, which may reflect a desire to avoid conception. However, studies also show an increase in coital frequency just after menstruation and near ovulation. The general conclusion seems to be that, like other mammals that exhibit a peak in sex drive (estrous behavior) around the time of ovulation (see Chapter 3), human females have retained some peak near ovulation. However, the actual expression of this increase in sex drive can be influenced greatly by psychological factors. Therefore, unlike in animals with a distinct behavioral estrous cycle, the human female can be sexually receptive at any time during the menstrual cycle.

Chapter 8, Box 2: Human Mating Systems

Animal mating systems can be of several types. *Monogamy* is when a single female and male pair within one breeding effort or for a lifetime. *Polygamy* is when an individual of one sex pairs simultaneously with more than one member of the other sex. Within polygamy, *polyandry* is when one female pairs with more than one male, and *polygyny* is one male paired with more than one female. *Promiscuity* is when no lasting pair-bonds are established and mating is more or less random.

Monogamous species tend to be *sexually monomorphic* (the sexes look alike), have offspring that require extended parental care from both parents, and evolve in environments in which resources (e.g., food) are spread out and predation is high. Also, individuals in the population tend to be widely scattered across the landscape. The idea is that the male stays

Continued on next page.

Chapter 8, Box 2 continued.

with one female to ensure viability of their offspring by helping care for the young, defending their young from other males (who could kill them) and predators, and obtaining food. Finally, monogamous pairs tend to exhibit more "affectionate" behavior to maintain the pair-bond. Monogamy is common in some animal groups; e.g., 92% of birds are monogamous. However, only 3% of mammalian species are monogamous, with the great majority being polygynous.

Polygyny (e.g., a dominant male with a harem of several females) is more adaptive when environmental resources (e.g., water, food) are patchy (clumped) and when females occur in groups. Thus, males compete for acquisition of a group of females by fighting with other males and displaying their prowess to females. Through a process Charles Darwin called *sexual selection* (see Chapter 17), this leads to *sexual dimorphism* (e.g., large male body size and "ornaments" of the male for display). The ovarian cycles of individual females in polygynous species tend not to be synchronized because if they were it would be difficult for a male with a harem to mate with all of his females at once! Furthermore, the male of polygynous species participates less in parental care, and the young tend to be born in a less helpless state.

Among primates, the group to which humans belong, 15% are monogamous, a higher percentage than for mammals in general. Among the great apes, our closest living primate relatives, there is an interesting variety of mating systems. The gorilla, for example, is polygynous. A large male acquires a harem, and he then mates periodically when a sexually receptive female in his harem presents herself to him. Gorillas are highly sexually dimorphic, with the males being twice the size of the females. Surprisingly, the male gorilla's testes and penis are relatively small; there is no need to produce a lot of sperm because mating is infrequent and the male knows that any pregnancies in his harem are his and there is no need to stimulate females or display (with a large penis) to females he already possesses in his harem. The chimpanzee, however, has a promiscuous mating system within a social group. Mating is frequent, and partners switch continuously. The sexes are about the same size (*monomorphic*), but the testes of a male chimp are relatively large (sperm for lots of matings). The penis of a chimp is also relatively large because it is used to display to the females. The female chimps advertise their receptivity by a swelling in the perineal region. The whole social group helps the mother care for the young, gather food, and defend against predators.

Humans have some signs of having evolved a form of monogamy in our ancestors. For example, the testes are relatively small (no need for a lot of sperm). However, the penis is relatively large (either for display or for giving sexual pleasure). The human female exhibits several unique traits among primates, and these traits probably evolved to ensure a prolonged pair-bond (to keep the male from wandering). Human breasts, unlike other primates, remain enlarged even when not nursing and serve as a sexual attractant. The breasts of other primates are not enlarged except when producing milk. Also, humans are the only primate to have coitus face to face, which allows more eye contact and affectionate interaction. Furthermore, human females have *concealed ovulation*, i.e., a male of our ancestors was unlikely to be able to tell when his mate was ovulating because she appeared in "estrus" at all times (this chapter discusses how present-day human females may feel more sexually receptive around ovulation, but they may not show it). Therefore, our ancestral male had to stay with his mate and copulate periodically to ensure that her pregnancy was his. This also kept him from wandering outside the pair-bond; if he knew for sure that he had fertilized his mate, it would have been advantageous for him to wander and fertilize another female so that he could pass on more of his genes to the next generation. Finally, the fact that groups of women tend to have synchronized ovulation (menstrual cycles; see Chapter 3) argues against a basic polygynous system in our past.

The fact that the average man is slightly taller, heavier, and stronger than the average women (moderate sexual dimorphism) may

Continued on next page.

Chapter 8, Box 2 continued.

A seemingly affectionate, perhaps monogamous pair of *Australopithecus afarensis* (of "Lucy" fame) strolling in the hot sun of the African savannah. This hominid primate, perhaps an African ancestor of *Homo habilis*, the first member of the genus of humans, had body features (small sexual size dimorphism, relatively small penis and scrotum, and continuously enlarged female breasts) different from the more polygynous primates such as modern gorillas. They also probably wandered far distances, using their erect posture, height, and free arms and hands, to search for food and water and to avoid the high temperatures of their environment. Their upright posture also suggests that they may have had a difficult birth (see Chapter 11) and may have borne infants that required both parents to feed and protect their child from predators during a somewhat prolonged parental care. Was this the advent of monogamy in the human ancestral tree? Perhaps.

Continued on next page.

Chapter 8, Box 2 continued.

not have evolved through sexual selection (i.e., as a signal for male–male competition or female attraction), although bigger males in our ancestors could have competed better with other males for mates. Another theory exists about sexual size dimorphism in humans; i.e., that the larger size of human males evolved through natural, not sexual, selection. In other words, it was beneficial for males to be large to protect their family from predators, build shelters, and gather food.

A survey of 853 present-day human cultures, from developed to hunter-gatherer, has shown in 99.5% of them that most women had only one man as a partner at one time. Although 84% of human cultures permit polygyny, only 10% of these 717 polygynous cultures *practice* polygyny. In a recent survey in the United States, 83% of men or women said that they were monogamous (married or cohabiting) and had 0 or 1 sexual partners in the past year. The same survey also reported that 75% of married men and 85% of married women reported that they were faithful to their mates. Thus, monogamy appears to be the main present-day human sexual strategy, with polygyny being a secondary choice. The driving evolutionary force behind this tendency for human monogamy was probably the fact that the human newborn is born in a relatively helpless state and needs extended parental care (from both parents). Humans might also have a tendency toward serial monogamy, i.e., changing monogamous partners when their children become more independent.

Recent studies of voles (mouse-like rodents in the genus *Microtus*) have suggested a difference in vasopressin and oxytocin-containing brain neurons (or their receptors) that could be a biological factor in whether a species is monogamous or promiscuous. Remember from Chapter 1 that both vasopressin and oxytocin (two closely related molecules) are made in the hypothalamus and are released from the pituitary gland into the blood as neurohormones. Oxytocin stimulates labor and milk ejection in women, whereas vasopressin causes water retention and constricts blood vessels in both sexes. Both hormones are also released during the sexual response cycle in both men and women. These two chemicals are also present in other neurons in the brain and act as neurotransmitters. In the monogamous prairie vole, the male is loyal to his mate for life, defends her with jealous rage, and helps care for their offspring. He also has more vasopressin and oxytocin neurons and fibers, plus a different distribution of brain receptors for these chemicals, than the closely related but promiscuous male montane vole; the latter shows no sexual loyalty to his mate, does not defend her against other males, and is not a good parent. It is thought, therefore, that vasopressin and oxytocin, which are released from neurons during mating, play a role in the sexual response in both species. In the prairie vole, however, a greater production of these chemicals leads to monogamy. If one administers a chemical antagonist to vasopressin or oxytocin to a male prairie vole, his affiliation toward his mate and pups ceases. Is it possible that a change in brain biology of vasopressin or oxytocin in promiscuous or polygamous primate apes has led to the pattern of monogamy seen in *Homo sapiens* and our earlier prehuman ancestors?

Homosexuality

Homosexual behavior is sexual contact with a member of the same sex. Persons who frequently or always choose a member of the same sex in their sexual relations, or usually are sexually attracted to the same sex, are called *homosexuals* (Greek *homo* meaning "same"). Male homosexuals are referred to as *gay*. Female homosexuals often prefer the term *lesbian*, a name derived from a Greek homosexual poetess named Sappho, who lived on an island of Lesbos (now Mytilene) in the Aegean Sea in 600 B.C. In contrast, a *heterosexual* person is attracted to and chooses to have sexual relations with the opposite sex (*heterosexual behavior*.)

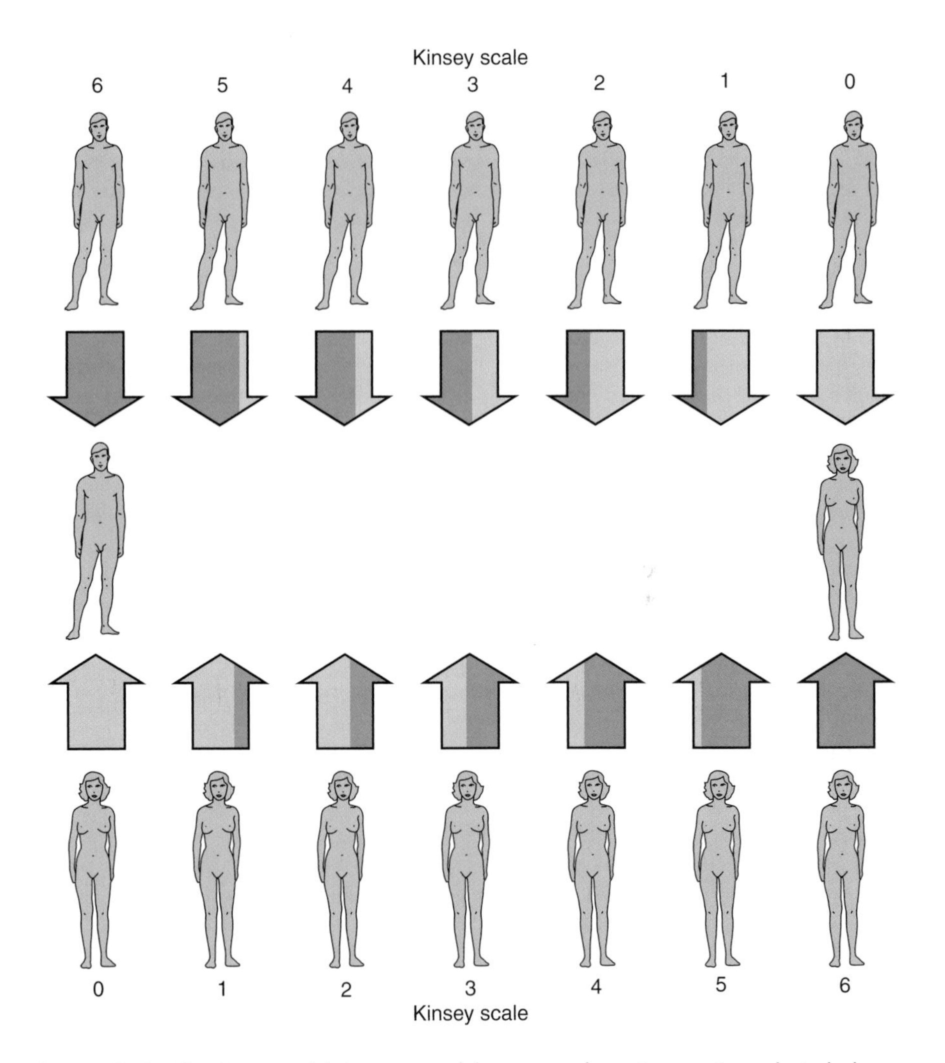

Figure 8-5 The Kinsey adult heterosexual–homosexual continuum: 0, exclusively heterosexual behavior; 1, mostly heterosexual, with incidental homosexual experience; 2, heterosexual, with substantial homosexual experience; 3, an individual with similar amounts of heterosexual and homosexual experience; 4, homosexual, with substantial heterosexual experience; 5, mostly homosexual, with incidental heterosexual experience; and 6, exclusively homosexual behavior. The proportion of each arrow shaded represents the percentage of heterosexual behavior. Individuals with scores of 1 through 5 could be termed *bisexual*.

Kinsey contributed greatly to our understanding and communication about homosexual behavior in pointing out that there is a continuum from heterosexuality to homosexuality. In his scale (Fig. 8-5), a person with a score of zero is exclusively heterosexual. At the other extreme are exclusively homosexual individuals, receiving a score of 6. These people always choose a sex partner of the same sex. Using this scale, about 2% of male Americans and 1% of female Americans have a score of 6 and are exclusively homosexual. About 63% of males and 87% of females have a score of 0 and are exclusively heterosexual. The remaining people, 35% of males and 22% of females, are *bisexual* in that they have had sexual experiences with their same and the opposite sex at one time or another in their adult life.

Surveys more recent than Kinsey's show that these frequencies of homosexual, heterosexual, and bisexual behaviors remain similar today. In the United States, about 1.3% of adult females had homosexual sex in the past year and 3.8% since the age of 18. Among American males, 2.7% had gay sex in the previous year and 7.1% since age 18. Slightly higher percentages reported being attracted to the same sex. It should be noted that a person may, for various reasons, never have expressed homosexual tendencies through behaviors, even though his or her feelings are almost exclusively homosexual. It should also be pointed out that there are many heterosexuals who have homosexual thoughts or fantasies at one time or another in their lives but have no tendencies to express these in their sexual behavior. Finally, there can be situational homosexual behavior. In prisons, for example, homosexual behavior is more common than in the general population. This is not because more homosexuals commit crimes and therefore end up behind bars. Rather, much of this behavior consists of heterosexuals switching to homosexual behavior because there is no heterosexual outlet for their sex drive. Thus, the fact that homosexual rape is common in prisons can be explained by the need to establish dominant and subordinate relationships in the prison. You can see from this discussion that it is difficult to classify a person as a "homosexual," and we really should talk about homosexual *behavior*.

Myths Some myths about homosexuals should be put to rest. One is that all male homosexuals are effeminate and all lesbians are masculine in appearance. The truth is that most homosexuals look and behave (except sexually) like heterosexuals. Only about 15% of male homosexuals and 5% of female homosexuals can be identified readily by appearance. Another misconception is that one person in a homosexual relationship always plays the "male dominant" role and the other the "female submissive" role. This certainly occurs in some homosexual couples, but in most couples the roles switch around or are more subtle than simply "masculine" and "feminine."

Another misconception is that contact with homosexuals will influence the sexual orientation of young children; e.g., that homosexuals in professions dealing with children (such as teachers and camp counselors) can introduce homosexuality to the children. The likelihood of this occurring is rare and can be compared to the possibility that a heterosexual teacher or counselor will influence the direction of sexual identity taken by children. In many cases, adolescents initiate other adolescents into homosexuality. Additionally, the vast majority of cases of child molesting occur when a heterosexual man sexually molests a young girl.

Some heterosexuals have a fear of being propositioned by a homosexual. In reality, gays are very good at limiting their sexual approaches to other homosexuals. Finally, homosexual behavior during preadolescence or earlier does not predict that such a person will be homosexual later in life.

Sexual Activity and Response The sexual behaviors of homosexuals are similar to those of heterosexuals except for, of course, vaginal coitus. Kissing, hugging, and caressing can be included in foreplay of homosexuals, as can fellatio and anal coitus in homosexual males, cunnilingus in lesbians, and mutual masturbation in both sexes. A study by Masters and Johnson indicates that the sexual response cycles of homosexuals and heterosexuals are physiologically identical regardless of the type of stimuli present.

Patterns of Sexual Behavior Male homosexuals tend to have more sexual partners in their life than heterosexual males. Lesbians tend to have fewer sexual partners than male homosexuals. Many homosexual couples, however, enter into long-lasting, stable relationships.

Psychoanalytic Theory What factors in a person's life predispose him or her to homosexual behavior as an adult? The answer is that we are not sure, but there are several theories. Psychoanalytic (Freudian) theory proposes that homosexuals experience an abnormal development of parental attachment. According to this theory, 3- to 6-year-old boys normally form an *Oedipus complex*, i.e., they fall in love with their mother and are jealous of their father. This is based on the Greek myth in which Oedipus killed his father and later married his mother—not knowing her true identity until it was too late. This conflict normally is resolved by the boy eventually identifying with his father and repressing his desire for his mother. According to this theory, male homosexuals retain a "negative Oedipus complex," i.e., they desire their father and are hostile to their mother. Some proponents of this theory suggest that a negative Oedipus complex could result from growing up with an overdominant or seductive mother and a weak, detached father. Thus, the child later fears relating to females because of the mother's jealous possessiveness and seductive behavior. Also, he may fail to gain the love of his father and seek it in other men.

Similarly, psychoanalytic theory suggests that young girls normally form an *Electra complex* in which they fall in love with their father and are jealous of their mother. Later, they repress their desire for their father and identify with their mother. This theory predicts that many lesbians were raised by rejecting mothers and distant or absent fathers. Thus, they have inadequate identification with their mother and also have not learned to relate to men.

One problem with these psychoanalytic theories is that they predict that a homosexual would have a gender identity of the opposite sex. However, most homosexuals have a normal gender identity but a reversed sexual object choice.

Learning Theory Learning theory also attempts to explain homosexual behavior. This theory assumes that all of us are born with both male and female potential. Our gender identity and sex object choice then are determined by what behaviors are rewarded or punished. If a boy, for example, is rewarded consistently for behavior that is typically exhibited by girls, he will develop a female-like gender identity and prefer sexual relationships with males as an adult. Evidence suggests that some male homosexuals have had a childhood pattern of cross-dressing, playing with dolls, and a tendency to play with girls more than boys. Similarly, lesbians would be rewarded as children when they exhibited behavior typically present in boys. Whether having play preferences like those of the opposite sex are causes or consequences of homosexuality is not clear. In this model, homosexuality is really a disturbance of sex role, with the individual being attracted to the opposite sex role. Another theory suggests that homosexual behavior in children may be rewarded by peers. A sociological aspect of this theory is that peer pressure can create homosexual behavior. For example, if a "sissy" boy and a "tomboy" girl are consistently labeled as "queer," they may be convinced that they are homosexual and will develop appropriate behavior for that role. One problem with these theories, as with many psychoanalytic theories, is that they predict disruptions in both gender identity and sexual object choice.

Hormones Do hormonal alterations during development cause homosexual behavior? Because early exposure to testosterone masculinizes the brain of male mammals, scientists theorize that homosexuals may have been exposed to abnormal levels of hormones as fetuses. Several lines of evidence support this theory, including the discovery of feminized brain areas in the brains of some gay men. We shall discuss these issues in Chapter 17.

Genetic Basis of Homosexuality Chapter 17 discusses how some parts of the brain of gay men appear to be feminized by hormonal exposure early in development or as an adult. Evidence also suggests a biological basis of homosexuality, i.e., there may be an inherited component to homosexuality. One kind of evidence in support of this comes from studies of identical and fraternal twins and their adoptive sisters. In females, if one identical twin is homosexual, there is greater than a 50% chance that her twin will be too, but only a 25% chance that her fraternal twin is a lesbian and only a 17% chance that her adoptive sister is a lesbian, being raised in the same household. Similarly, the numbers are 52% for identical male twins, 22% for fraternal twins, and 11% for adopted brothers. Even though the 17 and 11% incidences for adopted sisters and brothers are the lowest, they are still higher than the percentage of homosexuals in the general population. These studies therefore show a strong genetic basis, and a weak environmental influence, on both male and female homosexuality.

In 1993, a small region (*Xq28*) on the X chromosome was identified as a "gay gene" in males; it actually contains about 200 genes. First, it was discovered that only brothers, maternal uncles, and maternal male cousins exhibited an increased incidence of homosexuality in families of gay men. Then it was discovered that the X chromosomes of 33 out of 40 pairs of fraternal brothers who were both gay shared this gene cluster on their X chromosome. Only 25% of 12 heterosexual brothers shared this genetic cluster. This gene area is on the top of the long arm of the X chromosome and is on the mother's X. Finally, a similar study of 36 pairs of lesbian sisters lacked this gene cluster. The authors concluded that an inherited factor coded for on the maternal X chromosome is related to male homosexuality, but it is not the only determinant of male homosexuality as 7 of the gay brothers did not share this region. Other studies failed to confirm linkage between *Xq28* and male homosexuality so this continues to be an area of investigation. Female homosexuality may be controlled by different factors than male homosexuality.

There are indeed other observations that homosexuality, like heterosexuality, has a genetic, early developmental basis. For example, if one numbers the fingers on one hand from 1 (the thumb) to V (the little finger), the length ratio of fingers II and IV is informative. In heterosexual men, on average, II is shorter than IV, but in homosexual men, II is equal to IV. In heterosexual women, the two fingers are of equal length, but in lesbian women, II is shorter than IV on average. Differences in the growth of finger bones are thought to reflect androgen levels during development; whether there is a direct genetic basis is not known. Another indication that homosexuality has an early, neurodevelopmental basis is that a homosexual is 39% more likely to be left-handed than a heterosexual; this relationship between handedness and homosexuality is stronger in women than in men. Male homosexuals also have more older brothers than heterosexual men; each older brother increases the chance of homosexuality in a younger brother by 33%.

Another form of evidence that homosexuality has a biological rather than an environmental basis is that children raised by male or female homosexual couples are no more likely to be homosexual as adults than those raised by heterosexual couples. These children of homosexual couples were adopted, produced by artificial insemination (for lesbian couples), or were offspring of previous heterosexual sex.

If there is an inherited component to homosexuality, why have not the genes related to homosexuality disappeared? After all, homosexuals usually do not father or bear children. Maybe it confers evolutionary fitness in another way. For example, a recent study shows that female relatives of homosexuals have more children. Perhaps homosexuals in our ancestors' times were parenting "helpers," and thus gay gene(s) were retained because these individuals helped raise their relatives. One theory posits that male homosexuality is a polygenic trait involving several genes that feminize brain development. Heterosexual carriers of these alleles would tend to be more empathetic and sensitive, traits that make for more attractive mates and better fathers. This would keep the alleles in the gene pool.

Conclusions In summary, we do not know what causes homosexual behavior or, for that matter, heterosexual behavior. All human behavior is a complex interaction among inheritance, development, physiology, and learning experiences, and it is difficult to determine the relative influences of these factors. Furthermore, scientists who attempt to determine the causes of adult homosexual behavior often come up with differing conclusions. For example, remember that we discussed the general feeling that adult homosexuals do not always exhibit reversed gender identity. However, a recent study by the Kinsey Institute of Sex Research suggests that a disturbance of gender identity in childhood is very common in both male and female homosexuals. That is, male homosexuals exhibit feminine behaviors as young boys, whereas the reverse is true for female homosexuals. This study also suggests that homosexual behavior has its origins not as a result of a disturbed family environment, but as a deep-seated predisposition, perhaps biological in origin. Thus, perhaps there are several reasons why an adult may exhibit homosexual behavior, and we should speak of different kinds of homosexuality.

In 1973, the American Psychiatric Association removed homosexuality from its list of "personality disorders and certain other nonpsychotic mental disorders." Clinicians now agree that homosexuality is not an illness; it is just a sexual orientation. In recent years, homosexuals have become freer to express their sexual preference in public, a choice often called "coming out of the closet." Many homosexuals feel better about themselves because of support from "gay liberation" groups. Even so, there still is much prejudice against homosexuals.

Homophobia Fear of homosexuality in others or oneself is called *homophobia*. Often this phrase refers to repulsion, judgement, or prejudicial behavior against homosexuals. Why do people feel homophobic? One reason may be *xenophobia* (fear of differentness). A Freudian explanation is that we all begin as bisexual organisms during development, and as adults we fear our repressed "other sex" and are worried about our masculinity or femininity; we then project this fear on homosexuals. Yet another reason may be that we conform to a particular belief system, religious or otherwise, that condemns homosexuality. The AIDS epidemic (see Chapter 18) has brought about increased homophobia but also some openness about homosexuality in the general public, e.g., more open discussion of sexual practices, more awareness, and maybe even more sympathy (or empathy) among people. The recent scientific discoveries that at least some forms of homosexuality could have a genetic basis and are caused by developmental effects of hormones on the brain (see Chapter 17) should help people realize that sexual orientation is likely not a lifestyle preference, but a biologically predestined orientation of a person's personality. This, in turn, should bring about a lessening of homophobia in our society.

Treatment Should homosexuality be treated? A prevailing viewpoint is that a homosexual should seek professional help only if being gay makes him or her unhappy. Traditionally, psychologists or psychiatrists have found that it is difficult to "treat" a homosexual. Masters and Johnson, however, report success not only in treating sexual dysfunction in homosexuals, but also in eliminating homosexual tendencies in about two-thirds of those who wished to change their sex-object choice. Some critics of their findings feel that the patients studied were really mostly bisexuals, as a majority of the "homosexuals" treated by Masters and Johnson were rated scores of 3 or 4 on the Kinsey scale.

Transsexualism

A *transsexual* is a person who feels that he or she is trapped in the body of the wrong sex, i.e., the gender identity of these people does not fit with their biological sex. They may dress in the clothes of the opposite sex and want to be the opposite sex emotionally, psychologically, and sexually. About 60% of biologically male transsexuals are sexually attracted to men, 95% of biologically female transsexuals are attracted to women, and 10% of either sex are bisexual. We are unable to rule out exposure to abnormal hormone levels as a cause of sex-role reversal, even though adult transsexuals have the normal adult hormone levels of their biological sex (see Chapter 17). Chapter 17 discusses how the brain of transsexuals has a region that is typical of the opposite biological sex.

With proper hormonal treatment and/or surgery, the individual whose sexual identity is not in tune with his or her sexual anatomy need not remain so. A transsexual man can receive estrogen treatment, which can feminize his appearance. For example, his breasts may enlarge. Alternatively, he may choose to have a *sex-change operation*. This can involve castration and surgical removal of the penis. An artificial vagina can be created by making an opening in front of the anus and lining it with skin from the penis. Part of the scrotum can be used to form labia. Such "women" can have heterosexual intercourse and some adapt well to their new female role. Similarly, a transsexual woman can receive an ovariectomy and androgen treatment, which can produce a male-like body, facial hair, and a deep voice. There is little effect of this treatment on the external genitalia other than slight enlargement of the clitoris. A penis, however, can be constructed from skin and cartilage grafts. Sex-change operations for female transsexuals are much less common and more difficult than such operations for males.

Sexual Dysfunction

Sexual dysfunction occurs when an individual consistently fails to achieve sexual gratification. Such dysfunction is quite common. Masters and Johnson estimated that more than half of married couples have or will have a problem with sexual dysfunction. In one study of happily married, well-educated, predominantly white couples, 40% of the husbands reported that they had erection or ejaculation difficulties and 50% reported difficulties relating to a lack of interest in sex. Of the wives interviewed, 69% said that they had problems getting sexually aroused and 20% were unable to achieve orgasm. About 77% of the wives reported difficulties relating to a lack of interest in sex. Another study showed that 4 out of 10 women and 1 out of 3 men suffer from some kind of sexual dysfunction. About 10 to 20% of sexual dysfunctions have a diagnosed physical or physiological cause, with many of the remainder being caused by psychological conditions. About 40% of people who seek help from a sex therapist suffer from a lack of sexual desire. Obviously, physical dysfunction can influence a person's psychological well-being, and conversely, psychological dysfunctions can influence physiology. We must also remember that if a person is not attentive to the needs of a sexual partner and does not supply effective erotic stimuli, a man or woman may be sexually dissatisfied but not have a sexual dysfunction per se.

What are some of the psychological bases for sexual dysfunction? The availability of sex manuals may have led some people to expect too much from their sexual interactions. They read what "should" be experienced and therefore judge their own and their partner's behavior, placing unrealistic expectations upon themselves. This often leads to fear of failure and rejection and to performance anxiety. Also, failure to communicate one's needs often leads to a failure of a partner to supply effective stimuli. Another psychological cause of sexual dysfunction is anxiety relating to an earlier sexual experience. A person's previous sex acts may have been painful, unpleasant, or traumatic. He or she may have guilt about sex related to parental, religious, or moral training. Other factors inhibiting sexual enjoyment are choosing the wrong time for sex, too little foreplay, the inability to relax, too little variety in the stimuli or performance of coitus, hostility toward the sex partner, and not being sexually attracted to the sex partner.

We now review some of the more common kinds of sexual dysfunction. Many of these can be treated with psychological or behavioral therapy.

Vaginismus

Vaginismus refers to painful, spasmodic involuntary contractions of the outer third of the vaginal wall or its surrounding muscles. Vaginismus can cause severe pain during coitus or may even prevent intromission. It can be caused by such psychological states as fear of coitus or pregnancy or by frustration due to a partner with erectile dysfunction (see later). Also, it can be caused by the presence of scar tissue in the vagina or vulva due to childbirth, episiotomy, or infection of the vagina. It is not a common kind of female sexual dysfunction, occurring in only 9% of the women studied by Masters and Johnson. A woman has a difficult time stopping these contractions voluntarily, but many can learn to control them with therapy and biofeedback.

Dyspareunia

Dyspareunia is a term referring to difficult or painful coitus in either sex. In females, this can be due to sexual fears or inhibitions or to such physical conditions as vaginismus, irritation or damage to the clitoris, failure of the vagina to lubricate, displaced or prolapsed uterus, infection in the reproductive tract, or inflammation of the Bartholin's glands. In males, the glans may be hypersensitive because of an allergic response to spermicides or urethritis or the foreskin may be fused to the glans, a condition called *phimosis*. Male dyspareunia can also be caused by an accumulation of smegma in the uncircumcised penis. In some cases, the penis may be bent (*chordee*). This may be due to fibrous or calcified tissue in the top or sides of the penis, making the erect penis bend in some direction. This condition, called *Peyronie's disease*, can be the result of previous infection; it is most prevalent in older men. Also, scar tissue or infection in the male sex accessory ducts or glands can cause painful ejaculation.

Premature Ejaculation

Premature ejaculation, a common male sexual dysfunction, occurs when a man ejaculates too early. But what is too early? Masters and Johnson state that it is ejaculation that occurs at least 50% of the time before the woman reaches orgasm. Some men ejaculate immediately after intromission or even before intromission. Most studies suggest that premature ejaculation has a psychological basis. That is, it is a learned rapid response that can be unlearned. In rare cases, premature ejaculation is caused by prostatitis or a disease of the nervous system. In addition to being frustrating to the man, premature ejaculation can limit his female partner's sexual fulfillment, as women have a greater chance to achieve orgasm the longer intromission lasts before ejaculation.

Ejaculatory Incompetence

Ejaculatory incompetence is when a man is unable to ejaculate. He may never have been able to ejaculate or his problem may occur only in certain situations. A common occurrence is a man not being able to ejaculate into a vagina while having no such problem after oral or manual stimulation. Retrograde ejaculation and coitus reservatus can be cases of learned voluntary failure to ejaculate and are not considered to be cases of ejaculatory incompetence unless they have a physical basis. However, some men experience retrograde ejaculation after prostate surgery. It should be mentioned that a man can ejaculate without orgasm. This occurs most often when fear or boredom inhibits the orgasmic phase.

Erectile Dysfunction

Erectile dysfunction is the failure to gain or maintain an erection. It is often called *impotence*, but this is not an accurate or desirable term because it implies lack of fertilizing capacity or lack of power. Masters and Johnson divided erectile dysfunction into two types. The first, *primary erectile dysfunction*, is when a man has never had an erection. *Secondary erectile dysfunction*, the much more common type, is when a man has had erections before but now fails to have one more than 25% of the time. Actually, one out of eight men (about 10 million) in the United States will experience erectile dysfunction at some time in their lives.

About 60% of cases of erectile dysfunction have a physical basis. One such cause is unusually low blood pressure in the penis, which can sometimes be corrected surgically. Diabetes is often associated with erectile dysfunction. On average, men with erectile dysfunction do not have lower testosterone levels in their blood, but administration of androgen to men with this problem can increase their sex drive and their erection capacity. It has been shown that levels of the hormone prolactin are abnormally low in men with erectile dysfunction and premature ejaculation. In the past few years, *penile implants* have been used to treat more than 30,000 cases of erectile dysfunction in the United States. In this method, a silicone tube is implanted surgically into the penis of men with a physical basis of erectile dysfunction. The tube (and penis) can then be made erect by the man manipulating a pump in the scrotum.

Many cases of erectile dysfunction have a psychological basis. One way of telling if the problem is psychological is if a morning erection occurs; if so, the erection mechanisms are physiologically normal. Remember that erection requires parasympathetic dominance. Therefore, psychological conditions that activate the sympathetic system, such as fear, anxiety, or sudden intense stimuli, will tend to inhibit erection.

There are several related drugs now available that can support an erection in men with erectile dysfunction. The first such drug to appear on the market was Viagra (sildenafil). This pill causes an erection in a majority (80%) of men who have had difficulty obtaining this result. However, the drug works only if a man is also interested in sex psychologically. Viagra acts by slowing the breakdown of cyclic GMP, and the resultant increase in cGMP dilates the penile arteries and leads to an erection. Viagra takes at least 15 min to 1 h to work, and once male orgasm occurs there is a refractory period just as in the normal male sexual response cycle. Side effects can include flushing, headache, and nasal congestion, and a man should avoid this drug, and others like it, if he is taking nitroglycerine for heart problems. In fact, a man with heart problems should not use these drugs. Newer drugs that act in a similar manner to Viagra are Levitra (vardenafil) and Cialis (tadalafil). The effects of these two latter drugs are longer lasting (up to 36 h) and are more potent.

Orgasmic Dysfunction

Orgasmic dysfunction is the failure to achieve orgasm. It is the most common female sexual dysfunction. It is labeled as a "dysfunction," even though, as discussed earlier in this chapter, the variability of female orgasm could have a biologically evolved value. Masters and Johnson divided this condition into *primary orgasmic dysfunction*, when a woman never has had an orgasm, and *secondary orgasmic dysfunction*, when she fails to have orgasm only in selective situations. For example, some women can have an orgasm during manual or oral stimulation but not during coitus. In fact, about one in three women are unable to reach orgasm through vaginal intercourse. Studies of twins found that a woman's ability to reach orgasm is partly determined by her genes. According to the surveys of Hite and Masters and Johnson, about 10% of all women have primary orgasmic dysfunction and another 20% have experienced secondary orgasmic dysfunction. The unfortunate term *frigidity* is often used in reference to orgasmic dysfunction and is also used if a woman has no interest in sex. Periodic secondary orgasmic dysfunction is a normal part of the sex life of many women. However, it can interfere with the individual's happiness and sexual gratification, leading to residual sexual tension that could spill over into other aspects of a woman's life. A man should not necessarily feel guilty or "less manly" if his sex partner does not have an orgasm. However, more communication between the partners can often help both to achieve greater sexual gratification.

Psychological factors, such as those discussed at the beginning of this section, can inhibit orgasm. More rarely, orgasmic dysfunction has a physical basis. Illness or fatigue can be a cause, as can the absence of adequate levels of estrogens. Treatment of women who have orgasmic dysfunction with an estrogen, however, is not effective. Also, about 50% of women with diabetes exhibit a diminished capacity to have orgasm.

Women with orgasmic dysfunction sometimes, but not always, also exhibit a low sex drive. Several treatments for women with low libido are now on the market or under study. One of these is Viagra, best known for its effects on male erectile function, but which also increases blood flow to the clitoris. Hormonal therapies include estrogens in pill form or as topical creams or gels and androgens such as DHEA. The FDA has approved a device that creates gentle suction over the clitoris, increasing blood flow and sensation.

Drugs and Human Sexual Behavior

Therapeutic drugs (medicines) and nontherapeutic drugs (those taken for pleasure) can influence libido and the sexual response cycle. Before discussing some of these drugs, we should first discuss the legends about *aphrodisiacs*, substances that increase sex drive. Many of these seem to work only because one expects them to. A true aphrodisiac would be a substance that activates the part of the brain that controls sex drive, and to date there is no such substance, except, potentially, androgens and GnRH, as mentioned previously.

Many substances, however, have a reputation for being aphrodisiacs. Examples are artichokes, truffles, ginseng, and garlic. Oysters are also supposed to arouse one sexually, but there is no known chemical in oysters that has this effect. This idea probably came from the supposed resemblance of oysters to human genital organs. A similar argument goes for bananas! The Chinese have used powdered rhinoceros horn as an aphrodisiac for many years, much to the demise of the rhinoceros population. Perhaps this is the source of the term "horny." Recently, Indian scientists discovered ancient Hindu formulas for aphrodisiacs made from such exotic materials as crocodile eggs, elephant dung, burnt pearls, gold dust, and lizards' eyes. People in Shakespeare's time thought prunes were an aphrodisiac, and these fruits were handed out by brothels to the clients.

Nevertheless, it is true that some medicines and recreational drugs do influence sex drive and the sexual response, whereas others decrease sexual desire; the latter are called *anaphrodisiacs*. We know more about the effects of drugs on sexual behavior of men than women, primarily because the male sexual response is easier to observe.

Therapeutic Drugs

Some medicines used to treat human illness also influence sex drive. For example, a person with hypertension (high blood pressure) is often given an *antihypertensive drug*. These drugs lower blood pressure, and because blood pressure is important for erection, they can cause erectile dysfunction. *Spironolactone*, an antihypertensive drug, can lead to erectile dysfunction as well as to a decrease in sperm count and motility, gynecomastia, and a decrease in sex drive. This drug not only lowers blood pressure, but also inhibits the action of androgens on their target tissues. Secondary amenorrhea is seen in women using spironolactone. *Reserpine* is another antihypertensive drug that interferes with the action of brain neurons and can cause mental depression. Reserpine can also depress sex drive and reduce follicle-stimulating hormone (FSH) and luteinizing-hormone (LH) secretion in women. In men, it can cause erectile dysfunction, ejaculatory incompetence, and gynecomastia.

Drugs that influence the sympathetic or parasympathetic nervous systems can influence the sexual response. For example, *guanethidine* (also an antihypertensive) blocks the action of some sympathetic nerves and can inhibit ejaculation and possibly produce permanent damage to the neurons in the ejaculation center. Other antisympathetic drugs that can cause ejaculatory problems include ergot alkaloids, methyldopa, and rauwolfia alkaloids. Drugs that block the parasympathetic system can cause erectile dysfunction. One such substance is atropine. Some antihistamines also can have this effect.

Some minor tranquilizers that are popular prescription drugs in the United States, such as *diazepam* (Valium) and *chlorodiazepoxide* (Librium), can increase sex drive by removing inhibitions. However, chronic use of these drugs, as well as some antidepressants, can cause ejaculatory incompetence in males and menstrual irregularities in females.

Nontherapeutic Drugs

Nontherapeutic drugs are taken for their presumed pleasurable effects. Many users report that some of these drugs increase their sex drive, sexual performance, and enjoyment. Most of these effects are due to the drug acting (1) on the brain to remove sexual inhibitions, (2) on the peripheral nervous system to increase sensory acuity, or (3) on the sympathetic or parasympathetic nervous systems to influence the sexual response cycle. Also, as mentioned earlier, some of the effects of the drugs on sexual arousal occur simply because the person expects them to work.

Some of the *cantharides*, such as "Spanish fly," have reputations as aphrodisiacs. Spanish fly is derived from a beetle (*Lytta vesicatoria*) found in an area extending from southern Europe to western Siberia and in parts of Africa. It is used to heal blisters. If this drug increases libido, one explanation is that it can irritate the urinary bladder, urethra, and digestive tract. Prolonged use of Spanish fly, however, can cause *priapism* (prolonged abnormal erection) because it dilates blood vessels in the penis. This can lead to permanent penile damage and erectile dysfunction. Also, cantharides can cause vomiting and diarrhea and are poisonous at high dosages.

The active substance in marijuana and hashish from the hemp plant (*Cannabis sativa*) is tetrahydrocannabinol (THC). This hallucinogenic substance mimics the action of the sympathetic nerves. It first stimulates euphoria, causing increased sensory awareness and distorted perception. This is followed by sleepiness and a dream-like state. About one-quarter of marijuana and hashish users report an increase in sexual responsiveness and enjoyment; THC produces this effect by acting on the brain to relax sexual inhibitions and to increase sensory awareness. Some studies, however, indicate no effect of marijuana on sex drive. Other studies suggest that chronic use of marijuana (an estrogen; see Chapter 4) lowers the secretion of FSH and LH in young adult males and causes a decrease in testosterone levels, sperm count, and, in a few cases, erectile dysfunction. However, one study showed no effect of chronic marijuana use on the male reproductive system. Chronic THC use in females inhibits ovulation and shortens the luteal phase of the menstrual cycle if ovulation occurs.

Lysergic acid diethylamide (LSD) is a hallucinogen reported to increase sex drive due to general mood change. LSD affects the brain by acting like serotonin (a neurotransmitter) and also mimics action of the sympathetic nervous system. Mescaline has effects on sex drive similar to LSD.

Amphetamines are stimulants that decrease fatigue and activate the sympathetic nervous system. These include MDA (3,4-methylenedioxyamphetamine), TP (2,5-dimethoxy-4-methylamphetamine), and MMDA (3-methoxy-4,5-methylenedioxy-amphetamine). These drugs produce an aphrodisiac-like effect by increasing one's sensitivity to tactile and other stimuli.

Cocaine, a widely used drug in the United States, is said to increase libido, enhance enjoyment of sex, and lower inhibitions. This drug often is inhaled through the nostrils as a powder or is smoked, and some women apply it to the vaginal mucosa. Chronic users of cocaine (and amphetamines) tend to be more interested in drugs than sex.

Yohimbine is extracted from the bark of the Yohimbé tree of Africa. It is a reputed aphrodisiac and apparently acts by stimulating the parasympathetic nervous system, causing vasodilation and erection. This drug also acts on the brain and increases blood pressure, heart rate, and irritability and can cause sweating, nausea, and vomiting.

Amyl nitrite is a drug that dilates blood vessels. It is a chemical cousin of nitroglycerin and is used to relieve heart pain. This drug is called "poppers" because of the popping sound emitted when one breaks the capsule in which it is contained. Some inhale its fumes shortly before orgasm and claim it creates a more intense experience, but this drug also can cause dizziness, headache, fainting, and even death. *Butyl nitrite* has similar effects.

Saltpeter (*potassium nitrate*), which is the main ingredient in gunpowder and fireworks, is thought to decrease sex drive, but there is no scientific evidence for this.

Alcohol (ethyl alcohol) has an anesthetic, depressive effect on the brain and thus decreases sexual inhibitions. Alcohol in repeated dosages, however, suppresses the parasympathetic nervous system and can cause erectile dysfunction. At higher dosages it suppresses the sympathetic system and can cause ejaculatory incompetence. Chronic use of alcohol can destroy the neurons involved in the erection reflex. In addition, male alcoholics can be feminized by the drug, causing sterility and gynecomastia. Even a single dose of alcohol lowers testosterone levels in the blood by decreasing the secretion of luteinizing hormone and by increasing the activity of liver enzymes that destroy this hormone.

Narcotics such as heroin, morphine, and methadone generally depress the central nervous system and decrease sex drive and responsiveness. Low doses, however, remove sexual inhibitions. These narcotics lower testosterone levels in the blood of males. In women, morphine, heroin, and methadone can interrupt menstrual cycles by decreasing FSH and LH secretion.

Barbiturates such as *methaqualone* (Quaalude, Supor) are used as sleeping pills and hypnotic drugs. These drugs can decrease sexual inhibition, but there often is a reduced ability to perform sexually. The effects of barbiturates on the endocrine system are similar to those described for narcotics.

Chapter Summary

Our genetic sex determines our biological maleness and femaleness. Gender identity is the psychological awareness of one's biological sex. Sex (gender) role is the outward expression of one's gender identity (our masculinity or femininity). Sex role development results from an as yet unknown interaction between nature and nurture.

The human sexual response cycle is a sequence of physiological changes divided into four phases: excitement, plateau, orgasmic, and resolution. In general, this cycle in both sexes is characterized by vasocongestion of the external genitalia and skin as well as myotonia. In females, responses such as vaginal lubrication and changes in the orgasmic platform precede orgasm. In males, erection is caused by vasocongestion of the penis and is controlled by an erection center in the spinal cord. Ejaculation occurs during the male orgasmic phase and is controlled by an ejaculatory center in the spinal cord. In both sexes, orgasm involves the release of neuromuscular tension and is experienced as a highly pleasurable event. During the resolution phase in both sexes, the sexual systems return to an unexcited state. Women, unlike men, can experience multiple orgasms without going through resolution.

Hormones can influence sexual behavior by acting centrally (on the central nervous system) or on peripheral tissues. In men, testosterone, both centrally and peripherally, is necessary for male sexual behavior. In women, weak androgens secreted by the adrenal glands, along with ovarian testosterone and estradiol, maintain libido. A slight tendency for the female sex drive to increase near the time of ovulation in the menstrual cycle is masked by psychological variables so that the frequency of coitus during the cycle does not exhibit any consistent trend. The human male and female may produce pheromones that influence sexual behavior via sensing by the vomeronasal organs in the internal nasal cavities, but more needs to be learned about this topic.

Most of the adults exhibiting homosexual behavior are bisexual, and only about 2.7% of adult males and 1.3% of adult females are strictly homosexual. Sexual behavior and response in homosexuals are similar to those of heterosexuals, except for the absence of vaginal coitus. Although some (especially male) homosexuals have several sexual partners, other homosexual couples develop long-lasting relationships. Theories about what causes adult homosexuality are numerous and include retention of a negative Oedipus complex (psychoanalytic theory), reward for homosexual behavior as a child (learning theory), exposure to abnormal hormone levels (hormonal theory), the presence of a genetic influence on homosexuality (genetic theory), and situations in which no members of the opposite sex are available (situational theory). None of these theories fully accounts for the behavior of homosexuals.

Sexual dysfunction is present when a person's ability to receive sexual gratification is compromised consistently. Such dysfunction often has a physical basis, but it can also be caused by a psychological condition. Some of the more common sexual dysfunctions in women are vaginismus and orgasmic dysfunction, whereas men can develop ejaculatory incompetence, premature ejaculation, and erectile dysfunction. Both sexes can develop dyspareunia.

A substance that increases sexual desire is an aphrodisiac, whereas one that decreases sexual desire is an anaphrodisiac. In reality, no true aphrodisiac exists, but some therapeutic and recreational drugs can influence human reproduction and sexual behavior. Although some recreational drugs initially increase sexual drive and desire, most have long-term effects that lower sexual desire and performance as well as reproductive capacity.

Further Reading

Bancroft, J. (1984). Hormones and human sexual behavior. *J. Sex Marital Ther.* **10**, 3–21.

Buss, D. M. (1994). The strategies of human mating. *Am. Sci.* **82**, 238–251.

Concores, J., and Gold, M. (1989). Substance abuse and sexual dysfunction. *Med. Aspects Hum. Sexual.* Feb., 22–31.

Ezzell, C. (1994). Orgasm research: Struggling toward a climax. *J. NIH Res.* **6**(1), 23–24.

Fedler, H. H. (1984). Hormones and sexual behavior. *Annu. Rev. Psychol.* **35**, 165–200.

Goldstein, I. (2000). Male sexual circuitry. *Sci. Am.* Aug. 2000, 70–75.

Hite, S. (1976). "The Hite Report: A Nationwide Study of Female Sexuality." Macmillan, New York.

Hu, S., *et al.* (1995). Linkage between sexual orientation and chromosome Xq28 in males but not in females. *Nature Genet.* **11**, 248–256.

Jarrett, L. (1984). Psychosocial and biological influences of menstruation: Synchrony, cycle length, and regularity. *Psychoneuroendocrinology* **9**, 21–28.

Kinsey, A. C., *et al.* (1948). "Sexual Behavior in the Human Male." Saunders, Philadelphia.

Kinsey, A. C., *et al.* (1953). "Sexual Behavior in the Human Female." Saunders, Philadelphia.

Lemley, B. (2000). Isn't she lovely? *Discover Magazine*, Feb. 2000, 43–49.

Melman, A. (1999). Impotence in the age of Viagra. *Sci. Am.* **10**(2), 62–67.

Masters, W. H., *et al.* (1993). "Biological Foundations of Human Sexuality." HarperCollins College Publishers, New York.

Masters, W. V., and Johnson, V. E. (1966). "Human Sexual Response." Little, Brown, Boston.

Morris, N., *et al.* (1987). Marital sex frequency and midcycle female testosterone. *Arch. Sex. Behav.* **16**, 27–37.

Pennisi, E. (1995). Imperfect match: Do ideal mates come in symmetrical packages? *Sci. News* **147**, 60–61.

Seppa, N. (1998). Nailing down pheromones in humans. *Sci. News* **153**, 164.

Sherwin, B. B. (1988). A comparative analysis of the role of androgen in human male and female sexual behavior: Behavioral specificity, critical thresholds, and sensitivity. *Psychobiology* **16**, 416–425.
Siegel, R. K. (1982). Cocaine and sexual dysfunction: The curse of mama coca. *J. Psychoact. Drugs* **14**, 71–74.
Stahl, S. M. (1998). How psychiatrists can build new therapies for impotence. *J. Clin. Psychiatry* **59**, 47–48.
Taylor, R. (1994). Brave new nose: Sniffing out sexual chemistry. *J. NIH Res.* **6**(1), 47–51.
Thornhill, R., *et al.* (1995). Human female orgasm and mate fluctuating asymmetry. *Anim. Behav.* **50**, 1601–1615.
Thornhill, R., and Gangestad, S. W. (1996). The evolution of human sexuality. *Trends Ecol. Evol.* **11**, 98–102.
Weller, R. A., and Halikas, J. A. (1984). Marijuana and sexual behavior. *J. Sex Res.* **20**, 186–193.
Wojciechowski, A. P. (1992). Evolutionary aspects of mammalian secondary sexual characteristics. *J. Theor. Biol.* **155**, 271.

Advanced Reading

Argiolas, A., and Melis, M.R. (2003). The neurophysiology of the sexual cycle. *J. Endocrinol. Invest.* **26**, 20–22.
Bailey, J. M., *et al.* (1995). Sexual orientation of adult sons of gay fathers. *Dev. Psychol.* **31**, 124–129.
Baker, R. R., and Bellis, M. A. (1995). Human sperm competition: Ejaculate manipulation by females and a function for the female orgasm. *Anim. Behav.* **46**, 887–909.
Baskin, J. (1989). Endocrinologic evaluation of impotence. *South. Med. J.* **82**, 446–449.
Blanchard, R., and Bogaert, A.F. (2004). Proportion of homosexual men who owe their sexual orientation to fraternal birth order: An estimate based on two national probability samples. *Am. J. Hum. Biol.* **16**, 151–157.
Burnett, A. L., *et al.* (1992). Nitric oxide: A physiologic mediator of penile erection. *Science* **257**, 401–403.
Buss, D. M. (1989). Sex difference in human mate preferences: Evolutionary hypotheses tested in 37 cultures. *Behav. Brain Sci.* **12**, 1–49.
Campbell, B. C., and Udry, J. R. (1994). Implications of hormonal influences on sexual behavior for demographic models of reproduction. *Ann. N.Y. Acad. Sci.* **709**, 117–127.
Cutler, W. B., *et al.* (1986). Human axillary secretions influence women's menstrual cycles: The role of donor extract from men. *Horm. Behav.* **20**, 463–473.
Dennis, C. (2004). Brain development: The most important sexual organ. *Nature* **427**, 390–392.
Dewar, C.S. (2003). An association between male homosexuality and reproductive success. *Med. Hypoth.* **60**, 225–232.
DuPree, M.G., *et al.* (2004). A candidate gene study of CYP19 (aromatase) and male sexual orientation. *Behav. Genet.* **34**, 243–250.
Edgerton, M. T., *et al.* (1994). The surgical treatment of transsexual patients: Limitations and indications. *J. NIH Res.* **6**(1), 52–59.

Hamer, D. H., *et al.* (1993). A linkage between DNA markers on the X chromosome and male sexual orientation. *Science* **261**, 321–327.

Heller, J., and Gleich, P. (1988). Erectile impotence: Evaluation and management. *J. Family Pract.* **26**, 321–324.

Holden, C. (1992). Twin study links genes to homosexuality. *Science* **255**, 33.

Kadri, N. *et al.* (2002). Sexual dysfunction in women: Population based epidemiological study. *Arch. Women Ment. Health* **5**, 59–63.

Knussmann, R., *et al.* (1986). Relations between sex hormone levels and sexual behavior in men. *Arch. Sex. Behav.* **15**, 492–445.

Krane, R. J., *et al.* (1989). Impotence. *N. Engl. J. Med.* **321**, 1648–1659.

Lippa, R.A. (2003). Handedness, sexual orientation, and gender-related personality traits in men and women. *Arch. Sex. Behav.* **32**, 103–114.

LoPiccolo, J., and Stock, W. E. (1986). Treatment of sexual dysfunction. *J. Consult. Clin. Psychol.* **54**, 158–167.

McIntyre, M.A. (2003). Digit ratios, childhood gender role behavior, and erotic role preferences of gay men. *Arch. Sex. Behav.* **32**, 495–497.

Miller, E.M. (2000). Homosexuality, birth order, and evolution: Toward an equilibrium in reproductive economics of homosexuality. *Arch. Sex. Behav.* **29**, 1–34.

Muscarella, F., *et al.* (2001). Homosexual orientation in males: Evolutionary and ethological aspects. *Neuroendocrinol. Lett.* **22**, 393–400.

Mustanski, B.S., *et al.* (2002). A critical review of recent biological research on human sexual orientation. *Annu. Rev. Sex Res.* **13**, 89–140.

Preti, G., *et al.* (1986). Human axillary secretions influence women's menstrual cycles: The role of donor extract of females. *Horm. Behav.* **20**, 474–482.

Rice, G. *et al.* (1999) Male homosexuality: Absence of linkage to microsatellite markers at Xq28. *Science* **284**, 665–667.

Segraves, R. (1989). Effects of psychotropic drugs on human erection and ejaculation. *Arch. Gen. Psychiat.* **46**, 275–284.

Sokolov, J., *et al.* (1976). Isolation of substances from human vaginal secretions previously shown to be sex attractant pheromones in higher primates. *Arch. Sex. Behav.* **5**, 269–274.

Swerdloff, R. S., *et al.* (1992). Effect of androgens on the brain and other organs during development and aging. *Psychoneuroendocrinology* **17**, 375–383.

Weller, L., and Weller, A. (1993). Human menstrual synchrony: A critical assessment. *Neurosci. Biobehav. Rev.* **17**, 427–439.

CHAPTER **NINE**

Gamete Transport and Fertilization

Introduction

The process of *fertilization*, or *conception*, involves fusion of the nucleus of a male gamete (sperm) and a female gamete (ovum) to form a new individual. Because each gamete is haploid (N), fertilization restores the normal diploid (2N) chromosomal complement. Fertilization, however, is more than the simple fusion of gametes in that it is preceded by and requires a series of precisely timed events. Once sperm are deposited in the female reproductive tract, they travel a relatively long distance and overcome several obstacles before reaching the ovum. Similarly, the ovum travels through a portion of the female reproductive tract before it is fertilized. Not only do the gametes move to the appropriate regions of the female tract, but they undergo important physical and biochemical maturations that are a prerequisite for fertilization. Abnormalities in these maturational or transport processes, as well as in fertilization itself, can lead to infertility, spontaneous abortion (miscarriage), or birth defects.

Semen Release

After leaving the epididymides, sperm enter the vasa deferentia, which are long paired ducts serving as sperm storage and transport organs (see Chapter 4). Secretions of the male sex accessory glands (*seminal plasma*) mix with the sperm during ejaculation to form *semen* or *seminal fluid*. It has been theorized that the entire reserve of sperm in the epididymides and vasa deferentia would be depleted if an adult male had 2.4 ejaculations per day for 10 consecutive days. However, this normally does not occur, even with such Herculean ejaculation frequency because new sperm are produced continuously by the testes—about 200 million per day! Thus, frequent ejaculation is not an effective method of contraception.

Semen is released in three stages. Before male orgasm, a small amount of semen comes from the bulbourethral glands. In the second stage, the majority of semen is released; most of the seminal plasma of this stage comes from the seminal vesicles and prostate gland. In the third stage, another small amount of fluid produced by the seminal vesicles is exuded. Most of the sperm are expelled in the second stage, but some sperm are present in the semen of the first and third stages. Because sperm are present in the first stage, pregnancy can occur without male orgasm, which is one reason why *coitus interruptus* (withdrawal of the penis before ejaculation) is not an efficient method of birth control (see Chapter 14).

Contents of Seminal Plasma

Seminal plasma contains several substances, but the precise function of many of these components is not known. We do know, however, that some of them have roles in the maintenance, maturation, and transport of sperm. Water is present, which serves as a liquid vehicle for the sperm and seminal plasma constituents. Mucus from the bulbourethral glands serves as a lubricant for the passage of semen through the male reproductive tract. The prostate gland and the bulbourethral glands both secrete buffers, which neutralize the acidity in the male urethra and in the vagina. Some nutrients for sperm are present in the seminal plasma deposited in the vagina, the major ones being the sugar fructose and citric acid (from the seminal vesicles). *Carnitine*, concentrated from the blood by the epididymis, is also found in the seminal plasma. This chemical is involved in the metabolism of fatty acids, with the metabolites being used as another nutrient source for the sperm. Another constituent of seminal plasma secreted by the epididymis is *glycerylphosphocholine*. The enzyme *diesterase* in the uterus hydrolyzes (breaks down) this molecule, and the products of this digestion are used by the sperm as nutrients. Other enzymes secreted by the prostate gland and seminal vesicles are involved in the clotting and subsequent liquefaction of semen in the vagina. Human seminal plasma contains extremely high amounts of zinc (which may have antibacterial activity), and men with low zinc content tend to have a higher incidence of infertility. Finally, some kinds of prostaglandins are secreted into the seminal plasma, mostly by the seminal vesicles. Prostaglandins in seminal plasma may be involved in sperm transport. Finally, seminal plasma contains ATP, and men with low semen ATP levels tend to have lower fertility. Table 9-1 summarizes the sources of the major components of seminal plasma.

Table 9-1 Some Characteristics of Human Semen

General properties

Creamy texture: gray to yellow color

Average volume: 2.5–3.5 ml after 3 days of abstinence (range, 2–6 ml)

Fertility index (minimum qualifications for male fertility):

1. At least 20 million sperm/ml
2. At least 40% sperm must show vigorous swimming
3. At least 60% sperm must have normal shape and size

pH: 7.35–7.50 (slightly basic)

Sources and major components of seminal plasma

Epididymis (a slight amount)	**Seminal vesicles (about two-thirds of total volume)**	**Prostate gland (about one-third of total volume)**	**Bulbourethral glands (a few drops)**
Water	Water	Water	Water
Carnitine (a nutrient)	Fructose (a nutrient)	Bicarbonate buffers (neutralize vaginal pH)	Buffers (neutralize vaginal pH)
Glycerylphosphocholine (a nutrient)	Fibrinogen (clots semen)	Fibrinogenase (clots semen)	
	Ascorbic acid (a nutrient)	Fibrinolytic enzyme (liquifies semen)	Mucus (lubrication)
	Most of the prostaglandins (contract the vas deferens)	Citric acid (a nutrient)	
		A little prostaglandin	

Sperm Number and Structure

The number of sperm in a single ejaculate ranges from 40 million to 500 million (the average is about 182 million sperm). A male produces about 1 billion sperm (Fig. 9-1) for every ovum ovulated by a woman. Many ejaculated sperm (about 30%), however, are structurally or biochemically abnormal and are either dead or incapable of fertilizing; these are reabsorbed by the female reproductive tract or are lost through the vagina. For a male to be minimally fertile, his sperm count should be at least 20 million sperm/ml of semen; 40% of these sperm must swim and 60% should be of normal shape and size (Table 9-1).

Some evidence shows that human sperm count has declined over recent decades (see Chapter 4). One study suggests that the sperm count in healthy men has dropped 1% per year in the past 50 years. However, other studies contradict this idea, and whether there has been a worldwide decline in male fertility remains controversial. It is clear, however, that geographical differences in average sperm count exist. Differences in sperm production of men living in disparate regions of the world may reflect genetic, cultural, or environmental differences.

A healthy human sperm is 40 to 250 μm long and is composed of the following structures (Fig. 9-2): neck, midpiece, and tail. The *sperm head* contains an elongated haploid nucleus surrounded by a nuclear membrane. External to the nucleus is a membrane-bound vesicle called the *acrosome*. It fits closely over the tip of the sperm head like a cap, and the *inner acrosomal membrane* lies external to the nuclear membrane while the *outer acrosomal membrane* is just inside the sperm cell *plasma membrane*. The acrosome is filled with enzymes important in the penetration of the ovum. The short sperm neck is followed by the sperm midpiece, which contains mitochondria that generate energy for tail movement. The midpiece and sperm tail represent a flagellum, with the "9 + 2" arrangement of microtubules. This provides the propulsive force, allowing locomotion of the sperm cell as it moves toward the egg and during egg penetration. A human sperm cell is 60–70 μm in length.

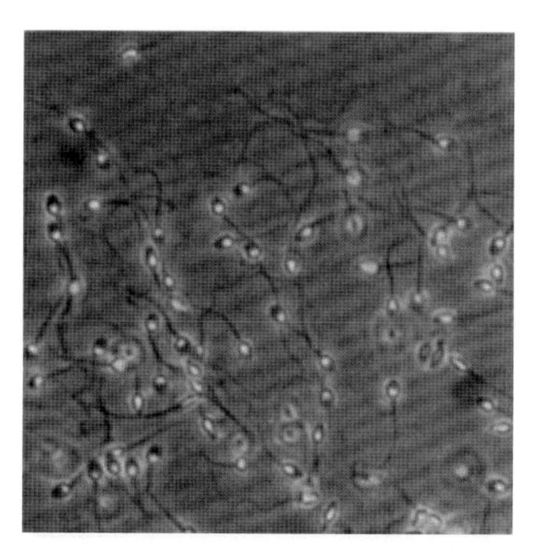

Figure 9-1 Photomicrograph of human sperm swimming in seminal fluid. The sperm heads shine because of a fluorescent dye.

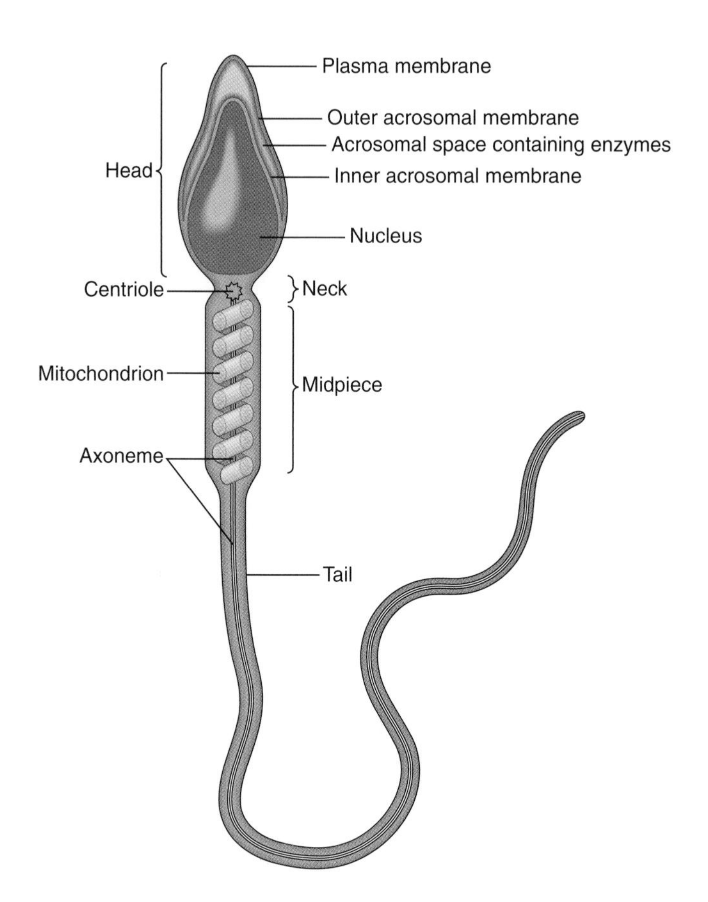

Figure 9-2 Sperm structure.

Sperm Transport and Maturation in the Female Reproductive Tract

Let us now follow the sperm cells on their journey through the female reproductive tract to the point of fertilization in the oviduct, a distance of about 15 cm (6 in.). The sperm are first deposited in the vagina; they then pass up this cavity and through the cervix into the uterus, up the uterus, through the junction between the uterus and oviduct (uterotubal junction), and up the isthmus of the oviduct to the usual area of fertilization in the oviduct: the ampullary–isthmic junction. Many of the millions of deposited sperm are lost during this journey, and only about 100 to 1000 reach the oviduct, with 20 to 200 reaching the egg itself. In addition, the sperm must undergo maturational processes during their journey, which give them the capacity to move and to fertilize an ovum.

Vaginal Sperm

About 1 min after deposition in the vagina, the semen becomes thicker and less liquid. This *semen coagulation* is brought about by the enzyme *fibrinogenase* in the seminal plasma, which converts the protein *fibrinogen* to *fibrin*, another protein. The major function of this coagulation may be to prevent sperm loss from the vagina. After about 20 min, however, the semen again liquefies. This *semen liquefaction*, which is caused by a *fibrinolytic enzyme* in the seminal plasma, stimulates some sperm to swim more rapidly and to reach the cervix. Even though semen liquefaction has not yet occurred, some sperm make it into the cervix and even into the uterus within a few minutes of deposition in the vagina.

The environment in the vagina is usually acidic (about pH 4.2), and this level of acidity inhibits sperm motility. The presence of semen in the vagina, however, increases the vaginal pH to a basic 7.2, which in turn increases sperm motility.

During coitus, female orgasm is accompanied by muscular contractions of the vaginal walls (see Chapter 8), and these contractions create a pressure in the vagina that is higher than that in the uterus. Sperm movement through the cervix may be aided by this pressure differential. Sperm, however, can move up the female tract without female orgasm.

Cervical Sperm

The cervical canal is lined by a complicated series of narrow folds and crypts and is blocked by a sticky mass of cervical mucus and tiny cervical fibers (see Chapter 3). In most stages of the menstrual cycle, the mucus is thick and fibers within it are densely packed. Shortly before ovulation, however, the rise in circulating estrogen levels causes the mucus to become more liquid and the gaps between the cervical fibers to widen. These gaps orient so that channels are formed. When the sperm enter the mucus, they line up in these channels almost in single file and pass through the cervix at a speed of about 1.2–3.0 mm per minute.

The cervical fibers may serve as a network upon which the sperm tails exert force, beating with a spiral motion and thus propelling the sperm upward. Also, these fibers may be of such dimension and length that they vibrate in rhythm with the tail beat frequency of normal sperm; this may allow normal sperm to move through the cervix, whereas sperm with abnormal or absent tail beats are detained. These latter sperm then die and are reabsorbed or lost from the body. Other sperm enter *cervical crypts* (deep recesses in the cervical wall), where they die or are lost, or they may remain alive as a reservoir of sperm that later may enter the uterus. Fewer than 1 million of the original 182 million sperm make it through the cervix.

Uterine Sperm

Upon leaving the cervix, the sperm travel up the uterus to the uterotubal junction. The uterine fluid is watery but sparse in humans, and the sperm essentially "climb" up the uterine lumen by beating their tails. The swimming rate of sperm (about 3 mm/min), however, cannot account for their traveling a distance of about 15 cm in the 30 min after ejaculation. Also, dead sperm reach the oviduct at about the same time as do live sperm. Thus, sperm tail beating probably is not important during sperm transport through the uterus so it must be the muscle contraction and movement of cilia in the female reproductive tract that facilitate sperm transport.

Mechanical stimulation of the cervix by the penis during coitus causes release of the hormone oxytocin from a woman's posterior pituitary gland. This hormone quickly travels via the blood to the uterus and increases the force of rhythmic uterine muscle contractions. These contractions act as waves to help the sperm move to the uterotubal junction. Prostaglandins in the seminal fluid may also cause uterine muscles to contract, but this is unlikely as very little if any seminal fluid enters the uterus through the cervix. The main function of the prostaglandins in seminal fluid is probably to contract the muscles of the vasa deferentia, thus aiding sperm passage during ejaculation.

The presence of sperm in the uterus initiates a massive invasion of white blood cells (*leukocytes*) into the uterine lumen. These cells then begin to engulf the dead or dying sperm that have not yet moved up to the uterotubal junction. No more than a few thousand sperm reach this junction.

The uterotubal junction is a muscular, tightly constricted region separating the uterus from the oviduct (see Chapter 2). Sperm enter the narrow opening of this junction and move through it at a relatively slow rate. Thus, the uterotubal junction allows the gradual entrance of sperm into the isthmus of the oviduct. About half of the sperm enter the wrong oviduct, and only a few hundred make it to the general proximity of the waiting egg.

Transport of the Sperm and Ovum in the Oviduct

Sperm tail beating is reduced, and the sperm "wait" in the isthmus for ovulation to occur. Other sperm previously residing in cervical crypts are also released around the time of ovulation. After ovulation, several sperm move up to the ovum, and fertilization by a single sperm usually occurs at the point where the isthmus joins the wider oviductal ampulla (ampullary–isthmic junction). Other sperm swim up the ampulla, through the infundibulum, and are lost in the body cavity.

Once ovulation has occurred, the infundibulum (funnel-shaped free end) of the oviduct moves to the ovary and envelops the ovulated ovum along with fluid derived from the ovulated follicle. Movement of the infundibulum is accomplished by the contraction of muscles in the membrane supporting the oviduct. Cilia are present in the wall of the fimbria (the edge of the infundibulum) and these beat toward the uterus. Thus, when the infundibulum envelops the ovary, the beating of the cilia moves the ovum into the ampulla of the oviduct. Cilia in the ampulla and isthmus of the oviduct also beat in a uterine direction, which sets up a flow of fluid toward the uterus.

The muscles of the oviduct also exhibit waves of muscular contraction after ovulation. These waves travel in the direction of the uterus and, along with the cilia, help the ovum move down the oviduct. Both ciliary beating and muscular contraction in the oviduct are influenced by ovarian sex hormones. Estrogens increase cilia number, and progesterone increases ciliary beating and egg transport.

A factor involved in the opposite movement of egg and sperm may be the direction of ciliary beating in the oviduct. Oviductal cilia exist in deep recesses in which cilia beat toward the ovary and on ridges where these cilia beat toward the uterus. Sperm may travel in these recesses, whereas the ovum may be propelled along the ridges. The presence of considerable amounts of mucus in the oviducts for 3 to 4 days after ovulation may serve as a medium for sperm transport. This mucus is gone when the fertilized ovum (embryo) travels down the oviduct to the uterus, as discussed in Chapter 10.

Sperm Capacitation and Activation

Freshly ejaculated human sperm are not capable of fertilization. A period in the female reproductive tract is necessary before sperm can fertilize an oocyte. Thus, during their journey, sperm gain the ability to fertilize an egg (a process called *sperm capacitation*). *Calmodulin*, a protein in seminal plasma, may also play a role in sperm capacitation. This protein (or another epididymidal secretion) may give the sperm the ability to be capacitated later on when they are in the uterus.

In general, the present scientific opinion is that capacitation involves removal or modification of molecules (glycoproteins) associated with the sperm head that stabilize the sperm plasma membrane. These molecules suppress the ability of sperm to fertilize. Alteration or removal of these inhibitory molecules allows the sperm to respond to signals that trigger the acrosome reaction, an important step in the fertilization process. Capacitation also increases the vigor or tail movements of the sperm (*hyperactivation*), propelling it toward the egg more effectively.

What substances in the female reproductive tract render the sperm capable of fertilization? One possibility is that molecules in follicular fluid escaping from the ovulating follicle play a role in sperm capacitation. Follicular fluid contributes only a small part of the oviductal fluid. Studies of mammals have demonstrated that two components of follicular fluid, progesterone and the protein albumin, facilitate the acrosome reaction. Calcium in follicular fluid increases the vigor of sperm tail beating. It is not clear if these substances are present in humans, although follicular fluid does activate human sperm. If operative in humans, they may have their effect when the sperm penetrates the cumulus oophorus (see later), which surrounds the ovum and is bathed in follicular fluid. A recent discovery is that follicular fluid, or the egg itself, produces a chemical that attracts human sperm (see HIGHLIGHT box 9-1). Another study suggests that mammalian sperm move toward the egg along a thermal gradient. The site of fertilization is slightly warmer than more proximal portions of the oviduct, and mature sperm have a preference for moving toward warmer fluid (*thermotaxis*). Sperm may be guided by temperature during most of their journey through the fallopian tube and then respond to chemical cues as they near the egg. In the future, we may expand our concept of sperm capacitation to include acquisition of the ability to detect chemical and/or thermal cues.

When Can Fertilization Occur?

Most references state that sperm live about 72 h and that an egg is fertilizable for 24 to 48 h. Thus, the fertile time in a menstrual cycle would be about 4 to 5 days,

Chapter 9, Box 1: Does the Human Egg Court Sperm?

Out of the millions of sperm deposited in the vagina during ejaculation, only 20 to 200 will reach close proximity to the egg in the oviduct, and yet the competition among sperm to be the one that fertilizes the ovulated egg is not finished. Part of the future of a sperm (to become part of a new embryo or to die as a haploid failure) could depend on its response to "courtship" by the egg and/or its surrounding fluid.

The sperm of algae, mosses, ferns, and some invertebrate animals are attracted to the egg chemically, but until recently only one case of such attraction has been described in a vertebrate animal. The egg of the herring (a teleost fish) is covered by a zona pellucida-like coat (the "chorion"), and the chorion cannot be penetrated by sperm swimming in the surrounding water unless it locates a small opening in it. This opening, the micropyle, secretes a chemical that activates and attracts sperm to it. However, until recently, there had been no evidence that the human egg attracts sperm. The pervading theory had been that the human sperm present in the vicinity of the egg bump into it by chance.

A new finding, however, suggests that the human egg produces a chemical that attracts sperm and influences their swimming motion. If follicular fluid from a large grafian follicle is placed at one end of a chamber, sperm will accumulate at that end, whereas they will not respond to a control fluid. The quantities of estradiol or progesterone in the fluid do not influence this response, but only some and not all follicles have fluid that works. A good correlation also exists between the fertilizability of an egg and the ability of its surrounding fluid to attract sperm. Control fluid previously containing an egg also attracts sperm, so it appears that this signal comes from the egg, not the surrounding follicular cells.

When sperm are exposed to the egg signal, they swim in a circle instead of in a straight line, which would increase their chances of contacting the egg. Interestingly, not all sperm are attracted to the egg; some could care less and some even swim away from the egg! A human sperm has about 20 chemical receptor molecules on its head, and maybe some sperm have not formed the receptor(s) used in this chemical orientation to the egg or perhaps they are abnormal in other ways. Nevertheless, they will not be the chosen one! Many questions still remain. What is the chemical that attracts sperm? Why do some eggs produce the chemical and some not? Why do some sperm respond whereas others do not? Does the chemical cause more sperm to move up the oviduct leading to the egg instead of the "empty" oviduct? Do X and Y sperm behave differently in response to this chemical? Could an inhibitor of this chemical be used as a new contraceptive agent? Only time will tell.

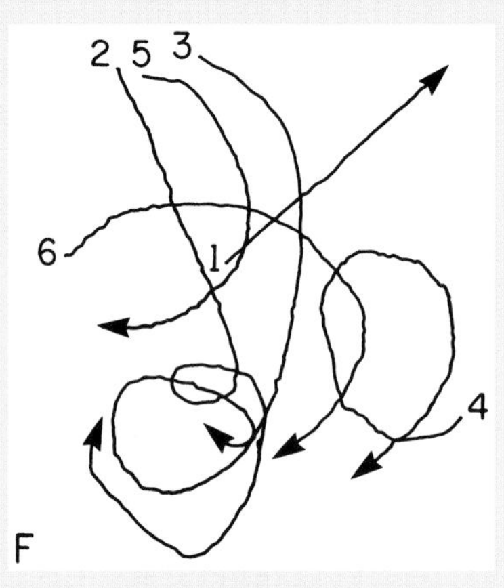

Six human sperm were placed in a fluid-filled chamber. Their starting position is represented by the numbers 1 through 6. Then, either follicular fluid from a human Graafian follicle or fluid exposed to a human egg was injected at the lower left-hand corner (F). Arrowed lines then indicate the path swum by each sperm. Note that sperm 2 through 6 turned and headed toward F. Sperm number 1, however, was not interested. (Adapted from Ralt *et al.* (1991).)

with ovulation occurring at about the middle of this time period. A recent study, however, has cast suspicion on this theory. This study found that conception can only occur in a 6-day period, i.e., during the 5 days before ovulation or on the day of ovulation. Therefore, some sperm live for 6 days and the egg lasts 12 to 24 h (or the change in cervical mucus after ovulation halts sperm transport).

The Process of Fertilization

Once a sperm and ovum are in the region of the ampullary–isthmic junction of the oviduct (Fig. 9-3), fertilization occurs. In the fertilization process, a sperm first penetrates between the cells constituting the cumulus oophorus and then through the zona pellucida and into the perivitelline space. The sperm then enters the oocyte through its cell membrane (the *vitelline membrane*). The following is a discussion of what happens during each of these processes, and Figs. 9-4 and 9-5 depict these processes. The entire process of fertilization takes about 24 h.

Sperm Passage through the Cumulus Oophorus

The ovulated ovum is surrounded by the cumulus oophorus, which is a sphere of loosely packed follicle cells (Fig. 9-4). Appropriately, cumulus oophorus means "egg-bearing little cloud." As a sperm enters the cumulus oophorus, the enzyme *hyaluronidase* on the sperm head dissolves hyaluronic acid, a major component of the cementing material found between the cells of the cumulus oophorus as well as between other cells in the body. Enzymatic dissolution of hyaluronic acid allows the swimming sperm to penetrate the cumulus oophorus and to reach the zona pellucida.

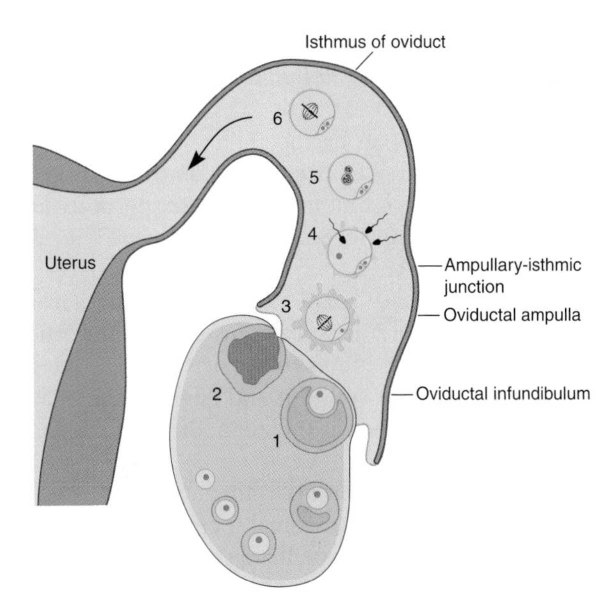

Figure 9-3 Diagram of the human ovary, oviduct, and part of the uterus showing fertilization: (1) Follicle in ovary is ready to ovulate; (2) new corpus luteum; (3) ovulated ovum is arrested in second meiotic division (note the first polar body); (4) formation of second polar body after fertilization; (5) fusion of egg and sperm pronuclei; and (6) beginning of first mitotic division of zygote.

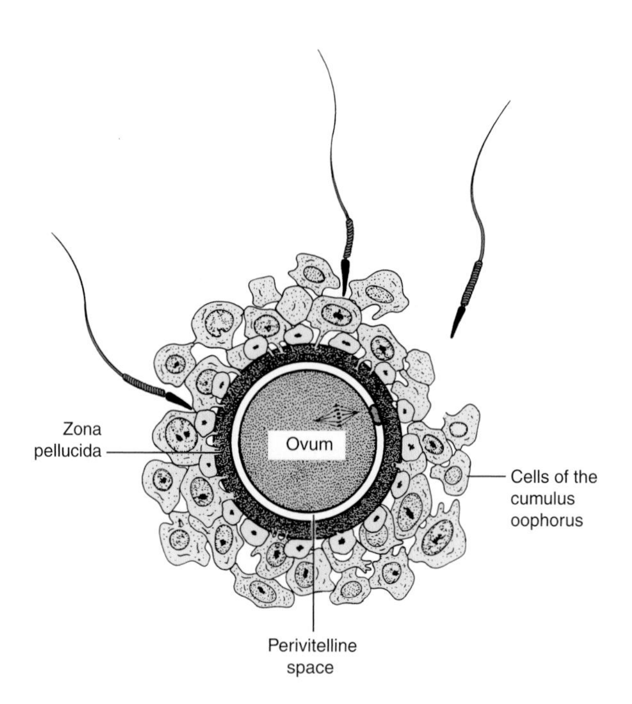

Figure 9-4 Illustration of the barriers around the recently ovulated ovum through which the capacitated sperm must pass to reach the perivitelline space and achieve activation and fertilization of the ovum.

Sperm Passage through the Zona Pellucida

The zona pellucida is an extracellular matrix composed of three glycoproteins termed ZP1, ZP2, and ZP3. Receptors on the sperm plasma membrane attach to ZP3. This ZP3 receptor binding allows the sperm to adhere to the zona pellucida and is a critical step in fertilization. It triggers the sperm head to undergo the *acrosome reaction.* An influx of calcium and a rise in pH and cAMP levels within the sperm head cause exocytosis of the acrosomal vesicle. That is, the plasma membrane of the sperm fuses with the outer acrosomal membrane, forming many small openings to the acrosome. Contents of the acrosome, which are hydrolytic enzymes, spill out and degrade the zona pellucida near the sperm head. This forms a tunnel in the zona, through which the sperm begins to move (Fig. 9-5).

Degradation of the sperm plasma membrane causes the loss of ZP3 receptors. However, now the inner acrosomal membrane is exposed, and it appears to have receptors for another zona pellucida glycoprotein called ZP2. This ZP2 binding maintains the contact between egg and sperm. The sperm tail continues to beat vigorously, helping the sperm penetrate through the zona pellucida and make contact with the plasma membrane of the egg. Once the sperm has penetrated the zona pellucida, it moves through a narrow, oblique path into the *perivitelline space* (the area between the zona pellucida and the vitelline membrane, see Fig. 9-4). Penetration of the human zona pellucida by a sperm takes less than 10 min under experimental conditions.

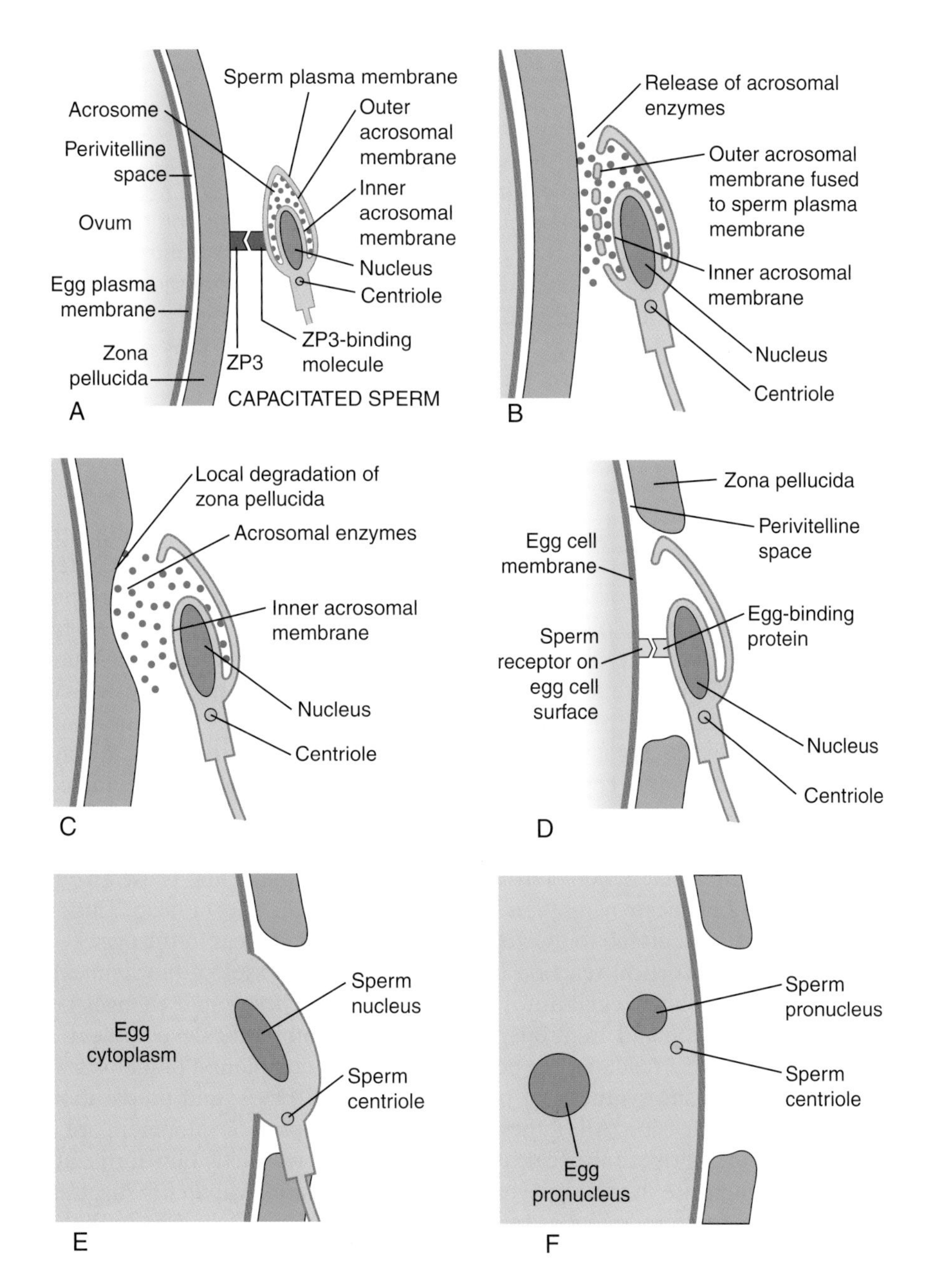

Figure 9-5 Stages of fertilization. Capacitated sperm have already passed through the cumulus oophorus surrounding the egg; for clarity, cumulus cells are not shown. (a) Proteins on the sperm plasma membrane bind to ZP3 molecules within the zona pellucida of the egg. (b) Zona binding triggers the acrosome reaction, in which the sperm plasma membrane fuses with the outer acrosomal membrane, causing exocytosis of acrosomal contents. (c) Acrosomal enzymes begin to dissolve a hole in the zona pellucida. This enzymatic degradation, accompanied by rapid sperm tail beating, moves the sperm through the zona. (d) Egg-binding proteins on the sperm cell surface bind to molecules on the egg cell membrane. (e) The sperm cell membrane fuses with the egg plasma membrane, allowing the sperm nucleus and centriole to enter the egg cytoplasm. (f) Egg and sperm pronuclei migrate toward each other in preparation for syngamy.

Sperm Attachment to the Egg Plasma Membrane

The sperm approaches the egg sideways instead of head on, and the sperm head now lies parallel to the egg cell surface within the narrow perivitelline space (Fig. 9-5). At this point, the posterior part of the sperm head attaches to the egg plasma membrane. The plasma membranes of sperm and ovum fuse, forming an opening into which the sperm nucleus, midpiece, and most of the tail sink into the egg cytoplasm. Scientists are actively investigating the molecules involved in egg–sperm adhesion and subsequent fusion. Finding the molecular basis of sperm–egg fusion may help us understand certain forms of infertility and could possibly lead to new contraceptives.

The Cortical Reaction

Once a sperm has entered the egg, it is imperative that no other sperm be permitted to fertilize it. If additional sperm were allowed to enter the egg, the extra genetic material they carry would disrupt normal development, and the resulting polyploid embryo would die. To prevent *polyspermy* (fertilization by more than one sperm), the egg now mounts a defense. Just underneath the plasma membrane of the egg lie small, membrane-bound vesicles called *cortical granules*. At fertilization, there is a sudden, dramatic burst in available free calcium in the egg cytoplasm as it is released from cytoplasmic storage. The rise in calcium causes cortical granule membranes to fuse with the adjacent cell membrane. Thus, the cortical granules open to the exterior and release their contents into the perivitelline space. Included in the cortical granule contents are enzymes that act on constituents of the zona pellucida. These enzymes alter ZP2 and ZP3, destroying their receptor sites for the sperm head. Thus, no additional sperm can attach to the zona pellucida to gain access to the egg.

The cortical reaction is the first step in a series of biochemical and physical changes in the egg known as *egg activation*. These rapid changes begin just after fertilization and are preparations for early embryonic development. In addition to the cortical reaction, egg activation involves completion of meiosis, increase in egg metabolism, synthesis of protein, RNA, and DNA, and preparation for the first mitotic division. All of these essential first steps in development are dependent on the initial rise in free calcium. We do not know exactly how fertilization initiates a calcium rise in the egg. One theory (the *receptor hypothesis*) suggests that binding of a sperm to an egg receptor induces biochemical changes in the egg cytoplasm that cause release of stored calcium. An alternative idea (the *cytoplasmic factor hypothesis*) is that as the sperm enters the egg cytoplasm, it carries a factor that causes free calcium to be released. Laboratory experiments lend support for each of these hypotheses, but the actual mechanism that occurs during normal fertilization remains unknown.

Completion of the Second Meiotic Division

The ovulated egg is arrested in the second meiotic division and still has a duplicated set of chromosomes. Before merging with sperm DNA, the egg must complete its second meiotic division and jettison one set of its chromosomes.

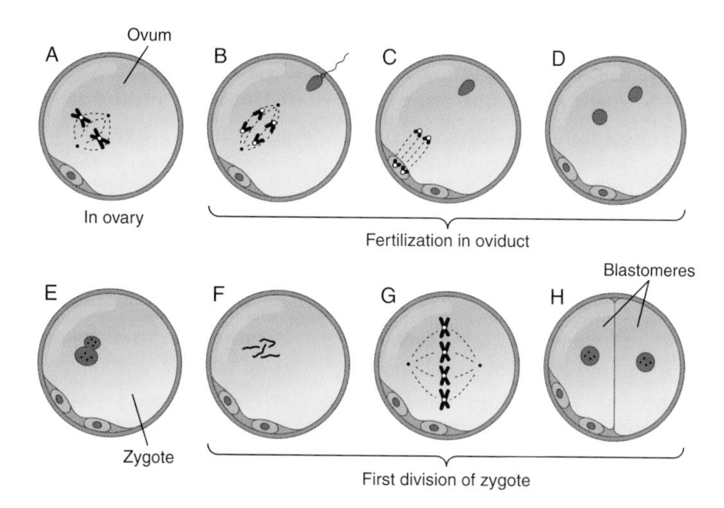

Figure 9-6 The nucleus of the ovulated egg is haploid and its chromosomes are arrested in the second meiotic division (1). The first polar body may divide into two small cells (1), one of which is pictured in further figures. Sperm penetration activates the egg so that the second meiotic division is completed (2, 3) and a second polar body is formed (4). The egg and sperm pronuclei then fuse (5) and the resultant diploid zygote now divides mitotically (6, 7) to form a two-cell embryo (8) consisting of two blastomeres. Note that only two chromosomes are shown in (1), even though there should be 23.

At fertilization, the rise in free calcium activates the egg nucleus to complete meiosis, and a second polar body is produced, removing the extra set of chromosomes from the egg. The second polar body can often be seen in the perivitelline space before it degenerates (Fig. 9-6).

Formation and Fusion of Sperm and Egg Pronuclei

Soon after the sperm nucleus enters the egg, its nuclear membrane breaks down. The sperm DNA decondenses as a result of exposure to factors in the egg cytoplasm. A new membrane then forms to enclose the *sperm pronucleus*. Sperm and egg pronuclei begin to migrate toward each other, replicating their DNA as they move. As they approach each other, their nuclear membranes break down and the two duplicated sets of chromosomes aggregate. Syngamy (merging of the two haploid genomes) has now occurred, and the fertilized egg (*zygote*) is the beginning of a new individual. In mammals, it takes about 12 h from the beginning of egg activation to pronuclear fusion. The centrosome contributed by the sperm organizes a mitotic spindle, and chromosomes now begin to line up at the metaphase plate. The zygote next divides mitotically, and two identical daughter cells, termed blastomeres, are formed (Fig. 9-6). Embryonic development has commenced.

We have seen that the sperm contributes its haploid chromosomes and centrosome to the zygote. The sperm tail disintegrates in the egg cytoplasm. What happens to the sperm mitochondria? It has long been known that the approximately 100 mitochondria brought by each sperm into an egg disappear soon after fertilization. Recent studies have demonstrated how this occurs. During spermatogenesis, sperm mitochondria are tagged with a protein called *ubiquitin*, a molecule used by all cells to mark proteins slated for destruction. These tagged paternal mitochondria are then destroyed and recycled by the egg after fertilization. Thus, all of our mitochondria are inherited from our mothers. Maternal inheritance of DNA-containing mitochondria has been a useful way to trace human origins.

Chapter 9, Box 2: Sperm Hitchhikers

Deprived of all but a scant amount of cytoplasm during the latter stages of spermatogenesis, the human sperm has, until recently, been considered to contribute nothing to the ovum except for its nuclear DNA and centriole. However, we know that factors carried by the sperm play active roles in the fertilization process. Some men with sperm apparently normal in shape, motility, and abundance still are infertile if they lack these biochemical factors. From the text, you know that some of these factors are enzymes such as hyaluronidase and acrosomal, enzymes necessary to break through the layers of cumulus cells and the zona pellucida before reaching the egg surface. Also necessary are zona-binding proteins on the surface of the sperm cell membrane and inner acrosomal membrane. However, these are not all of the players in the process of fertilization, and some sperm "hitchhikers" may also be important for normal development of the egg and embryo.

For example, the sperm head contains a protein, *fertilin-β*, on its surface. After the sperm penetrates the zona pellucida, the tip of its head approaches the vitelline membrane. Then the head turns laterally so that one side of the sperm head attaches to the vitelline membrane (see text). Fertilin-β appears to mediate this lateral attachment. If this protein is absent, fertilization does not occur because sperm–oocyte binding is inhibited. Fertilin-deficient sperm also have a reduced ability to bind to the zona, and our understanding of the normal action of fertilin is still evolving.

When the sperm penetrates the egg, waves of stored calcium ions are released in the egg cytoplasm. This sudden increase in calcium triggers egg activation (cortical granule release and reinitiation of meiosis). Scientists have long speculated that the trigger for calcium release is carried by the sperm. Researchers have found that the sea urchin sperm head contains an enzyme that can synthesize *nitric oxide*. This gas is injected into the egg at fertilization and can set off a calcium surge. It remains to be seen if a similar mechanism operates in humans. Study of human eggs has revealed that *phospholipase C* is carried by sperm into the egg. It also can cause the waves of calcium release and egg activation.

Ribonucleic acid (RNA) is produced when cells read the DNA sequences coded by the genes and transcribe these messages. During the later stages of spermatogenesis, sperm DNA becomes tightly compressed and gene expression ceases. However, scientists have found that sperm RNA is still present in the mature sperm even at fertilization. This is especially surprising because the sperm cytoplasm is virtually gone. Sperm cells contain an amazing repertoire of RNAs. It turns out that about 3000 of the 20,000–25,000 human genes are represented by sperm RNA. Some of the mRNAs represent known genes, others are unknown, and some of the RNAs do not code for proteins. Many types of mRNAs are found in the sperm cell nucleus.

Most of these 3000 transcripts are probably leftover RNA instructions for building the sperm cell during the process of spermatogenesis. However, scientists have identified six RNAs present in the spermatozoa but not in the unfertilized egg. They then asked if these transcripts are carried into the egg at fertilization. If the sperm delivers RNAs into the egg at fertilization, one would expect to find these RNA sequences in the sperm and in the zygote, but not in the unfertilized egg. Using cDNA probes, they found two sperm RNA sequences that are delivered to the egg at fertilization.

What happens to RNAs delivered by the sperm? Possibly they are simply destroyed by

Continued on next page.

Chapter 9, Box 2 continued.

the egg cytoplasm. However, it is also possible that these RNAs act as instructions for early embryonic development and that they are needed to launch the developmental program of the zygote. In fact, one of the mRNA transcripts delivered to the egg codes for *clusterin*, which has been implicated in cell–cell interactions, membrane recycling, and regulation of apoptosis (programmed cell death), processes central to embryonic development.

Preliminary evidence shows that the sperm of some infertile men lack some of the RNAs carried by sperm of fertile men. In fact, it is thought that a treatment that lowers or eliminates the RNAs from the sperm of fertile men may render them infertile, thus providing a potential male contraceptive method. Some scientists use the absence of male RNAs as a possible explanation of why embryonic development is so poor in most cases of cloning and in all cases of human parthenogenesis; neither process involves sperm. However, others cite the occasional success of cloning to argue against an important role for sperm RNAs.

As new sperm molecules are discovered, the role of the sperm has expanded from simply delivering a haploid genome to the egg to essential roles in the fertilization process and perhaps important roles in egg activation and early embryonic development as well.

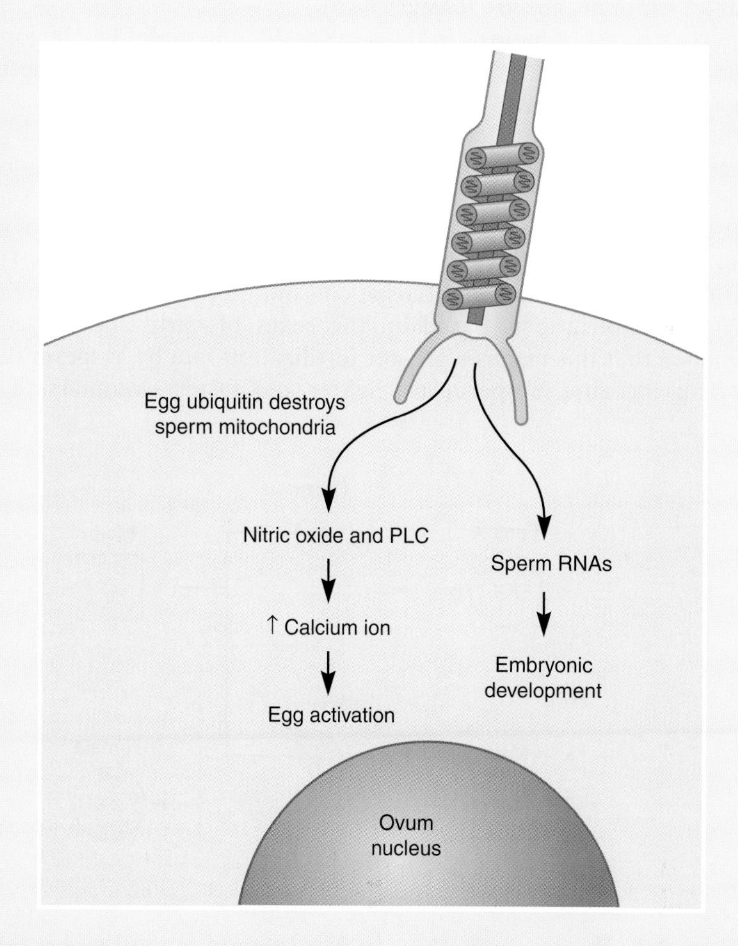

Possible influences of the sperm cell on the egg and/or early embryo in addition to the contribution of its haploid nucleus. For clarity, the sperm cell is shown oriented at right angles to the egg cytoplasm.

Chemical Inhibition of Fertilization

In the future, it may be possible to block fertilization by interfering with steps in the fertilization process. A search for vaccinations against sperm, egg, or the early embryo has long been underway. More recently, studies have focused on specific ways to thwart the actions of sperm cells, either by immobilizing them or by preventing them from undergoing the acrosome reaction, binding to the zona pellucida, or fusing with the egg cell membrane. Some of these potential future contraceptive methods are discussed in Chapter 14.

Sex Ratios

As discussed in Chapter 5, the normal chromosome number in humans is 46 (2N, diploid). Females have 22 pairs of autosomes and two X chromosomes. Males have 22 pairs of autosomes and an X and Y chromosome. The genes for male sex determination are carried on the Y chromosome. Thus, embryos without a Y chromosome are female.

As a result of meiosis in the adult testis, one diploid male germ cell (spermatogonium) gives rise to four haploid spermatozoa (see Chapter 4). Two of these spermatozoa will have 22 autosomes and a Y chromosome, whereas the other two will have 22 autosomes and an X chromosome. If a Y-bearing sperm (22Y) fertilizes an ovum (with 22 autosomes and an X chromosome), the embryo will be male; if an X-bearing sperm (22X) fertilizes an ovum, the offspring will be female. Thus, given an equal chance of X and Y sperm to fertilize, the sex ratio of embryos should be 100:100 (Fig. 9-7). However, the ratio of male to female embryos at conception (the *primary sex ratio*) is about 120:100. This ratio is based on the sexes of early aborted embryos. It is assumed that this means a greater fertilization rate by Y sperm than X sperm, perhaps because Y sperm are lighter and faster swimmers than X sperm.

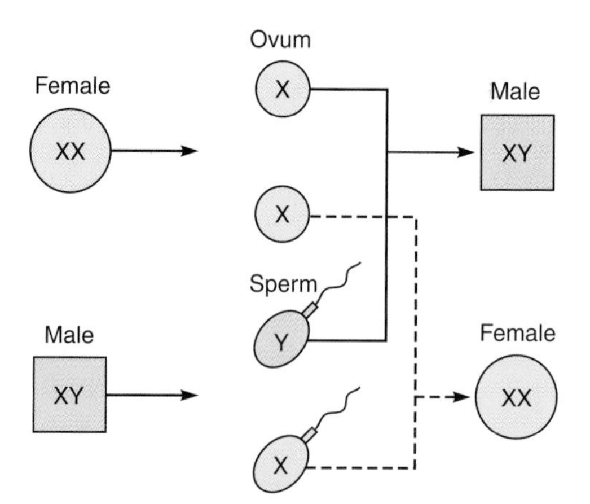

Figure 9-7 The chromosomal basis for the existence of an equal number of X and Y sperm, and thus a theoretical primary sex ratio of 100:100. As discussed in the text, this theoretical ratio is not borne out, and more embryos are male than female.

However, female embryos may die more frequently at an earlier age than male embryos or more X sperm may die in the female reproductive tract than Y sperm. The sex ratio of male births to female births (the *secondary sex ratio*) is 105:100. Thus, for reasons not yet understood, male fetuses suffer a greater mortality than female fetuses in the uterus.

Sex Preselection

Couples who desire to choose the sex of their baby may now do so. A relatively new technology (the *microsort method*) is the most effective procedure yet devised at separating X-bearing and Y-bearing sperm. It takes advantage of the fact that the large X chromosome has considerably more DNA than the tiny Y chromosome. A sperm sample is first collected from the prospective father. Then, the sperm cells are treated with a fluorescent dye that attaches to DNA and glows under laser light. Sperm with more DNA, scientists reasoned, would glow more brightly. Although X sperm have only 2.8% more DNA than those carrying a Y chromosome, the difference in brightness is sufficient to be distinguished by a light detector. The tagged sperm are sent through a very narrow tube with a diameter wide enough to allow only one sperm cell at a time. As sperm move through the tube, they are illuminated by a laser beam. An automated mechanical sperm sorter then separates the sperm, sending X sperm down one tube and Y sperm into another. The sorted sperm can then be placed in the woman's uterus (artificial fertilization) or used for *in vitro* fertilization. Approximately 91% of sperm cells in the X-bearing tube contain an X chromosome. This procedure is only about 74% effective in selecting Y-bearing sperm. Thus, the results are not foolproof, but this method does improve the chances of producing an embryo of the desired sex, especially if a couple wishes to have a girl.

A more accurate method of ensuring the sex of a baby is *preimplantation genetic diagnosis (PGD)*, which is available at a limited number of clinics. It involves *in vitro* fertilization followed by embryo selection. The mother's eggs and father's sperm are collected, and the eggs are fertilized in the laboratory. After 3 days of development, a cell is carefully removed from each embryo and chromosomes are examined. Those carrying a Y chromosome are separated from non-Y-bearing embryos. Only embryos of the desired sex are implanted into the mother's uterus. Although nearly 100% accurate, this procedure is more invasive, expensive, and controversial than sperm-sorting techniques.

Why would parents wish to preselect the sex of their offspring? One reason would be to avoid sex-linked genetic diseases, which are more likely to occur in males. Parents may also wish to balance their families or they may simply prefer to have a child of a given sex. Some have expressed concerns that the ability to select a baby's sex may be the first step to "designer children" chosen for other traits such as height, IQ, athletic, or musical ability. Others fear that widespread sex selection would lead to a gender imbalance in society and cause social problems. In fact, a preference for baby boys in China has led to a significant shift in the sex ratio in some areas of the country. In such cultures where boys are valued more highly than girls, the ability to select sex before fertilization could avoid costly and ethically controversial practices such as amniocentesis (genetic screening for sex), selective abortion, and even infanticide. In the United States, sperm selection likely would not lead to overall gender imbalance, as family preference for a girl or a boy baby is split more evenly. Finally, the ability to preselect a child's sex may help families limit their size. For example, using gender selection technology, a family with three boys could increase their likelihood of having a girl as their fourth and last child instead of continuing to have babies until a girl was conceived. However, the present high cost of the microsort and PGD methods likely will limit the practice of sex preselection in the foreseeable future.

Multiple Embryos

Twins occur in about 1 of every 80 or 90 pregnancies. When two ova are released and each is fertilized by a different sperm, *fraternal twins* are produced. These twins are *dizygotic* (the products of two different zygotes) and can be the same or different sex. Fraternal twins, which are *nonidentical* and are as different from each other as are nontwin brothers and sisters, account for two-thirds of all twins. The incidence of dizygotic twins is influenced by race and by inherited factors from the mother (not the father). Fraternal twins are more common in older mothers.

Identical twins, which are rarer than fraternal twins, usually occur when an early embryo divides into two. These twins are *monozygotic* (derived from one zygote) and are identical genetically. The incidence of identical twins is not related to race, inheritance, or age of the mother. Rarely, identical twins are *conjoined* (i.e., they fail to separate completely during embryonic development). These are called *Siamese twins*, after the first publicized Siamese twins, "Chang" and "Eng" (1811–1874), born in Siam of Chinese extraction. They were united at the chest by a thick mass of flesh. Some Siamese twins have been separated surgically after birth. For more on twin pregnancies, see Chapter 10.

When the number of embryos is greater than two (e.g., triplets, quadruplets), all are usually of multizygotic origin; in a few cases, some are multizygotic and some are monozygotic.

Parthenogenesis

Is it possible that an embryo can develop in a human female without previous fertilization? Embryonic development from an ovum not previously stimulated or penetrated by a sperm is called *parthenogenesis*. Such "virgin birth" is common in many insects, in some fish, amphibians, and reptiles, and in a strain of domestic turkeys. In addition, parthenogenetic mouse embryos can be produced in the laboratory, but they do not develop to term. There is no proven case of a parthenogenetic birth in humans. If parthenogenesis could occur, reduction division in the oocyte must not occur, the offspring would always be female, and the child would be genetically identical to the mother.

Chromosomal Aberrations

Errors of meiosis or fertilization can produce embryos with chromosomal aberrations. More than 90% of these embryos are aborted spontaneously, usually within the first trimester. In fact, 42% of embryos or fetuses that are aborted spontaneously have chromosomal abnormalities. A few fetuses with chromosomal defects, however, are born; about 1 out of every 100 newborns has such a defect. It must be emphasized that some of these disorders are not inherited in the strictest sense because the genes of the parents do not govern their occurrence.

In rare cases, one sperm will fertilize the ovum and a second sperm will fertilize the polar body. The two fertilized cells then form an embryo that is a genetic mosaic in that half of its cells will have a different genetic makeup from the other half. This condition also can occur when the haploid ovum divides into two cells and each cell is then fertilized by a separate sperm. If an X and a Y sperm were involved, half of the cells of an embryo would be male and half female, resulting in an intersex (see Chapter 5).

One kind of chromosomal aberration occurs when fertilization fails to activate the second meiotic division in the ovum. Thus, there is no egg pronucleus and the embryo develops with only one set of chromosomes (haploid) and genes of the male only. This process of embryonic formation is termed *androgenesis*. A similar situation occurs when the ovum pronucleus develops normally, but the sperm pronucleus does not form. In this case, called *gynogenesis*, the embryo also is haploid but has only the female's genes. Both of these conditions are lethal after only a few cell divisions in the embryo.

In contrast to the previously mentioned conditions, some embryos may develop with triploid cells (3N) that have 69 chromosomes (three complete sets). *Triploidy* can occur in at least three ways. First, sperm penetrating the ovum may be the product of a failure of reduction division during meiosis in the testis, and thus it has 46 instead of the normal 23 chromosomes. When this sperm fertilizes a haploid ovum, a triploid embryo develops. Second, even though mechanisms to prevent polyspermy are present, these mechanisms are not fail-safe. Thus, two haploid sperm can penetrate a single ovum (polyspermy) and both of their pronuclei then fuse with the haploid ovum pronucleus. Finally, reduction division (meiosis) may not have occurred in the oocyte, and the resultant diploid female pronucleus then fuses with a haploid sperm pronucleus to produce a triploid zygote.

The excess dosage of genes in triploid embryos tends to be less destructive than when there are too few genes, as in androgenesis or gynogenesis. Most triploid embryos develop to about the third month of pregnancy before aborting spontaneously. The very few triploid fetuses that survive to term are malformed and are stillborn or die soon after birth. Less than 1% of all human embryos are triploid.

Another error in fertilization results in embryos with either one too many (47) or one too few (45) chromosomes in their cells; these conditions are collectively called *aneuploidy*. This happens when there is aberrant chromosome movement during the first or second meiotic division in the testis or ovary or in the first cleavage division of the zygote. That is, a pair of hromosomes fails to separate during division, with both members going to one daughter cell (*nondisjunction*). The resultant cell has 47 chromosomes, and the cell coming up short has only 45. Thus, the aneuploid condition can be either *monosomic* (45 chromosomes) or *trisomic* (47 chromosomes).

Most monosomic embryos abort spontaneously early in their development. An exception, however, is when monosomy for a sex chromosome occurs. That is, each cell has only a single sex chromosome, either an X or a Y. About 98% of these embryos abort, but a few with one X (XO condition; Turner's syndrome) are born as sterile females with short stature and physical defects (see Chapter 5). Only 1 in 3500 living females has this syndrome.

Most trisomic embryos die in the second or third month of pregnancy and abort spontaneously; 20% of miscarried fetuses are trisomic. Some, however, are born with severe physical and mental defects. The most common trisomic condition in infants is *Down syndrome*, also called *Mongolism*, a condition in which the cells of the individual are trisomic for chromosome number 21. Children with Down syndrome exhibit abnormal body development and severe mental retardation.

For some as yet unknown reasons, the gametes of older men and women are more likely to produce trisomic embryos. The chances are 1 in 1000 for having a trisomic embryo for women under 35, but are 1 in 200 for 35-year-old women and 1 in 15 for 45-year-old women. Women over 35 have 15% of all babies but 50% of all Down's syndrome children. Therefore, it is recommended that women in their midthirties consider having the cells of their fetus examined by amniocentesis or chorionic villus biopsy (see Chapter 10) for evidence of chromosomal abnormalities. If certain chromosomal aberrations are found, induced abortion might be considered (see Chapter 15). It used to be thought that errors in meiosis in oocytes of older women were the main cause of trisomy. Recently, however, we have become aware that about one-fifth of trisomic infants are caused by chromosomal abnormalities in the sperm of older men.

As discussed in Chapter 5, nondisjunction of sex chromosomes can produce males with trisomic cells of an XXY or XYY makeup. In the former condition, Klinefelter's syndrome, males are sterile and have female-like breasts. About 1 out of 600 males is born with this condition. In the latter "supermale" condition (XYY), males are very tall and often have acne. These males tend to exhibit mental and social adjustment problems at a higher percentage than normal XY males. One in 2000 males has XYY cells. Some statistical evidence exists that the percentage of XYY males (1.8 to 12.0%) in penal institutions is greater than their percentage (0.14 to 0.38%) in the general population. Some controversy, however, surrounds these studies and it is not clear if the greater maladaptive behavior of XYY males is a direct result of their chromosomal abnormality or is due to social problems they had when growing up because of their unusual physical appearance. Apparently the elevated crime rate of XYY men is not related to aggression but may be related to low intelligence. Women with nondisjunction of the X chromosome have cells that are XXX. These women are female but sterile. Cases in which males have several X chromosomes (XXXY) are due to penetration of the ovum by more than one sperm.

Sometimes a gamete contains a chromosome with an extra piece from another chromosome attached to it; this is the result of *chromosomal translocation*. The chromosome from which the piece was taken thus suffers from *chromosomal deletion*. An example of a disorder resulting from chromosomal deletion is the *cri du chat* (French for "cry of the cat") syndrome, in which a piece of chromosome 5 is missing. These children are born with a small head, widely separated eyes, low-set ears, and mental retardation. When they cry, it sounds like a hungry kitten. Human kidney cancer has also been linked to an inherited chromosomal translocation in which a piece of chromosome 3 is hooked onto chromosome 8.

An inherited disorder of the X chromosome (*fragile X syndrome*) is the second leading cause of mental retardation. In these people, the X chromosome (in either sex) has an abnormally long, fragile arm. In this disorder, mental retardation is less severe in females than in males.

Chapter Summary

After sperm mature in the epididymides, they move down the vasa deferentia. Seminal plasma consists of secretions from male sex accessory glands. These secretions are added to the sperm to form semen (seminal fluid), which leaves the male urethra during ejaculation. Seminal plasma contains substances necessary for sperm movement, maturation, and maintenance.

About 66 million sperm are present in each milliliter of semen. Some of these sperm are abnormal and die. A healthy sperm is made up of a head (nucleus plus acrosome), neck, midpiece, and tail. After insemination of the female, the sperm move through the vagina, cervix, uterus, and into the oviduct. While in the uterus and oviduct, sperm acquire the ability to fertilize (capacitation) and are activated so that their tails beat more rapidly. Meanwhile, the ovulated ovum moves down the oviduct, and the sperm and ovum meet at the ampullary–isthmic junction of the oviduct, where fertilization occurs.

Before penetrating the ovum, a sperm moves first through the cumulus oophorus and zona pellucida. As it binds to the ZP3 glycoprotein on the zona pellucida, it undergoes the acrosome reaction, during which the sperm acrosome releases enzymes that help dissolve the zona. Once the sperm enters the ovum, it causes the completion of oocyte meiosis and the cortical reaction, which produces changes in the zona pellucida that act as a barrier to polyspermy. The haploid sperm pronucleus and egg pronucleus then merge, and a zygote is formed. In the future, certain chemicals may be used to block fertilization as a method of birth control.

Chromosomal sex is determined at fertilization, and couples may now be able to choose their baby's sex. Identical twins (monozygotic twins) are formed when a single sperm fertilizes a single ovum, after which the embryo divides into two. Fraternal twins (dizygotic twins) are formed by the fertilization of two separate eggs and sperm. Although several nonmammalian animal species can have offspring without fertilization (parthenogenesis), this has not occurred in humans.

Chromosomal errors that occur before or during fertilization can result in formation of an embryo that is haploid, triploid, aneuploid, or containing one or more chromosomes with added or deleted genetic material. Most embryos with serious chromosomal errors die early in development, but some genetic errors cause mild to severe disorders in humans.

Further Reading

Block, I. (1981). Sperm meets egg. *Sci. Digest* **89**(3), 96–99.

Fackelmann, K. (1998). It's a girl! Is sex selection the first step to designer children? *Sci. News* **154**, 350–351.

Hall, S. (2004). The good egg: Determining when life begins is complicated by a process that unfolds before a sperm meets an egg. *Discover* **25**, 30–39.

Ridley, M. (1993). A boy or a girl: Is it possible to load the dice? *Smithsonian Magazine* **24**(3), 113–124.
Travis, J. (2002). A man's job: A surprise delivery from sperm to egg. *Sci. News* **162**, 216–217.
Wassarman, P. M. (1988). Fertilization in mammals. *Sci. Am.* **259**(6), 78–85.
Wilcox, A. J., *et al.* (1995). Timing of sexual intercourse in relation to ovulation: Effects on the probability of conception, survival of pregnancy, and sex of the baby. *N. Engl. J. Med.* **333**, 1517–1521.

Advanced Reading

Davis, D. L., *et al.* (1998). Reduced ratio of male to female births in several industrial countries: A sentinel health indicator? *J. Am. Med. Assoc.* **279**, 1018–1023.
Evans, J. P.L., and Florman, H. M. (2002). The state of the union: The cell biology of fertilization. *Nature Med.* **8**(S1), S57–S63.
Garbers, D. L. (1989). Molecular basis of fertilization. *Annu. Rev. Biochem.* **58**, 719–742.
Ostermeier, G. C., *et al.* (2002). Spermatozoal RNA profiles of normal fertile men. *Lancet* **360**, 772–777.
Ralt, D., *et al* (1991). Sperm attraction to a follicular factor(s) correlates with human egg fertilizability. *Proc. Natl. Acad. Sci. USA* **88**, 2840–2844.
Roldan, E. R. S., *et al.* (1994). Exocytosis in spermatozoa in response to progesterone and zona pellucida. *Science* **266**, 1578–1581.
Schatten, H., and Schatten, G. (eds.) (1989). "The Cell Biology of Fertilization." Academic Press, San Diego.
Schatten, H., and Schatten, G. (eds.) (1989). "The Molecular Biology of Fertilization." Academic Press, San Diego.
Simon, C. (2003). The role of estrogen in uterine receptivity and blastocyst implantation. *Trends Endocr. Metab.* **14**, 197–199.
Wassarman, P. M. (1987). The biology and chemistry of fertilization. *Science* **235**, 553–560.

CHAPTER TEN

Pregnancy

Introduction

As discussed in Chapter 9, conception (fertilization) occurs when a sperm and ovum fuse to become a zygote. The zygote then divides mitotically to form two blastomeres. These cells, in turn, divide to produce four smaller blastomeres, and so on. It has been estimated that it takes only about 42 such sets of mitotic cell divisions to produce a newborn baby! Thus, the number of cells in the developing human increases exponentially. Not only proliferation but also cell differentiation takes place so that cells become disparate in form and function: some become liver cells, some nerve cells, some muscle cells, and so on. The zygote gives rise to the embryonic part of the placenta as well as to the embryo itself. This chapter describes the process of *pregnancy*, or *gestation*, during which the mother supports a developing human to a stage at which it can exist in the outside world.

What Is Pregnancy?

It takes about 38 weeks (9.5 months) for a developing human to grow from a zygote about the size of this period (·) to a birth weight of 7.0 to 7.5 lbs or so. Until it has implanted in the uterus by about 10 days after fertilization, the developing organism is a *pre*implantation *embryo*, or *preembryo*, and this time span is often termed the *embryogenic stage*. The developing organism is called an *embryo* during weeks 3 through 8 of development; after week 8, it is termed a *fetus*. The term *conceptus* is also used to refer to the products of conception (the embryo or fetus plus extraembryonic membranes).

Surprisingly, there is often confusion about the term "pregnancy" (see Table 10-1). In most textbooks and in legal rulings about induced abortion (see Chapter 15), pregnancy begins at conception: *We will also use that definition in this book*. Thus, pregnancy on the average lasts 38 weeks (8.75 months, or 266 days), with trimesters of approximately 3 months each. In most of the medical profession, however, due dates for birth are calculated from day 1 of the last menstrual period, and so are trimesters. Using this calculation, pregnancy lasts 40 weeks, or about 9.2 months (280 days on average). Finally, most of the general public feels that pregnancy begins "when you are pregnant," i.e., 2 weeks or so after conception, a day or two after your "missed menses," or when there is a positive pregnancy test. In this case, one would begin trimesters with a positive pregnancy test. Sometimes pregnancy is defined as beginning at

Table 10-1 Stages of "Pregnancy" as Determined by Various Methods

Pregnancy	The medical establishment	Developmental biologists and legal decisions	The general public (United States)
Beginning point	Day 1 of last menstrual period	Fertilization	First missed menstrual period (+pregnancy test)
First trimester (dated from fertilization)	−2–10 weeks	0–12 weeks	2–14 weeks
Second trimester (dated from fertilization)	11–22 weeks	13–24 weeks	15–25 weeks
Third trimester (dated from fertilization)	23–40 weeks	25–38 weeks	26–36 weeks
Length of pregnancy	9.2 months (40 weeks; 280 days)	8.75 months (38 weeks; 266 days)	8.3 months (36 weeks; 252 days)

implantation. A good practice when one is discussing this with physicians or other health professionals is to ask the precise way in which they are using the term "pregnancy." But remember, when these terms are used in this book, we do so in relation to age (time) since fertilization (formation of a diploid zygote).

Signs of Pregnancy

How does a woman know she is pregnant? *Presumptive signs of pregnancy* are possible indications of pregnancy. One presumptive sign is a missed menstrual period associated with coitus during the previous month. As discussed later, secondary amenorrhea is associated with pregnancy. Another presumptive sign of pregnancy is nausea, often after awakening. This is called *morning sickness* (see HIGHLIGHT box 10-3) and is due to a change in stomach function at this time. Morning sickness usually, but not always, goes away in a few weeks. Another presumptive sign that a woman is pregnant is an increase in the size and tenderness of the breasts, as well as a darkening of the areola surrounding the nipples.

Probable signs of pregnancy indicate that, in all likelihood, a woman is pregnant. These include an increase in the size of the abdomen and an increased frequency of urination (because the growing uterus presses on the urinary bladder). Also, the uterine cervix becomes softer by the sixth week of pregnancy; this condition, called *Hegar's sign*, is detected by a physician during a pelvic exam. Another probable sign of pregnancy is a positive pregnancy test, as discussed later.

Positive signs of pregnancy include detection of a fetal heartbeat, feeling the fetus moving, and visualization of the fetus by ultrasound or fetoscopy, methods reviewed later in this chapter.

Sometimes, a woman with either a great desire for, or a fear of, pregnancy can develop some of these presumptive or even probable signs of pregnancy. This is called *false pregnancy* (*pseudocyesis*) and is a good example of how our brain can influence our physiology. In some cases, a false pregnancy can last 9 months!

Pregnancy Tests

Pregnancy tests detect a hormone that is present in the blood and urine of a pregnant woman. This hormone is *human chorionic gonadotropin* (hCG), which is secreted by the placenta soon after pregnancy is established. In the past, bioassays were used to detect the presence of hCG. In these tests, a woman's urine was administered to such test animals as mice, rabbits, frogs, and toads. Because hCG acts like luteinizing hormone (LH), when present in the urine it causes ovulation in female animals or spermiation in male animals. One disadvantage was that most of these bioassays could not detect the presence of hCG until 2 to 4 weeks after the missed menstrual period. Another disadvantage was that many animals were killed. Therefore, newer, more effective pregnancy tests have been developed.

The current method of detecting pregnancy utilizes the fact that a monoclonal antibody to hCG (anti-hCG) can be obtained. When hCG is present in a solution, anti-hCG combines with it to produce a visible color reaction. In such an *immunoassay pregnancy test*, anti-hCG and urine are mixed in a test tube or on a glass slide, and the presence or absence of a certain color is noted. This kind of test takes about 2 h. A deficiency of this method is that in some cases a color forms when hCG is not present; this is due to the interaction of anti-hCG with other proteins in the urine and not to hCG. Also, a negative response can be obtained in a newly pregnant woman. This can occur because the test is done before hCG is secreted in high enough levels to be detected. Most urine tests are not sensitive enough to detect hCG until about 15 days after fertilization (about 1 day after the missed menses). Several home pregnancy test kits are available without prescription; they tend to be less accurate than laboratory results from a clinic or doctor's office. Therefore, if one of these kits is used, one should be aware of possible inaccuracies.

An extremely accurate pregnancy test uses radioimmunoassay of hCG in a woman's blood. With this method, hCG can be detected a few days after implantation. However, only a limited number of clinics have the facilities to perform this procedure and it is more costly than the previously described methods.

One problem with all these pregnancy tests is that they can be "tricked" (especially if done in the first 2 weeks of pregnancy). Not only do the immunoassay tests sometimes give false positives and negatives, but some abnormal kinds of embryonic tissue, such as hydatidiform moles as discussed later, can secrete large amounts of hCG when there is no embryo present. Also, a woman with an ectopic pregnancy (in which the blastocyst implants outside the uterus) may not have detectable levels of hCG in her blood. Failure to detect an ectopic pregnancy can be dangerous. Thus, a woman should consult with a physician in conjunction with any pregnancy test, positive or negative.

What to Do If You Are Pregnant

Pregnancy can be wanted or unwanted (or sometimes both). If a woman is pregnant but does not want to have a child, she has the options of induced abortion or adoption (see Chapter 15). If she wants her pregnancy to go to term, she should go to a physician, midwife, or clinic 2 or 3 weeks after her missed menstrual period. Once there, she will have her medical history taken, including information on the patterns of her menstrual cycles, previous pregnancies, miscarriages, surgery (especially of the pelvic region), childhood diseases, past or present venereal disease infections or other pelvic infections, and general health. Any genetically related disorder in her or in her partner's past should be communicated. The woman should also have a complete pelvic examination for evidence of infection and structure of her pelvis. A Pap smear should be done to check for cervical precancerous or cancerous growth. Her blood and tissue should be tested for sexually transmitted disease, blood type, and diabetes. She should find out about the fees and should discuss opinions about anesthetics, the role of her partner in delivery, and the availability of alternative birthing experiences (see Chapter 11). A woman who does not want her pregnancy to go to term should also visit a physician or clinic as soon as possible.

After the initial visit, most physicians advise a pregnant woman to be checked monthly for the first 5 or 6 months of pregnancy. Then, she is examined at least twice a month, until the last months, when she should be seen once a week.

The Process of Pregnancy

Implantation

As mentioned in Chapter 9, fertilization usually occurs in the ampullary–isthmic portion of the oviduct. Then the dividing ball of cells moves down the oviduct and into the uterus, where it implants in the uterine wall. We now describe this process.

After fertilization, the zygote undergoes several cleavage (mitotic) divisions to become a ball of approximately 32 cells by the third day after fertilization. This solid ball of cells is called a *morula* (Latin for "mulberry"), as shown in Fig. 10-1. The zona pellucida, which originally surrounded the oocyte within the follicle before ovulation, remains as a translucent membrane surrounding the morula. The morula continues dividing as it passes down the oviduct to the uterotubal junction (Fig. 10-2). It is assisted in its journey by the relative absence of oviductal mucus at this stage and by the beating of oviductal cilia in a uterine direction.

At about 3 to 4 days after fertilization, the preembryo enters the uterus (Fig. 10-2). Now, it is a larger mass of cells called a *blastocyst* (Fig. 10-1). The blastocyst looks like a solid ball of cells from the outside. If, however, it is cut in half, a fluid-filled cavity, the *blastocoel*, is revealed. A single layer of cells, the *trophoblast*, forms the outer layer of the blastocyst just inside the zona pellucida (Fig. 10-1). A clump of cells near one end of the blastocyst underneath the trophoblast layer is called the *inner cell mass* (Fig. 10-1). This group of cells gives rise to the embryo. It is also the source of *embryonic stem cells* (see Chapter 16).

The blastocyst rests freely in the uterine cavity for about 2 or 3 days, during which time it derives nutrients secreted by the uterine glands and increases slightly in size. On about the sixth day after fertilization, the uterus secretes an enzyme (protease) that dissolves the zona pellucida surrounding the blastocyst. Once the zona pellucida disappears, the inner cell mass end of the blastocyst

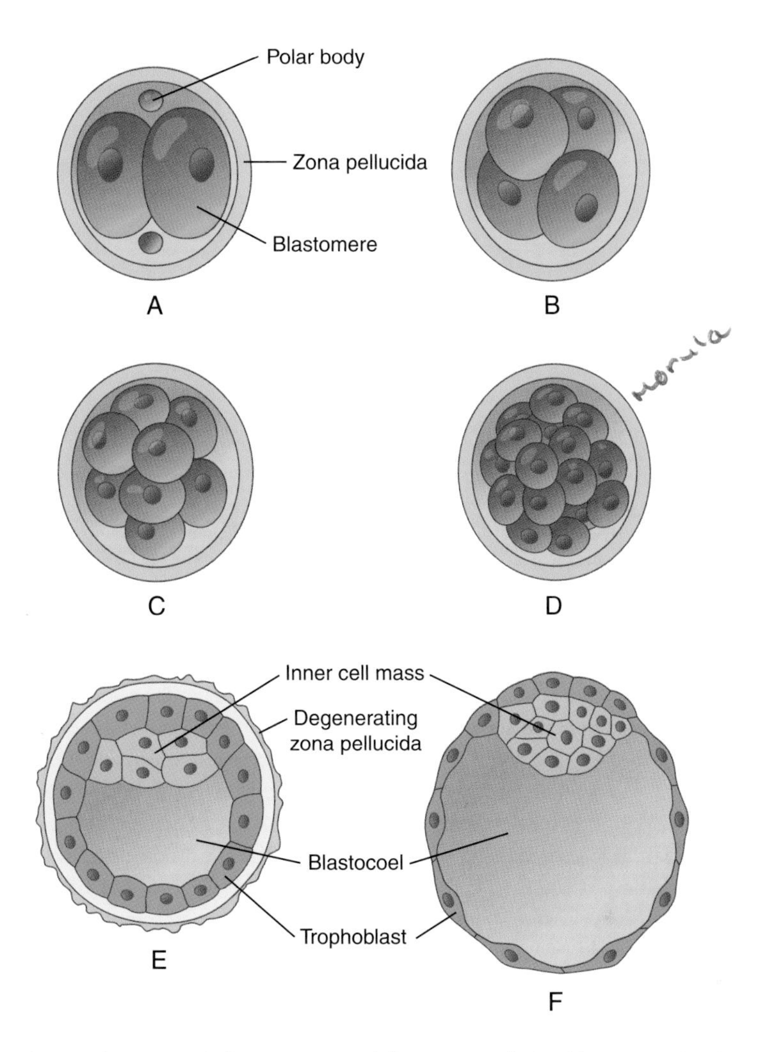

Figure 10-1 Cleavage of the zygote and formation of the blastocyst. (a–c) Various stages of cell division (cleavage). (d) A solid ball of cells called the morula. Note that as cleavage proceeds, cells become smaller. (e and f) Stages of the blastocyst.

attaches to the uterine wall. The blastocyst then begins to invade the endometrium, a process called *implantation*, or *nidation*. Implantation occurs 7 to 10 days after fertilization and usually occurs on the posterior wall of the fundus or corpus of the uterus (see Chapter 2).

During the early phases of implantation, the trophoblast differentiates into an outer *syncytiotrophoblast* and an inner *cytotrophoblast* (Fig. 10-3). The syncytiotrophoblast secretes proteases that break down cells of the uterine endometrium, thus allowing the blastocyst to penetrate into the uterine stroma. The syncytiotrophoblast acquires its name from the fact that it consists of a mass of cells that have lost their cell membranes and have communicating cytoplasm; such a structure is called a *syncytium*. Meanwhile, cells in the *uterine stroma* (the connective tissue framework of the uterus) multiply rapidly and form a cup

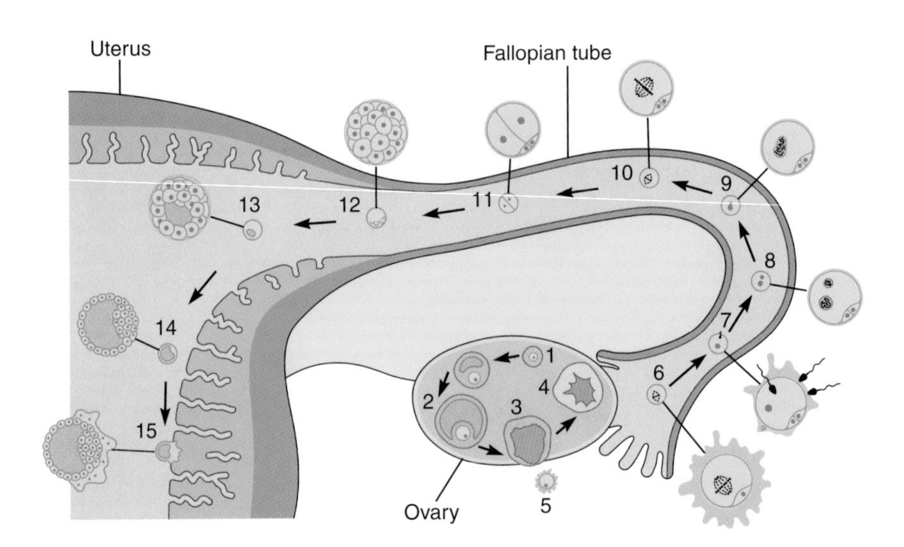

Figure 10-2 Internal fertilization of the human egg is one event in a complex process that begins when a primary follicle in the ovary (1) develops to become a mature follicle (2), which ruptures, releasing the egg (3). The follicle becomes a corpus luteum (4). (Follicular development is shown arbitrarily at different sites in the ovary for clarity.) The ovulated egg is swept from the surface of the ovary by the open end of the oviduct, or fallopian tube (5). The egg has just completed the first meiotic division, a division whereby its chromosomes are reduced to half the normal complement and a polar body is extruded to the margin of the egg; a second meiosis has begun (6). A sperm penetrates the egg cytoplasm, activating the egg to complete the second meiotic division, extrude a second polar body, and form the female pronucleus (7). The male pronucleus is released into the egg cytoplasm (8), and the two pronuclei fuse, mingling male and female chromosomes (9). The chromosomes replicate and divide (10), and the fertilized egg, or zygote, undergoes cleavage (11). Successive cleavages produce a morula (12), an early blastocyst (13), and, after some 4.5 days, a late blastocyst with an inner cell mass that gives rise to the embryo and an outer trophoblast (14). On the 7th to 10th day, the blastocyst implants in the uterine wall (15).

that grows over the blastocyst. This growth of uterine stromal cells is called the *deciduoma response*. Implantation is now complete. The conceptus has invaded its mother's endometrium and will now live off nutrients in the maternal bloodstream. In turn, the maternal tissue has cooperated to allow a controlled, limited incursion. As in all delicate negotiations, this complex, coordinated interaction between the conceptus and maternal tissues is possible only through communication. Researchers are now working to decipher the chemical signals (such as growth factors and other proteins) released by the conceptus and the maternal endometrium during implantation. Figures 10-2 and 10-3 summarize the process of implantation.

Implantation is a very important event. If it does not occur, the blastocyst will degenerate and pregnancy will be terminated. Implantation requires a uterus that has been exposed to just the right amounts of estradiol and progesterone at the right time. Remember from Chapter 3 that the corpus

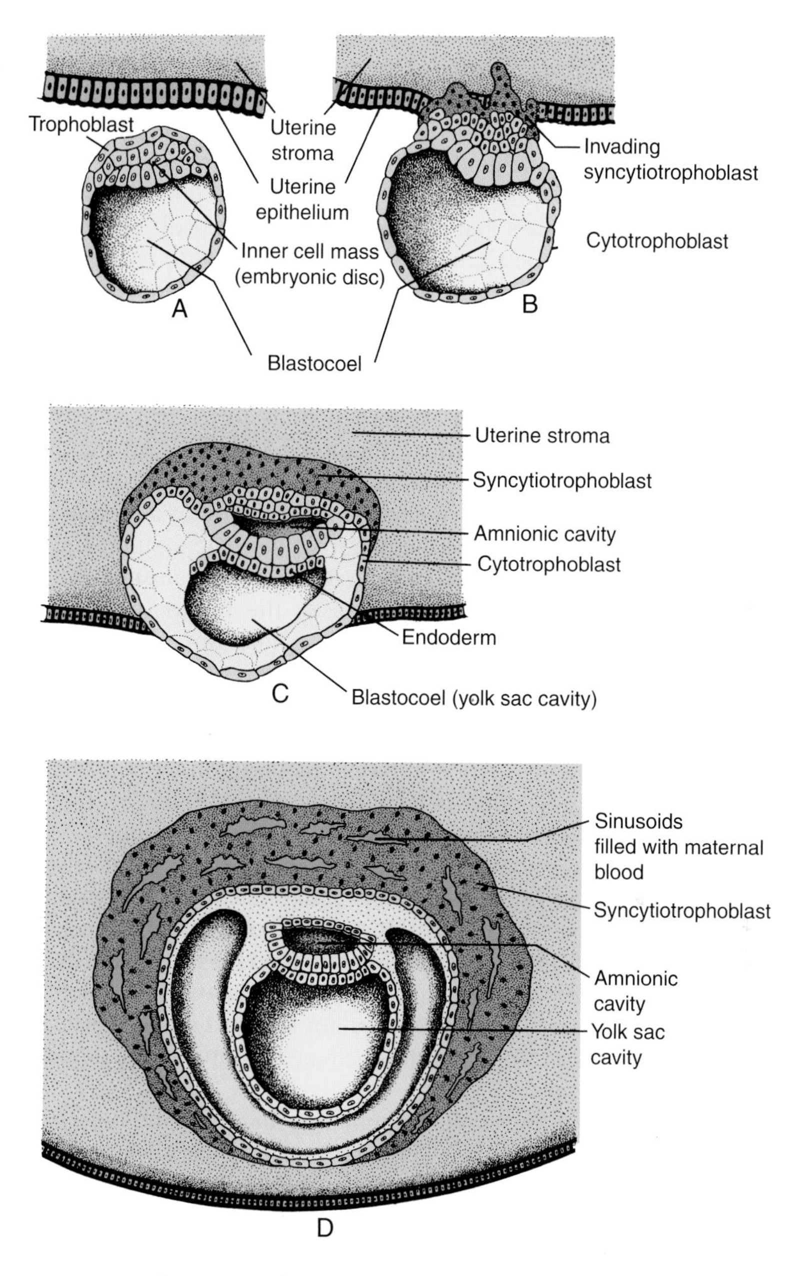

Figure 10-3 Implantation of the human embryo: (a) The blastocyst is not yet attached to the uterine epithelium; (b) the trophoblast has penetrated the epithelium and is beginning to invade the stroma; (c) the blastocyst sinks further into the stroma and the amnionic cavity has appeared; and (d) the uterine tissue has grown over the implantation site (deciduoma response), and irregular spaces, blood sinusoids, have appeared in the syncytiotrophoblast.

luteum formed after ovulation secretes moderate levels of estradiol and higher amounts of progesterone. This ratio of steroid hormones primes the uterus, making the endometrium vascular, secretory, and ready for implantation. Progesterone also causes the uterus to secrete the protease that dissolves the zona pellucida surrounding the blastocyst. Implantation also requires the active participation of the blastocyst. A blastocyst poised for implantation releases signaling molecules that participate in a reciprocal "conversation" between blastocyst and endometrium that allows the process to proceed normally.

A fetus's cells are genetically different from those of the mother. If doctors were to transplant a genetically different tissue or organ into a woman's body, her immune system would recognize the tissue as "foreign" and would attack and reject it. When she becomes pregnant, her fetus acts as an invading foreign parasite that one would expect her body to reject just as vigorously. However, for 9 months, the maternal immune system holds off the attack. How does the fetus avoid its mother's immune system assault? One possibility is that the fetus might somehow escape detection by the mother's immune cells. Alternatively, pregnancy may suppress the maternal immune system. The answer is not clear, but it may involve aspects of both of these mechanisms.

Recall that the external layer of the blastocyst is the *trophoblast*, which makes contact with the uterine epithelium as implantation begins. The *syncytiotrophoblast*, which develops as the outermost layer of the trophoblast, aggressively burrows into the uterine wall until the embryo is completely embedded in maternal tissue. This syncytiotrophoblast encloses the developing embryo and is the only embryonic tissue in direct contact with maternal tissue throughout pregnancy. It later forms the external layer of the placenta. Thus, the placenta is the front line of maternal/fetal interaction.

A woman's body can detect the presence of foreign cells because cell surface proteins called *histocompatibility molecules* (HLAs) "present" foreign antigens to her immune cells. These HLAs are found on most nucleated cells. However, placental cells largely lack these common HLAs. Instead, the surface of placental cells is studded with a unique molecule, HLA-G, not present in other cells of the mother or fetus. The mother may not recognize these particular histocompatibility molecules as "foreign," thus helping the fetus evade detection by the maternal immune system.

There is little evidence that the maternal immune system is suppressed overall during pregnancy. Pregnant women can successfully fight off colds, flu, and other common infections. However, those women with autoimmune diseases can experience more severe symptoms (e.g., lupus) or their symptoms can be alleviated during pregnancy (e.g., rheumatoid arthritis). This reflects a shift in the relative importance of cell-mediated and antibody defense systems during pregnancy. Changes in circulating levels of progesterone, prostaglandins, and other hormones likely induce alterations in the maternal immune system. Some evidence also shows local changes in the immune response within the uterus. Thus, a woman's immune system is altered but not disabled during pregnancy.

Fetal cells can occasionally slip across the placental barrier and make their way into the mother's bloodstream. For some reason, some of these cells escape being killed and can multiply and persist in the maternal circulation for decades (see HIGHLIGHT box 10-1). Whether the presence of these roaming fetal cells influences the mother's immune response to her fetus is not known.

Chapter 10, Box 1: Trading of Cells between Fetus and Mom during Pregnancy

We know that a mother and her offspring should remain close, but a new form of cellular closeness was discovered only recently. Normally the blood of the fetus and the pregnant woman does not freely cross the placenta. However, it turns out that there is some cellular trafficking of stem cells of the immune system across the placenta. Thus, fetal cells can end up in the mother's blood and tissues. The descendents of these cells can persist in the mother's body for decades and perhaps her lifetime, a condition called *fetal microchimerism* (FM). Similarly, maternal cells can reside in the offspring for many years, so-called *maternal microchimerism* (MM). Microchimerism is a surprisingly common phenomenon. The vast majority of women who have been pregnant (whether or not the pregnancy went to term) are thought to carry fetal cells, and at least 25% of us carry our mothers' cells within us. However, the number of foreign cells involved is very small. Women typically harbor about two fetal cells in a tablespoon of blood—less than one fetal cell per million maternal blood cells.

Most studies of the microchimerism of pregnancy have looked at mothers who have been pregnant with boys. The reason for this is that male cells can be detected in the maternal bloodstream by the presence of the Y chromosome. Fluorescent markers can be used to find individual Y chromosomes, and thus male cells. The presumption is that these could only have come from a male fetus, unless the woman had previously received a tissue transplant or blood transfusion from a male. The Y chromosome has been found on microchimeric cells called hematopoietic stem cells. These give rise to a wide variety of blood cells, including those of the immune system. Scientists find these stem cells by testing for the protein CD34, which is a marker for blood stem cells. Cells traded during pregnancy are not, of course, the same cells detected decades later. Rather, the foreign stem cells give rise to progeny that stay in the body, possibly for life. An intriguing possibility is that these cells may also be able to differentiate into other cell types. How they are tolerated by the host immune system is not understood.

Evidence suggests that both FM and MM may be involved in causing autoimmune diseases in the mother and offspring, respectively. Autoimmunity is when the immune system attacks one's own tissues. Many autoimmune diseases are much more common in women than in men and occur most often in the years and decades after pregnancy. This raises the hypothesis that some cases of autoimmune disease may be caused by descendants of fetal cells remaining in the mother's circulation. The maternal immune system may attack these foreign cells, or the foreign cells may launch an attack against the host maternal cells. What are the autoimmune diseases in which the presence of microchimerism has been implicated?

1. **Scleroderma and systemic sclerosis**. Excessive collagen (connective tissue) is deposited first in the skin (scleroderma) and then in internal organs (systemic sclerosis).
2. **Thyroid disease**. Disorders of the thyroid gland such as Hashimoto disease and Graves' disease, which lead to abnormal metabolism in the tissues.
3. **Primary biliary cirrhosis**. Progressive inflammatory destruction of the liver bile system.
4. **Sjögren's syndrome**. Lymphocyte infiltration into lacrimal (tear) glands and salivary glands, leading to chronic dry eyes and mouth.
5. **Multiple sclerosis**. Degeneration of the myelin sheath (fatty covering around nerve axons), causing severe sensory and motor nerve disorders.
6. **Rheumatoid arthritis**. The joints are attacked by one's own immune system.

Continued on next page.

Chapter 10, Box 1 continued.

7. **Systemic lupus erythematosus**. A generalized connective disorder causing inflammation of the skin, joints, blood cells, internal organs, and voluntary muscles.
8. **Juvenile dermatomyositis**. Inflammation of the skin and voluntary muscles.

Evidence linking microchimerism with scleroderma and systemic sclerosis is stronger than for other autoimmune diseases. For example, one study showed that systemic sclerosis patients who are mothers of sons have 30 times the number of fetal cells in their blood (as determined by the presence of the Y chromosome) as do unaffected women (fetal microchimerism). Evidence also implicates maternal microchimerism in systemic sclerosis. Individuals with this disease are over three times as likely to have their mothers' cells floating in their blood than nonaffected individuals. The numbers of foreign cells may be insufficient to accomplish the immune attack themselves. They may simply initiate an inflammatory response. Host immune cells attracted to the area would then carry out the immune response, leading to tissue damage.

Most of the evidence that autoimmune diseases are caused by microchimerism is circumstantial. At this point, the presence of foreign cells and development of an autoimmune disease is just correlation and does not prove a cause-and-effect relationship. In addition, because most women with male cells from their sons in their body do not develop these autoimmune diseases, there are other factors, perhaps genetic or environmental, involved in the risk of developing these diseases.

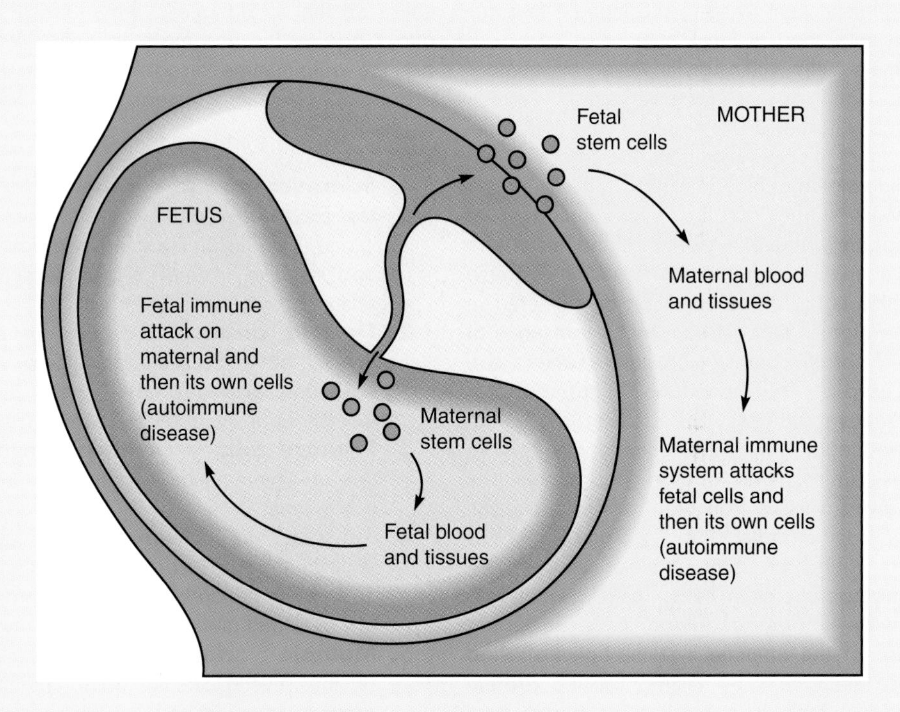

Maternal cells (red) occasionally cross the placenta and enter the fetal circulation. Very small numbers of their progeny may remain even after birth, a phenomenon termed *microchimerism*. A few cells from the fetus (blue) may also cross the placenta in the opposite direction, entering the maternal circulation. Descendants of these migrant cells can be detected in the maternal circulation years after she has given birth.

Continued on next page.

Chapter 10, Box 1 continued.

Some scientists have speculated a link between microchimerism and two pregnancy-related disorders that are not considered autoimmune diseases. These are polymorphic eruption of pregnancy, an outbreak of hives-like skin lesions, and preeclampsia. Both occur in the last trimester of pregnancy. Preeclampsia is the much more dangerous of the two conditions; it is the early stage of toxemia of pregnancy, characterized by high blood pressure, edema, and impaired kidney function. The risk of preeclampsia is correlated with unusually high levels of circulating fetal cells, and the severity of the disease increases with a greater presence of fetal cells.

As mentioned, most studies of microchimerism have focused on mothers and sons. Because cellular trafficking across the placenta between pregnant mothers and daughters must also occur, this type of microchimerism might also have a role in autoimmune disease. Microchimerisms can also be caused by blood transfusion or having a twin next to you in the womb, with sometimes a shared circulation. Theoretically, one could even possess cells from an older sibling passed first to one's mother during her previous pregnancy!

Considering that microchimerism is such a widespread phenomenon, is there a possible benefit for either mother or offspring? If the foreign stem cells are able to differentiate into a variety of cell types, they may be able to repair or replace damaged host tissue. In fact, two such cases have been discovered in which portions of one woman's thyroid gland and another woman's liver were discovered to be made up of entirely male cells, presumably from their sons. Could it be that one benefit of pregnancy is a "minitransplant" of stem cells from one's children?

Early Embryonic Development

Soon after implantation, the inner cell mass differentiates into two layers of cells, the *epiblast* and the *hypoblast* (Fig. 10-3). This two-layered structure is called the bilaminar embryonic disc, although only one of the layers (the epiblast) will give rise to cells of the embryo proper. The epiblast splits into three *germ layers*: ectoderm, mesoderm, and endoderm, pushing aside the hypoblast. Cells arising from the *ectodermal layer* will form the nervous system of the embryo, as well as the epidermis (outer layer) of the skin and related structures such as hair, nails, and tooth enamel. The *mesodermal layer* gives rise to many internal structures, including the skeleton, muscles, the circulatory system, the deep (dermal) layer of the skin, the kidneys and gonads, and the *notochord*, a longitudinal rod supporting the back of the early embryos that is later replaced by the vertebral column. The *endodermal layer* differentiates into the lining of two tubes: the digestive tube (and the liver, gall bladder, and pancreas that bud off the gut tube) and the respiratory tube, including the lungs. Thus, all of the cells in an individual's body originate in the epiblast layer. The hypoblast, along with cells from the epiblast and the trophoblast, contributes to the extraembryonic membranes, which are discussed next. These tissues are vital in sustaining the embryo during intrauterine development, but eventually degenerate or are shed at birth.

Extraembryonic Membranes

The inner cell mass produces three of the four *extraembryonic membranes*. The first of these to be formed is the *yolk sac*, which is an endoderm-lined membrane that surrounds the blastocoel; the blastocoel now is called the *yolk sac cavity* (Figs. 10-3 and 10-4). The yolk sac is not functional in humans, but remains as a vestige reflecting our evolution from ancestral reptiles that relied on stored yolk for embryonic nutrition. Most mammalian eggs have little or no yolk, and nutrients are instead transferred from maternal to embryonic

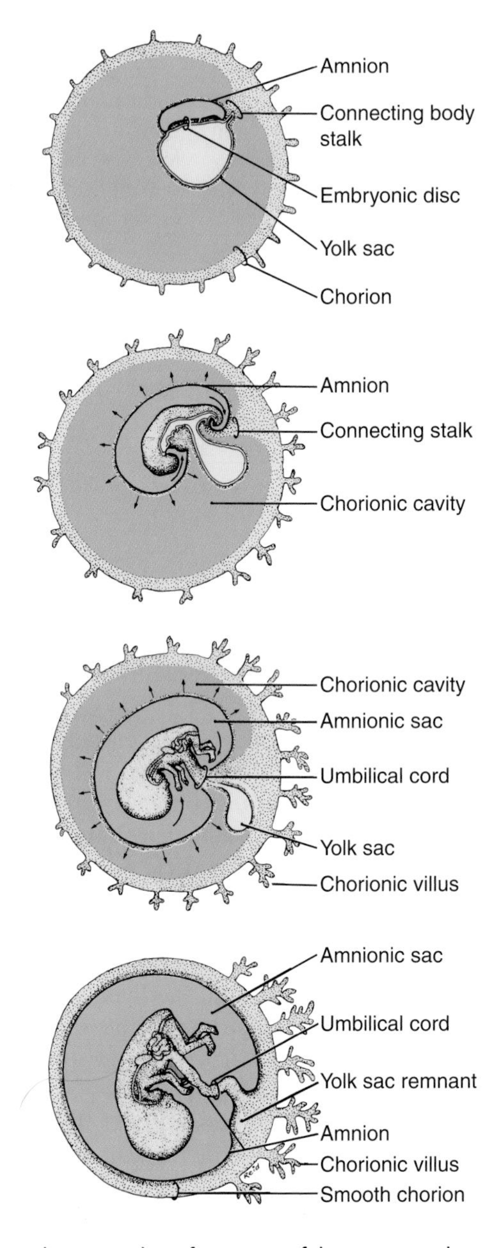

Figure 10-4 These drawings show formation of the amnion, chorion, and yolk sac. Note that the chorion eventually fuses with the amnion and that the yolk sac eventually degenerates. The allantois is not shown. The region of the extended chorionic villi in the lower figure is the placenta.

circulation via the placenta. The human yolk sac degenerates early in development. Before it does, however, it supplies the embryo with blood cells.

Another extraembryonic membrane formed from the inner cell mass, the *amnion*, then grows over the forming embryo (Fig. 10-4). The amnionic cavity becomes filled with *amnionic* (or *amniotic*) *fluid*. The amnion is an important extraembryonic membrane throughout development. The fluid in this sac supports and protects the fetus against mechanical shock and supplies water and other materials to the fetus. The amount of amnionic fluid is about 5 to 10 ml after 8 weeks of development, about 250 ml at 20 weeks, and increases to a maximum of 1000 to 1500 ml by the 38th week of pregnancy. The amount of amnionic fluid then declines to 500 to 1000 ml near the time of birth. Amnionic fluid is secreted and absorbed rapidly, at a maximum rate of about 300 to 600 ml per hour.

The third membrane is the *allantois*, which forms as a small pouch at the posterior (tail) end of the embryo. Although it has important functions in waste storage and gas exchange in many animals, the human allantois is a vestigial sac not used for these functions. Primordial germ cells originating in the proximal epiblast appear at the base of the allantois before migrating to the developing gonads (see Chapter 5). The allantois briefly extends into the umbilical cord of the early embryo. Blood vessels formed in the allantoic tissue contribute to umbilical cord development.

A fourth membrane, the *chorion*, is derived from the cytotrophoblast and surrounds the embryo after about 1 month of development; the chorion eventually fuses with the amnion (Fig. 10-4). The chorion forms an important component of the placenta, as discussed next.

The Placenta

The *placenta* is an organ vital to the developing fetus. Through this organ, the fetus receives important substances, such as oxygen and glucose, and it eliminates toxic substances, such as carbon dioxide and other wastes. The placenta is formed in the following way (Figs. 10-3 and 10-4). About the 14th day after fertilization, finger-like projections of the cytotrophoblast extend through the syncytiotrophoblast and toward the vascular uterine stroma. These projections are called *chorionic villi* because the trophoblast grows to surround the embryo and thus forms the chorion. The syncytiotrophoblast surrounding the chorionic villi secretes enzymes that dissolve the walls of small uterine blood vessels present in the stroma so that the mother's blood forms small pools (sinusoids) that actually bathe the villi. Thus, the human placenta is termed a *hemochorial placenta*.

It must be emphasized, however, that at no time does the mother's blood mix with that of the fetus. Each chorionic villus contains small blood vessels that are fed by blood coming from the fetus via the umbilical arteries present in the umbilical cord. Materials from the mother's blood, such as oxygen and glucose, then diffuse from the uterine sinusoids through the thin wall of each chorionic villus and into the fetal vessel within each villus. These materials travel to the fetus via the umbilical vein. In an opposite way, fetal waste products such as carbon dioxide leave the fetal blood and diffuse into the mother's blood to be excreted. The detailed structure of the placenta is shown in Fig. 10-5. The placenta serves as a nutrient, respiratory, and excretory organ for the fetus. As

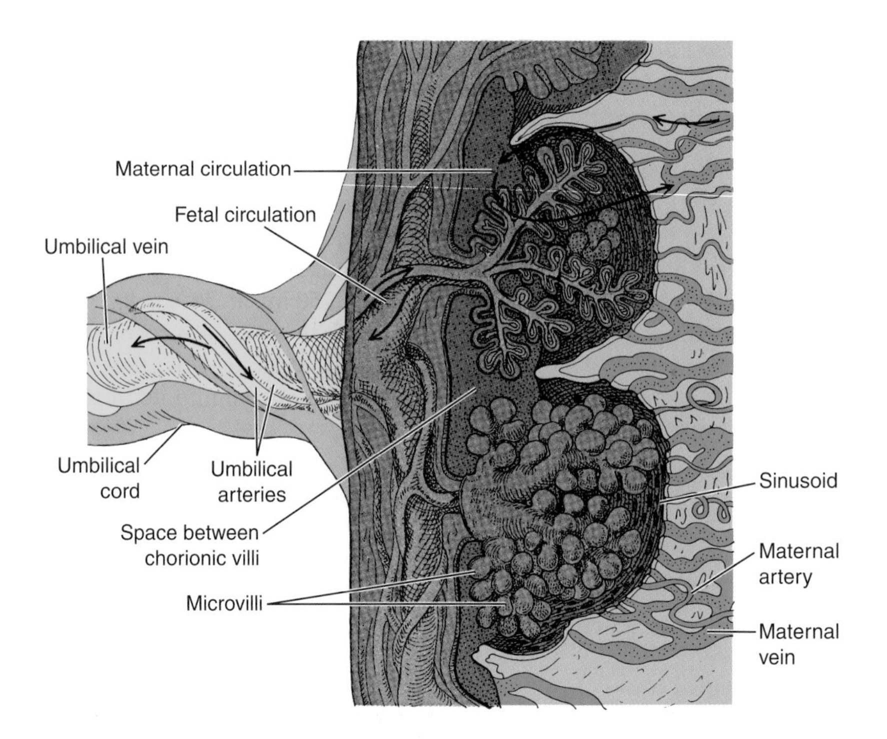

Figure 10-5 Structure of the human placenta in the vicinity of the umbilical cord. The human placenta is "hemochorial" because the chorionic villi are directly bathed by maternal blood.

discussed later, the placenta also secretes hormones that are vital to pregnancy. Table 10-2 summarizes the timing of events ending with implantation.

Molecules larger than about 500 molecular weight will not pass from the mother's blood into the fetal circulation because they are too large to diffuse through the chorionic villi and into fetal blood vessels. This means that most

Table 10-2 Summary of the Usual Timing of Early Events in Preembryonic Development, Considering Conception as Day 1

Day	Event
0	Ovulation
1	Conception (fertilization)
3	Morula
4	Early blastocyst
5	Late blastocyst
6	Blastocyst sticks to endometrium
7	Implantation begins
8	Amnionic cavity and embryonic disc form
9	Uterine sinusoids develop
10	Implantation complete
15	First missed menses (positive pregnancy test)

protein hormones of the mother do not reach the fetus; this is important because maternal pituitary hormones reaching the fetus could alter fetal development adversely. Actually, some large proteins from the mother do reach the fetus near the end of pregnancy. They are the maternal antibodies, which are actively pumped into the fetal circulation by placental cells, which is how the fetus is born with its mother's immunological protection against disease. Steroid hormones in the mother's blood are small enough to cross the placenta, but most do not because they are degraded by placental enzymes. Other maternal hormones, such as thyroxine, can enter the fetus. In fact, hypersecretion or hyposecretion of a woman's thyroid gland during pregnancy can harm fetal development. Bacteria are too large to cross the placenta, but many harmful viruses, such as HIV (see Chapter 18), as well as drugs such as cocaine, can cross.

The placenta continues to grow, like an expanding disk, throughout pregnancy. At the fourth week of pregnancy, this organ covers about 20% of the inner wall of the uterus, and by week 20 it covers about half the uterine wall. At this time, the placenta weighs about 200 g, and the fetus weighs 500 g. At term, the placenta is a disk-like structure that weighs about 700 g (about 1.5 lbs) and has a diameter of 20 cm (8 in.). Although the size of the term placenta varies among individuals, it is usually about one-sixth the weight of the term fetus. The placenta is an extremely vascular organ. Near the end of pregnancy, about 75 gal (285 liters) of blood pass through the placenta daily; this is about 10% of the total blood flow in the mother. This large organ, although supporting the fetus, has a life of its own; if the fetus dies or is removed, the placenta continues to flourish.

As the fetus and placenta grow, the stratum functionalis of the endometrium is transformed into the "decidua" of the pregnant uterus (Fig. 10-6). The maternal part of the placenta is the *decidua basalis*. The overgrowth of the endometrium during implantation (the deciduoma response) is now the *decidua capsularis*, and

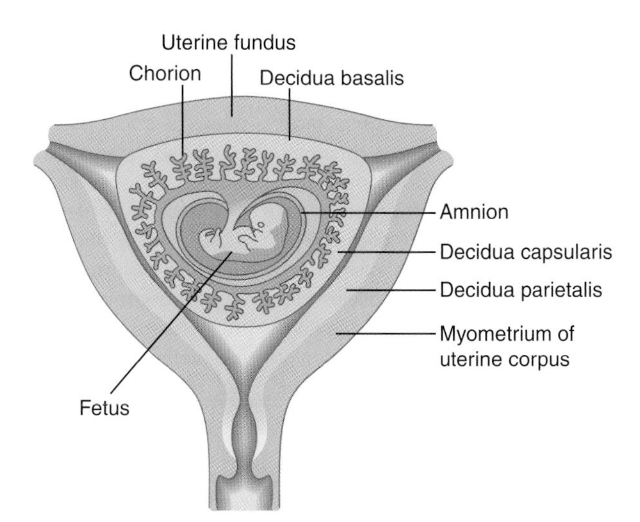

Figure 10-6 Diagram showing the three decidua of the pregnant uterus. The decidua basalis is the maternal part of the placenta. The decidua capsularis represents the deciduoma tissue described in the text. The decidua parietalis is the endometrium not associated with the fetus. Note that implantation could also have occurred on the corpus instead of the fundus of the uterus.

the endometrium away from the fetus is the *decidua parietalis* (Fig. 10-6). Note that the fetus actually resides within the uterine wall, not in the uterine cavity.

The *umbilical cord* connects the fetus with the placenta (Figs. 10-4 and 10-5) and is the lifeline of the fetus. It is derived from a structure connecting embryo and chorion, the *body stalk* (Fig. 10-4). At birth, this cord is 0.3 to 1.0 in. (1 to 2 cm) in diameter and 20 to 22 in. (50 to 55 cm) long. It is covered by the amnionic membrane and contains two *umbilical arteries* (which carry deoxygenated fetal blood to the placenta) and one *umbilical vein* (which carries oxygenated blood back to the fetus). Normally, arteries (vessels carrying blood away from the heart) contain oxygenated blood and veins (vessels carrying blood towards the heart) carry deoxygenated blood. In the umbilical circulation, however, the opposite is true. Vessels within the cord are cushioned by a gelatinous substance, *Wharton's jelly*. If one views a living umbilical cord, its pale color with its spiraling red and blue vessels explains the symbolism of the barber's pole, handed down by medieval barber–surgeons.

Chapter 10, Box 2: Maternal and Paternal Genetic Imprinting

If one looks through the lens of evolution, one could predict that human fathers would prefer big babies that are healthier and stronger. The mother, however, would favor having a slightly smaller fetus because carrying such a fetus to term and nursing it may be less of a drain on the mother. Thus, a "battle of the sexes" occurs over the size of a couple's offspring, and the weapons in this battle are the so-called "imprinted genes."

As you probably know, each gene of an individual comes in two copies, one from the male parent and one from the female parent. These two copies (alleles) of each gene most often operate independently of one another; although one allele can be dominant over the other, it does not matter which parent is the source of the dominant or recessive allele. With imprinted genes, however, only the allele from one of the parents is expressed; the allele from the other parent is silenced. That is, the two alleles of an imprinted gene behave differently depending on if they came from the mother or father. Some genes are expressed only when they are inherited from the mother (i.e., the ovum) and are repressed when they come from the father (i.e., the sperm), or vice versa.

The discovery of this phenomenon shed light on a previously unexplained problem in embryology. Diploid mouse embryos created in the laboratory from either two paternal pronuclei or two maternal pronuclei fail to develop properly. Mendelian genetics would predict that, as long as the embryo is diploid, the parental source of the two genomes should be irrelevant. Because both a sperm genome and an oocyte genome are required for normal development, it appears that, during gametogenesis, the genetic material in male and female gametes is treated differently. Thus, the maternal allele of a differentially altered or "imprinted" gene would behave differently from the paternal allele.

For imprinted genes, the expression of the alleles is controlled by methylation. Attachment of methyl groups (CH_3) to DNA usually acts to silence a gene. In this case, if the allele of a gene is methylated in the sperm during spermatogenesis but not methylated in the egg, only the maternal allele will be expressed during embryonic development. In other cases, methylation activates the allele. Thus, differential methylation of the maternal and the paternal alleles for a given gene is the "imprint" that we are discussing.

There are more than 40 known imprinted genes thus far discovered in mammals, and probably many more will be known in the future. Many of these imprinted genes influence the passage of nutrients from the mother to the fetus across the placenta. For example, in mice, a gene for insulin-like growth factor

Continued on next page.

Chapter 10, Box 2 continued.

2 (Igf2) causes growth of the placenta and delivery of more nutrients to the fetus, leading to larger newborns. Only the allele from the father is expressed; the maternal allele is silenced. Thus, expression of the father's Igf2 allele causes growth of the conceptus. At the same time, another gene (Igf2r) expressed only by the maternal allele encodes a receptor that limits the effectiveness of Igf2. This retards growth of the fetus, resulting in a smaller baby. As a result, these imprinted genes, one expressed when inherited from the father and one from the mother, have opposite effects on prenatal growth. Although most of the known imprinted genes were originally discovered in mice, many of the same genes are imprinted in the same direction in other mammals, including humans.

Chapter 8 discussed the general consensus that our species is primarily monogamous (one mate per life), but secondarily polygamous (more than one mate for life or at one particular time). One hypothesis about the evolution of imprinted genes assumes that these genes evolved in polygamous or at least sequentially monogamous species. In such species, a female would want smaller babies so that, although wanting to have a successful present pregnancy, she could save enough energy to carry the fetus of her next mate. However, the male would want his present mate to put all she has into "his" pregnancy. The existence of imprinted genes in humans suggests an evolutionary past of polygamy or sequential monogamy (having more than one mate per life, but not more than one at one time).

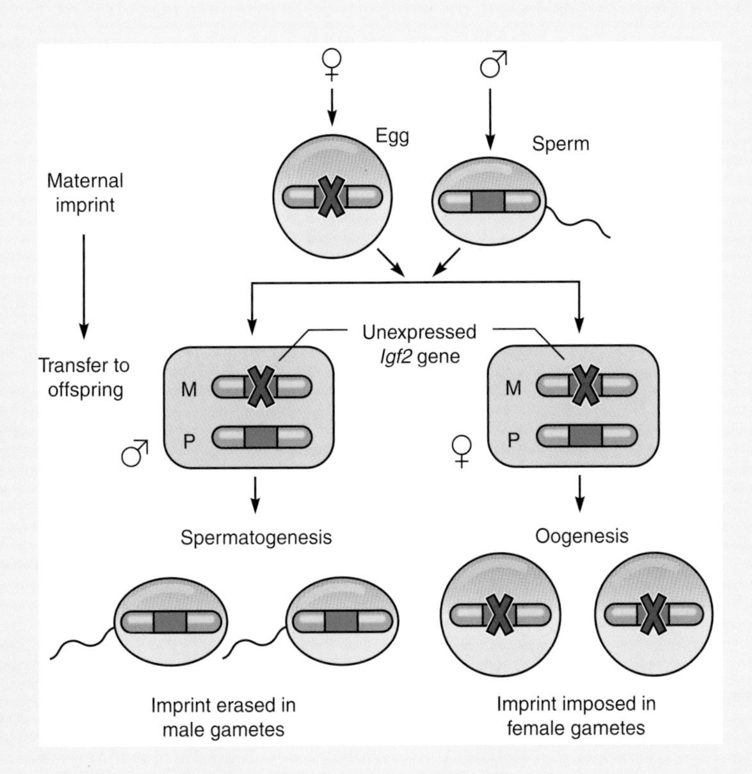

During oogenesis, the IGF2 allele becomes methylated (maternal imprinting). However, IGF2 alleles in the sperm are unmethylated and can be expressed. Thus, the fetus expresses only the IGF2 allele inherited from the father. Each generation, these genes become silenced during egg formation, and the imprint is erased during spermatogenesis.

Continued on next page.

Twin Pregnancies

As mentioned in Chapter 9, dizygotic (fraternal) twins develop from two separate zygotes. The two embryos usually implant separately, which results in two separate placentas, chorions, and amnions (Fig. 10-7). In some cases, however, the embryos implant close to one another, which results in a common placenta, fused chorions, and separate amnions (Fig. 10-7). In most cases, monozygotic (identical) twins result from the separation of a single inner cell mass into two. The two embryos share a common placenta and chorion but each is enclosed in its own amnionic sac (Fig. 10-8). In some cases, however, splitting of the early morula produces monozygotic twins with separate amnions and chorions and separate or fused placentas. In these cases, it is difficult to differentiate these monozygotic twins from dizygotic twins based on extraembryonic membranes alone.

The incidence of fraternal twin births in the United States is higher than for monozygotic births (Table 10-3). The rate of fraternal twins, but not identical twins, is higher in blacks than in whites (Table 10-3). Overall, the rate of twinning is highest in African blacks and lowest in Japan. Interestingly, the actual incidence of twin pregnancies is estimated to be up to four times the rate of twin births listed in Table 10-3. In many cases, this is because one of the twin embryos aborts spontaneously (unbeknownst to the woman) very early in development.

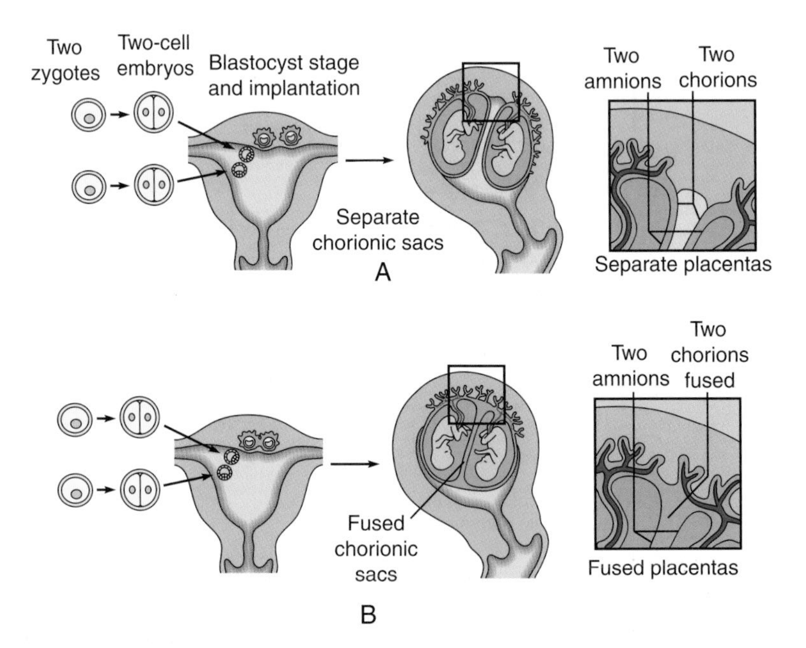

Figure 10-7 Pregnancy with dizygotic (fraternal) twins. Two zygotes are formed and two embryos implant in the uterus. When they implant separately, two separate placentas, chorions, and amnions appear (a). If they implant together, a single placenta, fused chorions, and two amnions are present (b).

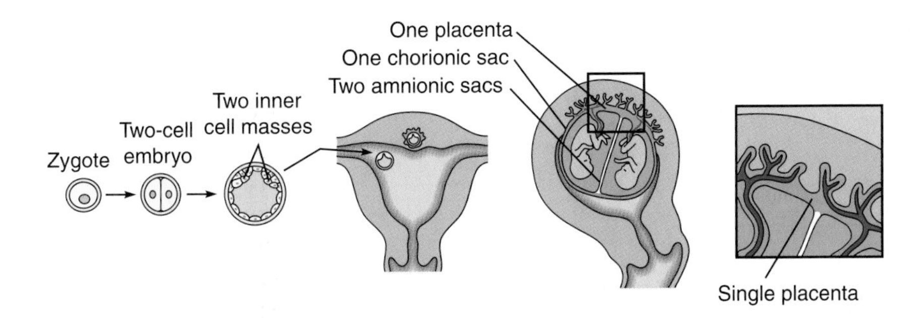

Figure 10-8 Pregnancy with monozygotic (identical) twins usually occurs when the inner cell mass of the blastocyst divides, producing two embryos with a single placenta and chorion, but two amnionic sacs.

Table 10-3 Rates of Twin Births Expressed as Number per Every 100 Births

Race	Overall rate	Fraternal	Identical
Whites (United States)	0.99	0.61	0.38
Whites (Europe)	1.22	0.86	0.36
Blacks (United States)	1.48	1.09	0.39
Blacks (Africa)	2.72	2.23	0.49
Asians (Japan)	0.64	0.23	0.41

Embryonic and Fetal Development

The Embryonic Period

By the end of the second week of postfertilization development, the preembryo has formed as a flattened disc consisting of the three germ layers: ectoderm, mesoderm, and endoderm. The next 6 weeks constitute the *embryonic period*, in which all of the major internal and external structures take shape. This is an extremely important period of development. It is also very sensitive to disturbances, and any alteration in development during this period may lead to death or major congenital malformations (Fig. 10-13). The stages of embryonic development, features of which are described later, are depicted in Figure 10-9.

During the third week of development, the flat, trilaminar embryonic disc begins to curl under to form a sausage-like shape. This movement places the ectodermal layer on the outside, the endoderm lining the inner tube (which will form the gut tube) and mesodermal tissue sandwiched between the other two layers. A *neural tube* develops along the embryo's back, and gives rise to the brain and spinal cord (central nervous system). The brain begins to enlarge in the future head end of the embryo. A series of lumps, the *somites*, form along either side of the neural tube. These will develop into the vertebrae, ribs, and muscles of the back.

At the beginning of the fourth week, the embryo is about 2 mm in length (about the diameter of the head of a pin). As the head and tail ends begin to curl in, the embryo assumes a C shape. In the head, the eyes begin to form and, more posteriorly, otic pits mark the position of the future inner ears. The neck region expands and a series of lumps can be seen; these are the *pharyngeal arches* that will

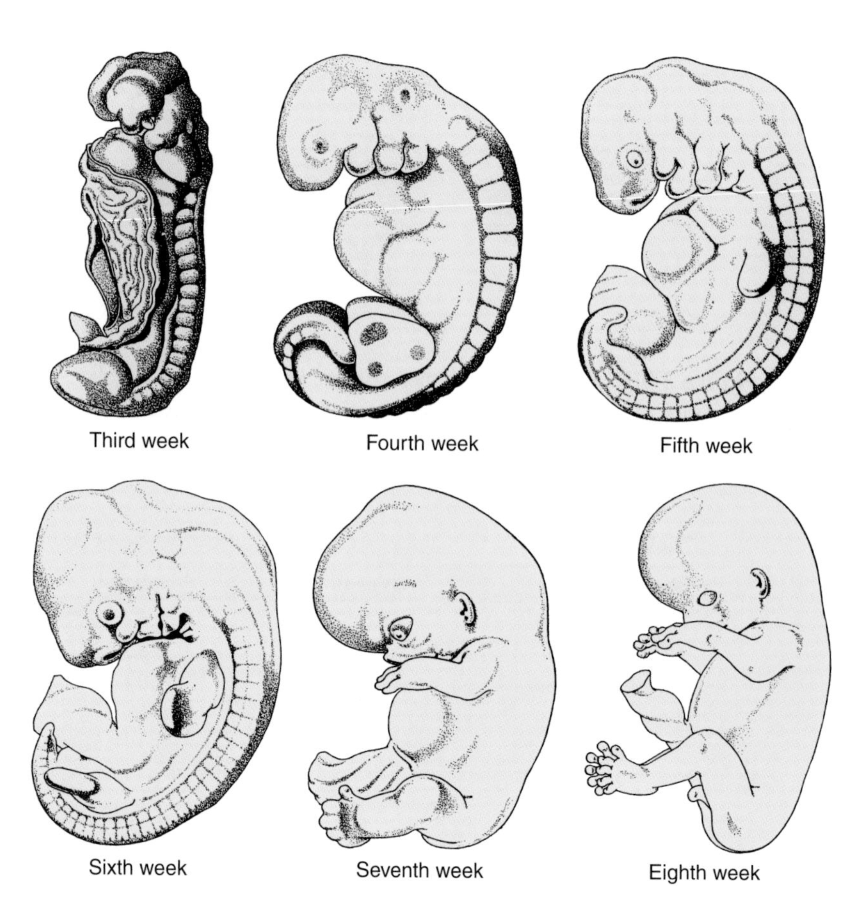

Figure 10-9 Development of the human embryo from the third week through the eighth week after conception. The embryo grows from about 2 mm (0.08 in.) in length after the third week to $1\frac{1}{4}$ in. at the end of the eighth week.

develop into jaws, parts of the ear, and other structures. Underneath the pharyngeal arches, the heart forms a prominent bulge and begins beating. Tiny arm and leg buds begin to swell. By the end of the week, the embryo has doubled in length.

Rapid growth of the brain occurs in the fifth week. In fact, this brain expansion is so extensive that the embryo's head bumps into its heart! The arm buds begin to flatten, and the hands becomes paddle shaped. The embryo is about 1 cm in length at the end of this week.

During the sixth week of development, the eyes become pigmented and more obvious. The external ears begin to form. Expansion of the brain and head continues, and the head now constitutes approximately half of the embryo. The leg bud becomes paddle shaped, and indentations in the hand bud called finger rays indicate the position of the digits. A distinct tail is still present. In the following (seventh) week, toe rays form. Rapid development of the gut tube causes intestines to protrude into the umbilical cord, which is evident as a swelling of the umbilical cord (*umbilical herniation*).

By the end of the eighth and final week of embryonic development, the embryo has grown to a length of $1\frac{1}{4}$ inch. The eyelids have grown to meet each other and fuse; thus the eyes are now closed. Fingers and toes can be seen clearly, but a thin sheet of webbing still exists between each digit. The tail bud has disappeared, and the embryo now begins to look distinctly human.

The Fetal Period

The fetal period begins at week 8 of postfertilization development and extends until birth. During this time, the organ systems established in the embryonic period continue to develop and differentiate. This is a period of rapid growth. By the end of the first trimester (12 weeks of development), the fetal heartbeat can be detected with a stethoscope. The fetus can react to stimuli and fetal movements begin, although the mother cannot yet feel these movements until the fourth or fifth month (halfway through pregnancy). By the end of the second trimester, the delicate fetal skin is covered with a protective layer of fatty secretions called *vernix caseosa*. In addition, the skin grows a layer of downy hair (*lanugo*), which usually is shed before birth. During the third trimester (weeks 25–38 of development), the fetus adds layers of fat and loses its wrinkled appearance. The lungs mature toward the end of this trimester. The fetus has a possibility of surviving if born after 26 weeks, although often only with advanced medical intervention.

Some of the major events during fetal development are presented in Table 10-4, and embryos and fetuses of different ages are depicted in Figs. 10-9 and 10-10. The position of the fetus in relation to the woman's body is depicted in Fig. 10-11.

Table 10-4 Changes Associated with Embryonic and Fetal Growth from Fertilization

End of month	Approximate size and weight	Representative changes
1	0.6 cm ($\frac{3}{18}$ in.)	Eyes, nose, and ears not yet visible. Backbone and vertebral canal form. Small buds that will develop into arms and legs form. Heart forms and starts beating. Organ systems begin to form.
2	3 cm ($1\frac{1}{4}$ in.) 1 g ($\frac{1}{38}$ oz)	Eyes far apart, eyelids fused, nose flat. Ossification begins. Limbs become distinct as arms and legs. Digits are well formed. Major blood vessels form. Many internal organs continue to develop.
3	7.5 cm (3 in.) 28 g (1 oz)	Eyes almost fully developed but eyelids still fused; nose develops bridge; and external ears are present. Ossification continues. Appendages are fully formed, and nails develop. Heartbeat can be detected. Organ systems continue to develop.
4	18 cm ($6\frac{1}{2}$–7 in.) 113 g (4 oz)	Head large in proportion to rest of body. Face takes on human features and hair appears on head. Many bones ossified and joints begin to form. Continued development of body systems.
5	25–30 cm (10–12 in.) 227–454 g ($\frac{1}{2}$–1 lb)	Head is less disproportionate to rest of body. Fine hair (lanugo hair) covers body. Rapid development of organ systems.
6	27–35 cm (11–14 in.) 567–681 g ($1\frac{1}{4}$–$1\frac{1}{2}$ lb)	Head becomes less disproportionate to rest of body. Eyelids separate and eyelashes form. Skin wrinkled. A 6-month fetus (premature newborn) is potentially capable of survival.
7	325–425 cm (13–17 in.) 1135–1362 g ($2\frac{1}{2}$–3 lb)	Head and body become more proportionate. Skin wrinkled.
8	40–45 cm ($16\frac{1}{2}$–18 in.)	Subcutaneous fat deposited. Skin less wrinkled. Testes descend into scrotum. Bones of head are soft. Chances of survival much greater at end of 8th month.
9	50 cm (20 in.) 3178–3405 g (7–$7\frac{1}{2}$ lb)	Additional subcutaneous fat accumulates. Lanugo hair shed. Nails extend to tips of fingers and maybe even beyond.

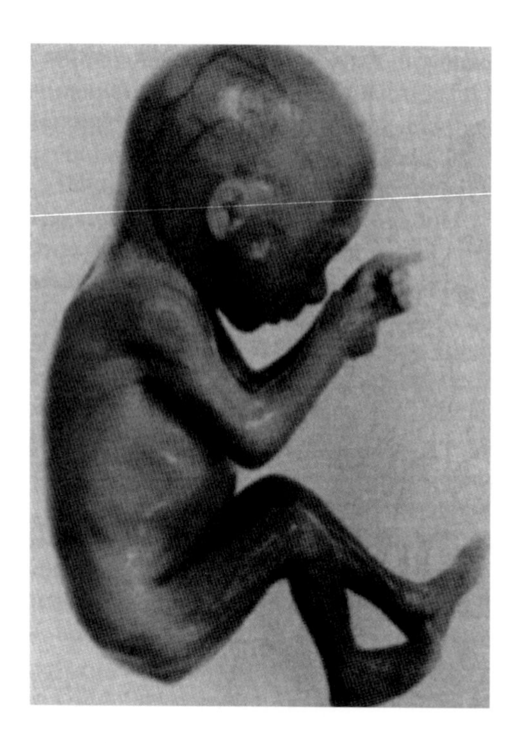

Figure 10-10 A 17-week fetus. Fetuses of this age are unable to survive if born prematurely, mainly because their respiratory system is immature.

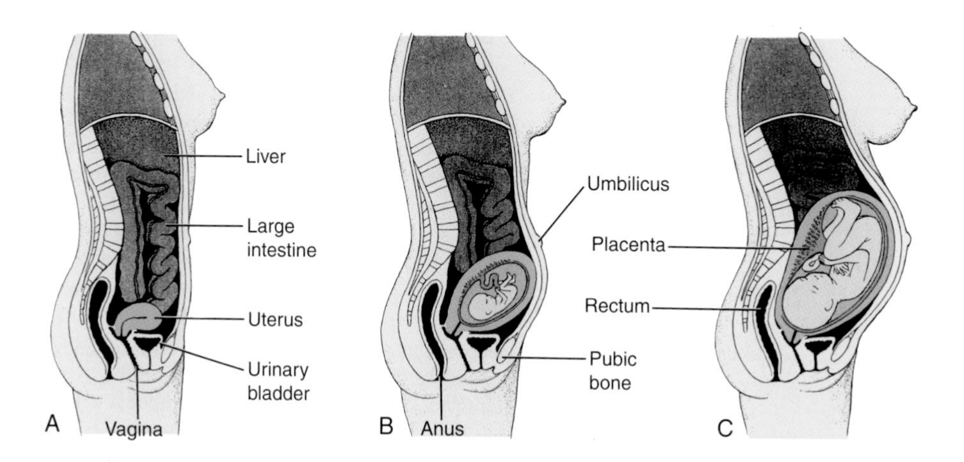

Figure 10-11 Diagrams of sections through a female: (a) not pregnant and (b) at 20 weeks of pregnancy. Note that as the fetus enlarges, the uterus increases in size. (c) Thirty weeks of pregnancy. Note that the uterus and fetus now extend above the umbilicus (belly button). The mother's abdominal organs are displaced, and the skin and muscle of her anterior abdominal wall are stretched greatly.

Digestive/Urinary Systems

The fetus derives nutrients from the mother's blood in the form of glucose, amino acids, fatty acids, vitamins, salts, and minerals. These nutrients pass to the fetus in the umbilical vein and are utilized by fetal tissues. Carbon dioxide and other wastes produced by fetal metabolism then pass back to the placenta via the umbilical arteries and are excreted by the mother. In late pregnancy, the fetus swallows about 500 ml of amnionic fluid each day, and this fluid contains water, salts, glucose, urea, and cell debris from the amnion and fetal skin. These ingested materials provide some nourishment for the fetus, and the waste products from this digestion combine with bile pigments to form feces in the large intestines. This fecal material, called *meconium*, is the first to be defecated by the neonate. The fetal kidneys are functional throughout pregnancy and produce about 450 ml of urine per day, which is excreted into the amnionic fluid.

Circulatory System

In the adult, deoxygenated blood (poor in oxygen and rich in carbon dioxide) enters the right side of the heart and then is pumped to the lungs via the pulmonary trunk and arteries. In the lungs, carbon dioxide is removed and oxygen is picked up. The oxygenated blood then returns to the left side of the heart via the pulmonary veins and is pumped to the dorsal aorta, which carries arterial blood to all other tissues.

The placenta, not the lungs, is the respiratory organ of the fetus, so the fetal circulatory system is different from that of the adult (Fig. 10-12). More specifically, after the oxygenated blood enters the right side of the fetal heart and is pumped into the pulmonary trunk, it is prevented from going to the collapsed lungs by a blood vessel shunt that goes from the pulmonary trunk to the dorsal aorta, the *ductus arteriosus*. The deoxygenated blood then reaches the placenta via the umbilical arteries, where it loses carbon dioxide and picks up oxygen. The oxygenated blood then travels via the umbilical vein back to the fetus, eventually reaching the right side of the fetal heart. Another shunt, the *ductus venosus*, shunts blood around the fetal liver (Fig. 10-12). The ductus arteriosus and ductus venosus close after birth.

The just-described fetal system, however, presents a problem. The volume of blood flowing through the left and right side of the heart must be equal or the system would become imbalanced. In the fetus, no blood flows from the lungs into the left side of the heart as in the adult, so how are the left and right sides of the heart kept in balance? The solution to this dilemma is a hole (covered by a flap) in the wall separating the left and right atria of the fetal heart, allowing blood to mix between the sides and to balance the heart. This hole, the *foramen ovale* (Fig. 10-12), closes after birth (see Chapter 11).

Nervous System

The nervous system of the fetus is formed very early in development. The central nervous system (brain and spinal cord) and peripheral nerves develop by the eighth week and influence development and function so that muscle sense and coordinated movements of the fetus appear very early. After the fetus begins to move on its own, some of its movements are well coordinated. The fetal sensory

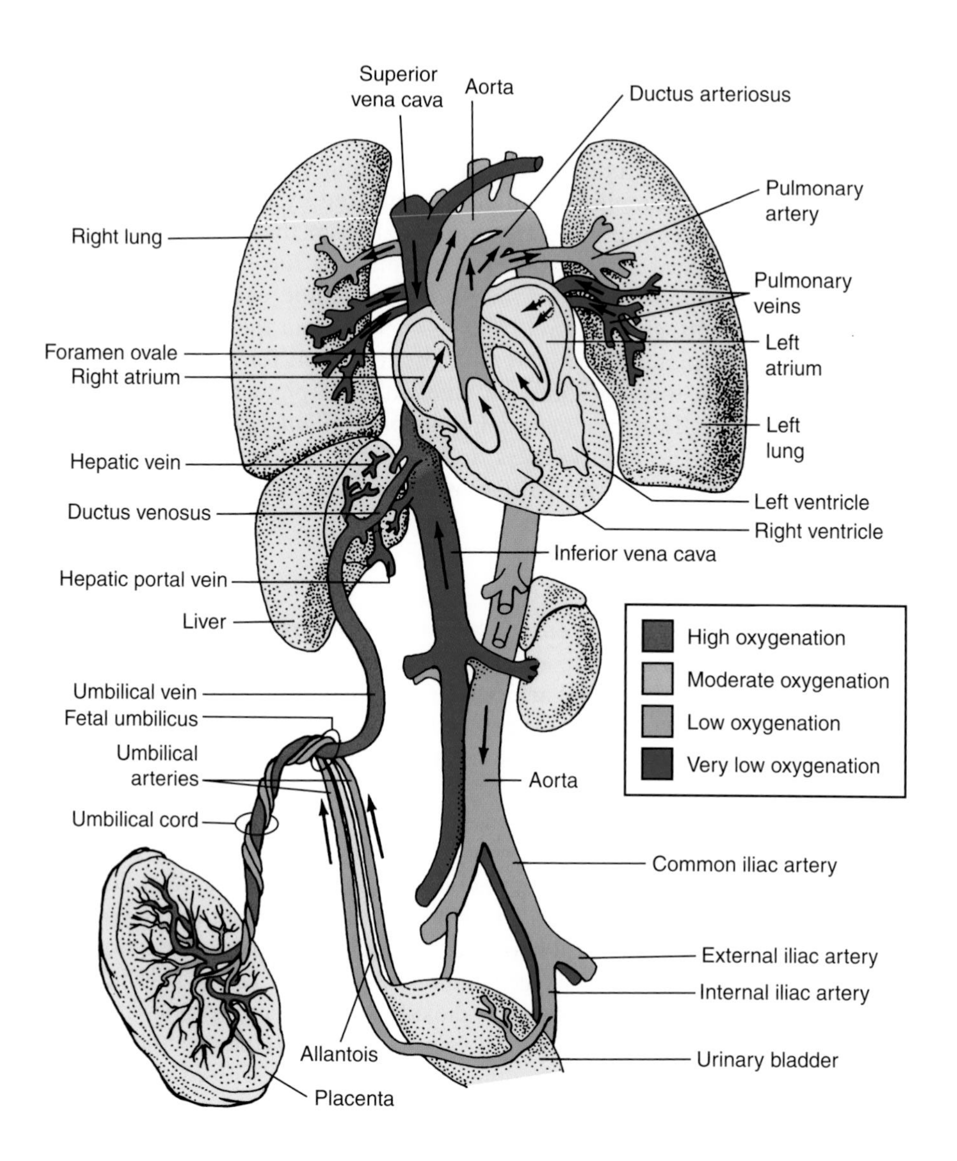

Figure 10-12 The fetal circulation. Note especially (1) the ductus arteriosus, which bypasses the fetal lungs; (2) the foramen ovale, which allows mixing of blood between the right and the left heart chambers; and (3) the ductus venosus, which shunts blood around the fetal liver.

nervous system also is functional, and the environment in the uterus is not totally devoid of stimuli. For example, the noise level within the amnionic sac is similar to that of a quiet room (about 50 decibels), and the light is like a dark room. The temperature of the amnionic fluid is about 0.5 °C higher than the mother's body temperature because the rapidly growing fetal tissues produce heat. It has been shown that a loud noise outside but near the woman's abdomen, a flash of light within the amnionic fluid, or pricking the fetal skin with a small instrument can evoke vigorous fetal reactions. Thus, the fetus, even in early stages, is not simply an inert, growing lump of tissue, but a moving, sensing organism that is capable of responding to changes in its environment.

Endocrine System

The endocrine system of the fetus is functional during most of pregnancy. The fetal anterior pituitary gland secretes the gonadotropins, FSH and LH, and these hormones may influence the development of the fetal gonads (see Chapter 5). In addition, some hCG reaches the fetus and may be involved in fetal gonadal function. The fetal pancreas secretes insulin, which allows fetal cells to use glucose supplied by the mother. As discussed later in this chapter, the fetal adrenal glands contain a special region that secretes steroid hormones that may play a role in the initiation of labor (Chapter 11).

As mentioned earlier, the fetus is surrounded by amnionic fluid, and there is a constant turnover of this fluid throughout pregnancy. Evidence shows that two hormones are involved in the regulation of amnionic fluid secretion and absorption. One is prolactin secreted by the placenta, which is found in high levels in amnionic fluid and is known to play a role in pumping sodium and therefore water across membranes. The second is a hormone secreted by the fetal neurohypophysis, *arginine vasotocin* (AVT). As discussed in Chapter 1, the adult neurohypophysis secretes oxytocin and vasopressin, not AVT. In some aquatic animals, AVT is an adult hormone that is involved in water transport across membranes. It is thought by some that ontogeny (the development of an individual) tends to recapitulate phylogeny (the evolutionary history of an animal). Because AVT is secreted by the fetus and is involved in the regulation of the amnionic "pond," fetal AVT secretion could be a recapitulation of the role of this hormone in aquatic lower vertebrates.

Fetal Disorders

As you have seen, the development of the fetus is a complex process beginning with fertilization and ending with birth. Thus, many errors in development are possible. It is estimated that about 50% of early embryos die within the first 3 weeks of development. Altogether, up to two-thirds of human embryos do not complete development. Usually the conceptus is lost before a woman recognizes that she is pregnant. Of confirmed pregnancies, about 15–20% end in spontaneous abortion (miscarriage). Most of these fetuses have chromosomal abnormalities. Fortunately, the vast majority of fetuses that survive through pregnancy are born as healthy infants. However, about 2% of newborns have serious birth defects (major *congenital* disorders). These may be caused by genetic mistakes or prenatal exposure to harmful substances, as discussed next. Damage due to perinatal (birth) trauma is discussed in Chapter 11.

Genetic and Chromosomal Disorders

The fetus can inherit genetic disorders. In fact, about 4000 human diseases are of genetic origin. Some of these disorders are mild, such as colorblindness. Others, however, can cause such troublesome handicaps as harelip, cleft palate, and club foot. Many others can kill the embryo or fetus. Chromosomal abnormalities account for 42% of spontaneously aborted fetuses and are present in 1 out of 200 newborns.

Rhesus Disease Commonly called "Rh incompatibility," *Rhesus disease* is an inherited phenomenon that damages not the present fetus but the fetus of a future pregnancy. This disease involves a gene with a dominant allele (*R*) and a recessive allele (*r*) for the Rhesus factor. Thus, cells of an individual can be Rh^+ (*RR* or *Rr*) or Rh^- (*rr*).

The cause for concern is when the pregnant woman is Rh^- and the father is Rh^+, which occurs in about 10% of marriages. In this situation, the Rh^- woman could be carrying an Rh^+ fetus. During labor and delivery, fetal blood cells could enter the maternal tissues because of broken blood vessels as the placenta detaches (see Chapter 11). Because the mother's tissues are Rh^- and the fetal blood cells are Rh^+, she will form antibodies to the fetal Rh^+ cells. If she then becomes pregnant again with an Rh^+ baby, these antibodies will enter the second fetus and destroy its mature red blood cells. As fetal red blood cells are destroyed, the fetus will develop *jaundice* (yellowish skin) due to the accumulation of *bilirubin* (a breakdown product of red blood cells) in its tissues. Bilirubin is toxic and can cause brain damage. Also, because mature fetal red blood cells are attacked, there will be many new immature red blood cells (erythroblasts) in the fetal blood, a condition termed *erythroblastosis fetalis*. Because these immature red blood cells are not efficient in carrying oxygen, the fetus is anemic and its tissues are unable to grow properly.

In the past, it was necessary to give a complete blood transfusion to a newborn with Rhesus disease. A new treatment, however, is much safer. An injection of Rhogam or Rho Immune (antibodies to Rh factor) is given to the mother within 2 or 3 days of delivery of the first infant or of a miscarriage or induced abortion (see Chapter 15). This drug destroys all Rh^+ fetal red blood cells that may have entered her blood, and thus she does not form antibodies that could harm her future fetus.

A fetus can also suffer from *ABO incompatibility* with the mother. If the fetus' ABO blood type is different from that of the mother (e.g., the fetus is type A or B and mother is type O), the mother will send antibodies across the placenta that will kill the fetus's red blood cells.

Teratogens, Mutagens, and Other Agents That Damage the Fetus

Exposure of the embryo or fetus to certain drugs, chemicals, or radiation can be mutagenic (damaging the genes or chromosomes of fetal cells) or teratogenic (affecting fetal growth and development). *Mutagen* means "mutant producing," and *teratogen* means "monster producing." These chemicals can also cause loss of a pregnancy before it is viable (*miscarriage*, or *spontaneous abortion*) or birth of a dead fetus (*stillbirth*). Many factors have been shown to be teratogenic or mutagenic in laboratory mammals and humans; a few of them are discussed here.

Viruses and Bacteria Some viruses can severely damage or kill the embryo or fetus. Examples are viruses causing AIDS, smallpox, chickenpox, mumps, and herpes. German measles (*rubella*) virus can produce heart defects, blindness (due to cataracts), deafness, microcephaly (small brain), mental deficiency, cleft palate, harelip, and spina bifida (exposed spinal cord). Exposure of the fetus to the rubella virus is most damaging from the 3rd to the 12th week of pregnancy. Before this time, little or no damage occurs. To prevent fetal exposure to rubella or other damaging viruses, a female child should be exposed to the virus to form antibodies, either by naturally contracting the disease or by vaccination.

Bacterial infections such as syphilis, pneumonia, tuberculosis, and typhoid can cause spontaneous abortion. Recent evidence suggests that about 16% of pregnant American women have a bacterial vaginal infection (*bacterial vaginosis*) that can cause premature birth (see Chapter 11).

Environmental Pollutants Many environmental pollutants can be teratogenic or mutagenic. Most of these agents have the most damaging effects during the fourth to the seventh weeks of pregnancy (Fig. 10-13). Before this time, they kill the embryo; after this time, they have less chance of harming the fetus. Examples of mutagens are mercury, lead, cadmium, arsenic, PCBs, DDT, benzene, and carbon tetrachloride. In Japan, mercury from fertilizers in factory effluent got into the fish population and caused *Minimata's disease*, which is characterized by damage to the fetal brain, resulting in abnormal muscle movement. *Cerebral palsy* (spastic muscle paralysis due to brain damage) is another name for this type of damage, which can be caused by fetal exposure to bacterial infection, oxygen deficiency, anemia, jaundice, and low blood sugar. Chapters 2 and 4 discussed how endocrine-disrupting contaminants could harm the reproductive system of fetuses.

Drugs, Alcohol, and Tobacco During pregnancy, especially in the first trimester, a woman should ask her physician about any medication, including nonprescription drugs. For example, taking the drug pseudoephedrine, a component of decongestant medicines and nasal sprays, can cause a baby to be born with *gastroschisis*, a condition in which there is a hole in the abdominal wall through which the intestines protrude.

Thalidomide is a mild tranquilizer that was used to treat morning sickness and to stop bleeding in pregnant women in the late 1950s and early 1960s, mostly in Europe and Canada. From 1958 through 1961, several thousand severely deformed infants were born to women who used this drug during pregnancy, especially during the fourth through the seventh weeks of pregnancy. Thalidomide causes the fetus to develop hands and feet but not arms or legs, a condition called *phocomelia*, or other abnormalities such as the total absence of limbs (*ectromelia*).

The drug *bendectin* has been part of a drama similar to that of thalidomide. At one time about 25% of pregnant women in the United States used this drug for morning sickness. In the 1970s, about 1.5 million pregnant women in 31 countries took the drug, despite evidence that bendectin could cause fetal abnormalities such as heart defects, hernia of the diaphragm, abnormal limbs, cleft palate, and stomach defects. The drug was withdrawn from the market in the United States in June 1983, and by July 1984 about $120 million had been awarded in lawsuits to victims of bendectin.

From the 1940s until the early 1970s, a synthetic estrogen, *diethylstilbestrol* (DES), was given to pregnant women to prevent miscarriage. About 2 million women in the United States were exposed to the drug during the first trimester of pregnancy. In the early 1970s, it was found that daughters born to these women exhibited an increased incidence of vaginal and cervical cancer, as well as an increase in miscarriage and premature births. Some of the sons born to these women also developed abnormalities in the male reproductive tract, including undescended testes and a low sperm count in their semen (see Chapters 2 and 4). Therefore, DES is no longer given to pregnant women.

Alcohol ingested by a pregnant woman can cross the placenta and affect fetal development adversely. Having at least two alcoholic drinks a week during

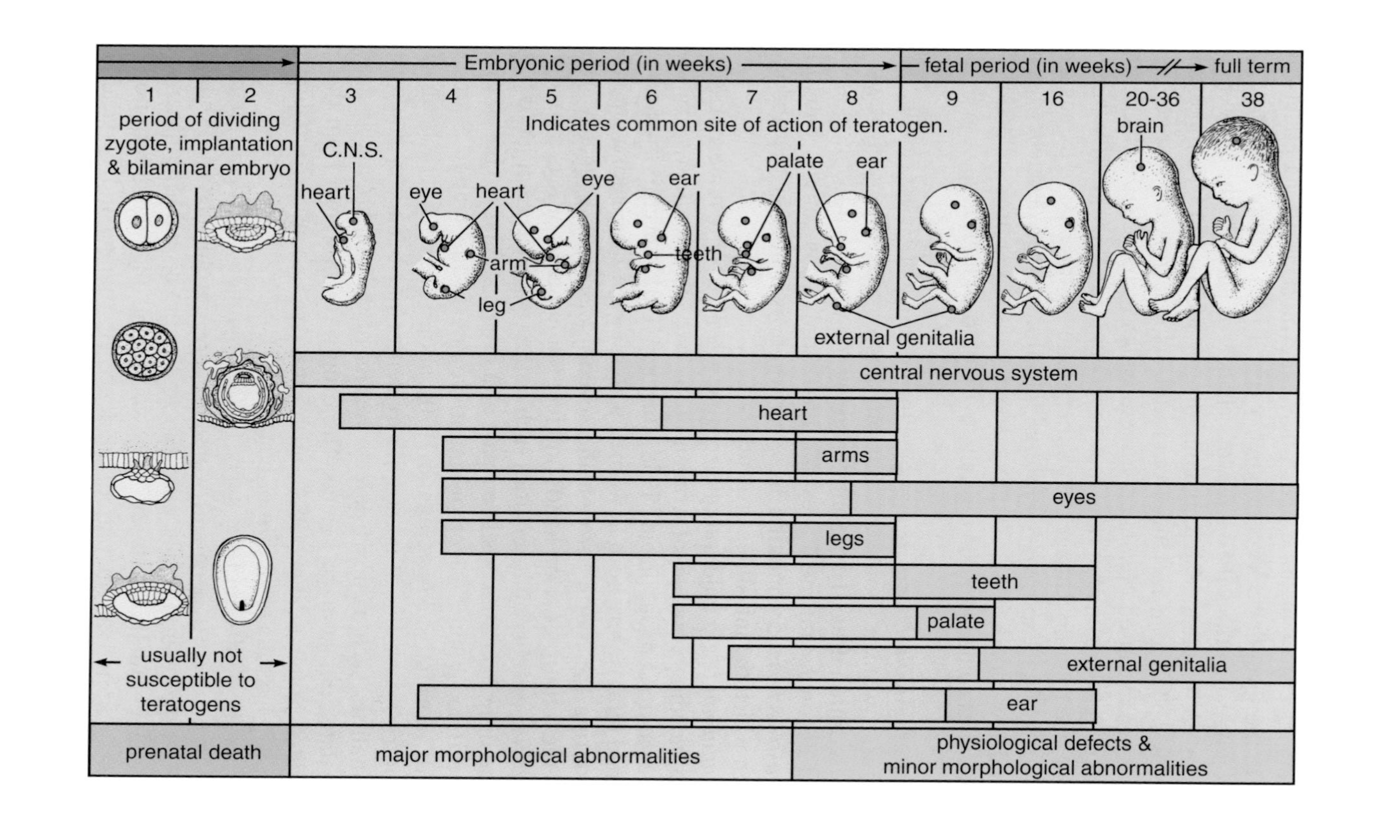

Figure 10-13 Critical periods of human development during which time teratogens are most effective. Yellow areas show the most critical spans of time, and the major morphological abnormalities most likely to occur are denoted.

Chapter 10, Box 3: The Adaptive Value of Morning Sickness

During the first trimester of pregnancy (and most often between the second and the eighth week), many women experience food aversions, nausea, and vomiting. This *morning sickness,* which often happens in the morning but can occur anytime, is usually blamed on abnormal stomach function. It occurs in about 75% of pregnant women. For years, the medical profession treated this "illness" with antinausea medications such as thalidomide and bendectin, but both drugs have been withdrawn because of their teratogenic effects on the developing embryo (see text), which is most sensitive to damaging agents in the first trimester. Today, morning sickness is simply not treated, which may be the correct choice after all.

Dr. Margie Profet of the University of California recently proposed a new theory about the adaptive value of morning sickness: that food aversions evolved in our ancestors to keep pregnant females from eating substances that could harm or abort their embryo and that the associated nausea and vomiting are responses to rid the body of such agents. Foods with strong or bitter tastes as well as pungent odors are avoided. As a result, females in early pregnancy tend to eat bland foods with very little smell or strong taste. Interestingly, Profet found that women with morning sickness have lower rates of miscarriage than women who did not have this symptom.

But why do certain foods contain such substances? The answer is they evolved these chemicals to inhibit ingestion by predators or plant eaters. In fact, most plants are not edible by people without their getting sick, and some of these have a long history of purposefully being used to induce abortion. Some of these plants contain estrogens (*phytoestrogens*), including clover, willows, and alfalfa. These phytoestrogens, which cause miscarriage in farm animals, have been used by some cultures to induce early abortions. Some Native American cultures in the American west, for example, would brew pine needle tea, which contains a phytoestrogen as well as other toxins that combine to induce abortion. The ancient Greeks discovered that stalks or seeds of plants in the genus *Ferula* (e.g., fennel, Queen Anne's lace) caused abortion when chewed or brewed as a tea. The active chemical in these plants blocks the synthesis of progesterone, which is vital for implantation as well as maintenance of pregnancy. These plants have a strong aroma (e.g., they are ingredients of some "steak sauces" sold in stores). Other plants known to cause abortion are pennyroyal, sage, myrrh, rue, papyrus, dates, and mustard.

Morning sickness may be a natural mechanism to avoid such plants that are used purposefully to cause miscarriage in some cultures. Some meats also induce nausea and vomiting in pregnant women, perhaps an adaptive response to avoid microbes in old or rotting meat. That morning sickness occurs in all cultures studied, both developed and underdeveloped, suggests its adaptive (evolutionary) origin. Profet correctly points out, however, that "modern" toxins in food, such as food additives, would not cause avoidance because they were not around thousands of years ago when humans were under the influence of natural selection. These "modern" artificial chemicals would be the ones to be most careful about!

Plants with pungent sap or seeds, such as those in the genus *Ferula* pictured here, will cause spontaneous abortion if ingested in the first trimester. In fact, they are employed to purposefully induce abortion by some women in central Asia as well as by a small number of women in the Appalachian Mountains of North Carolina. Adapted from Riddle *et al.* (1994).

early pregnancy can increase the risk of miscarriage. Chronic use of alcohol (six or more cocktails, or more than 3 oz of alcohol, daily) produces *fetal alcohol syndrome* 30 to 45% of the time. This syndrome is the third most common cause of mental retardation in infants in the United States today. It is characterized by the birth of relatively small infants with small heads. These children usually are retarded or have learning disabilities and behavioral problems. Milder alcohol intake (1 to 2 oz daily) can constrict umbilical blood vessels and cause miscarriage or the birth of an abnormally small infant. It has been shown recently that about three alcoholic drinks a day late in pregnancy can cause a reduction in IQ test performance in the resultant children at 4 years of age. Even small amounts of alcohol during gestation could harm a fetus, and no level of alcohol use during pregnancy has been proven safe.

Narcotics such as heroin and methadone, as well as cocaine, will cross the placenta and can cause addiction of the newborn. *Lysergic acid diethylamide* (LSD) can also cross the placenta and damage the fetal chromosomes, leading to deformities. In addition to reducing fertility, marijuana smoking (exposure to *tetrahydrocannabinol*) can decrease estrogen secretion from the placenta and cause miscarriage.

Tobacco smoking can have adverse affects on the fetus in several ways. First, the nicotine in tobacco smoke constricts blood vessels in the placenta and fetus, resulting in poor delivery of such blood-borne substances as oxygen and glucose to the fetal tissues. The carbon monoxide in tobacco smoke can bind to the hemoglobin of fetal red blood cells, thus preventing the oxygen from binding. Tobacco smoking during pregnancy can lower vitamin C levels in the fetus, impair fetal growth, damage the fetus and placenta (causing placenta previa), and lead to miscarriage, premature birth, or stillbirth. Smoking can also damage the part of the brain that will control respiration in the infant, and it has been estimated that some instances of crib death (see Chapter 12) are related to the mother's smoking during pregnancy. Some children of mothers who smoked during pregnancy have hearing difficulties and lower IQ tests. Obviously, then, it is recommended that a woman not smoke while pregnant.

Some evidence in laboratory mammals suggests that ingesting high levels of caffeine also could harm the fetus. In humans, studies of the possible effects of caffeine intake during pregnancy have yielded conflicting results. Although some studies have indicated no adverse effects of caffeine on human fetal development, several others demonstrate that caffeine consumption during pregnancy is associated with a higher risk of miscarriage. It is clear that caffeine crosses the placenta, and because caffeine is metabolized more slowly during pregnancy, a pregnant woman may have higher caffeine blood levels than a nonpregnant woman with the same caffeine intake.

Ingestion of aspirin, ibuprofen, or indomethacin by a pregnant woman can harm the fetal heart. The ductus arteriosus of the fetal heart is kept open by the secretion of prostaglandins, and these drugs are antiprostaglandins. Thus, exposure of the fetus can partially close the ductus arteriosus, resulting in babies born with poorly oxygenated blood and bluish skin (*cyanosis*). Also, because the ductus arteriosus is partially closed in these fetuses, too much blood is pumped into the vessels of the still-collapsed fetal lungs. This thickens the walls of the fetal lung blood vessels and can lead to *persistent pulmonary hypertension*, a condition in the newborn in which the arteries in the lungs have thick walls, and blood cannot pass through the lungs as well as it should. About two tablets of aspirin, ibuprofen, or indomethacin a day for 4 or 5 days in a row is too much during pregnancy. Acetaminophen, a nonprescription aspirin substitute, has only slight antiprostaglandin activity and therefore is probably safer than aspirin during pregnancy.

A recent study suggests that excess vitamin A ingested by a pregnant woman, especially in the first trimester, can increase birth defects of the face, head, brain, and heart. What is excess? Anything above the recommended adult daily intake of vitamin A (2700 IU/day) is too much, and some multiple vitamins contain 10,000 IU per pill!

Pregnant women may need to think twice about taking the antidepressant Prozac (fluoxetine), as its use has been shown recently to increase the risk of minor birth defects (such as fused toes) as well as premature delivery.

Radiation Radiation can also harm the fetus. X-rays and other radiation can be mutagenic in that they damage fetal genes or chromosomes. For example, infants born in southern Utah whose mothers were exposed to radiation fallout from atomic bomb tests in Nevada had twice the rate of birth defects as a control population. Also, before 1955, X-rays were used to test for pregnancy. Because of possible damage to the fetus, X-rays are now used only sparingly and in low dosages during pregnancy.

High Altitude Carrying a pregnancy at an altitude greater than 8000 feet can increase risks to both the mother and the fetus. Babies born at high altitude weigh about 2/3 lb less on average than babies born at sea level. Furthermore, there is an increased incidence of severe jaundice and neonatal fatality. High blood pressure (hypertension) is also more common in pregnant women at high altitude. If the mother smokes, all of these risks are increased two to three times.

Fetal Evaluation

For several reasons, it may be desirable to examine the condition of the fetus while it is still within the mother. For example, parents may be worried about inherited developmental abnormalities or there may be fear that the fetus has been exposed to teratogenic or mutagenic substances. At present, there are several ways that the condition of the fetus can be scanned. In the future, more attention may be paid to this matter, not only for health reasons but because physicians in the United States could be held liable if a deformed infant is born without previous fetal scanning.

Amniocentesis is usually done when pregnant women are over age 34 and/or have a family history of genetic problems. This procedure is performed most often between the 14th and the 16th week of pregnancy. A needle is inserted through the abdominal and uterine wall and into the amnionic fluid, using an ultrasound image as a guide (Fig. 10-14). A small sample of amnionic fluid is then withdrawn for analysis. Because fetal cells are present in the fluid, both these and the fluid itself can be examined. The cells can be cultured in dishes and their structure and function studied. More than 40 genetic abnormalities can be detected by amniocentesis. The sex of the fetus can also be determined by examining the sex chromosomes of the fetal cells. Although this procedure is relatively safe if done cautiously, some studies have shown an increase in uterine hemorrhage, miscarriage, and bone

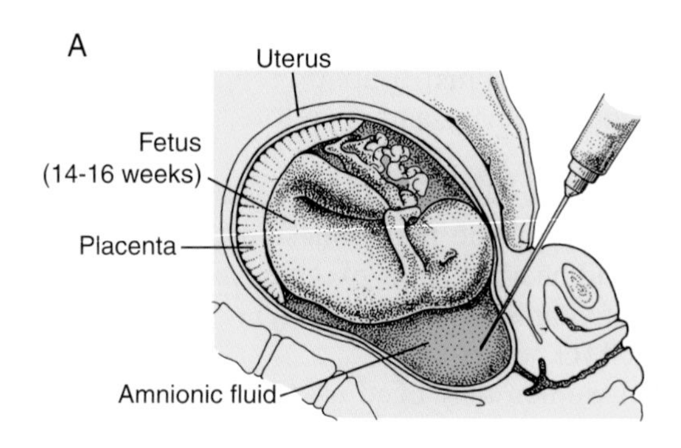

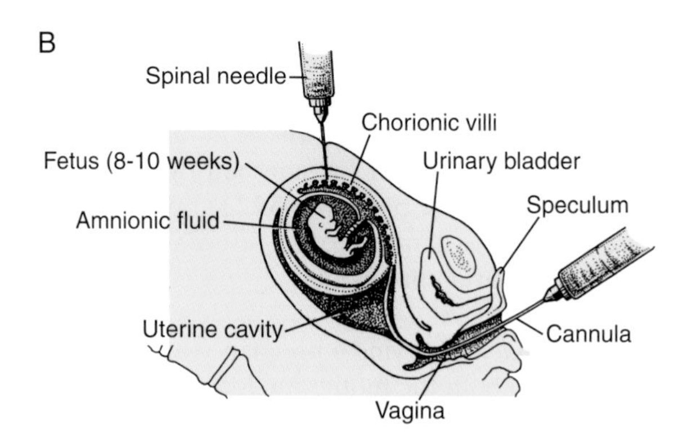

Figure 10-14 Two methods used to evaluate the genetic and developmental condition of the fetus. Amniocentesis (a) is done most often between the 14th and 16th week of pregnancy, whereas chorionic villus sampling, or CVS (b), is usually done between weeks 8 and 10. During CVS, cells are sampled from the chorionic villi via a tube going through the cervical canal (72% of the time) or by a needle inserted through the abdominal wall (28% of the time).

deformities in babies born to women who have had amniocentesis. In fact, one study from England suggests that amniocentesis kills 1.5% of the fetuses. Another disadvantage of amniocentesis is that it may take several weeks for the results to be known, and by that time an induced abortion, if desired, is more complex and dangerous (see Chapter 15). However, a new *fluorescent in situ hybridization technique* can give test results 3 days after amniocentesis. In certain high-risk pregnancies, the fetus is evaluated by *percutaneous umbilical blood sampling* or *fetal blood sampling*. A thin needle is guided through the mother's abdomen and into the umbilical cord, from which a sample of fetal blood is withdrawn. As you might expect, this procedure requires a high level of technical expertise. It is used to detect fetal infection, blood disorders such as hemophilia or anemia, certain metabolic disorders, or the presence of chromosomal defects. Because this procedure carries a higher risk of miscarriage than other fetal evaluation methods, its use is limited. Typically it is performed after the 18th week of pregnancy.

Chorionic villus sampling (CVS) is another way to monitor fetal condition. With this method, which is used between weeks 8 and 10 of pregnancy, a catheter is inserted through the vagina and cervical canal and into the uterine cavity, using ultrasound as a guide. Chorionic cells are then removed from the placenta and analyzed. Because this test can be performed earlier in pregnancy than amniocentesis and because one can obtain the results sooner (within 4 to 24 h), it has some advantages over amniocentesis. Disadvantages of CVS are that fewer tests can be done because there is no amnionic fluid to analyze and there is a slight increase in the risk of miscarriage caused by the procedure. The costs of the two procedures are similar.

Fetoscopy is a more involved and expensive method of fetal scanning. With this method, a small incision is made in the abdomen and uterus after injection of a local anesthetic. An optical viewer is then inserted into the uterus, and the fetus is viewed directly. Fetoscopy usually is done from the 15th to the 20th week of pregnancy. There is some risk, however, because fetoscopy causes miscarriage about 5% of the time.

Because the aforementioned fetal tests are invasive and carry the risk of miscarriage, a search is under way for noninvasive tests to screen for the presence of chromosomal abnormalities or developmental defects. *Ultrasound* has long been used to assess the fetal condition. With this method, a high-frequency sound source is applied to a pregnant woman's abdomen, and sound waves penetrate to the fetus. Dense fetal tissues, such as bone, reflect the waves, and these are detected by a receiver. In this way, fine measurements of the size and dimensions of the fetus can be made. Fetal heart rate can also be detected as early as the eighth week of pregnancy using ultrasound. The presence of twins can be confirmed. There is some concern but little evidence that ultrasound may affect fetal cells.

A noninvasive method used to evaluate the risk of Down syndrome is known as the *multiple serum marker test*. A sample of the mother's blood is drawn and levels of human chorionic gonadotropin (hCG), maternal serum a-fetoprotein (MSAFP), and unconjugated estriol are measured. A combination of high levels of hCG and low levels of the other two markers indicate a high risk of carrying a Down syndrome fetus, but the test is not definitive. If they so choose, women at high risk can then confirm the chromosomal abnormality by amniocentesis or chorionic villus sampling.

An alternative, noninvasive prenatal diagnostic test takes advantage of the fact that, in most pregnancies, fetal cells and free fetal DNA (not contained within cells) are found in maternal circulation (see HIGHLIGHT box 10-1). Blood is drawn from the mother, and fetal and maternal DNA are separated. The fetal DNA can then be screened for genetic abnormalities. This procedure is typically done when a male fetus is present, as male fetal cells can be distinguished from maternal cells based on presence of the Y chromosome.

The Pregnant Woman

Maternal Nutrition

During pregnancy, a woman not only must maintain her own health and well-being, but she faces the additional demands of a rapidly growing fetus and placenta. This requires her careful attention to diet, weight gain, and general health. Because the woman is supporting both herself and her fetus, her caloric intake must increase during pregnancy as well as during lactation. Also, special dietary supplements usually are prescribed by her physician. Specific nutritional requirements for pregnancy and lactation include extra protein, iron, calcium, folic acid, and vitamin B_6. Folic acid, for example, has recently been shown to reduce neural tube defects in newborns. Undernutrition can harm the fetus, resulting in a low birth weight or even miscarriage.

In the past, it was often recommended that a pregnant woman limit her weight gain. It is now felt, however, that she should gain about 25 lb during pregnancy. Of these 25 lb, about 11 lb should be fat. The increase in breast and uterine size adds about 3 lb, and the growing placenta another 2 lb. The amnionic fluid eventually weighs 1 lb, and the increase in maternal blood volume adds another 1 lb. The fetus itself weighs about 7 lb at term, bringing the total to 25 lb.

Exercise during pregnancy is recommended as long as a woman maintains her weight and consumes the specific nutritional requirements of the fetus. The amount of exercise recommended usually varies from physician to physician and depends a great deal on how the woman had exercised previously. In general, most forms of exercise are not harmful and are good for the mother's cardiovascular demands in supporting her growing fetus and in controlling excessive weight gain. In fact, burning 1000 and 2000 calories per week with exercise causes a 5 and 10% increase, respectively, in newborn weight.

A recent study of exercise in pregnant women showed that, of several kinds of activity, only standing for more than 8 h per day at work increased the risk of miscarriage. This effect of standing, however, was only true for women with a previous history of miscarriage. Other activities, such as heavy housework, caring for young children, number of hours at a job, commuting, stooping, bending, or lifting weights of greater than 15 lbs, had no influence on the miscarriage rate.

Physiological Changes during Pregnancy

Pregnancy makes great demands on the mother's body, and maternal adaptations to pregnancy are found in all organ systems. Increased cardiovascular function is needed to supply blood to the highly vascular placenta. Blood volume increases 45–50%. Because the increase in number of red blood cells (20–30%) is not as great as the increase in blood volume, a pregnant woman's hematocrit falls. This means that her blood is thinner, perhaps facilitating perfusion of the placenta. A woman's heart rate also increases during pregnancy. Thus, there is an increase in cardiac output (volume of blood per unit of time), which is needed to push blood into the placenta. The respiratory rate also increases by about 40%. This hyperventilation increases the ratio of O_2 to CO_2 in the pregnant woman's blood, facilitating the transfer of oxygen to the fetus and the removal of CO_2 from fetal to maternal blood. To filter this extra volume of blood, the kidneys enlarge and increase their filtration rate by 50%.

The Endocrinology of Pregnancy

The endocrine system of a pregnant woman operates differently from that of a nonpregnant woman. Many of the changes in hormone secretion are adaptations to maintain the fetus and to adapt the woman's body to her new nurturing role.

As mentioned in Chapter 3, the corpus luteum formed after ovulation secretes estradiol and progesterone during the luteal phase of the menstrual cycle, but this organ degenerates before menstruation in the nonpregnant woman. If implantation occurs, the corpus luteum does not die but continues to secrete high amounts of progesterone and low amounts of estradiol during the first trimester of pregnancy. This continued activity of the corpus luteum of a pregnant woman maintains the placenta in a functional condition. Steroids secreted by the corpus luteum of pregnancy also initiate the development of the mammary glands and inhibit ovulation by exerting negative feedback on pituitary gonadotropin secretion (see Chapter 3). Progesterone also increases fat deposition in early pregnancy by stimulating the appetite and diverting energy stores from sugar to fat.

What extends the life of the corpus luteum in a pregnant woman? The probable answer is that hCG prevents the corpus luteum from regressing and causes it to continue to secrete estradiol and progesterone. Secretion of hCG by cells of the cytotrophoblast begins as soon as 48 h after implantation. The saving of the corpus luteum by hCG is often referred to as the *maternal recognition of pregnancy*. The human placenta produces GnRH, which may regulate secretion of hCG as it regulates gonadotropin secretion from the anterior pituitary gland. Levels of hCG in the pregnant woman's blood rise steadily, reaching a peak in the second month of pregnancy (Fig. 10-15). The first hCG peak possibly is responsible for the increase in testosterone secretion in fetal males. hCG secretion then declines, exhibiting another, smaller rise in late pregnancy (at about 30 weeks). Even though hCG levels peak in the second month, secretion of steroid hormones from the corpus luteum begins to decline after the second month. What causes this slow decline of corpus luteum function in the second month, even while hCG levels are still high, is not known.

After about 5 weeks of pregnancy, the placenta begins secreting three estrogens: estradiol, estrone, and (predominantly) estriol. The placenta also begins

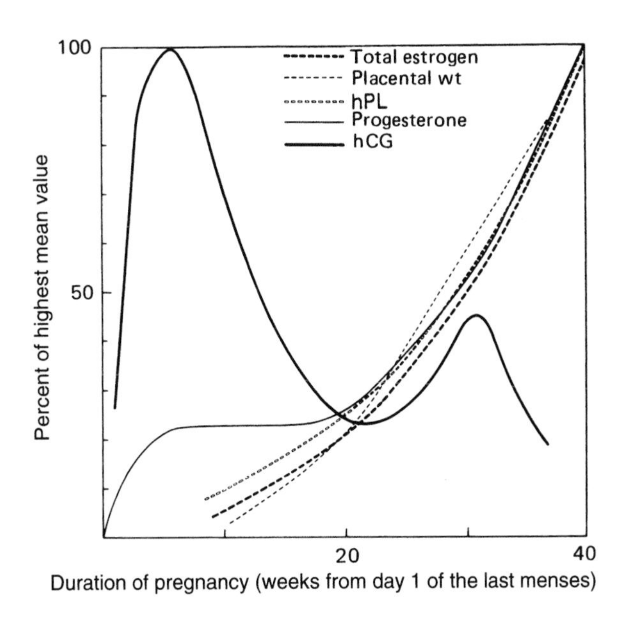

Figure 10-15 Hormone concentrations in a pregnant woman's blood and their relation to placental weight. Total estrogen includes estrone, estradiol, and estriol.

secreting progesterone. The secretion of estrogens and progesterone by the placenta is stimulated by hCG. The levels of these steroid hormones from the placenta increase throughout pregnancy (Fig. 10-15) and they continue to support the placenta and mammary glands and to inhibit ovulation. Can you now explain why removal of the ovaries before the seventh week of pregnancy causes miscarriage but does not after this time?

The fetus and placenta are both involved in estrogen and progesterone secretion from the placenta. This cooperative arrangement, the *feto–placental unit*, operates as follows (Fig. 10-16). First, the placenta converts cholesterol to progesterone, a conversion that the fetus is not capable of performing. Then, progesterone passes from the placenta to the fetus and reaches the fetal adrenal glands. These glands of the fetus contain a region called the *fetal zone*. This zone is very large but disappears soon after birth. The fetal zone converts progesterone to large amounts of the weak androgen dihydroepiandrosterone (DHEA), which is changed by the fetal liver to 16-OH DHEA-sulfate. Then, 16-OH DHEA-sulfate is carried in the fetal blood back to the placenta to be converted to the estrogen estriol. Thus, secretion of estrogens by the placenta requires the fetal adrenal glands. The fetal adrenals also secrete cortisol (a corticosteroid hormone), and this hormone can go to the placenta, where it influences estrogen and progesterone

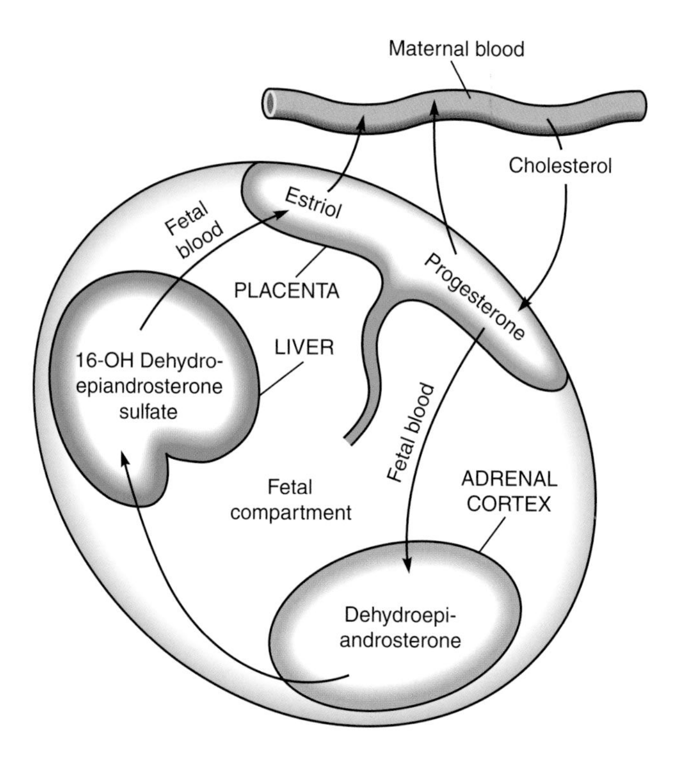

Figure 10-16 The human feto–placental unit showing how the mother provides cholesterol to the placenta, which converts it to progesterone for release into the maternal and fetal circulations. In the fetus, progesterone is converted to dehydroepiandrosterone (DHEA) by the fetal zone of the adrenal glands. DHEA is then converted to 16-OH DHEA sulfate in the fetal liver. This steroid then goes to the placenta and is converted to estriol, the major estrogen secreted by the placenta.

secretion. In fact, as shown in Chapter 11, cortisol secretion by the fetal adrenals near term may initiate labor in this manner.

In addition to secreting hCG, estrogens, and progesterone, the placenta also secretes a protein hormone that is similar in biological effect to pituitary growth hormone and prolactin. This hormone is called *human placental lactogen* (hPL), and it rises in the female's blood in late pregnancy (Fig. 10-15) and causes an increase in sugar in the mother's blood. Thus, hPL provides the fetus with additional glucose for its growth. This hormone (along with estrogens and progesterone) also helps prime the mammary glands for later milk secretion.

As mentioned earlier, the placenta secretes prolactin, which enters the amnionic fluid. The human placenta may also secrete chorionic corticotropin and thyrotropin, both similar to the pituitary hormones of the same name. In addition, prolactin secretion from the mother's pituitary gland increases during pregnancy. Another hormone, *relaxin*, is secreted by the corpus luteum and placenta. Levels of this polypeptide rise during pregnancy, and this substance relaxes the connective tissue connecting the two pubic bones (pubic symphysis) so that the fetus can pass through the birth canal with more ease during labor. Relaxin also helps efface the cervix and inhibits premature uterine contractions. As shown in Chapter 11, the cervix also dilates during early labor, and relaxin helps prepare the cervix for this event. It has been discovered recently that the placenta produces endorphins (opiate-like natural painkillers). This may mean that a woman in late pregnancy is less sensitive to pain.

Maternal Complications of Pregnancy

Some ailments of pregnancy are relatively common. These include constipation, headaches, and the development of enlarged veins that bulge under the skin (*varicose veins*). Many women also report changes in sleep patterns during pregnancy. A woman in the first trimester often requires more sleep than usual, but during late pregnancy, she may have difficulty sleeping.

Toxemia There are ailments of pregnancy that can be dangerous to the fetus and mother. For example, *toxemia* is a condition that develops in the last 1 or 2 months of pregnancy in 6 to 7% of all pregnancies in the United States. It is the leading cause of maternal and fetal death in the United States. It is more common in primiparous women and in multiparous women over 35 years of age than in other females. Other risk factors are preexisting obesity or high blood pressure, a pregnancy with twins or more, and having a close relative who has experienced toxemia. Black women have a higher occurrence of toxemia than whites, and a pregnancy with toxemia puts a woman at increased risk for the same condition in future pregnancies. The symptoms of toxemia are rapid weight gain, fluid accumulation in tissues (edema), high blood pressure (hypertension), and an increased excretion of proteins in the urine (proteinuria). Early toxemia is called *preeclampsia*. If this early condition is not controlled by diet, more severe toxicity (*eclampsia*) can develop, characterized by convulsions, coma, and death in about 15% of cases.

The cause of toxemia is not clear. It appears to involve the failure of trophoblast cells to fully invade deep uterine blood vessels, resulting in a poor maternal blood supply to the placenta. As the mother's cardiovascular system attempts to supply blood to the fetus through inadequate vessels, this may lead to maternal high blood pressure. There is some evidence that this abnormal placenta development may involve an immunological reaction against something in the father's semen. Toxemia is probably more common when a woman is newly exposed to the father's semen. This could occur if a woman becomes pregnant soon after beginning a new sexual relationship or if pregnancy occurs after a long period of condom use. (However, a study of women who conceived by *in vitro* fertilization showed that women using donor sperm had no greater rates of preeclampsia than those using husband's sperm.)

Toxemia is unique to humans; it has been difficult to study because there are no natural animal models of this disorder. However, a recent study has identified a key protein [*soluble fms-like tyrosine kinase 1* (sFlt1)] that is overexpressed in preeclampsia placentas in the third trimester of pregnancy. This protein inhibits blood vessel growth and may have a major role in this disorder. Injection of sFlt1 into pregnant rats elicits symptoms of a preeclampsia-like illness. Another recent finding showed that pregnant women at high risk for preeclampsia have low levels of placental growth factor and vascular endothelial growth factor in their urine. These factors act in concert to promote the growth of new blood vessels. Currently, no test is available to detect preeclampsia, but in the future a simple urine test for these growth factors may be able to identify women at risk. There is no known cure for this condition. Mild preeclampsia can be controlled by close monitoring, diet, and medication to control blood pressure. Severe eclampsia necessitates the premature delivery of the fetus.

Diabetes Mellitus About 1 out of 350 pregnant women in the United States develops *gestational diabetes mellitus*, a condition in which the mother's cells are not responding enough to insulin so that the maternal tissues can utilize glucose as energy. As a result, copious amounts of urine, containing glucose, are produced. This is not only damaging to the mother if not controlled, but also kills the fetus in 30% of cases. The greater the number of previous pregnancies, the greater the risk of diabetes.

Ectopic Pregnancies Normally, implantation occurs on the posterior wall of the uterus. If, however, implantation occurs outside the uterus, an *ectopic pregnancy* develops. In the United States about 1% of pregnancies are ectopic. About 96% of these are in the oviduct; they are called *tubal pregnancies*. The remaining 4% are in the abdomen (*abdominal pregnancies*), with implantation occurring on the gut, mesenteries, or ovaries. Tubal pregnancies are very dangerous to the mother because the embryo and placenta are growing in a restricted area and the oviduct walls are thin and vascular. These pregnancies are accompanied by pain and serious hemorrhage and require surgical removal of the embryo and placenta. Tubal pregnancies account for 10% of all maternal deaths in the United States; this is about 1 death per 1000 ectopic pregnancies. When ectopic pregnancies occur in the abdomen, the fetus dies and often is surrounded by calcium. These calcium deposits (*lithopedions*, or "stone babies") often are not discovered until later abdominal surgery for some other reason. In very rare cases, however, pregnancies can result in the birth of a healthy infant by cesarean section.

Ectopic pregnancies are more common in older, multiparous women, in nonwhites, and in women who have had a previous abortion, pelvic infection, or endometriosis than in other women. Birth control devices have some influence on the incidence of ectopic pregnancies. Barrier contraceptive methods (such as condoms and diaphragms) and use of the combination pill decrease the risk of ectopic pregnancy slightly. However, use of the minipill, intrauterine devices, and tubal sterilization can lead to an increased risk of ectopic pregnancy (see Chapter 11).

Hydatidiform Moles In rare cases implantation occurs, but the implanted body contains swollen chorionic villi and no embryo. Some of these *hydatidiform moles* become malignant and secrete large amounts of hCG. Therefore, they can be mistaken for a normal pregnancy. The cells of these moles are diploid, but the chromosomes are all from the father. A partial hydatidiform mole is one that contains a dead embryo; these moles are triploid (3N). Hydatidiform moles are removed surgically. These moles occur in about 1/1000 pregnancies.

Septic Pregnancy A *septic pregnancy* occurs when bacteria enter the uterus and cause severe infection. This condition is dangerous to both the mother and the fetus. It can cause maternal death, premature delivery and subsequent neonatal complications, death and spontaneous abortion of the fetus, or infection of the newborn.

Hemorrhage Excessive uterine bleeding during or immediately after labor can be dangerous to the mother if not controlled. One cause of excessive uterine hemorrhage during labor is *placenta previa*, a condition in which the placenta adjoins or covers the cervix so that the fetus is unable to be born properly. In addition to excessive bleeding, this condition sometimes requires a cesarean delivery (see Chapter 11). Excessive hemorrhage can also occur if the placenta detaches prematurely, a condition termed *abruptio placenta*. This is not only dangerous to the mother, but also means that the fetus can no longer receive life-giving materials from the mother.

Miscarriage About 50% of zygotes and preembryos undergo *miscarriage* (*spontaneous abortion*) before or right after implantation. In these cases the woman is not aware of being pregnant. This occurs because of genetic or chromosomal damage to the cells, immune responses, improper hormonal priming of the uterus, or exposure to drugs or pollutants.

After pregnancy is established, chromosomal or genetic errors account for 42% of miscarriages. About 15% of established pregnancies terminate by miscarriage, usually in the first trimester. Other factors that can induce spontaneous abortion after pregnancy is established include exposure to teratogens or mutagens, maternal stress, under- or overnutrition, vitamin deficiencies or excessive amounts of vitamins (especially vitamin A), thyroid and kidney disorders, diabetes, and uterine abnormalities. The rate of miscarriage is also higher in older women. When pelvic infection causes spontaneous abortion, it is called a *septic abortion*.

As discussed earlier, a critical stage in pregnancy is just when the corpus luteum is regressing and before the placenta is fully functional and secreting steroid hormones. Thus, weeks 10 to 14 are marked by a low level of estrogens and progesterone in the mother's blood, and in some women this drop may be low enough to cause placental detachment and miscarriage.

Many women experience a miscarriage and go on to have one or more successful pregnancies. (A woman's menstrual cycle usually resumes within 3 months of a miscarriage.) Other women, however, are functionally infertile because of continually losing pregnancies through miscarriage. This often is because the cervix is not capable of "holding" a fetus to term (known as *incompetent cervix*).

In the future, it may be possible to predict the danger of a miscarriage by detecting levels of certain chemicals in a woman's blood. One such chemical is *B_1-glycoprotein*, which is produced by the trophoblast (chorion). Abnormally low levels of this chemical after the ninth week of pregnancy have been shown to be associated with miscarriage.

Miscarriage is an emotionally draining experience for most couples. A woman may feel guilty over her inability to carry a fetus and may be fearful of future pregnancies. Many couples experience a profound degree of grief and sadness after a miscarriage similar to that experienced by those who have lost a child. Counseling is helpful to these people, and women who continually experience miscarriage should seek medical help at infertility clinics.

Even though the just-described conditions are potentially serious, they seldom result in the death of a pregnant woman. In developed countries such as the United States, about 14 per 100,000 pregnant women die of problems related to pregnancy. In underdeveloped countries, however, this number is about 740 per 100,000 because of poorer health care and nutrition.

Sex during Pregnancy

Coitus presents little danger to the fetus or mother. Coitus, however, should be avoided if there is any uterine bleeding during pregnancy or if the extraembryonic membranes have ruptured. For this reason, couples are often counseled to avoid coitus during the last few weeks of pregnancy. Oral–genital contact that involves air blown into the vagina could be dangerous in the last few weeks because of the possible introduction of air into the woman's bloodstream via the placenta. In general, the period of pregnancy requires sexual adjustments by both parents. A physician or counselor may suggest certain coital positions that can be used during this time (see Chapter 8).

Chances for a Successful Pregnancy

Given the complexity of pregnancy and its possible complications, you may be wondering how on earth a healthy baby is ever born! Well, the human body contains remarkable defenses against malfunction and disease. Of all newborns, about 87% will go home as healthy babies (Fig. 10-17). Of the remaining 13%, about 2% have congenital defects that kill them or result in a major handicap. Recently, however, a few successful operations on fetuses with defects, while they are still in the womb, have been performed. There is also a 1% chance of a baby dying in the first neonatal month. The other 11% are born with minor congenital defects, many of which are barely noticeable. The great chance that a healthy baby will be born and continue to be a healthy child should help alleviate the fear that some couples have about pregnancy.

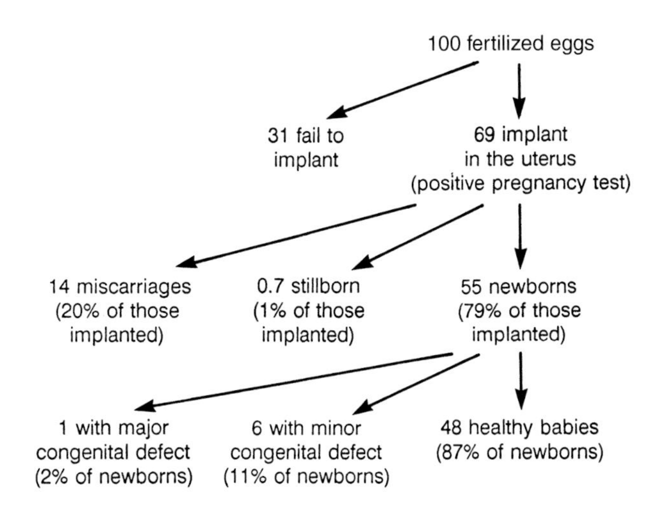

Figure 10-17 Average fate of 100 fertilized eggs in the United States. Losses of preembryos, embryos, fetuses, or newborns are due to maternal problems (such as endocrine imbalance, diabetes, and toxemia), as well as to problems of the embryo or fetus, including genetic disorders, chromosomal errors, and exposure to teratogens, mutagens, or other environmental toxins.

Chapter Summary

Pregnancy is the condition in which a developing human is nurtured within the uterus that begins at conception (fertilization) and lasts about 266 days. There are presumptive, probable, and positive signs of pregnancy. Pregnancy tests, which detect the presence of human chorionic gonadotropin in a woman's urine or blood, can utilize bioassay, immunoassay, or radioimmunoassay. If a woman believes she is pregnant, she should consult a physician or clinic within 2 to 3 weeks of her first missed menstrual period.

After fertilization, cleavage occurs to form a morula, which enters the uterus and develops into a blastocyst. Then, the blastocyst implants in the uterine lining. Implantation requires a uterus primed with estrogen and progesterone.

The embryo forms three primary germ layers (ectoderm, mesoderm, and endoderm), which give rise to all adult tissues. Four extraembryonic membranes (yolk sac, amnion, allantois, and chorion) form. The allantois and chorion contribute to the fetal portion of the placenta, which is a respiratory, excretory, and nutritive organ for the fetus.

Embryonic and fetal development is complex and is accompanied by the growth and development of all adult organ systems. Special adaptations of the fetal circulatory system reflect the use of the placenta (and not the lungs) for fetal respiration. Many fetal organ systems are functional early in development.

Although most fetuses are healthy, fetal disorders sometimes occur, leading to miscarriage or the birth of children with congenital defects. Some of these abnormalities have a genetic or chromosomal basis. Others, however, are caused by exposure of the fetus to teratogens or mutagens. Environmental pollutants, plant toxins, and nicotine can affect fetal health adversely, as can exposure to radiation. Because of possible fetal abnormality, some women choose to have their fetus examined by amniocentesis, chorionic villus sampling, ultrasound, or fetoscopy.

The body of a pregnant woman must support not only itself but the developing fetus. Under the influence of hCG secreted by the placenta, the corpus luteum life span is prolonged into the first trimester of pregnancy. After this time, the corpus luteum regresses, and the placenta secretes not only hCG but human placental lactogen, as well as estrogens, progesterone, and relaxin. The fetus interacts with the mother in controlling estrogen secretion from the placenta.

Some maternal complications of pregnancy include toxemia, diabetes mellitus, placenta previa, ectopic pregnancy, hydatidiform moles, septic pregnancy, and maternal hemorrhage. Nevertheless, the maternal death rate associated with pregnancy is low, at least in developed countries. About 20% of established pregnancies end in miscarriage. Of newborns, 2% have serious birth defects, whereas 11% have minor problems. The remaining 87% are perfectly healthy.

Further Reading

Barinaga, M. (2002). Cells exchanged during pregnancy live on. *Science* **296**, 2169–2172.

Beaconsfield, P., *et al.* (1980). The placenta. *Sci. Am.* **243**(2), 94–102.

Cibelli, J. B. (2002). The first human cloned embryo. *Sci. Am.* **286**(1), 44–51.

Couzin, J. (2002). Quirks of fetal environment felt decades later. *Science* **296**, 2167–2169.

Diamond, J. (1992). Our phantom children. *Nat. Hist. Magazine* **5**, 18–23.

Edwards, S. (1992). Use of coffee, alcohol, cigarettes raises risk of poor birth outcomes. *Family Planning Perspect.* **24**, 188–189.

Fackleman, K. (1997). The birth of breast cancer: Do adult diseases start in the womb? *Sci. News* **151**, 108–109.

Gura, T. (1998). How embryos avoid immune attack. *Science* **281**, 1122–1124.

Johnston, J. (1992). Chorionic villus sampling may cause rare fetal damage. *J. NIH Res.* **4**(7), 60–62.

McLaren, N., and Nieburg, P. (1988). Fetal tobacco syndrome and other problems caused by smoking during pregnancy. *Med. Aspects Hum. Sexual.* Aug, 69–75.

McLean, M., *et al.* (1995). A placental clock controlling the length of human pregnancy. *Nature Med.* **1**, 460–463.

Nilsson, L. (1990). "A Child Is Born." Delacorte Press/Seymour Lawrence, New York.

Pedersen, R. A. (1999). Embryonic stem cells for medicine. *Sci. Am.* **280**(4), 68–75.

Remez, L. (1989). Chorionic villus biopsy is useful alternative to amniocentesis, despite slightly higher risk. *Family Planning Perspect.* **21**, 188–189.

Riddle, J. M., *et al.* (1994). Ever since Eve: Birth control in the ancient world. *Archaeology* March/April, 29–35.

Rina, A. (1992). First-trimester chorionic villus sampling may raise risk of spontaneous abortion and limb abnormalities. *Family Planning Perspect.* **24**, 45–46.

Sapolsky, R. (1999). The war between men and women. *Discover Magazine*, **May, 1999**, 56–61.
Small, M. F. (2000) Gut instincts: Why do pregnant women get nauseated just when their bodies most need food? *Discover Magazine* **Sept., 2000**, 34–37.
Taussig, H. B. (1962). The thalidomide syndrome. *Sci. Am.* **207**(2), 29–35.

Advanced Reading

Albrecht, E. D., and Pepe, G. J. (1990). Placental steroid hormone biosynthesis in primate pregnancy. *Endocr. Rev.* **11**, 124–150.
Balat, O., *et al.* (2003). The effect of smoking and caffeine on the fetus and placenta in pregnancy. *Clin. Exp. Obstet. Gynecol.* **30**, 57–59.
Baker, M. E., (1995). Endocrine activity of plant-derived compounds: An evolutionary perspective. *Proc. Soc. Exp. Biol. Med.* **208**, 131–138.
Bianchi, D. W., *et al.* (1996). Male fetal progenitor cells persist in maternal blood for as long as 27 years postpartum. *Proc. Natl. Acad. Sci. USA* **93**, 705–708.
Clarren, S. K., and Smith, D. W. (1978). The fetal alcohol syndrome. *N. Engl. J. Med.* **298**, 1063–1067.
Cross, J. C., *et al.* (1994). Implantation and the placenta: Key pieces of the development puzzle. *Science* **266**, 1508–1517.
Dejmek, J., *et al.* (2002). The exposure of nonsmoking and smoking mothers to environmental tobacco smoke during different gestational phases and fetal growth. *Environ. Health Perspect.* **110**, 601–606.
Dlugosz, L., *et al.* (1996). Maternal caffeine consumption and spontaneous abortion: A prospective cohort study. *Epidemiology* **7**, 250–255.
Esplin, M. S., *et al.* (2001). Paternal and maternal components of the predisposition to preeclampsia. *N. Engl. J. Med.* **4**, 867–872.
Fenster, L. *et al.* (1997). Caffeinated beverages, decaffeinated coffee, and spontaneous abortion. *Epidemiology* **8**, 515–523.
Flaxman, S. M., and Sherman, P. W. (2000). Morning sickness: A mechanism for protecting mother and embryo. *Q. Rev. Biol.* **75**, 113–148.
Foster, W., *et al.* (2000). Detection of endocrine disrupting chemicals in samples of second trimester human amniotic fluid. *J. Clin. Endocrinol. Metab.* **85**, 2954–2957.
Galey, F. D., *et al.* (1993). Estrogenic activity in forages: Diagnostic use of the classical uterine mouse bioassay. *J. Vet. Invest.* **5**, 603–608.
Genbacev, O., *et al.* (2003). Disruption of oxygen-regulated responses underlies pathological changes in the placenta of women who smoke or who are passively exposed to smoke during pregnancy. *Reprod. Toxicol.* **17**, 509–518.
Govender, L., *et al.* (1996). Bacterial vaginosis and associated infections in pregnancy. *Intl. J. Gynecol. Obstet.* **55**, 23–28.
Hall, G. H., *et al.* (2001). Long-term sexual co-habitation offers no protection from hypertensive disease of pregnancy. *Hum. Reprod.* **16**, 349–352.
Hatch, M. C., *et al.* (1993). Maternal exercise during pregnancy, physical fitness, and fetal growth. *Am. J. Epidemiol.* **137**, 1105–1114.
Hillier, S. L., *et al.* (1995). Association between bacterial vaginosis and preterm delivery of a low-birth-weight infant. *N. Engl. J. Med.* **333**, 1737–1742.

Hughes, C. L., Jr. (1988). Phytochemical mimicry of reproductive hormones and modulations of herbivore fertility by phytoestrogens. *Environ. Health Perspect.* **78**, 171–175.

Infante-Rivard, C., *et al.* (1993). Fetal loss associated with caffeine intake before and during pregnancy. *J. Am. Med. Assoc.* **270**, 2940–2943.

Khattak, S., *et al.* (1999). Pregnancy outcome following gestational exposure to organic solvents. *J. Am. Med. Assoc.* **281**, 1106–1109.

Lambert, N. C., *et al.* (2001). From the simple detection of microchimerism in patients with autoimmune diseases to its implication in pathogenesis. *Ann. N. Y. Acad. Sci.* **945**, 164–171.

Makinen, J., *et al.* (1988). Causes of the increase in the incidence of ectopic pregnancy. *Am. J. Obstet. Gynecol.* **160**, 642–646.

Marchbanks, P., *et al.* (1988). Risk factors for ectopic pregnancy. *J. Am. Med. Assoc.* **259**, 1823–1827.

Mortensen, E. L., *et al.* (2002). The association between duration of breastfeeding and adult intelligence. *JAMA* **287**, 2365–2371.

Quinla, J. D., and Hill, D. A. (2003). Nausea and vomiting of pregnancy. *Am. Fam. Physician* **68**, 121–128.

Profet, M. (1992). Pregnancy sickness as adaptation: A deterrent to maternal ingestion of teratogens. *In* "The Adapted Mind: Evolutionary Psychology and the Generation of Culture" (J. H. Barkow, L. Cosmides, and J. Tooby, eds.), pp. 327–366. Oxford University Press, New York.

Rasch, V. (2003). Cigarette, alcohol, and caffeine consumption: Risk factors for spontaneous abortion. *Acta Obstet. Gynecol. Scand.* **82**, 182–188.

Robillard, P.-Y., *et al.* (1994). Association of pregnancy-induced hypertension with duration of sexual cohabitation before conception. *Lancet* **344**, 973–975.

Shaw, G. M., and Croen, L. A. (1993). Human adverse reproductive outcomes and electromagnetic field exposures: Review of epidemiologic studies. *Environ. Health Perspect.* **101**(Suppl. 4), 107–119.

Sibley, C. P., and Boyd, R. D. H. (1988). Control of transfer across the mature placenta. *In* "Oxford Reviews of Reproductive Biology," Vol. 10, pp. 382–435. Oxford University Press, Oxford.

Simpson, J. L., and Bischoff, F. (2004). Cell-free fetal DNA in maternal blood. Evolving clinical applications. *JAMA* **291**, 1135–1141.

Sleutels, F., and Barlow, D. P. (2002). The origins of genomic imprinting in mammals. *Adv. Genet.* **46**, 119–163.

Tycko, B., and Morison, I. M. (2002). Physiological functions of imprinted genes. *J. Cellul. Physiol.* **192**, 245–248.

Von Sydow, K. (1999). Sexuality during pregnancy and after childbirth: A metacontent analysis of 59 studies. *J. Psychosom. Res.* **47**, 27–49.

Vutyavanich, T., *et al.* (2001). Ginger for nausea and vomiting in pregnancy: Randomized, double-masked, placebo-controlled trial. *Obstet. Gynecol.* **97**, 577–582.

CHAPTER ELEVEN

Labor and Birth

Introduction

Chapter 10 showed that pregnancy brings about physiological changes and psychological adjustments in the future mother as well as remarkable development and growth of the fetus. The internal relationship between mother and fetus terminates in childbirth, or *parturition*. This chapter first discusses the timing of birth and the role of hormones in the birth process. Then we see how knowledge about the hormonal control of birth has allowed us to induce labor artificially. Next, we describe the stages of labor and birth and some aspects of premature, multiple, and difficult births. Finally, the use of medications during labor and birth and birthing methods are reviewed.

Time of Birth

The average length of pregnancy in humans is about 9 calendar months. The *due date* (or *term*) can be calculated by counting 280 days (40 weeks) from the first day of the last menstruation. This can be done quickly by adding 7 days to the first day of the last menstrual period and then counting back 3 months. The due date can also be determined by counting 266 days (38 weeks) from conception (see Table 11-1). Usually, ovulation and fertilization occur 13 to 15 days after the first day of the last menstruation, but conception can also take place at other lengths of time after day 1 of the cycle (see Chapter 3). This is one reason why babies often are not born on the due date if it is calculated by the date of conception. Female infants tend to be born a few days earlier than males, and women who exercise tend to give birth sooner. Also, women who have short menstrual cycles tend to have shorter pregnancies. The reasons for these differences are unknown. Birth within 2 weeks before or after the due date is considered normal.

Seasonal birth cycles occur in virtually all human populations, regardless of race, but interestingly the patterns of seasonality differ in various regions. In North America, the lowest number of births is in the spring, and there is a seasonal peak in the fall (Fig. 11-1). Furthermore, the magnitude of the seasonal change is greater in the southern (lower latitude) regions and lowest at more northern latitudes (Fig. 11-1). This North American pattern is also true for the Mideast, Asia, and Africa. However, Europe differs in that a seasonal birth peak occurs in the spring and a low in the winter (Fig. 11-2). Furthermore, the magnitude of the seasonal change in Europe is greater the higher the latitude.

Table 11-1 Day of Delivery, Measured from the Day of Conception[a]

Day of delivery	Percentage
Before day 212	12.7
Second week before due date (250–257)	12.3
First week before due date (258–265)	22.1
On day 266 (the due date)	2.7
First week after due date (267–273)	24.2
Second week after due date (274–280)	15.6
After day 280	9.4

[a]To measure from day of last menstrual period, add 14 days.

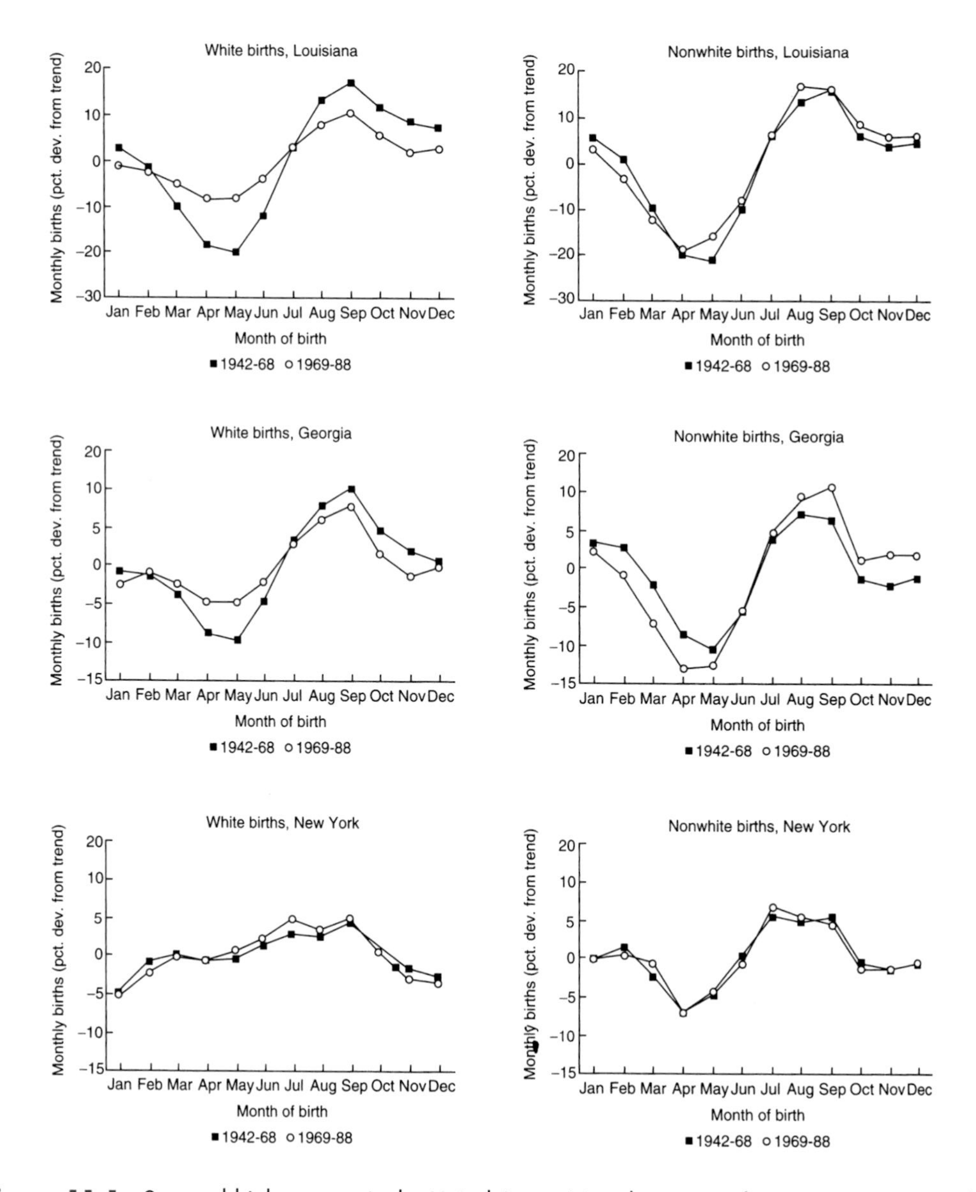

Figure 11-1 Seasonal birth patterns in the United States. Note that seasonality is more pronounced in southern states.

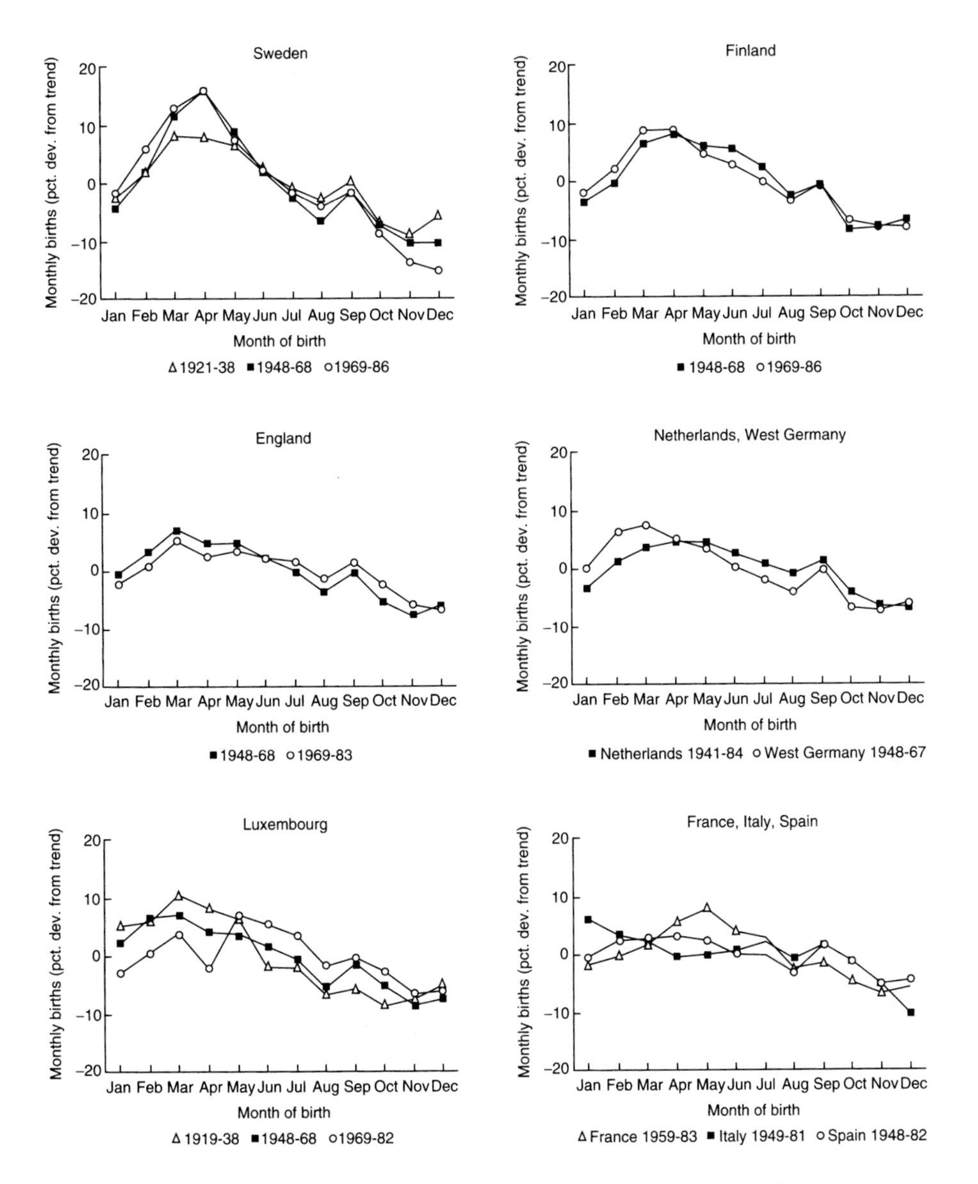

Figure 11-2 Seasonal birth patterns in Europe. Note that the spring high and the winter low in births increase in more northern countries.

In the North American pattern, the seasonal high in fall births corresponds to higher conception rates in winter, whereas the spring low in births corresponds to lower conception rates in the hottest months of the year. Because there is a lowering of sperm count and blood testosterone levels in American men around this same time (Chapter 4), the decrease in spring and early summer conceptions could be caused by an adverse effect of high temperature on sperm count or coital frequency. In Europe, however, it is thought that day length plays a greater role in causing the spring pattern of births. Whatever the causes, the seasonality in human births and conceptions may be evolutionary remnants of more pronounced seasonal cycles such as occur in wild primates as adaptations to seasonal food supplies. There is also a daily cycle in the frequency of human birth. More babies are born at night than during the day, with a small birth peak between 4:00 and 9:00 AM. This daily birth cycle may be an evolutionary vestige in that nocturnal birth in our ancestors may have offered protection against predators active during the day.

Hormones and Birth

During the 9 months after conception, the human fetus has undergone marked development and has increased greatly in size (see Chapter 10). Suddenly, it is expelled into the outer world. What factors determine when birth occurs? Remarkably, much of our present answer to this question originally came from observations of the food habits of sheep!

Delayed Birth in Sheep

Sheepherders in Idaho noticed that pregnant ewes grazing in certain pastures at specific times of the year failed to give birth to their lambs on time. Instead, extra-large lambs were born many days late, and the mothers often died in the process. In a search for the cause of this delayed birth, investigators discovered that, although the tissues of the mother appeared normal, the lambs had underdeveloped adrenal glands and malformed or absent pituitary glands. These fetal abnormalities occurred because the pregnant ewes ate the plant *Veratrum californicum*. This plant contains a chemical that passes across the placenta and harms the pituitary and adrenal glands of the fetus.

Further research has shown that birth was delayed in these sheep because the plant chemical reduced the secretion of corticotropin (ACTH) from the fetal pituitary gland. ACTH is needed for the fetal adrenal glands to develop and secrete adrenal steroid hormones (see Chapter 1). These steroid hormones play a major role in initiating birth. If the pituitary gland of a normal sheep fetus is removed (hypophysectomy), its adrenal glands are underdeveloped and pregnancy is extended 37 days past the normal due date. Injection of ACTH into hypophysectomized sheep fetuses results in normal fetal adrenal glands and delivery time is normal. Similarly, removal of the fetal adrenal glands (*adrenalectomy*) delays birth, and injection of adrenal steroid hormones into these fetuses reverses the effect of adrenalectomy. Finally, injection of either adrenal hormones or ACTH into the fetus any time after the second half of pregnancy results in early delivery. Normally, some factor causes the release of *corticotropin-releasing hormone* (CRH) from the fetal hypothalamus. This, in turn, causes secretion of ACTH, which induces secretion of fetal adrenal hormones (e.g., cortisol), and ultimately initiation of birth.

How does fetal cortisol initiate birth? The apparent answer is that this hormone travels to the placenta and affects placental secretion of sex hormones. Progesterone made by the placenta serves as a precursor for fetal cortisol (an adrenal steroid hormone) and an androgen (see Chapter 10), both of which are carried in the fetal blood to the placenta. The arrival of these steroid hormones at the placenta increases placental estrogen secretion and decreases placental progesterone secretion. Thus, the ratio of estrogen to progesterone (E/P ratio) in the mother's blood increases. In many mammals, this ratio in the blood of pregnant females increases near the onset of birth. Estrogens stimulate contraction of the uterine muscles, whereas progesterone inhibits uterine contractions. Therefore, the increase in the E/P ratio in the blood initiates uterine contractions (*labor*).

Hormonal Initiation of Human Birth

Is there evidence that a similar role of the fetal adrenal glands is present in humans as well as in sheep? The answer is "maybe." Recent information shows that there is an increase in the E/P ratio in the blood of women in the last 5 weeks of pregnancy. This elevated ratio may cause uterine contractions. Administration of a progestogen to women at risk for premature labor can delay labor initiation, and estrogen treatment can cause early labor. Significantly, women exhibiting early delivery tend to have newborns with relatively large adrenal glands, whereas women who have delayed delivery tend to have fetuses with relatively underdeveloped adrenal glands. The latter women also have relatively low levels of estrogen in their blood. *Postmature newborns* have low levels of cortisol in their blood. Thus, fetal adrenal glands may play a role in birth initiation in humans, but direct evidence is needed to substantiate this claim. If an increase in CRH secretion, and therefore fetal ACTH and cortisol secretion, plays a role in human birth initiation, the CRH likely is secreted by the placenta and not the fetus's brain. That is, placental CRH, levels of which rise during pregnancy, could be the "clock" that starts labor at the right time (see HIGHLIGHT box 11-1).

Chapter 11, Box 1: The Placental Clock

Although many babies are born at or near the "due date," you know that some are born prematurely and some after the due date. Given what you learned in this chapter about the hormonal mechanisms that initiate birth, one must question how these mechanisms can occur too early or too late. Most of our discussion of hormones and birth in this chapter revolved around sheep, with some mention that maybe the mechanisms are similar in humans. It turns out that there is a major difference between the reproductive endocrinology of humans (and other primates) and sheep.

In sheep, secretion of corticotropin-releasing hormone (CRH) from the fetal brain causes fetal ACTH secretion from the pituitary, which in turn increases fetal cortisol. You learned that fetal cortisol, by influencing the secretion of other hormones, initiates birth. Does this also hold true for human birth? Humans appear to differ from sheep in that most of the CRH that causes the fetal adrenal to secrete cortisol does not come from the fetus's brain but instead is secreted in large quantities by the placenta. Beginning at about the 16th week of pregnancy, levels of this placental CRH in the mother's blood increase exponentially as pregnancy advances. The increasing level of placental CRH has been hypothesized to serve as a "placental clock" that determines the timing of birth. According to this theory, the "clock" is set early in gestation, and its alarm goes off at birth.

In a 1995 study, researchers measured blood levels of CRH in 485 women during the midtrimester of pregnancy (16 to 20 weeks from the date of the last menstrual period). Then they recorded timing of births in this group. The 24 women who delivered prematurely (37 weeks or less) had almost four times

Continued on next page.

Chapter 11, Box 1 continued.

the amount of CRH in their blood in the second trimester as those women who delivered on time (37 to 42 weeks). The women who delivered late (more than 42 weeks) had slightly less CRH in their blood in early pregnancy. Two later studies confirmed the relationship between second trimester maternal CRH and premature delivery. Because CRH levels rise exponentially through the remainder of pregnancy, women who begin with higher levels early in pregnancy also reach an earlier CRH "peak."

Surprisingly, administration of cortisol to late-term women does not induce birth as it does in sheep. How then might CRH act to control timing of delivery? Here is a proposed explanation based on several recent studies. CRH synthesized in the placenta reaches the fetal circulation and causes hypertrophy (growth) of the part of the fetal adrenal called the *fetal zone*. When CRH reaches the fetal pituitary, it causes the pituitary to release ACTH, and the ACTH induces the fetal zone to release cortisol. Cortisol flows back to the placenta through the umbilical cord. In the brain, cortisol suppresses CRH through negative feedback. However, in the placenta, cortisol actually stimulates the production of more CRH. Thus, a positive feedback loop exists, which explains the sharp rise in both CRH and cortisol toward the end of gestation. Cortisol is important in promoting final organ maturation in the fetus prior to birth.

CRH also acts directly on the fetal zone to induce synthesis of DHEA-sulfate (an androgen). This steroid is the substrate for 80–90% of the estrogen synthesized in the placenta during late pregnancy. The effects of estrogen on the myometrium and cervix are important in increasing uterine contractility. Finally, CRH stimulates the production of prostaglandins (PGs) from the fetal membranes, and in turn PGs induce further production of CRH (this is another positive feedback loop). Prostaglandins play an important role in parturition by directly stimulating the uterus to contract.

Thus, CRH appears to play an important role in coordinating the events leading to birth, which are (1) an increase in the estrogen:progesterone ratio and the resultant effects on the uterus and cervix, (2) myometrial contraction, and (3) maturation of fetal organs in preparation for birth. Two positive feedback loops are involved, and they terminate at the moment of birth. It is important to realize that this is a "working hypothesis," a theoretical framework that can be tested experimentally. Further research will be needed to confirm the role of this CRH placental clock in the timing of birth.

This exciting work has the potential to help physicians predict if a woman is likely to deliver prematurely. It might also lead to the development of new therapies to prevent premature births.

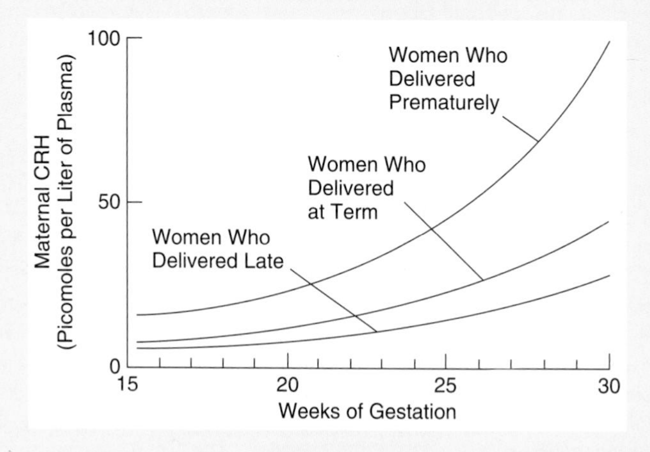

Levels of CRH in the blood of women during pregnancy. Note that the levels in weeks 16 to 20 can help predict the time of birth. (From Smith (1999))

In addition to the ratio of estrogen to progesterone in the maternal blood, other hormonal factors appear to be important in initiating or maintaining human labor. Prostaglandins are a group of fatty acids widely distributed in various tissues (see Chapter 2). Some prostaglandins stimulate the contraction of uterine muscles and now are used by physicians to induce labor at term. A high E/P ratio in the mother's blood appears to enhance the production of prostaglandins by the uterus, as does uterine contraction itself. But do prostaglandins play a role in normal human birth? We know that prostaglandin levels increase in the mother's blood a month before delivery and are very high during labor, with bursts in secretion 15 to 45 s after each uterine contraction. We also know that levels of prostaglandin in the amnionic fluid increase during labor. Certain drugs (such as aspirin) inhibit the synthesis of prostaglandins, and these drugs may delay labor if given routinely near term. In conclusion, prostaglandins do play a role in human labor, probably acting as paracrines (see Chapter 1).

It has long been known that a hormone secreted by the posterior lobe of the pituitary gland stimulates the contraction of uterine muscle (see Chapter 1). This hormone, oxytocin (in Greek, "quick birth"), also appears to be involved in human birth. In fact, a synthetic oxytocin (*pitocin*) is often used by physicians to hasten labor contractions. In humans, mechanical stimulation of the vagina, cervix, or uterus causes release of oxytocin; this reaction is called the *fetal ejection reflex*. Once oxytocin is secreted, it increases the intensity of uterine contractions, which in turn causes more oxytocin release. Recent research suggests that the placenta itself secretes oxytocin and that it is placental oxytocin, rather than pituitary oxytocin, that initiates labor. A high E/P ratio may increase placental oxytocin secretion.

In women, oxytocin levels in the blood are moderately low in the first stage of labor and are then higher and more variable in the second stage. Therefore, this hormone may simply increase the intensity of uterine contractions during later stages of labor. One theory suggests that even though oxytocin levels in the blood are low at the time of labor initiation, the number of oxytocin receptors in the uterus increases due to the increase in the E/P ratio in the woman's blood. Thus, oxytocin may actually initiate labor both directly and indirectly by causing the release of prostaglandins from the contracting uterus. Alcohol inhibits oxytocin release, and in the past it was not unusual for a physician to prescribe alcohol intravenously for women in premature labor. This practice is seldom followed today.

Scientists studying mice have found a way that the fetus might signal to the mother that it is time to give birth. In these animals, the concentration of surfactant protein-A (SP-A) rises sharply just before parturition. This protein synthesized in the fetal lung protects the lungs from microbial attack. Injecting SP-A into amnionic fluid of pregnant mice causes premature labor, whereas an antibody blocking the action of SP-A delays normal labor. The use of this signal ensures that the fetal lungs are developed sufficiently by the time of birth. Whether this mechanism operates during human labor is not yet known. It is possible that multiple signals exist to control the timing of parturition.

We have much to learn about the hormonal initiation of labor in humans. The sequence and relative importance of the processes just described are not well understood. Figure 11-3 presents a summary of how labor is initiated and maintained.

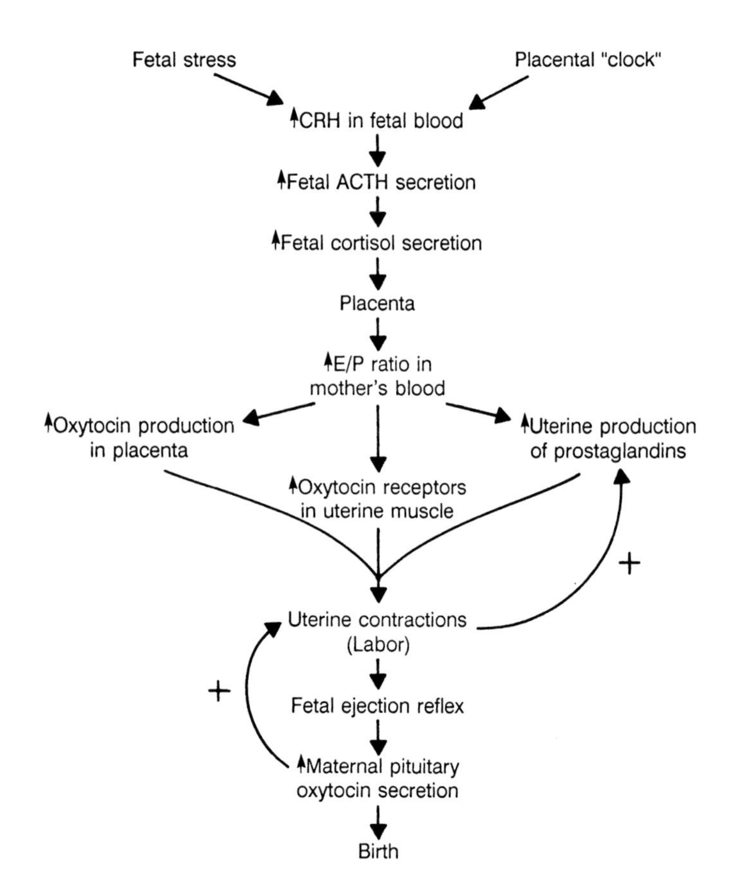

Figure 11-3 A summary of the hormonal events that interact to initiate and maintain labor. Note that corticotropin-releasing hormone (CRH), from the fetal brain and/or the placenta, is the signal for the beginning of labor. Note also two positive feedback systems (+), one involving the fetal ejection reflex and another involving the production of prostaglandins by the contracting uterus. ↑, increases; ↓, leads or travels to; ACTH, corticotropin; E/P ratio, relative amounts of estrogen (estriol) and progesterone in the mother's blood.

Induced Labor

Physicians can *induce labor* before term, near term, or after the due date by administering pitocin and/or prostaglandins, both of which are effective. First, however, the physician will break the amnionic membrane (see Chapter 10) because this alone often begins labor and hormones need not be administered. Pitocin is administered intravenously (by inserting a needle into a vein) in continuous slow drops monitored by a pump. Prostaglandins also can be given intravenously or are injected into the amnionic sac, given orally, or administered as a vaginal suppository.

One reason for inducing labor is if the birth is 2 weeks overdue (i.e., the fetus is *postmature*). It should be noted, however, that many "late" pregnancies are miscalculations of the due date, and there may be danger of delivering a premature baby if labor is induced in such cases. The physician usually can determine fetal maturity by physical and physiological characteristics. About 8 to 12% of babies are born postmaturely in the United States. Early labor can be induced artificially before 36 weeks of pregnancy, and babies born this way are *premature*, as discussed later. This can be done, for example, if the amnionic sac has burst 12 to 24 h beforehand and labor has not yet begun. A broken amnionic sac increases the danger of neonatal and maternal infection.

Preparation for Labor

About 2 or 3 weeks before labor, a woman could have a sensation of decreased abdominal distention produced by movement of the fetus down into the pelvic cavity. This is known as *lightening*; it is said that the baby has "dropped." Lightening occurs about 2 weeks before birth in a woman having her first baby but may not occur until labor begins in her subsequent pregnancies. A woman literally breathes easier after lightening because of less pressure on her diaphragm (a dome-shaped muscle under the rib cage that assists in breathing). Also, she may urinate more frequently because the fetus is now pressing on her bladder. A few hours to a week before labor begins, the part of the fetus that will exit first (usually the head) moves down into the pelvic girdle. This is termed *engagement of the presenting part*. Figure 11-4 illustrates how the female pelvis is adapted for passage of the fetus through the *birth canal* (cervix and vagina).

The Birth Process

The birth process can be divided into three stages: (1) *cervical effacement and dilation*, (2) *expulsion of the fetus*, and (3) *expulsion of the placenta*. The length of each stage varies among individuals and in the same individual between first and subsequent births. The entire process of labor and birth typically lasts from 8 to 14 h in women giving birth for the first time (*primiparous* women) but is shorter (4 to 9 h) in women who have had a child previously (*multiparous* women). Any length of labor up to 24 h, however, is considered normal.

Stage 1: Cervical Effacement and Dilation

Throughout pregnancy, especially in the last few weeks, a woman can experience mild, irregular uterine contractions. These *Braxton–Hicks contractions* have little rhythmic pattern. They sometimes cause the abdominal wall to become hard to the touch, but usually are not felt and subside after a few minutes. Some women have *false labor* contractions during late pregnancy, which they may experience as being moderately intense. These contractions can be rhythmic in nature, but they do not cause much cervical effacement or dilation. Also, they do not persist.

Eventually, true labor commences; the uterine contractions become more intense and they occur at regular intervals. These *effacement contractions* usually

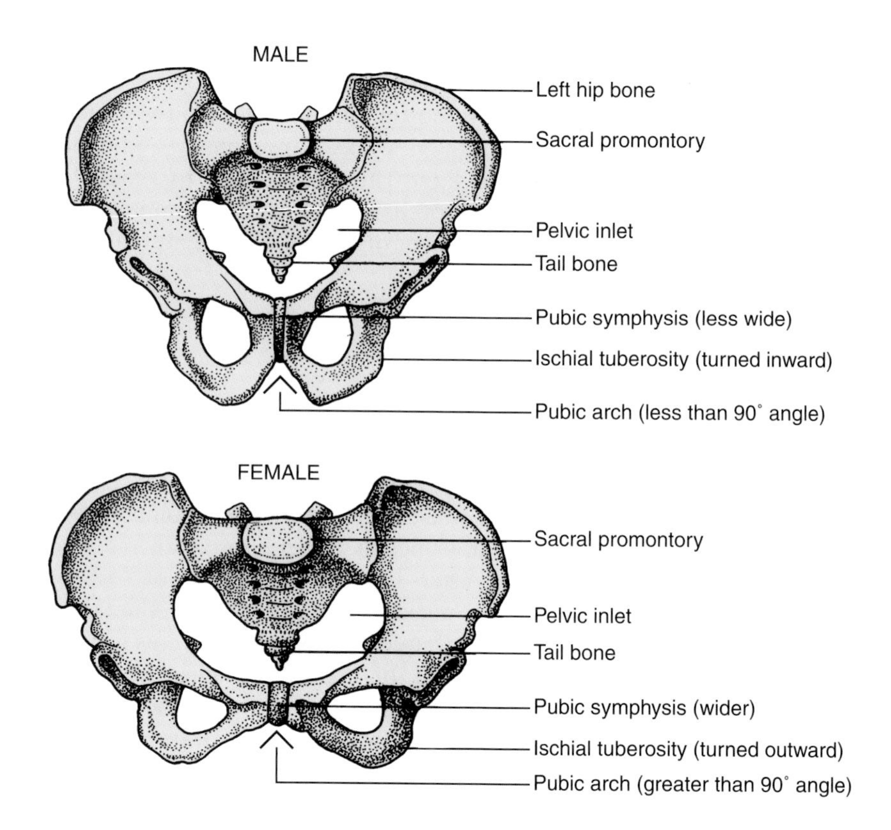

Figure 11-4 Sex differences in human pelvic structure. The female pelvis provides a wider passage for the fetus to move through the birth canal (cervix and vagina) in several ways, including (1) a larger and more oval pelvic inlet (space between the sacral promontory and the top of the pubic symphysis); (2) a wider pubic symphysis, even more so under the growth effects of relaxin during pregnancy (Chapter 10); (3) a wider pubic arch; and (4) outward turning of the ischial tuberosities. The last three characteristics produce a larger pelvic outlet (space between the tip of the tail bone and the bottom of the pubic symphysis). The fetus moves through the pelvic inlet and outlet (pelvic canal).

are felt in the back and then in the abdominal wall; they reach a peak and then relax. Some women may never be aware of them. The contractions can be felt and timed by placing a hand on the upper abdomen. Each contraction lasts about 30 to 60 s, with intervals between contractions of about 5 to 20 min (Fig. 11-5). The result of effacement contractions is *cervical effacement*, which means a thinning of the normally thick walls of the cervix and retraction of the cervix upward, making it easier for the fetus to pass into the *birth canal* (cervical and vaginal canals).

During pregnancy, the cervix is blocked by a *mucous plug*. At or immediately before the beginning of effacement contractions, mucus is dislodged along with a small amount of blood, and this *bloody show* (pinkish in color) exits through the vagina. Also at this time, or in the first stage, an enzyme weakens the amnion. A small tear then appears in the amnionic sac (actually made up of chorion on the outside and amnion on the inside), and clear amnionic fluid trickles or gushes from the sac and is expelled through the vagina. This bursting of the amnionic sac

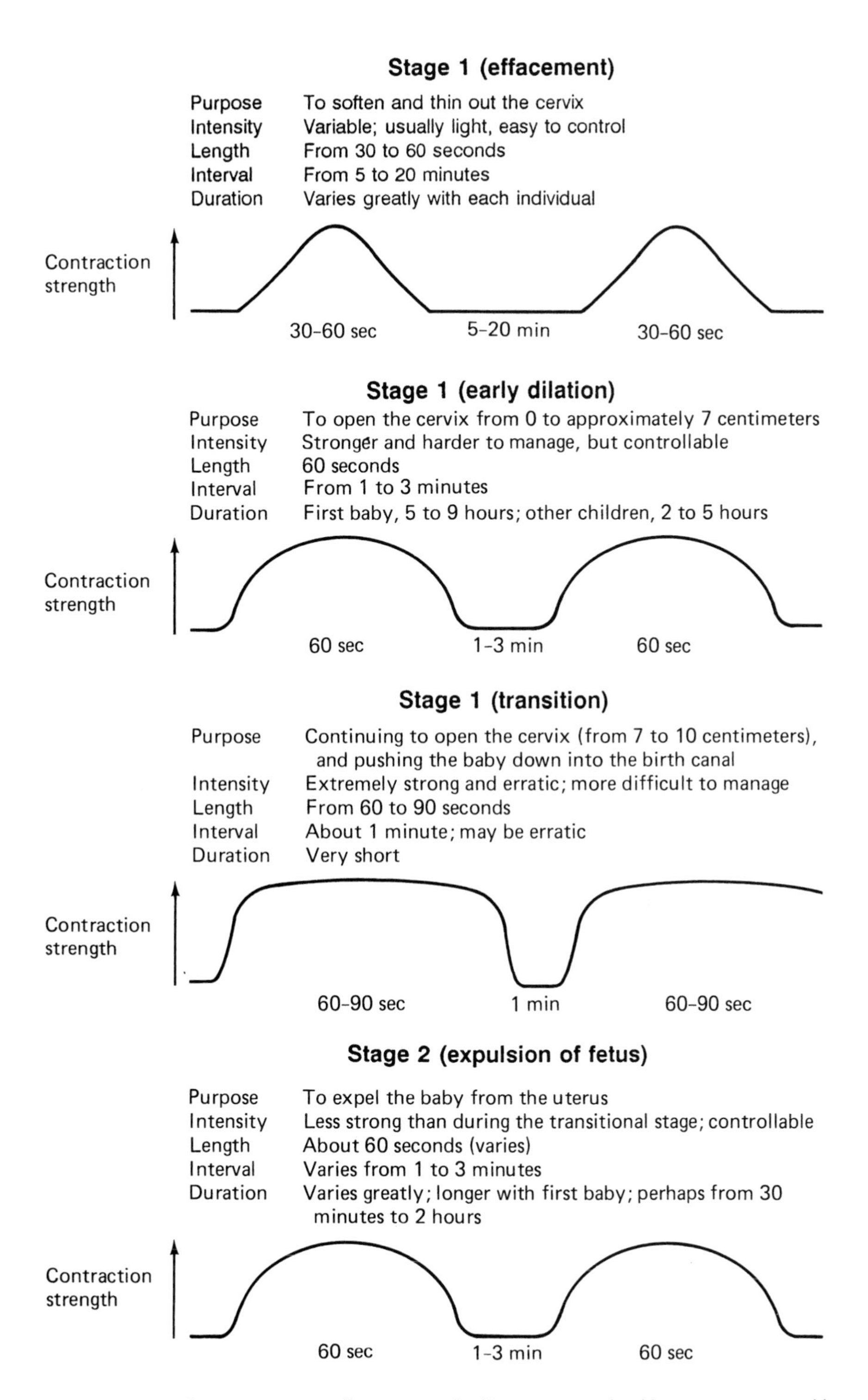

Figure 11-5 Characteristics and purpose of effacement, early dilation, transition dilation, and expulsion of fetus contractions. Times given are approximate. Stage 3 (expulsion of the placenta) contractions are not shown; these contractions are similar to those in early dilation, but the duration of the stage is shorter.

(*breaking of the bag of waters*) and the bloody show are sure signs that true labor is commencing. In about 12% of pregnancies, the amnionic sac breaks before labor begins. These "dry labors" proceed normally but often are shorter than usual. It is also common for the sac to remain intact after labor has advanced considerably, and in these cases the physician will puncture the amnion with an instrument (this does not hurt the mother, as there are no pain receptors in the amnion). It should also be noted that because the fetus resides in the uterine wall and not in the uterine cavity, a thin part of the uterine wall (the decidua capsularis; see Chapter 10) actually covers the chorion next to the uterine lumen. However, this thin uterine layer degenerates after 4 to 5 months of pregnancy.

A pregnant woman should notify her doctor or midwife of the advent of effacement labor, bloody show, or leaking of amnionic fluid. If she is planning a hospital birth, she may be advised to relax at home for a few hours, especially if this is her first baby. Leaving for the hospital before this time may not be necessary, even if the woman or her partner is nervous and anxious to get going. Meanwhile, they can time the contractions by feeling when the abdomen becomes hard and soft and can report to their doctor the duration of contractions and the interval between the initiation of successive contractions. The woman should not eat during this time, as digestion is inhibited during labor and vomiting should be avoided. An informed woman usually feels excited and confident during this stage of labor. Many physicians advise a woman to leave for the hospital (if she will use one) when contractions are about 5 min apart—from the beginning of one contraction to the start of the next. Of course, women who are a long distance from the hospital will be advised to leave earlier!

After a variable amount of time, the woman will notice that her contractions begin to last longer (about 60 s), are more intense, and occur more frequently, with rest intervals of only 1 to 3 min (Fig. 11-5). She has now entered the early dilation portion of stage 1; the result of these contractions is the beginning of *cervical dilation* (also called *cervical dilatation*). The external cervical os increases from its normal 0.3-cm diameter to a diameter of about 7 cm. The entire early dilation process lasts about 5 to 9 h in primiparous women, but is shorter (2 to 5 h) in multiparous women because the cervix is more pliable. The *dilation contractions*, because of their frequency and intensity, may become uncomfortable. If the mother is still at home, she should enter the hospital soon after these contractions begin. When she arrives at the hospital, she will be admitted, prepared for labor, and given a bed in the labor or birthing room. A nurse or her physician will then check for her degree of cervical dilation and often will rupture her amnionic sac if this has not occurred naturally. After washing the genital area, the nurse may shave all or some of the pubic hair to help protect the fetus against infection. An enema (induction of defecation) often is given, except if labor is advanced.

The final phase of stage 1 labor, during which the cervix dilates from about 7 to 10 cm in diameter, is called *transition dilation*. This phase lasts about 20 min to 1 h and tends to be shorter in multiparous women. Transition is characterized by very intense *transition contractions* that last longer (60 to 90 s) than those in earlier stages of dilation (Fig. 11-5). The interval between transition contractions is about 1 min, but often is erratic. In primiparous women, the cervix usually dilates after it effaces, but these two events occur together in multiparous women (Fig. 11-6). This is one reason why stage 1 of labor happens more quickly in multiparous women. During transition, the fetus descends into the pelvic basin, which puts pressure on the pelvic floor (Fig. 11-7). The woman has an urge

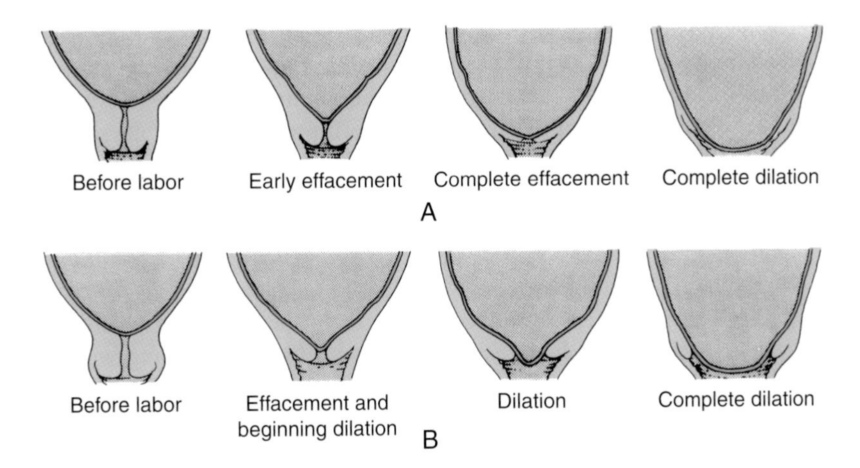

Figure 11-6 Degrees of cervical effacement and dilation. Note that in primiparous women (a), effacement occurs before dilation. In multiparous women (b), effacement and dilation occur together.

to push during these contractions, but she is advised not to do so. Pressure on the pelvic floor creates this pushing urge, which women say feels like an urge to defecate. Pushing before complete dilation to 10 cm will tire the mother but not move the fetus and could cause edema of the cervical tissues.

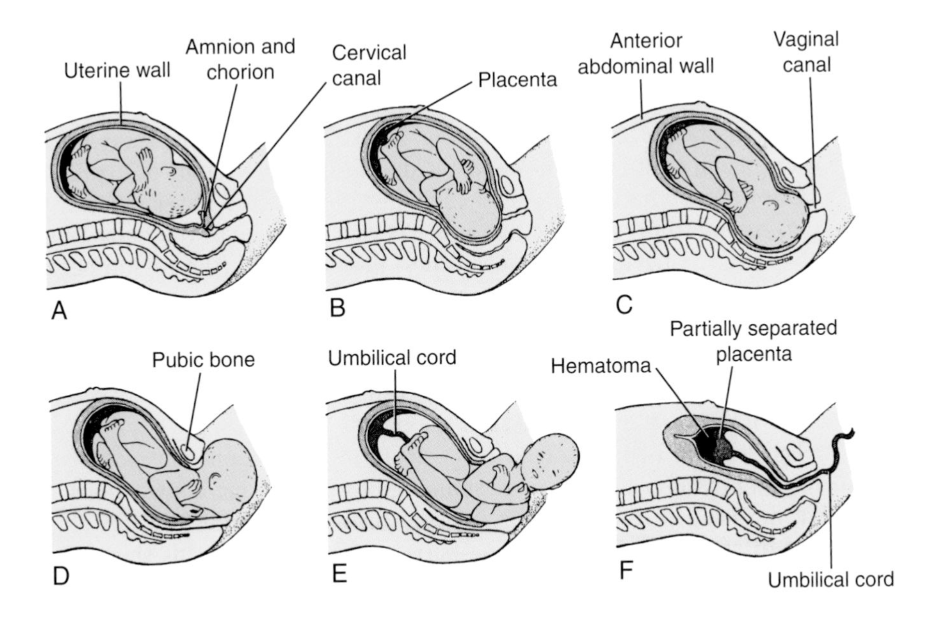

Figure 11-7 Diagrams illustrating the birth process. The cervix is dilated during the first stage of labor (a and b). The fetus passes through the cervix and vagina during the second stage of labor (c to e). Note that the fetal head rotates and first bends forward (flexion) and then back (extension). Then, as the uterus contracts, the placenta folds up and pulls away from the uterine wall (f). Separation of the placenta results in bleeding, forming a large blood clot (hematoma). Later, the placenta is expelled.

Transition is the most difficult part of labor, not only because of the severity of the contractions but because a woman may experience intense pain, nausea, vomiting, trembling, leg cramps, discouragement, and restlessness. Normally, the face of the fetus points toward the sacrum of the back, but sometimes the hard back of its head is toward the sacrum. When the latter situation is present, back pain (*back labor*) is felt during contractions. Transition does not last very long, and medications to relieve discomfort are often used during this phase (see later). A woman also can assume an upright or squatting position to allow gravity to help the labor process.

During transition the condition of the fetus can be monitored for signs of *fetal distress*. Fetal heart function is monitored by placing wire leads either on the mother's abdomen or on the fetus's scalp through the cervical opening. For some women, *fetal monitoring* procedures are an unwanted intervention. These procedures, however, can have several benefits, including early detection of fetal distress.

After full dilation (10 cm) is reached, the mother is moved to the delivery room. (If she is in a "birthing room," both labor and delivery will occur there.) A multiparous woman may be moved before full cervical dilation. In the delivery room, she moves onto a delivery table, where her legs are placed in stirrups with supports under her knees. Her inner thighs, abdomen, and genital regions are then cleansed with an antiseptic. The region is then covered partially with a sterile sheet.

Stage 2: Expulsion of the Fetus

Expulsion of the fetus begins when the cervix is dilated maximally and ends with delivery of the infant. The intensity of contractions in this stage is less than that in the transition portion of stage 1. Each contraction lasts about 60 s, with 1- to 3-min rest intervals between contractions (Fig. 11-5). Expulsion of the fetus happens in a shorter time in multiparous women; the duration for all women is 30 min to 2 h.

Once the woman is settled on the delivery table, the physician may perform an *episiotomy* to prevent tearing of the perineal tissues as the baby emerges. To do this, a local anesthetic is injected into the *perineum* (the region between the anus and the vagina) and a small incision is made in the perineal skin. This incision later will be sutured with absorbable material. About 70 to 80% of women in the United States have an episiotomy during labor; some feel that many of these operations are not necessary, especially since they can delay recovery and healing of the woman's tissues after the baby is born.

The woman is encouraged to push during expulsion contractions, and the increased pressure due to this pushing and to the uterine contractions begins to move the head of the fetus through the birth canal. Soon the top of the head begins to appear without receding between contractions. This show of the fetal head (*crowning*) signifies that the baby is about to be born. The bones of the infant's head are not yet fused and they overlap when the head is moving out. This facilitates passage of the fetus through the birth canal. The "pointed" appearance of the head disappears soon after delivery (see Chapter 12). Usually, the head emerges with the face down. Once the head is out, any mucus or amnionic fluid in the baby's nose or mouth is removed with a suction device. The infant rotates during labor so that the shoulders emerge in the up-and-down position that is the largest dimension of the birth canal (Fig. 11-7).

Once the shoulders have emerged, the physician or midwife determines if the umbilical cord is around the neck, a rather common occurrence with no special dangers if detected at this point. The infant then slides out (Fig. 11-8), takes his or her first breath, and usually emits an exhilarating cry. (The old tradition of slapping the back of the infant to stimulate its first breath usually is not necessary.) Mucus is removed from the baby's nose and mouth with suction. The umbilical cord is then clamped in two places about 3 in. from the baby's abdomen and is cut between the clamps. There are no nerve endings in the cord, and neither the mother nor the infant feels the procedure. Drops of penicillin or silver nitrate are then placed in the infant's eyes to prevent bacterial infection. This is required by law in all states because, without this treatment, the infant could be blinded by bacteria if the mother is infected.

Stage 3: Expulsion of the Placenta

During stage 3 of delivery, which lasts about 15 to 30 min, the placenta (*afterbirth*) is expelled. After the fetus is delivered, the next few contractions push the placenta, which has become detached from the uterine wall, through the birth canal. If it does not come out, the physician or midwife can massage the abdomen gently or pull on the umbilical cord to help expel the placenta. About 8 oz of blood normally is lost during delivery. Usually the uterus will be massaged through the abdominal wall to encourage contraction, which inhibits uterine blood flow. If uterine hemorrhage (excessive bleeding) persists, oxytocin or a chemical that constricts blood vessels is administered to contract the uterus and inhibit bleeding. Next, the episiotomy incision (if present) is sutured, and the mother is ready to attend to her infant.

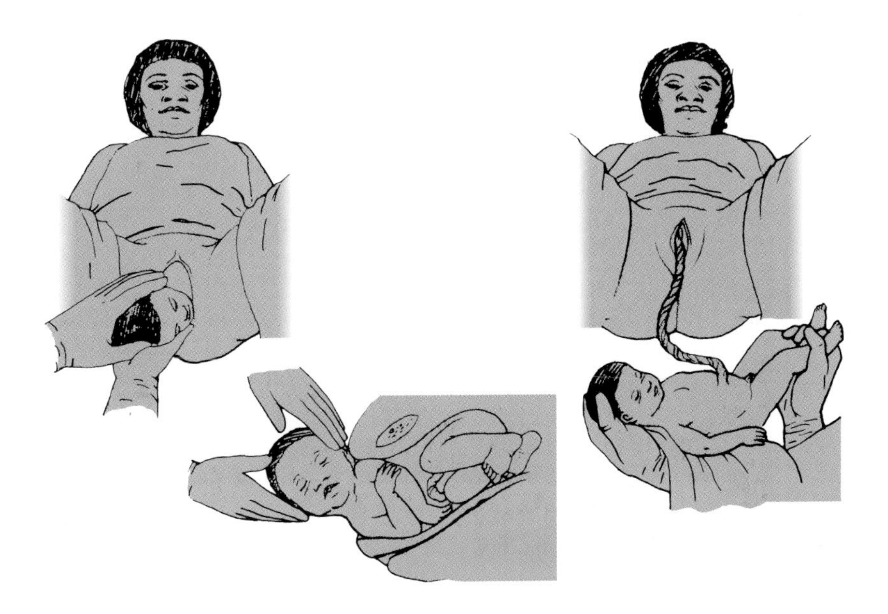

Figure 11-8 Stages in the final emergence of the infant.

In some societies, the expelled placenta is treated with reverence and is buried ceremoniously. Hospitals in the United States typically discard it unless it needs to be examined for medical reasons. However, appreciation for the placenta and the umbilical cord has grown recently because these tissues contain a small but valuable pool of stem cells. *Hematopoietic stem cells* found in "cord blood" have the potential to proliferate and differentiate into red blood cells, white blood cells, and platelets. Cord blood can be transplanted into leukemia and lymphoma cancer patients, where it can restore bone marrow destroyed by radiation and chemotherapy. These blood transplants are now being performed as an alternative to bone marrow transplants for certain patients. Both procedures require the immunological compatibility of donor and recipient, but donor cord blood is easier to harvest than bone marrow. Before giving birth, mothers can elect to donate placental and umbilical cord blood to a "cord blood bank." This blood can be frozen and stored for many years and could even be used to treat the infant donor in later years if this person became ill with a blood cancer.

Premature Births

A *premature* infant is usually defined as one born before 37 weeks of gestation and weighing less than 5.5 lb (2.49 kg). A newborn weighing between 1.0 kg (2 lb, 3 oz) and 0.5 kg (1 lb, 1.5 oz) is said to be *immature*. Below this lowest weight, the fetus is usually *stillborn* (born dead). The normal weight of a fetus 1 month before the due date is about 2.5 kg (5.5 lb), compared with the average weight of 3.40 kg (7.5 lb) at birth. A premature birth that occurs before the end of a full-term pregnancy is also called *preterm*; if the length of gestation exceeds 39 weeks, the infant is *postterm*, or *postmature*.

Preterm birth occurs in 12% of all U.S. pregnancies (one in eight pregnancies). It is responsible for 70% of all neonatal deaths and nearly 50% of all cases of congenital neurological disabilities. Earlier births carry a greater risk of death and disability such as mental retardation, blindness, deafness, and cerebral palsy. Births before 32 weeks of gestation, which make up 2% of all births, account for the vast majority of neonatal deaths. Premature delivery is widely considered to be the major problem currently faced in obstetrics in the United States. Although the infant mortality rate has declined in the past two decades, the incidence of preterm birth has actually increased during this time. For unknown reasons, black women have twice the incidence of premature infants as do whites.

The organs of premature newborns are not fully developed, and this and other possible fetal disorders that may have precipitated the premature birth in the first place reduce the chance for survival of these infants. In general, when development has reached 38 weeks, 99% survive, but this figure is 59% at 25 weeks and 18% at 23 weeks. Of infants born with less than 22 weeks of development, less than 1% survive. Unfortunately, about 85% of premature babies will face some developmental problems as they grow up. Premature babies often suffer from poor development of the brain. For example, infants born less than 5.5 lbs are four times less likely to graduate from high school by age 19. Premature babies also tend to score lower on achievement tests and are less likely to go on to college. Looking at this from another angle, for every 2.2 lbs additional weight at birth, boys score 4.6 points higher and girls 2.8 points higher on IQ tests at age 7, controlling for other relevant factors. Some aspects of the care of premature babies are discussed in Chapter 12.

We do not understand fully what causes premature labor. Almost one-half of twins are born prematurely. The average birth weight of twins (2.49 kg; 5.5 lb) is less than the average birth weight of single infants (3.40 kg; 7.5 lb) because each twin must share maternal nutrients. The increased incidence of prematurity in twins may be due to the greater distention of the uterus (fetal ejection reflex) or to a higher production of fetal adrenal hormones from two pairs of adrenals instead of one. Fetuses suffering from birth defects also tend to be born prematurely. Women who have had a prior premature baby are at especially high risk of preterm delivery. Maternal disorders (discussed in Chapter 10) such as severe malnutrition, infection of the uterus, diabetes, hypertension, toxemia, premature separation of the placenta, and bacterial vaginosis are other risk factors. As mentioned earlier, premature babies sometimes have oversized adrenal glands. Other evidence that premature labor is initiated by a hormonal change is that levels of estradiol in the mother's blood increase during premature labor. Ritodrine and nifedipine are drugs used commonly in the United States to prevent premature labor by relaxing the uterus, but these drugs have had limited effectiveness. A recent study found that weekly progesterone treatment can reduce preterm delivery in women who have previously had a preterm birth by 34%. The optimal mode of progesterone administration for high-risk pregnancies and its possible side effects are currently being evaluated.

In the near future, there may be better ways to detect if premature birth is going to occur. Recent research suggests that the fetal membranes produce an enzyme, *MMP-9*, that degrades the connective tissue holding the amnionic sac together and thus helps the sac to rupture. Measurement of the levels of this enzyme in the amnionic fluid could be used to predict premature labor. Another protein, *fetal fibronectin*, is the "glue" that holds the placenta onto the uterine wall. About 2 to 3 weeks before labor, this protein appears in cervical and vaginal fluid. Premature detection of this protein could indicate that premature birth is imminent. The most promising discovery, however, is that maternal blood levels of corticotropin-releasing hormone (CRH; see Chapter 1) at 14 to 18 weeks of gestation (early second trimester) are 3.64 times higher in women who later give birth prematurely than in women who give birth on time. Therefore, using CRH blood levels as an early predictor of premature birth could be very possible (see HIGHLIGHT box 11-1).

Multiple Births

Considering all births, the odds for having twins are about 1 in 71; for triplets, 1 in 6400; and for quadruplets, 1 in 512,000. However, heredity can influence the odds of having fraternal (but not identical) twins (see Chapter 10). Fraternal twins occur more commonly in women who have a family history of twins. Racial factors and age also appear to influence the incidence of fraternal twins; twinning is more likely to occur in blacks than in whites and is less frequent in Asians (Chapter 10). Women in their thirties have more fraternal twins than those in their twenties, and women who have been receiving certain fertility drugs tend to have multiple births (see Chapter 16). Delivery of multiple fetuses occurs about 22 days earlier, on the average, than single births.

In about 70% of cases, the presence of twins can be diagnosed before they are born. This can be done by ultrasound (discussed in Chapter 10), ascertaining excessive fetal movement, or detection of two heartbeats. When twins are born, both can come out head first or one can be head first and the other in a breech position (see the next section). One twin usually is expelled a few minutes to 1 h before the other, but there are records of the second twin being delivered up to 56 days after the first! A few of these cases in which a twin is born many days after its sibling may be due to *superfetation*, i.e., fertilization of a newly ovulated egg occurs while a previous fetus is developing in the uterus. This would have to occur before the fourth month of pregnancy, because after that time the amnionic sac obliterates the uterine lumen and would not allow sperm passage. There is, however, no direct proof that superfetation occurs in humans.

Difficult Fetal Positions

In 95% of all births, the fetus presents in the normal, head-down position. In 3 to 4% of births, the fetus is in a *breech presentation* at the beginning of labor, which means that the feet, buttocks, or knees rest against the cervix (Fig. 11-9). Actually, 50% of all fetuses are normally in the breech presentation before the seventh month of pregnancy; most then naturally turn 180° to the normal head-down position before the ninth month. Breech deliveries often occur with no difficulties (although labor is longer), but they sometimes require cesarean delivery. In 1 out of 200 births, the fetus is in a *transverse presentation*, the shoulders and arms emerging first (Fig. 11-9). In these cases, a cesarean delivery is necessary.

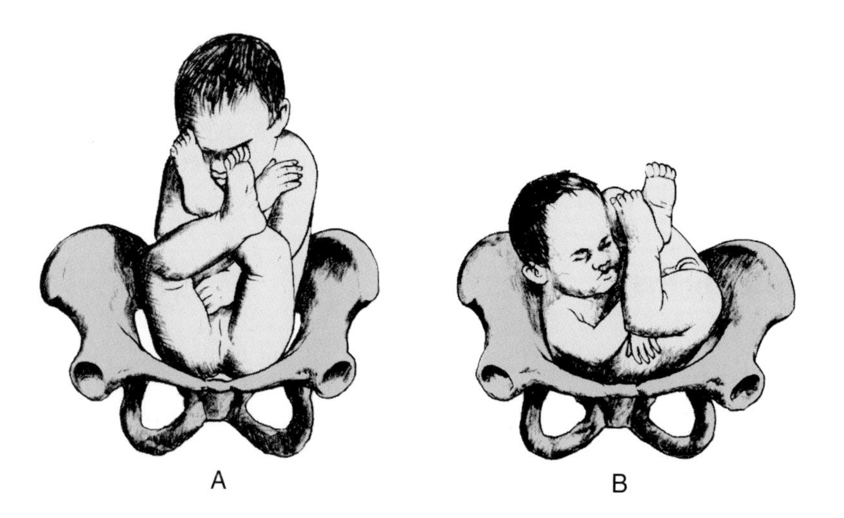

Figure 11-9 Breech (a) and transverse (b) fetal presentations. These relatively uncommon fetal positions may require forceps delivery or cesarean section. Other breech presentations can be variations of the one shown and include both knees folded (emerging feet first), one knee folded (one foot emerging first), or kneeling (knees first).

Handling Difficult Births

Forceps Delivery

If the fetus is not emerging easily, the physician can insert an instrument (forceps) into the birth canal and around the head to effect a *forceps delivery*. The forceps are inserted as two separate steel blades, the inner surfaces of which are curved to fit the fetus's head (Fig. 11-10). After both blades are in place, the handles are joined. The instrument is then pulled or twisted gently to assist in expulsion of the fetus. Forceps deliveries can be done if the head of the fetus is resting properly in the pelvis, the membranes are ruptured, and the cervix is dilated maximally (10 cm).

Some medical reasons for using forceps are (1) acute distress of the fetus, such as irregular or weak heartbeat and lack of oxygen caused by premature separation of the placenta, compression of the umbilical cord, or excessive pressure on the fetal head; (2) illnesses of the mother, such as heart problems, tuberculosis, or toxemia; (3) a previous cesarean section, as the wall of the uterus might tear; (4) presentation of the fetus in a breech position; and (5) an abnormally slow labor. Some physicians, however, use forceps as a matter of routine, a controversial practice. The fetus rarely is damaged by forceps if they are used when the fetus is crowning and not sooner. Anesthetics are required during a forceps delivery.

Vacuum Extraction

A newer method of extracting the fetus is used commonly in Europe. In this *vacuum extraction* method, a metal cup is placed on top of the fetus's head, negative pressure is applied to this cup, and the cup is attached firmly. The fetus is then pulled out. There are some reports, however, of damage to the fetus using this instrument, and the method is not used commonly in the United States.

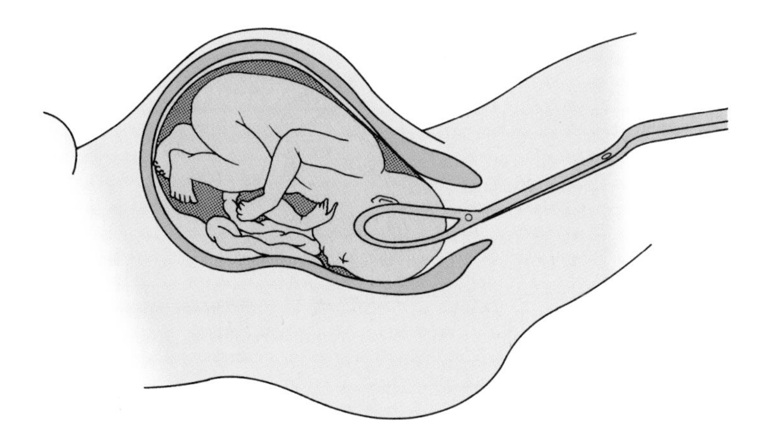

Figure 11-10 Use of forceps to deliver the fetus. Marks may appear on the infant's head where the forceps are applied, but these go away soon after birth. Forceps deliveries should be done only in specific situations (see text).

Cesarean Delivery

Cesarean deliveries (Latin *caedere*, meaning "to cut"; also known as cesarean sections, or C-sections) are performed in an operating room after a spinal or general anesthetic is given to the mother. (A few cesarean deliveries have been performed under *acupuncture*, which involves pain relief by inserting steel needles in specific body regions.) A cesarean operation usually lasts 20 to 90 min. Apparently, the name of this operation had its origin in an order by Emperor Julius Caesar of such an operation to be done on dying pregnant women in hopes of saving the unborn children.

In this procedure, an abdominal incision is made below the navel in the midline and through the uterine wall, and the baby is removed (Fig. 11-11). In Caesar's time, few if any of these operations were successful, but with today's modern surgical and antiseptic techniques, these operations are relatively safe for the mother and infant. In fact, women can have four or five babies or more by this method. Cesarean delivery is performed not only when the fetus is in a transverse presentation or less commonly in a breech presentation, but also when the pelvis of the woman is too small, the fetus is too large, or when the fetus shows signs of distress, such as abnormalities in heart function. Also, cesarean delivery is performed if the umbilical cord gets compressed between the head and the wall of the birth canal, if the placenta is coming out before the fetus (placenta previa), or if the placenta separates from the uterus prematurely. All of these situations are dangerous to the unborn child because the oxygen supply to the fetus is reduced, and prompt delivery is needed.

Some feel that too many cesareans are being done. This type of delivery accounted for 5.5% of deliveries in 1970 and has increased to about 28% today (over 1 in 4 births, or about one million annually in the United States). The World Health Organization recommends that cesarean sections make up less than 9.5% of all deliveries in wealthy industrialized nations such as the United

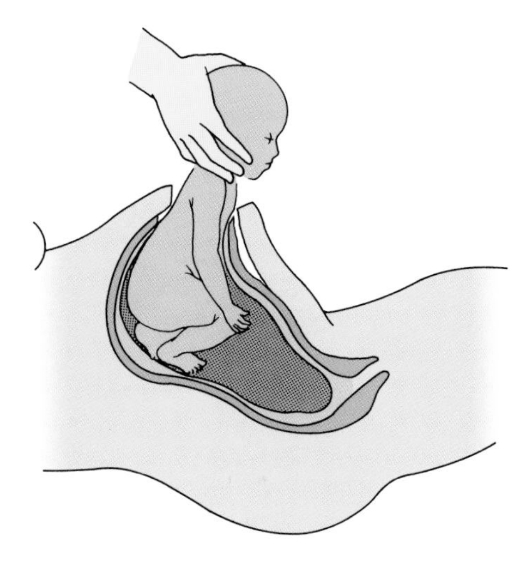

Figure 11-11 Removal of infant through incision in abdomen and uterus during a cesarean delivery.

States. Reasons given for the increase in cesarean deliveries include (1) the assumption that once a woman has a cesarean, she can never deliver a future child vaginally (this is not true in many cases); (2) using a cesarean delivery for a breech birth (this often is not necessary); (3) the increased use of fetal monitoring to detect fetal problems during labor; (4) the increasing pregnancies of older women; and (5) doctors' fear of liability lawsuits. Many physicians argue that cesarean deliveries decrease fetal and maternal death. Others, however, question the unnecessary use of this operation. Currently about 22% of all cesareans in the United States are performed purely for patient choice, not for any medical reason.

After a cesarean delivery, a woman usually stays in the hospital 2 to 4 days longer than women giving birth vaginally. This is because the stitched incision in the lower abdominal wall limits normal movement. Babies delivered by C-section tend to have more respiratory problems, possibly because they do not have a surge of epinephrine during delivery (see Chapter 12). Infants delivered with this operation can nurse normally, although the suckling response of some may be slow to develop. Milk production in mothers after this operation is normal once the infant's suckling pattern is established. Some women having this operation may feel inadequate in one way or another, but there appears to be no influence of cesarean delivery on the relationship between the mother and her child. Many women who have had a previous cesarean delivery can later deliver a child vaginally. However, the typical length of labor in the latter case is as long as for a primiparous woman.

Use of Medications during Labor

Anesthetics, which obliterate all sensations, can be used to relieve discomfort during childbirth. In general anesthesia, the anesthetic can be inhaled in the form of a gas or given intravenously. General anesthesia is usually not used during labor because of possible dangers to the baby and because it prevents a woman from experiencing or participating in the birth process.

Conduction anesthetics are given by injection, and there are several procedures for giving anesthesia in this way during labor. In an *epidural*, the anesthetic is injected into the outside membranes of the spinal cord. A *caudal* is injected lower in the back and into the central canal of the spinal cord. The epidural and caudal anesthetics numb sensations in the body below the point of injection. However, the woman still is able to use her muscles in this region to assist delivery. A *spinal anesthetic* is injected into the space just under the membranes surrounding the spinal cord near the center of the back, whereas a *saddle block* is injected into the same space lower in the back. Because the spinal and saddle blocks numb both sensory and motor nerves below the point of injection, the woman is unable to use the muscles of this region to help with delivery. In a *paracervical*, the anesthetic is injected into both sides of the cervix. Finally, in a *pudendal block* the anesthetic is injected into the pudendal nerve on each side of the vagina. The paracervical and pudendal blocks are used to ease the process of the baby passing through the birth canal.

Analgesics are sometimes used during labor to ease pain. They can be given intramuscularly (by injection into the muscle) or intravenously (by injection into a vein). Analgesics promote relaxation between uterine contractions and relieve discomfort and pain.

When any medication is used during labor and delivery, the choice of which to use and how it is given should be the result of communication between the physician and the patient. Often, a woman can consult with her physician before labor begins, and together they can decide on the choice of medication if needed. Questions to be asked about a medication are: "Is it safe for the mother?" "Is it safe for the fetus?" and "Will it lengthen the duration of labor?"

Although we do not have much information on the subject, it is possible that certain medications can affect the newborn adversely (Table 11-2).

Table 11-2 Medications for Pain Reduction in Labor and Delivery[a]

Analgesics	Effects on mother	Effects on baby[b]
Tranquilizers	Physical relaxation and reduced takes the edge off pain but does not eliminate it entirely anxiety;	Minimal
Barbiturates	Drowsiness and reduced anxiety; may slow the progress of labor	Can depress nervous system and breathing
Narcotics	Reduce pain and elevate mood, but may inhibit uterine contractions and cause nausea or vomiting	Can depress nervous system and breathing
Amnesics	Do not reduce pain but cause the woman to forget her experience after it is over; may cause physical excitation and wildness	Minimal, but possible danger of overdose when self-administered, with adverse effects on neonate
Anesthetics	**Effects on mother**	**Effects on fetus**
Local		
Paracervical	Blocks pain in the uterus and cervix, but relatively short-lasting and ineffective late in labor; can lower mother's blood pressure	Causes a slowing of fetal heart beat in about 20% of cases
Pudendal	Blocks pain from the perineum and vulva in about 50% of cases	Minimal
Regional		
Spinal, saddle block, epidural, caudal	Block pain from the uterus, cervix, and perineum; spinal and saddle block also block movement below point where administered; highly effective but can cause a serious drop in blood pressure or seizures. Can be used for cesarean deliveries	Generally does not affect fetus but requires forceps delivery more often than other methods
General	Woman unconscious; usually used only in the last few minutes of labor to eliminate pain completely, but may cause vomiting or other complications and is a leading cause of maternal death during delivery; used for cesarean deliveries	Can depress nervous system and breathing

[a]Adapted from W. H. Masters *et al.* (1993). "Biological Foundations of Human Sexuality," Harper Collins, New York.

[b]Some medications given during labor and delivery have been shown to possibly have long-term effects on the infant and child (see text).

For example, lack of oxygen (*anoxia*) can harm the newborn, and some anesthetics depress the newborn's respiration. Moreover, medications used during labor can cross the placenta into the fetus. When this occurs, the fetus is not able to inactivate the drugs efficiently because its kidneys and liver, two organs that are involved in breakdown and excretion of drugs, are not as functional as those of adults. Also, the fetal brain is not fully developed and may be susceptible to damage from the drug. There is disturbing evidence of adverse effects from drugs used during labor on the development and behavior of infants and even children up to 7 years after they are born. These adverse effects are seen on heart rate, suckling and feeding behavior, language development, alertness, mother–infant interactions, and cognitive development. However, controversy surrounds these findings, and we need to know much more about the potential effects of these medications on infant development. Certainly, medication is necessary in difficult birth situations. However, because the use of medications during labor can prevent the mother from totally experiencing the birth of her baby and because of possible adverse effects of these drugs on the mother and fetus, new methods of prepared childbirth have become popular in recent years.

Natural Birthing Methods

In the middle 1950s, most women requested medication during their labor. The father, instead of participating in the birth process, was usually banished to a waiting room where he paced the floor nervously and helplessly. In 1933, Dr. Grantly Dick-Read presented a new concept that he called "natural childbirth." He proposed that childbirth without drugs was better for the mother and infant. In the past, most women had been convinced that childbirth was filled with pain, suffering, and danger. Even the Bible says, "In sorrow thou shalt bring forth children" (Genesis 3:16). But labor means "hard work," not suffering, and it has been found that much of the childbirth pain is intensified by fear of such pain (but see HIGHLIGHT box 11-2). If the mother is well-versed in the biological events of the birth process and is physically and psychologically prepared for the process, the pain is greatly lessened and more manageable; often no medications are needed during "natural" labor.

Since Dr. Dick-Read's time, many birthing methods have appeared that can be called *prepared* (or "controlled," "natural," or "cooperative") *childbirth*. Of these, the *Lamaze* and *Bradley methods* have been the most popular in America. The former method was introduced to the Western world in 1951 by Fernand Lamaze (1890–1957), a French physician. It had been used previously in Russia. With these methods, the mother is taught not only the biology of the birth process, but also techniques of controlled breathing and relaxation to manage discomfort and pain. Her partner plays an important role in these exercises and in the delivery itself. Medications are used during natural births if the situation requires them, but the parents have an intelligent choice in the matter. In addition to the advantages of avoiding the use of medications, the mother is wide awake during the whole process, and both parents can more directly experience the birth of their child.

Some women choose to combine some prepared childbirth method with delivery in the comfortable surroundings of their own homes. The most common problems with home delivery are lack of oxygen for the fetus or newborn (usually due to compression of the umbilical cord) and hemorrhage of the mother. Those assisting at *home births* may not be prepared to cope with these and other unforeseen circumstances, which is why many physicians are opposed to home births. It is recommended that home births be attended by a physician or *nurse midwife* (a registered nurse with further education in midwifery). If home birth is planned, arrangements should be made for medical help in case of emergencies. Because of the potential dangers of home births, which now make up about 1% of births in the United States, many hospitals offer *birthing rooms* modeled after the home. Both labor and delivery can take place in these rooms.

Chapter 11, Box 2: Why Is Human Birth So Difficult?

Compared to our closest living relatives, the monkeys and apes, human labor and birth are much more risky, difficult, and often painful. Let us look at some of the differences in labor and birth between humans and other primates.

1. The relatively large fetal head size compared to pelvis size in humans. In other primates, the size of the fetal head is much smaller than the size of the pelvic canal, whereas in humans the fetal head is about the same size as the pelvic canal. This difference relates to two evolutionary trends in human evolution. First, *bipedalism* (two-legged walking) required narrowing of the pelvis to center the legs under the body. The second, *encephalization*, was the enlargement of the brain, perhaps due to the advent of tool use in early prehumans.

2. The necessity for the human fetus to twist and turn its head while moving through the birth canal (see Fig. 11-7). In apes, the front-to-back dimension of the pelvic canal is longer than the side-to-side dimension. The fetal skull is also longer from front to back. Thus, the fetus's head passes through with relative ease. In contrast, the human pelvic canal narrows part-way down. At the top it is widest side to side so the fetus enters with the head sideways. Lower down in the human pelvis, however, the canal is longest front to back so the fetus's head must turn 90°, but which way does it turn? The fetus's head is broadest at the back, but the dimensions of the pelvic canal are now broadest at the front so the fetus's head turns so that the head emerges face down. After it emerges, its head then turns sideways to allow the shoulders to emerge.

3. The risk of injury and infection to the human perineum resulting from too rapid delivery of the head and shoulders. Passage of the fetus through the cervix vagina and perineum is usually not a problem in apes, as it often is in humans. Lacerations and tearing of the soft tissues are not uncommon in women in the absence of an episiotomy, and one reason of course is the relatively large size of the human fetal head. Also, human mothers tend to want to push before complete cervical dilatation, which can cause laceration of cervical and perineal tissues, bleeding, infection, and perhaps damage to the fetus.

4. The usual presentation of the human fetus in the occipital position (see Fig. 11-7). In apes, the fetus usually emerges face up, with the occiput (back of the head) against the mother's sacrum. This allows the mother to use her hands to assist delivery and to clear mucus from her newborn's respiratory passages (see HIGHLIGHT figure). In contrast, most human babies are born in the occipital position (face down). This prevents a woman from catching her

Continued on next page.

Chapter 11, Box 2 continued.

newborn and assisting in delivery because pulling back on the emerging fetal head may damage the neck. Other problems with occipital delivery include increased possibility of soft tissue laceration, prolonged delivery because a broader part of the fetal skull must emerge first, and often severe backache ("back labor") in the mother.

5. The relatively prolonged labor and delivery in humans. The three stages of labor, together, last 50 to 75 min in baboons, about 150 min in orangutans, 21 to 175 min in gorillas, about 135 min in chimpanizees, but 40 to 2883 min in humans. Thus, human labor generally is three to four times longer! Not only does this increase the risk of fetal and maternal distress, but in ancient times would have more greatly exposed human mothers and their newborns to predation. The increase in delivery time is not only related to the fetal size and rotation and to pelvic canal dimensions, but also to the birth posture in some cultures, such as in the United States, where many women deliver on their backs. In other cultures, walking during labor, as well as sitting, squatting, or standing during delivery, is used to allow gravity to help expel the fetus.

As mentioned earlier, there seems to be general agreement that the evolution of the difficulty of human births related to the evolution of bipedalism and encephalization. Some also argue that once human birth became risky and difficult, a female in labor required help from other people to deliver and perhaps to help protect against predators, and thus supported the evolution of social empathy, communication, and cooperation. Apes and monkeys usually give birth in isolation and receive no assistance. In contrast, birth with assistance is almost universal in humans. Granted, some women do give birth alone, but the risks to neonates and mothers are higher. In a few hunter–gatherer cultures, women isolate themselves during birth. On the whole, however, birth attendants are used in most human cultures.

A controversy exists, however, about when birth became difficult in human evolution, an important question in relation to when bipedalism, encephalization, and social cooperation evolved. The apes are divided into families Pongidae (gibbons, orangutans, gorillas, chimpanzees) and Hominidae (prehumans and humans). The first hominids we know about evolved from apes in Africa about 4.2 million years ago. Prehumans are in the genus *Australopithecus*, and they probably were bipedal some of the time but also swung in trees. Their brains were relatively small (about one-third of modern human size). The first members of our genus (*Homo habilis*) appeared 2 million years ago and used tools; they were bipedal and their brain was somewhat larger than that of apes. *Homo erectus* appeared in Africa 1.8 million years ago. They were bipedal and their brain size was almost as big as ours. These prehumans lived in caves and used clothes, tools, and fire. Based on molecular and fossil evidence, our species (*Homo sapiens*) evolved in Africa about 200,000 years ago; they then migrated out of Africa and became differentiated into human races of yesterday and today.

Recent measurements of prehuman (*Australopithecus*) pelvises up to 3 million years old have shown that, unlike in apes, the pelvic canal was oval (wider side to side), probably an adaptation to bipedalism, meaning that the fetus's head may have come out sideways. Another study showed that *H. habilis* and *H. erectus* also had an oval pelvic canal. Meanwhile, brain size increased from 800 ml in *H. erectus* to 1200 ml in early *H. sapiens* (modern brain size averages 1400 ml). So the pelvis started to be "more difficult" in response to bipedalism before the genus *Homo* appeared, but then had to enlarge front to back to accommodate the larger fetal head about 200,000 years ago. Some argue that even in *Australopithecus* prehumans, the slight change in pelvis shape related to bipedalism caused more difficult births because the shoulders would require rotation to get out if the head first came out sideways. Therefore, attendance and social cooperation could have appeared in prehumans before the appearance of our species, *H. sapiens*.

Continued on next page.

Chapter 11, Box 2 continued.

Illustration of a monkey delivery. Note the squatting, sitting, and standing positions and that the fetus emerges face up.

Chapter Summary

The average length of pregnancy is 266 days, but parturition can occur before or after this due date. There are seasonal and daily cycles in the number of human births. The fetal adrenal glands secrete cortisol and androgen, both of which travel to the placenta and increase placental secretion of an estrogen and decrease placental secretion of progesterone. These high blood levels of estrogen in the female interact with oxytocin and prostaglandins to initiate labor. Physicians can induce labor artificially with synthetic oxytocin or prostaglandins.

Labor progresses in three stages: (1) cervical effacement and dilation, (2) expulsion of the fetus, and (3) expulsion of the placenta. Labor lasts longer in primiparous than in multiparous women. Causes of premature births can relate to hormonal factors in the fetus or mother, as well as to fetal or maternal diseases or abnormalities. Multiple births produce small-sized newborns, who often are premature. Forceps delivery can be performed if the fetus is in the breech presentation, if fetal distress exists, or if difficulties are encountered in delivery because of maternal disease or abnormality. Cesarean delivery can be performed when the fetus is in the breech or transverse position, when fetal distress is present, or if labor is not proceeding normally.

Medications used to reduce discomfort during labor include anesthetics and analgesics. The choice of drug or method used should result from communication between physician and patient. Methods of prepared childbirth incorporate education about labor and birth with controlled relaxation and breathing techniques and the assistance of a partner. Home births can be a joyful experience, but participants should be aware of potential dangers. The medical assistance of a physician or midwife is advisable in such births. The relative difficulty of human birth as compared to other primates, due to the evolution of bipedalism and a larger fetal head, could have led to the need for attendants to human birth.

Further Reading

Anonymous (1989). C-section rates remain high, but post-Cesarean vaginal births are rising. *Family Planning Perspect.* **21**, 36–37.

Anonymous (1989). Hospitals' cesarean rate reduced by one-third without adverse effects for mother and babies. *Family Planning Perspect.* **21**, 93–94

Fischman, J. (1994). Putting a new spin on the birth of human birth. *Science* **264**, 1082–1084.

Lam, D. A., and Miron, J. A. (1994). Factor contributing to the seasonality of human reproduction. *In* "Human Reproductive Ecology: Interactions of Environment, Fertility, and Behavior" (K. L. Campbell and J. W. Wood, eds.), Annals of the New York Academy of Sciences, New York.

Lagercrantz, H., and Slotkin, T. A. (1986). The "stress" of being born. *Sci. Am.* **254**(4), 100–107.

Lovejoy, C. O. (1988). Evolution of human walking. *Sci. Am.* **259**(5), 118–125.

MacDonald, P. C., and Casey, M. L. (1996). Preterm birth. *Sci. Am. Sci. Med.* **3**(2), 42–51.

Radetsky, P. (1994). Stopping premature births before it's too late. *Science* **266**, 1486–1488.

Rosenberg, K. R., and Trevathan, W. R. (2001). The evolution of human birth. *Sci Am.* **285**(5), 72–77.

Smith, R. (1999). The timing of birth. *Sci. Am.* **280**(3), 68–75.

Stupp, P. W., and Warren, C. W. (1994). Seasonal differences in pregnancy outcomes. *In* "Human Reproductive Ecology: Interactions of Environment, Fertility, and Behavior" (K. L. Campbell and J. W. Wood, eds.), Annals of the New York Academy of Sciences, New York.

Trevathan, W. R. (1987). "Human Birth: An Evolutionary Perspective." Adline de Gruyter, Hawthorne, NY.

Advanced Reading

Bronson, F. H. (1995). Seasonal variation in human reproduction. *Q. Rev. Biol.* **70**, 141–164.

Eason, E., and Feldman, P. (2000). Much ado about a little cut: Is episiotomy worthwhile? *Obstet. Gynecol.* **95**, 616–618.

Leake, R. D., *et al.* (1981). Plasma oxytocin concentrations in men, nonpregnant women, and pregnant women before and during spontaneous labor. *J. Clin. Endocrinol. Metab.* **53**, 730–733.

Liggins, G. C. (1994). The role of cortisol in preparing the fetus for birth. *Reprod. Fertil. Dev.* **6**, 141–150.

Lydon-Rochelle, M., *et al.* (2001). Risk of uterine rupture during labor among women with a prior Cesarean delivery. *N. Engl. J. Med.* **345**, 3–8.

McLean, M., *et al.* (1995). A placental clock controlling the length of human pregnancy. *Nature Med.* **1**, 460–463.

McLean, M., and Smith, R. (2001). Corticotropin-releasing hormone and human parturition. *Reproduction* **121**, 493–501.

Meis, P. J., *et al.* (2003). Prevention of recurrent preterm delivery by 17 alpha-hydroxyprogesterone caproate. *N. Engl. J. Med.* **348**, 2379–2385.

Nathanielsz, P. W. (1996). The timing of birth. *Am. Sci.* **84**, 562–569.

Wadhwa, P. D., *et al.* (1998). Maternal corticotropin-releasing hormone levels in the early third trimester predict length of gestation in human pregnancy. *Am. J. Obstet. Gynecol.* **179**, 1079–1085.

Weber, G. W., *et al.* (1998). Height depends on month of birth. *Nature* **391**, 754–755.

CHAPTER TWELVE

The Neonate and the New Parents

Introduction

The newborn child (known as a *neonate*) is expelled into an unfamiliar and, in some ways, harsh environment. A few hours before birth, it is resting comfortably in a warm, quiet bath of amnionic fluid and is receiving its nutrition and oxygen from its mother. Suddenly, it is forced out of this peaceful existence and must adapt to life outside. In other words, it now must breathe with its lungs, urinate and defecate, digest external nutrient sources, and combat the microorganisms and fluctuating temperatures of its new environment. Also, it must develop behaviors that elicit care from others. Similarly, the mother has undergone a joyful yet stressful experience. She now must adjust to the marked physiological and anatomical changes of the *postpartum* (after birth) period and must face the responsibilities, along with her mate, of caring for a helpless infant. This chapter discusses the condition and care of the newborn and the adjustments of the parents to this new arrival.

Treatment of the Newborn

Chapter 11 reviewed how the umbilical cord is cut, mucus is removed from the airways, and a substance is put in the newborn's eyes to prevent infection. The baby now is checked for possible bone breakage or dislocations due to its stressful passage through the birth canal, and its length, head circumference, heart rate and rectal temperature are recorded. Average newborn measurements are presented in Table 12-1.

Apgar Score

The *Apgar score* is named after Virginia Apgar, who invented the procedure. It is a rating of the general level of well-being of the newborn. Numerical values, from 0 to 2, are given to five responses of the newborn (Table 12-2): (1) heart rate, (2) respiratory effort, (3) muscle tone, (4) reflex irritability, and (5) color of the skin. Thus, a maximal score of 10 can be obtained. A baby with a score of 7 to 9 is normal or only slightly depressed, one with a score of 4 to 6 is moderately depressed, and a score of 0 to 3 reflects severe health problems. About 80% of newborns in the United States receive a score of 7 or above, and usually

Table 12-1 The Average Newborn in the United States

Measurements	Average
Weight	3400 g (7.5 lb)
Height/length	50 cm (20 in.)
Head circumference	33 cm (13 in.)
Chest circumference	30 cm (11.8 in.)
Heart rate	120–160 beats/min
Body temperature	35–36 °C (97–99 °F)

Table 12-2 Apgar Newborn Scoring System

	Score		
Sign	**0**	**1**	**2**
Heart rate	Not detectable	Below 100	Above 100
Respiratory effort	Absent	Slow (irregular)	Good (crying)
Muscle tone	Flaccid	Some flexion of extremities	Active motion
Reflex irritability	No response	Grimace	Vigorous cry
Color[a]	Pale	Blue	Pink

[a]If the natural skin color of the child is not white, alternative tests for color are applied, such as color of mucous membranes of mouth and conjunctiva, lips, palms, hands, and soles of feet.

no alarm is raised for scores of 6 or above. Newborns with low scores require intensive care immediately after delivery.

Leboyer Method

The *Leboyer method* of newborn care is moderately popular in the United States today. This method assumes that most hospital delivery rooms are stressful and harmful to the infant because of the presence of cold temperatures, loud noises, and bright lights. With the Leboyer method, the delivery room is kept at a warmer temperature than usual, noise is kept to a minimum, and lights are dimmed. The newborn is also immersed in a warm bath. Close physical contact between the mother and her baby is encouraged. There is no strong scientific evidence that this method improves the emotional or physical health of the newborn. In fact, early stress may be beneficial to the infant (see Chapter 6). Early mother–infant contact, however, has proven to be important to the development of the mother–infant bond.

Circumcision

Circumcision has been, and still is, a cultural and religious custom in many regions of the world. It is widely practiced among Jewish and Muslim populations for religious reasons; circumcision for social/cultural reasons is common mostly in the United States and Australia. This procedure involves surgical removal of the prepuce (foreskin) of the penis of a male infant; some traditions practice this procedure later in life. Females are also circumcised in some cultures; this involves removal of the prepuce of the clitoris, or even the entire clitoris, labia minora, and part of the labia majora. In Egypt, for example, a great majority of women have been circumcised. In the Jewish religion, male infants are circumcised on the eighth day of life as a symbol of the covenant between God and the Jewish people (Genesis 17:9–27). Choosing this day is medically sound as it is the first day that the infant boy produces his own vitamin K, which helps blood to clot.

About 65% of male infants born in United States hospitals are circumcised. The reasons parents choose this procedure are complex. Circumcision allows better hygiene of the glans penis; i.e., smegma and bacteria can be removed more easily (young boys, however, easily can be taught how to clean their penis). Circumcision increases the risk of penile infection in the first year, but then decreases the risk thereafter. Uncircumcised male babies have a higher risk of developing infections of the urinary tract in their first year of life and a higher risk of penile cancer (although the incidence of these disorders is rare in all males). As adults, circumcision reduces the risk of acquiring some sexually transmitted diseases (e.g., syphilis, HIV). However, sexual behaviors and condom use are far more important factors in the acquisition of sexually transmitted diseases than circumcision status. Some parents have their son circumcised because they feel he will be embarrassed if he is not and his peers are. Some people may feel that the absence of the prepuce will alter sexual pleasure during coitus, but there is no difference between circumcised and uncircumcised men in this regard.

In U.S. hospitals, male circumcision is done anytime in the first week of life, usually on the second day. A topical anesthetic is now often used because it does not harm the infant. After the baby is immobilized physically, the prepuce is removed with a knife or scissors, and the region is bandaged. The procedure is painful to the baby if an anesthetic is not used, and some feel it is an unnecessary trauma. There is also a small risk of bleeding and infection. In general, there is increasing concern about the value of circumcision in the United States. The American Academy of Pediatrics takes the position that the potential medical benefits of circumcision "are not compelling enough to warrant the AAP to recommend routine newborn circumcision." Furthermore, it calls for the "essential" use of pain relief if parents choose this elective procedure.

Adaptations of the Newborn

As mentioned earlier, the newborn is thrust from the uterus into a new environment, and now we discuss some ways that it adapts to this abrupt change.

The Respiratory System

The newborn takes its first breath immediately after delivery. The air sacs in its lungs are coated with *surfactant*, which is a lipoprotein that affects surface tension in the lungs. The presence of this surfactant is necessary so that the lung air sacs do not stay collapsed (by sticking together) after expiration. If little or no surfactant is present, the newborn's lungs are unable to fill with air, leading to *respiratory distress syndrome*. Chapter 11 showed how the fetal adrenal glands secrete cortisol soon before birth. In addition to playing a possible role in labor, this hormone causes secretion of the surfactant in the fetal lungs. Premature babies often have not secreted enough surfactant, but this condition can be alleviated by the administration of corticotropin (ACTH), which causes the adrenal glands to secrete cortisol.

When the fetus is stressed as it passes through the narrow birth canal, its adrenal glands produce large quantities of *epinephrine*, a stress hormone. Epinephrine in turn helps clear the newborn's lungs of fluid, supports the secretion of surfactant, and increases blood flow to the lungs and other organs. This is why babies born by cesarean section tend to have more respiratory difficulties; they do not experience as great a surge of epinephrine in their blood.

The newborn is quite capable of sneezing and hiccoughing, and will do so often, sometimes to the distress of parents. These phenomena, however, are normal reflexes that clear the respiratory tract of congestion. Newborns are also loud criers! In fact, this crying serves to elicit care-giving behavior by adults (see later) and increases the development of lung efficiency. It is noteworthy that analyses of the sounds of crying newborns can detect certain abnormalities. That is, sick babies cry differently from healthy ones.

The Circulatory System

Chapter 10 showed how the cardiovascular system of the fetus is adapted to permit use of the placenta as a respiratory, excretory, and nutritional organ. These adaptations include the ductus arteriosus, ductus venosus, and foramen ovale. Only a few days after birth do these structures close anatomically, but they are closed functionally soon after birth. As soon as the umbilical cord is tied, changes in blood pressure allow blood to flow to the newborn's lungs and to its liver, which causes functional closure of the fetal shunts. If the structures fail to close, the lungs will not receive enough blood, and thus the tissues of the infant will not have enough oxygen. When this fetal shunt fails to close, it is called *patent ductus arteriosus*. This can be due to a birth defect or premature delivery. Patent ductus arteriosus leads to a bluish color of the newborn skin, a condition called *cyanosis* ("blue baby"). Blue babies can be given antiprostaglandins to close the ductus arteriosus; if this fails, immediate heart surgery is necessary.

The Digestive Tract

The primary nutrient of the fetus is glucose supplied by the mother. After the baby is born, however, the placenta of course can no longer act as a source of this nutrient. Instead, the newborn's heart and liver release glucose into the blood for energy, but these sugar stores are depleted within a few hours. Then, fat is broken down into fatty acids and glycerol, which in turn are converted to more glucose. This energy should last until the baby receives milk from the mother or a bottle; if this nutrition is delayed, sugar water is sometimes fed to the infant to supplement energy and water requirements. Meanwhile, the newborn's digestive tract must begin to secrete digestive enzymes to process external food sources. A dark green, sticky feces (*meconium*) is defecated by the newborn up to the fourth day after birth. The feces change to a yellowish to brownish color when the infant receives milk.

Phenylketonuria (PKU) is an extremely rare, inherited condition that can lead to mental retardation. It is a disorder of protein metabolism that results in high levels of the amino acid *phenylalanine* in the newborn's blood and urine. Every baby is tested for PKU by pricking its heel and dabbing a drop of blood onto a special test paper. If the test is positive, the infant usually is put on a phenylalanine-free diet to control the disorder.

Thermoregulation

The temperature of the delivery room often is several degrees lower than the temperature of the womb, and the infant's capacity to maintain a normal body temperature is not yet fully developed. One way that the infant maintains its body temperature is by heat production from stores of *brown fat* around its vital organs and blood vessels. Brown fat differs from white fat in that its cells contain very high numbers of mitochondria, the energy-producing organelles. When brown fat is broken down immediately after birth, under the influence of epinephrine, the mitochondria release large amounts of heat that warm the newborn's body. Hibernating mammals use a similar mechanism to warm their bodies after their winter sleep.

The Nervous System and Behavior

The newborn can hear, see light and dark as well as movement, smell, and respond reflexively. Several reflexes can be tested to detect if the nervous system is functioning normally. One example is the *grasp reflex*. If you press an object on the infant's palm, near where the fingers join, the infant will flex its fingers and grasp the object. Similarly, its toes will flex if you stroke the sole of the foot; this is the *plantar reflex*. If the infant is subjected to a jolt or loud noise, it will respond with the *Moro* (or *startle*) *reflex*. That is, it will stiffen its body, draw up its legs, and fling its arm up and out. Then, it will bring its arms forward. These reflexes appear to be adaptations to hold onto the mother. If you rub a finger on the hungry infant's mouth or cheek, it will turn its head to keep in contact with the finger. This *rooting reflex* is used to stimulate the infant to find the breast. If you touch the infant's lips, it will display the *sucking reflex*. Walking reflexes also exist in the newborn, and exercise of these reflexes (by assisting them before they can walk on their own) leads to an earlier onset of walking.

There are five species-specific behaviors in human newborns: clinging, crying, smiling (beginning at week 5), following with eyes, and sucking. All of these elicit care-giving behaviors and increase survival. Babies at 2 months babble with speech-like sounds, and by 4.5 months can recognize their own names. Crying is especially obvious, and its function is to irritate the parents so that they will respond in a care-giving manner to ease the infant's needs (e.g., hunger, pain). Some say that the habit of Western cultures to ignore crying (to not "spoil" the child) is not natural. When comparing Western cultures to the !Kung San hunter-gatherers of South Africa, infant crying in both cultures peaks between the 6th and the 12th week and then declines. Whereas each crying bout lasts an average of 2–3 min in Western societies, it only lasts 5 s in the !Kung San tribes, in which mothers carry their infants with them and respond to their cries quickly. Is it possible that the relatively high incidence of child abuse and the low incidence of breast-feeding in Western societies are related to a delayed parental response to crying infants?

What a Newborn Looks Like

Some parents, upon seeing their new baby, wonder if it should have stayed in the womb longer! Its head often is egg shaped because the unfused skull bones were squeezed together as its head passed through the birth canal. However, the head rounds out in a few days, although the skull bones remain unfused for a year or more. The openings between the bones are "soft spots," or *fontanels*, on the head (Fig. 12-1). These are normal and slowly change to bone in the first year of life. Sometimes a thick, soft swelling occurs on the part of the scalp that once rested against the cervix. This *caput* (Latin for "head") normally goes away in a few days. Some newborns may also have a blood clot between the skull bone and its covering membrane due to pressure from the birth canal. This clot (*cephalohematoma*) often appears a day or two after birth and usually goes away in several weeks. It does not damage the brain. The newborn's soft nose frequently has been squashed during delivery, and its shape is not what it will look like in a few days. In addition, the use of forceps can temporarily alter head shape and appearance or can press on the facial nerves, which can lead to temporary paralysis of the eye and mouth muscles. This is quite disturbing to parents until they are reassured that normal function will develop soon.

The eyes of newborn whites are usually gray-blue, those of blacks, native Americans, and Hispanics are gray-brown or black, and those of Asians are green-blue. Only later will pigment change in the iris, resulting in the various eye colors. Some infants' eyes will be bloodshot, but this will disappear in a few days. Also, a pus-like secretion may exude from the eyes, especially in babies who received silver nitrate. This secretion will go away in about a day. Many parents are alarmed to see that their babies are cross-eyed. This is because the newborn is unable to focus its eyes properly, an ability that is gained over time. Also, the tear glands are not

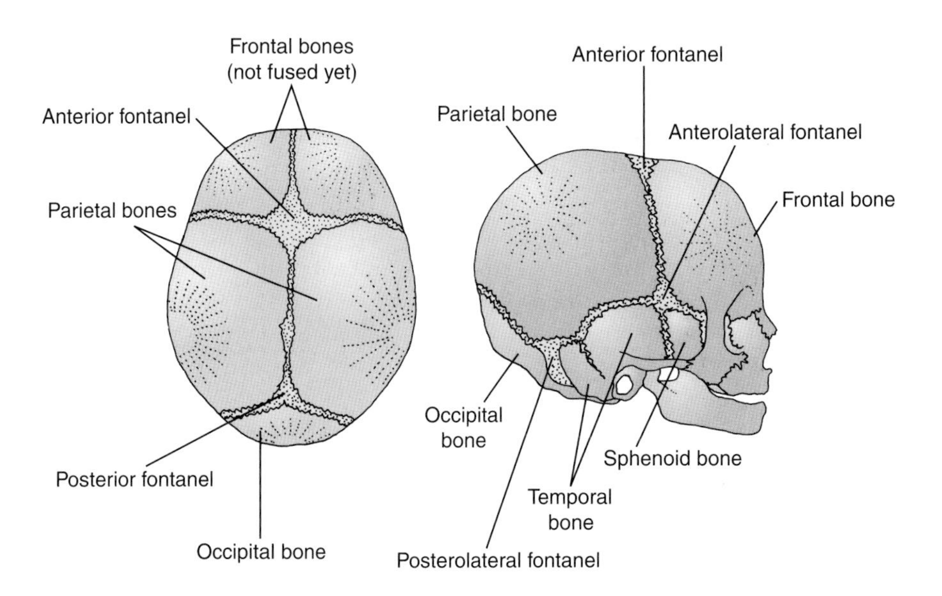

Figure 12-1 Fontanels of the newborn skull. These "soft spots" on the newborn's head are places where the skull bones have not yet fused.

functional until up to 3 months of age so the cries of a newborn have no tears. A disease called *retrolental fibroplasia* at one time afflicted 10 to 15% of premature babies, and this disorder caused partial or complete blindness. Now we know that it was caused by administration of too much oxygen to premature babies, and precautions against this practice have virtually eliminated this problem.

The delicate skin of newborn babies of all races and ethnicities is a pink, reddish or purplish color, revealing the color of underlying blood vessels. Later the true skin color will develop. Vernix caseosa, the cheese-like protective secretion of the fetal skin, may still be present, especially in premature babies. Fetal downy hair (lanugo) is present on the shoulders, back, forehead, and temples of some newborns. This hair is usually shed by the end of the first week. The skin of newborns is thin and sensitive and may exhibit temporary blemishes. Tiny white spots on the nose, chin, and cheeks are clogged pores, which go away soon. *Strawberry marks* are regions of dilated blood vessels in the skin and are common on the forehead, nose, upper eyelids, and nape of the neck. These usually disappear in a few months.

About one-third of babies develop physiological jaundice, which occurs on about the second to the fifth day postpartum. This condition, which appears as a transitory yellowish tint to the skin and eyes, is caused by excessive amounts of bilirubin. It is more common and severe in premature babies. Newborns with this condition can be treated with ultraviolet light in an incubator. Jaundice can also indicate erythroblastosis from Rh disease (see Chapter 10), but this jaundice develops on the first, not the second, day of life.

The reproductive tracts of male and female newborns are fully formed but inactive (see Chapter 5). The breasts of some males and females enlarge on the third or fourth day after birth. In fact, some secretion of fluid, termed *witch's milk*, can occur. During the medieval period, this fluid was believed to have wonderful healing powers. This breast enlargement, which goes away in a few days, is due to exposure of the fetus to placental hormones during late pregnancy. Sometimes a female newborn will have a little bloody vaginal discharge for the same reason.

Disorders of the Newborn

Premature babies are those born alive and weighing between 0.99 kg (2 lb, 3 oz) and 2.49 kg (5.5 lb). The chances of a 2-lb, 3-oz baby surviving, even with special intensive care, is about 50%. This percentage, however, increases to 90% in 4-lb babies. Elaborate hospital care, including incubation units that are almost like "artificial wombs," is increasing this survival rate.

Previous chapters showed how developmental abnormalities can occur in the embryo or fetus. These can be due to chromosomal anomalies as a result of fertilization errors, inherited disorders, the effects of teratogens or mutagens, or maternal disorders such as diabetes mellitus or toxemia (see Chapter 10). A newborn can also suffer damage during difficult births. For example, it could have been deprived of oxygen because of premature separation of the placenta or a kinked umbilical cord. Many fetuses suffering from developmental abnormalities are lost during pregnancy (spontaneous abortion), but some are born. Such newborns are said to have *congenital defects*, or *birth defects*.

Nevertheless, most babies born in the United States are healthy, with no defect (87%) or only a mild congenital disorder (11%). Of the remaining 2%, about half have a major defect and half die at birth or soon after. Babies born dead are *stillborn*; stillbirths occur in about 1 of every 200 deliveries. Other babies with severe defects suffer *neonatal death*, which means they die in the first month of life. Most of these die in the first 24 h after birth. Prematurity accounts for about one-half of the neonatal deaths in the United States. Other infants with severe birth defects can grow up to lead normal and happy, albeit handicapped lives, and some of these handicaps can be treated with medical and psychological procedures.

All states in the United States screen newborns for a variety of metabolic disorders; the specific screening requirements vary by state. A baby's heel is pricked to obtain a few drops of blood for laboratory analysis. Newborn screening tests include those for *phenylketonuria* (a deficiency of the enzyme needed to break down the amino acid phenylalanine), *galactosemia* (deficiency of an enzyme that breaks down the milk sugar galactose), *biotinidase deficiency* (this enzyme recycles the vitamin biotin), *sickle cell disease* (an inherited blood disorder), *congenital hypothyroidism* (inability to produce adequate levels of thyroid hormones), and *congenital adrenal hyperplasia* (deficient adrenal gland function). These disorders are not obvious in the newborn, but they can lead to serious physical and mental problems; some are even life-threatening.

Congenital hearing loss is a fairly common birth defect, occurring in 3–4 newborns per 1000. Infants whose mothers had a viral disease during pregnancy are at higher risk for hearing loss, as are preterm babies. Because hearing loss is difficult to detect in an infant, newborns are tested before they leave the hospital. A small microphone or earphone is placed by the infant's ear and the infant's sensory response to soft clicking sounds is recorded. Newborns who do not pass the hearing test should receive more extensive diagnostic testing to determine if a true hearing loss is present.

Sudden infant death syndrome (SIDS) is the sudden, unexpected death of an apparently healthy infant for which no known cause of death can be found. In the United States, this happens to over 2000 infants a year. It is most frequent between the ages of 1 month and 1 year, with a peak occurrence at about 3 months of age. Naturally, the risk of SIDS frightens parents, and it is unfortunate that we only have some vague inklings about what causes this disorder (see HIGHLIGHT box 12-1).

Chapter 12, Box 1: Back to Sleep

Sudden infant death syndrome (SIDS) is the sudden, unexpected death of a healthy infant for which no known cause can be found. This is also called "crib death" because most cases occur while the baby is sleeping in a crib. In the United States, about 2000–2500 babies currently die in this manner per year, most frequently between the ages of 1 month and 1 year (with a peak at age 3 months). SIDS is the third leading cause of infant death in this age range in the United States, after birth defects and prematurity/low birth weight. Although there have been a few cases of homicide that at first were thought to be crib death, a great majority of SIDS deaths are spontaneous and accidental.

A major theory has been that these infants suffer from bouts of *apnea* (periodic stoppage of breathing for at least 15 s) and are unable to arouse themselves after an episode of sleep apnea. In fact, several devices are available today to monitor apnea and automatically wake up a sleeping baby if it occurs. However, despite extensive research it has never been proven that sleep apnea is a precursor to SIDS, and in 2003 the American Academy of Pediatrics issued a policy that home apnea

Continued on next page.

Chapter 12, Box 1 continued.

monitors should not be used for the sole purpose of preventing SIDS.

What clues do we have about the cause of SIDS? There is no evidence that it is related to oxygen deprivation during birth, birth trauma, or the use of anesthetics or analgesics during birth. There are, however, several factors that correlate with an increased risk of SIDS.

When a pregnant woman smokes, or even when there is a smoker in the household after birth, there is an increased risk of SIDS. Other correlates include low birth weight, babies with type B blood, bacterial infection of the amnionic fluid, taking drugs and alcohol during pregnancy, maternal anemia, crowded housing, and high room temperatures. Also, bottle-fed babies are more likely to die from SIDS than breast-fed babies; the use of pacifiers decreases SIDS risk in bottle-fed babies. SIDS babies also tend to have relatively high thyroid hormone (thyroxin) levels in their blood. Newborns with an abnormal heartbeat (specifically, a prolonged QT interval) are much more likely to die of SIDS than babies without the abnormality. Interestingly, there may be a genetic aspect of SIDS. The brain chemical serotonin, which plays a role in the control of respiration, is attached in the blood to a carrier protein. The protein that carries serotonin has several genetic variations in different people, and the detrimental variants of this protein (coded by two alleles) increase the risk of SIDS. The frequency of these harmful alleles, and of SIDS itself, tends to run in families.

Although the actual cause of SIDS deaths remains obscure, we now know that a minor change in child care practices can reduce the risk of SIDS significantly. It has been shown that babies who sleep in the prone position (on their stomachs) are more likely to die from SIDS than those that sleep in the supine (on their backs) or lateral positions. Beginning in 1991, there has been a worldwide movement to convince people not to let their babies sleep on their stomachs. After this "Back to Sleep" campaign began in the United States, use of the prone sleeping position declined from 70% in 1992 to 14% today. This was associated with a 50% reduction in SIDS cases in the United States. Across the world, the incidence of SIDS has declined markedly as prone sleeping went from greater than 50% to less than 10%. The incidence of infant infections also decreases when babies sleep in the lateral or supine positions. Babies should be placed in the prone position to play during the day, as this helps develop upper body strength. Thus the recommendation is extended to "Back to Sleep, Tummy to Play."

When an infant's head or face is covered by bedding, the risk of SIDS goes up. Use of soft bedding, such as a pillow, fluffy blanket or quilt, or even a soft toy in a crib is to be discouraged. A firm mattress should be chosen, and only one infant should be placed in a crib. Avoiding hot room temperatures at night is also important; the room should be no warmer than that comfortable to a lightly clothed adult.

Even with the increasing practice of not putting infants on their stomach, studies show that about 30% of American mothers still switch their babies to the prone sleeping position at about 3 months of age. This practice, occurring within the most probable time for SIDS, should be stopped. Also, a higher percentage of low-income women living in inner city neighborhoods still put their infants to bed on their stomachs. Finally, being in child care increases the risk of SIDS greatly, often in the first week of child care. Why? Babies who are used to sleeping on their backs at home may be placed in the prone position when in child care; their arms and shoulders may not be strong enough to lift their heads to breathe. Therefore, "back to sleep" education must be continued in homes as well as day care facilities.

What about letting your infant sleep with you? Hunter-gatherers, regardless of their particular cultures, tend to cosleep with their infant in the same space (not necessarily in the same bed) for the first few years. The mother usually nurses the child to sleep and then again on demand throughout the night. The babies are usually placed on their back in a crib or on a mat, with no blankets, pillows, or toys. Communal sleeping is common, and sleeping while in a sling carried by the mother teaches the infant to sleep in a hectic and noisy environment. In contrast, many babies in

Continued on next page.

Chapter 12, Box 1 continued.

the industrialized world are put to sleep on their stomachs with soft, fluffy comforters, blankets, and pillows, sleep in their own room, are often not tended at night unless they cry, and are bottle-fed or breast-fed on a schedule.

Babies who sleep with their parents, either in a separate crib or in bed with them, get more rest, cry less, and experience more breastfeeding and parent–child bonding. However, it may be somewhat dangerous to sleep with an infant in the same bed because there have been cases of babies being suffocated by a rolling-over parent. In fact, in 2005, the American Academy of Pediatrics recommended putting babies to sleep in their own cribs instead of in their parents' bed. Perhaps cosleeping in a crib next to the parent's bed in the same room is the best bet. There is an increasing awareness of the benefits of cosleeping. One recent study in northeastern England showed that 65% of parents shared their bed with their infant, and 95% of these slept with both mom and dad.

Birth Weight and Adult Disease

The average birth weight in the United States is 7.5 lbs (3.4 kg). The normal range is 5.3 lbs (2.4 kg) to 9.7 lbs (4.4 kg). This is a 4.4-lb (2 kg) span of birth weights that falls within the normal range. Birth weight results from an interaction of genes that influence fetal growth with environmental factors that impinge on fetal development. It has been discovered recently that a baby's birth weight, even within the normal range, can influence its future health as an adult, decades after the fetus developed.

Apparently healthy newborns falling within the lower portion of the normal range can suffer a greater risk of acquiring cardiovascular disease (heart disease, stroke, hypertension) and diabetes when they become adults. One study revealed that newborns weighing about 5.1 lbs (2.3 kg) had double the risk of dying of heart disease than that of newborns weighing 9.9 lbs (4.5 kg). Smaller babies also had a significantly greater chance of having a stroke, hypertension (high blood pressure), and type II diabetes. In another study of 121,700 American nurses, women born weighing between 5.1 lbs (2.3 kg) and 5.5 lbs (2.5 kg) had a 23% higher risk of heart disease and an 80% greater risk of type II diabetes as adults when compared to those born weighing 7.1 lbs (3.2 kg) to 8.6 lbs (3.9 kg). Children born prematurely or with a low birth weight make 50% more insulin than normal birth-weight children. This "insulin resistance" can lead to adult-onset diabetes later in life.

A lower birth weight can be protective against some cancers. For example, women who were heavy newborns (over 8 lbs.) have a 30% increased risk of developing breast cancer as adults. That is, smaller newborns are somewhat more protected against this disease. Women with higher blood levels of estrogen (estriol) during pregnancy tend to have heavier babies, and it is thought that it is really these higher estrogen levels bombarding the mammary glands of the fetus that cause both a higher birth weight and the increased incidence of breast cancer; estrogens stimulate the division of breast cells and thus increase the risk of breast cancer. It is also thought that higher fetal estrogen exposure causes a greater risk of prostate cancer in adult men that were heavy at birth.

A lower birth weight could be due to genes, malnourishment, poor placental development, or to several factors that might influence fetal growth. The mechanisms that cause low birth weight, even if in the normal range, probably operate early in fetal development. However, there is a theory that the real cause of the low birth-weight correlation with several adult diseases may occur after birth, in childhood. This "catch-up growth" theory says that it is not the low birth weight that determines directly the increased incidence of adult disease, but actually it is the more rapid childhood growth of low birth-weight babies to catch up with their heavier counterparts. That is, something about the accelerated growth of low growth weight babies in childhood causes the higher incidence of adult disease mentioned earlier.

Condition of the New Mother

The new mother has just undergone the stress of delivery and now must face the responsibilities as well as the joys of motherhood. Also, her body is rapidly changing back to the nonpregnant condition following birth, i.e., during the postpartum period, or *puerperium*.

Physical Changes

After delivery, a woman's reproductive tract begins to return to the nonpregnant condition. Her uterus, which weighs about 2 lb after birth, regresses to about 3 oz in about 6 weeks. Most multiparous and a few primiparous mothers experience *afterpains* for about 3 days after delivery. These painful uterine contractions are often increased by nursing, which causes secretion of oxytocin (see later). During the postpartum period, most women have a vaginal discharge called *lochia*. This discharge is bright red for the first 3 or 4 postpartum days, turning pale pink or yellow–white a week or so after delivery. Lochia lasts about 10 days in nursing mothers and up to 30 days in nonnursing mothers. A woman, therefore, should wear a sanitary napkin (not a tampon) during this time. Frequency of urination is high in the first week after delivery because fluid is retained in a woman's body during late pregnancy. Normally, a woman can resume coitus about 6 weeks after delivery.

Psychological Changes

The human mother–infant bond is very important to the development of the child. In addition to providing warmth, protection, and nourishment, the mother's behavior provides visual and tactile stimuli that have important effects on the child's developing emotional and social behavior. Newborns form a specific attachment and preference for their mother's voice. A human mother can often identify her own infant by odor alone a few hours after birth; she may have learned this odor while she was still pregnant because her fetus's smell was in her own saliva and sweat! This "smell bonding" could influence maternal imprinting.

The period of pregnancy is a time during which the mother-to-be must come to accept the fact that she is pregnant. She must do this not only on an intellectual level, but also on all levels of mental functioning, especially the subconscious. The acceptance of pregnancy leads to acceptance of the neonate. If a woman has accepted pregnancy and is prepared for the appearance of the infant, she also is well prepared for the commencement of maternal behavior. However, problems can arise when the woman, for various reasons, has failed to accept her pregnancy. This may mean that she will be improperly prepared to accept the newborn. Many women have some negative feelings such as anger toward their newborn. Although most women eventually adjust to the demands of motherhood, some have a more difficult time doing so. These individuals experience emotional trauma, usually manifest some degree of denial or rejection of the infant, and may reject the father, who is held responsible. Hence, the entire relationship between the woman and her family may suffer.

During the first 2 days after delivery, a new mother is often elated but exhausted. Some women, however, experience depression and periods of crying on about the 3rd day postpartum. This depression, which can last a day or two or up to several weeks, is termed *postpartum depression*. Minor postpartum depression occurs in about 67% of women who have given birth. Severe mental disturbances, called *postpartum dysphoric disorder*, occur in about 1 out of 400 mothers. Most women who develop this syndrome do so during the first 10 days after birth.

There may be many causes for postpartum depression. The new mother may be unhappy because she is in an unfamiliar hospital environment, which is why some people favor home deliveries. She has just undergone the stress of labor, and she is undergoing marked physiological changes. Furthermore, she must now face, with her mate, the responsibilities of child care. A marriage also faces many adjustments to the new arrival.

Many women who suffer from postpartum depression have a history of premenstrual syndrome, and both premenstrual and postpartum periods are characterized by sudden drops in blood levels of progesterone. In fact, the administration of progesterone can alleviate the symptoms of both conditions. An interesting idea why human females are prone to postpartum depression is because, unlike many other mammals including primates, they do not eat their placenta, a rich source of progesterone. Another hormonal theory about the cause of postpartum depression is that the mother's brain has too little corticotropin-releasing hormone (CRH; see Chapter 1). This is because the high blood levels of cortisol in the mother's blood during birth (see Chapter 11) have suppressed, by negative feedback, CRH production in her brain. Low CRH brain levels are, in general, associated with depression in humans.

This discussion would not be complete if we did not mention *paternal behavior*, which means care-giving behavior by the father. In our species, certainly, the father participates in caring for the child. However, some fathers do not participate to any great extent, especially in the early stages of the infant's life. The usual reason these fathers give for not getting involved is related to the masculine cultural role, i.e., child care-giving is a woman's responsibility. During World War II, fathers who were absent when their first child was born or were absent during the child's first year later were more critical of its personality and were worried more about its eating, eliminating, and sleeping habits. Perhaps there is a sensitive period for paternal attachment in men, and maybe we should reevaluate the need for extensive early contact for the infant with both parents.

Breast-Feeding

Postpartum Endocrine Changes and Lactation

As noted in Chapter 10, the hormones of pregnancy cause growth of the mammary gland tissue and ducts. These hormones include estrogens and progesterone. During late pregnancy, the high levels of estrogens and progesterone in the mother's blood suppress the secretion of prolactin from her anterior pituitary gland so that the mammary glands secrete colostrum but do not undergo *lactation* (milk secretion). When the placenta is expelled, the levels of estrogens and progesterone drop rapidly to very low levels; this allows release of prolactin, which stimulates the alveoli of the mammary glands. This hormone then causes the primed mammary tissue to secrete milk. *Colostrum* is the first secretion of the mammary glands after birth, with milk appearing in 2 to 3 days.

When the infant *breast-feeds* (suckles the breast), sensory receptors in the nipples are stimulated and the nerve impulses travel to the woman's brain. These neural messages cause release of the hormone oxytocin from the posterior pituitary gland, which is carried by the blood to the mammary glands. Here, oxytocin causes contraction of the myoepithelial cells, squeezing the mammary alveoli and causing milk ejection from the nipples into the baby's mouth. This entire process is called the *milk-ejection reflex* (Fig. 12-2). Stressors such as distraction, embarrassment, fear, or anxiety can cause the nervous system to constrict blood vessels in the mammary gland. Thus, delivery of oxytocin is lowered and milk ejection slows or does not occur. The milk-ejection reflex can also be conditioned to the sight, smell, or sound of the baby so that after a while these stimuli alone can cause milk ejection. Nursing mothers often have been amazed that milk is ejected from their nipples when they hear their baby cry!

Suckling not only causes milk ejection, but nerve impulses also reach the hypophysiotropic area of the hypothalamus and inhibit gonadotropin-releasing hormone (GnRH) release (see Chapter 1). Suckling also increases the levels of a prolactin-releasing hormone (PRH) in the hypothalamo–hypophyseal portal vessels. This PRH, called vasoactive intestinal peptide, then helps increase prolactin secretion. Thus, suckling inhibits the secretion of follicle-stimulating hormone (FSH) and luteinizing hormone (LH) but stimulates the secretion of prolactin from the anterior pituitary gland. Blood levels of prolactin increase 2- to 20-fold within 30 min after suckling begins. Not only does nursing inhibit FSH and LH secretion, but prolactin also decreases the responsiveness of the ovaries to these gonadotropins. Therefore, nursing can inhibit ovulation.

Can nursing be a reliable contraceptive measure? Throughout human history, it has been the most commonly used and effective contraceptive measure, and even today its widespread practice can increase the average birth spacing in a population. However, an individual woman should not rely on lactation as a contraceptive because of its relatively high failure rate (see Chapter 14). Women who nurse on a regular schedule usually do not ovulate for 6 to 9 months postpartum compared with 1 to 4 months in women who do not nurse. After 6 to 9 months of nursing, however, menstrual cycles resume even if nursing is continued. Furthermore, the first postpartum ovulation occurs before the first menses. Therefore, it usually is recommended that nursing women use contraceptive measures beginning 3 to 6 months postpartum. It is unwise, however, to use oral contraceptives at this time as the estrogens in these pills can enter the breast milk and may have adverse effects on the infant. Also, use of a contraceptive pill containing a high amount of estrogen can decrease milk production and infant growth, especially if taken in the first 2 months after birth. Therefore, only injectable progestogens and progestogen-only pills, as well as condoms, IUDs,

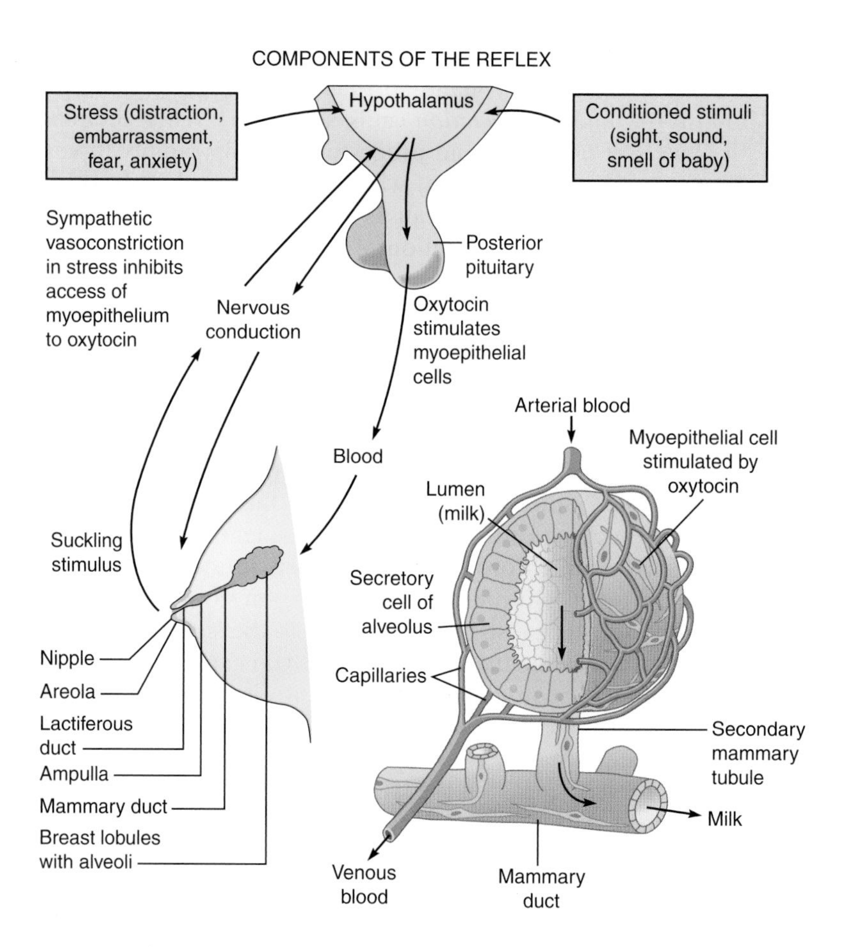

Figure 12-2 The milk-ejection reflex. When the baby suckles the nipple, nerve impulses reach the hypothalamus. Oxytocin is then released into the blood from the posterior pituitary gland and travels to the mammary glands. Here, oxytocin causes the contraction of myoepithelial cells surrounding each alveolus and milk is ejected. The entire reflex takes about 30 s. Stress can activate neurons of the sympathetic nervous system, which causes the constriction of mammary gland blood vessels, reduced delivery of oxytocin, and inhibition of milk ejection. Conditioned stimuli can replace suckling as an initiator of milk ejection.

spermicides, and diaphragms (see Chapter 14), should be used by nursing women.

The frequency of nursing may have a lot to do with its effectiveness as a contraceptive measure. For example, the !Kung San hunter–gatherers of Botswana and Namibia have an extremely long interval between births, averaging 44.1 months. This is despite the fact that nursing women have coitus frequently and do not use any other contraceptive measure. These women nurse briefly and frequently, about every 13 min during the day. It is this frequent nursing schedule, plus the fact that they do not wean their children until they are about 3.5 years old, that effectively keeps prolactin secretion high and FSH and LH secretion low, thus preventing ovulation (see HIGHLIGHT box 12-2).

Chapter 12, Box 2: The Duration of Breast-Feeding

The !Kung hunter-gatherers of Botswana have given us a fascinating and important view of the nursing biology of hunter-gatherers and possibly of our human ancestors in general. !Kung mothers nurse briefly and frequently throughout the day. Each bout of breast-feeding lasts only about 2 min. The average interval between nursing bouts is 13 min and is on demand, i.e., the baby cries and the mother responds by allowing the baby to nurse. This frequency of nursing raises blood prolactin levels in the mother, which causes more milk production and milk ejection. It also suppresses gonadotropin secretion and thus lowers estradiol and progesterone levels in the mother's blood. Because these hormones are at lower levels, the mother's uterus is not stimulated to grow once a month, and monthly ovulation and menstruation do not occur. The general hunter-gatherer pattern is (1) frequent and on-demand nursing, (2) almost continuous contact with the child, (3) a high level of touching and carrying of the child, and (4) suppression of maternal ovulation and menstrual cycles. Before being westernized, most hunter-gatherer women breast-fed their children for 3 to 4 years (the average today is 2.9 years). This served to be an effective contraceptive, leading to a long interval between babies, despite frequent coitus and no contraception.

What are the effects of a relatively late age of weaning on the hunter-gatherer child? Not long ago, there was a general feeling that breast-feeding beyond 12 months led to malnutrition of the child and suppression of its growth. It was even suggested that malnourished children should be weaned at 12 months. However, several recent studies of hunter-gatherers have shown that there may be a different interpretation. That is, lengthened breast-feeding is in response to an initial malnourished state of the child *before* 12 months of age. For example, a study in Senegal showed that children weaned after age 24 to 30 months had a lower height for age and greater stunting of growth than those weaned early, before age 18 to 24 months, but the former group was already more stunted at age 9 to 10 months. This study concluded that prolonged breast-feeding actually increases linear growth of initially malnourished children.

In the text, we saw how breast-feeding and breast milk have great benefits to both the child and the mother. Yet many mothers in modern, developed countries do not breast-feed their babies or do so for a relatively brief time. For example, about 99% of American women are physically capable of prolonged breast-feeding, but one-third of U.S. women do not breast-feed and the others usually stop before 6 months (average, 3.0 months). Hormonal contraceptives are then often used to suppress ovulation (menstrual) cycles because breast-feeding no longer serves that purpose.

What explains this pattern of reduction in nursing in modern industrialized societies? Commercial availability of milk substitutes (cow's milk, soy formula) encourages bottle feeding, especially in cultures where the mother often returns to work several months after giving birth. The use of pacifiers often is actually a substitute for on-demand nursing. Some women may also choose to bottle-feed their babies for psychological or social reasons. They may feel embarrassed about nursing or may think that it is not the right thing to do. In fact, women with ambivalent feelings about nursing and women who had an unplanned pregnancy tend to nurse for a shorter duration or not at all. Women who need to breast-feed their babies outside of the home may find it difficult to locate a socially acceptable place to nurse. A woman is often advised not to breast-feed her baby if she suffers from an infection, kidney impairment, heart trouble, or anemia. This is because nursing may put a dangerous stress on her diseased system. Some modern babies are unable to tolerate the milk sugar lactose and so parents have to switch to soy formula. Finally, some women think that they cannot make enough milk to support their infant so they begin to supplement with solid food or switch to a bottled formula.

Continued on next page.

Chapter 12, Box 2 continued.

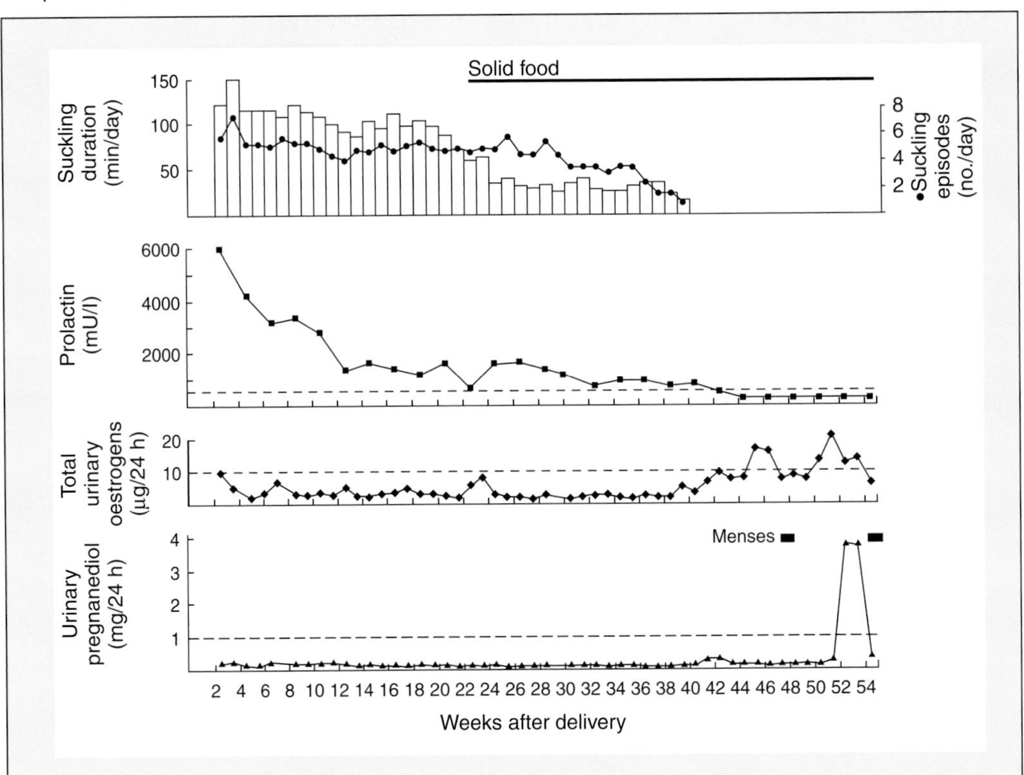

Delayed return of postpartum ovarian activity and ovulation in a breast-feeding mother. Note how suckling frequency correlates with blood prolactin levels. Ovulation (menstrual cycling) resumes soon after nursing ceases, as evidenced by the surge of pregnanediol, a metabolite of progesterone.

A recent study showed that there is no genetic difference in the ability of breast-feeding to suppress ovulation (menstrual cycles) when comparing hunter-gatherers with women in the United States. In both cases, 80 min of nursing per day in conjunction with a minimum of six nursing episodes will stop menstrual cycling up to 18 months after birth. More frequent nursing (on demand) would suppress ovulation for a much longer time. Compared to !Kung mothers, women in the United States experience a much faster restoration of menstrual cycling after giving birth because they nurse their babies less frequently and wean them much earlier. Women in the United States can return to fertility as soon as a month after giving birth, and it is not unusual for women to have two babies as little as a year apart in age. In contrast, the average birth spacing of !Kung women, using only breast-feeding for contraception, is about 3 years. Thus, the difference in natural fertility rate between hunter-gatherers and modern, civilized societies is not due to genetic differences but is cultural in origin.

In the best of possible worlds, more modern women would breast-feed more frequently and for a longer duration. As Roger Short, the well-known human reproductive biologist, said, "The benefits of breast-feeding for the baby are so great that it is a wonder in this litigious age that all bottle-fed babies do not sue their mothers, and the powdered milk companies, for denial of their natural birthright" (see Short, 1994). No one expects women in the developed world to breast-feed on demand for 3 years or more. Nevertheless, any change in the "hunter-gatherer direction" in the breast-feeding practices of modern women would be beneficial for both mothers and their infants.

Advantages and Disadvantages of Breast- and Bottle-Feeding

Breast-feeding offers several advantages to the infant. In addition to always being sterile and at the right temperature, human breast milk is a nutritionally complete infant diet. Table 12-3 compares the contents of human colostrum, human milk, and cow's milk (which is present in most infant formulas). In general, human milk contains all of the nutrients necessary for the infant in the first 6 months of life and most of those nutrients required after 6 months. Lactose sugar, fats (including fat-soluble vitamins), and proteins (casein, lactalbumins, lactoferritin) are present. Also, electrolytes such as sodium, chloride, calcium, and bicarbonate are abundant. Some nutrients in breast milk, such as certain amino acids, fatty acids, and cholesterol, are absent in cow's milk and infant formulas.

There is good evidence that breast-feeding has health benefits to the infant. Colostrum contains antibodies that protect the baby against bacterial infection of the digestive tract. In addition, the lactoferrin present in human milk binds to iron and prevents bacteria from multiplying in the newborn's gut. Breast-fed babies develop less atherosclerosis, hypertension, allergies, diarrhea, childhood diabetes, childhood lymphoma, and middle-ear and upper respiratory tract infections than bottle-fed babies. Artificial milk formulas contain more sodium and chloride than breast milk (Table 12-3), and some feel that this is one reason that the incidence of SIDS is higher in bottle-fed babies. Some infant formulas also contain more sugar than breast

Table 12-3 Approximate Concentration of the More Important Components per 100 ml Whole Milk Unless Otherwise Stated

	Human colostrum	Human milk	Cows' milk
Water, g	—	88	88
Lactose, g	5.3	6.8	5.0
Protein, g	2.7	1.2	3.3
Casein:lactalbumin ratio	—	1:2	3:1
Fat, g	2.9	3.8	3.7
Linoleic acid	—	8.3% of fat	1.6% of fat
Sodium, mg	92	15	58
Potassium, mg	55	55	138
Chloride, mg	117	43	103
Calcium, mg	31	33	125
Magnesium, mg	3	4	12
Phosphorus, mg	14	15	100
Iron, mg	0.09[a]	0.15[a]	0.10[a]
Vitamin A, μg	89	53	34
Vitamin D, μg	—	0.03[a]	0.06[a]
Thiamine, μg	15	16	42
Riboflavine, μg	30	43	157
Nicotinic acid, μg	75	172	85
Ascorbic acid, mg	4.4[b]	4.3[b]	1.6[b]

[a]Poor source.

[b]Just adequate.

milk, and the fact that bottle-fed babies tend to be more obese than breast-fed babies may relate to sugar craving. Another cause of the greater obesity in children who were bottle-fed may be that, unlike breast-fed babies who stop suckling when they are full, bottle-fed babies often are forced to drink more than they want. Unlike formula, breast milk contains *leptin*, which is thought to control obesity in adults. There may be other important components in breast milk that we are not aware of. For example, only recently has it been shown that *epidermal growth factor* is present in human milk and colostrum, and this substance may help the growth of infant tissues. Also, GnRH is present in human breast milk, but its role in the maturation of the infant reproductive system is not yet known. Human milk contains docosahexaenoic acid (DHA), a fatty acid not present in formula; DHA is required for normal vision development. Finally, delta sleep-inducing peptide is present in human milk; to tired parents, this may seem like the most important substance present!

Recent evidence strongly suggests that breast-feeding somehow increases intelligence of the recipient through childhood and even to adulthood. Breast-feeding increases scores on verbal and intelligence tests, and this effect is not due to the intelligence, educational level, or social status of the parents. The duration of breast-feeding is positively associated with intelligence, at least up to 9 months of age. A first guess is that there is something in breast milk, and not in formula or cow's milk, that stimulates an infant's brain development. Perhaps it is DHA found in breast milk and also in fatty fish such as salmon and mackerel; this omega-3 fatty acid is a major constituent of neuronal and glial cell membranes and plays a role in the transmission of neural signals. Alternatively, the effect on intelligence could be due to the physical and psychological effects of holding and nursing a child or there may be some unidentified factor that is related to the choice to breast-feed and leads to higher intelligence in the offspring.

One disadvantage of breast-feeding to the infant is that harmful substances ingested by the mother can pass into the breast milk. We already have seen that contraceptive estrogens are present in breast milk. Such things as nicotine, polychlorinated biphenyls (PCBs) and other pollutants, barbiturates, and anesthetics can also pass into the milk and affect the infant. Included in this list is alcohol, which causes excess hormone secretion from the infant's adrenal gland. A mother can pass active infections (such as tuberculosis and the HIV virus) to her baby through breast milk. Women who have silicone gel breast implants and breast-feed their babies could cause symptoms of autoimmune disease and gastrointestinal problems in their babies.

Breast-feeding also offers advantages to the mother. The oxytocin released during the milk ejection reflex causes the uterus to contract, limits uterine bleeding, and facilitates its postpartum recovery. The longer a woman breast-feeds, the greater her protection against breast cancer and type 2 diabetes. Nursing mothers tend to lose the weight they gained during pregnancy faster than mothers who bottle-feed. The increased secretion of prolactin and oxytocin in the nursing mother's blood could, as is true in other mammals, have a calming and "maternal" effect on the mother's brain. Some women even experience sexual pleasure during nursing, which is normal and part of the joy of the mother–infant contact. Resumption of sexual activities also occurs sooner in nursing mothers.

Finally, breast-feeding is more convenient than bottle-feeding because the mother does not have to purchase and prepare sterile formulas. Breast feeding is also less expensive than bottle-feeding.

There are, however, some disadvantages of breast-feeding to the mother. A nursing woman who is secreting 400 ml of milk a day is supplying 5 g of fat, 4 g of protein, and 28 g of lactose to her infant. Thus, breast-feeding is a nutritional drain on the mother, and she must watch her diet. In fact, nursing for at least 6 months can lead to loss of calcium and a risk of bone weakness in the mother; this lost bone mass usually is recovered quickly after nursing stops. Some women do not like the discomfort that nursing sometimes causes, as the breasts become engorged with milk and the nipples need care to avoid their becoming sensitive and painful during nursing. Finally, because a father can feed a bottle baby but not a breast-fed one, nursing mothers are often the ones getting up during the middle of the night. A father who can help bottle-feed his child can have quality time with the infant that one with a breast-fed infant does not have automatically. One solution is to have the father bottle-feed breast milk collected previously from the mother. There are special "breast pumps" to collect breast milk. Some of the disadvantages of breast-feeding could explain why *wet nursing* was common in early Europe; new mothers would hire women who recently gave birth to nurse their baby too.

From a biological point of view, breast-feeding is better for the infant and mother than bottle-feeding because breast milk is already mixed and is a complete nutritional package. In fact, feeding infants formula diluted (or overdiluted) with contaminated drinking water is a major cause of gastrointestinal problems and even death in infants in underdeveloped countries. That is why the World Health Organization (WHO) advises nursing for at least 2 years. The American Academy of Pediatrics recommends human milk as the preferred nutrition for virtually all infants, including premature and sick newborns, and (1) breast-feeding should begin as soon after birth as possible (within the first hour), (2) breast-feeding should be on demand, (3) water, sugar water, and formula supplements should be avoided unless there is a good medical reason to use them, and (4) breast milk should provide all of an infant's nutritional needs for at least the first 6 months of life. There are, however, many happy and healthy bottle-fed babies. In the final analysis, the choice to breast-feed or bottle-feed an infant is an individual one and should be based on the psychological and physical well-being of both the mother and the infant (see Table 12-4).

Chapter Summary

The newborn is expelled into a new environment to which it must adapt. The Leboyer method of childbirth attempts to reduce the stress of this environmental shock. The newborn's condition is checked by giving it an Apgar score, and body measurements are taken. Many young males (and, in some cultures, females) are circumcised in the first week of life.

The newborn has special adaptations. These include the secretion of lung surfactant, crying, and closure of the ductus arteriosus and venosus as well as

Table 12-4 Some Advantages and Disadvantages of Breast-Feeding as Compared to Bottle-Feeding with Formula

Advantages of breast-feeding and breast milk

1. Is sterile, convenient, and inexpensive.
2. Is at the correct temperature.
3. Has all nutrients necessary for first 6 months of life.
4. Contains maternal antibodies that increase newborn immunity and immune responses.
5. Contains secretory IgA, lactoferrin, and lysozyme, which are inflammatory/antimicrobial.
6. Protective effect against gastrointestinal and ear infections, obesity and resulting cardiovascular disease, diabetes, childhood cancers, meningitis, Crohn's disease, diarrhea, urinary tract infections, and allergies.
7. Reduces incidence of sudden infant death syndrome.
8. Evidence for higher IQ among breast-fed children.
9. Contains some chemicals (such as DHA, GnRH, epidermal growth factor, and delta sleep-inducing peptide) that could influence newborn growth and behavior.
10. Hormonal changes in the mother in response to nursing, such as secretion of oxytocin, enhance the healing of uterine tissue.
11. Speeds the mother's return to prepregnancy body weight
12. Aids in maternal–infant bonding.
13. Decreases the incidence of maternal ovarian and breast cancer.
14. Inhibits ovulation (but is at best an unreliable contraceptive measure).

Disadvantages of breast-feeding and breast milk

1. Breast milk can contain chemicals that could harm the baby such as environmental toxins or if the mother uses illegal drugs, alcohol, barbiturates, nicotine, or steroid hormones from contraceptive pills.
2. Breast milk can contain viruses that could infect the newborn, such as the AIDS virus (HIV); see Chapter 18.
3. An infant with galactosemia (a rare genetic disorder in which individuals lack the ability to break down galactose, a milk sugar) can be harmed by breast milk
4. The mother must watch her diet and caloric intake more carefully.
5. The father may have less opportunity to bond with the newborn early on.
6. Nursing causes mild to great discomfort (e.g., breast engorgement; nipple soreness).
7. Can cause temporary (reversible) loss of calcium in the mother's bones after at least 6 months of nursing.

the foramen ovale. The infant's cells release glucose for energy, and heat is released from stores of brown fat to maintain its body temperature. It has instinctive reflexes that are adaptations for maternal contact, nursing, and walking.

The newborn's appearance is alarming to some when they first view it. Its skull may be egg shaped, with soft fontanels. Swellings, blood clots, and strawberry marks can be present on its head. Its eyes, which can see light and dark, are not their final color and may be crossed and often exude a pus-like secretion. Newborn skin color varies in different races, but a bluish tinge (cyanosis) or yellowish tint (jaundice) may indicate a need for medical treatment. Its reproductive tract is formed but inactive, except that slight breast enlargement and secretion of fluid may occur.

Premature newborns need special intensive care, but many survive and are healthy. Some newborns exhibit abnormalities (birth defects). Congenital defects can cause stillbirth or neonatal death. Other congenital defects can be minor. Sudden infant death syndrome is a major cause of infant death during the first year of life.

The new mother has just undergone a sometimes joyous yet stressful experience and now must adapt to marked biological changes and the responsibility of raising (along with her mate) the new arrival. Some women experience postpartum depression, which may be related to endocrine and psychological factors. There is good evidence that early and frequent mother–infant contact is important to the development of maternal behavior.

Breast-feeding stimulates endocrine responses in a woman that cause lactation, milk ejection, postpartum infertility, and uterine changes. Although the widespread practice of breast-feeding within a population has the overall effect of wider birth spacing and slower population growth rate, breast-feeding for an individual woman is at best an ineffective contraceptive measure. Some women may choose not to breast-feed their babies for medical, sociological, or psychological reasons. Breast milk, however, is better for infants than most, if not all, artificial substitutes. Breast milk protects against digestive tract infections, atherosclerosis, hypertension, allergies, diabetes, middle-ear infections, and even sudden infant death syndrome. It is always sterile and at the right temperature. The nursing mother also benefits in that her physical recovery from childbirth occurs faster, and breast-feeding is more convenient and less expensive than bottle-feeding. Breast milk, however, can contain harmful substances ingested by the mother. Also, a father is unable to help the mother as much when she is breast-feeding her baby.

Further Reading

Anonymous (1982). More U.S. women breast feeding babies for a longer duration. *Family Planning Perspect.* **248**, 840–846.

Anonymous (1990). Duration of the contraceptive effect of human lactation. *Res. Reprod.* **22**(2), 1.

Ball, H.L., *et al.* (2003). Breastfeeding, bedsharing and infant sleep. *Birth* **30**, 181–188.

Carpenter, G. (1981). The importance of mother's milk. *Nat. History Magazine* **9**(8), 6–14.

Christensen, D. (2000). Weight matters, even in the womb: Status at birth can foreshadow illnesses decades later. *Sci. News* **158**, 382–383.

Delzell, J. E., Jr., *et al.* (2001). Sleeping position: Change in practice, advice and opinion in the newborn nursery. *J. Family Prac.* **50**, E2.

Ebomayi, E. (1987). Prevalence of female circumcision in two Nigerian communities. *Sex Roles* **17**, 13–52.

Erwin, P. C. (1994). To use or not use combined hormonal oral contraceptives during lactation. *Family Planning Perspect.* **26**, 26–30.

Flick, J. A. (1994). Silicone implants and esophageal dysmotility: Are breast-fed infants at risk? *J. Am. Med. Assoc.* **271**, 240–241.

Fogle, S. (1993). Darwin takes on mainstream medicine. *J. NIH Res.* **5**(5), 64–66.

Hanson, L. A. (1997). Breastfeeding stimulates the infant immune system. *Sci. Med.* Nov./ Dec., 1997, 12–21.
Haysson, V. (1995). Milk: It does a baby good. *Nat. History Magazine* **104**(12), 36.
Horgan, J. (1995). The mystery of SIDS. *Sci. Am.* **273**(2), 21–23.
Hrdy, S. B. (1995). Liquid assets: A brief history of wet nursing. *Nat. History Magazine* **104**(12), 40.
Hrdy, S. B. (1999). "Mother Nature: A History of Mothers, Infants, and Natural Selection." Pantheon Books, New York.
Hrdy, S. B., and Carter, C. S. (1995). Hormonal cocktails for two. *Nat. History Magazine* **104**(12), 34.
Itoh, H., *et al.* (2002). Hepatocyte growth factor in human breast milk acts as a trophic factor. *Horm. Metab. Res.* **34**, 16–20.
Janszen, K. (1980). Meat of life. *Sci. Digest* **122**, 78–81.
Kahn, A., *et al.* (2002). Sudden infant deaths: From epidemiology to physiology. *Forens. Sci. Intl.* **130S**, S8–S20.
Konner, M., and Worthman, C. (1980). Nursing frequency, gonadal function, and birth spacing among !Kung hunter-gatherers. *Science* **207**, 788–791.
Laumann, E. O. (1999). The circumcision dilemma. *Sci. Am.* **10**(2), 68.
Newman, J. (1995). How breast milk protects newborns. *Sci. Am.* **273**(6), 76–79.
Ostwald, P. F., and Peltzman, P. (1974). The cry of the human infant. *Sci. Am.* **230**(3), 84–90.
Restak, R. M. (1982). Newborn knowledge. *Science* **82**(3), 118–124.
Short, R. V. (1984). Breast feeding. *Sci. Am.* **250**(4), 35–41.
Short, R. V. (1994). Human reproduction in an evolutionary context. *In* "Human Reproductive Ecology: Interactions of Environment, Fertility and Behavior" (K. L. Campbell and J. W. Wood, Eds.). *Annals of the N. Y. Acad. Sci.*, **709**, 416–425.
Willinger, M. (1995). Sleep position and sudden infant death syndrome. *J. Am. Med. Assoc.* **273**, 818–819.

Advanced Reading

Abu Daia, J. M. (2000). Female circumcision. *Saudi Med. J.* **21**, 921–923.
Benini, F., *et al.* (1993). Topical anesthesia during circumcision in newborn infants. *J. Am. Med. Assoc.* **270**, 850–853.
Caldwell, J. C., *et al.* (1997). Male and female circumcision in Africa from a regional to a specific Nigerian examination. *Soc. Sci. Med.* **44**, 1181–1193.
Chapman, D.J., *et al.* (2001). Impact of breast pumping on lactogenesis stage II after Cesarean delivery: A randomized clinical trial. *Pediatrics* **107**, 1–7.
Cook, L. S., *et al.* (1994). Circumcision and sexually transmitted diseases. *Am. J. Public Health* **84**, 197–201.
Fergusson, D., *et al.* (1988). Neonatal circumcision and penile problems: An 8-year longitudinal study. *Pediatrics* **81**, 537–540.
Flick, L., *et al.* (2001). Sleep position and the use of soft bedding during bed sharing among African-American infants at increased risk for sudden infant death syndrome. *J. Pediatr.* **138**, 338–343.
Francis, D. D., and Meaney, M. J. (1999). Maternal care and the development of stress responses. *Curr. Opin. Neurobiol.* **9**, 128–134.

Goldberg, S. (1983). Parent-infant bonding: Another look. *Child Dev.* **54**, 1355–1382.

Golombok, S., *et al.* (1995). Families created by the new reproductive technologies: Quality of parenting and social and emotional development of children. *Child Dev.* **66**,285–298.

Hack, M., and Fanaroff, A. (1989). Outcomes of extremely-low birth-weight infants between 1982 and 1988. *N. Engl. J. Med.* **321**,1642–1647.

Herzog, L. (1989). Urinary tract infections and circumcision. *Am. J. Dis. Child.* **143**, 348–350.

Hopkins, J., *et al.* (1984). Postpartum depression: A critical review. *Psychol. Bull.* **95**, 498–515.

Howie, P.W. and McNeilly, A.S. (1982). Effect of breastfeeding patterns on human birth intervals. *J. Reprod. Fert.* **65**, 545–557.

Klonoff-Cohen, H. S., *et al.* (1995). The effect of passive smoking and tobacco exposure through breast milk on sudden infant death syndrome. *J. Am. Med. Assoc.* **273**, 795–798.

Klopper, P. H. (1996). Mother love revisited: On the use of animal models. *Am. Sci.* **84**, 319–321.

Lamb, M. E. (1983). Early mother–neonate contact and the mother–child relationship. *J. Child Psychol. Psychiat.* **24**, 487–494.

Lesko, S. M., *et al.* (1998). Changes in sleep position during infancy: A prospective longitudinal assessment. *J. Am. Med. Assoc.* **280**, 336–346.

Martorell, R., *et al.* (2001). Early nutrition and later adiposity. *J. Nutrit.* **131**, 8745–8805.

McNeilly, A. S. (2001). Lactational control of reproduction. *Reprod. Fertil. Dev.* **13**, 583–590.

Mortensen, L. *et al.* (2002). The association between duration of breastfeeding and adult intelligence. *J. Am. Med. Assoc.* **287**, 2365–2371.

Myers, B. (1984). Mother–infant bonding: The status of the critical-period hypothesis. *Dev. Rev.* **4**, 240–274.

Newcomb, P. A., *et al.* (1994). Lactation and a reduced risk of premenopausal breast cancer. *N. Engl. J. Med.* **330**, 81–87.

Sapolsky, R.M. (1997). The importance of a well-groomed child. *Science* **277**, 1620–1621.

Willinger, M., *et al.* (2003). Trends in infant bed sharing in the United States, 1993–2000: The national infant sleep position study. *Arch. Pediatr. Adoles. Med.* **157**, 33–39.

PART FOUR

Fertility and its Control

CHAPTER **THIRTEEN**

Human Population Growth and Family Planning

Introduction

In France, schoolchildren are given a riddle. They are told that a lily pond contains a single leaf. Each day thereafter the number of leaves doubles. They are then asked, "If the pond is full of leaves on the 30th day, at what point is the pond half full?" The answer is the 29th day! This riddle teaches about a reality of population growth; it dispels the myth that it will take as long as people have been on earth to double our present population. The previous chapters in this book presented what we know about human reproductive anatomy, physiology, and behavior and something about how science and medicine are providing us with the tools to influence these phenomena. This chapter deals with problems of choice in how we will go about our reproduction, which after all means "producing offspring."

The Biology of Population Growth

Basic Principles of Population Biology

The size of any given population of organisms is influenced by four factors: (1) *birth rate*, or natality, (2) *death rate*, or mortality, (3) *immigration*, or movement of new individuals into a population from another population, and (4) *emigration*, or movement of individuals out of a defined population. That is:

$$\text{Population size} = (\text{birth rate} + \text{immigration}) - (\text{death rate} + \text{emigration})$$

Birth Rate

Birth rate is the number of individuals born in a population in a given amount of time. Human birth rate is stated as the number of individuals born per year per 1000 in the population. For example, if 35 births occur per year per 1000 individuals, the birth rate is 35. Often this rate is expressed as a percentage, in this case 3.5 per 100, or 3.5%. Populations can be subdivided into *juveniles* (before puberty), *reproductive adults*, and *postreproductive adults* (those too old to have offspring). The younger a population, the faster that population grows because the birth rate is higher and the death rate is lower (Fig. 13-1). When

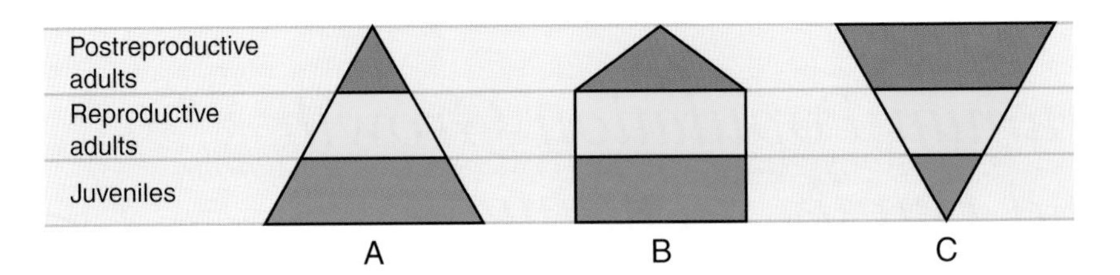

Figure 13-1 Three types of age pyramids. (a) Population is growing because of a large proportion of juveniles. (b) The population is stable; note the similar proportion of reproductive adults and juveniles and the relatively large proportion of postreproductive adults compared with (a). (c) Population is declining because the preponderance of postreproductive adults means that the death rate is high, while the birth rate is low.

birth rate is expressed per age group, it is called the *standardized birth rate*, as opposed to the *crude birth rate* of the total population.

Death Rate

Death rate is the number of individuals dying in a population in a given amount of time. As with birth rate, death rate is the number of humans in the population dying per year per 1000, and death rate can also be expressed as a percentage. The *crude death rate* can be broken down further into a *standardized death rate* for certain age groups. The higher the percentage of older individuals in a population, the higher the death rate and the lower the birth rate, and the more likely the population will decline (Fig. 13-1).

Emigration and Immigration

The balance between emigration and immigration of individuals determines the net addition or deletion of individuals in the population if birth rate and death rate are equal. We consider emigration and immigration to some extent when we talk about specific countries, but we ignore these when talking about the earth's population as there is no good evidence of anyone coming to or leaving our planet permanently!

Population Growth Rate

In general, and ignoring immigration and emigration, the population growth rate is the number of individuals added to or subtracted from a population in a given amount of time. Thus, growth rate is related to birth rate and death rate as described in the following equation:

$$\text{Population growth rate} = \text{birth rate} - \text{death rate}$$

The growth rate has a positive sign in a growing population and a negative sign in a declining population, depending on whether the birth rate exceeds or is less than the death rate. When the two rates are equal, the population is said to be *stable*.

Reproductive Potential

Theoretically, every species has a maximal birth rate defined by its genes. This maximal birth rate is the *reproductive potential* of the species. For example, humans usually have one child at a time with a minimal spacing between babies of about a year, assuming about 3 months of breast-feeding (which tends to inhibit ovulation, see Chapter 12) and no use of contraceptives. Thus, the reproductive potential of humans is about one child per year per reproductive female. Assuming a reproductive life span of women to be 12 to 50 years of age, a woman theoretically is capable of having 37 children!

Biotic Potential

The *biotic potential* of a population is the ability of that population to grow under optimal environmental conditions. When populations are expressing their biotic potential, the birth rate is maximal and deaths are few and are due to the aging process. Population growth under these conditions is at first slow but gets faster and faster. This kind of increase, called *exponential growth*, occurs because additions to the population are increasing like money does in a compound interest savings account. At a given amount of interest (percent population growth rate), the total amount of money in the account (population size) increases at a greater and greater rate because the interest is calculated on an ever-increasing amount in the account. The population growth curve of a population expressing its biotic potential looks like Fig. 13-2.

Environmental Resistance

Usually, natural populations do not express their biotic potential because environmental factors curb the birth rate and increase the death rate. This

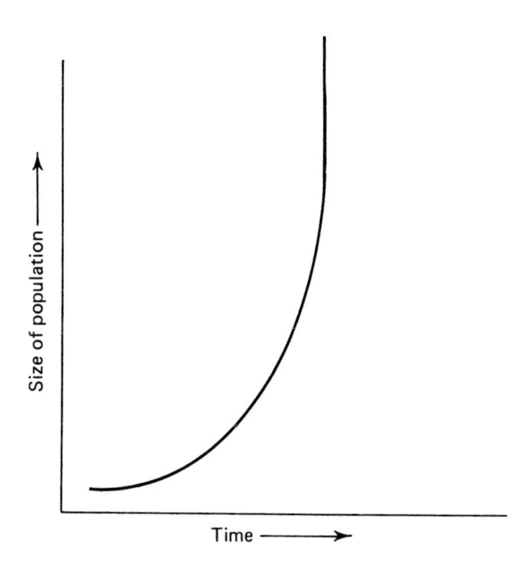

Figure 13-2 The exponential growth curve of a population under optimal environmental conditions.

environmental resistance to population growth can take the form of limited food and habitat for breeding and deaths due to disease, parasitism, and accidents. Many of these limiting environmental factors have a greater impact on a population the greater the population size. In other words, more individuals will suffer the higher the population density. In summary, populations are prevented from expressing their biotic potential as defined in this equation:

$$\text{Population growth} = \text{biotic potential} - \text{environmental resistance}$$

The *carrying capacity* of the environment for a given population is defined as the number of individuals that can be supported by that environment. Thus, environmental resistance usually causes a population previously expressing its biotic potential to level off at the carrying capacity; death rate increases, birth rate lowers, and population size stabilizes (Fig. 13-3). If the population is below carrying capacity, the opposite occurs. Thus, populations are regulated by a feedback system, and the carrying capacity is the "set point" (see Chapter 1 for a discussion of "set point"). In some nonhuman animals, behaviors such as aggressive defense of territories keep the population below the carrying capacity, but the extent to which this occurs in humans is not clear, except in the case of war.

Population Crashes

When a population grows beyond the carrying capacity of its environment, the results can be devastating. For example, a population may grow to the point at which an important environmental resource such as food is exhausted. The

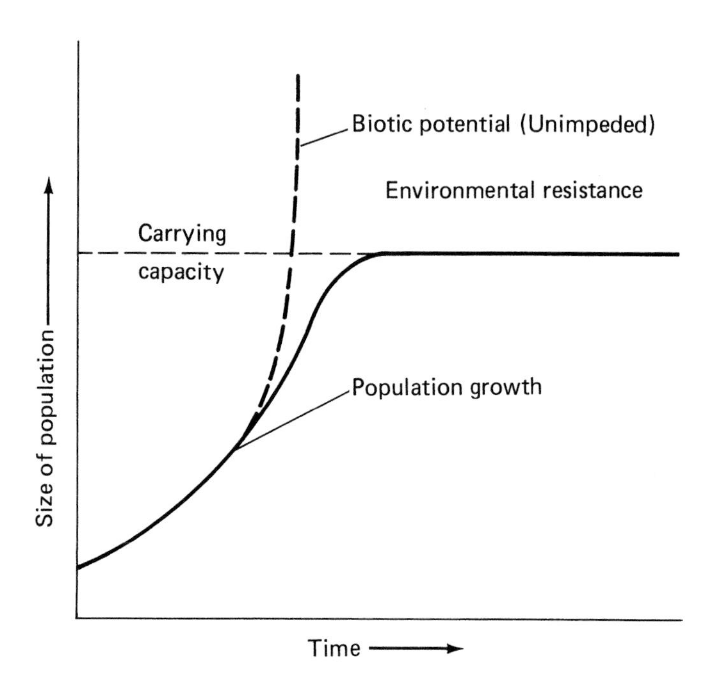

Figure 13-3 Environmental resistance causes a population previously expressing its biotic potential to level off at the carrying capacity of the environment.

result can be a population crash to a level far below the original carrying capacity (Fig. 13-4). After a crash and the subsequent recovery of the environment, the population would then express its biotic potential for a time but soon would be regulated at carrying capacity. If the environment has not recovered, the carrying capacity and stable population size would be lower. Some animal populations repeat this overpopulation error and exhibit cycles of population buildups followed by crashes.

Doubling Time

If one knows the birth rate and death rate of a population, the future change in size of that population can be predicted with some accuracy. For example, if the birth rate is 35 per 1000 per year and the death rate is 25 per 1000 per year, the growth rate of the population is +10 per 1000 per year. At this 1.0% rate of population increase, it will take 70 years for the population to double. Thus the number of years it will take for a population to double (*doubling time*) can be calculated by dividing the annual rate of population increase into 70:

$$\text{Doubling time in years} = \frac{70}{\text{Percent annual rate of increase}}$$

Thus, a 2.0% annual rate of increase would produce a doubling time of 35 years. Surprisingly, low growth rates still have relatively brief doubling times (Fig. 13-5).

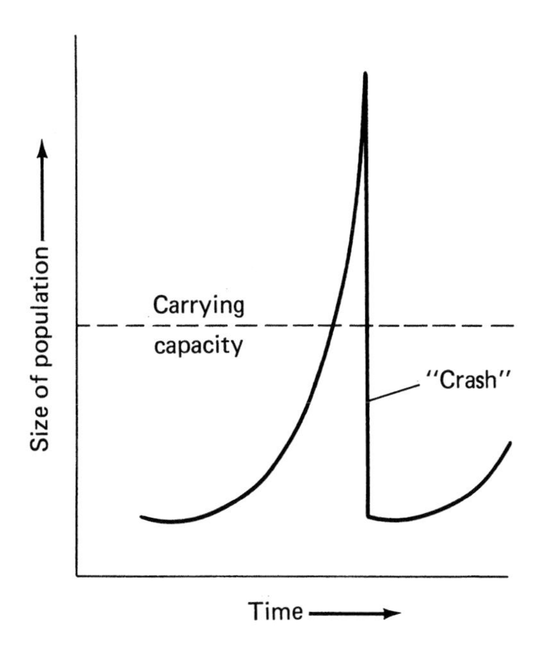

Figure 13-4 When population growth exceeds the carrying capacity of the environment, components of the environment are destroyed and the population may crash.

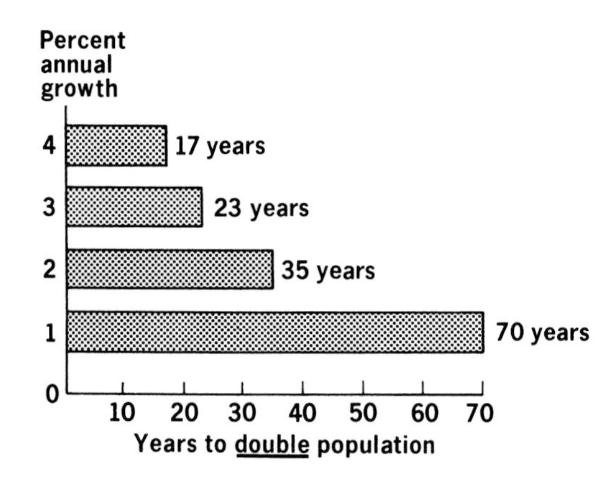

Figure 13-5 Even extremely small percentage annual growth rates produce rapid population increases.

Human Population Growth

The Prediction of Thomas Malthus

In 1798, Thomas R. Malthus, an English economist, historian, and clergyman, wrote an article titled "An Essay on the Principle of Population." In his discussion, he proposed that human populations have the potential to grow in an exponential manner, whereas the human agricultural food supply can increase only arithmetically. For example, suppose we begin with a human population of 1 (representing the initial size of a human population) and a food abundance of 1 (representing the initial size of the food source). According to Malthus, the growth of each would then proceed in the following manner:

$$\text{Exponential human population growth: } 1 \rightarrow 2 \rightarrow 4 \rightarrow 8 \rightarrow 16 \rightarrow 32$$
$$\text{Arithmetic increase in food: } 1 \rightarrow 2 \rightarrow 3 \rightarrow 4 \rightarrow 5 \rightarrow 6$$

Malthus concluded that "the power of population is infinitely greater than the power in the earth to produce subsistence of man." That is, the "passion of the sexes" will drive the human population to a level that exceeds the capacity of the environment to support it. The end result, in his words, will be "misery and vice."

Malthus failed to predict two things: (1) that people would be able to choose whether or not to express their reproductive potential and (2) that they would be able to increase artificially the carrying capacity, including food supply, of the earth's ecosystem. Nevertheless, there have been limitations to what we have done and can do about this Malthusian dilemma, and his prediction looms over the horizon.

Human Population Growth on Earth

Let us look at what has happened to the human population in the past and what may happen in the future. One way we can view this is by looking at the numbers of people on earth from ancient times to the present. About 25,000 years ago, people relied on hunting and food gathering for subsistence, and the total population on earth was about 3 million. At the beginning of the agricultural revolution (8000 B.C.), when people discovered how to grow food, the population was 5 million. Between that time and the birth of Jesus (A.D. 1), the population doubled about six to seven times, resulting in about 300 million people. In 1650, there were 500 million people, and it took only 180 years (to 1830) for this number to double again to 1 billion. By this time, the scientific–industrial revolution, which is still going on in developed countries and has not even begun or is just beginning in underdeveloped countries, allowed the population to double in only 80 years, to a population of 2 billion in 1930. By 1980, it had more than doubled again to 4.4 billion, which took only about 50 years. Thus, the human population has not only increased but has increased faster and faster in an exponential manner (Fig. 13-6). This history of human population growth and doubling time is summarized in Table 13-1.

The world's population (2005) is about 6.4 billion and is growing at a rate of over 80 million people a year. In the 1950s, the average couple in the world had 5.0 children, but in 2005 this had been reduced to 2.8. Despite the fact that the birth rate in the world has dropped recently, there are two reasons why the population is still growing at an alarming rate. First the world's death

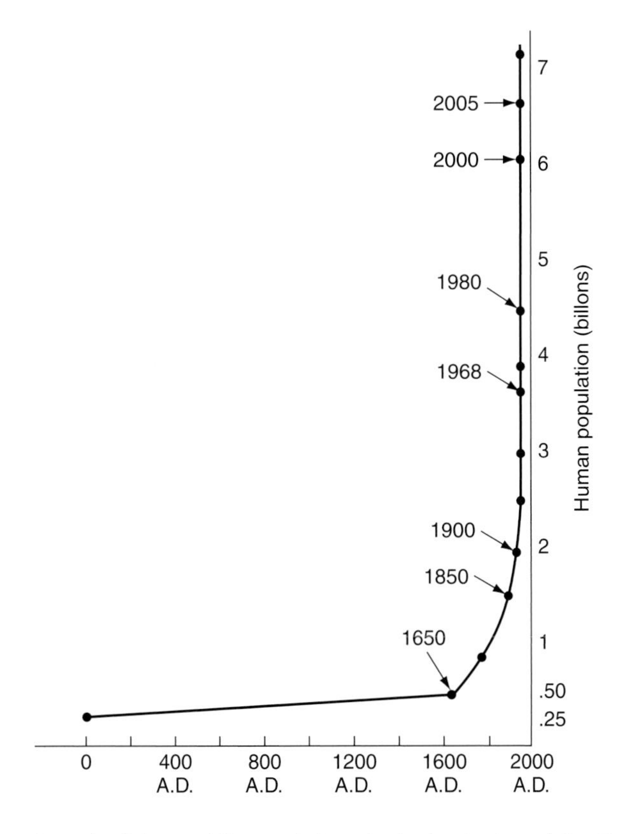

Figure 13-6 Growth of the world's population. At the beginning of the Christian era, the world population is estimated to have been 300 million. The 1650 population of 500 million had doubled by the year 1830 to 1 billion people. In the past 100 years, the growth rate has accelerated greatly. By the year 2000, the earth had approximately 6 billion inhabitants. Note the similarity of this curve to that in Fig. 13-2.

Table 13-1 Change in World Population and Doubling Time

Date	Estimated world population	Time it took for population to double
8000 B.C.	5 million	
400 B.C.	250 million	
A.D. 1650	500 million	2000 years
A.D. 1830	1000 million (1 billion)	180 years
A.D. 1930	2000 million (2 billion)	80 years
A.D. 1980	4000 million (4 billion)	50 years

rate has decreased because of medical advances and availability of health care. Second, the world's population still favors the young; about one-third of the population is under the age of 15. This means that even if all of the world's people immediately simply replaced themselves (had one child per person), the population still would grow for many years. It is projected that the world's population will be over 7.9 billion in 2025 and about 9.3 billion in 2050. As shown in Fig. 13-7, in the year 2025 about 6.7 billion will reside in countries that are currently less-developed and 1.25 billion in more-developed countries. Thus, the highest growth rates and shortest doubling times are in countries with the highest present populations (Table 13-2). Ninety percent of the increase in world population projected for 2050 will come from developing countries in Africa and Asia. Around the globe, 52 countries have doubling times of 30 years or less, and 32 of these are in Africa. Of the major regions of the world, only Europe is expected to experience a population reduction between now and the year 2050.

Age Distribution We have seen that the age distribution of a population is an important predictor of future population size. A greater percentage of young people in prereproductive or early reproductive years portends a higher future birth rate. Across the world, countries have great differences in age structure, and the youth population in developing countries far outpaces that in the more-developed world. For example, in sub-Saharan Africa, 44% of the population is under age 15; about 33% of the population is under 15 in Latin American and Asian countries (except for China). The youth population in developed countries tends to be much lower. Only 17% of the population in Europe and 14% in Japan are under 15 years of age, so these populations have relatively less potential for producing children than a population with a predominance of younger people (Fig. 13-1). Older people age 65 and above make up only 3% of the population in the sub-Saharan region, whereas 15% of Europeans are 65 and older. Thus, the age structure of a population is related to its growth rate. However, remember that even a very small population growth rate (e.g., 1.0% per year) can result in doubling of the population in a relatively short time (Fig. 13-5).

Birth Rate, Death Rate, and Fertility Rate As explained earlier in this chapter, yearly birth rate is expressed as number of births per 1000 individuals in the population for a given year. It can also be expressed as a percentage. Death

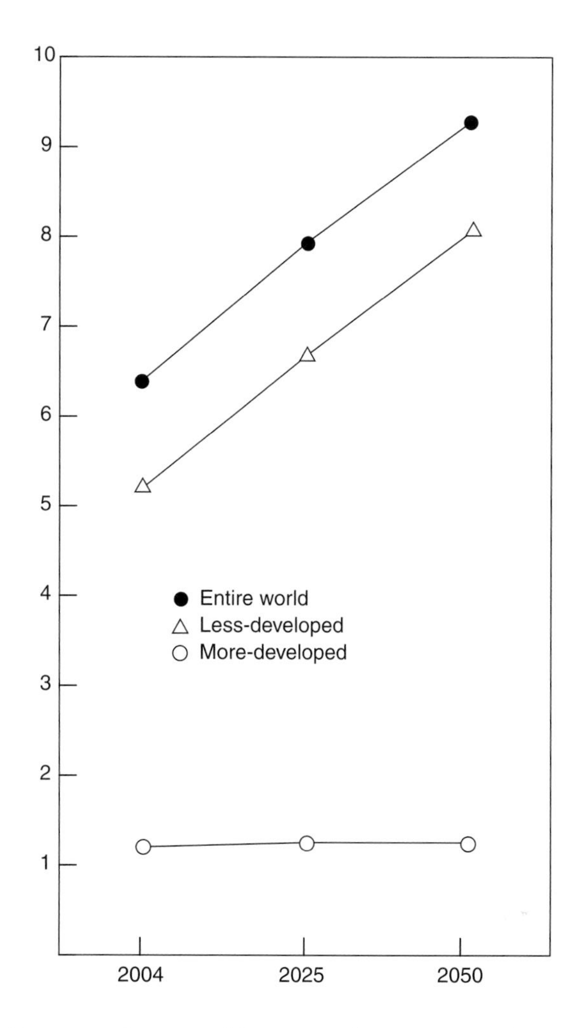

Figure 13-7 Present and projected future population sizes in the entire world and in countries of varying degrees of development (1994 to 2050). Data from the United Nations. More developed, Europe, North America, Australia–New Zealand, and Japan; less developed, Africa, Asia (except Japan), Latin America, and some areas of Oceania. Note that the most growth will occur in the less-developed countries that already have the greatest populations (e.g., China, India, Africa).

rates are described in the same way. To calculate the change in a population for a particular year (ignoring immigration and emigration), one would take the total number of births during that year and then subtract the total number of deaths. If, however, one is interested in predicting the size of the population in future years, it is important to ask questions like "How many infants born this year will die in childhood?" and "What proportion of new mothers are under age 20?" and "How many elderly individuals will die between the ages of 75 and 80?" Answers to these questions can be found by examining *standardized* birth and death rates, which break down birth and death data into age groups. A population with a high birth rate may

Table 13-2 Percentage Annual Growth Rates and the Resultant Doubling Times in Selected Global Regions[a]

Region	Annual growth rate (%)	Present population size (million)	Doubling time (years)
Entire world	1.3	6396	54
More developed	0.1	1206	700
Less developed	1.5	5190	47
Least developed (excluding China)	1.8	3890	39
Africa	2.4	885	29
Eastern Africa	2.3	270	30
Middle Africa	2.8	107	25
Northern Africa	2.0	191	35
Southern Africa	1.0	53	70
Western Africa	2.8	263	25
Asia	1.6	3875	44
Eastern Asia	0.6	1531	117
Southcentral Asia	1.8	1587	39
Southeastern Asia	1.5	548	47
Western Asia	2.0	209	35
Europe	−0.2	728	—
Eastern Europe	−0.5	299	—
Northern Europe	0.1	96	700
Southern Europe	0.1	149	700
Western Europe	0.1	185	700
Latin America and Caribbean	1.6	549	44
Caribbean	1.2	39	58
Central America	2.1	146	33
South America	1.5	365	47
North America	0.5	326	140
Canada	0.3	32	233
United States	0.6	294	117
Oceania	1.0	33	70
Australia	0.6	20	117
Melanesia	1.8	8	39
Micronesia	1.6	1	44
Polynesia	1.2	1	58

[a]2004 data from the Population Reference Bureau. "Annual growth rate" is rate of natural increase only (not counting immigration and emigration).

nevertheless have relatively slow future population growth if there is high infant mortality and a trend for women to delay childbearing. Also, a given number of deaths in the elderly members of a population has a much smaller impact on future population growth than the same number of deaths in juveniles or reproductive adults.

Birth data can also be described in terms of *fertility rate*, or the average number of children women have throughout their childbearing years. The average fertility rate in the world has decreased dramatically in the past 50 years, from 5.0 in the 1950s to 2.8 today. Despite the decline in fertility rate in

Chapter 13, Box 1: World Contraceptive Use and Fertility Rates

In the past two decades, there has been a decline in the fertility rate (number of children in a woman's lifetime) in the world. In 1985 to 1990, world fertility rates averaged 3.4 children per woman, but in 1996 this number fell to 3.0, and in 2005, the world fertility rate dropped below 3 children per female, to 2.8. In many more-developed countries, the fertility rate is at or below replacement value (around 2.0 children). In some areas, the fertility rate is well below replacement level (e.g., Australia (1.7), Canada (1.6), Western Europe (1.6), China (1.4) and Southern Europe (1.3). The fertility rate in the United States is at about replacement level (2.1). In Africa, the fertility rate fell from an average of 6.6 in 1975 to 4.8 only 20 years later. Countries whose fertility rates remain above 6 children per female are found in Africa (e.g., Nigeria, Congo, Chad, Uganda, and Burundi) and the Middle East (e.g., Yemen and Saudi Arabia).

Although the declines in the fertility rate in some world regions are encouraging, it certainly does not mean that the world's population problem is over. One only needs to be reminded that the doubling time of the present world's population (number of years for the population to double at the current fertility and death rates) is still only 54 years.

When the fertility rate of a population declines, one of the assumptions is that there has been greater use of contraception in that region. This has often been the case. In North America and western Europe, the percentage of married or cohabiting couples using contraception is about 75%. In some regions (e.g., China), use of contraception is nearly 84%. The fertility rates in these countries are at or below replacement value, whereas the reverse is true in many cases. That is, in some developing countries (e.g., Nigeria, Chad) the percentage of couples using contraception is between 10 to 15%, and the fertility rates are 6 to 7 in these countries. Clearly, availability of contraceptives gives women and families the ability to regulate their family size.

More couples in developing countries have used contraception recently because of greater availability of contraceptive methods and an increase in literacy, as well as government enforcement as in China. Better education allows couples to learn about the methods of contraception and the values of having a smaller family. Also, education, especially beyond the secondary level, results in a delay in marriage and childbearing.

In some regions (e.g., Nepal, Palestine), however, the fertility rate is fairly high (5 to 6), well above replacement value, even though contraceptive use is also moderately high (30 to 50%). This is because couples, for several reasons, choose to have a large family (see Chapter 14). Also, the relatively frequent use of contraceptive methods with high failure rates, such as the rhythm method or coitus interruptus, increases the fertility rates in these countries. In areas of the former Soviet Union, contraceptive use is well under 50% because modern birth control methods are often unavailable. The result is the frequent use of abortion to control family size (see Chapter 15).

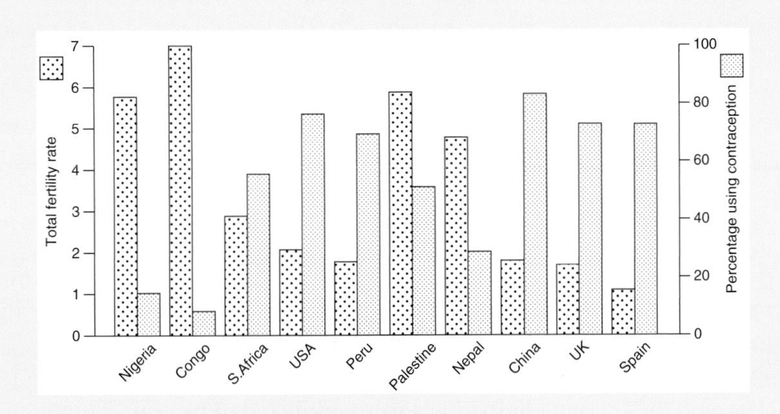

Contraceptive usage and fertility rates in selected countries.

North America (currently 2.0), Latin America (2.6), Europe (1.4), and Asia (2.6), the fertility rate remains high in Africa (an average of 5.1 children/woman), which also has the world's fastest population growth rate. Reasons for differences in fertility rate range from economic, cultural, educational and technological, to personal (see later). For example, the number of pregnancies a woman has often relates to the number of children she can expect to survive childhood. Each year an African infant is 13 times more likely to die than is an infant in Europe or North America.

Life Expectancy Another way of looking at the population situation is life expectancy, which relates to the death rate. Upper-class Egyptians 2000 years ago had a mean life expectancy of 25 to 30 years, as determined by the examination of mummies. Europeans in the Middle Ages lived an average of 35 years, which not long ago was the life expectancy of people in India. Tombstones suggest that only 18% of people in ancient Greece reached 40 years of age. Better nutrition and improved health care have now increased mean life expectancy to the late seventies in many developed countries such as the United States and United Kingdom. In these countries, most deaths occur in the very old or very young. The average infant mortality rate in Africa is nearly 10%, and in some areas of western Africa it is considerably higher. Although deplorable, this situation decreases the population growth rate more effectively than the mortality of postreproductive adults because these unfortunate children do not survive to reproduce.

Population Growth in Developed Countries

The population of the more developed regions of the world, currently at 1.2 billion, is not expected to change significantly in the next 50 years. Several developed regions, such as Europe, North America, and Japan, have reduced their population growth rate to well under 1% (excluding immigration), and Russia, Germany, Hungary, Romania, Italy, Austria, and several other European countries have stable or declining populations (Fig. 13-7). In Japan, abortion legalized in 1948 and supplied by the government resulted in about one-third of pregnancies being terminated in the first trimester (now decreased to about 22%), and the result has been a marked decline in Japan's population growth rate. Japan now has the largest proportion of elderly people in the world. The use of birth control methods such as IUDs, condoms, oral contraceptives, and sterilization has succeeded in lowering the rate of population growth in certain developed countries. Overall, the fertility rate in developed regions is 1.6. The United States has one of the highest fertility rates among developed countries, with an average of 2.1 children per woman.

U.S. Population Growth The birth rate in the United States exhibited a steady decrease until 1940. However, the so-called "baby boom" after World War II reflected an increase in the birth rate, from around 2.0% in 1940 to 2.7% in 1947. Since then, the birth rate again declined to a low in the late 1970s, but in the 1980s and 1990s it showed a slight increase (Fig. 13-8). The baby boom resulted in a bulge in the pyramid of U.S. age structure in the 1960 population in the 0–14 age class, and these young individuals were predicted to increase greatly the birth rate in the 1970s. However, this prediction has been only slightly borne out. This is because several factors prevented women from having as many children in the

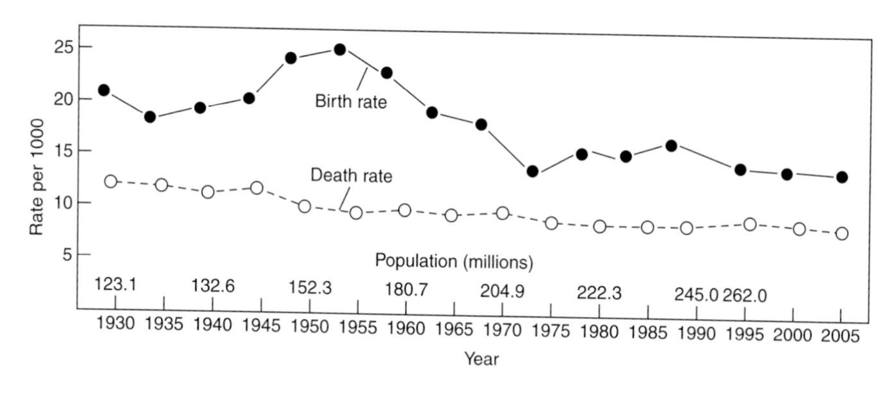

Figure 13-8 Birth rates and death rates in the United States from 1930 to 2005. Whenever birth rate exceeds death rate, a population is growing. Note that the death rate has been declining slowly, but the birth rate has fluctuated. The peak in births in the 1950s represents the post-World War II "baby boom."

1970s as in the past. The age of marriage increased, desired family size decreased, the use of birth control methods increased, abortion laws were liberalized, and there were changes in attitude about the role of women in our society.

The resident U.S. population in 1980 was about 222 million. At the rate of population growth at that time (slightly less than 1.0%), the doubling time was 70 years, and the U.S. Census Bureau predicted there would be about 300 million people in the United States by the year 2000. In 1994 the U.S. birth rate was 16 per 1000 per year, the death rate was 9 per 1000 per year, and the growth rate was 1.0% per year (a doubling time of 70 years). The U.S. population in 1994 was 260.6 million and is currently over 295 million. To reach zero population growth in the United States would require limiting family size to 2.0 children. (Actually the fertility rate needed to replace a generation is about 2.1 because some children will die before they are of reproductive age.) However, even if we averaged 2.1 children per female, the population would continue to increase because of the approximately 1 million legal immigrants and many additional illegal immigrants entering our country each year, who by most estimates account for at least one-third of the U.S. population growth. Stabilization of the U.S. population at this rate of immigration and with a family size of 2.0 children still would require at least 50 to 60 years (Fig. 13-9) because it would take that long for the age structure to shift to favor middle-aged instead of younger people. The average age in the United States in 1988 was about 32.3 years, and this would increase to about 37 years in these 50 to 60 years.

Population Growth in Underdeveloped Countries

The world population is growing the most rapidly in underdeveloped countries. Whereas the population in more developed regions is increasing at a rate of 0.25% per year, that of less-developed regions is increasing by 1.46% per year, nearly six times as fast (Table 13-2). The age structure of these countries is heavily weighted toward the 0–14 age group (Fig. 13-1). In fact, about one-third of the people in underdeveloped countries are under 15 years of age; in Africa the average is 42%.

Rapid population growth rate in the less-developed world is present even though many of these countries have introduced formal family planning

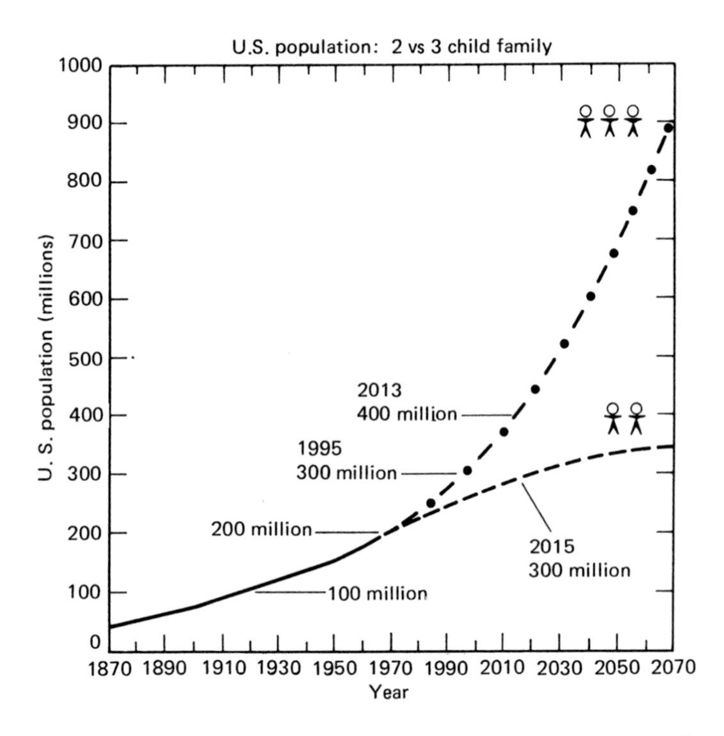

Figure 13-9 Differing predictions for population size in the United States, depending on a family size of two (– – –) or three (– • – • – •) children. These projections assume small future reductions in mortality and continuation of immigration at the present level.

programs. However, cultural, educational, and religious factors have been a barrier to the effectiveness of family planning programs in these countries. About 54% of married women throughout the world use some form of modern contraceptive method, including 69% in more developed countries. The use of these contraceptives is also relatively common in underdeveloped countries in Asia (58%), especially in China (83%), as well as Latin America and the Caribbean (62%). However, Africa is once again the exception, with less than 20% of married women using modern contraceptive methods (see Chapter 14).

Birth rates are declining in 14 of 15 underdeveloped countries (but not Latin American countries). The reasons for these declines are (1) women are marrying at a later age, (2) families want fewer children, and (3) there is greater utilization of birth control methods. Although the fertility rate among women in the developing world has dropped significantly in past decades, it is still well above the replacement rate. Life expectancy in many developing countries is increasing. Thus, because the birth rate has not decreased as rapidly as hoped for and the death rate has declined, birth rates remain higher than death rates and the end result has been rapid population growth. Over one billion people on earth still live in absolute poverty, and more than six million children under the age of 5 die of starvation each year.

Life expectancy and the death rate in areas of Saharan Africa have been impacted significantly by the devastating HIV/AIDS epidemic. In 38 highly affected countries of this region, the population projected for 2015 will be over

90 million fewer individuals, or 10% lower than it would have been without the presence of AIDS. Despite the tragic loss of human life from AIDS, high fertility rates will offset these deaths and continued growth of the population is probable.

Overall View of Human Population Growth

The world's population is increasing at an alarming rate (with a current doubling time of 54 years) due to an increase in "death control" without enough birth control. This phenomenon is most evident in underdeveloped countries, but is also present to a lesser degree in most developed countries, especially the United States. Even if couples worldwide immediately began to have 1.9 children each (below replacement), the world's population would still increase to a peak of 8 billion in 2050. This is because it would take many years for the age structure to equalize (see Fig. 13-1). Then, the population would decline to 5.6 billion by 2150. If the world remains at the present growth rate, it will take 280 years (seven doublings) for the earth's surface to be completely covered with a population density like that of a large urban city such as Chicago!

In contrast to the bleak picture depicted thus far, a recent population projection by the United Nations ("World Population Prospects, the 2002 Revision") gives us hope for the future if human fertility continues to move down toward replacement levels. This projection predicts that fertility in many developing countries will fall below 2.0 children per female at some point in the 21st century. In fact, by the year 2050 it is predicted that three-quarters of countries that are currently less developed will be below replacement levels of fertility. The AIDS epidemic is projected to continue for some time, increasing the death rate in several countries. With this decline in fertility and continued high mortality, it is predicted that the population of the world will be about 8.9 billion in 2050. The six countries accounting for half of the world's projected population growth are India, China, Pakistan, Bangladesh, Nigeria, and the United States.

Effects of Overpopulation

When a human population exceeds the carrying capacity of the earth to support it, as is happening now in some regions, many miseries are inflicted on each individual. These miseries take many forms, and it is beyond the scope of this book to discuss them in detail. You can read the book *The Third Revolution: Population, Environment, and a Sustainable World* by Paul Harrison or the more scholarly *AAAS Atlas of Population and Environment* (see the Further Reading list at the end of the chapter) for a more thorough discussion of the impact of population growth on our environment. We now discuss briefly some of these population-related problems.

Illiteracy

It may come as a surprise to you as you are reading this book that more than one-fifth of the people on earth are illiterate (cannot read or write). Two-thirds of these are women, which fact reflects social attitudes about women and their role in different societies. Ninety-eight percent of the world's illiterate inhabitants live in less-developed regions of the world. The rate is highest in Africa, where illiteracy averages 40%. Worldwide, a general trend is that the higher the birth rate in a country, the greater the illiteracy. There may be numerous and complex reasons for this association. High birth rate and low literacy may both arise as a result of dire economic conditions. A large number of young in a population puts a strain on the educational system, prohibiting full educational opportunities for all. Conversely, illiteracy and poor education in general can lead to an increase in the birth rate. For example, better-educated women tend to delay childbearing, and literate women have fuller access to birth control information. In addition to a high birth rate, accompanying widespread illiteracy are unemployment, low income, poor housing, and poor health care. One ray of hope is that world literacy is on the rise. In all regions of the world, more young people ages 15–24 can now read and write, compared to the general population.

Food Production and Hunger

Hunger and starvation are serious problems in many underdeveloped countries, whereas obesity is a major health concern in developed countries. The scientific–industrial revolution has not reached or solved food shortage problems in many areas in which malnourishment is predominant. Much of the land in these countries is overcultivated, which, along with overgrazing by domestic livestock and deforestation, has resulted in land erosion and destruction of land potentially available for crops. Aid from advanced countries, involving the shipping of food and technological assistance, has and is helping, but this aid has not worked as well as planned. This is because (1) this aid does not keep up with population growth, (2) such aid is only a stop-gap measure and usually does not motivate the people to produce more food for themselves or to decrease their birth rate, (3) the food and aid, once delivered to the country, are not distributed fairly, and (4) the availability (and cost) of food in developed countries is changing.

Unless population growth is curbed, the ultimate result in underdeveloped countries can be devastating famine. In fact, massive starvation and death have occurred quite recently in China, India, and Africa. Even without widespread famine, about half the people on earth are undernourished. Even in the United States, over 30 million people do not have the proper nutrition to live a healthy life.

Some people believe that more food production and better technology in underdeveloped countries would lower the birth rate in these countries. This is because some of the reasons for having large families are economic, such as a family wanting several children to help farm their land. Theoretically, if tractors or other technologies to help people farm were available, the need for large families would disappear. In countries such as Brazil and Mexico, however, rapid economic and technological development in the 1960s was not accompanied by a decline in birth rate.

Natural Resources and Energy

Although we all know that the earth's resources are finite, many people live in hope that resources yet to be discovered will sustain future human populations, and thus we should not be concerned about population growth. For example, there is a widespread belief that food from the sea will provide a major future food source. This hope is probably unfounded because most of the sea is a biological desert. Proper management may be able to double or even triple the yield of protein from the sea, but this will not solve food needs in the long run. Even now, ocean fisheries, which supply much protein to the world's population, are in trouble. The catch of fish, although tripling between 1950 and 1970, has declined in recent decades. As a result, the price of fish protein is rising. It is true that fish are a renewable resource, but harvesting must balance production if this resource is to remain stable. Fish have their own population problem due to excessive human predation and pollution.

Energy is at the forefront of world concern. We are already seeing that most of our current sources of energy are not renewable, and some are running out. Minerals and fossil fuels (oil, coal, natural gas) are becoming limited in some countries (such as the United States), and these countries must rely on imports from other countries or must develop alternative energy sources. Americans consume 35% of the fossil fuels of the world, yet make up less than 5% of the world's population. The struggle to control the world's remaining fossil fuels has led to economic pressures and political conflict among countries with and without these resources. Some countries, such as Japan, import *all* of their oil. Nuclear reactors are an important source of electricity and may become more so in the future. However, this source of energy has serious drawbacks, including high costs and the possibility of nuclear accidents. Alternative energy sources, such as solar energy (which potentially is unlimited), must be developed.

Pollution and Environmental Illness

Pollution of our ecosystems is a major problem that is related directly to both population number and lifestyle. Almost every waterway contains pesticides, insecticides, and industrial and human waste. Especially in large cities, the air is becoming heavily polluted with automobile and industrial emissions; many of these pollutants are not biodegradable. Some of these pollutants even affect our rainfall, making it more acidic, which could have profound effects on the earth's plant and animal life. Other pollutants are endocrine disrupters that cause abnormal reproductive development and cancers (see Chapters 2 and 4). Pollution with chlorofluorocarbons has depleted the ozone in our atmosphere and increased our exposure to ultraviolet radiation. The gases produced by industrial societies have entered our atmosphere and are causing global warming through the "greenhouse effect." This warming is already influencing global weather patterns and food production. Noise pollution is common in cities. It is clear that we must curb population growth and change our lifestyles in order to avoid poisoning ourselves. Much public health literature suggests that such severe human ailments as cancer and respiratory disease can be traced to environmental contamination.

Crowding and Stress

John Calhoun has studied the behavior of crowded rats and found that high densities cause, among other things, decreased reproduction, reduced maternal care, increased fighting, and aberrant sexual behaviors such as rape. In addition, crowding in animals produces physiological changes that cause stress-related conditions such as ulcers and cardiovascular disease. Although we have little reliable information on the effects of crowding on our own physiology and behavior, it is evident that the incidence of certain diseases and crime is higher in cities than in less densely populated areas. Humans are a gregarious species and tend to congregate in clumps. For example, most of the people in the state of California live on less than 10% of the land in the state and about 50% of the world population lives on 5% of the land surface. Because many Americans are moving away from cities, rural areas are also becoming crowded. As our total population increases, the effects on the environment will be accentuated by the tendency for people to clump together.

Quality of Life

It is a basic human right to be well nourished, have sufficient housing and clothing, live in a clean environment, and be able to commune with nature. In some underdeveloped countries, these rights are unfulfilled for a large proportion of their citizens. Even in the United States, the effects of overpopulation and the unequal distribution of resources are widespread. In this affluent society, millions of people are undernourished and poor. Air, soil, and water pollution are widespread, highways are clogged, schools and jails are overcrowded, and "urban blight" is common. Some natural areas are visited by reservation only, and some have been damaged or destroyed.

From a global viewpoint, about 40 million acres of tropical forest (an area twice the size of Austria) are being destroyed each year. Because of habitat destruction and pollution, about 27,000 animal and plant species are becoming extinct each year. Fifty percent of coral reefs on the oceans have been eliminated. Even if you are not bothered by the present situation, you must realize that, at the present growth rate, the population of the world will double in 54 years. Even if some countries improve the quality of life and decrease their rate of population growth, many regions of the world will not, and these regions will be of constant humanistic, political, and economic concern.

Will Science and Technology Save Us?

Through their creative genius, humans have been able to increase the carrying capacity of the earth to support an increasing population. The agricultural revolution, by increasing food production, and the scientific–industrial revolution, by providing new technologies, have greatly increased the earth's capacity to support more people in some regions, but these advances have reached only some people and many have reaped no benefits. Medicine and technology have decreased the death rate around the world due to mass vaccination and inoculation, improved sanitation, and the use of antibiotics, hormones, and other wonder drugs. Thus, scientific and technological advances have increased the capacity of the earth to support human life and, at the same time, have decreased the death rate so the population continues to grow. We are now at the limits of some components of the earth's carrying capacity and are beyond others.

There are a variety of suggestions as to how science and technology can save us in the future. For example, synthetic foods could be produced, we could farm the algae in the sea, we could desalinate seawater, we could develop better fertilizers, and we could cultivate unused land. It is true that we either now have or will have the capability to do many of these things, but the results will only be a temporary solution. For example, if the usable land that now is not cultivated is changed to crop land, the food produced may still not keep pace with the millions of people added to the earth each year. Similarly, increased food production from the sea may not keep up with population growth. Some have even suggested that we will be able to ship people to other planets in our solar system. However, we would need more than 2000 ships per day, each carrying 100 people and their baggage, to keep the earth's population stable! Even if this were possible, the planets in our solar system probably are uninhabitable.

Family Planning and Population Control

The messages of the previous discussion are that (1) the human population is subject to basic biological laws, (2) the human population is increasing at an alarming rate, especially in underdeveloped countries, (3) the carrying capacity of the earth's ecosystem for people is finite, and (4) unless our population growth is curbed, the earth will not be able to support our population, resulting, in the words of Malthus, in much "misery and vice." Thus, our goal must be to achieve zero population growth as soon as possible before nature does this for us and to achieve harmony of our activities with the earth's ecosystems. Two factors can decrease the rate of population growth: decreasing the birth rate and increasing the death rate. Since we do not want to increase the death rate, we must decrease the birth rate to about two children (or fewer) per couple. To do this, we must overcome educational, economic, cultural, psychological, religious, and racial barriers to curbing the birth rate.

Family Planning Programs

For individual freedom to continue, it is necessary for each person to have a choice as to how many children he or she will have. Twelve heads of state, in a United Nations Declaration in December 1966, declared that the opportunity to decide the number and spacing of children is a basic human right. This declaration inferred the right to limit offspring. Thus the goal of many *family planning programs* is to prevent unwanted pregnancies. Family planning programs can be supported by private, federal, state, or local governments and often are staffed by volunteers. Discussions with clients center on the age of marriage, the choice of being married, and the availability and use of birth control methods. Economic, educational, and health advantages of well-spaced and limited numbers of children are emphasized. Advice on pregnancy and child care often is provided. Complete physical examinations, blood tests, Pap cervical smear tests, and chest X-rays often are available.

Margaret Sanger, a nurse, is considered to be the instigator of family planning programs in the United States. Her interest stemmed from an experience, in 1912, of a young mother who nearly killed herself by attempting to abort her fourth child. Sanger's activities, in first disseminating birth control literature (which at that time was against federal and state law) and later setting up family planning clinics, got her arrested and jailed several times. In the end, however, her efforts had a great impact on the use of birth control, resulting in a reduction of the birth rate in the United States. Several agencies in the United States are now involved in family planning programs, including Planned Parenthood–World Population, The Family Planning Association, the Population Council, and Zero Population Growth.

In 1952, Ms. Sanger, along with Lady Ramu Rau of India and others, succeeded in forming the International Planned Parenthood Federation (IPPF), a combination of family planning groups in several nations. Many countries, most of which are underdeveloped, are now members of IPPF. Other agencies of the United Nations, such as the World Health Organization (WHO) and the United Nations Population Division, assist in governmental family planning programs, distribution of birth control devices, and advertising in the media or from village to village about the value of family planning. Family planning programs are active in every country that has more than 100 million people. Recent statistics show that in countries that have had formal family planning programs for 5 years or more, the birth rate has declined markedly.

Barriers to Family Planning

Limiting the number of children in a family is unacceptable to many people because of economic, social, and/or religious reasons. The family is the fundamental organizational unit of human societies, and the size of this unit varies from region to region and culture to culture. One must remember that family planning does not necessarily mean population control, as family planning simply implies avoiding having *unwanted* children, and the family may want and plan for more than two children. In fact, the average number of children desired in families of underdeveloped countries is four to five, as opposed to two in Europe and the United States. In less-developed countries, women are having an average of one more child than they wish. Even in the United States, almost half of all pregnancies are unintended. Thus, universal access to birth control would go a long way toward reducing the world's birth rate. However, even if unwanted births in underdeveloped countries were eliminated, it would not sufficiently reduce the marked rate of population growth in these countries. Family planning programs must change the attitudes of some societies and individuals about desired family size if the overall goal of curbing population growth is to succeed. This, however, is a "tough nut to crack."

Cultural Barriers Cultural traditions have a strong influence on birth rate in some regions. Early age of marriage and the status achieved by having a large family (especially of boys) are factors leading to higher birth rates. Some women or their spouses find contraceptive use objectionable. Nevertheless, evidence suggests that family planning programs are causing the birth rate to decline in some underdeveloped countries.

Some people want large families for economic security. Consider a couple in rural India. They may want five or more children to ensure enough people to work their small plot of land. In addition, when they get old and cannot work, they want children around to take care of them, taking into account the fact that some of their children may die from malnutrition or disease. How would you go about convincing this couple to have only two children? The population of India in the late 1970s was about 516 million and was growing at 2.5% per year despite a governmental family planning program that began in the 1950s. In 2004, the population in India (the second most populous country behind China) was almost 1.1

billion and the annual growth rate was 1.7 (doubling time of 41 years!). Consider that there are 550,000 Indian villages with varying languages and customs, and you begin to see the real problem in reducing population growth in this country.

Religious Barriers Religious attitudes and beliefs may also limit the effectiveness of family planning programs. Consider this quote from the Bible (Genesis 1:26–28, King James version):

> So God created man in his own image, in the image of God created he him; male and female created he them.
>
> And God blessed them, and God said unto them, Be fruitful, and multiply, and replenish the earth, and subdue it: and have dominion over the fish of the sea, and over the fowl of the air, and over every living thing that moveth upon the earth.

The Roman Catholic Church has proclaimed that any "artificial" means of birth control (i.e., anything except rhythm methods or abstinence) is a sin. Even with this position, there is recent evidence that the usage of all forms of birth control in the United States is similar among Catholics and Protestants, although family size is slightly higher among U.S. Catholics as opposed to Protestants. The Anglican (Episcopal) Communion in England and the United States approved all methods of contraception in 1958, as did the Central Conference of American Rabbis (Reformed) in 1960 and the National Council of Churches in 1961. Religious beliefs, however, still influence the usage of contraception and family size in many other countries.

Other Barriers Racial and economic factors sometimes influence population control. For example, many family planning programs in the United States are attempting to reach the poor, and there is a misconception that most poor are racial minorities and that most minorities are poor, which is not true. However, some minority groups believe that attempts to limit family size in the poor are a form of genocide in that it permits whites to reproduce at the expense of other races. What is true is that the poor in the United States comprise many races, including whites. About 42% of families with greater than five children are poor in the United States, whereas only about 10% of families with one or two children are poor. The incidence of unintended pregnancies is also higher in women living below the poverty line (61%) compared with the general population (49%).

Other factors influencing family size include established patterns of sexual behavior, tolerance of women working, the social status of women, and the cost of raising and educating children. All of these factors should be of interest to family planners. Tables 13-3 and 13-4 summarize the reasons people have for limiting or not limiting family size. You may want to discuss these with your instructor, classmates, and friends. Another major barrier to effectively reaching zero population growth is that humans tend not to believe in something that is not at present having a serious effect on their lives. Thus, education about the present and future effects of the population explosion should be at the forefront of family planning programs. Not only must these programs deal with the practices and attitudes of a target population about contraception, but they must educate people, especially young men and women, about the potentially disastrous consequences of population growth.

Table 13-3 Some Reasons Why a Family Would Want to Have Several Children

Health	
	Children often die. It is necessary to have large families in order to get living children who grow to adulthood.
Economic	
	Children are an economic advantage. They are needed or are useful in helping the family earn a living. They pay for themselves by working as they grow. If you have many children, one or more will be able to take care of you and provide security in old age.
Family welfare	
	Children can help with work around the house. Older children help the younger ones.
	Big families are happy families. Family life is more enjoyable; they have a good time together.
	Children from big families are better adjusted, better able to get along with other people, not so spoiled or egotistical.
	Children continue the family name. It is necessary to have many children to be sure to have a son to carry on the family name.
	Strength of the clan. The family is stronger; sons can help you fight your battles; family can be upheld.
Marriage adjustment	
	Large families promote good marriage adjustments. Couples get along with each other better, marriage is better.
Personality needs	
	Ego support. Children are a demonstration of virility and manliness.
Community and national welfare	
	Large families are good for the community or nation. They promote population growth.
Moral and cultural	
	Large families are the will of one's god. It is against religious belief to limit fertility.
	Large families promote morality. They help prevent divorce or infidelity.
	Tradition. The community, village, family, and clan expect large families.
	You have high status in the community. If you have a big family, you are more important, are looked up to.
Dislike for contraception	
	Contraception is disliked because of aesthetic or health reasons or because it interferes with sex.

Family Planning in the United States

In 1969, the Congress of the United States established the Commission of Population Growth and the American Future. This commission recommended several changes for controlling population size in the United States. For example, the government should develop a national plan for population stabilization and schools should educate children about the population problem. Additional recommendations included that child care services be facilitated so that more women could work, that adoption procedures should be simplified, and that women should have equal rights compared with men. Finally, minors' access to contraceptives should be greater, there should be no restrictions on sterilization services, and development of better contraceptives should be emphasized in government-sponsored research programs. Another recommendation, that abortion laws be liberalized, was accomplished by the U.S. Supreme Court (see Chapter 15).

Many of these recommendations have been put into action, resulting in smaller families and a reduction in the number of unwanted births in the United States. However, a significant portion of unmarried young American men and

Table 13-4 Some Reasons Why a Family Would Want to Limit Their Offspring Number

Health
Preserve health of mother.
Ensure healthy children.
Reduce workload of father and mother.
Economic
Everyday, general expenses are less.
Avoid worsening present (poor) economic condition.
Gain a higher standard of living, more comfort, afford better house.
Permit saving for future, for retirement.
Desire to avoid subdividing property or savings among many children.
Family able to have money for recreation, vacation.
Family welfare
Improve children's lot in life, give them good education, help them get started in a career.
Opportunity to do a better job of rearing children; able to devote more time to each, better able to socialize with child.
Avoid overcrowding of house; more opportunity for individual expression.
Easier to find a more desirable house or apartment.
Marriage adjustment
Provides husbands and wives more leisure opportunity to enjoy each other's companionship.
Improves the sexual adjustment by eliminating or reducing fear of unwanted pregnancy.
Personality needs
Facilitates realization of ambitions. Permits husband and/or wife to pursue occupational or vocational objectives.
Facilitates self-development. Permits a wife to express herself outside the home, yet have a normal family and married life.
Facilitates realization of social needs. Permits the person (especially wife) to have contacts and friendships outside the home and participate in neighborhood activities.
Reduces worry of the future. Avoids danger of childbearing when one is really too old—danger of dying and leaving behind orphan children.
Community and national welfare
Helps avoid overpopulation, overcrowding.
Helps community meet demands for education, other community services.
Helps nation with economic development.
Helps keep down delinquency, social problems of youth.
Helps reduce welfare burden of the community.

women are still sexually active and many of them do not practice effective contraception. A survey by the Alan Guttmacher Institute indicates that two-thirds of the 5 million sexually active teenage females and 7 million sexually active teenage males in the United States do not use effective contraceptive protection. In this same survey, 51% of the females said that they did not think they could get pregnant. In sexually active teenagers, 58% of those not using contraception get pregnant as opposed to only 6% of those who do use contraception. And the U.S. population still is growing at a rate of about 0.6% a year.

It is important that Americans, as well as citizens of other countries around the world, become aware of the impacts of their personal and lifestyle choices on the population and the environment. Choices made today will greatly affect the quality of life for future generations. An average middle-class American uses about 26,000 gallons of water, 21,000 gallons of gasoline, and 10,000 pounds of beef in his or her lifetime. As noted earlier, Americans use 35% of the world's supply of minerals and fossil fuels but make up only 5% of the world's population.

It may be time for a voluntary change in lifestyle before such a change is required for survival. The ideas that "I'll get mine while I'm here" and "Why should I save it for somebody to use and enjoy after I'm dead?" must give way to a broader sense of responsibility as inhabitants of a shared planet.

Individual Freedom and Family Planning

To respect individual freedom, family planning should be on a voluntary basis. However, if voluntary reduction in family size is not achieved, governments could initiate measures to control population size. This governmental intervention has already been realized in China (see HIGHLIGHT box 13-2). Table 13-5

Chapter 13, Box 2: Government-Enforced Population Control in China

The People's Republic of China was the first country to embark on a deliberate campaign to reach zero population growth by the year 2000. As the Communists came to power (in 1949), Chairman Mao Zedong stated that "The absurd theory that increases in food cannot catch up with increases in population, put forth by such western bourgeois economists as Malthus and company, has not only been refuted by Marxists in theory, but has also been overthrown in practice in the post-revolutionary Soviet Union and in the liberated China." In 1957, however, in a reversal of marked importance, he stated that "population growth must be controlled" in the face of vast unemployment and food shortages. Since then the Chinese government has called for "birth planning" responsibility at all political and social levels, educational and motivational campaigns about such planning, and wide and free availability of contraception and abortion. The goal was to decrease the birth rate, which indeed has gone from 34/1000/year to the present 12/1000/year. In 1970, the average Chinese family had six children and now it is down to about two, but population growth continues. At present, China (the largest country in the world) has 1.3 billion people and is adding nearly 8 million people each year! The present natural growth rate (births minus deaths) of China is 0.6% per year (the same as the United States), with a doubling time of 117 years. The growth rate has slowed down as a result of tough and controversial measures taken by the government.

In 1971, China adopted the "*wan xi shao*" ("later, longer, fewer") policy. "Later" meant waiting to marry and have children, "longer" meant spacing successive children at least 3 years apart, and "fewer" meant limiting the number of children per couple to two. The government soon realized that this was not enough, so in 1979 they made the policy "*wan xi shao you,*" or "later, longer, fewer, and better." "Better" meant having only one child per couple and raising him or her better. The "one-child only" policy is voluntary, but there are great benefits to signing up for it. After the initial child is born, a couple acquires a "one-child" certificate, which entitles them to higher income, lower-cost health care, better housing, larger old-age pensions, and preferential schooling and employment for that child. Part of this policy also involves free, easy access to a wide variety of contraceptives as well as abortion. In China, 84% of women use birth control and 36% use IUDs, several of which were invented in China (such as the "Canton Flower" and the "Human Beetle"). Also, 34% of women and 8% of the men have been sterilized, and several hormonal and barrier contraceptives are also available. Abortion is free and on request.

The "one-child only" policy has had an effect on birth rates in China in some local areas and not others. For example, an allowance of two children in rural areas resulted in an increase in

Continued on next page.

Chapter 13, Box 2 continued.

the birth rate in some regions (see Highlight figure). If a couple chooses not to use contraceptives after having a child, there can be political pressures. These include "militant propaganda teams" that visit the home, apply pressure, and even exercise control over the family's drinking water, food, or work permits. Couples with a male child also have been "mobilized" to undergo sterilization or abortion. Penalties occur if a couple that has a "one-child" certificate later has another child. These include returning all stipends or work points received, a reduction in wages if the couple has more than two children, no subsidies, poorer health care, no preferential treatment related to schooling, and no job benefits. If a single child dies, however, a couple is permitted to have another.

One problem with the "one-child only" rule is the traditional Chinese desire to have a son to support a couple in their old age. Thus, if the first child is a girl, a couple would want another until they had a son. The Chinese government has attempted to help this situation by making daughters as well as sons legally responsible for their parents, and a daughter as well as a son can now take over a father's job when he retires. In addition, job preferences are now given to the daughters of one-child families. Still, there are rumors, some verified, of terrible things happening to first-born daughters in China. Selective abortion (after determining fetal sex) may be occurring, as well as even infanticide of baby girls. Orphanages are showing a rise in abandoned, mainly female, infants.

Because of the increase in male children in Chinese families in recent years, it is ironic that now the Chinese government is concerned about the increasing gender gap. About 119 boys for every 100 girls are being raised in Chinese families, and in rural regions this gap is about 133 boys:100 girls. New regulations aim to close the gender gap by the year 2101. The government is making it a crime to selectively abort female fetuses and is prohibiting doctors from telling parents the sex of their fetus. Still, these practices will continue as long as a strong cultural bias for male children remains.

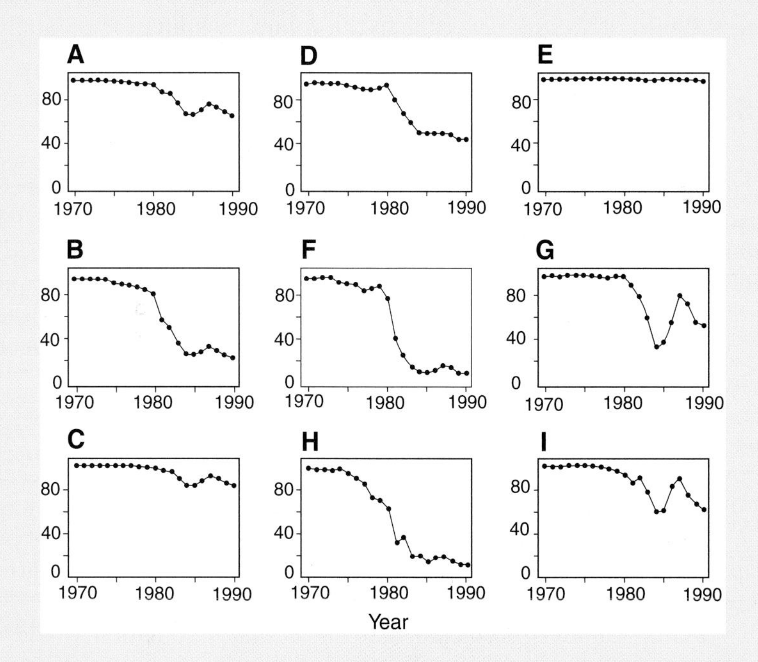

Proportions of women having a second child in China as a whole (A), in urban (B), and rural areas (C) of China as a whole, and examples of specific urban (D, F, and H) and rural (E, G, and I) regions. Note the marked drop in second children, especially in urban areas, after the 1979 "one-child only" policy was established. Used with permission from Feeney (1994).

Table 13-5 Possible Proposals Whereby a Government Could Control Population Growth

Birth control	Taxing	Marriage	Social and political
1. Sterilize poor people 2. Liberalize abortion and sterilization laws 3. Sterilize men or women who have more than two living children 4. Temporary sterilization of women; reversal only by government consent 5. More contraceptive devices available without prescription 6. Payment for cost of abortion 7. Mandatory abortion of pregnancies outside marriage 8. Selective placing of contraceptive chemicals in water supplies	1. Reduce taxes for single people 2. No tax exemptions for parents 3. Tax families with more children 4. Free schooling for up to two children 5. Tax benefits for working wives	1. Charge a high fee for a marriage license 2. Incentive payments for late marriages 3. Issuance of two kinds of marriage licenses: (a) family marriage, licensed for children; more difficult to dissolve; expensive (b) couple marriage, licensed for no children; easy to dissolve; inexpensive 4. Issue no or one-child marriage certificates, with rewards	1. More support of contraceptive research 2. Pressure on religious and other groups 3. Reduce new housing, transportation, and water supplies to certain regions 4. Make foreign aid inversely proportional to population growth rate in a given country 5. Reward families of no or few children with job benefits, higher standard of living, and educational support.

lists some possible ways that governments could attempt to limit our reproduction. Look at each suggestion, think about it, and discuss its advantages and disadvantages. Which of these is favorable to the voluntary limitation of family size?

Chapter Summary

The subject of this chapter has been human population growth, the effects of this growth upon us and our environment, and the importance of family planning and population control. Population size and growth rate are determined by interactions among birth rate, death rate, emigration, and immigration. Rapidly growing populations are characterized by a predominance of the young, whereas stable populations have similar numbers of all ages except the very old. Declining populations have a predominance of older individuals.

A population expresses its biotic potential when the individuals are exhibiting their reproductive potential, which is their maximal rate of reproduction. However, populations do not reproduce at their biotic potential because of environmental resistance. Population growth must stabilize eventually at the carrying capacity of the environment. Populations expanding far above carrying capacity will crash. Populations with a high rate of population growth have a short doubling time.

Thomas Malthus predicted that the human population would outgrow its environmental requirements, which may well occur. The earth's population, currently about 6.4 billion people, has exploded in the past 100 years. At the current growth rate (1.3%/year), the population will double again in 54 years. In underdeveloped countries, death rates have lowered and birth rates have remained stable or decreased only slightly so the population growth rate in these countries is still alarmingly high (1.5% on average, although some individual countries have growth rates of up to 3.5%). In many developed countries, the growth rate has decreased to 1.0% or less because of a marked decrease in the birth rate. In fact, some developed countries have stable or even declining populations. However, in the United States, it will still take 50 to 60 years to reach a stable population even if each couple only has two children because the population has a disproportionate number of young people and a strong immigration rate.

Many human populations are plagued by illiteracy, hunger and starvation, depletion of natural resources and energy supplies, pollution and environmental illness, stress due to crowding, political conflict, and an overall reduction in the quality of life. These problems can be exacerbated by further population increases. Science and technology, which in the past have increased the earth's carrying capacity for people, will make further advances. These improvements, however, are not likely to keep pace with the population growth if it continues at the present rate.

The goals of family planning programs are to decrease the rate of birth of unwanted children and to limit family size. These programs deal with such subjects as age of marriage, availability and use of contraceptive methods, and the economic, educational, and health advantages of well-spaced children and small families. The International Planned Parenthood Federation and the United Nations offer assistance to family planning programs in various countries. Family planning programs in the United States and some other developed countries, as well as in several underdeveloped countries, have helped reduce family size. Economic, social, cultural, psychological, racial, and religious barriers still confront the proponents of family planning, and these barriers vary in different regions. Many people, especially in underdeveloped countries, want more than two children. If voluntary limitation of the birth rate does not occur in the future, governments may seek to control family size through legislation, such as the current situation in China.

Further Reading

Anonymous (1990). Natural family planning in today's world. *Res. Reprod.* **22**(3), 1–2.

Ausubel, J. H. (1996). Can technology save the earth? *Am. Sci.* **84**, 166–178.

Boland, R. (1992). Selected legal developments in reproductive health in 1991. *Family Planning Perspect.* **24**, 178–185.

Brown, L. (1995). Facing food scarcity. *World Watch* **8**(6), 10–20.

Caldwell, J. C., and Caldwell, P. (1990). High fertility in sub-Saharan Africa. *Sci. Am.* **262**(5), 118–125.

Chege, N. (1995). Kenya's plans for its children. *World Watch* **8**(1), 20–22.

Chen, P., and Kols, A. K. (1982). Population and birth planning in the People's Republic of China. *Popul. Rep. Ser. J.* **25**, 579–618.

Greenhalgh, S. (1986). Shifts in China's population policy 1984–86: Views from the central, provincial, and local levels. *Popul. Dev. Rev.* **3**, 491–515.

Doyle, R. (2000). The U.S. population race. *Sci. Am.* **283**(2), 26.

Gleick, Pl H. (2000). Making every drop count. *Sci. Am.* **284**(2), 40–45.

Harrison, P. (1993). "The Third Revolution: Population, Environment, and a Sustainable World." Penguin USA.

Henricksen, D. (1994). Putting the bite on planet earth. *Int. Wildlife* Sept./Oct., 36–45.
Keyfitz, N. (1989). The growing human population. *Sci. Am.* **261**(3), 118–127.
Klitsch, M. (1995). U.S. fertility rates fell during 1992, even among teenagers, women over 30. *Family Planning Perspect.* **27**, 91–92.
"Managing Planet Earth: Readings from *Scientific American*" (1990). Freeman, New York.
Mukerjee, M. (1998). The population slide. *Sci. Am.*, **279**(6), 30–32.
Nelson, T. (1996). Russia's declining population. *World Watch* **9**(1), 22–23.
Pan, S. (1993). China: Acceptability and effect of three kinds of sexual publication. *Arch. Sex. Behav.* **22**, 59–71.
Platt, A. E. (1995). Dying seas. *World Watch* **8**(1), 10–22.
Potts, M. (2000). The unmet need for family planning. *Sci. Am.* **282**(1), 88–93.
Postel, S. (2002). Growing more food with less water. *Sci. Am.* **284**(2), 46–51.
Robey, B., *et al.* (1993). The fertility decline in developing countries. *Sci. Am.* **269**(6), 60–67.
Wilken, E. (1995). Assault on the earth. *World Watch* **9**(1), 22–23.

Advanced Reading

Feeney, G. (1994). Fertility decline in East Asia. *Science* **266**, 1518–1523.
Harrison, P. *et al.* (2000). "AAAS Atlas of Population and Environment." University of California Press.
Lee, R. (1994). Human fertility and population equilibrium. *In* "Human Reproductive Ecology: Interactions of Environment, Fertility and Behavior" (K. L. Campbell and J. W. Wood, eds.). New York Academy of Sciences, New York.
Raloff, J. (1996). The human numbers crunch: The next half century promises unprecedented challenges. *Sci. News* **149**, 396–397.

CHAPTER **FOURTEEN**

Contraception

Introduction

Advances in reproductive science and medicine have made it possible to manipulate human fertility. For couples who desire to avoid or terminate an unwanted pregnancy, contraceptive measures and induced abortion (discussed in Chapter 15) are available. This chapter reviews current methods of contraception, including their usage, advantages and disadvantages, mechanisms of action, effectiveness, and major and minor adverse side effects. In addition, possible future methods of contraception are discussed.

Contraception literally means "against conception," and many contraceptive devices prevent fertilization of an ovum by a sperm. For our discussion, however, we also include contraceptive methods that prevent the transport of a developing preembryo through the oviduct or implantation of a preembryo in the uterus. Figure 14-1 summarizes how present contraceptive methods prevent fertility. Contraceptive use has increased throughout the world as people have begun to realize the problems of overpopulation (see Chapter 13) and as couples come to understand the effects of an unwanted pregnancy on their lives.

About 4 million births occur each year in the United States. One-third of these are to unmarried women. Approximately 55% of these births to unmarried women result from unplanned or unwanted pregnancies. Even in married U.S. women, about 30% of pregnancies resulting in live births were unintended. Fully half of all women in the United States will have had an unplanned pregnancy between the ages of 15 and 44, but 9 out of 10 women at risk for an unwanted pregnancy report that they use contraception. Even when contraceptives are used, about one-half of the unintended pregnancies in the United States are due to inconsistent or incorrect use of contraceptives. The rates of unplanned pregnancies, teenage pregnancies, and abortions in the United States are the highest in the industrialized world. This is mainly due to failure to use contraceptives or failure to use them correctly.

In the United States today, about 62% of women in their childbearing years (between the ages of 15 and 44) use some form of contraception (Table 14-1). Of those not using birth control, most are pregnant or postpartum, wish to be pregnant, or infertile. Sterilization (female and male) is the most popular birth control method overall and among married couples, whereas the combination birth control pill is the method of choice of never-married women. Use of the birth control pill is greater among younger women and those with a higher educational level. As a woman ages, she is less likely to take the pill and more likely to use sterilization as a contraceptive.

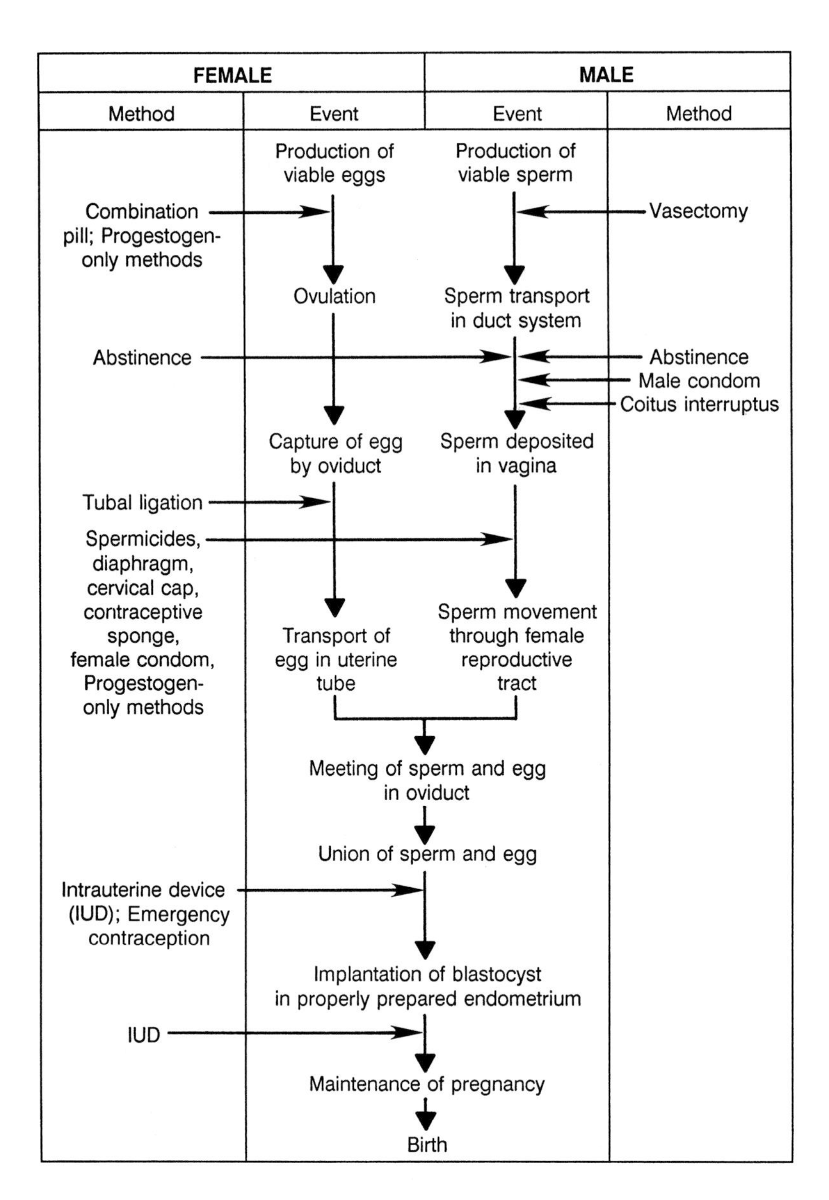

Figure 14-1 The main mechanism of action of major contraceptive methods in men and women.

Patterns of contraceptive use vary widely around the world. Some examples of the most popular contraceptive methods in various countries are: Japan, the male condom; Hungary, Sweden, and much of Africa, the pill; Canada and Brazil, female sterilization; Egypt and Norway, the IUD; Indonesia, hormonal injections and implants; Spain and Jamaica, the diaphragm/cervical cap; Turkey, Romania, and Italy, withdrawal; and Poland, the rhythm method. In general, countries in more developed regions favor the pill, followed by the condom and female sterilization. In underdeveloped countries, female sterilization is the most common method, followed by the IUD and the combination pill. We now discuss each of the major contraceptive methods in detail.

Table 14-1 Percentage Use of Various Contraceptive Methods by People Ages 15 to 44 in the United States

	1982	1995	2002
Using contraceptives	55.7	64.2	61.9
Tubal sterilization	12.9	17.8	16.7
Vasectomy	6.1	7.0	5.7
Combination pill	15.6	17.3	18.9
IUD	4.0	0.5	1.3
Diaphragm	4.5	1.2	0.2
Condom	6.7	13.1	11.1
Three-month injectable	NA	1.9	3.3
Withdrawal	1.1	2.0	2.5
Other methods	4.8	3.5	2.3
Not using contraception	44.3	35.8	38.1

Source: Vital and Health Statistics, U.S. Center for Disease Control.

Combination Pill

Scientific research on the possible use of hormones for human contraception began in 1897 when John Beard, an anatomist at the University of Edinburgh, surmised that the corpus luteum inhibited ovulation during early pregnancy. In 1898, an endocrine function of the corpus luteum was proposed by August Prenant of France. In 1931, the Austrian physiologist Ludwig Haberlandt, while doing research on the role of the corpus luteum in reproduction of mice and rabbits at the University of Innsbruck, suggested that hormonal sterilization could be an effective contraceptive method in humans. Soon after, research on laboratory mammals clarified that a certain ratio of the levels of estrogen and progestogen in the blood during the luteal phase of the estrous cycle blocks ovulation (see Chapter 3). This was an exciting revelation that led to discussions about the possible use of these steroid hormones for human contraception. The three natural estrogens (estradiol-17β, estriol, and estrone) were identified in 1929 and 1930, and the natural progestogen (progesterone) was identified in 1934. In 1937, Sir Charles Dodd synthesized an artificial estrogen (stilbestrol), which was followed by the introduction of more synthetic estrogens such as ethinyl estradiol and synthetic progestogens such as ethisterone and norethynodrel. An important development in the 1940s was the finding by Russel Marker in the United States that the sweet potato was an inexpensive source of the material needed to make synthetic steroid hormones. Thus, the idea of human steroidal contraception, as well as an inexpensive source of synthetic steroid hormones, was available by 1950. In the early 1950s, Margaret Sanger, a pioneer in the establishment of family planning in the United States, along with scientists Gregory Pincus and M. C. Chang, began serious discussions about the possibility of steroidal contraception.

By the mid-1950s, Gregory Pincus, as well as John Rock and Celso Ramon-Garcia, began clinical investigations on the effectiveness of the progestogen norethynodrel as a contraceptive agent. Then, in 1956 in Puerto Rico, John Rock, Gregory Pincus, and Edris Rice-Wray, a feminist and birth control advocate, began clinical trials of norethynodrel in pill form. The original preparation of norethynodrel contained small amounts of mestranol, an estrogen, and it was discovered that the presence of this estrogen increased the effectiveness of the pill. Thus, in 1960, the Food and Drug Administration (FDA) approved the use and marketing of the *combination pill* containing an estrogen *plus* a progestogen. This first combination pill contained 10 mg of norethynodrel and 150 μg of mestranol. By 1963, the progestogen content was reduced to 2 mg or less and the estrogen to 100 μg or less. The combination pill became the most popular contraceptive measure in the United States as well as in many other parts of the world. Although its use has dropped because of concern about possible side effects (to be discussed later), the combination pill is still the most popular reversible contraceptive method in the United States and in several other countries.

Ingredients

Today, there are over 30 different brands of combination pills. All of these contain a synthetic estrogen and a progestogen. The brands differ in the relative and absolute amounts of these compounds in each pill. The estrogen is either mestranol or ethinyl estradiol, and the progestogen can be one of eight types (ethynodiol diacetate, norethindrone, norethindrone acetate, norethynodrel, norgestrel, desogestral, gestodene, or norgestimate). Some synthetic estrogens and progestogens are more potent than others, and the choice of what brand to use depends on each individual woman's physiological makeup.

How the Combination Pill Works

The combination pill prevents conception by inhibiting ovulation. The combination of estrogen and progestogen inhibits tertiary ovarian follicle growth and prevents the surge of follicle-stimulating hormone (FSH) and luteinizing hormone (LH) secretion from the pituitary that normally causes ovulation during the middle of the menstrual cycle (see Chapter 3). In essence, this pill mimics the negative feedback effects of estrogen and progestogen present during the luteal phase of the menstrual cycle and in pregnancy. Even if the pill fails to prevent ovulation, conception does not occur because the pill renders the cervical mucus hostile to sperm transport. Even if some sperm reach the egg and fertilization occurs, the pill causes the uterine endometrium to be unreceptive to the embryo because the estrogen:progestogen ratio, although effective in blocking ovulation, is not quite correct to support implantation.

Use of the Combination Pill

The pills come in 21- or 28-day cycle packages. A woman takes one hormone-containing pill each day, usually beginning on the Sunday after her last menses began. She then continues for 21 days. During this time, ovulation is suppressed. She then either stops taking any pills for 7 days or switches to an inert "reminder" pill for the remaining 7 days of her cycle. After stopping the hormonal pill, menstruation begins. A woman is unlikely to get pregnant during the 7-day period when she is not taking the hormonal pills because she is menstruating for a few days and an embryo would not implant. Furthermore, 7 days is not long enough for follicular growth and ovulation to occur. If she misses 1 day of taking a hormonal pill, she can take two the next day with no loss of protection. If, however, she misses 2 or more days, she should stop taking the hormonal pills for 7 days and then begin a new pill cycle; other contraceptive protection (foam, condom, diaphragm) should be used during this 7-day waiting period, as pregnancy certainly is possible.

It should be mentioned that other forms of combination pills are available. The birth control pill discussed earlier is called monophasic because all of the active, hormone-containing pills have the same dose of estrogen and progestogen. Biphasic and triphasic pills were designed to more accurately mimic the changing hormonal levels in a natural cycle. Each package contains two 14-day or three 7-day sequences of pills with changing amounts of estrogen and/or progestogen. By 1998, triphasics made up 35% of the prescriptions for combination pills in the United States. They are as effective as the monophasic combination pill in preventing pregnancy. In 2003 the FDA approved Seasonale, a 91-day pill sequence that delivers a fixed dose of estrogen and progestogen for 84 days, followed by 7 days of inactive pills during which time menstruation occurs. Thus, a women taking the 91-day pill has a menstrual flow only once every 3 months.

Failure Rate

Because the combination pill prevents ovulation, conception, and implantation, it is a very effective contraceptive method. In this discussion, we refer to failure rates of contraceptive measures in 100 woman-years, or 100 WY (*Pearl's formula*). A *woman-year* is the use of a contraceptive device by a woman for 1 year (12 potential menstrual cycles). Thus, 100 woman-years of use could be, for example, the use of a contraceptive by 10 women for 10 years, or 100 women for 1 year. A failure rate of 1/100 WY means that 1 woman out of 100 would become pregnant using this device for 1 year. Failure rates of a contraceptive measure are given in a range, with the lower figure representing its theoretical effectiveness if used properly and the larger figure representing the actual failure rate. The latter figure is higher because people sometimes do not follow instructions or are not conscientious in their use of the device. The failure rate of the combination pill is 0.1 to 3.0/100 WY, with the higher figure present because some women forget to take, or choose to not take, a pill every day. This failure rate is very low in relation to rates of many other contraceptive devices. Contrary to popular belief, the use of common antibiotics does not decrease the effectiveness of the combination pill or vice versa.

Side Effects

Mild Side Effects Some women experience mild adverse side effects of the combination pill, usually during the first few months of use. Estrogen-related side effects can include nausea or (rarely) vomiting, bloating, fluid retention and slight weight gain, irritability, mood changes, headaches, breast tenderness, dysmenorrhea, uterine bleeding between periods (spotting), changes in sex drive, increased blood sugar, minor blood clotting, increased blood pressure, and suppression of lactation if taken while nursing. Some women develop a persistent yeast infection in the vagina due to the increase in blood sugar levels. Progestogen-related adverse side effects can be erratic menstrual bleeding, amenorrhea, breast shrinkage, breakthrough bleeding, and depression. Taking estrogen in birth control pills for more than 2.5 years can elevate the risk of developing gallstones. Not all women develop these symptoms, but a woman experiencing one or more of them may want to consult with her doctor about switching to a brand with a different dosage or a different kind of estrogen or progestogen.

About 70% of women who stop taking the combination pill are fertile in 3 months or less, although some require 18 months or more to become fertile again. There is no evidence that use of the combination pill produces infertility. Also, there is no need for women on the pill to periodically "take a break" from pill use.

Serious Side Effects Much has been written in the popular press about possible serious side effects of the combination pill, which probably explains the decrease in its use in the United States in recent years. Does the combination pill cause cancer? The answer is that there may be a small increased risk for breast cancer and also for cervical cancer. A few recent studies suggest that combination pill use increases the risk of breast cancer, especially in younger women. For example, one study found that pill use for greater than 6 months doubled the risk of early onset breast cancer in women under age 35. However, a recent study by the U.S. Centers for Disease Control on a larger number of women showed no increased risk of breast cancer associated with pill use. A Norwegian study of over 100,000 women found a 25% increased risk of breast cancer related to the estrogen dosage and duration of use of combination pills. Finally, a worldwide analysis of 54 studies involving 150,000 women found a slightly elevated risk of breast cancer in oral contraceptive users; this risk disappeared 10 years after pill use was discontinued. Long-term use (10 years or more) of the combination pill increases the risk of cervical cancer. This risk may decrease when women discontinue use of the pill. Finally, there is some evidence that combination pill use may slightly increase the risk of cancerous liver tumors.

Studies of the cardiovascular effects of the birth control pill have yielded mixed results. Use of the combination birth control pill can pose a risk to women who are predisposed to cardiovascular disease. This is the most serious potential complication of pill use. Cardiovascular disease can lead to deaths from heart attacks, blood clots in the lungs (pulmonary embolism), or bursting of a blood vessel in the brain (cerebral hemorrhage). Two of the rare but serious side effects of the combination pill are hypertension (a rise in blood pressure) and increased risk of blood clots in the legs, heart, lungs, and brain. Thus it is recommended that women who have a history of blood clots or vein inflammation, a history of heart disease or stroke, or those who have uncontrolled high blood pressure should not be on the pill. Some evidence suggests that the combination pill can raise the levels of triglycerides and cholesterol in the blood. Side effects of the pill related to cardiovascular disease were more pronounced in the early birth control pill formulations; they are now less of a concern with the newer, lower-dosage pills. In fact, a recent massive study of 162,000 women concluded that birth control pills can actually reduce the overall incidence of heart attacks. However, smoking cigarettes, especially in women over 35, increases adverse cardiovascular effects greatly in users of birth control pills. Despite warnings on pill containers about smoking and pill use, approximately 30 to 40% of women in the United States who take the combination pill also smoke.

The steroid hormones in the pill enter breast milk, but there is no evidence that consumption of these steroids by the infant harms its development. Nevertheless, a lactating woman should not take the combination pill because it suppresses milk production. Furthermore, although there is no evidence that taking the combination pill before or during the first few days of an unknown pregnancy harms the fetus, pregnant women should avoid use of the combination pill as a precaution.

Because of the potential minor and serious adverse side effects of the combination pill, women should consider some other contraceptive measure if they smoke or have a history of blood clotting, liver disease, high blood pressure, epilepsy, diabetes, heart attack, or stroke.

Beneficial Side Effects Despite these potentially harmful effects of the combination pill, the death rate from pill use is still less than that of pregnancy and birth. In fact, the combination pill has beneficial side effects other than preventing conception. For example, women using the pill have a decreased incidence of ovarian cancer, with a 50% lower risk after 5 years of use. The protective effect is increased with longer pill use, and the protective effect continues after use of the pill is discontinued. The combination pill also lowers the risk of endometrial cancer and, as with ovarian cancer, the protective effect increases with length of pill use and protection continues after a woman stops using the pill. Other beneficial side effects of the combination pill are alleviation of acne, excess body hair, breast cysts, heavy and irregular menstrual periods, and symptoms of premenstrual syndrome. The monthly menstrual flow is more regular and averages less in volume in pill users than in nonusers. Thus, there is often less severe cramping and the pill can also protect against anemia in women who experience heavy loss of blood in the menstrual flow.

Costs and Benefits

One must consider the costs and benefits of every contraceptive device (see Table 14-2). The combination pill has the advantage of being convenient in that its use does not interfere with ongoing sexual activity. Another benefit is its very low failure rate. It has been well tested for many decades. The yearly cost of the combination pill use is relatively low, but more expensive than some other means. On the negative side, the combination pill has potential minor and serious adverse side effects. It must be used daily and conscientiously. Use of the pill confers no protection against sexually transmitted diseases. A couple must weigh such costs and benefits when choosing any contraceptive method, including the combination pill.

Table 14-2 Summary of Common Contraceptive: Methods in the United States

Method	Approximate ideal failure rate (per 100 WY)	Approximate actual failure rate (per 100 WY)	Some advantages	Some disadvantages
Hormonal Combination pill (estrogen and progestogen)	0.1	3	Regular and lighter menses; decreased uterine cramping; no interruption of sex; protects against ovarian and endometrial cancer, pelvic inflammatory disease, premenstrual syndrome, ovarian cysts, benign breast lumps, heavy uterine bleeding, anemia, and ectopic pregnancy	Prescription; daily use; fertility often does not return for several months after use; cannot be taken when nursing; should not be used by women over 35 who smoke; does not protect against sexually transmitted diseases
Minipill (progestogen)	0.5	10	Can use if combination pill use is ill-advised (e.g., previous reproductive cancer, nursing); no interruption of sex	Prescription; daily use; may take awhile to become effective; irregular uterine bleeding; weight gain, no protection against STDs
Progestogen implant (e.g., Norplant)	0.3	0.3	Works up to 5 years; no interruption of sex; continual protection; no daily pills	Prescription; weight gain; may be infertile for awhile after implants are removed; often difficult to remove because of tissue encapsulation; no longer on market in U.S.; no protection against STDs
Progestogen injections (e.g., Depoprovera)	0.3	0.3	Continual protection; no interruption of sex;	Prescription; must get injection every 3 months; may be infertile 3–18 months after last injection; weight gain; no protection against STDs
Estrogen + progestogen injections (e.g., Lunelle)	<1.0	<1.0	Continual protection; no interruption of sex	Prescription; must get injection every month; cannot be taken when nursing; should not be used by women over 35 who smoke; no protection against STDs
Postcoital estrogen (or + progestogen) pill (emergency contraception or "morning-after pill")	1	25	Can be taken after the sexual act (before implantation)	Prescription; must take within 120 h of coitus; no protection against STDs

Barrier male condom (no spermicide)	2	12	Nonprescription; latex and polyurethane protect against sexually transmitted diseases; male participates; decreased risk of ectopic pregnancy	Unpleasant to some; interrupts sex; can be used only once; less penile sensation in some men, easy to forget; must use water-based lubricant; may slip off during withdrawal; can age or have structural defects
Male condom (with spermicide)	1	5	See information on male condom alone and on spermicides	See information on male condom alone and on spermicides
Female condom (with spermicide)	5	20	Nonprescription; protects against sexually transmitted diseases	Unpleasant to some; interrupts sex; can use only once; must be removed after coitus; see also information on spermicides
Diaphragm (with spermicide)	3	18	No hormonal alterations; reusable; immediately reversible	Prescription; requires fitting; unpleasant for some; can interrupt sex; easy to forget; must insert within 2 h of coitus and be left in 6–8 h after coitus, but no longer than 16 h; incomplete protection against STDs; see also information on spermicides
Cervical cap (with spermicide)	6	18	No hormonal alterations; reusable; immediately reversible	Prescription or nonprescription; may require fitting; can be difficult to insert; can irritate the cervix; can be dislodged; must be left in at least 6–8 h after coitus, but removed 24–48 h after coitus; should have Pap smear after cap is inserted; see also information on spermicides
Sponge (with spermicide)	6 (before childbirth) 9 (after childbirth)	18 (before childbirth) 28 (after childbirth)	No prescription; simple to use; effective for several acts of coitus; some protection against sexually transmitted diseases	Unpleasant for some; must expose to water; can tear during removal; must be inserted up to 24 h before coitus and left in at least 6 h after coitus, but removed within 30 h; can use only once; see also information on spermicides
Intrauterine device	0.8 (with copper) 2 (with progestogen)	4 (with copper) 6 (with progestogen)	No need to remember to use; no sex interruption; remains effective for several years	Prescription; painful to insert; can be expelled; must check string to see if it is in place

continued

14-2—*Continued*

Method	Approximate ideal failure rate (per 100 WY)	Approximate actual failure rate (per 100 WY)	Some advantages	Some disadvantages
Spermicides alone (foam; cream; jelly; suppositories; film)	3	21	No prescription; acts as lubricant; can kill some sexually transmitted organisms	Unpleasant for some; must be inserted properly; can interrupt sex; easy to forget; can sting or burn; must have sex within 30 min to 6 h of insertion (depending on brand) and must not douche for 6–8 h after coitus
Surgical tubal ligation	0.2	0.4	Continuous protection; protects against ovarian cancer; no sex interruption	Expensive; should be considered irreversible; no protection against STDs
Vasectomy	0.1	0.15	Continuous protection; no sex interruption	Expensive; can be irreversible; no protection against STDs;
Natural Natural family planning (rhythm)	1.9	20	Acceptable to religions having moral concerns about birth control	Requires high motivation and periodic abstinence; unreliable, takes time to learn; no protection against STDs
Withdrawal (coitus interruptus)	4	18	Acceptable to religions having moral concerns about birth control	Requires high motivation; reduces sexual pleasure; unreliable; no protection against STDs
Breast-feeding	15	50	Acceptable to religions having moral concerns about birth control; protects against breast cancer	Unreliable; no protection against STDs
No contraceptive use	85	85		

Minipill

The *minipill* first was marketed in 1973 as another reversible oral contraceptive measure. This pill contains only a small amount of progestogen and is taken daily, even during menstruation; regular menstruation often occurs every 25 to 45 days when on the minipill. It is recommended that a woman use a spermicide during her first 1 or 2 months of minipill use because the contraceptive action of this pill may take a while to be effective. The minipill can block ovulation; its major effects, however, are to render the cervical mucus hostile to sperm transport and to disrupt transport and implantation of the early embryo. One advantage of this pill over the combination pill is that there are fewer adverse side effects, although there is an increase in breakthrough bleeding, some variation in cycle length, amenorrhea, and ectopic pregnancy in some women. Also, the failure rate of the minipill (0.5 to 10/100 WY) is higher than that of the combination pill. For this reason, use of the minipill is not widespread. In situations in which the use of estrogens is ill-advised, however, a prescription for this pill may be indicated.

Chapter 14, Box 1: Where Is the Male Birth Control Pill?

Advances in our knowledge of the reproductive biology of women have led to several methods of female hormonal contraception, designed to inhibit ovulation or prevent implantation. Male reproduction also depends on hormonal controls, as learned in Chapter 4. So why is there no male birth control pill available? Part of the reason is that men produce sperm continuously, unlike females who release eggs on a cyclic basis. Therefore, men are constantly fertile, while women are fertile during only a short time each month. It is relatively easy to give females hormones that mimic pregnancy, lactation, or the luteal phase and thus block ovulation. More of a challenge is to shut down ongoing sperm production in men by hormonal manipulation. Also, some feel that the lack of robust research on a male contraceptive pill stems from the idea that men are not as committed to taking responsibility for birth control as are women. However, surveys indicate that many men in committed relationships are willing to take a contraceptive pill. The idea of a hormonal birth control pill for men has been around since the mid-1950s. More recently, research has also focused on immunological control of sperm production. We now look at the "state of the art" in male contraception.

First, the hormonal approaches: One of the initial ideas that comes to mind is to give extra testosterone to a male. This would lead to infertility (lack of mature sperm) because the extra testosterone would inhibit LH secretion, and FSH secretion to a lesser extent, by negative feedback on GnRH secretion and action. Although testosterone produced by the testes would drop, the contraceptive dose of testosterone in the man's blood would be sufficient to maintain his normal sex drive and abilities. Unfortunately, natural testosterone is degraded if given as a pill or by injection. However, a synthetic ester of testosterone, called testosterone enanthate, has proven to be effective as an injection or by skin patch.

However, several clinical trials of testosterone enanthate have shown that, although it lowers gonadotropin secretion, suppression of sperm production to the point of infertility (below 3 million sperm per milliliter of ejaculate) occurs in only 40 to 70% of Caucasian men and in 90% of Chinese and Indonesian

Continued on next page.

Chapter 14, Box 1 continued.

men. These are unacceptably high failure rates, especially when compared to those of the female birth control pills. Furthermore, there is fear that high levels of testosterone enanthate could damage the liver and also increase the risk of prostate cancer. Other side effects of extra androgen include acne, weight gain, and a reduction in high-density lipoprotein (HDL) in the blood, which is the "good" kind of cholesterol carrier. A new synthetic androgen, 7α-methyl-19-nortestosterone, halts spermatogenesis in rats without stimulating the prostate, so we may see this androgen as a possible substitute for testosterone enanthate in future human trials.

Because testosterone or testosterone enanthate alone probably are not sufficiently effective for male contraception, why not give an androgen with another hormone or hormone antagonist that also suppresses sperm production, thus lowering the failure rate to acceptable levels? This in fact has been one of the most popular approaches in recent clinical trials. For example, 24 men were given testosterone by skin patch while taking a daily pill of a progestogen (desogestrel). After 24 weeks, 23 of the men had sperm counts below the acceptable level for fertility. Another approach is to give a GnRH antagonist with testosterone; this has been shown to be an effective method, but it probably would not be used by many men because the antagonist is expensive and must be given as a weekly intramuscular injection. Finally, a combination of testosterone with an estrogen has been suggested; this would work, but not many men would agree to take it.

A good possibility for a male contraceptive method is similar to the intradermal progestogen implants (e.g., Norplant) used by many women. Clinical trials in the Netherlands have been conducted using such a method. In one of these trials, 120 male volunteers received hormonal implants. These tiny rods, inserted under the skin of the arm, released a constant amount of etonogestrel (a synthetic progestogen). These implants did block sperm production. However, the blood levels of natural testosterone were reduced significantly with this treatment, a side effect not acceptable to most men. Thus, the men in the trials also received injections of testosterone enanthate every 4 to 6 weeks to maintain their male secondary sex characteristics and sexual ability.

All of these "testosterone plus" methods have drawbacks. Will the side effects of excess androgen in the blood be similar to those during the use of anabolic steroids (Chapter 4)? What about the fact that it takes so long (about 3 months) for such methods to completely suppress sperm levels in ejaculate, and 3 to 4 months to recover from such suppression when the method is stopped? These problems and the fear of lawsuits from the drug companies that are developing these methods point to other, nonhormonal methods as better possibilities for male contraception.

A major effort recently has centered around developing contraceptive methods that use the immune system to attack the egg, the sperm, or the zygote. One potential male contraceptive involves a calcium channel found in the sperm tail. An influx of calcium through these channels allows the sperm tail to beat vigorously enough to penetrate the zona pellucida and reach the egg surface. Mice lacking the gene encoding this calcium channel are infertile because their sperm are very poor swimmers; otherwise, these mutant mice are healthy. Would disruption of sperm tail calcium channels provide contraception in men? There is hope that inducing formation of antibodies to these calcium channel proteins, by administering the protein into a man's body, would render his sperm incapable of fertilizing an egg. Because this protein is not expressed in any other place in the body besides sperm tails, there probably is a low risk of side effects. One problem, however, is the question of reversibility; how long will a man's sperm be rendered incapable of fertilization after antibodies to this protein are induced? An interesting sidelight is that calcium channel blockers commonly prescribed for high blood pressure and heart disease can leave men infertile. However, men taking these drugs have normal sperm motility, so the blockers do not seem to affect the sperm tail calcium channels. Instead, the drugs act to coat sperm with cholesterol, inhibiting their ability to fertilize an egg.

Continued on next page.

Chapter 14, Box 1 continued.

So the search goes on for a male birth control method that ideally would be effective (low failure rate), safe, reversible, self-administered, and would not require long periods of abstinence nor affect male traits. Some estimate that it will be 10 to 20 years before such a male method will appear. Until then, men who wish to take responsibility for contraception are limited to abstinence, coitus interruptus, condoms, spermicides, and vasectomy.

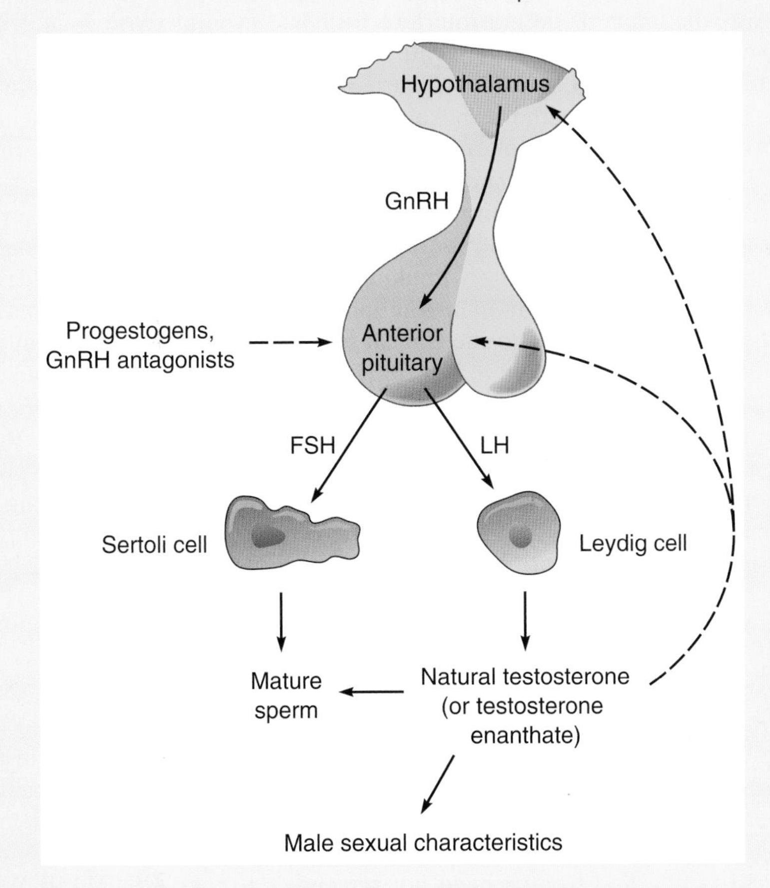

Summary of how hormonal male birth control methods would have contraceptive effects. Solid lines, stimulation; dashed lines, inhibition.

Intradermal Progestogen Implants

Intradermal ("within the skin") *progestogen implants* have been used by women as a contraceptive measure in various parts of the world and were approved in the spring of 1991 for use in the United States (under the brand name Norplant). Until it was taken off the market in 2000, 1 million American women used Norplant. Six matchstick-sized tubes were surgically inserted under the surface of the skin of the inner, upper forearm. The hormone (levonorgestrel) was then released in small amounts for up to 5 years. The mechanism of action of the progestogen implant is similar to that of the minipill. Side effects include irregular or prolonged uterine bleeding, amenorrhea, weight gain, acne, headaches, and increased risk of ectopic pregnancy in some women. Also, use of the implants may not be confidential as they might be visible under the skin. These implants, however, have some distinct advantages. They are convenient, and some evidence suggests that they lower blood cholesterol and triglycerides. Also, the implants do not interfere with, and even increase, milk production in lactating women. The failure rate of these implants is about 0.3/100 WY. After stopping Norplant, a woman will usually be fertile again in a few months.

Norplant was withdrawn from the market primarily because the intradermal progestogen implants often proved difficult to remove. In fact, removal can result in pain, scarring, and even muscle or nerve damage; lawsuits have been filed in this regard. Additionally, some women experienced more severe neurological side effects of intradermal progestogen implants, including dizziness, depression, moodiness, fuzzy thinking, and numbness in arms and legs, as well as weight gain, month-long menstruation, and an increase in fluid pressure within the brain. This had led to the clinical trials of implants that are smaller, fewer in number, and do not need to be removed because they dissolve in the body; these changes may alleviate the aforementioned problems. The FDA may soon approve an intradermal implant called Implanon, which is used in Australia, the United Kingdom, and elsewhere around the world. This is a single progestogen-containing rod that releases the drug for up to 3 years after being implanted under the skin. The lower level of progestogen could mean fewer side effects compared to Norplant.

Injectable Hormones

Injectable progestogen ("Depo-Provera, or the shot") has been used in 64 countries since the late 1950s but was approved only recently for use in the United States. The progestogen is injected every 90 days and works by blocking the LH surge, thus preventing ovulation. A combination estrogen–progestogen monthly shot (Lunelle) is also available. The failure of these injections is only about 0.3/100 WY, and women need not remember to take a daily pill. Women who cannot take estrogen or who are breast-feeding can use the progestogen-only form of injectable birth control. The disadvantages and side effects of Depo-Provera are similar to those of the intradermal progestogen implant. If a woman stops this method, she will usually be fertile 3–18 months after the last injection. The advantages and side effects of Lunelle are similar to those of the combination pill.

Transdermal Hormone Delivery

Estrogens and progestogens can be delivered across the skin (transdermally) through the *contraceptive patch*, which is currently marketed as Ortho Evra. The thin plastic patch is worn on the skin, where it provides a constant flow of hormones into the bloodstream. A woman places the patch on her torso, stomach, buttocks, or upper arm. Because it is highly adhesive, women can bathe, swim, and carry on normal activities while wearing the patch. It should be applied once a week for 3 weeks. It is important to remember to apply the patch on the same day for the 3 consecutive weeks. The fourth week is patch free, and menstruation occurs at this time. The patch appears to be as effective as the combination pill in preventing pregnancy. Like other combination hormonal contraceptive methods, ovulation is prevented and thickening of the cervical mucus inhibits sperm movement. Advantages include convenience and ease of use; there is no daily pill to take, no injections, and a woman can apply the patch herself. Side effects are similar to those of the combination birth control pill. Finally, another method of delivering contraceptive hormones is Nuring, a plastic ring containing estrogen and progestogen that is placed within the vagina. Hormones released from the ring are transported across the vaginal epithelium and enter the bloodstream.

Emergency Contraception

Emergency contraception (also called postcoital contraception or the "morning-after pill") involves the administration of estrogens and/or progestogens soon after having unprotected sex. There are over 20 kinds of pills used for emergency contraception, but only two have been marketed specifically for emergency contraception in the United States and approved by the FDA. Plan B is the first progestogen-only postcoital pill. One pill is taken as soon as possible after having sex; a second pill is taken 12 h later. The first pill can be taken as late as 120 h (5 days) after unprotected sex, but the contraception is most effective when started within 24 h. This treatment can inhibit ovulation, interfere with transport of the preembryo down the oviduct, or alter the endometrium so that implantation does not occur. Emergency contraceptive pills are not effective if implantation has already occurred. Other emergency methods, including Preven, are combination pills (and estrogen plus a progestogen), taken as described for Plan B except that two to five pills are taken first and the same dosage is taken again 12 h later. Methods using combination pills are simply a new use for some of the combination pills already on the market. After taking these combination pills, the preembryo dies and then is expelled, sometimes with slight uterine bleeding. Side effects, which are fewer with Plan B when compared with the combination pill methods, include nausea and vomiting. Failure rates of emergency contraception are fairly high. If started within 72 h of unprotected sex, Plan B is 89% effective, and the combination pill methods are 75% effective. Currently, in six U.S. states and 33 countries emergency contraceptive pills are available without prescription. The FDA is now considering over-the-counter sale of Plan B in all U.S. states. This important decision has the potential to prevent as many as half of the 3 million unplanned pregnancies in the United States each year.

Intrauterine Devices

An *intrauterine device* (IUD) consists of a T-shaped piece of flexible plastic placed through the cervical canal into the uterine cavity. Some IUDs have a wrapping of copper wire around the plastic or contain a progestogen. Probably the first IUDs were pebbles placed in the uterus of camels by Arabs and Turks. One of the first IUDs made for human use was invented by the German physician Richard Richter in 1909. It was ring shaped and made of silkworm gut. In the 1920s, Ernst Grafenberg (another German) introduced the first widely used IUD, a ring of gut and silver wire.

After a pelvic exam, the physician prescribes an IUD and places it in the uterus through the cervix. Some IUDs have an attached string that protrudes through the external cervical os. The woman should feel for this string after each menses to check if the IUD is in place. The string usually is not felt by the male during coitus. An IUD can be inserted at any time, although many physicians prefer to insert it immediately after cessation of menstruation.

We are still not absolutely sure how an IUD works. However, we now know that the primary effects occur prior to fertilization. Presence of the IUD reduces sperm motility and viability, resulting in reduced sperm counts in the uterus and cervix. It also affects the development and maturation of the ovum. Secondarily, there is some evidence that the presence of an IUD, as a foreign body, causes inflammation and thus increases the number of white blood cells in the uterus; these cells then may block implantation. *Copper IUDs* (e.g., ParaGard) have been shown to be more effective than plastic-only IUDs. Copper IUDs can be left in place for 10 years and appear to cause less uterine bleeding. The copper may have a toxic effect on spermatozoa and/or interfere with their capacity to fertilize an egg or it may interfere with the role of zinc in implantation. Copper IUDs are also employed postcoitally to block implantation; they are highly effective when used as emergency contraception. Another type of IUD (Mirena) contains progesterone. This device not only has the effects of other IUDs, but the progesterone prevents implantation and inhibits sperm transport through the cervical mucus. It can be left in place for 5 years. The failure rate of IUDs is very low (Table 14-2). Worldwide, it is by far the most popular reversible method of contraception. About two-thirds of the world's IUD users live in China.

IUDs have several side effects. Insertion can be painful. A woman receiving an IUD may suffer from abnormally high menstrual flow and spotting, especially in the first weeks after insertion, and the average monthly menstrual flow in IUD users is more that in non-IUD users. This bleeding and spotting is less in women using the progestone-treated IUDs. Abdominal cramps, especially in the first few weeks after insertion, are common. Also, women who have had repeated or severe venereal disease infections in the past should not use an IUD. If a vaginal infection is present during insertion of the IUD, the procedure could introduce microbial contamination into the uterus, where it can cause a dangerous pelvic infection. Of most concern are women who have asymptomatic chlamydia or gonorrhea infections. The most appropriate users of the IUD may be women in mutually faithful relationships.

In the early 1970s, an IUD called the Dalkon shield was used widely. It was withdrawn by the FDA in 1974 because of a higher than usual incidence of pelvic infections during pregnancies. Severe pelvic infection can take the form of *pelvic inflammatory disease*, with the danger of scarring of the uterus and oviducts, resulting in infertility (see Chapter 16). In June 1980, a woman in Denver was awarded \$6.8 million in a lawsuit against the company that manufactured the Dalkon shield. The woman used this IUD but later became pregnant and suffered a serious, almost fatal, spontaneous abortion caused by uterine infection. In all, over 50,000 women filed successful lawsuits against the company. The resulting public concern about the safety of the IUD significantly reduced the use of the IUD in the United States (less than 1% of U.S. women use the IUD, compared to about 12% in Europe). However, recent careful studies show a low risk of pelvic inflammatory disease among users of the newer types of IUDs and no increased risk at all for women in a monogamous relationship. Risk of infection is greatest in the first 20 days after IUD insertion, but after that time it remains low. Thus, IUD users have a slightly higher risk of pelvic infection than nonusers, but the absolute risk is very low. The incidence of pelvic infection is lower using the progesterone-containing IUD when compared with the copper-containing IUD. Recent studies show no apparent risk of tubal infertility associated with IUD use.

In about 7% of women, the IUD is expelled from the uterus, often because of a poor fit. Current IUD use does not increase the incidence of ectopic pregnancies, although past IUD use may elevate the risk slightly. If a uterine pregnancy occurs with an IUD in place, there is a chance that a miscarriage will occur because of uterine infection (*endometritis*). Thus an IUD should be removed once pregnancy is verified, even though IUD removal itself carries with it a minor risk of miscarriage.

Spermicides

The word *spermicide* means "killing of sperm." Actually, spermicides provide a mechanical barrier to sperm transport as well as having adverse effects on sperm. In the past, such substances as gum, animal dung, and various acidic compounds were used as spermicides. In late 19th century England, a combination of quinine sulfate, cocoa butter, and lactic acid was used widely. Present-day spermicides consist of an inert base and an active ingredient. The base provides a mechanical barrier to sperm movement and suspends the active ingredient. The active ingredient is usually nonoxynol-9 or octoxynol in present-day spermicides; these substances are surfactants that disrupt the structure of the sperm membrane.

Spermicides are placed high in the vagina next to the external cervical os. They can come in the form of creams and jellies, aerosol foams, and foaming tablets or suppositories. Most creams, jellies, and foams should be applied 1 h or less before coitus, although an "extra-strength" jelly can be used up to 6 h before coitus. Tablets and suppositories should be inserted between 15 min and 1 h before coitus. Douching should not be done until 6 to 8 h after coitus to allow the spermicide to be effective. A new application of spermicide should precede subsequent sexual activity.

The failure rates of spermicides range from 3 to 21/100 WY. There are several reasons for the high failure rates of spermicides under actual use, e.g., failure to use sufficient amounts of spermicide or not placing the substance high enough into the vagina. Also, spermicides should be applied before any penetration by the male because sperm can be present in the small drop of semen during stage 1 of ejaculation (see Chapter 9). Other reasons for the failure of spermicides are ejaculation before the spermicide has dispersed (sperm can reach the cervix in 90 s or less), coitus without reapplication of the spermicide, and lack of use during the "safe" period of the menstrual cycle.

From a cost–benefit point of view, spermicides are inexpensive and do not require a prescription; they can also prevent the spread of some sexually transmitted diseases (see Chapter 18). These benefits, however, are balanced by the relatively high failure rate and the fact that their use can interfere with sexual interaction. Spermicides are most effective when used with other contraceptive devices such as a condom or diaphragm. Use of a spermicide more than once a week increases the risk of urinary tract infection in the female. This is probably because surfactant spermicides also damage beneficial bacteria normally found in the vagina, allowing those that cause urinary tract infections to thrive. Nonoxynol-9 may increase the risk of a sexually transmitted infection by irritating the vaginal epithelium. There is no good evidence that spermicides affect an embryo in the case of pregnancy while using spermicides. Several new spermicides are undergoing clinical trials (see HIGHLIGHT box 14-2).

Douching (bathing the vagina with an acidic solution) is not an effective contraceptive measure because it usually is done after sperm have already entered the cervical canal. In fact, persistent douching may lead to vaginal irritation or infection (*vaginitis*) and can even lead to pelvic inflammatory disease.

Chapter 14, Box 2: Future Contraceptive Methods

The ideal contraceptive would (1) be highly effective, (2) be virtually free of side effects, (3) not require conscientious adherence to the method, (4) be long-acting, (5) be reversible, (6) help prevent transmission of sexually transmitted disease, (7) be inexpensive, (8) be used by either sex if possible, and (9) not interfere with sexual satisfaction. As noted in this chapter, there presently is no contraceptive method that satisfies all of these criteria, and scientists are continuously attempting to invent better methods. However, many drug companies are not interested in supporting these efforts because they feel that the market is well served, because they are worried about product liability and regulatory restrictions, and because it takes a tremendous amount of money and years of testing before a new method is approved for use. Nevertheless, some of the new contraceptive methods that could appear in the future are summarized here.

It is estimated that over 450 plant species are known to contain natural chemicals that have a contraceptive effect, and many ancient or present cultures use these as "folk-medicine" birth control. For example, tribal peoples in the state of Rajasthan in northwest India use the bark, oil, roots, leaves, and seeds of 18 different plants as male and female contraceptives or abortifacients. The Chinese are making a great effort to discover plant-containing contraceptive chemicals. For example, an extract from the roots of a twining Chinese vine, *Tripterygium wolfordii*, has been shown to reversibly inhibit sperm production in human males without effects on tesosterone, FSH, or LH secretion as well as on sex drive and the ability to have an erection. Could this be a future contraceptive for men? Perhaps, but one must approach this with caution. *Gossypol*, an extract of the cotton plant, was developed and tested on 100,000 Chinese men as a possible male contraceptive, but has been rejected because of its possible irreversibility and side effects, especially paralysis due to excessive amounts of potassium in the blood!

Continued on next page.

Chapter 14, Box 2 continued.

Possible New Contraceptive Methods

Category	Name	Male or female method	How given	How it works	Present status
Vaccine	Anti-hCG antibody	Female	Injection lasts 6 months; booster shot lasts 5 years	Blocks implantation, causes early miscarriage	Clinical trials
Vaccine	Antizona pellucida antibody	Female	Injection	Blocks part of zona that binds to sperm	Animal testing
Vaccine	Antisperm antibody	Female	Injection	Destroys sperm	Animal testing
Vaccine	Anti-FSH antibody	Male	Injection	Halts spermatogenesis with no effect on testosterone	Clinical trials (now used in India)
Hormonal	Melatonin plus progestogen (β-Oval)	Female	Daily pill	Blocks ovulation	Clinical trials (now used in Europe)
Hormonal	Androgen (testosterone enanthate)	Male	Weekly injections	Blocks spermatogenesis while maintaining androgen levels in the blood	Clinical trials
Hormonal	Androgen (testosterone enanthate) plus progestogen	Male	Implants, pills, patch, or injections	Blocks spermatogenesis while maintaining androgen levels in the blood	Clinical trials
Hormonal	GnRH antagonist plus androgen (testosterone enanthate)	Male	Daily pill	Blocks spermatogenesis while maintaining androgen levels in the blood	Clinical trials
Hormonal	Androgen (testosterone enanthate) plus estrogen	Male	Daily pill	Blocks spermatogenesis while maintaining androgen levels in the blood	Clinical trials
Hormonal	Intracervical progestogen ring	Female	Placed around cervical os	Inhibits sperm movement through the cervix	Clinical trials
Spermicides	E.g., PRO2000, cellulose sulfate, polystyrene sulfate	Female	Cover vaginal and cervical lining	Kill sperm and various STDs	Clinical trials
Spermicides	E.g., Buffergel, Acidoformgel	Female	Cover vaginal and cervical lining	Increase vaginal acidity	Clinical trials

Diaphragm

A *diaphragm* is a shallow cup of thin rubber stretched over a flexible wire ring; this device is placed in the vagina so that it covers the external cervical os (see Fig. 14-2). The diaphragm thus prevents sperm from entering the cervical canal. A spermicide should be used with the diaphragm for complete protection. In the distant past, diaphragms consisted of gums, leaves, fruits, seed pods, and sponges. Casanova (1725–1798) recommended using the squeezed half of a lemon as a diaphragm; it may have worked in some cases because of the acidity of lemon juice! Use of the diaphragm decreased in the United States between 1965 and 1974 but now is increasing again among women who wish to avoid the drawbacks of the combination pill or IUD.

A diaphragm is fitted and prescribed by a physician. A refitting should be done after delivery of an infant or if the woman's weight changes by 10 lb or more. Diaphragms can be one of three types. The coil-spring type can be

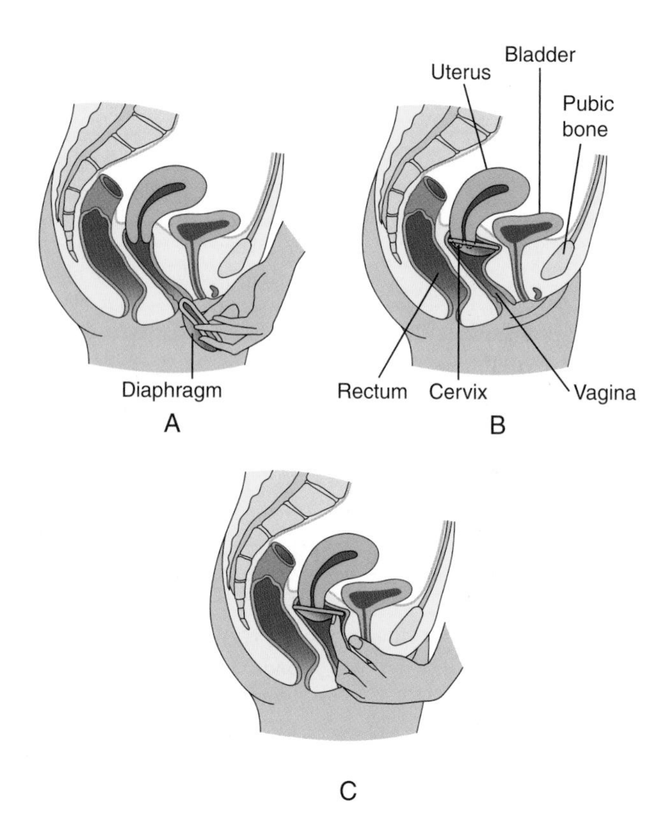

Figure 14-2 Insertion, proper position, and removal of a diaphragm. (a) The diaphragm is inserted manually into the vagina (inserters can also be used with some kinds of diaphragms). (b) The diaphragm is in proper position, fitting snugly between the posterior aspect of the pubic bone and the posterior vaginal fornix, completely covering the cervix. (c) The diaphragm is removed by hooking a finger under the forward rim (the edge behind the pubic bone).

compressed anywhere along its circumference while being inserted, either manually or with an inserter. The flat-spring type can only be compressed in one place before inserting manually or with an inserter. The third type, called an *arcing spring*, forms an arc when squeezed and can only be inserted manually. Most women find manual insertion easier with all types of diaphragms. Spermicidal jellies or creams should be applied to the inner side of the rubber dome and rim. The diaphragm should be inserted less than 2 h before coitus and left in at least 6–8 h after coitus to allow the spermicide to work. It must be removed, however, within 16 h after coitus to avoid irritation of the vagina or risk of toxic shock syndrome.

A diaphragm has several advantages: (1) It does not affect a woman's natural hormones; (2) it is relatively safe, although some users develop bladder infections; (3) it is easily reversible; (4) it usually is not felt by either partner during coitus; and (5) a diaphragm can be used by a lactating woman. A diaphragm does not protect against sexually transmitted diseases.

The relatively high failure rate of a diaphragm with a spermicide (3 to 18/100 WY) is due to several factors. For example, use of the diaphragm requires high motivation because it may disrupt sexual interactions. In some couples, the man inserts the diaphragm as part of foreplay, which may alleviate some of this disruption. Use of the diaphragm may also be avoided by some women because it is messy, the genitals must be handled, or it is difficult to insert. Some failures may occur because of not applying the spermicide or because of removing the diaphragm too soon. Finally, the penis may dislodge a diaphragm during coitus.

Cervical Cap

The *cervical cap* is a small silicone cup that blocks the cervix. It is held in place by suction. A cervical cap should be used with a spermicide. It must be left in place at least 8 h after coitus and must be removed within 48 h. Disadvantages when compared with the diaphragm are that it is often more difficult to insert and it should not be worn during menstruation. Because of possible effects of cervical caps on the cervix, a Pap smear should be done within a few months of beginning its use. Its actual failure rate is 6 to 18/100 WY in nulliparous women and 9 to 28/100 WY in multiparous women. Cervical caps are approved for general use in the United States, marketed as FemCap and Lea's shield. They last about 1 year, but should be examined regularly for holes or weak spots. Most caps need to be fitted and prescribed by a physician, but one, a disposable silicone rubber cap, can be obtained without a prescription.

Sponge Contraceptive

A form of *sponge barrier contraceptive* is available without prescription. One version of this device is a collagen sponge, about 2.5 by 1 in. in size. This sponge is moistened and placed in the vagina over the cervix. The sponge holds up to 3 oz of semen (the average amount of ejaculate is 0.2 oz), and it catches the semen and prevents penetration of sperm into the uterus. The sponge contains a spermicide. It is inserted up to 24 h before coitus and must be removed 30 h after coitus. The developers of this sponge say that there is no danger of toxic shock syndrome (see Chapter 3) with this device; however, because of a possible association, women should not use a sponge during menstruation. The actual failure rate is high (6 to 18/100 WY), and even higher if a woman has had a child (Table 14-2). This contraceptive, marketed as the "Today Sponge" has recently been approved by the FDA for return to the U.S. market after an 11-year hiatus.

Male and Female Condoms

Male Condom

Condoms (from Latin *condus* for "receptacle") are sheaths of latex rubber, polyurethane plastic, or (less commonly) animal cecum (a pouch on the intestine) that are placed over the penis to prevent semen from entering the vagina. Condoms of animal intestine or silk have been used for hundreds of years. After the discovery of vulcanization of rubber in 1944, latex rubber condoms were introduced. Condoms come in a standard 7-in. length; they can be dry or prelubricated with silicone or treated with a spermicide. One can even purchase condoms of different colors or those with rubber bristles or flaps ("French ticklers"). These novel condoms, however, often break and are not good contraceptive agents.

A condom is placed on the erect penis before coitus. About 0.5 in. is left at the end for collection of semen. After ejaculation, the penis should be withdrawn while it is still erect, holding the base of the condom with fingers to prevent loss of semen into the vagina. Petroleum jelly should not be used as a lubricant because it could damage the condom. Instead, water-soluble lubricants such as surgical jelly can be used with dry condoms, if desired.

Condoms have many advantages. They are inexpensive, disposable, and easily available without prescription. They are an easily reversible method of contraception and need be used only at the time of coitus. They allow men to take part in preventing pregnancy and the spread of sexually transmitted diseases. Condoms are manufactured with rigorous governmental inspection for damage, but they can develop cracks or holes if very old or if kept at a high temperature (like in the glove compartment of a car). A definite advantage of condoms is that they are a prophylactic against the transfer of sexually transmitted diseases including HIV; the use of condoms, however, does not totally eliminate the possibility of this transfer (see Chapter 18). The high failure rate of condoms (2 to 12/100 WY) is due to the fact that they might not be used in the passion of the moment or that they are put on or taken off improperly. The use of a spermicide with a condom reduces the failure rate to 1 to 5/100 WY. Rubber condoms may decrease sensation by the penis; this is less likely with the use of the newer, thin plastic types.

Female Condom

A *female condom* is available without prescription; it should be used with a spermicide. This is a 7-in.-long, soft, loose-fitting plastic (polyurethane) pouch that lines the vagina. A soft ring at each end holds it in place; the inner end is

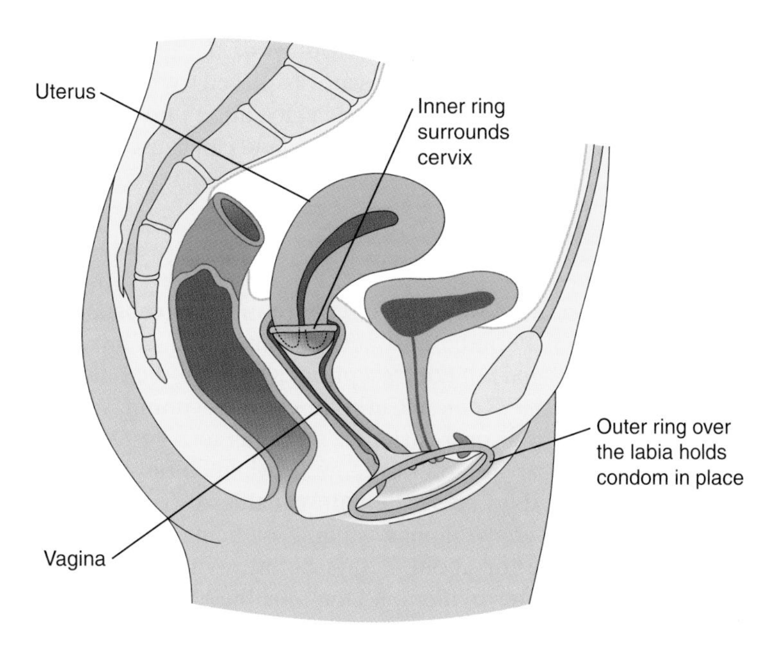

Figure 14-3 A female condom in place.

closed (covering the cervix) and the outer open ring partially covers the labia (Fig. 14-3). A woman can insert the condom up to 8 h before sex and it should be removed after coitus and before the woman stands up. It is disposable. The failure rate (5–20%) is relatively high. Use of the female condom, however, does lower the transmission of sexually transmitted diseases, including the AIDS virus (see Chapter 18).

Coitus Interruptus

Coitus interruptus, or withdrawal of the penis before ejaculation, is a time-honored contraceptive method and still is a commonly practiced method. It is more widely used in developed regions of the world, especially in Europe. Withdrawal is condemned by some religions, which may stem from the book of Genesis 38:8–9, where Onan was killed for "spilling semen on the ground." Disadvantages of this method include the fact that it requires high motivation and is highly frustrating to some couples. Another disadvantage is that any sperm deposited before withdrawal or left on the vulva wall during withdrawal could reach the cervix. These factors account for the high failure rate of coitus interruptus (4 to 18/100 WY).

Coitus Reservatus and Coitus Obstructus

More rare is the practice of *coitus reservatus*, which allows loss of an erection while the penis still is inserted, with no ejaculation. This is an inefficient method, however, because it takes a high degree of motivation, and pregnancy may occur from sperm present in the initial drop of the ejaculate. Even less common is *coitus obstructus*, in which pressure is placed, by squeezing, on the base of the spongy urethra, causing retrograde ejaculation into the bladder. This is relatively ineffective as a contraceptive measure and can cause bladder infection.

Natural Family Planning

Obviously, one sure way to avoid conception is to abstain from coitus, a practice called *celibacy*. Periodic abstinence is also used in *natural family planning* (or the *rhythm method*) of contraception, which are the only contraceptive methods condoned by the Roman Catholic church without special permission. These methods are based on restriction of coitus to a "safe period" of the menstrual cycle. The "fertile" period of the cycle, when coitus is avoided, is determined by the number of days sperm remain capable of fertilization (6–7 days) and the time that the ovum remains capable of being fertilized (1 day). Therefore, a woman is fertile for about 5 days before ovulation, the day of ovulation, and possibly 1 day after ovulation. One must be able to estimate the time of ovulation to use this method, and there are several ways that couples can do this. Before reviewing these methods, be aware that the use of rhythm methods of contraception requires instruction, daily monitoring, and a strong commitment from both sexual partners.

Calendar Method

Use of the *calendar method* of predicting the safe period began with studies by H. Knaus in Austria and K. Ogino in Japan in the 1930s. This method is based on the predicted time of ovulation (14 days before day 1 of the next cycle), taking into account individual variation in menstrual cycle length. To use the calendar method, a woman should keep track of her cycle lengths for at least a year. Then, 18 days are subtracted from the shortest cycle recorded and 11 days from the longest cycle recorded. Thus, the fertile period is defined. For example, suppose a woman, after recording her cycle lengths for a year, had a longest cycle of 35 days and a shortest cycle of 25 days. Her possible fertile period then would be from days 7 to 24 of her cycle. This obviously does not leave much time for sex! The failure rate using the calendar rhythm method (14 to 47/100 WYS) is very high (Table 14-2) due to the inaccuracy of the method, nonadherence by couples, and the possibility of coitus-induced ovulation (discussed in Chapter 3).

Basal Body Temperature Method

There is a 0.3 to 0.5 °C (0.5–1.0 °F) rise in *basal body temperature* immediately after ovulation during the menstrual cycle (see Chapter 3). A special basal body thermometer with a 96° to 100 °F range is used. Temperature should be measured soon after awakening in the morning, before becoming active or eating. The most effective way to use basal body temperature measurement as a contraceptive method is to limit coitus to a time beginning 3 days after the temperature rise and extending to day 1 of the next cycle. Failures (1 to 20/100 WY) occur because it is difficult to detect such a small rise in body temperature and because there can be variation in the pattern of this rise in one individual in different cycles.

Cervical Mucus Method

The amount and consistency of the cervical mucus change throughout the menstrual cycle (see Chapter 3). During the "wet days," immediately before and after ovulation, the mucus becomes more abundant, becomes clear and slippery (like raw egg white), and has a high degree of threadability as detected by placing it between two fingers. A woman is most fertile at this time. Before this stage, it is cloudy yellow or white and sticky. After the wet days, the mucus volume decreases and it is cloudy. Coitus should be avoided from the time that the wet days begin until the fourth day after the wet days end. This *cervical mucus method* is very inefficient (3 to 86/100 WY).

Sympto-Thermal Method

The *sympto-thermal method* (sometimes called "natural family planning") uses several indicators to detect the fertile period. These include basal body temperature, cervical mucus, breast tenderness, vaginal spotting, Mittelschmertz (pain of ovulation), and the degree of opening of the external cervical os. Because this method combines several indicators of impending or recent ovulation, its failure rate (about 1.9 to 20/100 WY) is less than for methods using basal body temperature or cervical mucus alone.

Is Breast-Feeding a Contraceptive Measure?

Breast-feeding can inhibit ovulation and produce postpartum amenorrhea if done frequently and consistently (see Chapter 12). The inhibition of ovulation by even a rigorous breast-feeding schedule, however, can become ineffective after about 6 to 9 months. After this time, ovulation and pregnancy can occur even if breast-feeding continues. In addition, ovulation occurs in 3 to 10% of breast-feeding women *before* their first postpartum menstruation. Thus, a breast-feeding woman should use another form of contraception. Combination pills, however, are not recommended for lactating women because they suppress milk production and the steroids can enter the breast milk. Other devices such as an IUD, diaphragm, or condom (with spermicide) are better for lactating women. The failure rate for breast-feeding as a contraceptive measure is about 15 to 50/100 WY.

Surgical Sterilization

Surgical sterilization is the leading method of contraception in the United States (Table 14-1) as well as worldwide. It is used by one in every four couples in the world (21% female sterilization, 4% male sterilization). More than one in four U.S. women using contraception and about one-third of all married women in China and India have been sterilized. Male sterilization by vasectomy is considerably less complicated and less expensive than female tubal sterilization. Despite this, five women around the world are sterilized for every man vasectomized; in the United States, nearly three times as many women as men have been surgically sterilized.

Even though some methods of surgical sterilization are reversible, they should be considered to be a permanent, irreversible form of contraception. Most physicians in the United States will not surgically sterilize anyone under 21 years of age for contraceptive reasons. A person contemplating surgical sterilization should ask: What are the possibilities of wanting children in the future? If divorced or your partner dies, would you want to remarry and have children? If you lost a child, would you like to keep the option to become pregnant again? Would *in vitro* fertilization (see Chapter 16) or adoption be acceptable alternatives?

Tubal Sterilization

Tubal sterilization involves surgery during which the oviducts are excised, plugged, cauterized, or tied so that gametes cannot pass through them. This type of sterilization is commonly called tubal ligation, which means "tying the tubes." Originally, the tubes were tied with thread; the first such operation was performed in England in 1823. Today, several methods of tubal blockage are performed, as well as several methods of exposing the tubes surgically. In the United States, tubal sterilization is the leading method of contraception in married women and in women over 30. More than 700,000 such operations are performed annually in the United States.

Several methods have been used to expose and view the oviducts before the actual sterilization procedure. One method uses a 10-cm abdominal incision, an operation called a *laparotomy*. This operation is done under a spinal or general anesthetic. After sterilization, the woman must be hospitalized for 2 to 5 days, and recovery takes several weeks. In 1961, a variation of this operation, termed a *minilaparatomy*, was introduced. In this method (also called the "band-aid" method), a small (2.5 cm) incision is made in the abdominal wall just above the pubic bone using local anesthesia. The complete operation (exposure of the oviducts as well as sterilization) takes about 10 min, and the patient, after a brief rest, goes home. Hospitalization usually is not required unless there are complications. In both laparotomy and minilaparotomy, gas (carbon dioxide) is pumped into the abdominal cavity using a syringe to render the tubes more visible to the surgeon so that the tubal sterilization can be performed. The gas is removed later.

Laparoscopy is another technique used to expose the oviducts (Fig. 14-4). This method, introduced in 1967, involves insertion of an optical tube (laparoscope) into the abdominal cavity to view the oviducts after gas is injected. Attachments to manipulate, anesthetize, and block the tubes can be contained in the optical instrument. Laparoscopy and subsequent sterilization take about 20 min and require a brief (1 to 2 nights) hospital stay.

Two other methods for surgical approach to the tubes also use an optical instrument. In one, an instrument is inserted through a 3- to 5-cm incision through the posterior vaginal fornix and into a blind peritoneal pouch between the anterior wall of the rectum and the posterior uterine wall. The oviducts

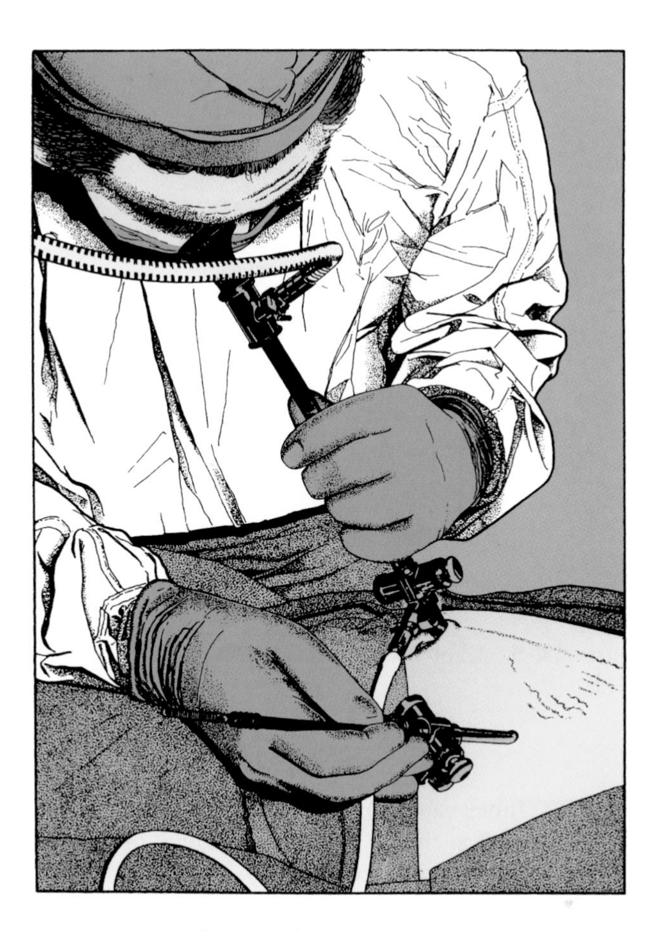

Figure 14-4 Laparoscopy offers a safe and simple way to perform tubal sterilization. Light from an external source is transmitted through glass fibers (enclosed in the tube in the upper part of the illustration). The surgeon has a clear view of the abdominal organs. The abdomen is distended with an inert gas to give the surgeon room to maneuver the instruments. In the surgeon's right hand is a specially designed forceps to manipulate the tissue. Another type of laparoscope contains an optical viewer, forceps, anesthetizer, and a clip or band applicator, all in one instrument.

are then pulled from their natural location into the region of incision and sterilization is performed. This procedure is called *culdoscopy*. A local anesthetic is used, and no exterior scar remains. A *colpotomy* is like a culdoscopy except that the tubes are not brought through the incision with forceps.

Once the tubes are exposed, they can be blocked in several ways and in several places. The infundibulum of the oviduct can be excised, buried, plugged, or capped. The ampulla and isthmus can be tied, cut, excised, clipped, banded, or buried. Near the uterotubal junction, the oviducts can be clogged or blocked with chemicals or plugged with an inert material such as silicone. The most common method is that of Pomeroy, first developed in 1930. In this method, each tube is cut and the free ends are folded back on themselves and tied

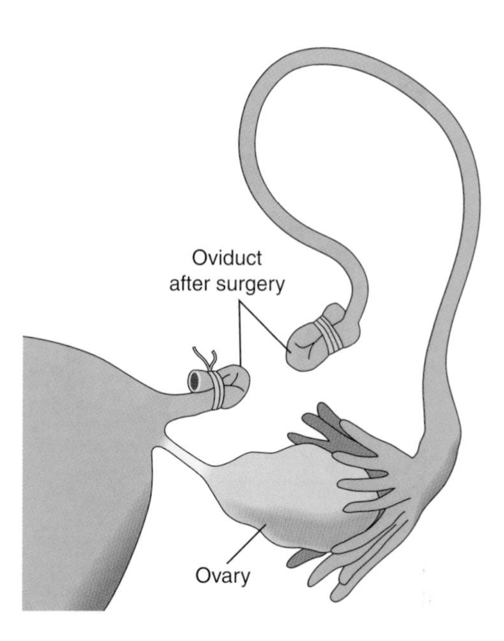

Figure 14-5 Tubal sterilization using the Pomeroy method. Each oviduct is cut and tied off.

(Fig. 14-5). Tubes can also be burned (electrocautery) or compressed by the use of clips or bands such as a silastic band called the "falope ring."

A no-incision method, called *hysteroscopic tubal sterilization* (Essure), is now available. A physician inserts small, soft metallic coils through the vagina and uterus, into the ends of the oviducts nearest the uterus. Here the coils induce formation of scar tissue, which closes the tubes. Three months later, the tubes must be tested for complete blockage by doing a *hysterosalpingogram*. In this procedure, the physician inserts a dye through the vagina into the uterus and then visualizes movement of the dye by X-ray. No dye should be seen flowing into the oviducts. During this 3-month waiting period, a woman must use another form of contraception. This method is safer and has a faster recovery period than other female sterilization techniques. No anesthesia is necessary, and the procedure can be performed in a physician's office. There is a risk that the coils may dislodge before scar tissue closes the oviducts completely.

Although tubal sterilization should be considered permanent, in reality it is reversible in some cases. The success of reversibility is close to 50% and is most successful when the original operation left a sufficient length of the oviduct intact to allow it to be reconnected. A woman may desire reversal, for example, because of divorce or remarriage, the death of her children, or an improvement in her economic situation. However, this microsurgery is considerably more involved and expensive and requires longer recovery time than the original sterilization procedure. Women undergoing reversal of tubal sterilization are at high risk for an ectopic pregnancy.

Advantages of tubal sterilization as a contraceptive measure are that its failure rate is low (0.2 to 0.4/100 WY) and it can be permanent, but disadvantages include that it is expensive and there are some dangers to health. The major risks are pain, discomfort, bleeding, and infection related to the incision, as well as possible side effects of anesthesia, if used. Another problem is a small risk that the tubes will partially reopen. Ectopic pregnancy could occur in women with incompletely blocked tubes. Rarely, sperm may enter a small gap in the oviduct and fertilize an ovum, with the embryo then implanting in the oviduct. As discussed in Chapter 10, tubal ectopic pregnancies can be dangerous. However, tubal sterilization has been shown to be associated with a reduced risk of ovarian cancer.

Hysterectomy

Hysterectomy (surgical removal of the uterus) is performed for various medical reasons, such as uterine cancer. In many instances in the past, however, hysterectomy was used as a method of sterilization. This is considered inadvisable now because it is an operation carrying all the risks of major surgery. Tubal sterilization is safer and less expensive. One advantage of a hysterectomy is that its failure rate is zero, although some women continue to use contraceptives to avoid ectopic pregnancies.

Vasectomy

A *vasectomy* is sterilization of the male by excising and tying the vasa deferentia. It is permanent birth control method for men who wish to enjoy sex without the risk of impregnating their partners, who have completed their families and want no additional children, or who wish to avoid passing on a hereditary illness. Such operations were first done in England and Sweden in 1894 and now are becoming a popular form of contraception throughout the world. About 500,000 vasectomies are performed each year in the United States, and more than 6 million per year are done in India.

Vasectomies are done in a doctor's office or clinic, using a local anesthetic. A small incision is made in the middle or on each side of the scrotum to expose the vasa deferentia. Once exposed, a small section of each vas is removed, and the loose ends are cauterized (burned) and tied back on themselves (Fig. 14-6). Trials of newer methods of occluding the vasa, such as the use of clips, plugs, or valves, have met with only limited success. After the vasa are blocked, the incision is sutured and the patient can leave. It may take several months, however, for the sperm present in the vasa to leave the body. Thus, a man should abstain from unprotected coitus until his sperm count (checked after about 15 ejaculations) is very low or zero.

About 2% of men receiving a vasectomy experience some minor side effects of the operation, such as skin discoloration, bruising, swelling, a blood clot in the area of incision, or a slight infection of the scrotum or epididymides. These discomforts usually are gone in 2 weeks.

A method of "no-scalpel" vasectomy was developed in China in 1974; since then, it has been performed on more than 10 million men in developing countries. In this method, one tiny puncture instead of two larger incisions (one on each side) is done, leading to fewer complications and quicker recovery.

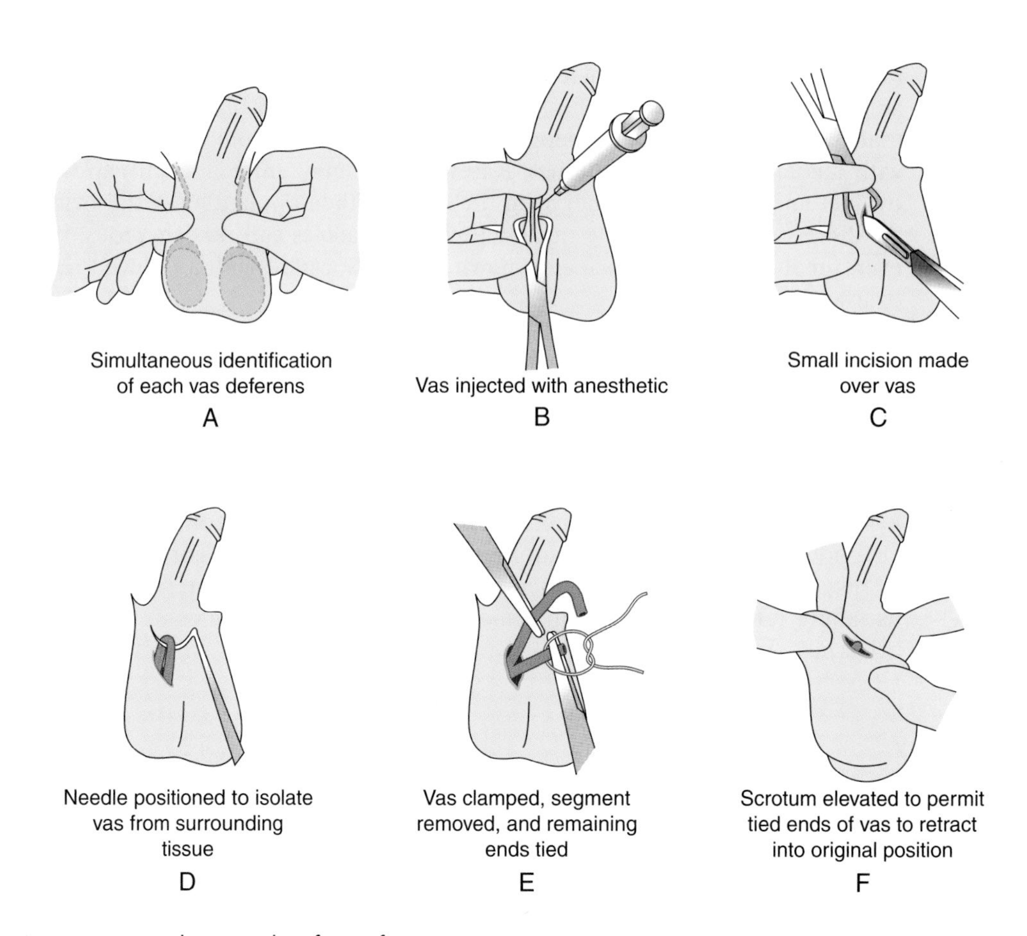

Figure 14-6 The procedure for performing a vasectomy.

Most evidence suggests that a vasectomy either has no effect on sexual motivation or actually improves it; there is no effect of vasectomy on the levels of testosterone in the blood. Also, semen volume and the experience of orgasm are virtually unchanged in vasectomized men. The failure rate of vasectomy is very low (about 0.1 to 0.15/100 man-years); failure is due to not cutting the vas or to anastomosis (growing together) of the cut ends of the vasa (Fig. 14-6).

After a vasectomy, the testes continue to manufacture new sperm, which then enter the vasa deferentia but are blocked from passing from the body during ejaculation. The blocked sperm die and are resorbed. In some men, however, sperm (or sperm fragments) may leave the vas deferens and enter the abdominal or scrotal cavity. In these cases, a man can form antibodies to his own sperm. Antisperm antibodies are found in one-third to two-thirds of vasectomized men; they are also present in about 2% of nonvasectomized men. If sperm leave the male tract at the site of the incision, a man could develop *autoallergic orchitis*, an allergic reaction at the point of incision. The lump formed (*sperm granuloma*) may have to be removed surgically. It is not known if there are general adverse effects of the presence of antisperm antibodies on a man's future health. Most evidence suggests no effect of vasectomy on later risk of testicular or prostate cancer.

For reasons similar to those discussed for women desiring reversal of sterilization, a man may decide to have his vasectomy reversed. Even though a vasectomy should be considered permanent, about half are reversible. The more extensive the pieces of vasa removed, the less likely that reversal can occur. Future development of valves to block the vasa may render this operation more reversible. Of men who have had successful reversal of their vasectomies, however, only about one-half are able to impregnate their partner. Whether or not this is because they possess antibodies to their own sperm is not clear.

For a summary of the contraceptive methods described in this chapter, see Table 14-2.

Psychology of Contraceptive Avoidance

Considering the wide (albeit less than completely satisfactory) range of contraceptive choices available, what explains the fact that half of the pregnancies in the United States are unplanned? Some couples experience an unwanted pregnancy even though they use a contraceptive measure consistently. This may be due to the inherent imperfection in the measure itself. Others, however, experience an unwanted pregnancy because of misuse or avoidance of contraception. This can happen because a couple is not well informed about contraceptive use and availability. In other instances, however, psychological patterns are responsible for contraceptive misuse or avoidance. For example, the man or woman may deny the possibility of pregnancy, uttering statements such as "It can't happen to me." A couple could skip the use of contraception in the heat of passion or they may deny responsibility for a pregnancy. Love may even be an excuse for an unwanted pregnancy in that a person may feel that giving one's partner a child is a "gift of love."

Some couples may not use contraceptive measures because of guilt. They may feel that contraceptive use labels them as promiscuous, and they may even accept pregnancy as a punishment for their sexual activities. Others, perhaps with certain religious beliefs, feel that having a child is the only natural and condoned result of coitus and that to use contraception is a sin. Some couples may avoid the use of contraceptives out of shame. For example, they may avoid birth control counseling because it would expose their sex life or their ignorance about sexual matters. Other people may relate their fertility to their self-esteem; for example, a man may feel that his fertility is a reflection of his virility and masculinity.

Unwanted pregnancies can also result from hostility in that pregnancy can be used as revenge against a partner or parents, or to recapture a lost partner. Women or men may be coerced into sex that results in a pregnancy. People even avoid or misuse contraception for the thrill of risk-taking because the passion of the moment is the only important experience. Finally, couples may not use contraception because they fear the possible adverse side effects on their physical health or sex life.

Although induced abortion is available when unwanted pregnancies occur, it is psychologically a much more difficult event than contraceptive use. Abortion puts more responsibility on the woman, medically it can be more involved than the use of contraceptives, and it is more expensive. Induced abortion should never be considered a form of contraception.

Choosing a Contraceptive

The ultimate contraceptive would be one that could be used by either sex, is free, never fails, is completely reversible, does not interfere with sexual activities, does not require conscientious adherence to the method, and has no adverse side effects. Although there is an ongoing search for new methods, no such contraceptive exists at the present time, with each measure having its unique portrait of advantages and disadvantages. When a person chooses to use a certain contraceptive, he or she must consider failure rate and possible adverse side effects. When a measure interferes with comfortable sexual activity, the problem often can be overcome by caring communication between sexual partners. The ability to protect against sexually transmitted diseases may be an important factor in choosing a contraceptive method. Couples may decide to use a combination of methods. For example, the best protection against pregnancy and sexually transmitted disease is the use of a highly effective method such as the combination pill or IUD plus a condom. Although cost may interfere with the use of certain measures, the potential financial costs of a pregnancy are even more burdensome. The contraceptive needs of an individual or a couple may change over their reproductive lives, especially the decision to use a reversible or a permanent method of contraception. Finally, a couple must realize that adherence to conscientious use of the method they choose is important if avoidance of pregnancy is a primary concern.

Chapter Summary

Contraceptive measures either prevent fertilization or disrupt preembryo transport or implantation. These measures are used to prevent unwanted pregnancies. The combination pill is the most popular reversible contraceptive measure in the United States. This pill contains a synthetic estrogen and progestogen, and it inhibits ovulation, sperm transport, and implantation. It has a very low failure rate. Minor side effects of this pill are common, but the only proven serious side effect relates to a higher incidence of cardiovascular system disease and deaths in pill users, especially those who smoke and are over 35 years of age. Some evidence shows that long-term pill use could increase the risk of cervical and breast cancer, but more studies are needed to confirm this. The minipill, intradermal progestogen implants, injectable progestogen or estrogen + progestogen, and transdermal estrogens and progestogens are other methods of delivery of these steroid hormones. Postcoital estrogen and/or progestogen provides a high dose of these hormones, which interferes with implantation.

Intrauterine devices are flexible plastic devices placed within the uterus; most contain copper or a progestogen. They probably work by causing an inflammatory response in the uterus, thus preventing implantation and/or causing "hostile" cervical mucus. Side effects such as abdominal cramping, increased menstrual bleeding, and pelvic infection are seen in some women who use IUDs. Also, IUDs are spontaneously expelled in a few women. The use of IUDs in the United States has declined because of the perceived threat of pelvic infection. However the IUD is the most commonly used reversible contraceptive method throughout the world.

Spermicides are placed in the vagina and act both as barriers to sperm transport and as sperm-killing agents. The failure rates of spermicides are relatively high, and they are best used with other contraceptive measures such as the diaphragm, cervical cap, sponges, or condoms.

The diaphragm is a shallow cup of thin rubber stretched over a flexible wire ring and is placed in the upper vagina to block the cervix. The effectiveness of this barrier to sperm transport is enhanced by the simultaneous use of a spermicide. Failure usually is the result of improper use. A major disadvantage of the diaphragm is that its insertion can disrupt sexual interaction. Advantages, disadvantages, and failure rate of the cervical cap are similar to that of the diaphragm. Another barrier method is the use of a male condom, a sheath of latex rubber, animal cecum, or polyurethane, placed over the erect penis before penetration. A female condom is also available. Failure of condoms is high, mainly because people do not use them properly or consistently. The use of a spermicide with condoms increases their effectiveness.

"Natural" contraception can involve inhibition of ejaculation by coitus interruptus (ejaculation outside the female) and coitus obstructus (retrograde ejaculation into the male bladder). Rhythm methods of contraception utilize the prediction of a safe period for coitus and avoidance of coitus during the fertile period of the menstrual cycle. Means for predicting the fertile and safe periods include the calendar method, basal body temperature method, cervical mucus method, and sympto-thermal method. Breast-feeding can be another form of natural contraception. All of these natural contraceptive measures, however, have very high failure rates.

Surgical sterilization is an effective, popular contraceptive measure. In females, the oviducts can be exposed by surgery using laparotomy, minilaparotomy, laparascopy, culdoscopy, or colpotomy. The tubes can then be blocked by cutting, electrocautery, tying, or using clips or bands. Hysterectomy is no longer used as a common contraceptive measure. In males, the vasa deferentia can be cut, burned, or tied in a procedure called vasectomy. Although surgical sterilization may be reversible in both sexes, it still should be considered a permanent contraceptive measure. Side effects of these procedures are minor, and the failure rates are very low.

New contraceptive measures may be developed in the future. These include the use of antibodies to reproductive hormones (hCG; FSH) or structures (e.g., zona pellucida), hormonal administration (melatonin plus progestogen; testosterone enanthate alone or with a GnRH antagonist, estrogen, or progestogen), or barrier (intracervical progestogen ring).

An unwanted pregnancy can result from the inherent failure of a contraceptive measure. More often, however, it is the result of misuse or lack of use of contraception for many, often subtle, psychological reasons. A couple must weigh the advantages and disadvantages of the various contraceptive measures, with special attention to failure rates and adverse side effects, before choosing.

Further Reading

Alexander, N. J. (1995). Future contraceptives. *Sci. Am.* **275**(3), 136–141.

Alexander, N. J. (1996). Barriers to sexually transmitted diseases. *Sci. Am. Sci. Med.* **3**(2), 32–41.

Alexander, N. J. (1999). Beyond the condom: The future of male contraception. *Sci. Am.* **10**(2), 80–85.
Althaus, F. A. (1990). Women who use the pill may be at increased risk of primary liver cancer. *Family Planning Perspect.* **22**, 137.
Althaus, F. A., and Kaeser, L. (1990). At pills' 30th birthday, breast cancer question is unresolved. *Family Planning Perspect.* **22**, 173–176.
Anonymous (1989). Pill users face increased risk of cervical cancer, but decreased risk of other genital cancer. *Family Planning Perspect.* **21**, 33.
Anonymous (1989). No added risk of strokes or heart attack is found among pill users. *Family Planning Perspect.* **21**, 381.
Beck, J., and Davies, D. (1987). Teen contraception: A review of perspectives on compliance. *Arch. Sex. Behav.* **16**, 337–368.
Christensen, D. (2002). What activates AIDS? The body's immune reaction to HIV is a double-edged sword. *Sci. News* **161**, 360–361.
Crust, M. P., *et al.* (1989). Administration of steroids by skin patches. *Res. Reprod.* **21**(4), 1–2.
Damaris, D. (2000). Male choice: The search for new contraceptives for men. *Sci. News.* **158**, 222–223.
Edwards, S. (1994). Women who have undergone a tubal sterilization have a reduced risk of contracting ovarian cancer. *Family Planning Perspect.* **26**(2), 90–92.
Ezzell, C. (1995). Sperm-stoppers: Researchers target the sperm cell membrane in male contraception. *J. NIH Res.* **7(3)**, 43–47.
Fackelmann, K. (1998). Stamping out syphilis: Can the United States finally vanquish this sexually transmitted disease? *Sci. News* **154**, 202–204.
Ferrier, V. (2001). A unisex contraceptive drug target. *Nature Cell Biol.* **3**(11), E249.
Harvey, S. M., *et al.* (1989). Factors associated with use of the contraceptive sponge. *Family Planning Perspect.* **21**, 179–183.
Jones, E. F., and Forest, J. D. (1992). Contraceptive failure rates based on the 1988 NSFG. *Family Planning Perspect.* **24**(1), 12–19.
Klitsch, M. (1988). FDA approval ends cervical cap's marathon. *Family Planning Perspect.* **20**, 19–40.
Klitsch, M. (1992). The new pills: Awaiting the next generation of oral contraceptives. *Family Planning Perspect.* **24**(5), 226–228.
Levinson, R. A. (1986). Contraceptive self-efficacy: A prospective on teenage girls' contraceptive behavior. *J. Sex Res.* **22**, 347–369.
McCormack, M., and LaPointe, S. (1988). Physiologic consequences and complications of vasectomy. *Can. Med. Assoc. J.* **138**, 223–225.
Mitwer, M. (1989). Oral contraceptive use linked to chlamydial, gonococcal infection. *Family Planning Perspect.* **21**, 282–283.
Mitwer, M. (1995). Vasectomy appears unlikely to raise men's chances of developing either prostate or testicular cancer. *Family Planning Perspect.* **27**, 95–96.
Pleck, J., *et al.* (1988). Adolescent males' sexual behavior and contraceptive use: Implications for male responsibility. *J. Adolesc. Res.* **3**, 275–284.
Potts, M. (1996). The myth of a male pill. *Nature Med.* **2**, 398–399.
Raloff, J. (1995). Drug of darkness: Can a pineal hormone head off everything from breast cancer to aging? *Sci. News* **147**, 300–301.
Reinisch, J. M., *et al.* (1995). High-risk sexual behavior at a midwestern university: A confirmatory study. *Family Planning Perspect.* **27**(2), 79–82.

Riddle, J. M., *et al.* (1994). Ever since Eve: Birth control in the ancient world. *Archaeology* March/April, 29–35.
Schmidt, K., and Hitchcock, P. J. (1999). HIV infection and AIDS. *In* "Encyclopedia of Reproduction," Vol. 2, pp. 624–636.
Tanfer, K., *et al.* (1992). Determinants of contraceptive choice among single women in the United States. *Family Planning Perspect.* **24**(4), 155–161.
Travis, J. (2001). Sperm protein may lead to male pill. *Sci. News* **160**, 228.
Trussell, J., and Grummer-Strawn, L. (1990). Contraceptive failure of the ovulation method of periodic abstinence. *Family Planning Perspect.* **22**, 65–75.
Trussell, J., and Stewart, F. (1992). The effectiveness of postcoital hormonal contraception. *Family Planning Perspect.* **24**, 262–264.
Trussell, J., *et al.* (1992). Emergency contraceptive pills: A simple proposal to reduce unintended pregnancies. *Family Planning Perspect.* **24**, 269–273.
Turner, R. (1995). No overall pill and breast cancer link, but young users' risk may be higher. *Family Planning Perspect.* **27**, 45–46.
Whitley, B. E. (1990). College student contraceptive use: A multivariate analysis. *J. Sex Res.* **27**, 305–313.
Wilcox, A. J., *et al.* (1995). Timing of sexual intercourse in relation to ovulation: Effects on the probability of conception, survival of pregnancy, and sex of the baby. *N. Engl. J. Med.* **333**, 1517–1521.

Advanced Reading

Aldous, P. (1994). A booster for contraceptive vaccines. *Science* **266**, 1484–1486.
Althuis, M. D., *et al.* (2003). Hormonal content and potency of oral contraceptives and breast cancer risk among young women. *Brit. J. Cancer* **88**, 50–57.
Amory, J. K., and Bremmer, W. J. (2000). Newer agents for hormonal contraception in the male. *Trends Endocrinol. Metabol.* **11**, 61–65.
Amory, J. K., and Bremmer, W. J. (2003). Regulation of testicular function in men: Implications for male hormonal contraceptive development. *J. Steroid Biochem. Mol. Biol.* **85**, 357–361.
Anderson, R. A., and Wu, F. C. W. (1996). Comparison between testosterone enanthate-induced azoospermia and oligospermia in a male contraceptive study. II. Pharmacokinetics and pharmacodynamics of once weekly administration of testosterone enanthate. *J. Clin. Endocrinol. Metabol.* **81**, 896–901.
Brachen, A., *et al.* (1990). Conception delay after oral contraceptive use: The effect of estrogen dose. *Fertil. Steril.* **52**, 21.
Brinton, L. A., *et al.* (1995). Oral contraceptives and breast cancer risk among younger women. *J. Natl. Cancer Instit.* **87**, 827–835.
Comhaire, F. H. (1994). Male contraception: Hormonal, mechanical, and other. *Hum. Reprod.* **9**, 586–590.
Cox, B. C., *et al.* (2002). Vasectomy and risk of prostate cancer. *J. Am. Med. Assoc.* **287**, 3110–3115.
DeRossi, S.S., and Hersch, E.V. (2002). Antibiotics and oral contraceptives. *Dent. Clin. North Am.* **46**, 653–664.
Dieckmann, K. P. (1994). Vasectomy and testicular cancer. *Eur. J. Cancer* **30A**, 1039–1040.
Dumeaux, V., *et al.* (2003). Breast cancer and specific types of oral contraceptives: A large Norwegian cohort. *Int. J. Cancer* **105**, 844–850.

Forinash, A.B., and Evans, S.L. (2003). New hormonal contraceptives: A comprehensive review of the literature. *Pharmacotherapy* **23**, 1573–1591.

Glasier, A. (2002). Contraception: Past and future. *Nature Med.* **8**(S1), S3–S6.

Grabrick, S. M., *et al.* (2000). Risk of breast cancer with oral contraceptive use in women with a family history of breast cancer. *J. Am. Med. Assoc.* **284**, 1791–1798.

Hankinson, S., *et al.* (1992). A quantitative assessment of oral contraceptive use and risk of ovarian cancer. *Obstet. Gynecol.* **80**, 708–714.

Healy, B. (1993). News from NIH: Does vasectomy cause prostate cancer? *J. Am. Med. Assoc.* **269**, 2620.

Herndon, E.J., and Zieman, M. (2004). New contraceptive options. *Am. Fam. Physician* **69**, 853–860.

Hubacher, D. (2002). The checkered history and bright future of intrauterine contraception in the United States. *Persp. Sex. Reprod. Health* **34**, 98–103.

Inter, T., *et al.* (1992). Oral contraceptive use and the incidence of cervical intraepithelial neoplasia. *Am. J. Obstet. Gynecol.* **167**, 40–44.

Jick, H., *et al.* (1981). Vaginal spermicides and congenital disorders. *J. Am. Med. Assoc.* **245**, 1329–1332.

Kinniburgh, D., *et al.* (2002). Oral desogestrel with testosterone pellets induces consistent suppression of spermatogenesis to azoospermia in both Caucasian and Chinese men. *Hum. Reprod.* **17**, 1490–1501.

McLachlan, R. I., *et al.* (2002). Effects of testosterone plus medroxyprogesterone acetate on semen quality, reproductive hormones, and germ cell populations in normal young men. *J. Clin. Endocrinol. Metabol.* **87**, 546–556.

Mieusset, R., and Bujan, L. (1994). The potential of mild testicular heating as a safe, effective and reversible contraceptive method for men. *Int. J. Androl.* **17**, 186–191.

Mills, J. L., *et al.* (1982). Are spermicides teratogenic? *J. Am. Med. Assoc.* **248**, 2148–2151.

Nager, C., and Murphy, A. A. (1989). Antibiotics and oral contraceptive pills. *Sem. Reprod. Endocrinol.* **7**, 220–223.

Qian, S. Z. (1987). *Tripterygium wilfordii*, a Chinese herb effective in male fertility. *Contraception* **36**, 335–345.

Orenstein, J. M., *et al.* (1997). Macrophages as a source of HIV during opportunistic infections. *Science* **276**, 1857–1860.

Ray, S., *et al.* (1991). Development of male fertility-regulating agents. *Med. Res. News* **11**, 437–472.

Ray, S. C., and Quinn, T. C. (2000). Sex and the genetic diversity of HIV-1. *Nature Med.* **6**, 23–25.

Schlesselman, J. J., *et al.* (1988). Breast cancer in relation to early use of oral contraceptives: No evidence of a latent effect. *J. Am. Med. Assoc.* **259**, 1828–1833.

Schwartz, J.L., and Gabelnick, H.L. (2002). Current contraceptive research. *Persp. Sex. Reprod. Health* **34**, 310–316.

Service, R. D. (1994). Barriers hold back new contraception strategies. *Science* **266**, 1489.

Service, R. F. (1994). Contraceptive methods go back to the basics. *Science* **266**, 1480–1481.

The Cancer and Steroid Hormone Study of the Centers for Disease Control and the National Institute of Child Health and Human Development (1986).

Oral contraceptive use and the risk of breast cancer. *N. Engl J. Med.* **315**, 405–411.

Trussell, J., and Vaughn, B. (1992). Contraceptive use projections: 1990 to 2010. *Am. J. Obstet. Gynecol.* **167**, 1160–1164.

Turner, R. (1994). Pill users may experience elevated risk of Crohn's disease, ulcerative colitis. *Family Planning Perspect.* **26**, 281–282.

Turner, L., *et al.* (2003). Contraceptive efficacy of a depot progestin and androgen combination in men. *J. Clin. Endocrinol. Metabol.* **88**, 4659–4667.

Tyrer, L. B. (1993). Current controversies and future direction of oral contraception. *Curr. Opin. Obstet. Gynecol.* **5**, 833–838.

Vessey, M. P., *et al.* (1983). Neoplasia of the cervix uteri and contraception: A possible adverse effect of the pill. *Lancet* **ii**, 930–934.

CHAPTER **FIFTEEN**

Induced Abortion

Introduction

Induced abortion is the termination of an implanted embryo by artificial measures. This chapter discusses the history of induced abortion in the world and, more specifically, in the United States. In addition, this chapter reviews reasons why a woman or couple would want an abortion and some of the controversies surrounding this procedure. Finally, the medical procedures used to induce abortion and some benefits and problems of these procedures are summarized.

Induced abortions have been used to terminate unwanted pregnancies for centuries. Many, often dangerous, "folk methods" have been used to induce abortion and some are still used in certain parts of the world. These include ingestion of certain plant compounds (see Chapter 10), physical trauma to the abdomen, and introduction of chemicals or sharp objects into the uterus. Throughout history, induced abortion, along with *infanticide* (killing the newborn), was practiced in various parts of the world to adjust sex ratios, control population pressure, eliminate deformed children, or terminate pregnancies resulting from incest or adultery. In more recent times, the laws of different governments concerning abortion have ranged from permissive to highly restrictive. In many countries, abortion has changed in recent decades from a largely hidden and disputed practice to a medically and socially acceptable procedure. In other countries, abortion is practiced in secrecy and peril. The worldwide yearly incidence of abortion is 35 per 1000 women. Today, 26% of all known pregnancies end in abortion.

Abortion laws vary widely among different countries. About one-third of the world's population live in countries with nonrestrictive abortion laws, allowing abortion on request up to the 10th to 24th week of pregnancy (defined in this chapter as the time since conception), with the usual time being the 12th week. The United States, as shown later in this chapter, is such a country at present. "Abortion on request" means that one can obtain an abortion simply by asking for one, regardless of the reasons. Another third of the world's population live in countries with moderately restrictive abortion legislation. Abortion on request is not allowed in these countries, but abortion is permitted for a wide range of medical, psychological, and socioeconomic reasons. Finally, about one-third of the people on earth live under restrictive abortion laws. That is, abortion is illegal unless there is a threat to the woman's health or life. In countries where abortion services are limited, illegal abortions are frequent and present a major health hazard, resulting in 100,000 deaths of women each year. In fact, approximately 44% of all abortions worldwide are performed illegally. However, in many countries that outlaw abortion, "menstrual regulation" is allowed. This is a procedure to induce a menstrual period in a woman who has missed one or (less often) two periods. It involves evacuating the uterus using a flexible cannula and a hand-held suction device. Typically it is allowed only if a pregnancy has not yet been confirmed. Most menstrual regulation procedures, however, are actually early first-term abortions. Although abortion rates vary greatly among countries, abortion is most common in areas experiencing high rates of unwanted pregnancies, regardless of whether or not abortion is legal. Thus, access to effective contraception is key to lowering the incidence of abortion around the world.

Induced Abortion in the United States

History of Abortion Legislation in the United States

The Comstock Law of 1873 made induced abortion illegal in the United States, and this was the situation until the late 1960s. Then, because of the increasing concern about women's rights, overpopulation, and the high frequency of dangerous illegal abortions, some states liberalized their abortion laws. From 1967 to 1970, about a dozen states enacted legislation allowing abortion in cases in which pregnancy posed a substantial risk to a woman's mental or physical health, when the fetus had grave physical or mental defects, or if the pregnancy had resulted from rape or incest. In 1970, four states enacted even more liberal laws, allowing early abortion simply on request.

In 1973, the U.S. Supreme Court (in the cases of *Roe vs Wade* and *Doe vs Bolton*) legalized abortion in the United States under certain conditions. Under the court's decision, pregnancies in the first trimester (first 12 weeks) can be performed on request, after communication between the woman and her physician. Second trimester abortions (the 13th through the 24th week), however, are restricted to cases of danger to the woman's physical or mental health, and states can regulate the qualifications of the people doing the procedures and the facilities where these abortions occur. States, however, must allow second trimester abortions related to fetal defects. In the case of third trimester abortions (after the 24th week), individual states can regulate if and how abortions are performed, except that these late abortions must be permitted in all states if there is danger to a woman's physical or mental health.

In 1977, the Hyde Amendment (sponsored by Representative Henry Hyde) was supported by the U.S. Supreme Court. This amendment prohibits the use of federal funds (Medicaid) for abortions except to preserve a woman's life. In early 1980, however, U.S. District Judge J. F. Dooling declared this amendment unconstitutional, primarily because it made abortion less available to poor people. About one-third of the more than 1.3 million legal abortions performed each year in the United States are done on women receiving welfare. On June 30, 1980, the Supreme Court ruled against Dooling's ruling and supported the Hyde Amendment by a 5-to-4 vote. The court also allowed Medicaid abortions only in cases of rape or incest or when at least two physicians agree that childbirth would endanger a woman's life or cause severe and long-lasting danger to her physical health. Thus, abortion is now legal (under the conditions noted earlier) in the United States but is available more readily to those who can pay for it.

At present, 32 U.S. states have laws that restrict the access of minors to abortion; these laws, which require parental notification or consent or a court order, are being tested in several courts. In 1989, the U.S. Supreme Court (*Webster vs Reproductive Health Services*) upheld a Missouri law that restricted the involvement of state personnel or hospitals in abortions and banned the abortion of viable fetuses (those that could survive outside the mother's body). By doing this, the Supreme Court allowed states to enact their own laws concerning abortions.

More recently (*Planned Parenthood of Southeastern Pennsylvania vs Casey*; 1992) the U.S. Supreme Court upheld (5 to 4) *Roe vs Wade*, but deleted the references to "trimesters." Instead, they ruled that (1) abortions are legal on request prior to the ability of a fetus to survive on its own (about 22 weeks from conception); (2) after this time, individual states can prohibit abortion except if there is danger to a woman's health or life; and (3) states can regulate abortion throughout pregnancy as long as there is no "undue burden" on a woman's right to terminate her pregnancy.

Needless to say, the legal aspects of induced abortion could change in the future. In 1983, the U.S. Congress debated the "Human Life Amendment" to the Constitution, which said that "the Congress finds that present-day scientific evidence indicates significant likelihood that actual human life exists from conception." It therefore would have banned all abortions. This amendment failed in the Senate, the vote being 18 shy of the two-thirds majority needed. If such an amendment passes in the future, any contraceptive measure that could act after conception could be banned, including IUDs, the combination pill, the minipill, intradermal and injectable progestogen, emergency contraception, mifepristone, and anti-hCG methods.

Present-Day Abortion Statistics in the United States

In America today, induced abortion is a common choice to end unwanted pregnancies among married and unmarried people. Each year, about 6 million pregnancies occur in the United States. Half of these are unplanned. Almost half of the unplanned pregnancies end in induced abortion. Currently, more than one in five pregnancies in the United States are terminated by abortion. Abortions in the United States account for a small proportion (3%) of the abortions performed worldwide each year.

Since abortion became legalized in 1973, medical records reveal changes in the incidence of this procedure in the United States. In 1973, there were 616,000 induced abortions; this number was about 1.2 million in 2004. The abortion rate (abortions per 1,000 women) reached a peak in 1980, and since then has decreased overall (Fig. 15-1). At the same time, the tendency has been to perform abortions earlier in pregnancy (Fig. 15-3). American women of all ages have abortions, but induced abortion is especially common in younger women (Fig. 15-2). About 1 in 12 teenaged women in the United States becomes pregnant each year. Over half of all abortions performed are on women under 25. Fortunately, the rate of teen pregnancy has decreased since 1990, and teen abortion is also on the decline. In addition to being young, women seeking abortions in the United States tend to be poor. Three of every four abortions are

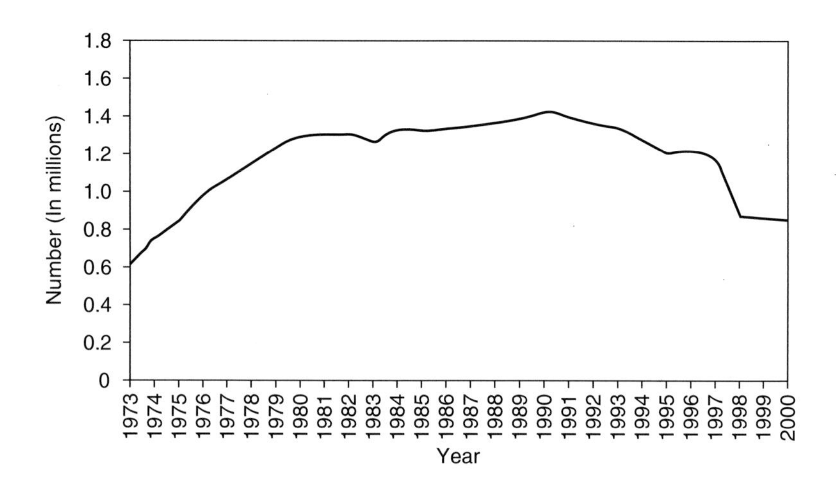

Figure 15-1 Number of abortions in the United States for the years 1973–2000. Note that the number of abortions has declined in recent years.

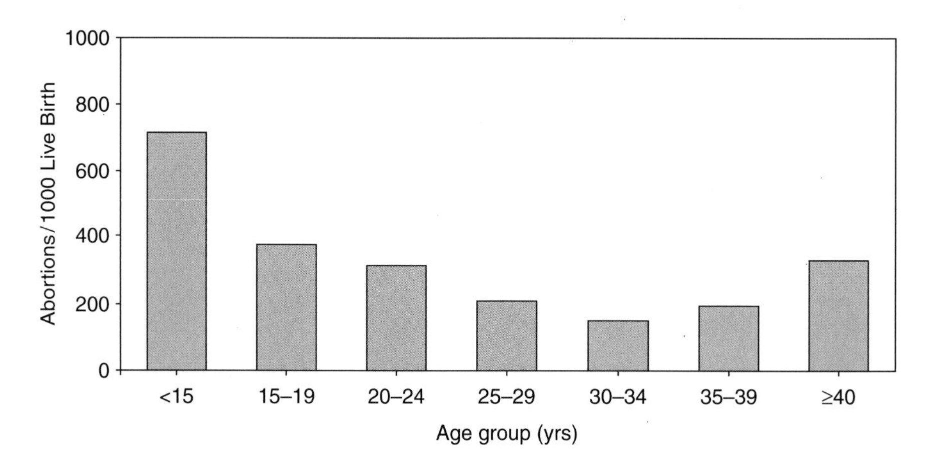

Figure 15-2 Abortions per 1000 live births in American women of different ages in 2000. Note that the most abortions occur at younger ages.

performed on unmarried women, and nearly two-thirds of women seeking abortions already have one or more children.

Why Women Have Abortions

There are several reasons why a woman who desires a child would want her pregnancy terminated by induced abortion. First of all, her physical health may be in danger because of being pregnant so the abortion is needed because of the threat of severe illness or death. Maternal disorders that threaten a

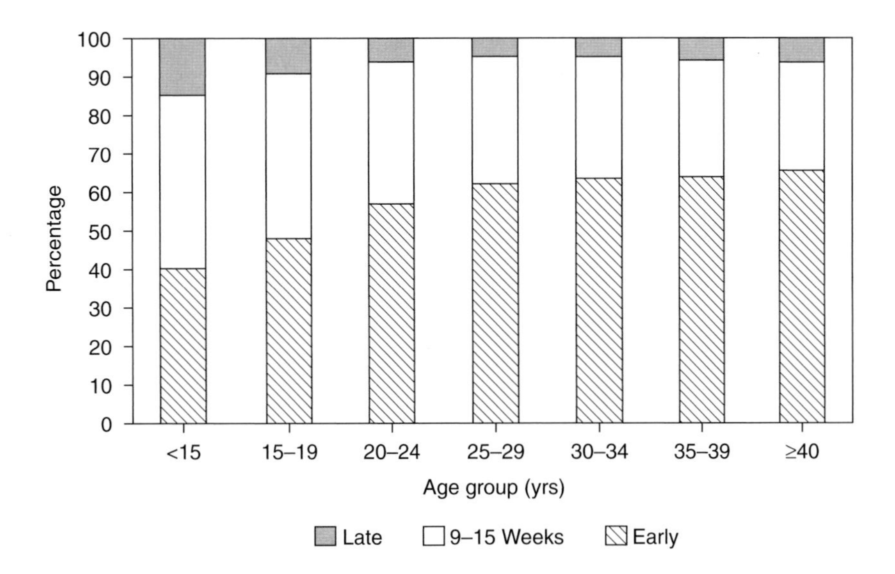

Figure 15-3 Percentage of American women of various ages who had an early (before 8 weeks), midterm (9–16 weeks) or late (after 16 weeks) abortion in 2000. Note that older women tended to have earlier abortions.

pregnant woman include diabetes, kidney disease, and disorders of the circulatory system. *Therapeutic abortion* is an induced abortion to prevent harm to the woman.

Second, amniocentesis, fetoscopy, ultrasound, or fetal blood tests may have indicated grave fetal abnormalities, and a woman or couple may choose to terminate the pregnancy to avoid giving birth to a severely handicapped infant. Amniocentesis is performed most effectively around the 16th week of pregnancy (see Chapter 10), and it can take several weeks for the results of some of the tests to be known. Therefore, in these cases, induced abortion in the late second trimester or early third trimester must be done at some discomfort or even risk to the woman, as discussed later. The results of chorionic villus biopsy, however, can be obtained earlier in pregnancy (see Chapter 10).

Although some feel that a deformed or handicapped child has a right to life and happiness, others feel that the impact of the birth of a grossly deformed child on parents' lives and finances, on siblings, and on the social institutions that care for such children justifies pregnancy termination. Inherent in the difficulty of choosing abortion in these cases is determining the potential severity of the fetal problem. Some defects are minor and some severe, with many somewhere in between.

In another category is the termination of unwanted pregnancies. Why would a pregnancy be unwanted? One reason could be because it occurred "by accident" as a result of coitus between a couple with no long-term relationship or, in extreme cases, it could have been the result of incest or rape. Another reason could be that a couple that originally wanted a child changed their minds; perhaps their financial situation worsened or a divorce is imminent. An unwanted pregnancy often occurs because a contraceptive measure failed or was not used properly or simply not used at all. Thus, termination of a potentially healthy fetus carried in a physically healthy woman can occur because of threats to the emotional health and happiness of the potential parent(s).

In considering an abortion, a woman or couple must weigh the various reasons for and against this choice. These reasons can be religious, psychological, moral, legal, or medical. With respect to medical considerations, the maternal death rate due to childbirth and that due to abortion both increase with the age of the woman. In general, abortion before the 17th week of pregnancy carries with it a lower maternal death rate than childbirth. The maternal death rate for abortions after the 17th week, however, is greater than for childbirth; both death rates are very low. An abortion is expensive, but the cost is low compared with the cost of childbirth at a hospital or birthing center. Abortion allows a woman to choose to be a parent or not after pregnancy has occurred and gives her a choice about raising a child she may not want.

Chapter 15, Box 1: The History and Ethics of Induced Abortion

As is known, people have different beliefs about the morality of induced abortion. Much of this controversy relates to the concept of *ensoulment*, or when in human development an embryo or fetus acquires a soul and becomes a person. The "event" of ensoulment (or in nonreligious terms, the acquisition of "humanhood") is morally important because killing a fetus before this time would simply be preventing a potential human life, whereas pregnancy termination after this time would be, in some people's opinion, murder. Different religious and other groups have defined ensoulment at various points in the cycle of pregnancy: (1) at the moment of conception, (2) at implantation (about day 7), (3) during the 9th week of pregnancy (when the fetus begins to look like a person), (4) at the 16th to the 20th week of pregnancy (when "quickening" occurs), (5) at the 24th week (when about 15% of fetuses can survive with special care if born), or (6) at the 28th week (when most fetuses can live with special care if born). Some people believe that a human does not acquire a soul until it interacts with another human after birth and some do not believe there is a soul.

The difficult decisions and controversies about induced abortion have a long history. Chapter 10 discussed how natural *abortifacients* (substances that induce abortion) contained in certain plants have been used for centuries. A work by the Chinese Emperor Shen Nung, 4600 years ago, contained a recipe to induce abortion using mercury. Greek, Hebrew, and Roman law generally protected the fetus only after it had recognizable features. The famous Greek physician, Hippocrates of Cos (late 5th century B.C.), is generally thought of as the "father of medicine." The "Hippocratic oath" has been the ethical guide for the medical profession to the present day. In its original form, this oath swore not to administer an abortive pessary (suppository); Hippocrates said nothing about oral pills or surgical methods. In fact, he recommended violent exercise to terminate a pregnancy. The Greek philosopher Aristotle (384 to 322 B.C.) wrote, "If conception occurs in excess of the limit so fixed . . . have abortion induced before sense and life have begun in the embryo." Later, Scribonius Largus, a Roman physician (1st century A.D.), interpreted the original Hippocratic oath to mean that physicians should condemn any method to expel the embryo or fetus. Roman stoic philosophers, however, believed that the soul first appeared when the newborn was first

Continued on next page.

Chapter 15, Box 1 continued.

exposed to cool air. Hebrew religious law did not consider pregnancy to occur until 40 days after conception.

The present Roman Catholic Church and other conservative sects of Christianity, as well as some branches of Judaism, are restrictive about induced abortion. In contrast, such religions as Buddhism, Islam, and Hinduism are more tolerant of induced abortion. In the mid-1800s, the Roman Catholic Church allowed abortion at any time up to day 40 of pregnancy, a point where it was believed ensoulment occurred. In 1869, however, Pope Pius IX declared that ensoulment occurs at conception and thus disallowed abortion at any stage of pregnancy. Pope Paul VI reaffirmed this opinion in his encyclical, *Humanae vitae*, in 1968. In this statement, he proclaimed that "the direct interruption of the generative process already begun and, above all, abortions, even for therapeutic reasons, are to be absolutely excluded as lawful means of controlling the birth of children."

In reality, it is difficult if not impossible to describe a point in human development when an embryo or fetus becomes a person. Indeed, it is more accurate to speak of the potential for human life, which is present in the egg and sperm as well as at all stages of embryonic and fetal development. Thus, many abortion decisions weigh the potential life and rights of the fetus against the potential length and quality of life, as well as the right to choose, of the woman. These always are difficult decisions, regardless of one's beliefs.

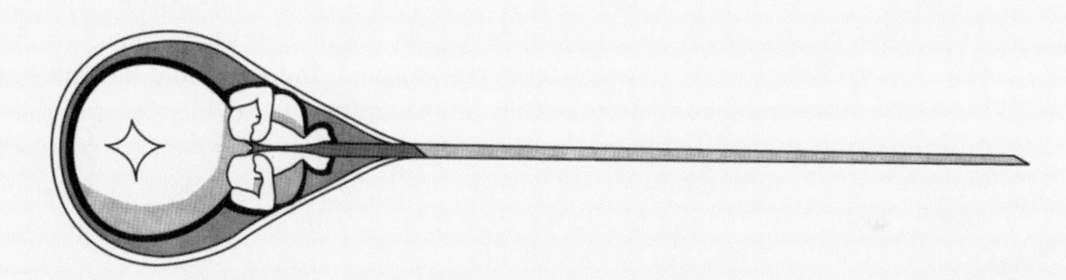

Based on anatomical observations of miscarried fetuses, Aristotle decided that the male fetus became "human" at 40 days after conception and the female at 80 days. He also believed that, before this time of acquisition of the human soul, the embryo and early fetus first went through a "vegetative soul" and then an animal "vital principle." Aristotle's teachings led many religions to accept abortion in the early first trimester. This view, however, was later contradicted by the belief of some embryologists, in the 1600s and 1700s, that they saw a fully formed little human (called a "homunculus") inside each sperm! Until it was disproven by better microscopes, this led to much indecision by religious groups as to when human life began and if and when induced abortion should occur.

First Trimester-Induced Abortions

Induced abortion during the first trimester of pregnancy (up to and including the 12th week) is relatively safe and simple from a medical viewpoint. In the United States, about 87% of all induced abortions are performed during the first trimester. We consider here termination of pregnancy during the embryonic period (up to the 8th week) and the early fetal period (9th to the 12th week) of the first trimester. Two major methods are used to terminate a first trimester pregnancy. A *surgical abortion*, the most common choice among U.S. women, uses aspiration and/or surgical instruments to empty the contents of the uterus. In a *medication abortion*, women are given drugs, either singly or in combination, that disrupt implantation, interfere with embryonic/fetal development, or cause uterine contractions.

Before any induced abortion procedure, a complete medical history of the woman should be taken. Certain previous or present conditions such as diabetes, circulatory disorders, epilepsy, venereal disease, uterine infection, excessive uterine bleeding, or previous cesarean section may require hospitalization for even the most simple abortion procedures. In addition, tests for pregnancy and the Rhesus factor should be performed. If the woman is Rh^- and the partner is Rh^+, the fetus may be Rh^+. In these cases, fetal blood could mix with the mother's blood during the abortion procedure, and the antibodies formed by the mother could damage the fetuses of future pregnancies. Therefore, such a woman is given Rhogam or a similar product to prevent antibody formation (see Chapter 10).

Vacuum Aspiration

Vacuum aspiration (or *vacuum curettage*) is the most frequently used method to perform first trimester abortions in the United States. It is commonly done from the 5th to the 11th week after conception (Fig. 15-4). An anesthetic is usually injected into the cervical wall, and a tube (*vacurette*) is placed into the uterus. This tube is connected to a suction device attached to a collection bottle (Fig. 15-5). After the tube is inserted so as to touch the amnionic sac, the suction is turned on and the collection bottles are examined for evidence of embryonic and placental tissue. If this material is not completely removed by suction, the endometrium can then be scraped with a curette (a metal scraper). The procedure usually lasts 3 to 10 min. Some women (about 3 in 1000) experience medical

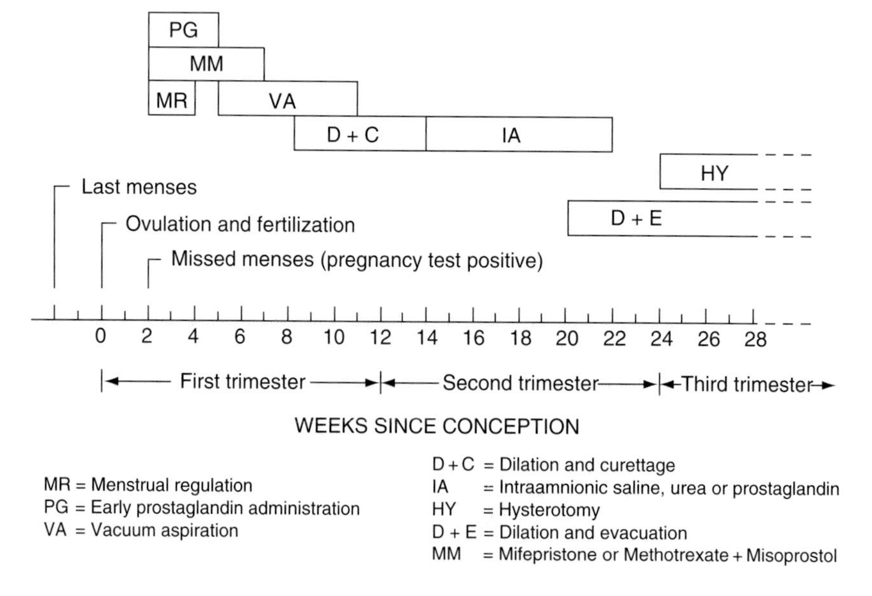

Figure 15-4 The usual periods of pregnancy during which various methods of induced abortion are utilized.

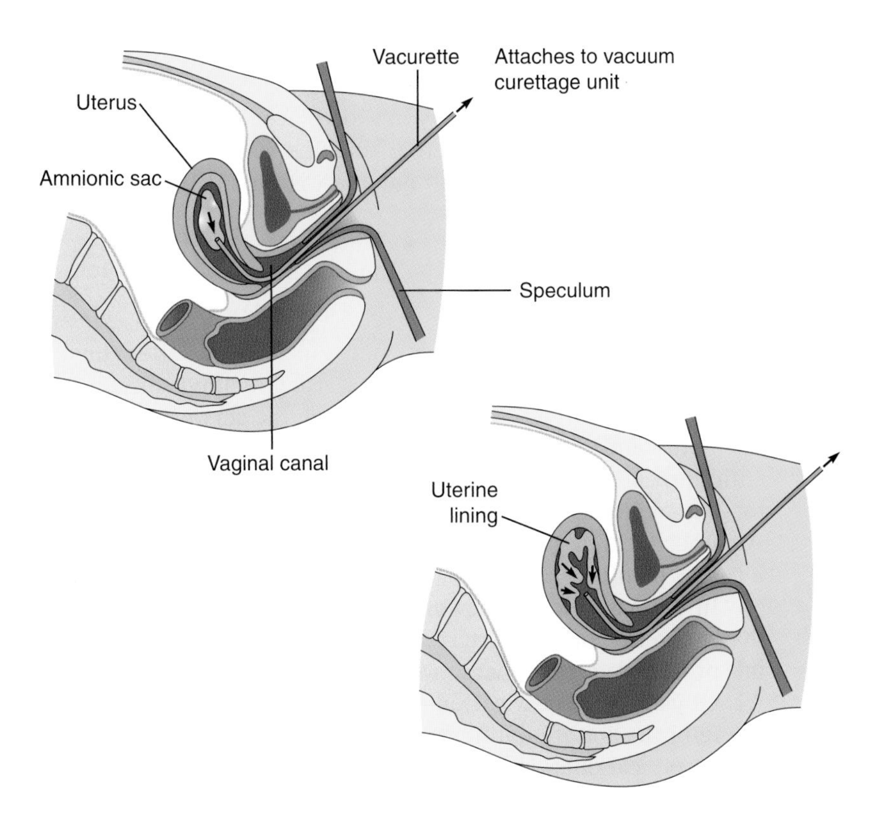

Figure 15-5 A vacuum aspiration abortion. A tube (vacurette) is inserted through the cervix and into the uterus, after the vagina is opened using a speculum (top). The embryo and uterine lining are then suctioned into a bottle (not shown) using a suction pump (bottom).

complications such as uterine cramping, bleeding, or infection after vacuum aspiration, but these complications occur with less frequency than after a *dilation and curettage* (D&C; see later). This method is replacing dilation and curettage as a safer and easier method of first trimester abortion. It is quicker, causes less uterine bleeding, is less painful, and has a lesser risk of uterine infection.

Menstrual regulation (or *menstrual extraction*) is a term used to describe an early uterine evacuation for women who have delayed menses, typically performed with a manual suction device. It is a relatively simple method to induce early abortion of the embryo. This method usually is done within 2 weeks of a woman's missed menstrual period, i.e., 4 weeks after conception (Fig. 15-4). No positive pregnancy test is needed for menstrual regulation, which is why this procedure is practiced commonly in countries that outlaw abortions of recognized pregnancies. Menstrual regulation is the least painful of all abortion procedures. It involves placing a flexible plastic tube (cannula) through the vagina and cervical canal and into the uterus. The other end of the tube is attached to a hand-held syringe, and the embryo and endometrial tissue are sucked into the cannula using this syringe. Many women do not require an anesthetic during this procedure, although some request one. After a 10- to 30-min rest in the clinic or physician's office, the woman can leave for home.

Some women experience minor adverse reactions to menstrual regulation, including uterine infection, uterine cramping, or excessive uterine bleeding. In a few cases, some of the embryonic or placental products are not removed or the embryo is missed altogether and the pregnancy continues. The maternal death rate due to menstrual regulation is only 0.4/100,000, making this a very safe procedure.

Dilation and Curettage

Dilation and curettage usually is done 8 to 14 weeks after conception (Fig. 15-4). In this procedure, the cervix is dilated using methods discussed later and then a metal scraper (*curette*) is inserted through the cervical canal and into the uterus. The fetus and endometrium are then scraped and removed (*curettage*). The entire procedure takes about 10 to 15 min. Some women require hospitalization for this method, whereas others have a D&C in a clinic or physician's office. A paracervical or general anesthetic often is used during this procedure because it can be painful.

In the past, metal instruments (*dilators*) were used to expand the cervix so that a curette could be inserted into the uterus. In about 5% of cases, however, this procedure caused laceration of the cervix or perforation of the uterine wall. Also, this type of cervical dilation can lead to an incompetent cervix, wherein future births are premature because the cervix is unable to support the growing fetus. Cervical dilation with metal instruments can lead to a higher incidence of premature births, stillbirths, and miscarriages in future pregnancies.

For these reasons, cervical dilation is now commonly done in other ways. In the past, items such as sponges, slippery elm bark, and inflatable rubber tubes were used to dilate the cervix. Prostaglandins can also be used to dilate the cervix. Today, however, the safest and most common method of cervical dilation is the use of *laminaria tents*. These are cylinders of dried and sterilized seaweed (*Laminaria japonica* or *Laminaria digitata*) that absorb water and swell to three to five times their original size when placed in the cervix. They are inserted in the cervical canal (Fig. 15-6) for a minimum of 3 to 5 h before curettage. The

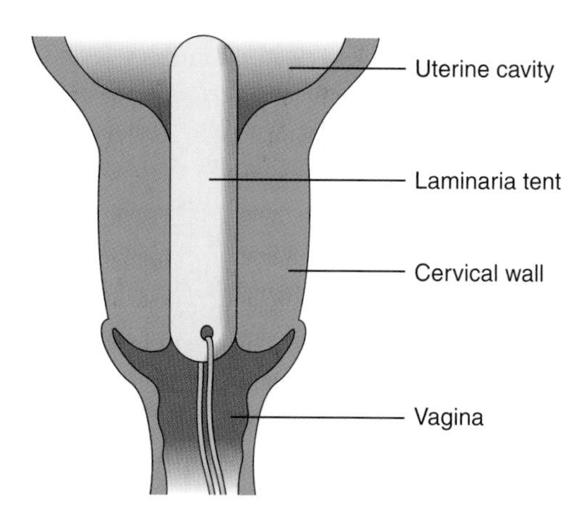

Figure 15-6 Laminaria tent inserted properly in the cervix.

woman is usually free to go home during this waiting period. The tent must be inserted properly or it could be expelled or enter the uterus. Some women experience uterine cramping while the tent is in place. However, laminaria tents are safe, with no danger of cervical laceration, uterine perforation, or incompetent cervix. They also are less painful and cause less bleeding than metal dilators. Laminaria tents are used not only for cervical dilation before curettage abortion, but also for other abortion procedures and for medical therapy related to dysmenorrhea or infertility because of a blocked cervical os. It should be mentioned that a new method of dilating the cervix could replace the use of laminaria tents in the future, especially for second trimester abortions. This device involves the use of a rubber catheter connected to an inflatable, pear-shaped balloon.

Prostaglandins

Another method of early first trimester abortion involves the use of prostaglandins. These drugs can be administered orally or as a vaginal suppository up to 3 weeks after a missed menstrual period (Fig. 15-4). Prostaglandins cause contraction of the uterine muscle and abortion of the fetus. Some advantages of this method are that it is nearly 100% effective if done less than 5 weeks after conception and it has little risk to the woman. No pregnancy test is required for this procedure. One disadvantage of this method is that prostaglandins can cause contraction of smooth muscle in the digestive tract, leading to nausea and diarrhea, but pretreatment with a digestive tract relaxant can alleviate this problem. Another disadvantage in some women is that uterine bleeding begins about 3 to 6 h after the procedure and can last 10 to 14 days. Also, the abortion can take several hours. However, newer and more potent prostaglandin analogs are now being tested. Prostaglandins are also used in combination with other drugs, as described next.

Medication Abortion: Mifepristone, Methotrexate, and Misoprostol

In 1988, an antiprogesterone compound, *mifepristone* (also called mifeprex, RU486, or "the abortion pill"), became available commercially in France. This chemical prevents progesterone from acting by blocking progesterone receptors. Therefore, it inhibits any progesterone function in the body and thus can block ovulation (because progesterone helps estradiol to cause an LH surge), prevent implantation (which depends on high levels of progesterone), and cause abortion (because the intact placenta depends on progesterone). Because its efficiency as an ovulation inhibitor or postcoital contraceptive method is too low (about 20/100 WY failure rate) to favor its use as a contraceptive, mifepristone is used primarily to induce abortion. An abortion performed using medications only (not surgical methods) is called a *medication abortion* (or *medical abortion*).

Mifepristone has been used to induce early abortion in France, Germany, England, Sweden, and China. In 1994, the French company owning the patent on the drug donated its rights to the Population Council, a nonprofit contraceptive research group in New York City. The approval of mifepristone in the U.S. led to other clinical uses in addition to abortion, as evidence shows that it can be used to treat ovarian and endometrial cancers, some types of breast cancers, fibroid uterine tumors, endometriosis, glaucoma (an eye disease), and serious psychotic depression, as well as in assisting labor induction.

Methotrexate, a drug used to treat cancer, ectopic pregnancy, and arthritis, can be used instead of mifepristone in combination with misoprostol to induce early abortion. Methotrexate inhibits DNA synthesis in rapidly dividing cells, in this case those of the trophoblast. This drug is injected, and then prostaglandin is administered as a pill or vaginal suppository 2 days later. Abortion usually then occurs within 24 h, accompanied by minor uterine cramping and bleeding. This method is about 96% effective. Mifepristone, methotrexate, and misoprostol reproduce many of the same physiological changes that occur during spontaneous abortion (miscarriage). Although medication abortions currently account for less than 10% of all abortions in the United States, they are becoming more common (see HIGHLIGHT box 15-2).

Chapter 15, Box 2: The Early Abortion Pill in the United States

In September 2000 the U.S. Food and Drug Administration approved the use of mifepristone (formerly called RU-486; brand name Mifeprex) to terminate very early pregnancy. Soon after approval, antiabortion activists predicted that this drug would be unsafe and would increase the number of abortions greatly in the United States. Prochoice activists, in turn, praised the approval of this drug in that it offered a more private and accessible early abortion method. Five years later, only a moderate number of women (about 360,000) had used this method in the United States, representing less than 10% of all abortions. Women have been slow to accept "medication abortion" as an alternative to surgical abortion, but its use is increasing as this method of ending a pregnancy becomes more widely understood.

The FDA approved the use of RU-486 for only the first 49 days of pregnancy, counting since the first day of the last menstrual period. This is a shorter period of time than the 63 days of pregnancy allowed in European countries. If a pregnant woman chooses this method, she first sees a physician, who determines her stage of pregnancy and prescribes medications. She then takes RU-486 (by pill) at home. Two days later, she goes to the doctor's office or clinic and takes a pill or inserts a vaginal suppository containing misoprostol. The RU-486, an antiprogestogen, causes the early embryo to detach from the uterine wall, and the misoprostol, an analog of prostaglandin E1, softens the cervix and stimulates uterine contractions. A miscarriage occurs, usually within 24 h after using the suppository; the uterine contents appear as a few large blood clots, which are expelled into the toilet. The woman then returns to the doctor's office 2 weeks later to determine if abortion was complete. This combination of two medications induces abortion about 95–97% of the time.

There are several advantages of this chemical method over early surgical abortion. It can be used at an earlier stage of pregnancy than surgical methods so women do not have to experience weeks of anxiety and nausea while waiting. This method can be done privately at home. Whereas 84% of U.S. women live in counties where there are no surgical abortion clinics, this medication abortion method would be

Continued on next page.

Chapter 15, Box 2 continued.

readily available to these women; it can be prescribed by a physician (OB/GYN, family doctor, etc.) in one's home town. Some women feel that this method of abortion is more "natural" than surgical abortion; it is similar to spontaneous abortion (miscarriage). With medication abortion, there is less risk of uterine infection than with surgical abortion. No surgical instruments are used, and women may feel more in control of the experience.

There are, however, some drawbacks to the use of mifepristone for early abortion. This method takes longer than the 1 day needed for a surgical abortion and requires two or more visits to the hospital or clinic. Also, side effects (uterine bleeding and cramping) are common; bleeding and spotting last 9 to 16 days. In 1% of cases, bleeding is severe and must be treated surgically. Nausea, diarrhea, and vomiting may occur, although they are generally mild. In a minority of women it may take 2 weeks or more for abortion to occur. In the 5% of women who do not experience complete abortion after taking these medicines, surgical abortion then needs to be done. The prescribing physician must be able to provide surgical intervention if needed. Also, this chemical method does not work on an ectopic pregnancy. Some women choose a surgical abortion because they would prefer to have a doctor or nurse present during the procedure and to know when and where it will occur. Finally, because chemical abortion using mifepristone has to be done within the first 49 days from the first day of the last menstrual period, a woman has only 21 days from a positive pregnancy test to decide if this method is right for her and to arrange the medical appointments.

Second Trimester Induced Abortions

Induced abortions in the second trimester of pregnancy (13th to 24th week) are more complicated and riskier to the woman. After the 19th week of pregnancy, the maternal death rate due to abortion is greater than that for childbirth. Still, more than 100,000 second trimester abortions are performed in the United States each year.

Why a Second Trimester Abortion?

Why would a woman wait until the second trimester to have an abortion? One answer may be that she did not have the money for a first trimester abortion or perhaps she lacked information on the availability of abortion. Also she may not have realized that she was pregnant, may have denied this fact, or thought the problem would "go away." More important, she may have developed some maternal disorder related to her pregnancy that did not appear until this time or it may have been diagnosed during midpregnancy that the fetus had a severe abnormality. New ways of testing the DNA of a fetus could identify abnormal genes, which in turn could cause difficult choices about terminating pregnancies. Finally, she, alone or with her partner, may have decided against childbirth for various emotional or psychological reasons.

From the 12th through the 15th week of pregnancy, the uterine wall is thin and susceptible to perforation. Also, the placenta is now highly vascular, and a

D&C or vacuum aspiration could cause excessive uterine hemorrhage. Some physicians will do a D&C or vacuum aspiration abortion up to the 15th week (Fig. 15-4), although this requires careful training and experience.

Intraamnionic Saline

One method used to induce abortion in the second trimester is injection of *intraamnionic saline*. In this method, usually done in a hospital using a spinal or general anesthetic, the cervix first is dilated with a laminaria tent. Then, a long needle is inserted through the abdominal and uterine walls and into the amnionic sac. After about 250 ml of amnionic fluid is removed, a syringe is used to inject saline solution into the amnionic sac (Fig. 15-7). The saline solution kills the fetus and induces delivery within 24 to 48 h. In some cases, a solution of urea or glucose is injected instead of saline. Saline, urea, and glucose cause uterine contractions by inducing uterine secretion of prostaglandins. Oxytocin can also be administered to facilitate a saline-, urea-, or glucose-induced abortion.

There are several disadvantages to this method of induced abortion. In 10 to 16% of saline-induced abortions, the placenta is not delivered with the fetus and must be removed using a curette. In other cases, the fetus is born alive but soon dies of respiratory failure. Also about 34 per 1000 women develop complications due to misinjection of saline into a blood vessel, uterine infection, or uterine hemorrhage. Delivery of the fetus after saline, urea, or glucose administration can be similar to normal labor with its risk of maternal complications. The maternal death rate after saline abortions is about 12 to 18 per 100,000.

Prostaglandin Injections

Prostaglandin injections were introduced in 1970 as a method for second trimester abortions and are now the most widely used procedure in the United

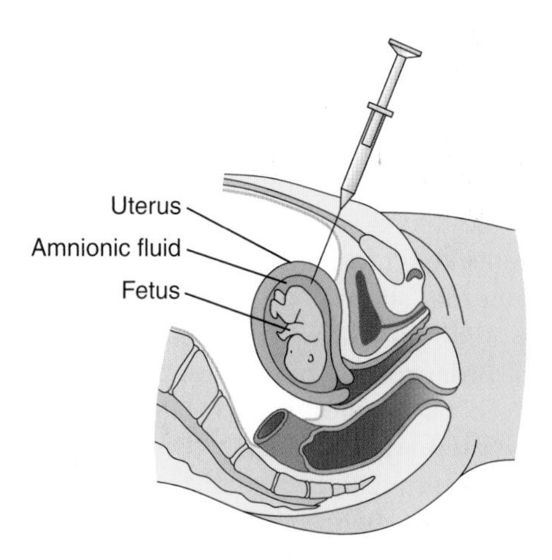

Figure 15-7 Diagram of administration of intraamnionic saline, urea, glucose, or prostaglandin to induce second trimester abortion.

States at this stage of pregnancy. These drugs work best from the 15th to the 20th week of pregnancy. The procedure is similar to that used during saline-induced abortions, except that no amnionic fluid is withdrawn and only a small volume of fluid is injected. The prostaglandin can be injected into the amnionic fluid or between the fetal membrane and the uterine wall. The latter "extraamnionic" route, using a needle passed through the cervical canal, requires a lower amount of prostaglandin and has fewer side effects than the "intraamnionic" route. Vaginal suppositories containing prostaglandins are also being used for second trimester abortions and are considered 100% effective. Urea, saline, and oxytocin can be used with prostaglandins. The combination of mifepristone plus prostaglandins is also being used for second trimester abortions.

The use of prostaglandins for second trimester abortions has certain advantages over the use of saline. Primarily, it is faster, with delivery occurring after 6 to 8 h if the cervix is dilated with a laminaria tent. Also, there are fewer complications. However, some women develop diarrhea and nausea because the prostaglandins cause contraction of smooth muscle in the digestive tract. Another disadvantage of prostaglandin-induced second trimester abortion is that the fetus may be born alive.

Dilation and Evacuation

Dilation and evacuation (D&E) is performed after the 12th week of development. In this abortion procedure, the cervix is dilated using a laminaria or other method and then the uterine contents are removed using a combination of suction and instruments (curettes and forceps). The procedure can take up to 30 min. Anesthesia and/or analgesics are often used. D&E abortion is a safe and effective method when performed by trained, experienced providers. It is used in the United States for the majority of abortions at greater than 12 weeks since conception.

Dilation and Extraction

Intact dilation and extraction, or D&X ("partial birth abortion"), is usually performed after 20 weeks gestation (in the late second trimester and early third trimester). The fetus is partially delivered feet first and then its brain is removed by suction while its head is still in the birth canal. The entire fetus is then removed. This procedure is done if the fetus is already dead, if it has severe hydrocephaly, or if continuing a pregnancy would put the mother's life in danger. It has been used very rarely as an elective procedure. After much legal and public debate, dilation and extraction abortions were banned by the United States Congress and president in late 2003.

Third Trimester-Induced Abortions

Third trimester abortions (after the 24th week of pregnancy) are rare in the United States, mainly because after this time the fetus, with special care, has the potential to survive on its own. *Hysterotomy* is the usual procedure used to induce abortion in the third trimester, although saline, mifepristone, and/or prostaglandins, as well as dilation and extraction, are also used. Hysterotomies can be performed from the middle of pregnancy to the early part of the third trimester. After the fetus can survive on its own (24th week on), induced abortion by any means is rare. A hysterotomy is major surgery done under anesthesia of the pelvic region or general anesthesia. It often is called a "minicesarean" because the procedure is similar to a cesarean delivery. Hospitalization is required for 4 to 6 days. Minor or major complications of this surgery occur in about 23 to 51% of women, and the mortality rate is about 45 to 271 per 100,000. One negative aspect of these abortions is that the fetus may be delivered alive and then must be left to die of respiratory failure, a disturbing task for all persons involved.

Folk Abortifacients

As mentioned in Chapter 10, there are herbal substances that can induce abortion. For example, pennyroyal (an aromatic member of the mint family) is at present used as an "underground" method to induce abortion; it is brewed as a tea or taken as a concentrate in pill form. Because this herb induces uterine bleeding, it can abort an embryo or fetus. However, it can also cause hemorrhage, seizures, hallucinations, kidney failure, liver damage, dizziness, cramping, fainting, and even be fatal to the woman. Abortion can be incomplete, especially in the later stages of pregnancy.

Around the world, women have long used a variety of substances and techniques in attempts to end their unwanted pregnancies. In addition to a wide variety of herbs and other natural substances, common abortifacients include high doses of drugs such as aspirin, alcohol, or birth control pills; injecting caustic substances such as bleach, formaldehyde, and gasoline into the vagina or uterus; inserting sharp objects into the uterus; physical trauma to the uterus such as beating the abdomen or intentionally falling a long distance; and using prayer and magic to end an unwanted pregnancy.

Safety and Consequences of Induced Abortion

Induced abortion is one of the most common surgical procedures performed in the United States. Modern surgical and medication abortion procedures have a low rate of deaths and complications. Currently, less than 1 death results from 100,000 abortions performed in the United States. The death rate has dropped dramatically since abortion was legalized in 1973. Medical complications are also relatively low. A recent report of 170,000 women receiving abortions found that 0.7 out of 1000 had complications resulting in hospitalization. Improved methods and a trend toward earlier abortions have contributed toward a reduction in deaths and complications from abortion procedures. Unfortunately, deaths and complications from illegal abortions remain a serious health risk to women in some countries.

An induced abortion using modern methods does not affect a woman's future fertility, risk of premature birth, ectopic pregnancy, or miscarriage. Some studies have shown an increased risk for placenta previa, whereas others have not. Ovulation usually occurs 2 or 3 weeks after a first trimester abortion. This delay is longer after second or third trimester abortions and is similar to the delay after a normal delivery in a nonnursing woman (about 3 months).

Abortion does not increase a woman's risk of cancer. Early studies led to claims of an increased risk of breast cancer among women who had abortions. However, more recently large studies of women who obtained abortions show no effect on breast cancer risk, and the National Cancer Institute has concluded that there is no evidence for an association between abortion and breast cancer.

One might ask whether experiencing an abortion has significant psychological effects on a woman. As with other major decisions faced by a woman in her life, the decision to have an abortion can have both short- and long-term consequences. Numerous studies examining psychological effects of an abortion find that the most common short-term response is relief, and that in subsequent years, the majority of women continue to feel that they made the right decision. Adverse psychological reactions such as depression occur in a minority of women and occur more often in women with a prepregnancy history of depression. Younger women and those having more children are also at higher risk of negative postabortion symptoms.

Chapter Summary

Induced abortion is the termination of a pregnancy by artificial measures. Governments can be permissive or restrictive in their legislation regulating abortion. Induced abortion is legal in the United States today, where over one in five pregnancies end in induced abortion. Although women of all types use abortion services, women seeking abortions in the United States tend to be young, poor, unmarried mothers.

A woman or couple with a wanted pregnancy may choose an abortion because of danger to the woman's physical or mental health or because the fetus is gravely deformed or genetically at risk. Also, some women simply do not wish to be pregnant and terminate pregnancy as a means of fertility control or to avoid the birth of an unwanted child. A moral controversy about induced abortion relates to the concept of "ensoulment." Choosing to have an abortion usually is a difficult decision and involves various religious, moral, psychological, legal, and medical considerations.

Induced abortion procedures are safer and simpler at earlier stages of pregnancy. During the embryonic period (early first trimester), abortion can be induced by menstrual regulation or administration of prostaglandins or mifepristone and misoprostol. During the early fetal period (late first trimester), pregnancy can be terminated by dilation and curettage or, more commonly, by cervical dilation using a laminaria tent followed by vacuum aspiration. Common methods of inducing abortion during the second trimester are administration of intraamnionic saline, urea, glucose, and prostaglandins, alone or in combination. Extraamnionic prostaglandin is also used, as well as dilation and evacuation. Induced abortion of a third trimester fetus is rare; when done, it usually involves hysterotomy. Intact dilation and extraction (so-called "partial birth abortion") in the late second and third trimester have been banned in the United States.

Most abortion procedures do not influence future fertility. Menstrual cycles resume about 2 or 3 weeks after a first trimester abortion and in about 3 months after a second or third trimester abortion. Folk abortifacients have long been used by women in attempts to end unwanted pregnancies.

Further Reading

Adler, N.E., *et al.* (2003). Abortion among adolescents. *Am. Psychol.* **58**, 211–217.

Anonymous. (1992). Court reaffirms *Roe* but upholds restrictions. *Family Planning Perspect.* **24**, 174–177, 185.

Elam-Evans, L.D., *et al.* (2003). Abortion surveillance – United States, 2000. *MMWR Surveill. Summ.* **52**, 1–32.

Fackelmann, K. A. (1994). Beyond the genome: The ethics of DNA testing. *Sci. News* **146**, 298–299.

Finer, L.B., and Henshaw, S.K. (2003). Abortion incidence and services in the United States in 2000. *Perspect. Sex. Reprod. Health* **35**, 6–15.

Gammon, M. D., *et al.* (1996). Abortion and the risk of breast cancer: Is there a believeable association? *J. Am. Med. Assoc.* **274**, 321–322.

Gold, M., *et al.* (1997). Medical options for early pregnancy termination. *Am. Family Physician* **56**, 533–537.

Grobstein, C. (1982). When does human life begin? *Science* **82**, 3(2), 14.

Henshaw, S. K. (1990). Induced abortion: A world review, 1990. *Family Planning Perspect.* **22**, 76–89.

Hollander, D. (1995). Mifepristone and vaginal misoprostol are effective, acceptable, and inexpensive medical abortion regimen. *Family Planning Perspect.* **27**, 223–224.

Klitsch, M. (1992). Abortion experience does not appear to reduce women's self-esteem or psychological well-being. *Family Planning Perspect.* **24**, 282–283.

Riddle, J. M., *et al.* (1994). Ever since Eve: Birth control in the ancient world. *Archeology* March/April, 29–35.

Stone, R., and Waszak, C. (1992). Adolescent knowledge and attitudes about abortion. *Family Planning Perspect.* **24**, 52–57.

Torres, A., and Forest, J. D. (1988). Why do women have abortions? *Family Planning Perspect.* **20**, 158–168.

Ulmann, A. (1990). RU 486. *Sci. Am.* **262**(6), 42–49.

Advanced Reading

Baulieu, E. (1989). Contragestion and other clinical applications of RU 486, an antiprogesterone at the receptor. *Science* **245**, 1351–1357.

Couzinet, B., *et al.* (1986). Termination of early pregnancy by the progesterone antagonist RU 486 (mifepristone). *N. Engl. J. Med.* **315**, 1565–1569.

Creinin, M. D. (1997). Medical abortion with methotrexate 75 mg intramuscularly and vaginal misoprostol. *Contraception* **56**, 367–371.

Finer, L.B., and Henshaw, S.K. (2003). Abortion incidence and services in the United States in 2000. *Persp. Sex. Reprod. Health* **35**, 6–15.

Frank, P., *et al.* (1993). The effect of induced abortion on subsequent fertility. *Br. J. Obstetr. Gynaecol.* **100**, 575–580.

Goldberg, A.B., *et al.* (2004). Manual versus electric vacuum aspiration for early first-trimester abortion: A controlled study of complication rates. *Obstet. Gynecol.* **103**, 101–107.
Grimes, D., *et al.* (1988). Early abortion with a single dose of the anti-progestin RU-486. *Am. J. Obstet. Gynecol.* **158**, 1308–1312.
Hausknecht, R. (2003). Mifepristone and misoprostol for early medical abortion: 18 months experience in the United States. *Contraception* **67**, 463–465.
Jones, R.K., *et al.* (2002). Mifepristone for early medical abortion: Experiences in France, Great Britain, and Sweden. *Persp. Sex. Repro. Health* **34**, 154–161.

CHAPTER SIXTEEN

Infertility

Introduction

Couples who use contraception wish to avoid unwanted pregnancies. In contrast, many couples who wish to bear children are unable to conceive. Current estimates place couples suffering from *infertility* in the United States at about 2.4 million, or about 1 out of 6 married couples of childbearing age. A couple is considered infertile if they have participated in unprotected coitus for a year without becoming pregnant. In about 35% of these couples, the problem is with the female, in about 35% with the male, in 20% with both partners, and in 10% the cause is unknown. A couple can also be infertile because of *pregnancy wastage*, or the inability to maintain a pregnancy once established.

Hope is present for infertile couples because the cause of infertility in 85–90% of cases can be diagnosed, and about 50–60% can be treated successfully. In those situations in which treatment fails, a couple may want to consider adopting a child or to seek help in other ways. We now look at some causes and treatments of infertility.

Seeking Medical Help for Infertility

When infertile couples seek medical help, the first step is to diagnose the cause of their problem by doing an "infertility work-up" on the woman, the man, or both partners. In the case of the female, several tests can be administered. Her menstrual cycle, if occurring, can be followed using the sympto-thermal method and/or hormone assay. If it is found that she is ovulating, the administration of dyes, gas, or laparoscopy can be used to check if her uterus or oviducts are blocked. An endometrial biopsy can assess the condition of her uterine lining, and her reproductive tract can be tested for infection. In the *postcoital test* (*Huhner's test*), a woman's cervical mucus is checked for sperm number and condition within 2 h of coitus. This test can be used to evaluate semen quality as well as the survival of sperm in the woman's mucus. There also are tests to see if the woman is producing antibodies to her husband's sperm.

In the case of the male, semen can be collected and a sperm count done. If a man's sperm count is in the fertile range (greater than 20 million sperm/ml of semen), the sperm will be examined to determine whether motility and structure are normal. The semen will also be checked to see if it is ejaculated in normal amounts (1 to 6 ml) and if it contains enough fructose (greater than 1200 μg/ml). If the male's sperm count is low, levels of hormones [luteinizing hormone (LH); androgens] will be assayed in his blood or urine, and the blood supply to his testes may be examined. Also, a sample of testicular tissue may be taken to see if spermatogenesis is normal; this is called a *testicular biopsy*. Finally, the condition of the sex accessory glands and ducts can be assessed and immunological tests done to see if he is producing antibodies to his own sperm.

Infertile couples can suffer from psychological and emotional problems related to their condition. These problems can be caused by feelings of inadequacy, loss, and frustration over their situation. Therefore, it is often wise for infertile couples to seek counseling about their problem. For a discussion of the psychological response of couples to infertility, see the reference by B. E. Menning in the Further Reading section. Organizations such as RESOLVE and the American Fertility Association can provide information and support to infertile couples.

Female Infertility

Age is an important factor in female infertility. Women are most fertile in their early twenties. In the United States, about 4.1% of women between the ages of 20 and 24 suffer from infertility; this increases to 5.5% at ages 25 to 29, 9.4% at ages 30 to 34, and 19.7% at ages 35 to 39. Thus, the older a woman becomes, the more difficult it is for her to become pregnant, even though she may become more equipped emotionally and financially to care for a child as she matures beyond her midtwenties. Although men do not experience a similar dramatic decrease in fertility, age-related changes in sperm production and sperm delivery can also impair male fertility.

Failure to Ovulate

Failure to ovulate is the leading cause of infertility in females. Irregular or absent menstrual periods can be a sign of ovulation problems. A woman may not ovulate because her hypothalamus or pituitary gland is not fully functional, such as resulting from the presence of a pituitary tumor. Severe stress, excessive exercise, or low body fat can cause failure to ovulate (Chapter 3), or the hypothalamus may not be secreting enough gonadotropin-releasing hormone (GnRH) to stimulate a LH and follicle-stimulating hormone (FSH) surge that results in ovulation. In this case, administration of GnRH stimulatory agonists (Chapter 1) in small dosages has been effective in inducing ovulation. A small pump can be placed under the skin, which slowly releases GnRH into the bloodstream. Sometimes a woman's pituitary gland is not capable of secreting enough LH and FSH even though adequate levels of GnRH are present. In this case, one common treatment is the drug *clomiphene* (or Clomid), which can be administered orally. Clomiphene is an antiestrogen that inhibits the negative feedback action of estrogens on FSH and LH secretion. Thus, FSH and LH levels increase when clomiphene is given, and 30 to 50% of infertile women so treated will ovulate and become pregnant. One effect of clomiphene is that the chance of having twins is 5% compared with 1% in untreated women. This is because the increased gonadotropin levels may hyperstimulate the ovary, causing ovulation of two eggs and the resultant dizygotic (fraternal) twins. For some unknown reason, twins that result from clomiphene treatment are also sometimes monozygotic (identical). Other side effects can include hot flashes, mood swings, headaches, visual disturbances, and ovarian cysts. Although women with certain kinds of infertility may be at increased risk for ovarian cancer, the use of clomiphene does not appear to affect this risk.

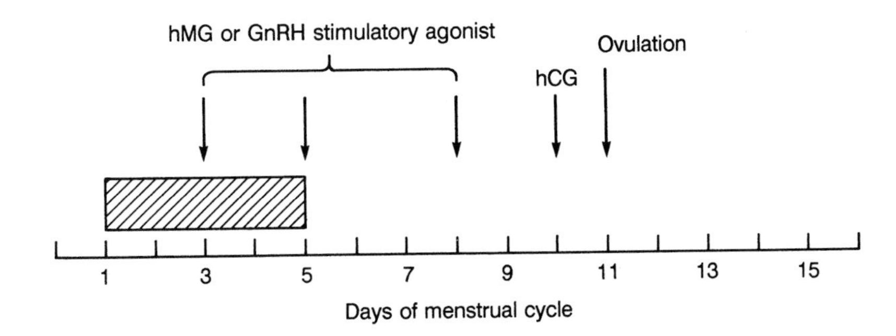

Figure 16-1 Treatment of an infertile woman with human menopausal gonadotropin (hMG) or a GnRH stimulatory agonist followed by human chorionic gonadotropin (hCG) to induce ovulation. The shaded area represents menses.

Another treatment for a woman who is secreting insufficient FSH and LH is the administration of human menopausal gonadotropin (hMG) (e.g., Pergonal), which contains FSH and LH, followed by hCG (Fig. 16-1). These gonadotropins are administered as injections. This treatment causes ovulation in 50 to 70% of cases so it is highly effective. The incidence of twins, however, is 15% and of triplets or more is 5%. For these reasons, gonadotropin treatment is used only as a last resort unless, of course, twins or triplets are desired. Some women using hMG develop abdominal pain caused by enlargement of the ovaries.

Another cause of infertility in relation to malfunction of the pituitary gland is excess pituitary secretion of prolactin. Some infertile women have high prolactin levels in their blood. That this condition causes their problem is demonstrated by the fact that the drug *bromocriptine*, which inhibits prolactin secretion, restores fertility in many of these women.

In other women, failure to ovulate may not be caused by malfunction of the hypothalamus or pituitary, but instead the ovary may be incapable of responding to an LH and FSH surge. This may be due to the presence of endometriosis, ovarian cysts, tumors, or scars caused by ovarian infection. In these cases, ovarian surgery may be needed to restore fertility. Some women, however, may have permanently malfunctioning ovaries. The ovaries of women with Turner's syndrome (XO), for example, lack follicles (see Chapter 5).

Tubal Blockage

In some infertile women, ovulation occurs but the sperm fail to reach the egg or the fertilized egg cannot reach the uterus because of a blockage in one or both oviducts. Tubal blockage is the second leading cause of infertility in females, occurring in 30 to 35% of infertile women. Tubal blockage can be caused by a kink in the tube or by scarring as a result of past venereal disease infection, especially *Chlamydia* (see Chapter 18). Also, a piece of the uterine endometrium may have become displaced from the uterus and lodged in the oviduct. This endometriosis (see Chapter 2) causes sterility because the uterine tissue grows and blocks the tubes. Endometriosis can also cause infertility when the tissue does not block the tubes but occurs outside the tubes or in the abdominal cavity, especially around the ovaries.

Tubal blockage sometimes can be repaired by introducing a fluid or gas (CO_2 or air) into the tubes. Also, a small balloon can be inserted into the obstructed tube, and when it is expanded the tube is unblocked. This is called *transcervical balloon tuboplasty*. If this fails to open the tubes, it may be possible to repair them surgically.

Absence of Implantation

In some women, the preembryo may reach the uterus but implantation does not occur. Priming of the uterus by estrogen and progesterone is needed for implantation to occur (see Chapter 10), and this priming may be inadequate in some women. Some of these cases can be treated by the administration of steroid hormones (estrogens or progestogens) to render the uterine endometrium more receptive to the blastocyst.

Other cases of infertility may be due to damage to the endometrium. Perhaps fibroids or scars from pelvic infection are present (see Chapter 2) or a previous unsafe abortion (see Chapter 15) may have damaged the uterine lining.

Reduced Sperm Transport or Antibodies to Sperm

The female tract may not allow transport or survival of the male's sperm. The woman's vagina, for example, may be highly acidic, which can be treated using alkaline douches. Alternatively, her cervical mucus may be hostile to sperm movement, a condition perhaps alleviated by estrogen administration. If the cervix has been damaged, as sometimes occurs as a result of infection, this damage may be corrected surgically. In some cases, a woman produces antibodies to her husband's sperm, and in this case the couple may choose to become pregnant using donor artificial insemination (see later).

Table 16-1 summarizes the causes and possible treatments of female infertility.

Table 16-1 Possible Causes and Treatments of Infertility in Women[a]

Cause	Possible treatment
Maternal age	Start family earlier
Excessive exercise	Decrease exercise
Low body fat or obesity	Gain or lose weight
GnRH insufficiency	GnRH stimulatory agonists
Gonadotropin insufficiency	Perganol; clomiphene
Prolactin oversecretion	Bromocriptine
Other endocrine disorders	Hormonal therapy; surgery
Ovarian disorders, e.g., dysgenesis[b], cysts, cancer	Microsurgery; donated egg; surrogate pregnancy
Tubal blockage, e.g., dysgenesis[b]; scarring	IVF; GIFT; ZIFT; microsurgery; balloon tuboplasty
Uterine abnormalities, e.g., dysgenesis[b]; abnormal endometrial growths	Microsurgery; gestational carrier
Endometriosis	GnRH antagonist; danazol; microsurgery
Cervical abnormalities; e.g., dysplasia; cancer; hostile mucus	Microsurgery; estrogen treatment; gestational carrier
Pelvic inflammatory disease	Antibiotics; safe sex
Antisperm antibodies	Immunosuppressant drugs; AID; ICSI
Antibodies to own egg	Immunosuppressant drugs; surrogate pregnancy; donated egg

[a]See Fig. 16-3 for other "high-tech" solutions.

[b]Abnormal development.

Chapter 16, Box 1: Seasonal Changes in the Ability to Conceive

Even though a couple may be fertile throughout the year, there is a certain season when they may have a harder time conceiving and carrying a pregnancy to term. A seasonal cycle in fecundity (the ability to conceive) exists in both hemispheres, and of course it is related to the seasonal cycles of birth mentioned in Chapter 11. For example, births peak in the late summer and early fall in the United States. This means that the peak in fecundity is in late fall and early winter. However, the birth trough in the United States (spring) is preceded 9 months earlier by a time (summer) when it is more difficult for a couple to conceive. The birth peaks and troughs in Europe are different from those in the United States, but the same rules hold there as well as in the southern hemisphere where the seasons are 180° out of phase with those in the northern hemisphere.

Research has shown that the low fecundity of a couple around 9 months before the birth trough in the population could be related to reduced ovum quality and sperm quality or quantity, as well as other seasonally fluctuating factors. A moderate reduction in coitus frequency in the summer months, especially in warmer regions, contributes only partially to the lower conceptions at that time. It is as if there are inherent, subtle mechanisms that interfere with sperm or egg quantity or quality, embryo quality, or the ability of the embryo to implant in the uterine wall and establish and maintain a pregnancy. According to Pl. H. Jongloet, an expert on human seasonal reproduction at the University of Nijmegan, The Netherlands, eggs ovulated at the wrong time of year produce babies, and later adults, who experience more medical problems. According to this "seasonal overripenesss ovopathy" hypothesis, eggs ovulated from the ovary at a time on either side of the seasonal peak of conception in a population remained in the ovary too long and are aged. Pregnancies resulting from conception in the trough of the fecundity cycle (e.g., with aged ova) are prone to a slight increase in gestation length, birth of preterm babies, ectopic pregnancies, preeclampsia, miscarriage, and stillbirth.

This seasonal resistance to conceive occurs even during *in vitro* fertilization, when infertile couples have their sperm and egg (or someone else's sperm and/or egg) brought together in the laboratory, and then the embryos formed are implanted in the woman's uterus. There is a reduction in success rate of this procedure if done during the season of low fecundity. Even when a woman is impregnated by a donor's sperm (artificial insemination) the results are poorer if done 9 or so months before the trough of the seasonal birth cycle in that population. These problems are related to a lower rate of fertilization, poorer embryo quality, and a lower implantation rate in the uterus. Low sperm quality and number are also contributors to this seasonal change in the success rate of *in vitro* fertilization.

Being conceived during the low season of fecundity can lead to a slightly greater risk of certain problems when the babies become adults. In relation to reproduction, women have an increased risk of early menarche, longer menstrual cycles, heavy menstrual bleeding, stillbirths, shorter breast-feeding, and early menopause. There are conflicting reports, but a few studies show that being conceived in the fecundity trough increases one's risk of getting breast cancer later in life. People conceived in the fecundity trough also increase their chances as adults of developing schizophrenia, bipolar (manic-depressive) disorder, epilepsy, panic disorder, and major depression. They also have a better chance of becoming alcoholics and to use illicit drugs.

Thus, the males and females of today's industrialized world may have internal, biologically evolved mechanisms that say "Don't conceive now." Although these mechanisms probably evolved in our hunter-gatherer ancestors to prevent pregnancy or birth when food and water were not abundant, the ghost of their presence still haunts us. However, births still occur, even though at a lower rate, from pregnancies beginning in

Continued on next page.

Chapter 16, Box 1 continued.

the season of low fecundity. These effects of birth season are subtle. Children conceived in the trough of the fecundity season (summer in the United States) have only a small increased risk of the aforementioned problems (usually under 10%) and the great majority of infants born in these months grow into healthy adults.

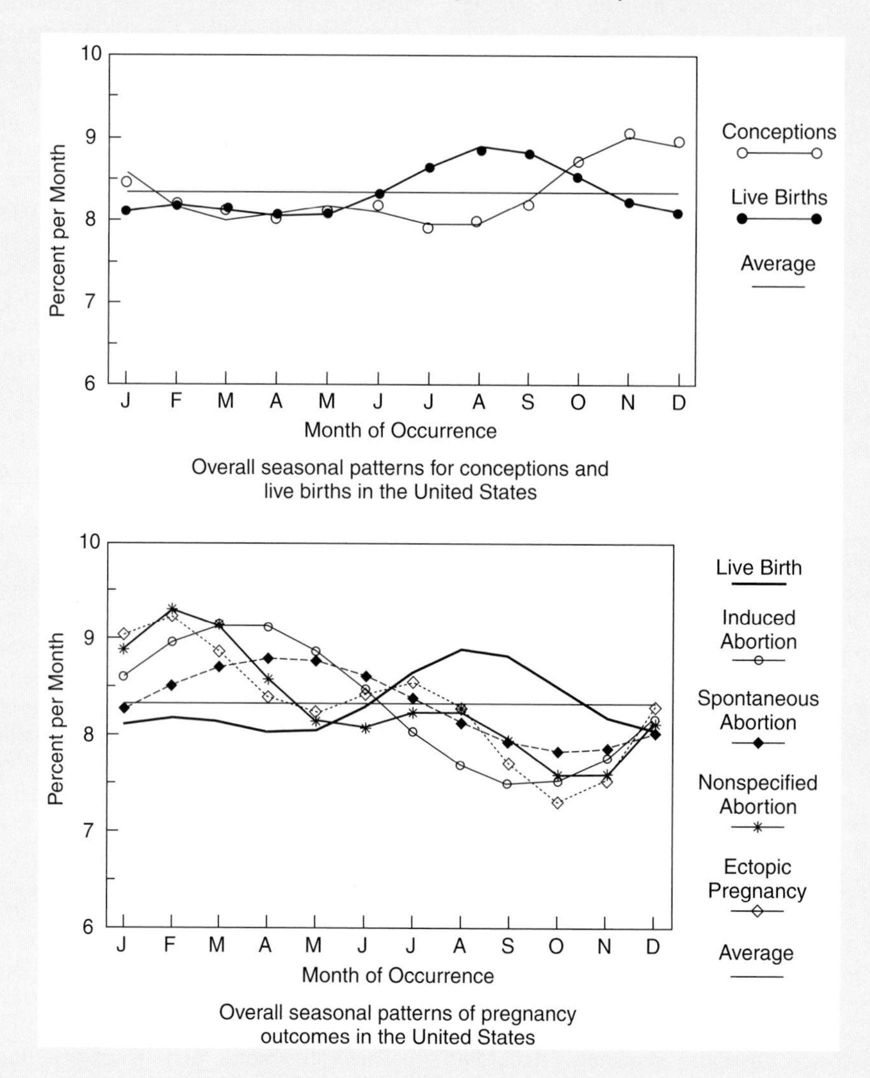

(Top) Overall seasonal patterns of conceptions and live births in the United States. (Bottom) Seasonal patterns of pregnancy outcomes in the United States.

Pregnancy Loss (Miscarriage)

About 15% of known pregnancies end in miscarriage. However, it is estimated that up to 50% or even 75% of all fertilized eggs do not complete development; these very early pregnancy losses often occur before implantation and before

Table 16-2 Some Causes and Possible Treatments of Miscarriage

Cause	Possible treatment
Embryonic chromosomal abnormalities, e.g., maternal age; errors of fertilization	Surrogate pregnancy; donated egg; preimplantation genetic diagnosis
Uterine developmental abnormalities, e.g., inherited; diethylstilbestrol-induced	Microsurgery; gestational carrier
Uterine disorders, e.g., fibroids; polyps; hyperplasia; cancer	Microsurgery; gestational carrier
Uterine prolapse	Surgery
Uterine infection or scarring	Antibiotics; microsurgery; gestational carrier
Cervical abnormalities, e.g., dysplasia; cancer	Surgery; gestational carrier
Cervical incompetence	Surgery
Low secretion of progesterone and/or estrogens	Administer progestogens and/or estrogens
Drugs, e.g., tobacco; alcohol; some prescription drugs; some illegal drugs	Stop using them
Teratogens, e.g., environmental pollutants	Avoid
Immunological rejection of embryo	Paternal leukocyte immunization; surrogate pregnancy

a woman realizes that she is pregnant. Thus, miscarriage is the most common disorder of pregnancy. Unfortunately, some women are infertile because they experience recurrent miscarriages (usually defined as three or more miscarriages). Most early miscarriages occur because of chromosomal abnormalities of the fetus. A woman who has had repeated miscarriages may simply have the misfortune to have carried fetuses with unrelated chromosomal anomalies. Women over age 35 have an increased risk of miscarriage, probably because chromosomal errors accumulate as the egg ages. However, some miscarriages do not have an apparent genetic cause. Recently, attention has focused on possible immunological causes of miscarriage. Recall from Chapter 10 that the mother usually does not reject the fetus as foreign tissue because of several adaptations of the maternal immune system. It was suspected that couples who shared certain *major histocompatibility* (MHC) *antigens* were at higher risk for miscarriage, but there is no clear evidence to support this idea. Current research is focused on one MHC gene (HLA-G) that is expressed mostly in fetal tissues of the placenta, the cytotrophoblast. Because it is found where maternal and fetal cells interact, this antigen may play a role in the support of pregnancy. The presence of certain HLA-G variants in the male and/or female may increase a couple's risk of having a miscarriage.

Other possible causes and treatments of pregnancy loss are summarized in Table 16-2.

Male Infertility

Low Sperm Count

A low sperm count is the leading cause of infertility in males. An infertile man may have a low sperm count (*oligospermia*) or the absence of sperm (*azoospermia*) in the semen because his hypothalamus or pituitary gland is functioning below normal levels. Treatment with GnRH stimulatory agonists or gonadotropins may restore fertility in these men. Also, giving the antiestrogen clomiphene to infertile men can increase their GnRH secretion (by reducing the negative feedback of estradiol on GnRH) and increase fertility. Some infertile men have high prolactin levels in their blood, and treatment with bromocriptine can make them fertile.

Sometimes the testes themselves may be incapable of responding to gonadotropins. For example, there may be structural abnormalities or permanent damage to the testes. When testes are damaged, the treatment may be artificial insemination of the woman with donor sperm. Some structural problems, however, can be corrected by testicular surgery. Permanent damage could occur because of exposure to radiation or chemotherapy from cancer treatment. Men at risk of infertility from these therapies might consider cryopreserving their sperm before undergoing treatment. In the future, it may be possible to remove spermatogonial stem cells prior to the cancer treatment; these may be kept alive in the laboratory and reimplanted into the man at a later time, restoring his fertility.

A varicose vein in the scrotum (a condition called *varicocele*) can raise testis temperature and cause infertility. Men with *retrograde ejaculation* have low sperm counts because semen flows into the bladder rather than into the urethra during ejaculation. This condition can arise from previous surgery, diabetes, or certain medications. It can be treated surgically or reversed by stopping the medication. *Orchitis*, or inflammation of the testis, can cause temporary or permanent infertility. It can result from infection associated with a sexually transmitted disease or other bacterial or viral (usually mumps) infection. *Cryptorchid* (undescended) testes often are unable to produce sperm. When only one testis is impaired, however, usually the unaffected testis is sufficient to maintain normal fertility.

In about 8 to 13% of infertile men, the problem is caused by the production of antibodies to their own sperm. This occurs because some sperm inadvertently enter a man's body outside the reproductive tract. Treatment with adrenal hormones has been shown to alleviate this problem in some men. Finally, about 1 in 1000 infertile men have a low sperm count because they are missing part of the Y chromosome.

Sperm Transport

In some infertile men, the secondary accessory ducts or glands are not functioning properly. The vasa deferentia, for example, may be occluded by an enlarged testicular vein pressing on it (varicocele). The vasa can also be blocked by scar tissue caused by venereal disease infection (see Chapter 18). Many of these cases can be corrected with surgery. Finally, a sex accessory gland may be malfunctioning or inactive. If the glands are simply underdeveloped, this condition can be treated by the administration of an androgen. Infection of the prostate can lead to sterility in some men. *Impotence*, the inability to gain or maintain an erection long enough to ejaculate, can also result in infertility. Some types of erectile dysfunction can be treated (see Chapter 8).

Environmental Factors

Some men may be temporarily infertile because of environmental factors. Smoking, for example, can decrease sperm motility and increase the number of structurally abnormal sperm in the ejaculate. Testosterone levels in the blood

Table 16-3 Possible Causes and Treatments of Infertility in Men[a]

Cause	Possible treatment
Recreational drugs, e.g., cocaine; tobacco; marijuana; alcohol	Stop using them
Pollutants, e.g., endocrine disrupting contaminants; carbon monoxide; lead	Avoid
Anabolic steroids	Stop using them
GnRH insufficiency	GnRH stimulatory agonists; clomiphene
Gonadotropin insufficiency	Perganol
Prolactin oversecretion	Bromocriptine
Other endocrine disorders	Hormonal therapy; surgery
Testicular disorders, e.g., dysgenesis; varicocele; cancer	Surgery; AID
Orchitis	Antibiotics, anti-inflammatories
High testicular temperature	Loose underwear
Low testosterone secretion	Androgen administration
Sex accessory duct or gland disorders, e.g., dysgenesis; scarring	Surgery; androgens; IVF; AIH; AID
Prostate disorders	Surgery; hormone therapy; chemotherapy; radiation
Sexually transmitted diseases or other infections	Antibiotics; safe sex
Low zinc or magnesium	Administer
Antibodies to one's own sperm	AI with ICSI; AID
Missing part of Y chromosome	AID

[a]AI, artificial insemination; AIH, artificial insemination with husband's or partner's sperm; AID, artificial insemination using donor sperm; ICSI, intracytoplasmic sperm injection; IVF, *in vitro* fertilization.

are also lower in men who smoke. Accumulation of some environmental pollutants and endocrine disrupters, such as xenoestrogens, can also reduce sperm count (see Chapter 4). Chronic use of some drugs, such as alcohol, marijuana, and anabolic steroids, can also reduce fertility. Certain medical conditions, including obesity, diabetes, and thyroid disease, can affect a man's fertility, and a gradual decline in fertility often occurs after age 35.

Table 16-3 summarizes some causes and possible treatments of infertility in men.

Assisted Reproductive Techniques (ART)

For couples whose infertility cannot be reversed by treating one or both partners with medication, surgery, or other means, several "high-tech" procedures are available to bring together the egg and sperm. These include new methods to retrieve and store gametes, control and monitor the process of fertilization using the couple's gametes or those donated from others, and test an early embryo for the presence of genetic abnormalities.

Gamete Storage and Artificial Insemination (AI)

When sperm are introduced into a woman's reproductive tract by means other than coitus, it is called *artificial insemination*. The sperm can come from a "donor" man who is not the woman's husband or partner ["artificial insemination donor"

(AID)] or from the husband ["artificial insemination husband" (AIH)]. In both cases, the man deposits semen into a vial by masturbating or into a specially designed condom during intercourse. Several ejaculations, which are collected and frozen, are required to pool enough sperm to be effective. If a man's sperm count is very low, sperm can be retrieved directly from the testis by testicular biopsy. A sperm sample can also be taken from the epididymis using a tiny syringe or by aspirating epididymal fluid using microsurgery under a microscope.

If the quantity of motile sperm is sufficient, the pooled sperm sample is "washed" (processed to concentrate motile sperm in a small amount of fluid) and then loaded into a thin, flexible catheter and delivered into the woman's uterus. This usually involves only slight discomfort. The timing is critical; sperm should be introduced into the uterus within a few hours of ovulation. Fertility drugs may be used to induce ovulation of several oocytes, thus improving the chances for fertilization.

If the male partner cannot produce sperm of sufficient quantity or quality or if a single woman wishes to become pregnant without a partner, donor sperm from a "sperm bank" can be used. Donors are classified, anonymously, by their physical and even personality characteristics, and some sperm banks screen semen for hereditary diseases and chromosomal abnormalities. In the United States, no cases of AIDS (see Chapter 18) have resulted from the use of donor artificial insemination. Nevertheless, such transmission is possible, and for this reason many sperm banks now test semen samples for the presence of AIDS antibodies. More than 60,000 babies are born each year in the United States as a result of artificial insemination, about two-thirds of them a result of AIH. Usually, the sperm are chilled before use, but they also can be frozen at −196 °C (−321 °F) for up to 10 years. About 80 sperm banks in the United States, run by hospitals or private businesses, store chilled or frozen sperm. Artificial insemination can be useful for infertile couples because of some problem with the male partner such as a low sperm count or other kinds of infertility. In such cases, semen can be used from the husband or a donor male. Sperm storage can also be useful for a man who is going to have a vasectomy, just in case he may later want to father a child.

Anyone who is contemplating artificial insemination should be aware, however, that there are biological questions to be answered. For example, how does freezing affect sperm? After being frozen and thawed, many sperm lack vigor and motility, and up to 85% are shaped abnormally. This may explain why the rate of spontaneous abortions after artificial insemination is higher than the normal rate. More male babies than usual are born by this method.

There are also moral, ethical, and legal considerations about artificial insemination. There are no federal regulations for artificial insemination, and individual states vary in their legal attention to this matter. Questions can arise, such as: Who is the legal father of a child born using semen from a male other than the husband? If a couple having a child by this method is divorced, is the ex-husband responsible for child support? If a child conceived in this manner has a birth defect, can the physician be charged with malpractice? What are the inheritance rights of a child conceived by sperm from a man other than its legal father? Should semen be used for selective breeding for certain inherited traits, and what are the moral implications of selective breeding? An example of such an attempt at selective breeding is a sperm bank in California using sperm donated by Nobel prize winners. What happens when sperm from a different race are used by mistake? This situation has already occurred.

In 1977, the first scientific report appeared showing that ovulated ova (oocytes) in the laboratory mouse could be collected from the oviduct and stored at −196 °C for 6 years. The eggs were then thawed, fertilized in a dish, implanted into the uterus of a female mouse, and developed into normal mouse pups. This is now possible with human ova as well, and many babies have already been born using frozen eggs. "Ovum banks," however, are not likely to become as common as sperm banks because egg retrieval is usually much more expensive and difficult than sperm collection.

In Vitro *Fertilization (IVF)*

On July 25, 1978, a 5 lb, 12 oz baby girl named Louise Joy Brown was born to Mr. and Mrs. Gilbert J. Brown of Bristol, England. Louise, a perfectly normal little girl, was the world's first "test-tube baby." That is, her father's sperm fertilized her mother's ovum in a dish, and the early embryo resulting from this *in vitro* (Latin for "in glass") *fertilization* was then put back into Mrs. Brown's uterus to develop and be born. Thus, Louise was conceived in a dish in the laboratory. This remarkable achievement was the culmination of a long history of research in mammalian reproduction. The first human egg to be fertilized by *in vitro* fertilization was accomplished by J. Rock and M. Menkin in 1944. Drs. P. C. Steptoe and R. G. Edwards of England then refined the technique and eventually determined how to implant the embryo into the mother's uterus. This procedure is used as a treatment for infertile couples with conditions such as a fallopian tube blockage or very low sperm count.

Ovarian Stimulation and Egg Retrieval First, "fertility drugs" (clomiphene citrate or gonadotropins) are administered to cause several large follicles to mature in a woman's ovary. GnRH agonists or antagonists may be used additionally to prevent premature ovulation. Growth of the follicles over the next 8–14 days is monitored using ultrasound. Circulating hormone levels are also checked. During follicular growth, estrogen levels should rise but progesterone levels should remain low. When several large follicles have grown, an injection of hCG is usually given. This mimics LH and causes the eggs to mature and also triggers ovulation. Before ovulation occurs, however, oocytes (ova) are removed from the large follicles in the ovary. In most cases, several (up to 12) ova are removed to ensure a sufficient number of embryos to transplant. The operation is done by inserting a long needle through the vaginal wall toward the surface of the ovary. Guided by ultrasound, the needle contacts each enlarged follicle and the ovum is removed by suction.

Fertilization and Embryo Transfer Before the ova are retrieved, the husband produces (by masturbation) fresh sperm, which are placed in a petri dish. The dish contains a fluid that nourishes and capacitates the sperm. The ova are then added to the dish with the sperm, and fertilization, as evidenced by the presence of two pronuclei, usually occurs within 12 to 14 h. Here it would be possible to use a donor ovum instead of the ART patient's own ovum, donor sperm, or even a donor ovum and donor sperm.

After fertilization, the embryos are transferred to a new culture dish for incubation, and their development is monitored closely. It takes 2 days for an embryo to reach the 2- to 4-cell stage and in 3 days the embryo has 6–10 cells. By the 5th day, the embryo becomes a blastocyst. Embryos that exhibit normal development are transferred into the uterus between 1 and 5 days of development. The preembryos are placed into a tiny tube, which is inserted into the woman's uterus through her cervix. Finally, the embryos are released into the uterus, and implantation can now occur. It takes about 2 weeks to know if one or more embryos have implanted. Most ART clinics transfer only two or three embryos in one *in vitro* fertilization cycle. The remaining eggs or embryos can be frozen for possible use in future IVF cycles.

Of 79 women first accepted for this procedure in England in the late 1970s, 68 underwent laparoscopy, the method formerly used to retrieve eggs. Of the 68 women, 44 produced ova in the correct state, and 32 of the ova were fertilized and developed into 8- to 16-cell embryos, which were then transplanted back into the mothers. Only 4 of the 32 women became pregnant, meaning that only 4 had successful embryonic implantation, and only 2 of these pregnancies produced normal infants at term. The other two fetuses were aborted early in pregnancy and were found to exhibit chromosomal abnormalities.

With recent refinements of the external fertilization technique, *in vitro* fertilization, as well as other assisted reproduction methods (see later), have been more successful. One of these improvements is *intracytoplasmic sperm injection* (ICSI), in which a single sperm is injected directly through the zona pellucida and into the ovum. Currently, just over half of the ART cycles in the United States use ICSI. Another is *assisted hatching* or *zona drilling*, in which part of the zona pellucida is removed from the ovum surface. This may facilitate hatching of the embryo prior to implantation. In an effort to rejuvenate aging eggs, cytoplasm from a young donor egg is injected into the egg of an older ART patient, a process called *cytoplasmic transfer*. This highly experimental treatment has had some limited successes, which may be explained by the introduction of healthy mitochondria into the oxidatively weakened eggs. Embryos resulting from this treatment, however, would have mitochondria from two different females instead of inheriting mitochondria from the mother only, as in normal development. The long-term consequences of this are unknown.

Sperm can be removed directly from a man's testes and injected into an ovum *in vitro* and then transferred into the uterus. Even immature spermatogenic cells (spermatids, spermatocytes) have been used for intracytoplasmic transfer. In Korea, a few births have occurred after removing an immature oocyte from a woman's ovary, maturing it in a dish containing gonadotropins, fertilizing the mature egg with sperm, and then implanting the embryo in the uterus. This method would allow a woman whose ovaries fail to grow large follicles to have a child. Additionally, frozen embryos have been used successfully. In fact, there were about 400,000 frozen embryos in United States fertility clinics in the year 2002. The success rate with frozen embryos, however, is lower than with fresh embryos.

Currently over 115,000 cycles per year are performed in the United States. The U.S. Center for Disease Control reports that in 2002, the woman's own fresh eggs were used in 74% of cycles and her frozen eggs in 14%. Donor eggs (fresh or frozen) were used in 11% of ART cycles.

Risks of IVF One of the major "risk factors" associated with IVF is the risk of failure. In 2002, the most recent year for which data are available, the success rate of cycles in which women used their own, nonfrozen eggs was 28.3% (Fig. 16-2). The success rate is better for women under age 35 (36.9%), but it decreases with every additional year (30.6% from age 35 to 37, 20.5% from age 38 to 40, and 10.7% for women age 41 to 42). Using a woman's own previously frozen eggs, the success rate is 25% per cycle; using donor frozen eggs, it is 29%. Consider a 36-year-old woman (the median age for an IVF patient) who undergoes an IVF cycle and freezes extra eggs or embryos from that cycle. If she does not become pregnant on the first try, she is likely to use her frozen eggs for a subsequent attempt. Each succeeding IVF cycle has a lower probability of success because of the greater failure rate with frozen eggs and also because she is getting older. Smoking, especially in the male partner, also reduces the success of IVF cycles, probably because it can damage DNA in the sperm.

In an effort to increase the chance of pregnancy, IVF clinics typically transfer multiple embryos into the uterus. However, this practice increases the risk of multiple births. Currently, 35% of births resulting from *in vitro* fertilization using a woman's own, nonfrozen eggs are multiples (twins or more). Using donor eggs, this rises to 42%. Multiple births pose greater potential problems for mother and infants alike. Women carrying multiple fetuses have a greater probability of requiring a cesarean delivery. Infants born from multiple gestations have a higher risk of prematurity, low birth rate, developmental delays, disabilities, and infant death. The medical, emotional, and societal costs of these complications can be very high. Although the rate of multiple births from ART has decreased slightly in recent years, complications resulting from multiple-fetus pregnancies remain a serious risk for families undergoing IVF. National guidelines in the United Kingdom now stipulate that no more than two embryos be transferred per cycle except for unusual circumstances. There are no federal laws or agencies in the United States that regulate ART procedures used in infertility clinics, although some states have regulations. The 1992 Fertility Clinic Success Rate and Certification Act requires all U.S. clinics performing ART to report annual data on IVF, GIFT, and ZIFT procedures to the CDC, but not all clinics currently comply.

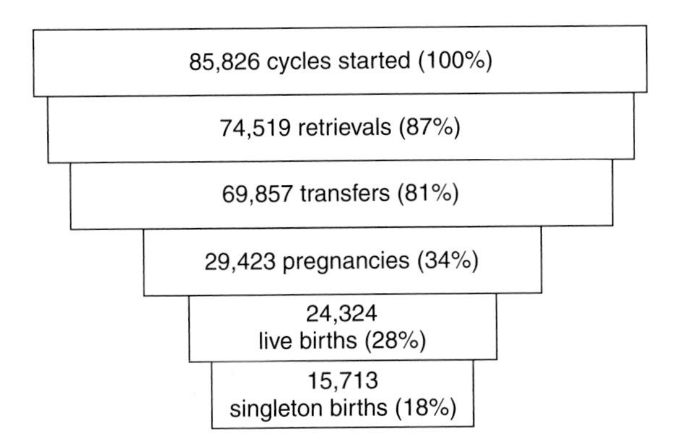

Figure 16-2 Outcome of IVF cycles in the United States using fresh nondonor eggs. Numbers in parentheses are percentages of cycles started that reached each stage. 2002 data from the CDC.

Although there is no clear evidence overall of an increase in congenital defects in babies conceived through *in vitro* fertilization, some studies have reported a higher risk. There is a small increased rate of genetic abnormalities in infants conceived using intracytoplasmic sperm injection. This technique is often used when the male partner's sperm count is very low or the sperm lack normal motility. Defective sperm are often cause by genetic mistakes, and men using ICSI could pass these genetic errors and their fertility effects to their sons. Pregnancies using IVF have an increased rate of miscarriage. Whether the high miscarriage rate is an effect of the IVF treatment itself or a reflection of fertility problems already present in women seeking IVF is not known.

About 30% of women undergoing ovarian stimulation with fertility drugs will develop *ovarian hyperstimulation*. The ovaries swell, and fluid accumulates in the abdomen. Symptoms include abdominal pain, bloating, intestinal disorders, and nausea. Most cases are mild and can be treated with rest and nonprescription painkillers; rarely, medical intervention is required.

Chapter 16, Box 2: Costs of Assisted Reproductive Technologies

In 2002, the latest year for which data are available, 85,826 women in the United States underwent assisted reproductive technology cycles (IVF, GIFT, ZIFT) and there were well over 200,000 cycles in Europe. The numbers almost certainly are higher today. A great majority of these procedures used *in vitro* fertilization, about half used intracytoplasmic sperm injection, and many also used frozen sperm and/or frozen preembryos. Nevertheless, despite manipulating variables such as type of culture medium and age of preembryos before implantation, whether frozen or not, the overall success rate in the United States (having a live birth from one attempt) is 28%. This means that a determined couple will repeat the procedure an average of four cycles until a successful pregnancy is established. The cost of an IVF cycle averages $12,400 in the United States, according to the American Society of Reproductive Medicine. (The fee is much lower in Europe.) The average expenditure for IVF per live birth of a baby is about $60,000. Added to this is the cost of delivery and postnatal hospital care. Because pregnancy rates decline and miscarriage rates increase with age, IVF costs per live birth are three times higher for women over 40 than for those 30 or younger. Insurance coverage of "high-tech" infertility treatments varies in each state.

The high failure rate of each IVF cycle, and the consequent need to repeat the procedure in most cases to achieve pregnancy, adds considerably to the overall costs. What is the reason why fewer than 20% of preembryos transplanted into the uterus implant successfully? A major factor is that many of the preembryos produced by IVF have chromosomal abnormalities such as aneuploidy. However, even in natural fertilization, it is estimated that at least 50% of embryos fail to implant or are lost to early miscarriage because of chromosomal abnormalities. Couples attempting to conceive naturally have only a 25% chance of becoming pregnant within a month of trying. Thus, the poor implantation results of IVF may simply be a reflection of the low natural reproductive capacity in humans. New tests for chromosomal abnormalities (*preimplantation genetic diagnosis*) in 3-day-old preembryos will raise the success rate of artificial implantation. Other reasons for loss of implanted embryos or fetuses (miscarriage) include uterine problems related to poor hormonal exposure and the basic difficulty of a human uterus to carry multiple pregnancies.

Another factor driving up the cost of IVF babies is the fact that 35% of IVF births in the United States are of multiple infants. Until recently, three, four, or even more preembryos

Continued on next page.

Chapter 16, Box 2 continued.

were routinely implanted into the uterus in these procedures. Often, more than one will implant. This is the reason that IVF and similar techniques produce about 31.6% twins and 3.8% triplets or more. This can be compared to 1.4% for fraternal twins and 0.016% for triplets born through natural means. The disturbingly large proportion of multiple gestations leads to more complications during or after pregnancy. The average hospital cost for the birth of twins is four times higher than for a singleton birth; for triplets, it is 11 times higher. These expenses do not include the monetary and emotional costs related to the long-term complications frequently associated with multiple births. Each additional fetus shortens a 40-week pregnancy by about 3.5 weeks, so many babies born by artificial reproductive technologies are born prematurely and have low birth weight. Even singleton newborns conceived through IVF have a greater chance of prematurity and low birth weight. These infants are more likely to suffer from mental retardation, learning disabilities, lung disorders, and cerebral palsy. Furthermore, low birth weight children tend to have high blood pressure (hypertension), heart disease, and osteoporosis when they become adults.

Women carrying multiple fetuses can elect to selectively abort one or more of the fetuses to increase the likelihood of a positive outcome of her pregnancy. Opting for *multifetal pregnancy reduction* is certainly a difficult decision for a woman highly motivated to become a mother. Furthermore, the procedure carries the risk of miscarriage of all of the fetuses. Sometimes the loss of a fetus in a multiple gestation occurs naturally. To minimize the possibility of multiple pregnancies, there is a recent tendency for IVF clinics to culture preembryos for 5.5 to 7.5 days and then implant blastocysts instead of earlier preembryos. However, only about 25% of blastocysts implant, a rate similar to that when earlier preembryos are implanted. Finally, there is a worldwide tendency at

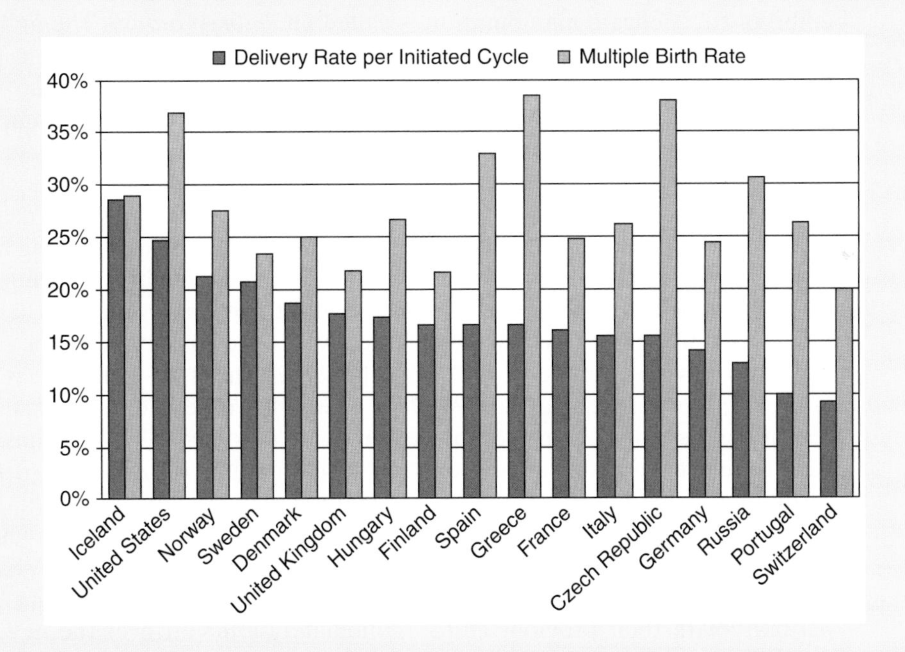

Comparison of IVF delivery rates and multiple birth rates in various countries.

Continued on next page.

Chapter 16, Box 2 continued.

present to reduce the number of embryos implanted to two. This is because using two preembryos has the same success rate as using three or more, but has a decreased risk of multiple births.

The monetary cost of having a baby through assisted reproductive technologies includes the costs of the IVF procedures, hospital delivery costs, and the possibility of extensive additional expenditures for complications such as multiple births, low birth weight, and prematurity. Added to this can be significant emotional costs, psychological stress, and loss of work time during and after the treatments. The psychological considerations can also be significant for couples who attempt one or more IVF cycles but ultimately fail to conceive. The high cost of ART inevitably leads to a disparity between those infertile couples who can afford the treatments and those for whom it is an economic impossibility. The inequality of access to these infertility treatments is an important ethical consideration regarding ART.

Artificial reproductive technologies have produced, since 1978, about one million new babies throughout the world. These procedures have given infertile couples priceless happiness and the ability to satisfy the powerful desire for procreation. Nevertheless, anyone choosing these procedures should be aware of the financial costs and potential risks to their future babies.

Preimplantation Genetic Diagnosis (PGD)

Remarkably, it is possible to remove a single cell from an early (approximately 8-cell) embryo without jeopardizing the future development of the embryo. This delicate manipulation is called an *embryo biopsy*. The ability of an embryo to *regulate* and compensate for the loss of one cell makes it possible for scientists to remove a cell to examine the genetic make-up of an embryo. Thus, embryos formed by *in vitro* fertilization can be screened for possible inherited disorders. For example, if a woman and her partner each carry a recessive mutation for a genetic disease, their child has a 25% chance of inheriting both mutations and having the disease. Couples can use IVF with preimplantation genetic diagnosis to eliminate the risk of passing a serious or fatal familial disease to their child. Only those embryos found to be free of genetic errors would be transferred into the woman's uterus. There are now over 1000 children who were genetically screened as embryos. The procedure may have prevented devastating illness and suffering in these children. However, the potential to use PGD to screen for other nonlife-threatening conditions has raised considerable controversy. For example, PGD has been used to select the sex of a baby, even by fertile parents who used *in vitro* fertilization for that purpose only. Several couples who already have a child with an inherited blood disease have used this method to "design" a baby who can be a stem cell donor for their sick child. Embryos obtained by IVF were tested genetically for the inherited disease, as well as for immunological compatibility with their previous child. Immediately after birth of the newborn, umbilical cord blood cells were transplanted into the affected sibling. Such preimplantation genetic testing that has no direct value to the newborn carries serious ethical concerns. Should parents be able to select for other heritable characteristics of their offspring, such as height and eye color? With a cost of over $20,000 for each IVF/PGD cycle, will only the wealthy be able to have "designer babies?"

Egg Donation

If a woman's ovaries are incapable of producing large, mature follicles or if she has a genetic disorder that she does not wish to pass on to her offspring, she could use the eggs produced by another woman (an *egg donor*). This could be a friend or family member or an anonymous donor. Egg donation is more complicated and expensive than sperm donation because the donor must undergo ovarian stimulation with fertility drugs and egg retrieval, as described previously. If fresh eggs are to be used, the infertile woman must also be treated with medication so that her cycle is synchronized with that of the egg donor. This prepares her uterus to receive the donor egg after it is fertilized. Eggs can also be frozen, either before or after being fertilized, but the success rate is lower when using frozen eggs and embryos. Because the quality of a woman's eggs diminishes with age, the likelihood of implantation decreases and the chance of miscarriage increases. Thus, women in older age groups (over age 39) are more likely to use donor eggs. Several babies have been born using this method, including some to postmenopausal grandmothers! The success rate using fresh donor eggs is around 50% regardless of the age of the infertile woman. Of course, an infant produced from a donor egg will be genetically unrelated to the mother (unless the donor is a relative).

Scientists are now refining techniques that allow a woman to be her own egg donor. Women undergoing chemotherapy and radiation for cancer treatment can suffer damage to their ovaries, resulting in infertility. Several years ago, when a Belgian woman faced chemotherapy for Hodgkin's lymphoma, her doctor made a radical suggestion. Why not preserve a piece of her healthy ovary before undergoing the destructive treatment? A section of her ovary was removed and frozen. Five years after the end of her chemotherapy, the piece of tissue was implanted under her ovary, and she conceived less than a year later. Although there is a slight possibility that the egg grew from her damaged ovaries, this transplant procedure may provide hope for fertility renewal of cancer patients and could also potentially prolong fertility in older women (see also Chapter 7).

Gamete or Zygote Intrafallopian Transfer

Gamete intrafallopian transfer (GIFT) is used for infertile women who are ovulating but have blocked oviducts (fallopian tubes) or for infertile couples who, for religious reasons, wish to avoid fertilization outside the human body. Mature ova are removed from large follicles in the infertile woman's ovary, as in the IVF method. Then the ova are inserted by laparoscopy into one of the woman's oviducts below the point of blockage and the husband's sperm is also placed into the oviduct with the ova. Fertilization and implantation then can occur; steroid hormones (estrogen, progesterone) may be administered to assist implantation and prevent miscarriage.

In a variation of this method, ova are removed and fertilized with the husband's sperm (AIH) *in vitro*. Then, zygotes (single diploid cells) are inserted into the oviduct and allowed to travel down the oviduct before implantation. This is called *zygote intrafallopian transfer* (ZIFT). Success rates for ZIFT and GIFT are about the same as for IVF. These techniques are not widely used and account for a very small fraction of the ART cycles performed in the United States.

Surrogate Mothers and Gestational Carriers

A woman who is infertile because she cannot produce eggs or cannot carry a pregnancy to term can hire another woman to carry her child. A *surrogate mother* is artificially inseminated with the sperm of the infertile woman's partner. The surrogate then becomes pregnant, delivers the child, and gives the child to the original couple. She is the biological mother of the child. Usually, a contract is written between the couple and the surrogate mother. Stipulations can include that the surrogate be married with children of her own, have a physical and psychological examination, be willing to surrender the child, and not use tobacco, alcohol, and other drugs while she is pregnant. Her fee for this favor can be $13,000 or more. Legal questions about this procedure, including who is the mother of the child and visiting rights of the surrogate mother, are complex and have resulted in several recent court cases. Gay men who wish to father children may use a surrogate mother.

An infertile woman may be able to produce eggs but her uterus cannot carry a pregnancy, e.g., because of a hysterectomy. After ovarian stimulation, her ova can be retrieved and fertilized *in vitro* by her partner's sperm, and the resulting embryo(s) placed in another woman's uterus. The woman who "hires out" her uterus is a *gestational carrier*. She is not related to the child genetically, but the legal, financial, and psychological issues are similar to those surrounding surrogate pregnancies.

Figure 16-3 summarizes some of the "high-tech" methods that can be used by infertile couples to have children.

Cloning and Ethical Issues in Assisted Reproduction

The development of techniques used in assisted reproductive technologies has made possible the manipulation of human eggs and embryos to achieve goals scarcely imagined until recent years. This includes skills that enable scientists to use *cloning* techniques on human cells. As discussed in Chapter 9, development of a human embryo begins after fertilization, when maternal and paternal genes combine and begin to send instructions to activate the waiting egg cytoplasm. Thus, the two components needed to initiate human development are a diploid complement of chromosomes and the egg cytoplasm equipped with specialized cellular machinery to respond to developmental instructions from the DNA. Cytoplasm from an egg cell is essential to initiate embryonic development in animals, but the genetic complement doesn't necessarily have to come from a sperm and an egg. In the 1950's Briggs and King developed a method to remove the nucleus from (*enucleate*) a frog's egg and replace the egg nucleus with a diploid nucleus harvested from one of the frog's *somatic cells* (i.e., a non-germ cell). Subsequent experiments proved that nuclei from highly differentiated cells such as those in the skin of a frog's foot can be "reprogrammed" by placing them in the egg cytoplasm, and that they can give rise to a wide variety of specialized cells. Tadpoles that develop from these recombined cells are genetically identical

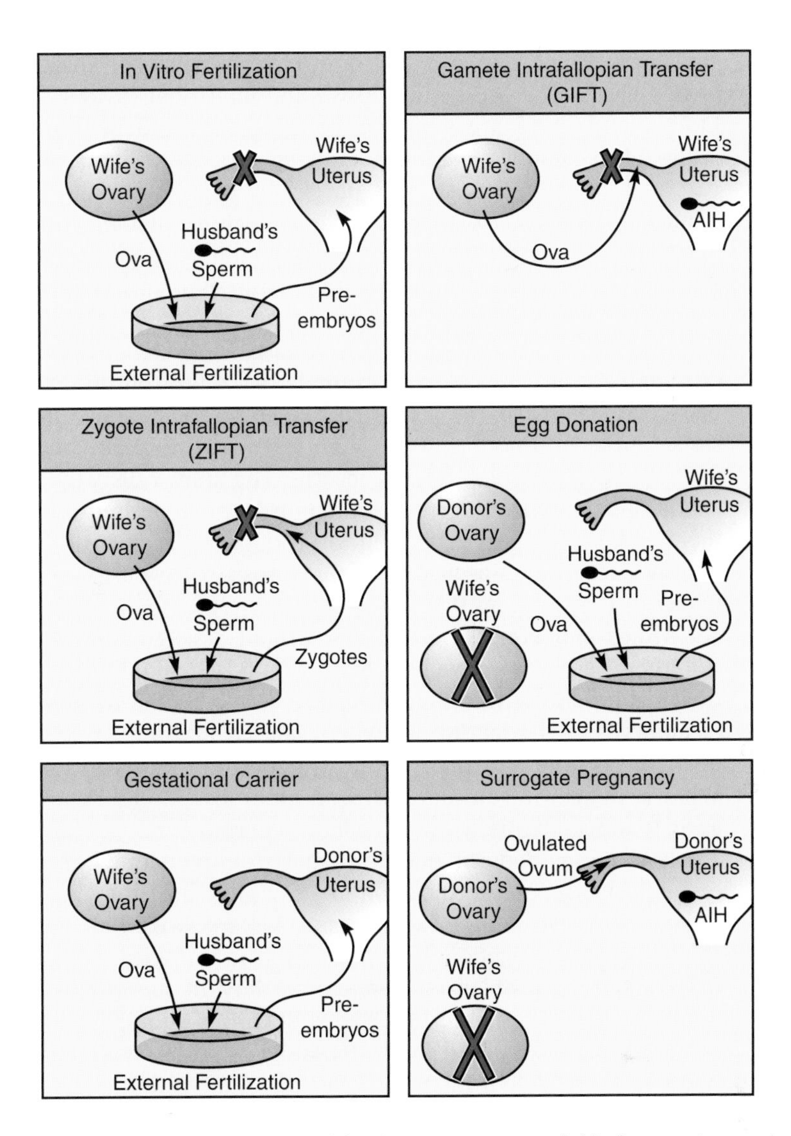

Figure 16-3 Some of the advanced fertility treatments available for couples in which the woman has ovarian malfunctions, tubal blockage, or uterine problems. X, abnormal function. AIH, artificial insemination with the husband's sperm. Donor sperm could be used for each procedure if the husband is infertile or if the woman is unmarried.

clones of the frog donating the nuclei. These nuclear transfer experiments first demonstrated the potential of animal cloning.

The first mammal to be cloned from adult DNA was a sheep, Dolly, by Scottish scientists at the Roslin Institute. A single cell from the udder of a donor sheep was placed adjacent to the enucleated egg cell from another sheep. An electrical pulse caused the two cells to fuse, placing the somatic cell nucleus inside the egg cell. After allowing the resulting hybrid cell to divide and develop for a few days *in vitro*, the preembryo was placed into the uterus of a third sheep, who gave birth to the cloned lamb. Since the birth of Dolly in 1997, many other mammals including goats, cows, mice, pigs, cats and dogs have been cloned through nuclear transfer technology. This process is expensive and inefficient (it took 276 tries to produce Dolly), and there are concerns about the overall health of cloned animals. Although techniques exist to potentially produce a human through cloning, the vast majority of scientists and physicians express strong ethical opposition to human *reproductive cloning*. However, many hope that *therapeutic cloning* will produce new tools to treat some of the most common, serious human diseases.

Therapeutic cloning involves the synthesis of preembryos for research and medicine. The objective is to make embryonic stem cells that are a genetic match for a patient suffering from a particular disease. These stem cells, which can potentially differentiate into any cell type in the body, come from the inner cell mass of the blastocyst (see Chap. 10). Using techniques described above, human cloned blastocysts can be produced and the stem cells extracted. The blastocyst is destroyed in the process, raising ethical concerns. The stem cells would then be grown in conditions that encourage differentiation into the desired cell type. In the future, therapeutic cloning may be used to produce replacement cells to treat disorders such as cancer, Parkinson's, and heart disease, or even to synthesize replacement tissues or whole organs that would not pose a risk of immunological rejection.

Moral and ethical questions surround assisted reproduction technologies. Will they lead to selective breeding? Would rich people pay poor people to carry their babies and, if this happened, who would be the parents? Should the death of embryos in the laboratory be considered murder? A recent case involved a divorcing couple and their battle over custody rights to their frozen embryos. What is done with unwanted frozen embryos? Could some clinics sell or give away frozen embryos, without a couple's permission?

It is these kinds of questions that prompted the U.S. government to ban research on external fertilization in 1975 while an Ethics Advisory Board of the Department of Health, Education and Welfare looked into the matter. In March 1979, this board sanctioned such research in America with the ultimate goal of overcoming the problem of human infertility. In June 1979, the board approved of research on external fertilization with certain stipulations, some of which were (1) no embryos fertilized in the laboratory can be maintained for over 14 days (the maximum time it takes for a normal embryo to implant fully in the uterus); (2) people donating eggs or sperm must be notified about, and must approve of, the purposes and uses of their cells; (3) if the embryo is to be implanted back into the uterus, the gametes must be obtained from married couples; and (4) the public must be informed about incidences of abnormal offspring resulting from these procedures.

The U.S. government then again banned federal support of any research involving human preembryos or embryos in 1980. In 1993, however, Congress lifted the 1980 ban on government funding for such research. In 1994, two committees of the U.S. National Institutes of Health supported the funding of studies of very early preembryos, including those developed solely for research. Then, in late 1994, the U.S. federal government banned the support of research on human embryos created for such a purpose, but left open the possibility of supporting research on human embryos discarded by fertility clinics. In 2001, President Bush signed an executive order banning government funding for research that would involve the destruction of stored embryos.

The achievement of Drs. Steptoe and Edwards was indeed a landmark of reproductive medicine, and their method has great promise in helping couples who otherwise are infertile to have children. More than one million children around the world have been conceived by ART. However, the success rate remains low, the procedure is prohibitively expensive for many couples, and the risk of multiple births remains high. Improvements in the methods and technologies of ART are rapidly bringing new hope for infertile couples, along with additional medical and ethical concerns among the public.

Adoption

Infertile couples, of course, have the option of adopting a child. Adopted children usually are the product of unwanted pregnancies or conditions in which the biological parents want their child but are unable to care for it. In traditional, "closed" adoptions, the people who give up a child for adoption are not told who adopted their child. A person (single man or woman) or a couple who desires to adopt should first contact an adoption agency. Then, the agency determines if the couple or person fits their standards. What are common standards? For many agencies, a couple must be infertile and not more than 40 years older than the child they adopt. Additionally, a home study is done to determine if the family will provide adequate physical and emotional care for the adopted child. If the couple is approved, they may have to wait up to 5 years for a healthy American infant; this wait is often less for a foreign child or a "special-needs child" who may be older, non-Caucasian, or have physical, developmental, or psychological challenges. Recently, private "open" adoptions, in which the adopting parents know the birth mother, are becoming more frequent, especially for couples who would not be approved by an agency. Adoption of an infant is more expensive than pregnancy and delivery.

Adopted children can be as happy and healthy as children raised by their biological parents. Some adopted children, especially when over 2 years of age when adopted, have adjustment problems that may be long-lasting. Adopted parents also have adjustments to make. For the most part, however, families with adopted children are no different from those with biological parents and offspring.

Chapter Summary

A woman can be infertile for several reasons. If her pituitary gland is not secreting enough FSH and LH to cause ovulation, she can be treated with GnRH, clomiphene, or hMG followed by hCG. If her ovaries are not responding to FSH and LH because of the presence of cysts, tumors, or endometriosis scars, ovarian surgery can help. Blocked oviducts can sometimes be opened by forcing gas into them or by surgery. If a woman's uterus does not support implantation, treatment with an estrogen and a progestogen may allow implantation. If her cervix is hostile to her husband's sperm, artificial insemination may be necessary.

An infertile male with a low sperm count can be helped by the administration of gonadotropins or GnRH. If his testes are damaged or if he is producing antibodies to his own sperm, artificial insemination may be the answer.

Nonfunctional male sex accessory ducts or glands can be made functional with androgen treatment or surgery.

Artificial insemination utilizes sperm from the husband (AIH) or from a donor (AID). One problem with this procedure is the introduction of nonmotile or abnormally shaped sperm. *In vitro* fertilization is the most frequently used high-technology solution to infertility. After a woman's ovaries are induced to produce several mature ova, the eggs are removed and fertilized with the husband's sperm in the laboratory. Embryos are then placed back into the woman's uterus. In many IVF procedures, a sperm cell is injected directly into the egg. Other alternatives for infertile couples include gamete intrafallopian transfer, zygote intrafallopian transfer, and the use of frozen sperm, eggs, or embryos. Some infertile couples hire another woman to bear a child for them, utilizing artificial insemination with the husband's sperm and even embryo transfer to a surrogate mother. Infertile couples may also be able to adopt a child.

Further Reading

Aitken, R. J., and Irvine, D. S. (1996). Fertilization without sperm. *Nature* **379**, 493–495.

Chen, I. (1990). Egg glitch: Too cold or old. *Sci. News* **138**, 263.

Collins, J. A. (1994). Reproductive technology: The price of progress. *N. Engl. J. Med.* **331**, 270–271.

Edwards, R. C. (1990). A decade of *in vitro* fertilization. *Res. Reprod.* **22**(1), 1–2.

Fackelmann, K. A. (1990). Zona blasters. *Sci. News* **138**, 376–379.

Fackelmann, K. A. (1994). Cloning human embryos. *Sci. News* **145**, 92–95.

Fackelmann, K. A. (1994). Embryo research panel ignites debate. *Sci. News* **146**, 212.

Fackelmann, K. A. (1994). Test-tube diagnosis: Analyzing embryos for genetic flaws. *Sci. News* **146**, 286–287.

Fackelmann, K. A. (1994). Germ cell transfer boosts fertility. *Sci. News* **146**, 356.

Hall, S. S. (2004). The good egg. *Discover* **25**, 30–39.

Lovell-Badge, R. (1996). Banking on spermatogonial germ cells: Frozen assets and foreign investments. *Nature Med.* **2**, 638–639.

Menning, B. E. (1987). "Infertility: A Guide for the Childless Couple." Prentice-Hall, Englewood Cliffs, NJ.

Phipps, W., *et al.* (1987). The association between smoking and female infertility as influenced by cause of infertility. *Fertil. Steril* **48**, 377–382.

Plumez, J. H. (1982). "Successful Adoption." Harmony Books, New York.

Rensberger, B. (1994). Human embryo clones: Dividing fact and fiction. *J. NIH Res.* **61**, 26–27.

Schwartz, L. (1989). Surrogate motherhood III: The end of a saga? *Am. J. Family Ther.* **17**, 67–73.

Seibel, M. (1988). A new era in reproductive technology. *N. Engl. J. Med.* **318**, 828–834.

Sherins, R. J. (1995). Are semen quality and male fertility changing? *N. Engl. J. Med.* **332**, 327.

Witkin, S. S., and David, S. S. (1988). Effect of sperm antibodies on pregnancy outcome in a subfertile population. *Am. J. Obstet. Gynecol.* **158**, 49–62.

Advanced Reading

Auger, J., *et al.* (1995). Decline in semen quality among fertile men in Paris during the past 20 years. *N. Engl. J. Med.* **332**, 281–285.
Callahan, T. L., *et al.* (1994). The economic impact of multiple-gestation pregnancies and the contribution of assisted-reproduction techniques to their incidence. *N. Engl. J. Med.* **331**, 244–249.
Dodson, W. C. (1989). Role of gonadotropin releasing hormone agonists in ovulation induction. *J. Reprod. Med.* **34**(1), Suppl., 76–80.
Howards, S. S. (1995). Treatment of male infertility. *N. Engl. J. Med.* **332**, 312–317.
Infante-Rivard, C., *et al.* (1993). Fetal loss associated with caffeine intake before and during pregnancy. *J. Am. Med. Assoc.* **270**, 2940–2943.
Morell, V. (1995). Attacking the causes of "silent" infertility. *Science* **269**, 775–777.
Neumann, P. J., *et al.* (1994). The cost of a successful delivery with in vitro fertilization. *N. Engl. J. Med.* **331**, 239–243.
Skakkebaek, N. E., *et al.* (1994). Pathogenesis and management of male infertility. *Lancet* **343**, 1473–1479.

PART FIVE

Special Topics in Human Reproductive Biology

CHAPTER SEVENTEEN

Brain Sex

Introduction

This chapter is about differences between the male and the female human brain from fetus through adulthood. The differences discussed are structural (e.g., size of certain brain areas), chemical (e.g., amount of neurotransmitters in certain brain regions), physiological (e.g., metabolic activity in certain brain regions during a specific state; differences in reproductive hormone control systems), psychological (aggression; sex role; gender identity), and cognitive (e.g., verbal fluency; visual-spatial ability). For our purposes, we discuss humans; much is known about this topic in other mammals such as the laboratory rat, but it turns out that other mammals may differ somewhat from humans in this regard.

This is a fascinating and yet sensitive and controversial topic. First, a few of the brain sex differences between women and men are described in some scientific studies but are not verified by others. This chapter emphasizes those sex differences that appear to be valid and observed repeatedly. Second, some of the sex differences are not large, and there can be much overlap between the sexes, with just the average of the two sexes being different. Third, and most important, is the "nature versus nurture" question. That is, is a sex difference in human brain structure or function a biologically predetermined genetic or hormonal sex difference (like having a uterus or not) or is a brain sex difference the result of experience, e.g., the result of treating boys and girls or men and women differently, with these experiences in turn causing the developing or adult brain to change its structure or function? It is possible that some brain sex differences can be the result of pure "nature" or pure "nurture," but in reality most probably result from the interaction of biology *and* experience, as is true for many human traits. Because this book is about biology, however, the "nature" component of brain sex differences is emphasized, but remember that "nurture" affects the brain also. Finally, any sex differences in brain function should not be used to judge that one sex is superior to the other or to discriminate against either sex.

Biological Causes of Brain Sex Differences

Most animal studies suggest that biology can influence brain sex differences in two major ways. Let us use a sex difference in the structure of a certain brain region as an example, with this structural difference causing a sex difference in adult behavior. One way that this adult brain sex difference can appear is by an *organizational effect* of a hormone early in life. For example, remember from Chapter 5 that testosterone secretion from the fetal testis produces masculinization of the genitalia, i.e., testosterone (via conversion to DHT), and organizes the genitalia in a male direction, forming a penis and scrotum. Absence of testosterone results in feminization of the genitalia (i.e., formation of a clitoris and labia). An organizational effect of testosterone on the brain would masculinize the brain in a similar fashion, whereas the absence of testosterone would feminize the brain. Four more important points about the organizational effects of testosterone on the brain are (1) they must occur during a *critical period* (a specific time during development, possibly between weeks 16 and 28 of development in humans); (2) they are permanent and irreversible; (3) there can be different critical periods for different organizational effects; and (4) they can be anatomical and/or physiological.

There is good evidence that, in organizing the brain in a male direction, testosterone is actually converted by the enzyme aromatase (in the brain itself) to estradiol so it is actually a "female sex hormone" that masculinizes the male brain! Why does the brain of fetal or juvenile females not become masculinized by placental or ovarian estrogen? In rats, there is a blood protein (α-fetoprotein) that binds to estrogen and prevents it from entering the brain. In human fetuses, however, although there is α-fetoprotein, it does not bind to estrogen so we do not know why the human female brain is not masculinized by estrogen. Perhaps it is because there is not enough estrogen circulating in the female fetal blood during the critical period.

Some brain sex differences could require only organizational effects of hormones. In many cases, however, even though the brain of males and females has been organized by testosterone to be different in the fetus, the functional aspects of this difference (e.g., physiological, psychological, or behavioral) may not appear, or not be maximally expressed, until this brain area is acted upon by a certain hormonal condition at puberty and in adulthood. Thus, adult hormones can have an *activational effect* on an organized sex difference in the brain. For example, let us say that a brain area controlling aggression is organized (masculinized) by testosterone activation early in development. The adult expression of the function of this brain area will also require high blood levels of testosterone. Most evidence suggests that, unlike in other mammals, the activational effects of testosterone in humans do not work through conversion to estrogen, but rather the testosterone acts directly on the brain or via conversion to dihydrotestosterone (DHT; see Chapter 5). Therefore, orchidectomy of an adult male (thus lowering testosterone levels) would reduce aggression. In addition, the administration of testosterone to this male should again increase his aggression. Classically, however, administration of testosterone to an adult female with an organized sex difference in the aggression area of the brain should not cause her to be more aggressive because her "aggression brain area" is different from that in a man. In summary, organizational effects of a hormone can cause a permanent sex difference in a region of the adult brain, but the maximal expression of this difference often, but not always, needs the presence of a specific activational effect of the same or another hormone (Fig. 17-1). Both organizational and activational effects of gonadal steroid hormones require specific receptors in brain neurons.

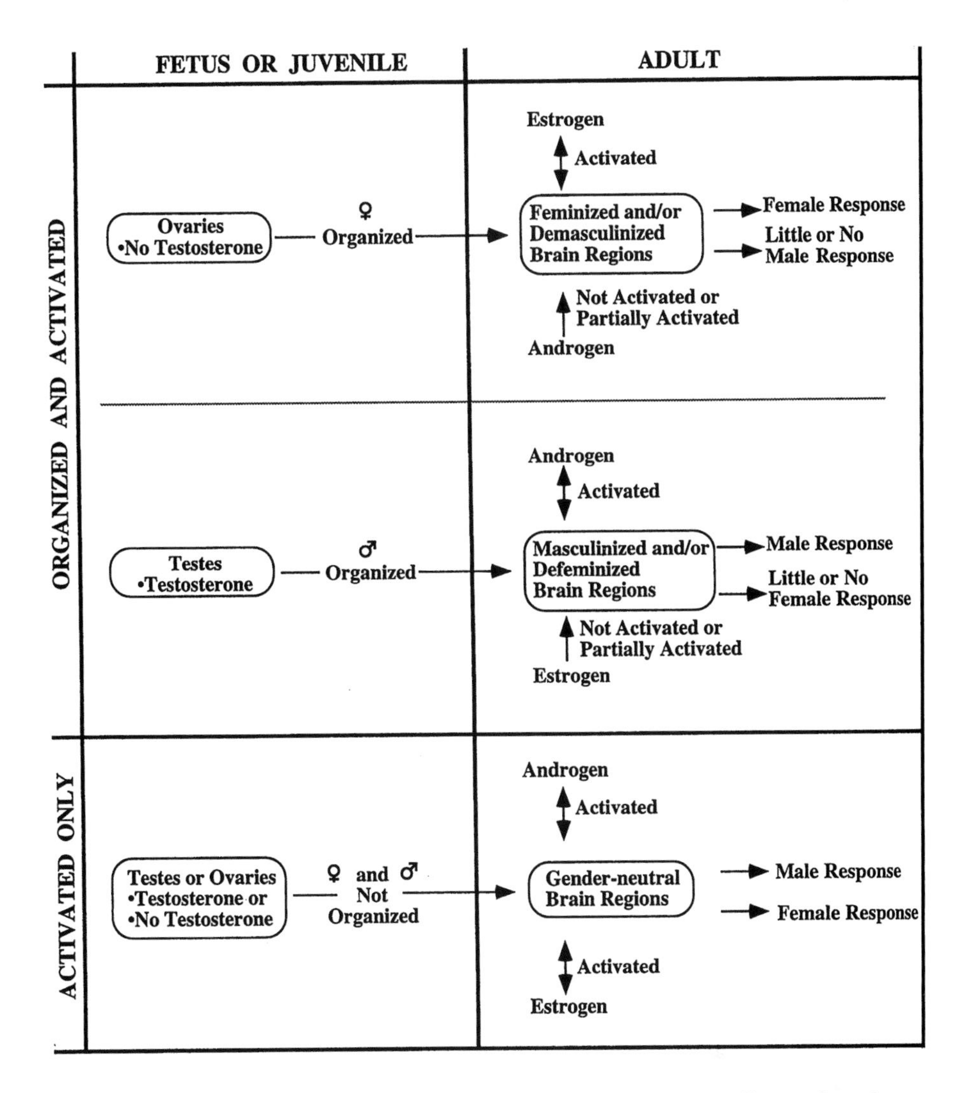

Figure 17-1 (Top two rows) Organizational and activational effects of androgen (e.g., testosterone) on the brain. A brain organized by androgen will be activated by androgen in the adult to produce a male response but will not respond to estrogen with a female response. A brain that has not been organized by androgen will respond in the opposite manner. Some organized brain areas do not need hormonal activation. (Bottom row) Activated-only effects of hormones on the brain. Regardless of the biological sex, the brain will respond in a male manner if androgen is present and in a female manner if estrogen (e.g., estrodiol) is present. A double arrow means that the activational effect is reversible. It is feasible that a sex difference could be solely the result of an organizing effect of a hormone, with no activation required. (The authors thank Ned Clark for his help with this illustration.)

There is a second possible biological cause of a sex difference in brain function. That is, the sex difference is *only* due to the activational effect of a hormone(s). In this scenario, no permanent organized differences exist in the brain of adult females and males. The presence of testosterone, therefore, would cause a masculinized pattern of behavior in either sex, whereas the absence of testosterone would demasculinize the behavior (cause it to be reduced or absent). The activational effect of testosterone in this case would be direct or after conversion to DHT, and not by conversion to estrogen. The presence of high blood levels of estrogen would feminize the behavior in either sex, whereas the absence of estrogen would defeminize the behavior. Activated-only hormonal effects are reversible, i.e., they disappear when the hormone is not present. The same is true for the activational effects on organized sex differences described previously. A good analogy of an "activational only" control of a sex difference relates not to the brain but to the mammary glands. Normally, milk production in these glands requires a certain combination of hormones (e.g., an estrogen, progesterone, and prolactin). This combination usually only is present in the blood of nursing women following birth (see Chapter 12). However, if a man is treated with this hormone combination, his mammary glands not only can produce milk, but he can nurse a baby! Thus, lactation is an "activated-only" action of hormones, and there is no organized sex difference in the mammary gland tissue itself.

A final possible biological cause of a sex difference in the human brain is genetic. That is, female cells have two X chromosomes (one inactive), whereas males have one X and one Y. Perhaps a gene(s) on the Y chromosome, for example, could result in a sex difference in the brain. There is little evidence of this in relation to the brain, although one study in rats showed that sex differences in the structure and function of certain dopamine-containing neurons appeared in a culture dish without previously being exposed to sex hormones (i.e., before gonadal differentiation). Nevertheless, we shall assume here that any biological sex differences in the brain are related to hormone differences.

Although it is probable that early organizational effects of hormones do cause some of the sex differences in brain structure and function in humans described later, it has been common dogma that the adult human brain (and in fact the adult vertebrate brain in general) is unable to grow new neurons. It has also been assumed that solely activational effects of hormones on the human brain described in this chapter (e.g., on the surge center and cognition) do not cause brain structural changes. However, it is well known that certain brain nuclei that control song production in birds exhibit hormonally controlled seasonal changes in the number of cells in certain brain regions. That is, there is neuronal cell death and then production of new neurons seasonally. We now know that new neurons can appear in the adult human brain also. Evidence shows that one brain area in the male and female human hypothalamus, the vasopressin-containing cells of the *suprachiasmatic nucleus* (SCN; see later), has more cells in autumn than in summer. Thus, some of the activational effects on the human brain could actually change neuron number, synaptic connections, and neurotransmitter function (see HIGHLIGHT box 17-2).

A final clarification: the brain can be masculinized or feminized. If the degree of development of a part of the brain controls both male and female aspects of a sexual behavior, depending on early exposure to testosterone, for example, then a heterosexual male's brain would be masculinized by testosterone and a heterosexual female's would be demasculinized by the lack of testosterone. If, however, there are separate brain areas, one causing male sexual behavior and the other female sexual behavior, then the brain of males would be masculinized and defeminized, and the brain of females would be feminized and demasculinized. Thus, a person's brain could potentially be masculinized but not defeminized, or feminized but not demasculinized, which may be the case for some brain functions in homosexual or transsexual individuals. To simplify matters in the remainder of this chapter, however, only the terms *feminized* and *masculinized* are used.

Sex Differences in Neonatal Behavior

There are sex differences in neonatal behavior that could be due to genetic or hormonal effects on the fetal brain. However, they could also be due to differences in the treatment of newborn boys and girls in the first few hours after birth. At any rate, newborn boys tend to sleep less, be more irritable, are more distressed when separated from their mother, have greater muscular strength, are sturdier and larger, and exhibit greater reflexive startle during sleep. Newborn girls, however, are more advanced in skeletal and neural development, are more sensitive to tactile, oral, and visual stimuli, are more responsive to sweet tastes, have more rhythmic mouth movements, are more orally receptive, and begin nursing earlier.

Sex Differences in Childhood Behavior

There are clear sex differences in the psychology and behavior of children. The left cerebral hemisphere develops a dominance over the right earlier in girls, which relates to the fact that girls, on average, learn to talk earlier and exhibit earlier verbal fluency than boys. Boys, however, have visual-spatial and mathematical skills that exceed, on the average, those of girls. In addition, young boys tend to spend more energy in aggressive ("rough and tumble") play than girls, who in turn tend to spend more time with fantasies and play related to mothering.

About 10% of human children are extremely inhibited (shy). There is a sex difference in what happens in these shy children when they mature. Boys who were shy at 21 months of age tend to lose their shyness beginning around age 7, but shy girls do not become even a little less shy until about age 16. Although extensive shyness has a genetic basis, we do not know if these sex differences in shyness are due to brain differences or to learning and/or experiential differences between boys and girls as they grow up.

Some evidence suggests that testosterone secretion from the testes of male human fetuses could be responsible for some of the early sex differences in neonatal and childhood behavior. That is, testosterone might have organizational effects on the human brain early on that could influence sex differences in the behavior of children. Unfortunate "natural experiments" occur when female human fetuses are exposed to androgens [e.g., during congenital adrenal hyperplasia (CAH); see Chapter 5] or if the mother is exposed to androgenic progestogens during pregnancy to prevent miscarriage. The behavior of these androgen-exposed young females after birth is interesting. Although these girls are masculinized physically, they usually are raised as girls. Studies indicate, however, that they are "tomboyish" in their behavior, attitudes, and preferences. They usually exhibit relatively less maternal fantasy and play and are more aggressive than their sisters. Similarly, female fetuses of Rhesus monkeys exposed to androgens exhibit male-like play and aggressive behavior after birth. Concluding, however, that exposure of human females to androgen when they are fetuses produces male-like childhood behaviors is premature. Rather, it is possible that their early postnatal experience, especially the behavior of parents toward the infants, could have produced their tomboyish behavior.

Sex Differences in Adult Cognition and Motor Skills

There are no sex differences in the general intelligence of men and women, and when corrected for body size, the brain weight of men and women is similar (see HIGHLIGHT box 17-2). However, when psychologists measure a few specific kinds of cognition (problem-solving abilities) and motor skills, there are average differences between men and women, even though there is considerable overlap between the sexes. These average sex differences in cognition (see HIGHLIGHT box 17-1) include that women are better at perceptual speed (e.g., grouping together similar objects; identifying landmarks), verbal fluency (e.g., spelling, grammar), precision manual tasks (e.g., manual dexterity), and mathematical calculation (e.g., adding and subtracting). Men, however, are better at visual-spatial tasks (e.g., mentally rotating or manipulating an object; navigating a maze or map), target-directed motor skills (e.g., dart throwing), and mathematical reasoning (problem solving with numbers). At least some of these sex differences are present in children, although this childhood gender gap in cognition may be narrowing; see Holden (1994) in Further Reading section. It is not yet clear if these sex differences in cognition and motor skills are present in all human cultures.

In humans, there appears to be lateralization of function between left and right sides of the cerebral cortex. For example, the left side is more important for verbal processing, whereas the right is more important for visual-spatial function. Also, there appears to be a sex difference in the degree of lateralization. The right and left male brain is more lateralized, i.e., the two sides of the female brain are more similar, and there appears to be more communication between the right and the left cerebral cortex in women.

Techniques such as *functional magnetic resonance imaging* and *positron emission tomography* can visualize the metabolic activity of certain brain regions during specific activities. These techniques have demonstrated additional sex differences in cognition in the human brain. For example, during attempts to rhyme words, men use a region in the lower frontal lobe of their left cerebral cortex (a speech-related area), whereas women use this region equally on both right and left sides (Fig. 17-2).

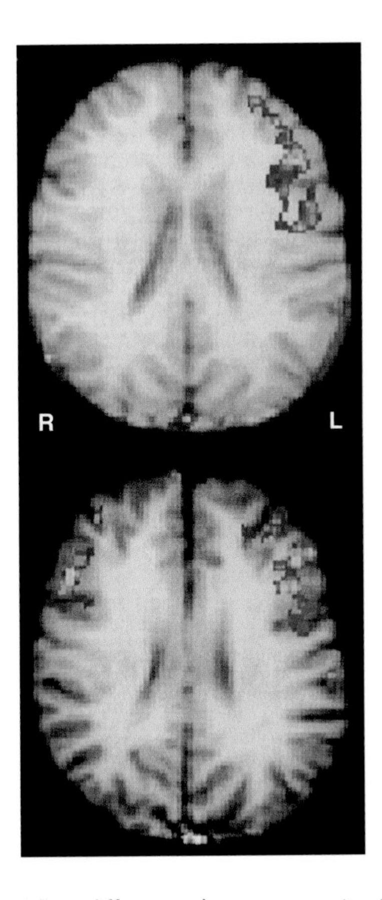

Figure 17-2 Brain blood flow differences between men (top) and women (bottom) trying to decide if two words rhyme. The yellow and red areas indicate that a brain region has more blood flow (is more active) than in the blue areas. Note that men use the left side of their cerebral cortex only, whereas women use both sides.

Sex differences in the structure of the human brain (Table 17-1) could relate to the previously mentioned differences in cognition between the sexes. In general, some nerve fiber tracts that connect right and left sides of the brain are larger in women, perhaps relating to the greater communication between the two sides. These include the massa intermedia, splenium of the corpus callosum, and anterior commissure (Fig. 17-3; Table 17-1). Other structural sex differences could also relate to the greater verbal fluency in women, (e.g., larger planum temporale of the cerebral cortex; Table 17-1). The sex difference in lateralization of the visual area of the cerebral cortex (Table 17-1) could relate to the greater visual-spatial ability in men.

Are some of these cognitive sex differences in humans the result of early organizational effects of hormones? At least one of the previously mentioned

Table 17-1 Some Known Sexually Dimorphic Areas in the Adult Human Central Nervous System[a]

Structure	Brain region	Sexual dimorphism in heterosexuals (size)[b]	Homosexual males[a]	Homosexual females[c]	Transsexual males[d]	Function of structure
Suprachiasmatic nucleus	Hypothalamus	Sex difference in shape, not cell number	Enlarged	?	No difference	Controls daily and seasonal rhythms
Sexually dimorphic nucleus, or INAH-1 (first interstitial nucleus of the anterior hypothalamus)	Hypothalamus	M > F	No difference	?	No difference	Male mating behavior
INAH-2 (second interstitial nucleus of the anterior hypothalamus)	Hypothalamus	M > F	No difference	?	?	Some aspect of male sexual behavior?
INAH-3 (third interstitial nucleus of the anterior hypothalamus)	Hypothalamus	M > F	Reduced	?	?	Some aspect of sexual orientation
Bed nucleus of the stria terminalis (central region)	Hypothalamus	M > F	No difference	No difference	Reduced	Connects amygdala (which influences sexual behavior) with hypothalamus; gender identity?
Massa intermedia	Thalamus	F > M	?	?	?	Connects right and left thalamus (a sensory relay region)
Anterior commissure	Cerebrum	F > M	Enlarged	?	?	Communication between right and left cerebral cortex; connects olfactory bulbs on either side

Planum temporale in temporal lobe of cerebral cortex	Cerebrum	Larger on left than on right in both sexes, but less lateralization (right larger too) in females	?	?	?	Auditory speech functions; melodies and speech tones
Splenium (posterior end of corpus callosum)	Cerebrum	F > M	Enlarged	?	?	Communication between right and left visual cortex
Cuneate area in occipital lobe of cerebral cortex	Cerebrum	Larger on right in F; larger on left in M	?	?	?	Vision
Caudate nucleus of basal ganglia	Cerebrum	Larger on right in M; larger on left in F	?	?	?	Unconscious skeletal muscle movement
Spinal nucleus of bulbocavernosus muscle (ejaculatory center)	Spinal cord	M > F	?	?	?	Contracts bulbocavernosus muscle during orgasm

[a]For the location of some of these areas, see Fig. 17-4.
[b]M, male; F, female.
[c]As compared to heterosexual individuals of the same genetic sex; differences present regardless of the presence or absence of AIDS before death.
[d]As compared to homosexual and heterosexual individuals of the same genetic sex.

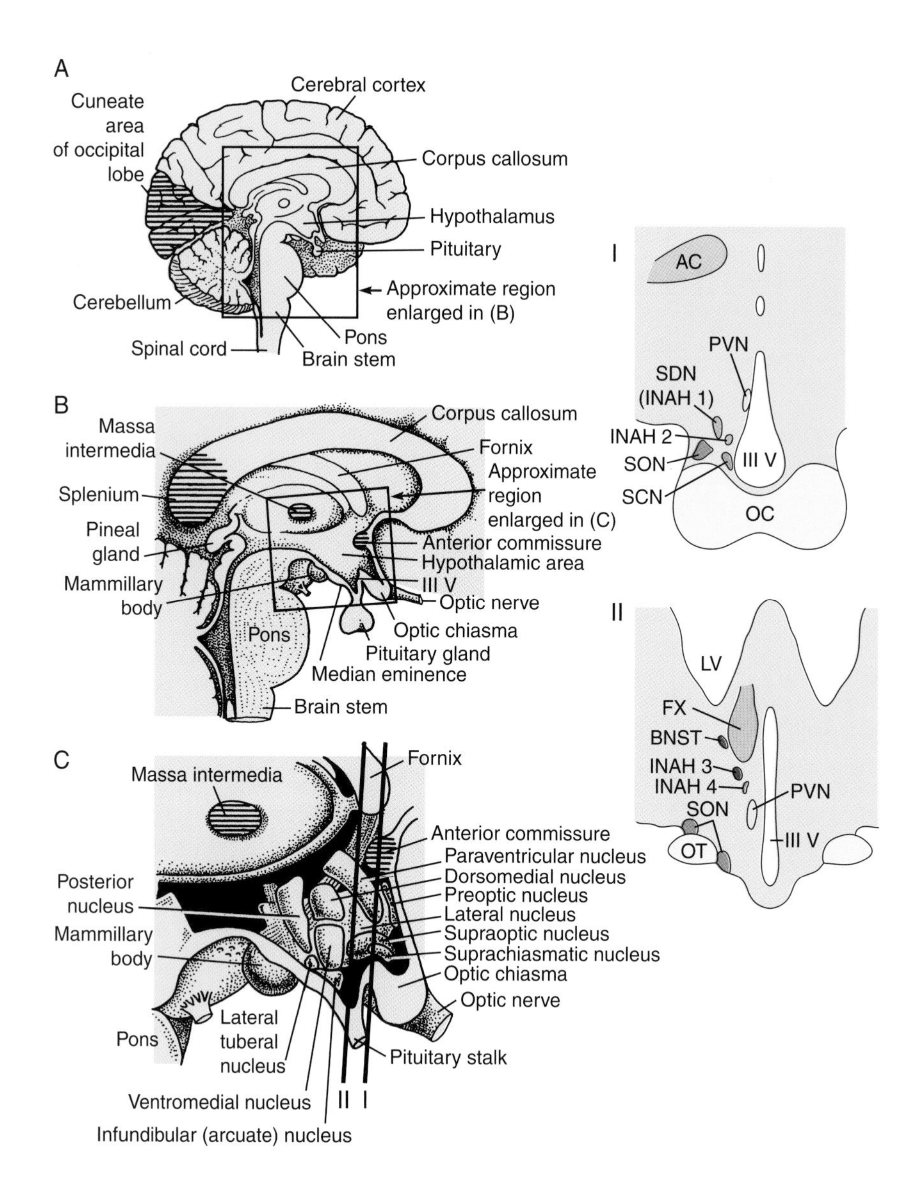

Figure 17-3 Schematic diagrams of the locations of some human brain regions (striped areas) that exhibit a size difference between men and women (see also Table 17-1 and Fig. 17-6). (A, B, and C) Side views in the middle of the brain (anterior to the right). Note that B is an enlargement of the boxed region in A and that C is an enlargement of the boxed region in B. In C, several hypothalamic nuclei are shown. I and II are frontal sections through C in the medial preoptic area of the anterior hypothalamus; see lines in C. Sexually dimorphic nuclei in I and II are also shown as stripes. See Table 17-1 for a list of which is larger in women or men. In I and II, AC, anterior commissure; BNST, bed nucleus of the stria terminalis; FX, fornix; INAH, interstitial nuclei 1, 2, 3, and 4 of the anterior hypothalamus; LV, lateral ventricle; OC, optic chiasma; OT, optic tract; PVN, paraventricular nucleus of the hypothalamus; SCN, suprachiasmatic nucleus; SDN, sexually dimorphic nucleus or INAH 1; SON, supraoptic nucleus; IIIV, third ventricle. Sexually dimorphic regions in the human brain not shown (planum temporale in the temporal lobe of the cerebral cortex; caudate nucleus; spinal nucleus of the bulbocavernosus of the spinal cord) are indicated in Table 7-1.

Chapter 17, Box 1: Evolution of Sex Differences in Problem-Solving Skills

As discussed in the text, on average there are sex differences in problem-solving skills between men and women, even though there is substantial overlap. On average, women are better at verbal fluency (spelling, grammar), perceptual speed (speed at which objects with common appearance are matched), verbal and item memory (such as listing objects of the

Problem-Solving Tasks Favoring Women

Women tend to perform better than men on tests of perceptual speed, in which subjects must rapidly identify matching items–for example, pairing the house on the far left with its twin:

In addition, women remember whether an object, or a series of objects, has been displaced:

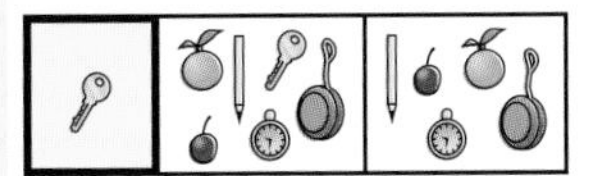

On some tests of ideational fluency, for example, those in which subjects must list objects that are the same color, and on tests of verbal fluency, in which participants must list words that begin with the same letter, women also outperform men:

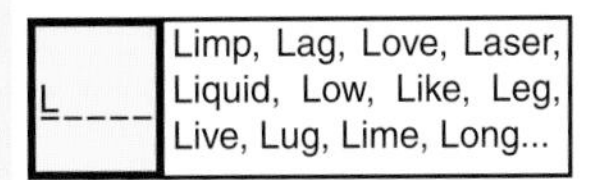

Women do better on precision manual tasks; that is, those involving fine-motor coordination such as placing the pegs in holes on a board:

And women do better than men on mathematical calculation tests:

77	$14 \times 3 - 17 + 52$
43	$2(15 + 3) + 12 - \frac{15}{3}$

Problem-Solving Tasks Favoring Men

Men tend to perform better than women on certain spatial tasks. They do well on tests that involve mentally rotating an object or manipulating it in some fashion, such as imagining turning this three-dimensional object

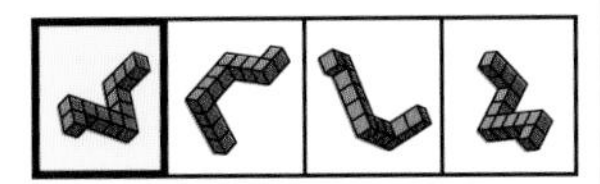

or determining where the holes punched in a folded piece of paper will fall when the paper is unfolded:

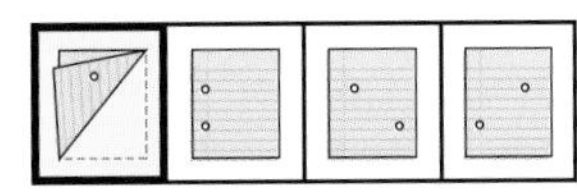

Men also are more accurate than women in target-directed motor skills, such as guiding or intercepting projectiles:

They do better on disembedding tests, in which they have to find a simple shape, such as the one on the left, once it is hidden within a more complex figure:

And men tend to do better than women on tests of mathematical reasoning:

1,100	If only 60 percent of seedlings survive, how many must be planted to obtain 660 trees?

Sex differences in cognition and motor skills between men and women.

Continued on next page.

Chapter 17, Box 1 continued.

same color or words beginning with the same letter), plotting or recall of direction between objects (so-called spatial mapping), and fine motor skills (such as finger dexterity). Finally, women excel in mathematical calculations (adding subtracting, multiplying, dividing).

Men, on average, excel on some spatial tasks (such as the ability to imagine rotation of a figure), spatial-motor targeting ability and accuracy (such as in throwing or catching objects), perception of vertical and horizontal (such as finding a given shape embedded in a more complex background), mathematical reasoning (problem-solving), and map reading and maze learning.

Many may assume that these sex differences are mainly caused by culture; that is, how we raise boys versus girls determines adult sex differences in these skills. However, there are three lines of evidence suggesting that inherited sex differences in brain biology are mainly responsible for differences in these cognitive abilities between men and women. First, these sex differences are present in young children. Second, these differences are influenced greatly by sex hormones, which have both organizational and activational effects (see text). Third, the differences between the sexes are consistent across geographic regions, among social classes and ethnic groups, and when comparing a wide range of cultures, from hunter-gatherer people to populations in developed countries. Certainly, nurture interacts with nature, but these sex differences are produced regardless of how a child is raised; nurture probably merely influences the average degree of difference.

Many scientists that study sex differences in problem-solving abilities (and the related brain differences and effects of hormones) hypothesize that these differences evolved under natural selection favoring the division of labor in our hunter-gatherer history. Until about 10,000 years ago, when the agricultural revolution occurred, all human beings had some form of hunter-gatherer culture. In these small groups of related people, there was a division of labor in that men typically did most of the long-range travel and hunting for game, and their abilities listed earlier would help in tracking, spotting, and killing of animals. Women, however, often foraged for food and attended to family and camp. Here, they could use their enhanced abilities in foraging and childcare, as well as for navigation that emphasized familiar landmarks.

This is not an argument that men should go to work ("hunt") and women should go shopping and take care of home and children. On the contrary, even though our brains may have evolved sex differences related to sex roles in hunter-gatherers, there is much overlap between the sexes in these abilities, and learning and choice can play a major role in defining sex roles in more modern societies.

adult brain structural differences is present in the fetus (the splenium of the corpus callosum), suggesting an early hormonal influence. In addition, abnormal circumstances suggest an early hormonal influence on cognitive and motor sex differences. For example, females born with congenital adrenal hyperplasia, a genetic disorder, have abnormally high androgen levels in their blood during development. These females, who often have masculinized genitalia (Chapter 5), are usually raised as girls and often have surgery to "feminize" their genitalia. Nevertheless, they tend to exhibit greater visual-spatial ability and less verbal ability than their sisters, suggesting masculinization of cognition. Similarly, when mothers were given androgenic progestogens in the past to prevent miscarriage, they tended to masculinize cognition in their daughters. Interestingly, the synthetic estrogen diethylstilbestrol (DES), banned in 1971, was also given in the past to pregnant women to prevent miscarriage, and it tended to masculinize cognition in their daughters. DES also caused a male-like greater asymmetry in verbal processing in these daughters. Some sons of DES-treated mothers as adults have brains that are less lateralized, and these sons are relatively less adept in spatial ability, both feminized traits. All of these conditions (CAH, DES, androgenic progestogen, as well as maternal stress) tend to masculinize play behavior in daughters, i.e., they cause more "tomboyish behavior." In contrast, biological males with testicular feminization syndrome (or androgen insensitivity), who have high testosterone levels as fetuses but no testosterone receptors (Chapter 5), exhibit feminized cognitive function. Why would estrogen masculinize cognition? Because testosterone normally has organizational effects after being converted to an estrogen in the brain.

Evidence also suggests that sex differences in cognition are also strongly influenced by the activational effects of hormones, even perhaps overcoming any organizational effects of hormones or sex differences in experience. Table 17-2 summarizes the results of experiments on the cognitive changes in transsexuals after "sex-change" treatment. As discussed in Chapter 8, these individuals feel that their gender identity (see later) is the opposite of their biological sex. Females do not differ from heterosexual biological females in adult reproductive hormone levels or secondary sexual characteristics, but they feel that they are, and want to change to, the opposite sex. These individuals then request a sex-change operation (castration and genital surgery) and treatment with the dominant sex hormone of the opposite sex. A summary of these results (Table 17-2) is that (1) before the sex-change treatment, the cognitive abilities of female to male (FM) transsexuals

Table 17-2 Relationships of Sexual Orientation, Gender Identity, and Sex Reversal of Transsexuals on the LH Surge Center, Cognitive and Motor Skills, Sexual Arousability, and Aggressive Tendencies

Adult biologic sex	LH surge?[a]	Cognition[b]	Sexual arousability[c]	Aggression[d]
Female				
Heterosexual	Yes	♀	♀	♀
Homosexual	Yes	♀	?	?
Transsexual (untreated)	Yes	♀	♀	♀
Transsexual (treated with androgen)	?	♀ → ♂[e]	♀ → ♂[e]	♀ → ♂[e]
Transsexual (ovariectomized and treated with an androgen)	Yes → No[e]	?	?	?
Male				
Heterosexual	Yes in one-third (with low testosterone)[f]	♂	♂	♂
Homosexual	In half (with low testosterone)[f]	♀	?	?
Transsexual (untreated)	No	♂	♂	♂
Transsexual (treated with an estrogen or an antiandrogen)	?	♂ → ♀[e]	♂ → ♀[e]	♂ → ♀[e]
Transsexual (orchidectomized and treated with an estrogen)	No[e] → Yes	?	?	?

[a]Is there an LH surge in response to high estrogen levels?

[b]Relates to cognitive and motor skills that exhibit sex differences (see text).

[c]Sensitivity to erotic stimuli.

[d]Based on answers to test questions.

[e]Changed in the same person before and after treatment.

[f]Low testosterone secretion in response to administration of hCG.

are feminine, whereas those of the male to female transsexuals (MF) are masculine; (2) after the FMs receive an androgen for 3 months, their cognitive function is masculinized (increase in visual-spatial ability; decrease in verbal fluency); and (3) after the MFs receive an estrogen or an antiandrogen (a chemical that blocks androgen receptors) for 3 months, their cognition is feminized (increase in verbal fluency; decrease in visual-spatial skills). These results occur without castration and suggest that hormones have a predominantly activational effect on sex differences in cognition. As you know, testosterone is predominant in adult male blood, whereas estradiol is predominant in female blood.

Another indication that at least some of the sex differences in cognition and motor skills are controlled activationally by sex hormones are the changes in these abilities during the menstrual cycle. As discussed in Chapter 3, levels of estradiol, progesterone, and testosterone change in a woman's blood during the menstrual cycle. In the menstrual phase, all three hormones are low; in the midfollicular phase, estradiol dominates; around the time of ovulation, estradiol and testosterone dominate; and during the luteal phase, progesterone and estradiol dominate. High levels of estrogen, such as occur near ovulation and during the midluteal phase, correlate with relatively low visual-spatial ability, but higher verbal and fine-motor ability, as well as a sense of well-being and sensory perception in adult women. Low levels of estradiol (e.g., during the early follicular phase) correlate with an increase in visual-spatial ability. The relative influence of estradiol and testosterone in these changes is not clear. Recent studies, however, indicate that men with relatively low testosterone in their blood (but still much higher than in women) score better on visual-spatial tests and mathematical reasoning than men with high blood testosterone. Women with relatively high testosterone score better on visual-spatial tests (but not mathematical reasoning) than women with very low testosterone. Perceptual speed does not appear to be related to testosterone levels in men or women.

Brain Sex Differences in the "Surge Center"

During the menstual cycle, high blood estrogen (estradiol) levels a day or two before ovulation activate the "surge center" in the female hypothalamus, thus causing a positive feedback effect on gonadotropin-releasing hormone secretion. This, in turn, causes a surge of luteinizing hormone (LH) (and, to a lesser extent, follicle-stimulating hormone) in the blood, which causes ovulation (Chapter 3). Early organization of the surge center is dependent on exposure to testosterone. Because the surge center in males has not been organized to respond to estrogen, if high estrogen is administered to an adult male rat, an LH surge will not occur. Thus, the surge system in rats is defeminized by the organizational effects of testosterone early in life, whereas the feminized surge center in the hypothalamus of rat females is activated by high estrogen. Is the same true for humans? The answer, surprisingly, is no. One study demonstrated that when an estrogen (Premarin) was injected into 18 women, an LH surge occurred. This was true in heterosexual, homosexual, or transsexual women, indicating that only biological sex, not gender identity or sexual orientation, correlates with an active surge center in females (all of these women had normal menstrual cycles). In this same study [see Gooren (1986a) in Advanced Reading section), 44 men were also given Premarin: 11 of 23 gay men responded with an LH surge, 5 of 15 heterosexual men had an LH surge, and none of the 6 MF human transsexuals males responded. Males with a responsive surge center also had the lowest amount of testosterone secretion by their testes if they were given a gonadotropin (hCG). The conclusion was that a surge center is active when blood testosterone is low and is not active when blood testosterone is high, regardless of male gender identity or sexual orientation. In conclusion, it appears that there is *not* an organized sex difference in the presence of a functional surge center in humans as there is in rats.

In fact, if FM transsexuals are ovariectomized (as part of a sex-change operation) and also given an androgen for 30 to 37 months, the surge center changes from responsive to nonresponsive. In addition, if MF transsexuals are orchidectomized and given an estrogen for the same time period, the surge center changes from nonresponsive to responsive. Thus, in the same person, the surge center can be feminized or masculinized depending on the hormonal environment, a solely activational control system [see Gooren (1986b) in Advanced Reading section].

Brain Sex and Human Sexual Behavior

As discussed in Chapter 5, our sex chromosomes determine our biological sex (female or male). Gender identity is the awareness or belief about one's femaleness or maleness. Gender role (or sex role) is the public manifestation of one's gender identity, i.e., one's femininity or masculinity, including appearance, behavior, and sexual orientation.

Sexual Orientation

As discussed in Chapter 8, male homosexuals (gay men) can be said to have a feminized sex role in terms of sexual orientation, whereas female homosexuals (lesbian women) have a masculinized sex role in terms of sexual orientation. Other aspects of sex role, appearance, or behavior in homosexuals can be typical or atypical of heterosexual individuals of the same biological sex, depending on the individual.

Do homosexuals have specific brain areas or functions that are similar to those of their opposite biological sex? Most studies along these lines have been of the brains of gay men, so the question that has some answers is, at present, are the brains of gay men feminized in some way? Recent evidence suggests that this is the case.

In 1984, B. A. Gladue and associates published a paper in the respected journal *Science* (see Further Reading section) that had great impact. They showed that high amounts of estrogen given to 12 heterosexual females (Kinsey scale of 0) caused an LH surge (as expected; see Chapter 3), which did not occur in 17 heterosexual men (Kinsey scale 0). The surprise was that estrogen given to 14 male homosexuals (Kinsey scale 6) produced a moderate LH surge (Fig. 17-4). The authors suggested that the LH response to estrogen represented a "biological marker" of homosexuality in men. Although some initially interpreted this as meaning that the brains of gay men were organized early on in a female direction, more recent studies, as discussed earlier, have shown that this difference between gay and heterosexual men, and indeed the difference between heterosexual men and women, depends *only* on an activational substance from the testes of the adult. That is, the testes of

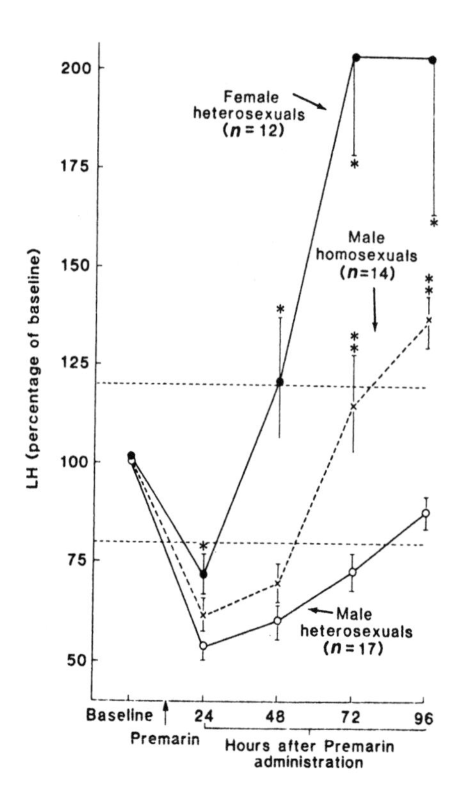

Figure 17-4 Changes in blood levels of LH in response to the positive feedback effects of Premarin (an estrogen) on the surge center of the brain. Note that an LH surge occurs in heterosexual women but does not occur in the heterosexual men tested. Note also that a moderate LH surge also occurred in the homosexual men tested.

heterosexual males secrete something (perhaps testosterone) that inhibits the response of the surge center in their brain to estrogen. Heterosexual women, having little or none of this factor, have a reactive surge center. Gay men, therefore, may have only moderate levels of this factor. Another more recent study demonstrated that some heterosexual men (5 of 15) exhibit an LH surge in response to estrogen and that these men have lower testosterone in their blood after the estrogen injection than heterosexual men without an active surge center. In contrast, 11 of 23 homosexual men tested exhibited a female-like LH surge. Other studies have found that some homosexual women have a reduced (but still present) LH surge in response to estrogen. Finally, transsexual males and females who undergo a sex change (castration plus hormone treatment of the opposite sex) develop a surge center characteristic of their "new" sex (see Table 17-2).

The general feeling now is that males (whether heterosexual or homosexual) with blood testosterone levels on the lower end and/or blood estrogen levels on the higher end tend to have a responsive, female-like, surge center. Therefore, in the study indicated in Fig. 17-4, the homosexual men must have represented the "responsive" type of male as shown in later studies, and the heterosexual men the "nonresponsive" type.

Do the blood androgen or estrogen levels of adult male homosexuals differ from their heterosexual counterparts, as one would predict if there is an overall difference in the responsiveness of their surge centers to estrogen? One study found significantly lower testosterone levels and impaired testicular function in male homosexuals rated 5 or 6 on the Kinsey scale. Another study, however, failed to find a difference in hormone levels in male homosexuals. If the blood levels of sex hormones are the same in homosexual and heterosexual men and women, how does one explain the intermediate state of sex differences in the surge center and cognition seen in homosexual men and women? Even if hormone levels in the blood of homosexuals differ from those in heterosexuals, there is no evidence that these differences influence the direction of sexual orientation. Rather, these differences probably could affect the level of sex drive by hormones acting either centrally or peripherally (see Chapter 8). For example, if one administers testosterone to a male homosexual, his sex drive may increase, but it still would be directed toward other males.

Table 17-1 indicates several sex differences among heterosexual men and women in the size of certain cell groups (nuclei) in the hypothalamus. The hypothalamic areas where these cell groups are located (the preoptic/anterior hypothalamic area; Fig. 17-3) may be involved in the male sexual response and sexual motivation. For example, the *sexually dimorphic nucleus* (SDN), also called the first *interstitial nucleus of the anterior hypothalamus* (INAH 1) in humans, is important for male sexual responses and coitus in rats, whereas another brain area, the *amygdala* (not shown in Fig. 17-3), controls male rat sexual motivation. The SDN is in the preoptic area of the anterior hypothalamus. Testosterone activates both areas behaviorally, at least in the rat. The male rat SDN is five to eight times larger (more neurons) than that of the female, a result of an organized effect of testosterone (converted to estrogen) on the brain during the newborn period.

There is also a sex difference in the sexually dimorphic nucleus of humans. This brain area is 2.2 times as large in adult human males as in adult females. This sex difference appears at about 4 years of age and results from a reduction in the size of the SDN in females (Fig. 17-5). In 2-month-old male but not female human infants, there is a surge of testosterone in the blood that lasts until the fourth month. It is not known if this postnatal surge, or the earlier fetal testosterone surge (Chapter 5), maintains the SDN in males after their fourth year of life. The behavioral effects of this sex difference in humans are not clear, but note that the SDN of gay men in Table 17-1 is not feminized. This supports the knowledge, discussed previously in Chapter 8, that the male and female human sexual response cycle is basically similar.

Two other hypothalamic nuclei, located near the SDN, are larger in the human male than in the human female (Fig. 17-3; Table 17-1). One of these (INAH 2) is not feminized in gay men, but the other (INAH 3) is smaller (feminized) in gay men regardless if they did or did not have AIDS. Some believe that the INAH 3 functions in male sexual orientation. It is not known if the SDN, INAH 2, or INAH 3 are organized and/or activated by sex hormones in humans.

The suprachiasmatic nucleus (SCN) of the hypothalamus is involved in the control of daily and seasonal rhythms. The shape of the human SCN is sexually dimorphic; it is oval in women and spherical in men, but the number of its vasopressin-containing cells is the same between heterosexual men and women.

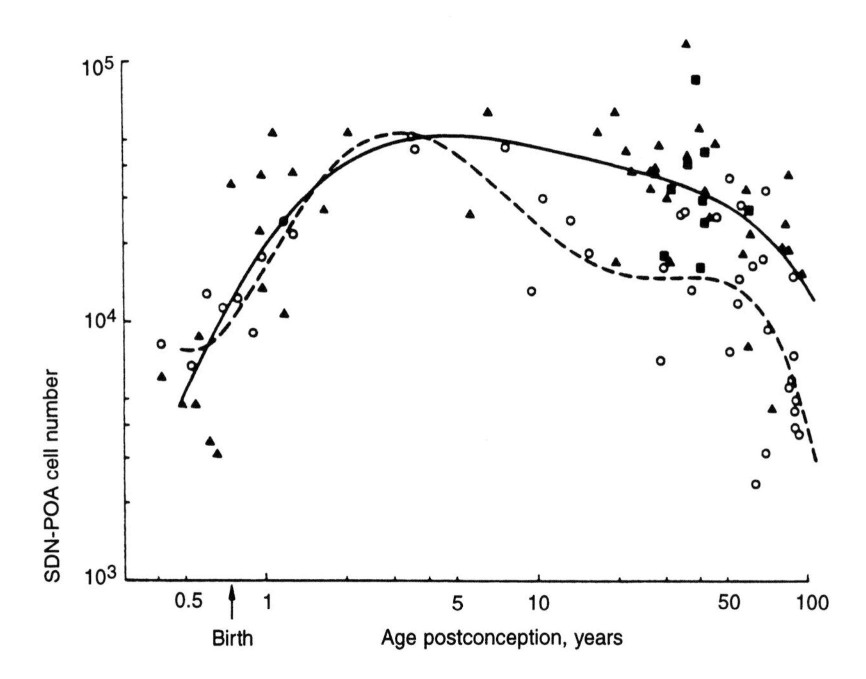

Figure 17-5 Development of the sexually dimorphic nucleus (SDN, or INAH 1) in the human hypothalamus. Note that the number of cells in this nucleus is similar in children up to age 4. Then, more cells die in the female than in the male nucleus, producing a sex difference (male > female) in the size of this nucleus in adult humans. This difference is thought to be the result of organizational effects of testosterone preventing neuron death in the male brain. ▲, heterosexual males; O, heterosexual females. The SDN of homosexual men (■) does not differ from that of heterosexual males (see also Table 17-1 and Fig. 17-6).

Interestingly, the total volume of the SCN, and the number of vasopressin-containing cells, is about twice that in gay men when compared to heterosexual men or women (Fig. 17-6). The significance of this enlargement in gay men is unknown, but it is not related to the presence of AIDS before death.

A group of neurons in the human spinal cord, called the *spinal bulbocavernosus* (or *Onuf's nucleus*), is larger in human males. This nucleus (the ejaculatory center) innervates the bulbocavernosus muscles that connect to the base of the penis and play a role in ejaculation (Chapter 8). In rats, it is organized and activated by testosterone (converted to DHT), but there is no knowledge about this in humans.

The *anterior commissure*, which is a main nerve fiber connection between the cerebral cortex (especially the olfactory region) on each side of the brain (Fig. 17-3), is larger in heterosexual women than in heterosexual men and is also enlarged (feminized) in gay men (Table 17-1). Also, the *splenium* (posterior end of the corpus callosum; Fig. 17-3), which is larger and more bulbous in heterosexual women than in heterosexual men, is enlarged (feminized) in gay men (Table 17-1). Overall, these brain feminizations in gay men (regardless of previous HIV infection) could help explain the higher degree of similarity between function of the two sides of the cerebral cortex (a feminine trait) in gay men. It is significant that gay

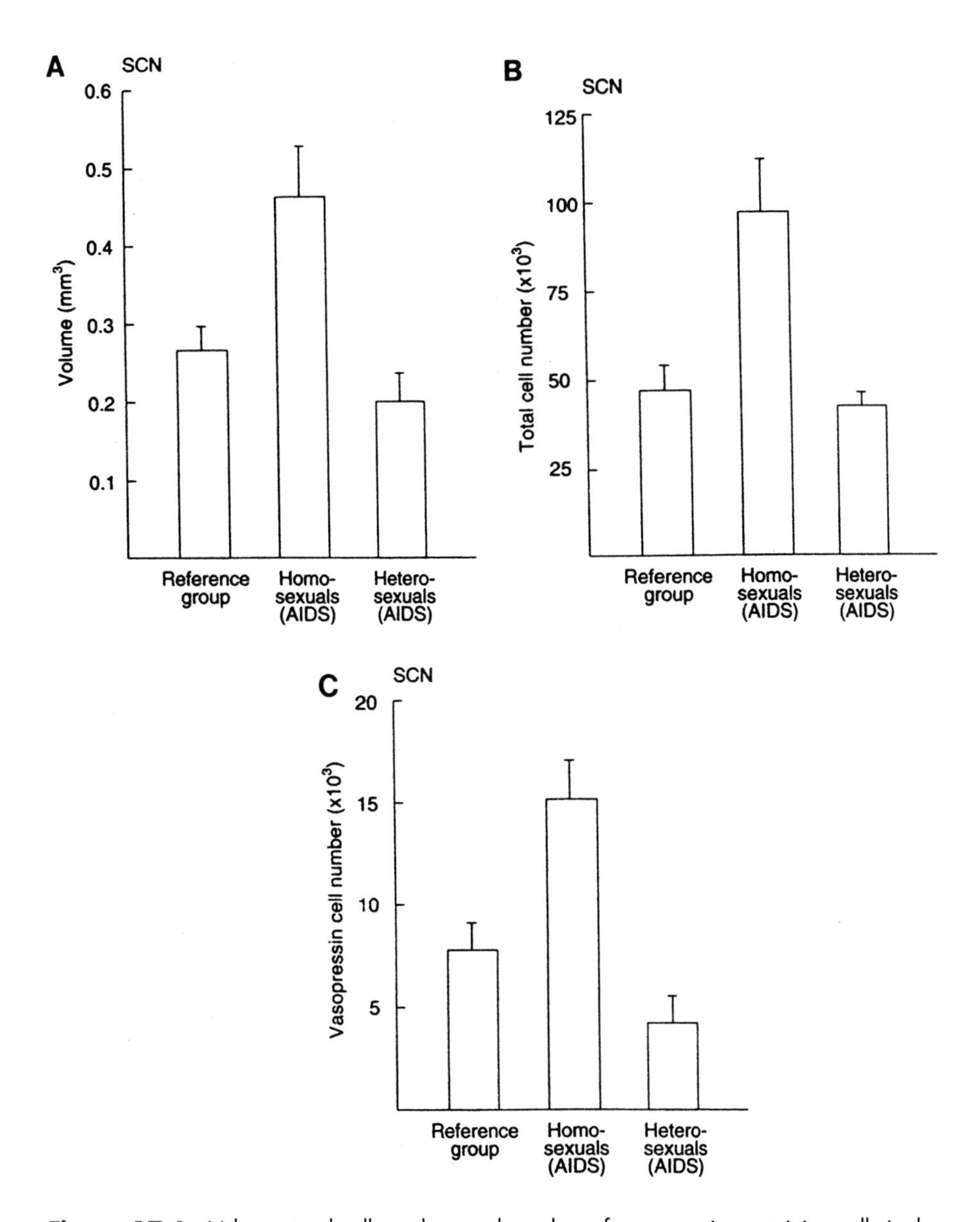

Figure 17-6 Volume, total cell number, and number of vasopressin-containing cells in the suprachiasmatic nucleus of heterosexual men with or without AIDS (see Chapter 18). Note the increased size of this nucleus in gay men; this nucleus is different in shape but not size between heterosexual men and women. Reference group: men without AIDS.

men and homosexual women both exhibit cognitive functions and motor skills intermediate between those of heterosexual men and women.

Is the absence of organizational effects of testosterone in gay men responsible for feminization of certain regions of their brain? Perhaps this is the case, for several reasons. First, there is a strong inherited component to at least male homosexuality (see Chapter 8). Second, one of the feminized brain regions in gay men, the corpus callosum, appears as a sex difference even in the late fetus.

Third, some gay men have reduced visual-spatial and targeting motor skills and increased verbal skills, and this difference appears early in life. Fourth, abnormal hormonal conditions in the fetus or child can increase the incidence of homosexuality. For example, sons of pregnant mothers exposed to stress (in some studies) exhibit an increased evidence of homosexuality as adults; stress could reduce testosterone production by the fetal testes. Also, biological males with testicular feminization syndrome (Chapter 5) tend to exhibit sexual preference toward males. Men who have many older brothers are more likely to be homosexual; the "maternal immune hypothesis" posits that mothers of sons produce anti-H-Y antibodies during pregnancy that interfere with the masculinizing effects of H-Y antigen on future sons. Finally, "Guevedoces," biological males with normal testosterone levels as fetuses, are born with feminized genitalia and are raised as girls (Chapter 5). Yet at puberty, as their genitals and body become masculinized, they exhibit male heterosexual orientation. Thus, their experience of being raised as girls did not override the organizational imprint of testosterone on their sexual orientation early in life.

Do abnormal early organizational effects of androgens increase the incidence of homosexuality in women? The answer also could be "yes." For example, daughters of DES-treated mothers exhibit a markedly increased incidence of bisexual or homosexual fantasies and behavior as adults (remember, DES is an estrogen, and testosterone organizes the fetal brain by being converted to an estrogen). Also, over one-third of females with congenital adrenal hyperplasia (high androgen exposure as fetuses) are lesbians as adults, a much higher percentage than in the normal population.

The brains of heterosexual men and women respond differently to certain odors. When men are presented with an estrogen-like chemical distilled from women's urine, believed to act as a human pheromone, the hypothalamus is activated. This region of the brain does not respond to a testosterone-related chemical derived from male sweat. However, women respond in the opposite way, with a hypothalamic response to the male pheromone but not to the female compound. When gay men were exposed to the substances, their brain responses were similar to those of heterosexual women. Whether the feminization of this physiological brain response in gay men has a biological basis or is the result of experience and learning is not clear.

Brain Sex and Gender Identity

As mentioned earlier, transsexuals have a disturbed gender identity, i.e., they feel that they are trapped in the body of the wrong sex. This is true even though they have normal gonads and genitalia, adult hormone levels, and secondary sexual characteristics of their biological sex. About 60% of MF transsexuals are attracted to men, 10% of transsexuals are bisexual, and 95% of FM transsexuals are attracted to women. Furthermore, most homosexuals and offspring of DES-treated, progestogen (androgen)- treated, and stressed mothers or women born with CAH have a normal gender identity, even if their sex role and orientation may be atypical. These examples show that gender identity is not always related to sex role and orientation.

What causes transsexuality? Perhaps a disturbance in brain development early in life, as transsexuals report feeling the way they do even as small children. One study showed that the central part of the *bed nucleus of the stria terminalis* (BNST) of the hypothalamus (Fig. 17-3), which in rats is essential for sexual behavior, is normally 2.5 times larger in human heterosexual males than in human heterosexual females, and it is feminized (reduced in size) in 44% of MF transsexual males (Table 17-1). In this study, some of these transsexuals were homosexual, some were bisexual, and some were heterosexual; thus, feminization of the BNST was not influenced by sexual orientation. Also, this same work [see Zhou *et al.* (1995) in Advanced Reading section] found no feminization of the SDN or SCN in these transsexuals. Therefore, gender identity may relate somehow to the BNST.

Sex Differences in Aggression

Higher levels of testosterone appear to activate aggressive tendencies. For example, blood concentrations of testosterone have been positively correlated with aggressiveness in pubertal males and females, as well as in males in prisons who have committed violent crimes. Furthermore, aggression scores increase in FM transsexuals that are treated with an androgen and decrease in MF transsexuals treated with estrogen or an antiandrogen [see Van Goozen *et al.* (1995) in Advanced Reading section]. Activational effects of androgen on aggression may be preceded by early organizational influences of androgen. For example, some daughters of mothers treated with DES or androgenic progestogens during pregnancy, as well as 75% of girls with CAH, exhibit increased "rough-and-tumble" play at a level similar to the higher level normally shown by boys as compared to girls. Brain areas controlling aggression include the hypothalamus and amygdala, but there have been no brain studies of people with histories of high androgen exposure to see if their "aggression brain areas" are different.

Chapter 17, Box 2: The Meaning of Brain Measurements

Do human males or females have bigger brains? To answer this "emotional powder keg" of a question, past scientists would weigh fresh or preserved brains, fill empty skulls with a fluid to measure brain volume, or measure head shape to estimate brain size. Men, being on the average larger than women, do have a 10 to 15% larger brain. To answer this question, however, one should correct brain measurements by dividing brain size by something like body height or weight. However, if you divide by body weight, women have a larger brain; the opposite occurs if you divide by height.

Overall, most studies have concluded that *corrected* brain size is similar between the sexes. However, even if one sex had a bigger brain, would it mean anything in relation to intelligence? Our ancient human relatives, the Cro-Magnon people, had bigger brains than modern humans, but were they smarter?

More modern approaches to looking at sex differences in the human brain count neuron number in the brain of men and women. In 1987, an anatomist from Germany demonstrated that women have 4000 more neurons per mm^3 in their cerebral cortex than men. Because the absolute brain size of women is 10 to 15% smaller, it was concluded that the total number of neurons in male and female brains is the same. However, in our examples in this chapter, sex differences in the number of neurons are corrected for brain size, so they are truly different. But what does cell number mean, anyway?

Continued on next page.

Chapter 17, Box 2 continued.

More important than cell number might be the number of synaptic connections between cells. The more complex the connections, the more complex or different the function. For example, there are more dendritic connections in the male versus female rat preoptic area of the hypothalamus. Similarly, there is a sex difference in synaptic organization in the amygdala of the rat. Both areas are involved with sexual behavior and are sensitive to androgens and estrogens. In fact, a given nucleus within the brain could exhibit no sex difference in neuron number but still could exhibit such a difference in synaptic connections.

Even more importantly, the neurotransmitters delivered to, or emanating from, the neurons of a particular nucleus could exhibit a sex difference, even with no sex difference in neuron number or synaptic connections. There are several such examples in male and female rats, but less is known for humans. The number of vasopressin-containing neurons in the suprachiasmatic nucleus of heterosexual women is greater than in heterosexual men. Another finding is that cells containing vasoactive-intestinal peptide in the bed nucleus of the stria terminalis are more numerous in heterosexual men than in heterosexual women.

Finally, the number of brain receptors for hormones and neurotransmitters, and the amount of enzymes utilized for enzymatic conversion of hormones (e.g., aromatase; 5α-reductase), can differ between the sexes, with no other sex difference being obvious. As a result, we must be careful in our studies of biological sex differences in the human brain, and we probably have only seen the tip of the iceberg!

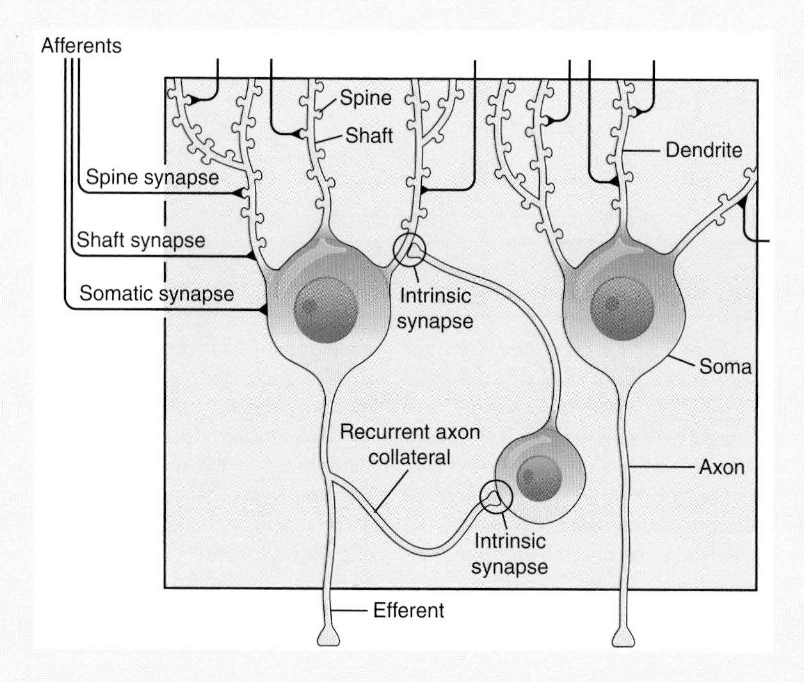

This diagram illustrates the kinds of neuronal networks found in brain areas that are sexually dimorphic. Hormones could influence, through organizational or activational effects, factors such as number of soma (cell bodies), numbers of spines on dendrites, numbers of afferent (incoming) or intrinsic (with the nucleus) synapses, and kinds and amounts of neurotransmitters, as well as number and kinds of hormone receptors in the neurons.

Conclusion

It is clear that there are at least a few structural sex differences in the adult human brain and that some of these are feminized in male homosexuals and MF transsexuals. In the rat brain, over 15 morphological sex differences have been identified, as well as more than 10 sex differences in neurotransmitter production and about 10 sex differences in receptors to hormones and neurotransmitters. More brain sex differences will likely be discovered in the human brain in the future. For example, a hypothalamic structure, the *ventromedial hypothalamic nucleus*, is necessary for *lordosis* (the female mating reflex) in the rat, but this nucleus has never been studied in men or women. Some of the human brain sex differences could be organized, activated, or both, and their roles in gender identity, sexual motivation, the sexual response cycle, and sexual orientation, as well as sex differences in cognition and motor control, remain elusive. Furthermore, we have little information about the influence of environment and learning on brain structure and function in human males and females.

One final point. There are two kinds of genetic selection during evolution: natural selection and sexual selection. *Natural selection* results in differential reproductive success as organisms interact with their environment. *Sexual selection* is a special kind of natural selection for characteristics that enhance mating success. It involves competition for mates among individuals of the same sex. Reproductive success is increased by the presence of display characters (signals) that give an advantage to certain individuals in *courtship behavior* and mate choice, often for males competing for females and/or environmental resources. Sexual selection leads to the evolution of secondary sexual characteristics (*sexual dimorphisms*) that are preferred by the opposite sex. Other sexual dimorphisms are used not in courtship but during *mating behavior* to transfer gametes, to divide resources between males and females, or to separate roles of the sexes in, for example, raising of offspring or food gathering. These sexual dimorphisms are not secondary sexual characteristics and are the result of natural, not sexual, selection. Of course, some sexual dimorphisms that originally evolve through natural selection (e.g., large body size) could later have become a sexual signal. Sexual selection, as Charles Darwin discussed, depends not only on male or female display behaviors, but also on perception and response by the receiving individual. It is possible that some of the sex differences in human brain structure and function were the product of sexual selection (i.e., were involved in courtship) and that some are the product of natural selection (i.e., involved in the sexual response cycle or division of labor between the sexes) in our hunter-gatherer ancestors [see Jacobs (1996) in the Advanced Reading section].

Chapter Summary

Sex differences in human brain structure and function result from an interaction between "nature" (genes, hormones, etc.) and "nurture" (learning, stress, etc.). This chapter emphasizes biological influences. Early (fetal) exposure to sex hormones irreversibly organizes brain sex differences, but adult hormones activate these organized differences. In addition, some adult brain sex differences are the result of activational hormone effects only. There are neonatal and childhood behavioral sex differences that could have a strong biological component. Adult women tend to be better at perceptual speed, verbal fluency, precise manual tasks, and mathematical calculation, whereas adult men on average are better at visual-spatial tasks, target-directed motor skills, and mathematical problem solving. Although these sex differences are undoubtedly influenced by experience, they are also affected by early and adult hormone levels. Several anatomical sex differences between the brains of men and women, in both the cerebrum and the hypothalamus, have been demonstrated. Some of these might explain behavioral and physiological sex differences and some appear to be the result of early or adult hormone exposure. Feminization of some of these brain areas in homosexual and transsexual men suggests a role of these brain regions in sexual orientation and gender identity.

Further Reading

Barinaga, M. (1991). Is homosexuality biological? *Science* **253**, 956–960.

Breedlove, S. M. (1992). Sexual differentiation of the brain and behavior. *In* "Behavioral Endocrinology" (J. B. Becker, S. M. Breedlove, and D. Crews, eds.), pp. 39–70. MIT Press, Cambridge, MA.

Byne, W. (1994). The biological evidence challenged. *Sci. Am.* **270**(5), 50–55.

Carter, C. S. (1992). Hormonal influences on human sexual behavior. *In* "Behavioral Endocrinology" (J. B. Becker, S. M. Breedlove, and D. Crews, eds.), pp. 131–142. MIT Press, Cambridge, MA.

Ellis, L., *et al.* (1988). Sexual orientation of human offspring may be altered by severe maternal stress during pregnancy. *J. Sex Res.* **25**, 152–157.

Fausto-Sterling, A. (1992). "Myths of Gender: Biological Theories about Women and Men." Basic Books, New york.

Gibbons, A. (1991). The brain as "sexual organ." *Science* **253**, 957–959.

Gladue, B. A., *et al.* (1984). Neuroendocrine response to estrogen and sexual orientation. *Science* **225**, 1496–1499.

Hampson, E., and Kimura, E. (1992). Sex differences and hormonal influences on cognitive function in humans. *In* "Behavioral Endocrinology" (J. B. Becker, S. M. Breedlove, and D. Crews, eds.), pp. 357–400. MIT Press, Cambridge, MA.

Hooper, C. (1992). Biology, brain architecture, and human sexuality. *J. NIH Res.* **4**(10), 53–59.

Kelley, D. B., and Gorlick, D. L. (1990). Sexual selection and the nervous system: The neurobiology of courtship and aggressive displays. *Bioscience* **40**, 275–283.

Kimura, D. (1999). Sex differences in the brain. *Sci. Am. Presents* **10**(2), 26–31.

LeVay, S., and Hamer, D. H. (1994). Evidence for a biological influence in male homosexuality. *Sci. Am.* **270**(5), 44–49.

Mustanski, B. S., *et al.* (2002). A critical review of recent biological research on human sexual orientation. *Annu. Rev. Sex. Res.* **13**, 89–140.

Pillard, R. C., and Bailey, J. M. (1998). Human sexual orientation has a heritable component. *Hum. Biol.* **70**, 347–365.

Richardson, S. (1995). S/He-brains. *Discover* June, 36.

Swaab, D. F., and Hofman, M. A. (1995). Sexual differentiation of the human hypothalamus in relation to gender and sexual orientation. *Trends Neurosci.* **18**, 264–270.

Swaab, D.F., *et al.* (2003). Sex differences in the hypothalamus in the different stages of human life. *Neurobiol. Aging* **Suppl. 1**, S1–S16.

Touchette, N. (1993). Estrogen signals a novel route to pain relief. *J. NIH Res.* **5**(April), 53–58.

Advanced Reading

Adkins-Regan, E., *et al.* (1997). Organizational actions of sex hormones on sexual partner preference. *Brain Res. Bull.* **44**, 497–502.

Allen, L. S., and Gorski, R. A. (1992). Sexual orientation and the size of the anterior commissure in the human brain. *Proc. Natl. Acad. Sci. USA* **89**, 7199–7202.

Allen, L. S., *et al.* (1989a). Sex difference in the bed nucleus of the stria terminalis of the human brain. *J. Comp. Neurol.* **302**, 697–706.

Allen, L. S., *et al.* (1989b). Two sexually dimorphic nuclei in the human brain. *J. Neurosci.* **9**, 497–506.

Allen, L. S., *et al.* (2003). Sexual dimorphism and asymmetries in the gray-white composition of the human cerebrum. *Neuroimage* **18**, 880–894.

Archer, J. (1991). The influence of testosterone on human aggression. *Br. J. Psychol.* **82**, 1–28.

Arnold, A. P., and Breedlove, S. M. (1985). Organizational and activational effects of sex steroid hormones on vertebrate behavior: A re-analysis. *Horm. Behav.* **19**, 469–498.

Bailey, J. M., *et al.* (1991). A test of the maternal stress theory of human male homosexuality. *Arch. Sex. Behav.* **20**, 277–293.

Berenbaum, S. A., and Resnick, S. M. (1997). Early androgen effects on aggression in children and adults with congenital adrenal hyperplasia. *Psychoneuroendocrinology* **22**, 505–515.

Byne, W., *et al.* (2000). The interstitial nuclei of the human anterior hypothalamus: An investigation of sexual variation in volume and cell size, number and density. *Brain Res.* **856**, 254–258.

Byne, W., *et al.* (2001). The interstitial nuclei of the human anterior hypothalamus: An investigation of variation with sex, sexual orientation, and HIV status. *Horm. Behav.* **40**, 86–92.

Cooke, B., *et al.* (1998). Sexual differentiation of the vertebrate brain: Principles and mechanisms. *Front. Neuroendocrinol.* **19**, 323–362.

Diamond, M. C. (1991). Hormonal effects on the development of cerebral lateralization. *Psychoneuroendocrinology* **16**, 121–129.

Dittman, R. W., *et al.* (1990). Congenital adrenal hyperplasia. I. Gender-related behavior and attitudes in female patients and sisters. *Psychoneuroendocrinology* **15**, 401–420.

Eals, M., and Silverman, I. (1994). The hunter-gatherer theory of spatial sex differences: Proximate factors mediating the female advantage in recall of object arrays. *Ethol. Sociobiol.* **15**, 95–105.

Ehrhardt, A. A., *et al.* (1985). Sexual orientation after prenatal exposure to exogenous estrogen. *Arch. Sex. Behav.* **14**, 57–77.

Goldstein, J. M., *et al.* (2001). Normal sexual dimorphism of the adult human brain assessed by *in vivo* magnetic resonance imaging. *Cereb. Cortex* **11**, 490–497.

Gooren, L. (1986a). The neuroendocrine response of luteinizing hormone to estrogen administration in heterosexual, homosexual, and transsexual subjects. *J. Clin. Endocrinol. Metabol.* **63**, 583–588.
Gooren, L. (1986b). The neuroendocrine response of luteinizing hormone to estrogen administration is not sex specific but dependent on the hormonal environment. *J. Clin. Endocrinol. Metabol.* **63**, 589–593.
Gooren, L. J., and Kruijver, F. P. (2002). Androgens and male behavior. *Mol. Cell. Endocrinol.* **198**, 31–40.
Hall, J. A. Y., and Kimura, D. (1995). Sexual orientation and performance on sexually dimorphic motor tasks. *Arch. Sex. Behav.* **24**, 395–407.
Hamer, D. H., *et al.* (1993). A linkage between DNA markers on the X chromosome and male sexual orientation. *Science* **261**, 321–327.
Hamer, D. H., *et al.* (1999). Genetics and male sexual orientation. *Science* **285**, 803.
Hampson, E. (1990). Variations in sex-related cognitive abilities across the menstrual cycle. *Brain Cognit.* **14**, 26–43.
Hines, M. (2003). Sex steroids and human behavior: Prenatal androgen exposure and sex-typical play behavior in children. *Ann. N.Y. Acad. Sci.* **1007**, 272–282.
Hines, M., and Shipley, C. (1984). Prenatal exposure to diethylstilbestrol (DES) and the development of sexually dimorphic cognitive abilities and cerebral lateralization. *Dev. Psychol.* **20**, 81–94.
Hofman, M. A., and Swaab, D. F. (1992). Seasonal changes in the suprachiasmatic nucleus of man. *Neurosci Lett.* **139**, 257–260.
Hu, S., *et al.* (1995). Linkage between sexual orientation and chromosome Xq 28 in males but not in females. *Nature Genet.* **11**, 248–256.
Imperato-McGinley, J., *et al.* (1979). Androgens and the evolution of male gender identity among male pseudohermaphrodites with 5-alpha reductive deficiency. *N. Engl. J. Med.* **300**, 1233–1237.
Imperato-McGinley, J., and Gautier, T. (1986). Inherited 5α-reductase deficiency in man. *Trends Genet.* **2**, 130–133.
Jacobs, L. F. (1996). Sexual selection and the brain. *Trends Ecol. Evol.* **11**, 82–86.
Joseph, R. (2000). The evolution of sex differences in language, sexuality, and visual-spatial skill. *Arch. Sex. Behav.* **29**, 35–66.
Kendler, K. S., *et al.* (2000). Sexual orientation in a U.S. national sample of twin and nontwin sibling pairs. *Am. J. Psychiat.* **157**, 1843–1846.
Kerr, M., *et al.* (1994). Stability of inhibition in a Swedish longitudinal study. *Child Dev.* **65**, 138–146.
Kimura, D. (1996). Sex, sexual orientation and sex hormones influence human cognitive function. *Curr. Opin. Neurobiol.* **6**, 259–263.
Kruijver, F. P., *et al.* (2000). Male-to-female transsexuals have female neuron numbers in a limbic nucleus. *J. Clin. Endocrinol. Metab.* **85**, 2034–2041.
Lasco, M. S., *et al.* (2002). A lack of dimorphism of sex or sexual orientation in the human anterior commissure. *Brain Res.* **936**, 95–98.
LeVay, S. (1991). A difference in hypothalamic structure between heterosexual and homosexual men. *Science* **253**, 1034–1037.
Lippa, R. A. (2003). Handedness, sexual orientation, and gender-related personality traits in men and women. *Arch. Sex. Behav.* **32**, 103–114.
Loraine, J. A., *et al.* (1971). Patterns of hormone excretion in male and female homosexuals. *Nature* **234**, 552–553.

Macke, J. P., *et al.* (1993). Sequence variation in the androgen receptor gene is not a common determinant of male sexual orientation. *Am. J. Hum. Genet.* **53**, 844–852.

Mann, V. A., *et al.* (1990). Sex differences in cognitive abilities: A cross-cultural perspective. *Neuropsychologia* **28**, 1063–1077.

Meyer-Bahlburg, H. F. L., *et al.* (1995). Prenatal estrogens and the development of homosexual orientation. *Dev. Psychol.* **31**, 12–21.

Money, J., *et al.* (1984). Adult heterosexual status and fetal hormonal masculinization and demasculinization: 46, XX congenital virilizing adrenal hyperplasia and 46, XY androgen-insensitivity syndrome compared. *Psychoneuroendocrinology* **9**, 405–414.

Morris, J. A., *et al.* (2004). Brain aromatase: Dyed-in-the-wool homosexuality. *Endocrinol.* **145**, 475–477.

Muscarella, F., *et al.* (2001). Homosexual orientation in males: Evolutionary and ethological aspects. *Neuroendocrinol. Lett.* **22**, 393–400.

Neava, N., *et al.* (1999). Sex differences in cognition: The role of testosterone and sexual orientation. *Brain Cognit.* **41**, 245–262.

Nishizuka, M. (1992). Sexual differentiation and neuronal plasticity of synaptic organization during and after the brain development in the rat. *In* "Brain Control of the Reproductive System" (A. Yokoyama, ed.), pp. 21–48. Japan Scientific Societies Press, Tokyo.

Nopoulos, P., *et al.* (2000). Sexual dimorphism in the human brain: Evaluation of tissue volume, tissue composition and surface anatomy using magnetic resonance imaging. *Psychiat. Res.* **98**, 1–13.

Pearcey, S. M., *et al.* (1996). Testosterone and sex role identification in lesbian couples. *Physiol. Behav.* **60**, 1033–1035.

Pilgrim, C., and Hutchison, J. B. (1994). Developmental regulation of sex differences in the brain: Can the role of gonadal steroids be redefined? *Neuroscience* **4**, 843–855.

Quinsey, V. L. (2003). The etiology of anomalous sexual preferences in men. *Ann. N.Y. Acad. Sci.* **989**, 105–117.

Rabinowicz, T., *et al.* (2002). Structure of the cerebral cortex in men and women. *J. Neuropathol. Exp. Neurol.* **61**, 46–57.

Reinisch, J. M., *et al.* (1991). Hormonal contributions to sexually dimorphic behavioral development in humans. *Psychoneuroendocrinology* **16**, 213–278.

Resnick, S. M., *et al.* (1986). Early hormonal influences on cognitive functioning in congenital adrenal hyperplasia. *Dev. Psychol.* **22**, 191–198.

Rice, G., *et al.* (1999). Male homosexuality: Absence of linkage to microsatellite markers at Xq28. *Science* **284**, 665–667.

Robinson, S. J., and Manning, J. T. (2000). The ratio of 2nd to 4th digit length and male homosexuality. *Evol. Hum. Behav.* **21**, 333–345.

Roper, W. G. (1996). The etiology of male homosexuality. *Med. Hypotheses* **46**, 85–88.

Savic, I., *et al.* (2005). Brain response to putative pheromones in homosexual men. *Proc. Natl. Acad. Sci. USA* **102**, 7356–7361.

Sharma, U. R., and Rissman, E. F. (1994). Testosterone implants in specific neural sites activate female sexual behavior. *J. Neuroendocrinol.* **6**, 423–432.

Silverman, J., and Eals, M. (1992). Sex differences in spatial abilities: Evolutionary theory and data. *In* "The Adapted Mind: Evolutionary Psychology and the

Generation of Culture" (J. H. Barkow *et al.*, eds.), pp. 533–554. Oxford University Press, New York.

Swaab, D. F., *et al.* (2002). Sexual differentiation of the human hypothalamus. *Adv. Exp. Med. Biol.* **511**, 75–100.

Swaab, D. F., and Hofman, M. A. (1990). An enlarged suprachiasmatic nucleus in homosexual men. *Brain Res.* **537**, 141–148.

Weiss, E., *et al.* (2003). Sex differences in brain activation pattern during a visuospatial cognitive task: A functional magnetic resonance imaging study in healthy volunteers. *Neurosci. Lett.* **344**, 169–172.

Wrase, J., *et al.* (2003). Gender differences in the processing of standardized emotional visual stimuli in humans: A functional magnetic resonance imaging study. *Neurosci. Lett.* **348**, 41–45.

CHAPTER **EIGHTEEN**

Sexually Transmitted Diseases

Introduction

Sexually transmitted diseases (STDs), also called *sexually transmitted infections* (STIs), are those that are transmitted from one person to another during coitus or other genital contact. They sometimes are called *venereal diseases*, after Venus, the Roman goddess of love. The incidence of these approximately 25 diseases is disturbingly high. It is estimated that one in four Americans between the ages of 15 and 55 will acquire some form of STD, and 15.3 million Americans will contract an STD each year. Half of these will remain infected for the rest of their lives. The incidence of STDs in the United States is considerably higher than in most other industrialized nations.

Organisms causing STDs usually do not live and reproduce on dry skin surfaces. Instead, they require the moist environments of membranes in the so-called "transitional zones" of the body—those that occur at openings between the external and the internal body surfaces. These transitional zones include the vulva, vagina, and urethra of the female; the penis and urethra of the male; and the mouth, oral cavity, eyes, and anus of both sexes. These zones usually are where the STDs first gain a foothold; from there they can invade other body tissues. Although the body forms antibodies to many of the STD organisms, immunities are slow to develop or may never occur.

This chapter discusses the incidence, causes, transmission, symptoms, diagnosis, and treatment of the more common STDs (such as chlamydia, gonorrhea, genital warts, herpes, syphilis, and AIDS), as well as some of the less common ones. Although each disease is discussed separately, a person frequently has more than one STD at the same time. The present yearly incidence of several sexually transmitted diseases in the United States and worldwide is summarized in Figures 18-1 and 18-2.

Bacteria and Viruses

Bacteria are single-celled, microscopic organisms. Most have a cell membrane and all lack membrane-bound organelles, including a nucleus. The bacterial genetic material is a single, circular molecule of DNA not arranged into a chromosome. Bacteria can have several shapes (e.g., rod shaped; filamentous; spiral shaped). Many bacteria cause disease by producing toxins. Bacterial infections that cause human illness can be prevented by vaccines or can be cured by antibiotics. A virus is a tiny, noncellular particle composed of a nucleic acid core (DNA or RNA) and a protein coat. Viruses are parasitic and reproduce only

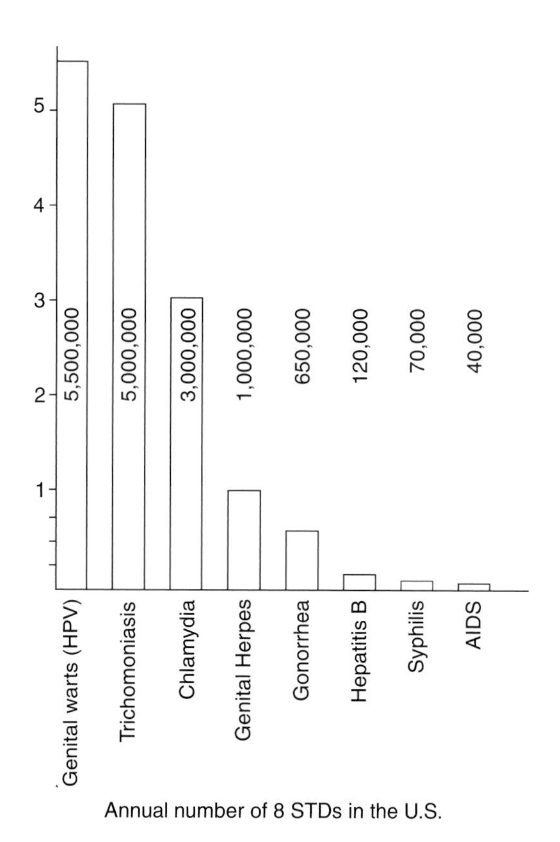

Figure 18-1 Yearly incidence (new cases/year) of eight common sexually transmitted diseases in the United States (2003).

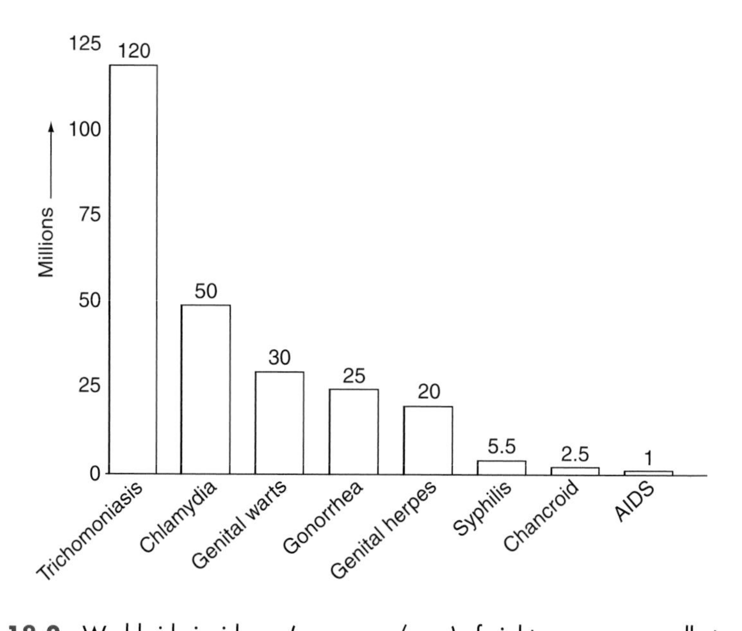

Figure 18-2 Worldwide incidence (new cases/year) of eight common sexually transmitted diseases.

within a host cell. Some viral-caused human illnesses can be prevented by vaccination, but viruses are not harmed by antibiotics. First we discuss some of the sexually transmitted diseases caused by bacteria and then those caused by viruses. Other kinds of organisms causing STDs, such as fungi, protozoa, and invertebrates, are also mentioned.

Gonorrhea

Gonorrhea is an STD that has been afflicting humans for centuries. References are made to this infirmity in ancient Chinese and Hebrew writings. In the early 1960s, the frequency of gonorrhea in the United States exhibited an epidemic increase that peaked in the late 1970s and then decreased until it reached a low in 1997. Since then, the incidence has risen slightly. Currently, about 335,000 new gonorrhea cases are reported each year in the United States. This is an underestimate of the actual extent of the disease in the United States because only about half of the cases are reported, and the Centers for Disease Control (CDC) estimates that about 650,000 new cases actually occur each year (Fig. 18-1). Most new cases occur in the 15- to 29-year-old age group, although this disease can occur in people of any age. The incidence of gonorrhea is especially high in two groups—sexually active individuals under age 24 and men who have sex with men (MSM), especially those who live in high-density urban areas and who have unprotected sex with multiple partners. It is about 20 times higher among African Americans than among American Caucasians.

Cause of the Gonorrhea Epidemic

What caused the gonorrhea epidemic that occurred in the 1960s and 1970s in the United States? One factor was the increase in sexual activity and in the number of sexual partners of young people. Furthermore, many sexually active teenagers were ignorant of precautions against, and symptoms of, gonorrhea. The condom, diaphragm, and spermicidal foams, creams, and jellies offer some protection against transmission of this disease, but the increasing preference for the combination pill over these birth control methods in recent years (see Chapter 14) has decreased the use of these measures. The combination pill may actually increase a female's susceptibility to gonorrhea infection because it increases the moisture and pH of the vagina, conditions favorable to the gonorrhea-causing bacterium. There is no good scientific evidence, however, that the incidence of gonorrhea infection is increased by use of the pill alone. Another factor influencing the gonorrhea epidemic in the United States was the Vietnam War. After every war, there is an increase in STDs because people carry new strains of the organisms to other geographic regions. Finally, as shown later in this chapter, new strains of gonorrhea appeared that were resistant to traditional treatment with penicillin.

Cause of Gonorrhea

Gonorrhea is caused by the bacterium *Neisseria gonorrhoeae*, named after the scientist Albert Neisser, who identified it in 1879. The term *gonorrhoeae* is derived from a Greek word that means "flow of seed." *Neisseria gonorrhoeae* is a gram-negative, diplococcus bacterium. In 1883, Christian Gram of Denmark invented a stain that differentiated gram-positive from gram-negative bacteria. Gram-positive bacteria stain more darkly because of differences in their cell wall structure. "Diplococcus" means that these bacteria occur in pairs, with their adjacent sides flattened. A common name for the bacterium *N. gonorrhoeae* is *gonococcus*, and gonorrhea is often referred to as a "gonococcal infection."

Transmission

There are 16 different strains of *N. gonorrhoeae*, some more damaging than others. Because all strains die quickly when exposed to dry air and sunlight, it is virtually impossible to catch this disease by touching toilet seats and only very rarely can it be transmitted by moist towels, clothes, or hands. The main way this bacterium is transmitted is during heterosexual or homosexual coitus. The bacteria can be transmitted to the mouth or anus during oral or anal sexual contact. Nonoxynol-9, an ingredient in spermicides, can reduce gonorrhea transmission by 60%. The gonorrhea bacteria thrive in the moist membranes of the urogenital tract, as well as in the mouth and oral cavity, anus, and eyes. Once bacteria are introduced into one of these regions, the *incubation period* (the time it takes before symptoms appear) is usually 2 to 5 days, but it can be as little as 1 day or as long as 8 days.

Symptoms in Females

About 75% of females who acquire gonorrhea are asymptomatic (show no symptoms). There are, at present, about 800,000 females in the United States with undiagnosed gonorrhea. This presents a problem because they are carriers of the disease without knowing it, and the disease can reach an advanced stage before a female knows that she has the affliction. In the remaining 25% of women who exhibit symptoms, the first sign usually is the appearance of a clear or whitish fluid discharge from the vagina (infection or inflammation of the vagina, called *vaginitis*). This discharge soon changes to a yellowish or greenish color, i.e., it becomes a *purulent discharge* (a pus-filled discharge). The vaginal wall can become quite irritated at this time, and bleeding may occur with vaginal intercourse. Eventually, the infection can reach the cervix; infection and inflammation of this organ (*cervicitis*) contribute to the purulent discharge. A discharge can also come from the urethra (*urethritis*). Urination can become difficult and painful when urethritis is present. Also, the bacteria can reach the urinary bladder, causing infection (*cystitis*). Because symptoms such as vaginitis and urinary tract infection can have other causes, women with a gonorrheal infection can be misdiagnosed.

Complications in Females

If left untreated, the bacteria can infect the uterus (*endometritis*) and can reach the oviducts 2 to 10 weeks after the initial infection. Inflammation and infection of the oviducts (*salpingitis*) can lead to infertility (see Chapter 16). The bacteria can spread to other pelvic and abdominal organs, resulting in a dangerous condition called *pelvic inflammatory disease*, which is discussed later in this chapter. The bacteria can also cause inflammation of the heart, brain, spinal cord membranes, eyes, skin, and joints. As mentioned earlier, oral coitus with a person carrying the bacteria in his or her genital region can lead to infection of the oral cavity. Anal coitus can transmit the disease into the anus and cause inflammation of the rectum (*proctitis*). Both women and men infected with gonorrhea are more likely to acquire HIV infection and to transmit HIV to their sexual partners.

Pregnancy and Gonorrhea Pregnant women who have gonorrhea can pass the disease to the fetus. The bacteria can enter the fetal blood across the placenta. If this happens in the first trimester, there is an increased risk of miscarriage. In addition, *N. gonorrhoeae* in the birth canal can infect the eyes of a newborn. Untreated, the newborn's eyes develop a thick discharge within 21 days of birth, and the eyes can eventually be destroyed. Treatment of the newborn's eyes with silver nitrate or an antibiotic (see Chapter 11) prevents this from occurring.

Symptoms in Males

Most men (70 to 90%) develop recognizable symptoms of gonorrhea, but the symptoms may take as long as 30 days after infection to appear. The first signs usually are a purulent discharge from the urethra and redness of the glans of the penis. Urination can become painful and difficult, and scar tissue can form in the urethra. Men can also have painful erections, pain in the groin region, and a low fever. If not treated, the infection can spread in about 3 weeks to the urinary bladder and prostate gland and can infect these organs. The epididymides also can become infected, and in some cases, the testes themselves become infected and inflamed (*orchitis*), sometimes leading to infertility.

Diagnosis

The symptoms of gonorrhea are not especially useful in diagnosing the disease. This is because they are similar to symptoms of some other kinds of STDs and because many females and some males are asymptomatic. Unfortunately, there is as yet no reliable blood test for gonorrhea. A sample of the discharge from the penis or cervix can be stained for microscopic observation. This reveals the organisms only in about half of infected women, although the test is more accurate for men. Alternatively, swabs of the urethra, cervix, and/or rectum can be made and cultured in a special (Thayer–Martin) medium. The colonies of bacteria growing in this medium are then examined with a microscope for the presence of *N. gonorrhoeae*, and several biochemical tests are used to confirm the presence of this organism. This culture test takes about 24 to 48 h. Unfortunately, in about 15 to 20% of people who have the disease, gonorrhea bacteria are not visible in a culture. Therefore, it is a good idea to have repeat cultures done. A new test can detect the genetic material of gonococcus bacteria.

Treatment

Treatment with an injection of penicillin has been the standard treatment for gonorrhea since the 1940s. However, strains of *N. gonorrhoeae* have appeared that produce an enzyme (penicillinase) that destroys penicillin. These penicillin-resistant strains, called "supergonorrhea," are becoming more common in the United States and produce fewer and milder early symptoms. With the alarming increase in penicillin-resistant strains, the CDC stopped recommending penicillin as the treatment of choice. Instead, since 1993, the CDC has recommended the use of fluoroquinolones (i.e., ciprofloxacin, ofloxacin, or levofloxacin) for treatment of the disease. However, fluoroquinolone-resistant *N. gonorrhoeae* have been found recently in people who had been infected in Asia, the Pacific Islands, and California and among men who have sex with men. Thus, in 2004, the CDC recommended that fluoroquinolones no longer be used to treat these gonorrhea cases. The list of alternative antibiotics for these patients is limited, and treatment often is in the form of an injection rather than oral medication.

A follow-up culture should be done 1 week after antibiotic treatment. Preliminary trials with a gonorrhea vaccine and antibody treatment have been encouraging.

It is common for persons infected with gonorrhea to also be infected with *Chlamydia* (discussed later in this chapter). Therefore, a combination of antibiotics, such as ceftriaxone and doxycycline, is often prescribed. Sexual partners of the infected individual should be tested and, if found to be infected, treated with antibiotics even if they are asymptomatic.

Syphilis

Syphilis is a serious sexually transmitted disease caused by a bacterium, *Treponema pallidum*. About 5.5 million people in the world are diagnosed with syphilis each year (Fig. 18-2). In the United States, its incidence has been decreasing since reporting began in 1941, with the exception of a sharp, transient rise around 1990. The number of new cases in the United States reached an all-time low in 2000.

Control of the disease has been especially successful among African Americans and people living in the South, where syphilis has been historically prevalent. There is even hope that the opportunity now exists to eliminate the disease in the United States. Currently, about 34,000 new cases are reported each year. As with gonorrhea, the actual incidence is higher because many cases are not reported (Fig. 18-1).

Theoretical Origins

Columbian Theory There are at least two theories about the origin of syphilis. The "Columbian theory" proposes that Christopher Columbus and his crew contracted the disease from indigenous people on their first voyage to the West Indies in 1493. They then introduced the disease to Europe. The first documented epidemic of syphilis occurred in western Europe at the end of the 15th century. People in one country usually blamed foreigners for introducing the disease. Thus, it was called the "Neopolitan disease" in France, the "French pox" by Italians, and the "French or Spanish disease" by the English. In 1520, after the epidemic was over, an Italian physician and philosopher, Hieronymus Fracastorius, wrote a poem in which the people of the earth were given a horrible disease by the sun god, Apollo, because a shepherd encouraged his people to worship the king instead of Apollo. The shepherd's name was Syphilis.

Evolutionary Theory The "evolutionary theory" of the origin of syphilis proposes that this disease is related to other nonvenereal diseases such as yaws and nonvenereal syphilis. *Yaws*, a tropical disease of the skin, is the most primitive of all these diseases. It is caused by a bacterium closely related to *Treponema pallidum*, called *Treponema pertenue*. When people migrated to cooler, drier northern climates, their skin became drier, and *Treponema pertenue*, favoring moist regions, migrated to the axilla, mouth, nostrils, crotch, and anus and caused *nonvenereal syphilis*. This disease (also called *endemic syphilis*) is a childhood infection in arid regions of the world. It is transmitted by direct (nonsexual) body contact, in drinking water, or by eating utensils. The bacterium that causes this disease is not distinguishable from *T. pallidum* and probably evolved from *T. pertenue*. Later, *T. pallidum* began to favor the even moister areas of the genitals and became the sexually transmitted affliction we now know as syphilis. Since it was recently proven that both yaws and syphilis were present in the Americas before Columbus, both theories appear to be true.

Syphilis as Distinct from Gonorrhea

For years, it was thought that syphilis and gonorrhea were caused by the same organism. An English physician, John Hunter (1728–1793), once attempted to prove that these diseases were caused by different organisms by injecting pus from a gonorrhea patient into himself. Unfortunately, and unbeknownst to him, the patient also had syphilis. Hunter contracted both diseases, perpetuating the misconception that they were caused by the same organism, and he died of his self-inflicted syphilis. In 1838, Phillipe Ricord of France determined that gonorrhea and syphilis were caused by different bacteria.

Transmission

Treponema pallidum is a corkscrew-shaped (spirochete) bacterium. It thrives in moist regions of the body and will survive and reproduce only where there is little oxygen present. It is killed by heat, drying, and sunlight. Therefore, one cannot catch syphilis from contacting toilet seats, bath towels, or bedding. It can, however, live in collected blood for up to 24 h at 4 °C and thus, in rare cases, is transmitted during blood transfusion. Nine out of 10 cases of syphilis transmission occur during sexual intercourse, although it can also be introduced into an open wound in the skin. Fortunately, only about 1 in 10 people exposed to the bacterium develops syphilis. Persons infected with syphilis can acquire the HIV virus more easily and transmit it to others.

Stages of the Disease

Primary Stage The symptoms of untreated syphilis occur in four stages. The *primary stage of syphilis* usually appears as a single sore called a *chancre* (pronounced shang'ker) at the place where the bacteria first entered the body. This is a round, ulcer-like sore with a hard raised edge and a soft center. It looks like a crater, about $\frac{1}{2}$ to 1 in. in diameter. This chancre, which for all its awful appearance is painless, appears 10 to 90 days after entry of the bacteria. Because the chancre is painless and may be in a location not readily noticed, a person may not realize that he or she is infected. In males, the chancre usually occurs on the glans or corona of the penis, but it can occur anywhere on the penis or on the scrotum. In females, it usually appears on the vulva, but sometimes can occur on the cervix or vaginal wall. After oral coitus with an infected person, it can appear on the lips, tongue, or tonsils, and it can appear in the anus after anal coitus with an infected person. Lymph nodes enlarge a few days after the sore appears. The chancre heals in 1 to 5 weeks, and the primary stage is then over. Meanwhile, the bacteria travel in the blood or lymphatic system to other parts of the body and, in one-third of cases, they will eventually cause the secondary stage of syphilis if the person is not treated.

Secondary Stage The *secondary stage of syphilis* occurs 2 weeks to 6 months after the primary stage. This stage is characterized by a rash that appears on the upper body, arms, hands, and feet and can spread to other skin regions. In light-skinned people, the rash appears as cherry-colored blemishes or bumps that change to a coppery-brown color. In dark-skinned people, the blemishes are grayish blue. Larger bumps can develop and burst, especially in the inguinal region. The rash does not itch and is painless, but the syphilis bacteria are present in great numbers in these sores, and contact with the sores is very infectious to other people. Other symptoms of the secondary stage include hair loss, sore throat, headache, loss of appetite, nausea, constipation, pain in the joints and abdominal muscles, a low fever, and swollen lymph gland. The symptoms are minor and cause little inconvenience in about 60% of untreated individuals in the secondary stage and thus can be completely overlooked. The secondary stage goes away in 2 to 6 weeks but can recur over the next 1–2 years, after which the untreated individual then enters the latent stage of syphilis.

Latent Stage During the *latent stage of syphilis*, which can last for years, a person exhibits few or no symptoms. An individual in the latent stage can no longer transmit the bacteria (except to a fetus, as discussed later). About half of the people who enter the latent stage never leave it, even if not treated. The other half eventually enters the tertiary stage of syphilis if not previously treated with antibiotics.

Tertiary Stage Entrance into the *tertiary stage of syphilis* occurs because the bacteria have invaded tissues throughout the body, although people in the tertiary stage are not infectious. The tertiary stage is characterized by large, tumor-like sores (*gummas*) that form on tissues of skin, muscle, the digestive tract, liver, lungs, eyes, nervous system, heart, or endocrine glands. Infection of the heart (cardiovascular syphilis) can cause severe damage to the heart and its valves. Invasion of the bacteria into the central nervous system causes "neurosyphilis," and the brain and spinal cord can be severely damaged. People with neurosyphilis can develop partial or total paralysis, blindness, or psychotic and unpredictable behavior. Damage from tertiary syphilis can cause death.

Congenital Syphilis

As noted earlier, a person is not infectious in the latent or tertiary stages of syphilis. This is true except in the case of an infected pregnant woman, who can pass the bacteria to her fetus at any stage of syphilis. The placenta protects the fetus against invasion of syphilis bacteria up to the sixth month of pregnancy, after which time the *T. pallidum* organism passes through the placental membranes into the fetal bloodstream. Then the fetus can contract the disease from the mother. If this happens, about 30% of the fetuses miscarry and 70% are born with *congenital syphilis*. The latter children are contagious in their first and second year and go through all the stages of syphilis if left untreated. About 23 in 100 such cases develop tertiary syphilis in 10 to 20 years. Symptoms of tertiary congenital syphilis include damage to the eyes, deafness, flattening of the bridge of the nose ("saddle nose"), and central incisor teeth that are spread apart and notched ("Hutchinson's teeth"). Many of these individuals die from this affliction.

Diagnosis

Several of the symptoms of syphilis can be confused with those of other sexually transmitted diseases. Also, it has proven difficult to grow cultures of *T. pallidum* in the laboratory. Therefore, other tests are necessary to see if a person has contracted the disease. Diagnosis of syphilis can be accomplished in several ways. A blood test for the presence of antibodies to the disease can be done a week or so after the primary chancre appears. Several such tests are available, including the Venereal Disease Research Laboratory (VDRL) test, the Rapid Plasma Reagin (RPR) test, and the Syphla–Chek test. All are equally sensitive, but Syphla–Chek seems to be the best for primary syphilis. The older "Wasserman test" for syphilis has been displaced by these newer methods. False positive results occur in 1 out of 3000 of these blood tests and, more importantly, false negatives occur about 25% of the time. Because of this error factor, an individual's tissues should also be checked for the presence of live *T. pallidum*, based on their characteristic shape and movement, using a dark-field microscope to confirm the diagnosis.

Treatment

Once it has been determined that a person has syphilis, treatment with one of several antibiotics is effective. Most commonly, benzathine penicillin G is given as a single injection each day for 8 days if the person is in the primary stage and for 3 to 4 weeks at higher dosage if the person is in a more advanced stage of the disease. Tetracycline or erythromycin can be used if a person is hypersensitive to penicillin. It must be emphasized that syphilis, like gonorrhea, is a curable disease. However, individuals with tertiary syphilis, though treated, still may suffer permanent tissue damage.

Chlamydia

Nonspecific urethritis has, in the past, referred to any sexually transmitted *urethritis* (urethral infection) not caused by *Neisseria gonorrhoeae*. We know now that the leading cause of this affliction is the gram-negative, bacterium-like microorganism *Chlamydia trachomatis*. This is a member of a group of very small bacteria that, unlike other bacteria, live inside cells and were once thought to be viruses. About half the cases of nonspecific urethritis in men and women are caused by *chlamydia*, which is a sexually transmitted organism. Chlamydia can also be passed to the eyes by touching infected regions. At present, there are about three million new cases of chlamydia each year in the United States (Fig. 18-1).

In females, the cervix is the main site of chlamydia infection, leading to spotting between periods, a yellowish vaginal discharge, and frequent urination. In addition, the cervix, vagina, urethra, and vulva become reddened and irritated. There is an association between chlamydia infection and cervical cancer. About 75% of infected women have no symptoms, which makes early detection difficult. The result of an untreated infection could be pelvic infection and even infertility. It is estimated that about 11,000 women in the United States become infertile annually as a result of chlamydia infection. A chlamydia infection in a pregnant woman can be passed to her child during delivery, and the result can be lung and eye infections in the newborn.

Male partners of chlamydia-infected women usually have chlamydia in their bodies. Although 50% of men are asymptomatic, half of infected men develop frequent and painful urination. For men, the incubation period is 7 to 28 days. In both men and women, chlamydial infection increases by three- to five-fold the risk of infection by HIV for those exposed to the virus. Correct use of latex condoms can reduce the risk of passing chlamydia to a sexual partner.

When cases of chlamydia were first reported by state health departments in 1984, there were 7500 known cases in the United States. Ten years later, the reported incidence had jumped to 450,000, and in another decade, this number doubled again. Because of the dramatic spread of this infection and because 50% of men and 75% of women are asymptomatic, this disease is referred to as a "silent epidemic." Nearly half of all chlamydia cases in women occur in adolescents ages 15–19. The CDC now recommends that all sexually active females under the age of 25 be screened for chlamydia at least once a year.

Chlamydia is detected by tissue culture, an enzyme test, and a new DNA assay. For both sexes, chlamydia can be cured by treatment with antibiotics such as azithromycin or doxycycline. Both partners should be treated, even if only one has symptoms. Women whose partners are not treated are at high risk for reinfection. Other causes of nonspecific urethritis include *Ureaplasma urealyticum*, which is related to chlamydia. *Haemophilus vaginalis*, a bacterium, also produces some cases of nonspecific urethritis. In women, these infections can also cause pelvic infection and even infertility, and they can cause infertility in men through scarring of sex accessory tubes or by damaging sperm.

Genital Herpes

Herpes genitalis is a very common STD. Each year, about one million people in the United States will contract this disease (Fig. 18-1). Throughout the world, genital herpes is more common in industrialized nations and in urban areas, and prevalence is also high in sub-Saharan Africa and in the Caribbean. Because herpes is, at present, incurable, it is estimated that there are about 45 million sufferers in the United States today (one out of five adolescents and adults in the United States). It is most prevalent in teenagers and young adults, and more common in women than in men, but it can infect anyone who is sexually active. Once infected with the virus that causes genital herpes, the affected person retains the virus for the rest of his or her life.

Cause

This disease is caused by *herpes simplex type 2* (HSV2) *virus*. There are 25 herpes viruses, which cause such diseases as fever blisters and cold sores (*herpes simplex type 1 virus*), chicken pox in children or shingles in adults (*varicella–zoster virus*), infectious mononucleosis (*Epstein–Barr virus)*, and cytomegalic inclusion disease, which affects the fetus and newborn and results in enlargement of the liver and spleen (*cytomegalovirus*). The HSV2 virus usually affects the body below the waist (e.g., the genitals, thighs, and buttocks), whereas type 1 usually invades areas above the waist. About 20% of herpes infections of the genital region, however, are caused by herpes simplex type 1, usually the consequence of oral coitus with an infected person. Similarly, HSV2 occasionally is isolated from mouth sores. Transmission of the virus usually occurs during sexual contact with someone who is having a herpes "outbreak," i.e, when the virus has caused visible sores. However, infection can also occur when the sexual partner has no obvious symptoms. Condoms are not 100% effective in preventing transmission of the herpes virus. It should be emphasized that herpes genitalis can be transmitted by nonsexual contact with an infected person. Because herpes viruses can survive a few hours on moist toilet seats, gloves, in tap water, and on plastic surfaces in spas, a person at least theoretically could acquire the virus from these surfaces.

Symptoms

Once a person contracts the herpes simplex virus, usually through genital, oral, or anal contact, clusters of tiny blisters develop that change to painful round sores in 4 to 7 days. Two or 3 days later, these take the form of multiple small, round, itchy ulcers. Severe ulcers occur only in about 10% of infected people; in the remaining 90% the sores are minor and often go unnoticed. In males, the sores occur mainly on the penis (shaft, foreskin, glans, urethral orifice), especially in uncircumcised males. The primary symptoms usually are more painful in females, with sores appearing on the labia, clitoral hood, cervix, vaginal introitus, urethral orifice, or perineum. Urination and coitus can be painful. More severe but less common symptoms in both sexes include fever and enlargement of the inguinal lymph nodes. In general, the symptoms are more severe in people who have never been exposed to any herpes virus. If a person touches an open sore and then his or her eyes, he or she can develop a virus infection that can lead to blindness. The sores, if they develop, heal in 1 to 6 weeks. Individuals can infect other people most when sores are present but can also be infectious after the sores and scabs disappear.

After the herpes sores have healed, the virus migrates up sensory nerves to clumps of nerve cells near the spinal cord. They lie dormant there for several days, weeks, or months. Then they become active again, migrate back to the skin, and cause recurrence of the symptoms. Such recurrence can be on different parts of the penis or scrotum in males or on the vulva, vagina, or cervix in females. The recurrent attack often is accompanied by enlargement of the lymph nodes in the groin, as well as fever and headaches. The symptoms go away in 1 to 4 weeks. Recurrence of symptoms can occur frequently (e.g., once a month) or there may be several months between attacks. The recurrences may be associated with times of stress or, in females, with menstruation. Eventually, antibodies are formed that alleviate or stop recurrences, and a few people may never have a second attack. Nevertheless, HSV probably stays with people for their entire lives. There is a disturbing positive correlation of females developing cervical cancer after having contracted genital herpes.

Herpes simplex type 2 virus in the blood of a pregnant female can cross the placenta and damage the fetus. This, however, is rare. More often, the fetus is exposed to the virus during birth, especially if the woman has open sores during delivery. However, about a third of cases of neonatal herpes occur during deliveries in which the woman had no open sores. About 25% of newborns exposed to herpes virus type 2 develop blindness or brain damage, and another 25% of exposed newborns develop skin lesions. For this reason, the fetuses of women with herpes infection often are delivered by cesarean section (see Chapter 11).

Diagnosis

In addition to the previously described symptoms, genital herpes can be diagnosed by taking a smear from the cervix, culturing it with tissue cells, and examining the cells for the presence of the virus. Also, a blood test is now available that can differentiate between antibodies to herpes simplex type 1 and type 2 viruses.

Treatment

There is no reliable cure for herpes genitalis, which explains why the actual number of people having the virus is high. Currently the only treatment available is an antiviral medication such as acyclovir, which can limit the frequency and length of outbreaks. Antiviral medications are available as a topical cream, but they appear to be more effective when taken orally. Suppressant topical agents can help limit the transmission of the virus to a sexual partner. As with any viral infection, the development of a vaccine is highly desirable. Clinical trials of a genital herpes vaccine are now under way in the United States.

Genital Warts

Genital warts (*Condyloma acuminata*) can occur in the genital region because of the presence of a *human papilloma virus* (HPV). This virus often is transmitted sexually, which is why the condition may be called "venereal warts." These warts can also appear spontaneously. (The kind of wart that occurs on the skin in other body regions is caused by a different virus.) About 20% of sexually active 14- to 18-year-old females have the virus, and most of their sex partners do too. This amounts to about 5.5 million new cases in the United States annually (Fig. 18-1). At least 20 million Americans are carriers of HPV, which makes HPV one of the most prevalent STDs.

After the HPV virus is contracted, the warts appear in 3 weeks to 8 months. They are moist, soft, cauliflower-like bumps occurring singly or in groups. They can be pink, red, or dark gray. Females often get them on the cervix, labia, vulva, or perineum, whereas in males they appear on the prepuce, glans, or coronal ridge of the penis, as well as in the urethra and on the scrotum.

Genital wart infection can be serious. In women, there is a strong association between the presence of the virus and abnormal growth (dysplasia; cancer) of the cervix. There are about 100 types of HPV, and 4 are commonly associated with cervical abnormalities. About 95% of women with cervical cancer have one or more types of HPV virus (compared with 20% of women without cervical cancer). It takes about 2.5 years after acquiring the virus for the resultant cervical abnormalities to appear. Although a Pap smear does not detect the virus, any cervical abnormalities detected by a Pap smear should be followed by an antibody test for HPV. Women with a positive HPV test should get a Pap smear every 6 to 12 months. Men who are positive for HPV should be aware that some types of cancers of the penis appear to be associated with the virus. Women are 5 to 11 times more likely to have HPV and cervical cancer if their sexual partner frequents prostitutes or has numerous other sexual partners. There also is an association of the virus with cancer of the anus in men who have sex with men.

Genital warts can be treated with the medications podophyllin (an antimitotic agent), imiquimod (an immunosuppressant), trichloroacetic acid, dry ice, or liquid nitrogen; they usually dry up and fall off a few days after being treated. If this does not work, the warts can be removed by laser surgery or heat cauterization. However, there is no cure for HPV infection. The search for a vaccine against HPV is focused on preventing transmission of the virus, in addition to preventing women already infected with the virus from developing cervical cancer.

Trichomoniasis

Trichomoniasis is a very common but curable sexually transmitted infection. It is caused by a flagellate protozoan, *Trichomonas vaginalis*. This single-celled organism is pear shaped and has fine flagella that it uses to swim with (Fig. 18-3). Trichomoniasis does not survive in the mouth or rectum, and it is transmitted most commonly during coitus. It is estimated that over five million men and women contract the infection each year in the United States (Fig. 18-1). *Trichomonas* is commonly present in the vagina (see Chapter 2), and about one in five women who have this protozoan in their vagina do not have trichomoniasis. It is only when the organisms multiply in large numbers that symptoms of vaginitis occur. There is a profuse, watery, frothy, odorous (stale or musty) discharge, ranging from grayish

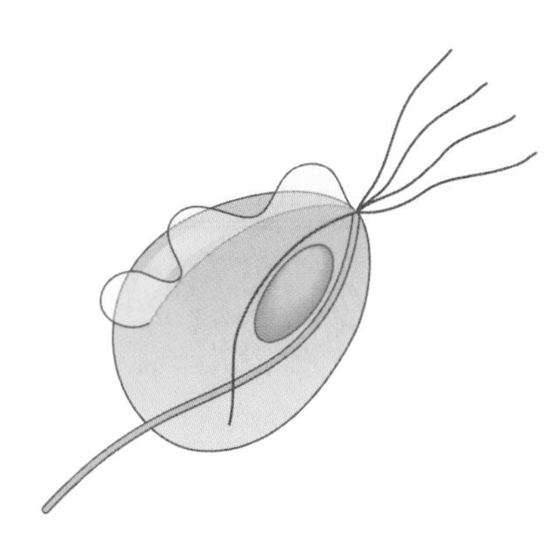

Figure 18-3 *Trichomonas vaginalis*, a flagellated protozoan that causes vaginitis.

white to yellowish brown, that develops in 4 to 28 days. The vulva can swell, become red and itchy, and the swelling and redness can spread to the cervix, urethra, and urinary bladder. Most men who have trichomoniasis have no symptoms, but it is responsible for 5 to 15% of nonspecific urethritis and cystitis in men. In this case, a fluid discharge will be exuded from the urethra.

The presence of *T. vaginalis* can be detected by examining some of the discharge under a microscope. It can be treated with oral administration of a single dose of metronidazole (Flagyl). Side effects of Flagyl can include an upset stomach, loss of appetite, diarrhea, hives, dizziness, and a bad taste in one's mouth.

Viral Hepatitis B

Hepatitis virus B (*viral hepatitis, type B*) is a sexually transmitted virus that causes liver disease. It often is transmitted by the use of an infected hypodermic needle. It can also be transmitted during sexual contact or during other close contact with the blood or body fluids of infected people. The virus is present in saliva and semen and can be transmitted during kissing and anal or oral intercourse. It can cross the placenta. People at high risk for hepatitis B infection include injection drug users, persons who have multiple sex partners, men who have sex with men, and infants born to infected mothers. There are about 120,000 new cases of hepatitis B in the United States each year (Fig. 18-1), and 1 to 2% of these are fatal. Symptoms include an inflamed liver (*hepatitis*), liver cancer, fever, weakness, headache, and muscle pain. A hepatitis B vaccine has been available since 1981. The CDC recommends routine vaccination of all children ages 0 to 18, as well as those in high-risk groups.

Pediculosis Pubis

Pediculosis pubis, or "crabs," is caused by a tiny, parasitic, blood-sucking crab louse by the name of *Phthirus pubis* (Fig. 18-4). This organism can be transmitted by direct body contact and also by contact with hair, clothing, or bedding. The organism can be seen at the base of hairs or as black spots visible on underwear. It lives in pubic, axillary, eyebrow, eyelash, and facial hair, but never in scalp hair.

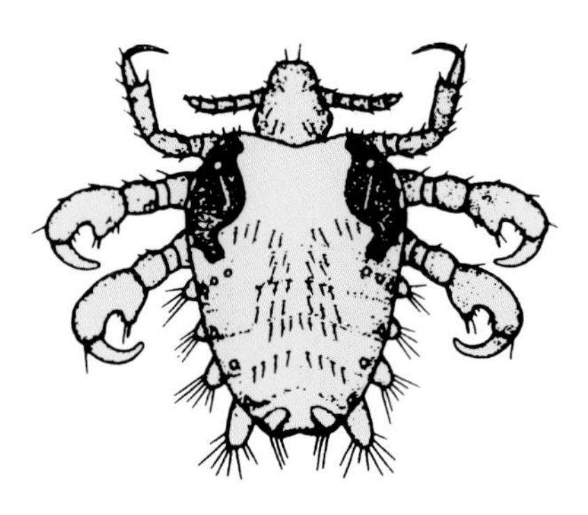

Figure 18-4 *Phthirus pubis*, a crab louse that causes pediculosis pubis (crabs).

It needs the environment of human hair, as it dies in 24 h if removed. Female parasites lay tiny white eggs at the base of hairs, and the eggs hatch into larvae in 7 to 9 days. Adults or larvae cause itching, and scratching can lead to secondary infection of the skin or hair follicles. In World War II, parasitized people were shaved and a 10% DDT powder was used for 24 h. Now, an insecticide called gamma benzene hexachloride (Kwell) is applied as a cream, lotion, or shampoo. If one member of a household has crabs, all members should be treated. All underclothing and linens should be washed in hot water with bleach.

Scabies

Scabies is caused by the mite *Sarcoptes scabiei* and is transmitted by close contact, sexual or otherwise. The female mites burrow under the skin and then lay their eggs. This leads to itchy rashes on the abdomen, flanks, thighs, genitals, buttocks, or forearms—anywhere except the face. All members of a household must be treated if one member has it. Treatment for crabs also kills scabies mites.

Lymphogranuloma Venereum

Lymphogranuloma venereum is a tropical sexually transmitted disease that is rare in the United States, but is most prevalent in Africa, southeast Asia, Central and South America, and Caribbean countries. It is caused by *Chlamydia trachomatis*, the same organism (but a different strain of it) that is one cause of nonspecific urethritis. Although most people catch this disease after close sexual contact, it can also be transmitted on clothing. A newborn's eyes can also be infected with this organism if it is present in the vulva of a delivering mother. After an incubation period of 7 to 21 days, a small, painless blister forms. In males, this blister occurs on the penis or scrotum, and in females occurs on the vagina, vulva, cervix, or urethra. The anus, tongue, or lips can also be infected. Other symptoms can include swelling of inguinal lymph nodes, fever, chills, backache, abdominal and joint pain, and loss of appetite. In a few cases, the spreading organisms can cause a serious complication, *Reiter's syndrome*, which is characterized by rheumatism, arthritis, conjunctivitis, and heart valve defects. Diagnosis of this disease is by a skin test (Frei test) or by a test for antibodies. Both tests, however, only become positive 3 to 4 weeks after the onset of the disease. Treatment is with tetracycline or the sulfonamides.

Chancroid

Chancroid (or "soft chancre") is a sexually transmitted disease caused by a very small gram-negative bacterium, *Haemophilus ducreyi.* It is rare in the United States, being most common in the tropical Far East. However, it has been appearing more frequently in the United States. Once the bacteria enter a cut in the skin or invade the mucous membranes of the genital region, a small *papule* (small elevation on the skin) appears in 12 to 24 hr. This papule, commonly found on the penis or vulva, then bursts and an ulcer forms in 1 to 2 days. This ulcer looks like the chancre of primary syphilis except that it has soft edges. It also is painful, whereas the hard chancre of primary syphilis is painless. Multiple soft chancres can develop as the organism spreads, even to the thighs. Another common symptom is swollen lymph glands in the groin region. Many females do not show symptoms but are carriers of the bacteria. Anal coitus can lead to soft chancres in the anus. Chancroid can be diagnosed by examining a smear or culture under a microscope and looking for *Haemophilus ducreyi*. Also, the presence of *H. ducreyi* can be detected by a skin test (Ducrey's skin test), which becomes positive 1 to 2 weeks after a person is infected. Standard treatment is with a sulfa drug such as sulfonamide.

Molluscum Contagiosum

Molluscum contagiosum is a disease caused by a virus related to the chickenpox virus. This virus can be transmitted sexually, but it can also be transmitted by skin contact. The symptoms are small, pink, wart-like growths on the face, arms, back, umbilical region, or buttocks. This disease is relatively harmless, and the growths can be removed by freezing or burning.

Granuloma Inguinale

Granuloma inguinale is caused by a gram-negative bacterium by the name of *Calymmatobacterium granulomatis* (formerly called *Donovania granulomatis*). This is an extremely rare sexually transmitted disease, with about 100 cases occurring annually in the United States. It is more common in south India and New Guinea. From one to several weeks after the bacteria enter the body, a tiny blister occurs on the penis or vulva, which then develops into an open, bleeding sore. The skin around the sore becomes swollen and red. One can diagnose this disease by looking for bacteria in a small piece of skin taken from the edge of the sore, using a microscope. Treatment is with tetracycline.

Acquired Immunodeficiency Syndrome (AIDS)

Statistics

Acquired immunodeficiency syndrome is a usually fatal disease caused by two types of the *human immunodeficiency virus* (HIV), HIV-1 and HIV-2. These viruses probably arose from similar viruses in African primates (see HIGHLIGHT box 18-1). The first known HIV infection was found in a man from the Democratic Republic of Congo in 1959. Undoubtedly, you are well aware that HIV infection and AIDS are now a global epidemic (Fig. 18-5). At the end of 2004, an estimated 40 million people in the world were infected with HIV, and there have been about 20 million deaths worldwide from AIDS. Each day, 13,000 people are newly infected with the virus. About two-thirds of all HIV-positive people in the world live in sub-Saharan Africa; thus, the HIV epidemic is much more severe in this area than in any other region of the world. In seven countries of southern Africa, at least one in five adults is HIV positive. The highest level of HIV infection is in Botswana, where 37% of the adults are infected. More women than men are infected with HIV in sub-Saharan Africa. Half a million children in this region are HIV positive, about 90% of whom were infected by

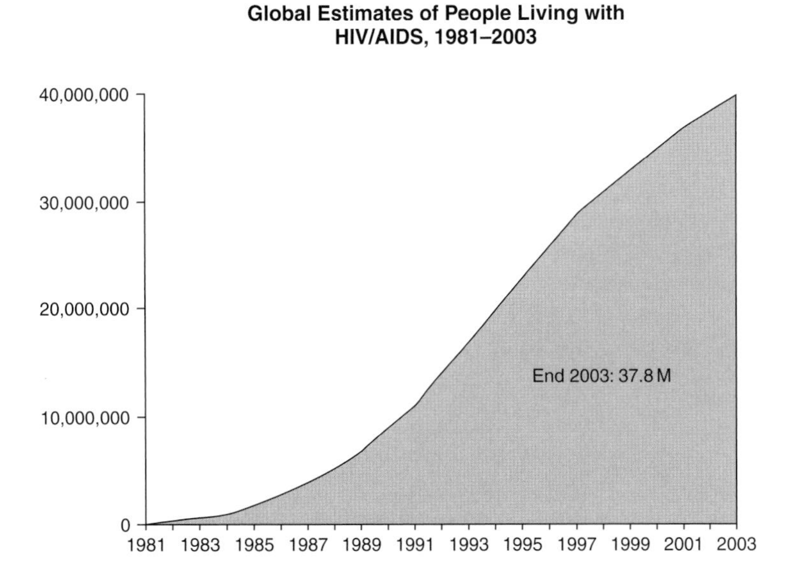

Figure 18-5 Growth of the HIV/AIDS epidemic. The number of individuals living with HIV/AIDS from the beginning of the epidemic (1981) until 2003 is shown. About 40 million individuals are infected with HIV worldwide. (Data from UNAIDS.)

their mothers during pregnancy, childbirth, or breast-feeding. The virus affects millions of other children through the deaths of their parents, teachers, and doctors, disruption of family and community life, and increased social and financial responsibilities placed on children in AIDS-affected communities.

Asia has been less affected by the HIV epidemic than Africa. However, over eight million people in this region are living with HIV. The huge size and high mobility of the population portend the potential for the epidemic to have a devastating effect on this region. China has only recently recognized the scope of its HIV-infected population, and a widespread campaign to prevent transmission of the virus will be necessary to avoid further expansion of the epidemic. Latin America and the Caribbean are also affected by the HIV/AIDS epidemic, although the prevalence differs greatly among countries. One of the highest HIV infection rates in this region is found in Haiti, where it is estimated that over 200,000 children have lost one or both parents to AIDS. Another region of concern is eastern Europe. Per capita, the fastest-growing HIV/AIDS epidemic is occurring in the countries of the former Soviet Union, fueled mainly by illicit drug use.

The first group of AIDS cases in the United States appeared in 1981. By October 1995, over 500,000 AIDS cases had been reported. This number has risen to 900,000 total reported AIDS cases in the United States from 1981 to 2005, and over half a million Americans have died from this disease. Today, about 950,000 people, or 1 in 170 adults in the United States are HIV positive. Nearly one-third of these are unaware of their HIV status. Over 400,000 Americans are living with AIDS. Therefore, the number of HIV-infected people without AIDS exceeds the number with AIDS (Fig. 18-6).

Is AIDS presently increasing or decreasing in the United States? Note in Fig. 18-7 that a peak in the annual rate of newly diagnosed AIDS cases occurred in the early 1990s. This peak followed about 10 years after the peak in new HIV infections (Fig. 18-8). Following these peaks, the rates of HIV infection and new AIDS cases have declined and then leveled off. Since the late 1990s, the number of new AIDS cases has stabilized at about 40,000 new cases per year.

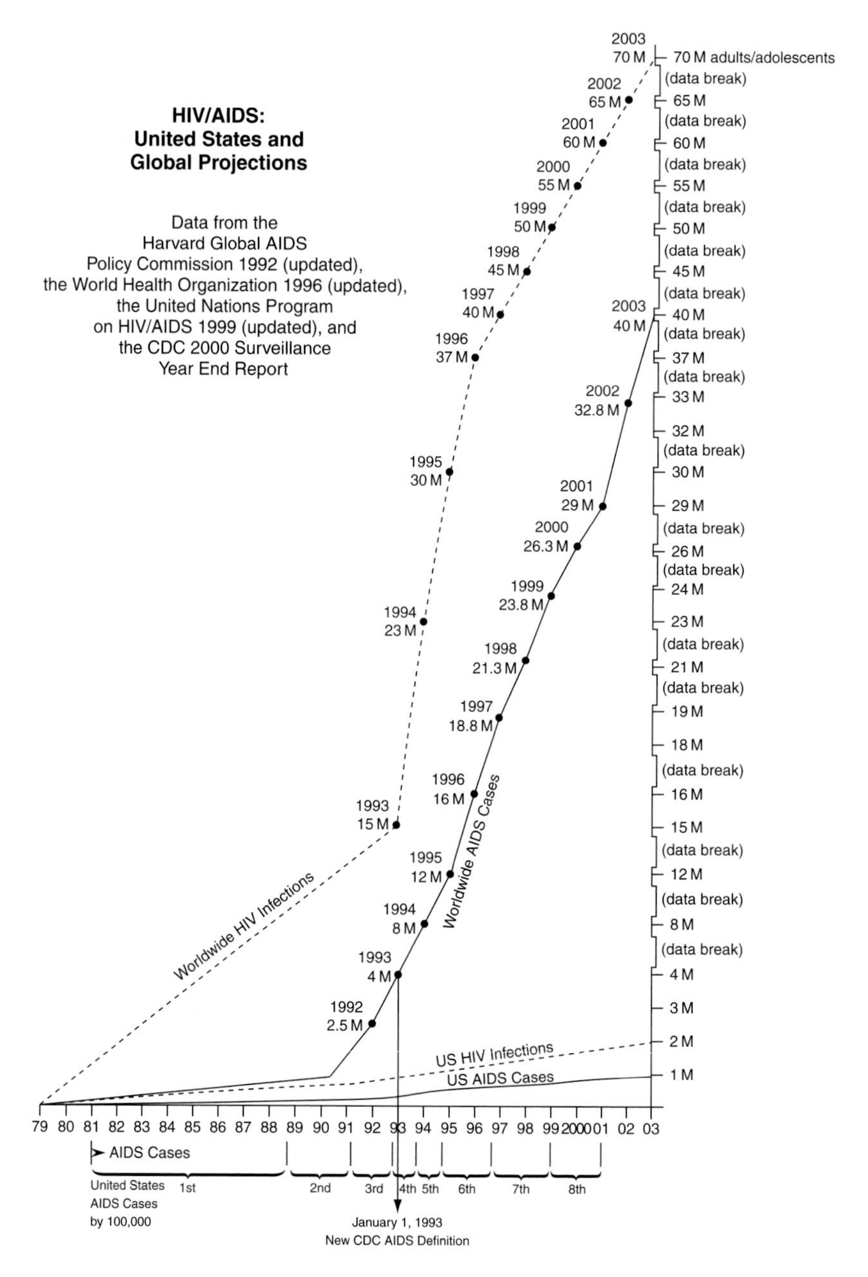

Figure 18-6 United States and global estimated total (cumulative) number of HIV infections and AIDS cases. If prevention does not work and a cure is not found, about one billion people in the world will be infected with HIV by the year 2025.

The face of HIV/AIDS is changing, both in the United States and throughout the world. In the 1980s, an American AIDS patient was typically white, male, and homosexual. More recently, transmission of the virus has been shifting from homosexual to heterosexual contact. More women are being infected, and now one-quarter of those living with the HIV virus in the United States are females (worldwide, women account for 46% of HIV-positive people). African American and Hispanic populations in the United States bear a higher burden of the disease. In fact, about 75% of new AIDS cases currently occur among racial minority groups (Fig. 18-9). Half of new AIDS cases are diagnosed in African Americans, even though they make up only about 13% of the U.S. population. The average age at infection has decreased, resulting in more HIV-positive adolescents and young adults.

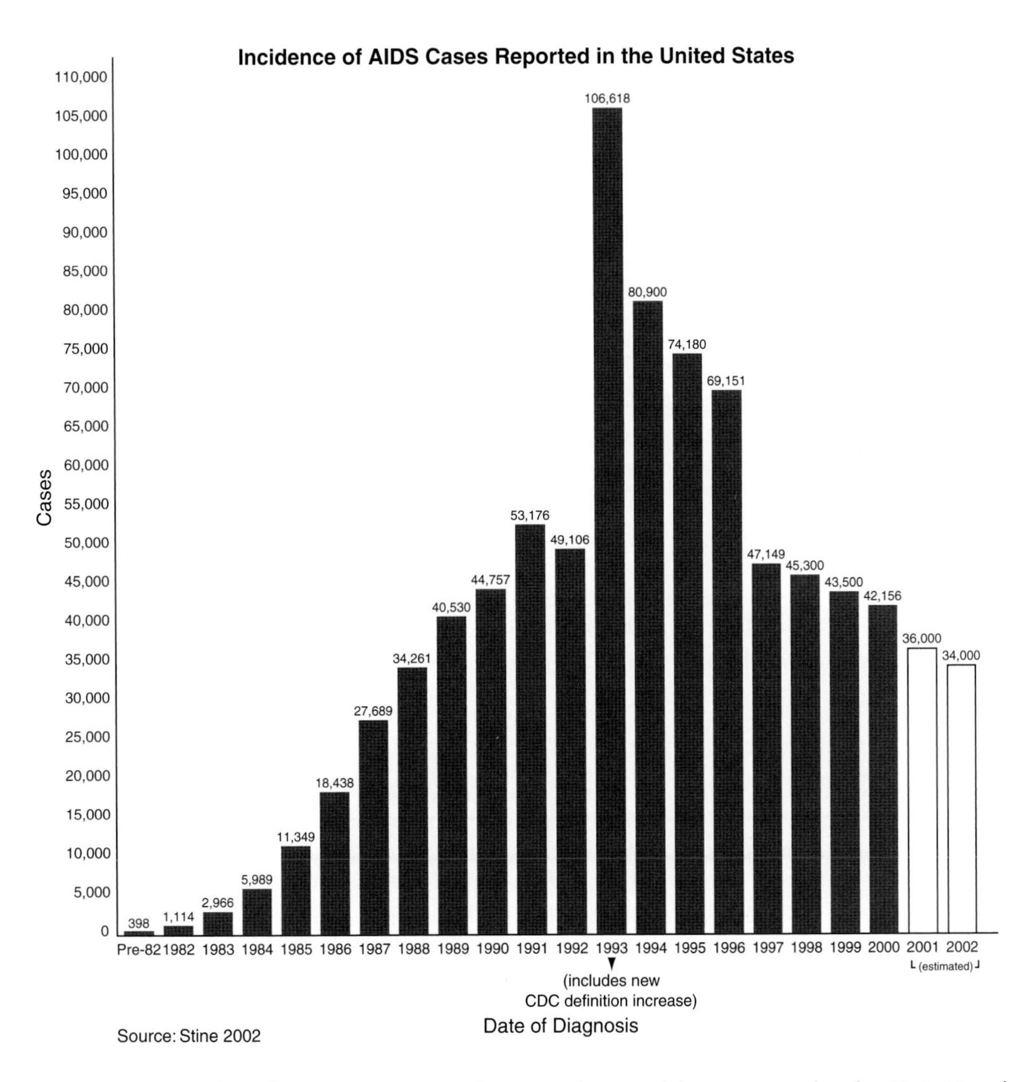

Figure 18-7 Number of new AIDS cases each year in the United States reported to the CDC. Note the dramatic increase in 1993, when the CDC definition of AIDS was expanded.

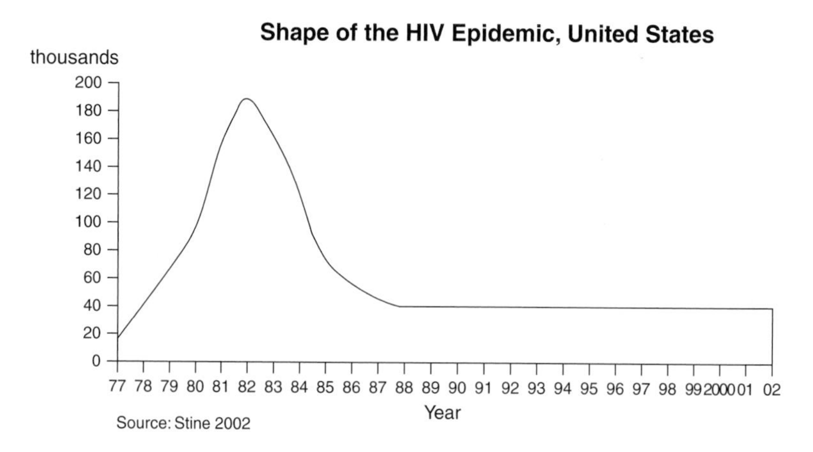

Figure 18-8 Number of individuals newly infected with HIV in the United States for the years 1977–2002. (Data from the CDC.)

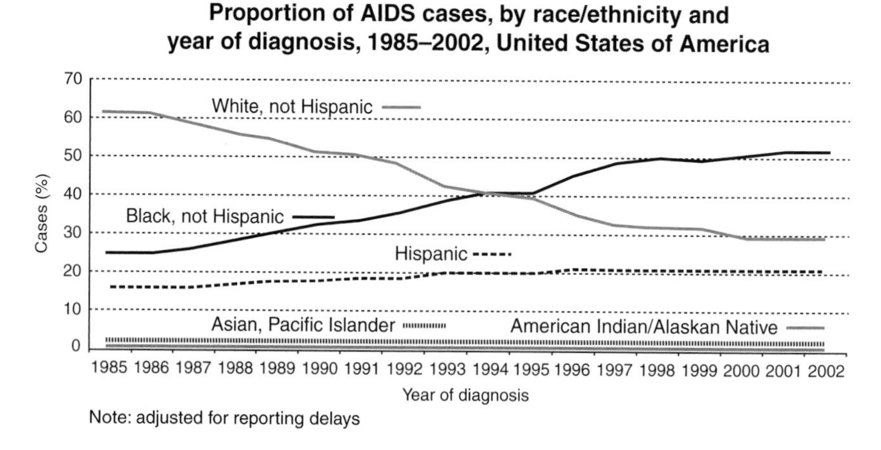

Figure 18-9 Change in incidence of AIDS by race/ethnicity in the United States for the years 1985–2002. Today, African Americans and Hispanics make up a disproportionate number of people living with AIDS. (Data from the CDC.)

Chapter 18, Box 1: Origins of the HIV Viruses

Because human immunodeficiency viral infection is a fairly recent phenomenon, there were several initial rumors about its origin. The press in the former Soviet Union speculated that the United States created HIV as a germ warfare weapon. But who would want a weapon that would take several years to kill the enemy? Another rumor was that white South Africans hybridized the human T-cell leukemia virus with a sheep virus to create HIV, with the purpose of using it to eradicate black South Africans. This is now known to be not only unthinkable but also impossible, as the gene sequence of HIV is not compatible with that of either "parent" virus. Another rumor with no evidence was that HIV came from a similar virus in cats; the cat virus has never been shown to be transmitted to humans.

Continued on next page.

Chapter 18, Box 1 continued.

The most probable origins of the HIV viruses were closely related viruses in wild primates in Africa. Many African (but not Asian) primates carry 1 of over 30 types of *simian* immunodeficiency virus (SIV).

Computer modeling of the DNA sequences of HIVs and SIVs suggests that HIV-1, which causes most of the world's human AIDS cases, originated from the SIV of the African chimpanzee of central Africa in the mid-1940s. Once passed into humans, this chimp SIV then mutated into HIV-1. In contrast, HIV-2, which is mostly confined to AIDS cases in west Africa and islands off its coast (but is also present in the United States), is very close to the SIV of the west African sooty mangabey monkey; some think that this mangabey SIV in humans is HIV-2. SIVs have been transmitted to humans accidentally in laboratories, but as yet there are no reported cases of human AIDS derived from SIV cross-infection. HIV-2, however, can cause AIDS in baboons, and sooty mangabey SIV can cause AIDS in Asian Rhesus monkeys (which normally do not have an SIV).

How and why did SIVs begin to infect people and change to HIVs? Some feel that the original cross-species transmission of SIV occurred when people butchered primates for meat. As humans began encroaching more and more on rain forests and wilderness

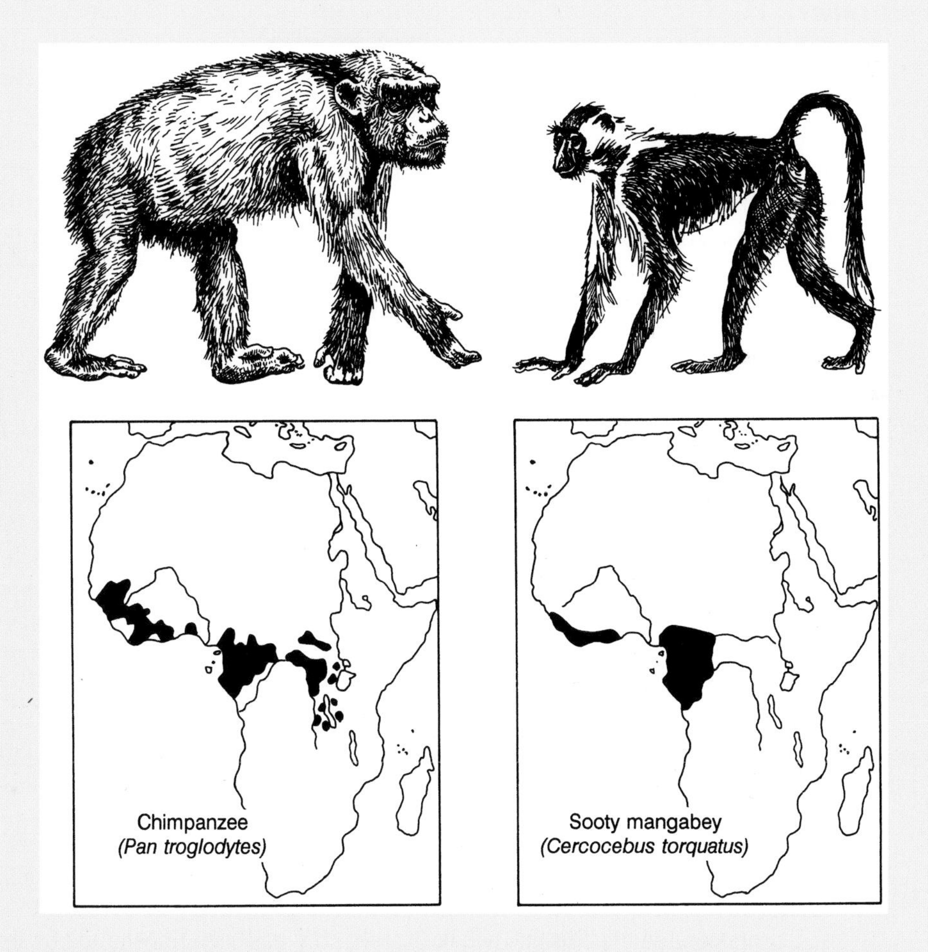

The SIV of the chimpanzee (upper left) and sooty mangabey monkey (upper right) gave rise to HIV-1 and HIV-2, respectively. The distribution of those two West African primates is shown.

Continued on next page.

Chapter 18, Box 1 continued.

areas, they placed themselves in closer contact with primates and SIVs. Viruses could be transmitted by a bite. Another theory is that the monkey kidney cultures used to produce the oral polio vaccine carried SIV, but there is no evidence of this occurring. Whatever the method of transfer, it is clear that HIVs evolved from similar viruses of our African primate relatives. Since HIV-1 and HIV-2 appeared, several subtypes of each have evolved in humans. For example, there are at least seven kinds of HIV-1, all evolving from an original viral type around 1960. The rapid evolution of the HIV viruses makes a vaccine more difficult to develop.

The African primates carrying SIV do not get AIDS. During evolution of a virus and its host, viral strains will be favored by natural selection that replicate efficiently without causing cell or tissue destruction in the host, and the host will eliminate parts of its immune response that may influence host survival negatively. In time, will HIV reach such a steady state in humans? That is, will HIV and humans coevolve such that AIDS does not occur, as is true with SIV in wild African primates?

Transmission

The HIV virus is present in blood, as well as such bodily fluids as semen, cervical and vaginal secretions, breast milk, and (at low levels) in saliva and tears. It is transmitted when the contaminated fluid enters another's body through the internal membranes of the mouth, urethra, anus, and vagina, especially if there are small tears or wounds in these membranes. Transmission can occur during oral, vaginal, or anal intercourse. If there are sores in the genital region due to another disease, it makes it easier for the HIV viruses to enter the body. The virus in semen can be transmitted to men or women, and the virus in cervical or vaginal fluid can be transmitted to men or (rarely) women.

During anal coitus (between homosexual males or heterosexuals), the virus can be transmitted more easily from the insertive than from the receptive partner. The same is true for vaginal coitus in heterosexuals. This is because there is a much higher concentration of HIV in semen than in anal or vaginal–cervical secretions and also because penile insertion and movements cause abrasions of the lining of the rectum or vagina. Furthermore, the anus is more likely to tear than the vagina, and rectal cells have receptors to which the virus can attach. This helps explain (along with an often large number of sexual partners) why anal coitus produces a high incidence of HIV transmission in men who have sex with men (MSM), i.e., homosexual or bisexual men. Anal coitus is responsible for 90% of new HIV infections in MSM. Thus, homosexual sex (among men) is more likely to transmit the virus than heterosexual vaginal sex. During heterosexual sex, it is about 18 times more likely for an HIV-infected male to pass the virus to a female than the other way around. Uncircumcised men who are HIV infected can pass the virus more easily.

Transmission can occur when contaminated fluid contacts a wound on the skin surface of another person, though this is rare—it has happened to fewer than 100 U.S. health workers. The virus can also be transferred when one uses a previously used, contaminated hypodermic needle. Rarely, HIV has been transmitted by sharing a razor blade. Transmission can also occur from an infected mother to her fetus, across the placenta (especially during early pregnancy), by the fetus swallowing amnionic fluid, or during delivery (because the virus is present in cervical and vaginal cells and secretions). The virus is also present in breast milk and can be passed to newborns during nursing. Finally, the virus has been transmitted in blood transfusions or in donated semen during artificial insemination (Chapter 16), especially from 1978 to 1985. Since March 1985, a blood test for the presence of antibodies to the HIV-1 and HIV-2 viruses has been used to screen donated blood or semen (in some states) before it is used. Carriers of the virus, however, can lack antibodies to HIV in their blood for up to 18 weeks or even (rarely) up to 3 years after the virus enters their body. In this way, so-called "silent transfusions" could occur, during which one contracts the virus during the standard transfusion of blood that was screened. Presently, however, testing for the virus itself in potential blood donors has lowered the risk of getting HIV from a blood transfusion to 1 in 20 million.

The good news is that the virus does not live long outside the body, and one cannot contract HIV from classroom activities, bathrooms, swimming pools or hot tubs, or by touching a person with the virus. Although present in saliva and tears at low levels, there is no evidence that transmission can occur from these fluids during even open-mouth kissing, sharing food, coughing, sneezing, sweating, sharing cups or utensils, or by contacting tears or urine. Finally, one is unable to catch the virus by *giving* blood, and there is no evidence that mosquitoes or other insects, spiders, or ticks can transmit the virus to humans. Table 18-1 summarizes the usual modes of HIV transmission.

The HIV virus is not transmitted as readily as are many other STIs. It is estimated that a person's chance of acquiring the virus in a single exposure to an infected sexual partner is about 1 in 500, as shown in Table 18-2. Therefore, on average, numerous unprotected sexual encounters with an HIV-positive partner are required for a person to become infected. (However, it is important to note that infection can occur at any time and that unprotected sex with an infected individual should be avoided completely.) This may be one reason why HIV/AIDS is so prevalent in sub-Saharan African heterosexuals. In this region it is fairly common for men (and possibly women) to have more than one simultaneous long-term sexual relationship. Condoms may be used at the beginning of a relationship, but

Table 18-1 HIV Transmission

Transmission routes
Blood inoculation
Transfusion of HIV-infected blood
Needle sharing among injection drug users
Needlesticks, open cuts, and mucous membrane exposure to (or rarely from) health care workers
Use of HIV-contaminated, skin-piercing instruments (ears, acupuncture, tattoos, surgical, or dental instruments)
Injection with unsterilized syringe and needle
Sexual contact: Exchange of semen, vaginal fluids, or blood
Homosexual, between men
Lesbian, between women (rare)
Heterosexual, from men to women and (less so) women to men
Perinatal
During pregnancy (across the placenta or by the fetus ingesting amnionic fluid)
During birth
Breast-feeding

Table 18-2 Risk of HIV Infection for Heterosexual Intercourse in the United States[a]

	Estimated risk of infection	
Risk category of partner	**1 sexual encounter**	**500 sexual encounters**
Do not know if partner is infected with HIV		
Not in any high-risk group		
Using condoms	1 in 50,000,000	1 in 110,000
Not using condoms	1 in 5,000,000	1 in 16,000
High-risk groups[b]		
Using condoms	1 in 100,000 to 1 in 10,000	1 in 210 to 1 in 21
Not using condoms	1 in 10,000 to 1 in 1,000	1 in 32 to 1 in 3
Partner has had a negative HIV antibody test		
No history of high-risk behavior[c]		
Using condoms	1 in 5,000,000,000	1 in 11,000,000
Not using condoms	1 in 500,000,000	1 in 1,600,000
Continuing high-risk behavior[c]		
Using condoms	1 in 500,000	1 in 1,100
Not using condoms	1 in 50,000	1 in 160
Partner has had a positive HIV antibody test		
Using condoms	1 in 5,000	1 in 11
Not using condoms	1 in 500	2 in 3

[a]Adapted from Stine, GJ (2002). *AIDS Update 2002*. Englewood Cliffs, NJ: Prentice Hall.

[b]High-risk groups with prevalences of HIV infection at the higher end of the range include homosexual or bisexual men, injection drug users from major metropolitan areas, and hemophiliacs. Groups with prevalences at the lower end of the range include homosexual or bisexual men and injection drug users from other parts of the country, female prostitutes, heterosexuals from countries where heterosexual spread of HIV is common (including Haiti and central Africa), and recipients of multiple blood transfusions between 1983 and 1985 from areas with a high prevalence of HIV infection.

[c]High-risk behavior consists of sexual intercourse or needle sharing with a member of one of the high-risk groups.

are often discontinued as a sign of trust. If many men and women have multiple long-term sexual partners, a web of sexual relationships exists throughout the culture. This would favor the widespread transmission of the HIV virus more rapidly than in heterosexual cultures where the practice of serial monogamy (consecutive long-term sexual partners) is more common.

Who Is Most Susceptible to HIV Infection?

In the United States, certain segments of the population are particularly susceptible to HIV infection (Table 18-3). Of the present U.S. HIV/AIDS cases, 73% are males and 27% are females. Of males, about 63% are men who have sex with men, 14% are injection drug users, 5% are MSM who are also injection drug users, 17% acquired HIV through heterosexual sex, and less than 1% are from blood transfusions or other causes. About 79% of infected women obtained the virus through heterosexual contact and 19% by injection drug use. Young men under age 25 account for 15% of HIV-positive men; young women

Table 18-3 Adult/Adolescent Behavioral Risk Groups, Race, and Sex: Percentage of Total AIDS Cases in the United States in 2001

Group	Number of AIDS cases	% of cases
Exposure group		
Men who have sex with men	16,280	44
Injection drug user (IDU), heterosexual	8,140	22
Heterosexual contact	3,700	10
Homosexual/bisexual IDU	1,480	4
Transfusion related	111	0.4
Hemophiliac	37	0.2
None of the above	7,400	20
Total		100
Race/ethnicity		
Black, non-Hispanic	17,020	46
White, non-Hispanic	11,840	32
Hispanic	7,030	19
Other	1,110	3
Sex (adults only)		
Male	28,490	77
Female	8,510	23
Age group		
13–19	185	0.5
20–24	1,332	3.6
25–29	5,254	14.2
30–39	16,946	45.8
40–49	9,620	26
50–59	2,701	7.3
60 and above	962	2.6

in this age group make up 27% of affected women. A fraction of a percentage of AIDS cases are newborns of HIV-infected mothers. About 20 to 30% of HIV-infected mothers pass the virus to their fetus.

Thus, in the United States about 17% of men and 79% of women Who are HIV positive contracted HIV from infected people of the opposite sex during heterosexual coitus, resulting in about one-third of the HIV/AIDS cases in the United States. Is heterosexual AIDS increasing? In 1991, only 4% of HIV/AIDS cases involved heterosexual transmission so there has been a major increase in heterosexual transmission of HIV in the United States. In the rest of the world, it is estimated that heterosexuals make up fully 90% of the HIV-infected people at present. Sex workers play a role in this heterosexual transmission, but with the increase in infected women, heterosexual transmission of HIV now occurs largely via spouses and other long-term sexual partners.

Life Cycle of the HIV Virus

HIV infection leads eventually to failure of the immune system. Immune cells (*lymphocytes*) reside in lymphatic organs (lymph nodes, spleen, tonsils) as well as in the blood and membranes lining or covering surfaces of the body. In discussing the life cycle of HIV, we will be concerned with B lymphocytes, T lymphocytes, and macrophages. B lymphocytes make an *antibody* (protein in the blood) in response to a foreign substance (*antigen*). The antigen can be a foreign molecule, a portion of a molecule, or a whole microorganism such as a virus or bacterium. An antibody binds to and helps destroy a specific antigen. T lymphocytes are of several types. *Cytotoxic (killer) T lymphocytes* seek out cells infected with viruses and destroy them. *Helper T lymphocytes* increase the number of cytotoxic T cells and increase antibody production by B cells. Finally, *macrophages* capture foreign antigens and "present" them to the other lymphocytes, thus assisting the immune reaction. There are also memory T and B cells that "remember" the specific foreign invader if it appears again, and thus the immune system responds more quickly (i.e., one is "immune" to that antigen). A *vaccine* (or vaccination) artificially introduces a harmless antigen, such as a dead virus or piece of a virus, so that the body will form a specific antibody that will defend against the real virus if it invades.

The mature HIV virus consists of a core, or "capsid," with two strands of ribonucleic acid (RNA) plus viral proteins and a surrounding envelope (Fig. 18-10). The HIV viral envelope has protein "spikes" on its surface; each spike consists of two glycoproteins, gp120 and gp41. The gp120 protein attaches (binds) to a cell surface receptor protein (CD4) on the surface of only some immune cells, especially helper T lymphocytes. When gp120 binds to CD4-bearing lymphocytes ("CD4 cells" or "T4 cells"), the three-dimensional structure of gp120 changes. This allows binding sites on gp120 to attach to chemokine c-receptors on the surface of the host cell, usually the CCR5 receptor or the CXCR4 receptor (also called *fusin*). Binding of the gp120 protein to the coreceptor allows the virus to fuse with the cell membrane and enter the host cell. The viral RNA is then reverse transcribed into strands of viral deoxyribonucleic acid (DNA) using the enzyme *reverse transcriptase* and is then made double stranded. The viral DNA then inserts itself into the DNA in the chromosomes of the host cell using the enzyme *integrase*; the viral DNA spliced into the host cell DNA is now called *proviral DNA*. Eventually, the proviral DNA is transcribed into viral messenger RNA and translated into new capsid and envelope proteins. Thus, the virus has replicated (multiplied), and new viral particles burst free ("bud") from the host cell as the latter dies. This is how the virus multiplies and, at the same time, kills CD4 lymphocytes. Without enough CD4 T cells, the cytotoxic T cells and B cells are unable to defend against the virus as it continues to multiply. Eventually, the entire immune system of the individual is weakened, opportunistic infections and cancers arise, and the person eventually dies of AIDS. The life cycle of the HIV virus is summarized in Fig. 18-10.

Symptoms of HIV Disease and AIDS

Infection with the HIV virus causes a slowly progressing *HIV disease*, the last stage of which is AIDS itself. The virus can be transmitted to another person at any stage, but is transmitted most readily when the infected person's viral load is high. Once the HIV virus enters a person's body, the virus replicates rapidly and the amount of virus in the blood increases dramatically. During this *window period* (or incubation period), there are no symptoms or detectable antibodies to HIV in the blood; CD4 lymphocyte counts in the blood are high during the window period. However, a person can transmit the virus with greatest risk at this time; this is important because a person still would not know that he or she is HIV positive unless he or she had a test for the virus itself.

After the window period, antibodies to HIV appear in a person's blood, usually within 1 to 18 weeks of initial infection. Almost all (99%) of HIV-infected

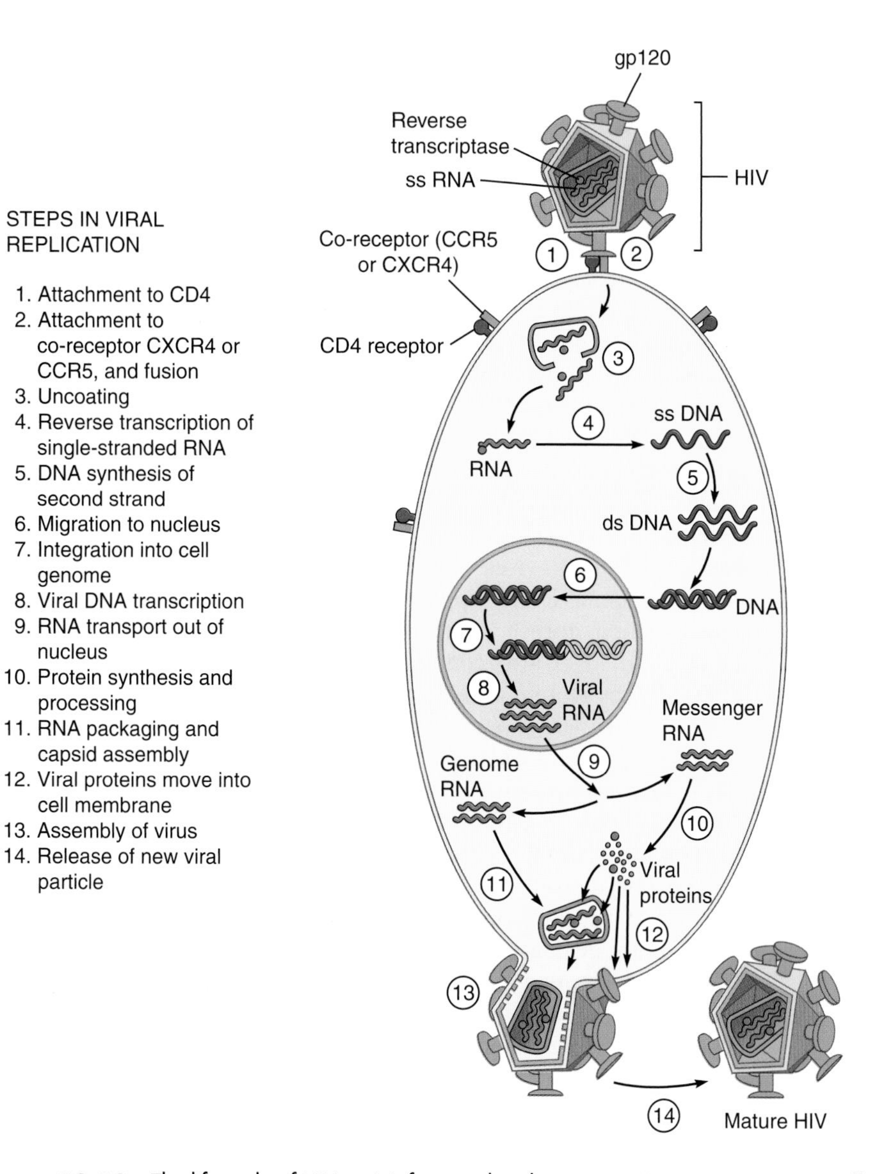

Figure 18-10 The life cycle of HIV as it infects and replicates in a CD4 immune system cell. Note that the mature HIV particle is actually *much* smaller than the CD4 cell that it infects.

people have detectable antibodies by 6 months, but a few can be antibody free for up to 3 years! The appearance of HIV antibodies in the blood is called *seroconversion*. The presence of the HIV virus can be determined by testing for HIV antibodies. When HIV antibodies are detected, a person is considered "HIV positive." A home test for HIV antibodies is available. One lances a finger and places a drop of blood on a card that is mailed to be tested anonymously, using a code number. Results are available in 7 days by telephone. If the test is positive, it should be repeated. A newer, rapid test for HIV antibodies in saliva can be performed using a swab of the oral cavity. This test is 99% effective, but the results should be verified. Antibody tests are not used before 6 to 12 weeks of infection, but a test for viral RNA can detect the infection before seroconversion.

In about 70% of HIV-infected people, seroconversion is accompanied by a flu-like acute illness (fever, swollen lymph nodes, muscle ache, weight loss). This *acute phase of HIV disease* usually lasts 2 to 4 weeks and then most people enter an *asymptomatic phase of HIV disease*. Although people in the asymptomatic phase exhibit few or no external symptoms and low virus levels in the blood (Fig. 18-11), the HIV virus is present, replicating, and destroying CD4 lymphocytes in the lymph nodes. As a result, the CD4 count in the blood is slowly falling (from a normal of 900 to 1200 cells/μl of blood to around 500/μl; see Fig. 18-11). It is estimated that the HIV virus kills about one billion lymphocytes per day!

After several asymptomatic years, a person then enters the *symptomatic phase of HIV disease*. The CD4 count is still falling, as are the numbers of cytotoxic T lymphocytes in the lymph nodes and blood (Fig. 18-11). HIV antibody levels in the blood also begin to fall as B lymphocytes decrease their antibody production (Fig. 18-11). Now begin the so-called *opportunistic infections* of HIV disease (Fig. 18-12). These infections are not directly caused by the HIV virus, but instead are caused by other viruses, bacteria, fungi, or cancer-causing factors because the HIV virus has destroyed the immune system that usually fights these invaders. Early opportunistic diseases can include shingles, bacterial skin infection, thrush, gastrointestinal infections, and tuberculosis (Fig. 18-12; Table 18-4). It is also common for some HIV-infected people to have other STDs, such as syphilis and herpes.

In 4.2 to 15.0 years (average 7.8 years) from being infected, the HIV disease usually progresses to AIDS itself. In 1987, AIDS was defined as having evidence of HIV infection (a positive HIV antibody test and/or microscopic visualization of the virus itself) plus at least 1 of 23 opportunistic infections (see Table 18-4). In 1993, AIDS was redefined by the following CDC criteria (Table 18-4). To have AIDS, one must have positive evidence of HIV infection plus a CD4 lymphocyte count of less than 200 cells/μl of blood (or less than 14% CD4 cells of total blood lymphocytes) and/or evidence of HIV infection plus at least one of the conditions listed in Table 18-4. Common AIDS opportunistic infection symptoms include fever, weight loss, fatigue, pneumonia, diarrhea, and neurological problems (meningitis, headaches, seizures, and dementia partial blindness), as well as cancers of the skin (Kaposi's sarcoma), cervix, and lymph nodes (Table 18-4). Once full-blown AIDS develops, without aggressive treatment death follows in about 2 years. Figure 18-13 summarizes the progression of HIV disease, ending in AIDS itself.

Long-Term HIV Survivors

Despite the fact that the average time between HIV infection and the onset of AIDS is 7.8 years, there are a few HIV-infected people who are still AIDS free without medication after more than 15 years. Furthermore, these "long-term survivors" are generally healthy and their CD4 count remains only moderately reduced (about 500 cell/μl) and steady. In addition, some babies born with HIV are free of the virus within months. No one knows why, but some possibilities are beginning to emerge. Several Australian men who were infected with HIV-1 via blood transfusions before 1985 never developed a lowered CD4 count, HIV-related diseases, or AIDS and all have survived at least 15 years (two died of other causes). Furthermore, the gay man who donated the blood remained disease free.

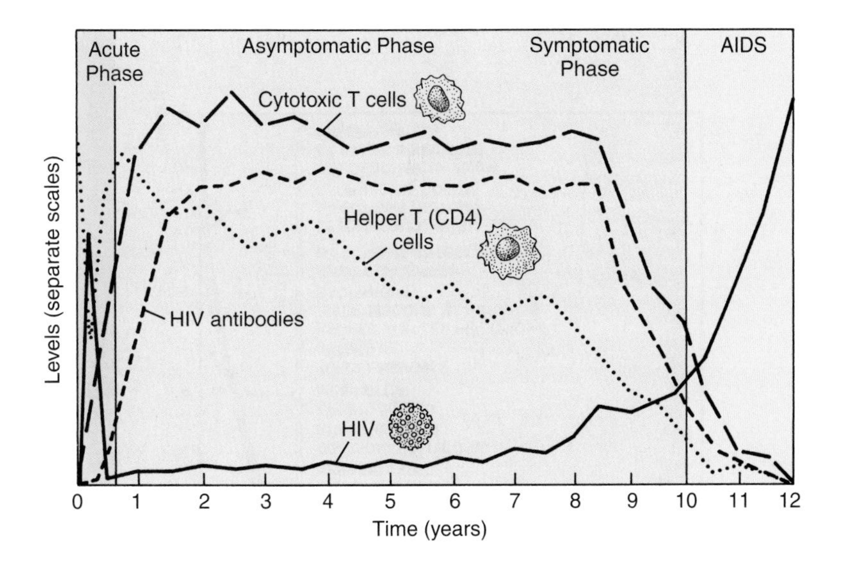

Figure 18-11 Progress of the HIV disease. After an initial window period, when CD4 cells in the blood are reduced and before antibodies appear, an acute phase is characterized by flu-like symptoms, high HIV antibodies, high blood CD4 counts, and low blood virus levels. This is followed by an asymptomatic phase in which the person is relatively healthy but blood CD4 cells are declining gradually. During this phase, however, there is a war going on between the HIV virus and immune cells in lymphatic tissue, with the latter being destroyed gradually. When the CD4 count falls below 500 cells/μl of blood, the symptomatic phase of HIV disease appears, in which opportunistic infections can occur. This is accompanied by a lowering of blood cytotoxic T-cell count and HIV antibodies, as well as a rise in HIV virions in the blood. AIDS then usually appears, with several opportunistic infections and death in usually about 2 years.

The viral counts of these people were very low, indicating poor viral replication. It turns out that the strain of HIV-1 that infected these people had an abnormal gene (*nef*) that is involved in viral replication. In fact, there is some hope that one could genetically engineer normal HIV to have a deficient *nef* gene and use it as a live vaccine. More recently, it has been discovered that some people are resistant to HIV infection because their CD4 cells lack the protein CCR5, one that helps the virus bind to the immune cell.

Treatments of AIDS

There is, at present, no cure for AIDS, but several drugs are presently approved for use to fight HIV infection by inhibiting viral replication. They are also used to prevent the transmission of the virus from an HIV-infected mother to her fetus during pregnancy. One class of medications is known as *nucleoside reverse transcriptase inhibitors* (NRTIs). They include AZT (zidovudine), ddC (zalcitabine), ddI (didanosine), 3TC (lamivudine), d4T (stavudine), and seven other drugs currently approved for use in the United States. All of these drugs inhibit viral replication by interfering with reverse transcriptase (see Fig. 18-10 and HIGHLIGHT box 18-2). As the reverse transcriptase enzyme reads the

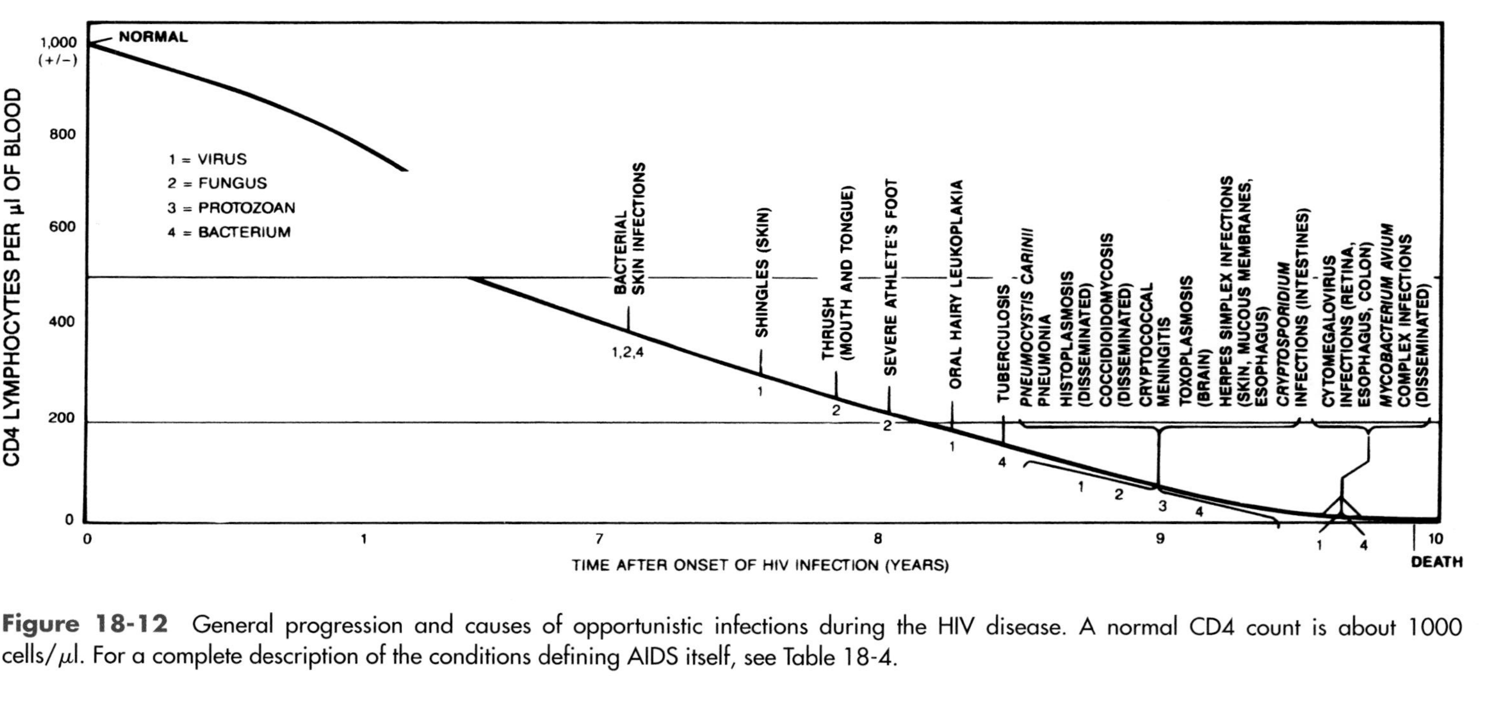

Figure 18-12 General progression and causes of opportunistic infections during the HIV disease. A normal CD4 count is about 1000 cells/μl. For a complete description of the conditions defining AIDS itself, see Table 18-4.

Table 18-4 List of 26 Conditions in the 1993 AIDS Surveillance Case Definition[a,b]

Candidiasis of bronchi, trachea, or lungs
Candidiasis, esophageal
Cervical cancer, invasive[c]
Coccidioidomycosis, disseminated or extrapulmonary
Cryptococcosis, extrapulmonary
Cryptosporidiosis, chronic intestinal (>1-month duration)
Cytomegalovirus disease (other than liver, spleen, or nodes)
Cytomegalovirus retinitis (with loss of vision)
HIV encephalopathy
Herpes simplex: chronic ulcer(s) (>1-month duration), bronchitis, pneumonitis, or esophagitis
Histoplasmosis, disseminated or extrapulmonary
Isosporiasis, chronic intestinal (>1-month duration)
Kaposi's sarcoma
Lymphoma, Burkitt's (or equivalent term)
Lymphoma, immunoblastic (or equivalent term)
Lymphoma, primary in brain
Mycobacterium avium complex or *M. kansasii*, disseminated or extrapulmonary
Mycobacterium tuberculosis, disseminated or extrapulmonary
M. tuberculosis, any site (pulmonary[c] or extrapulmonary)
Mycobacterium, other species or unidentified species, disseminated or extrapulmonary
Pneumocystis carinii pneumonia
Pneumonia, recurrent[c]
Progressive multifocal leukoencephalopathy
Salmonella septicemia, recurrent
Toxoplasmosis of brain
Wasting syndrome due to HIV

[a]Adapted from Stine, GJ (2002). *AIDS Update 2002.* Englewood Cliffs, NJ: Prentice Hall.

[b]To have AIDS, one must have a positive HIV antibody test plus a CD4 count of less than 200 cells/μl of blood (or less than 14% CD4 cells of total lymphocytes in the blood) and/or a positive antibody test plus at least one of the conditions listed here.

[c]New AIDS opportunistic diseases as of 1993.

viral genetic code from RNA to synthesize proviral DNA, NRTIs present in the cell act as "dummy building blocks" substituting for DNA nucleotides, disrupting the construction of a proviral DNA strand.

Another class of antiretroviral medications is *nonnucleoside reverse transcriptase inhibitors* (NNRTIs). Like NRTIs, these drugs target reverse transcriptase, but they do so directly, probably by binding to and disabling the enzyme. Three of these drugs are currently approved for treating HIV/AIDS patients. *Protease inhibitors* (PIs), a third class of drugs, inhibit the viral enzyme protease. This enzyme cuts up the large viral proteins into smaller, functional ones. PIs thus prevent synthesis of the proteins needed to produce new viral particles. Eight protease inhibitors are available in the United States. Finally, a *fusion inhibitor* (T20, or enfuvirtide) is now available. It inhibits the fusion of the viral envelope with the host cell membrane, thus preventing infection of the cell.

Additional drugs in each of these four categories are in the development phase or in clinical trials. For several reasons, there is a continued need for the discovery of new antiviral drugs. One is that the HIV virus is capable of rapid mutation,

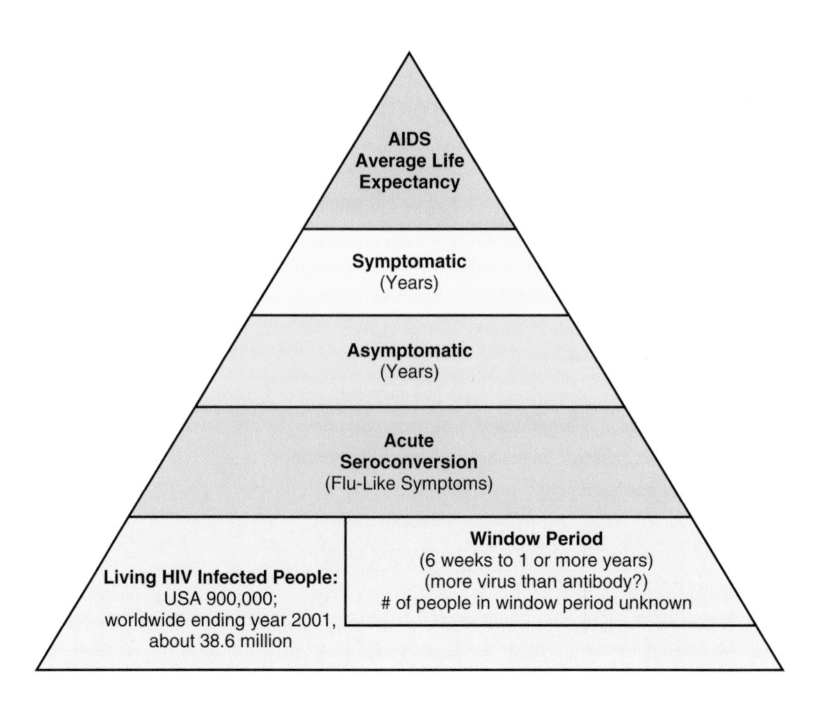

Figure 18-13 The HIV/AIDS disease progression pyramid.

becoming resistant to certain drugs. Second, none of the current therapies is successful in completely shutting down viral replication. In addition, each carries significant side effects (such as nausea, rash, insomnia, vomiting, headaches, anemia, muscle abnormalities) that can be tolerated by individual patients to varying degrees. Because different drugs work in different ways, and at different stages of the viral life cycle, a combination of drugs has been found to be more effective than single medications in slowing progression of the disease and prolonging life. Typically, a "cocktail" of two NRTIs plus either an NNRTI or a PI is most effective. These drugs are expensive and must be taken in a strict regimen. Some possible future treatments for HIV infection are discussed in the HIGHLIGHT box 18-2.

Chapter 18, Box 2: Attacking the HIV Virus

Soon after the appearance of AIDS in the world, intensive research began to find medicines that would either prevent HIV infection (e.g., vaccines) or attack the HIV virus itself. After several decades, research did produce drugs that attack the HIV virus after it has invaded a person's body. Many of these FDA-approved drugs fall into a general category of *reverse transcriptase inhibitors*, which reduce transcription of viral RNA into viral DNA. Others are *protease inhibitors*, which inhibit viral protein processing. The approach today is to give two reverse transcriptase inhibitors in combination with one protease inhibitor, so-called *highly active antiretroviral therapy* (HAART). However, some evidence suggests that HAART therapy is not as effective as once thought (see text) so the search continues for effective anti-HIV drugs.

There are at present clinical trials of drugs that block entrance of the HIV virus into the host cell or prevent integration of the HIV proviral

Continued on next page.

Chapter 18, Box 2 continued.

DNA into the host DNA. Additionally, there are at least 39 clinical trials using several different HIV vaccines. That is, a portion of the HIV virus is injected, and the recipient's immune system then produces antibodies that kill any living HIV that later enters the body. However, researchers estimate that an effective HIV vaccination could take many years to develop. Why so long? First, the virus hides within cells, where antibodies cannot go. Also, the virus mutates rapidly, even in one person as the disease progresses, making it difficult for one specific vaccine to be effective. Third, one must stimulate cytotoxic T-cell activity to fight the virus, whereas a vaccine would only increase antibody formation. Fourth, HIV kills the very immune cells that support antibody production. Fifth, clinical trials must use several thousand people at high risk

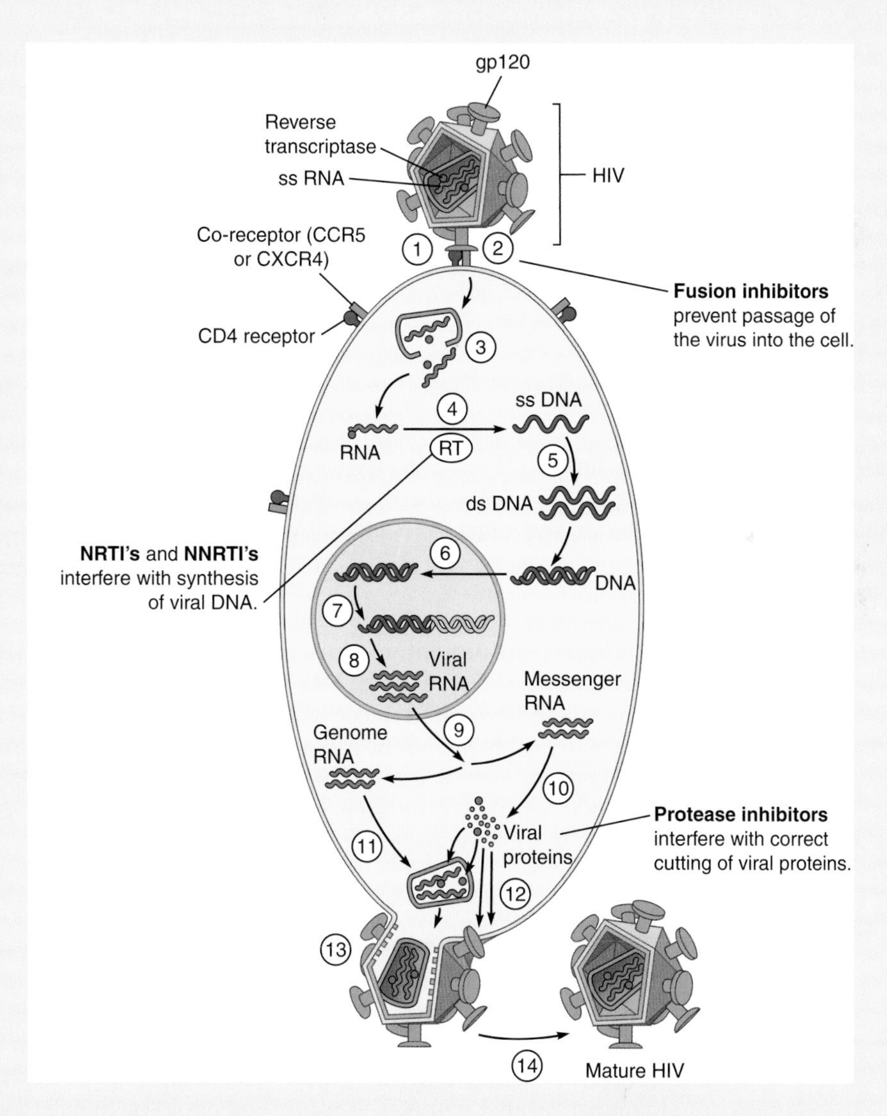

How HIV/AIDS treatments interfere with the life cycle of the HIV virus. See Fig. 18-10 for descriptions of the numbered steps in viral replication.

Continued on next page.

Chapter 18, Box 2 continued.

for contracting AIDS, as well as several more thousand for a control group, and then wait several years for the results. Finally, a lack of funding for AIDS research has inhibited progress toward an effective HIV vaccine.

The points of action of already approved or future HIV drugs and vaccines (numbers refer to points of action in the HIV life cycle listed on the left of the figure).

Approved anti-HIV drugs

(4) Reverse transcriptase inhibitors, often used in combination with other drugs, block reverse transcriptase and thus transcription of viral RNA into viral DNA.

(10) Protease inhibitors, often used in combination with other drugs, inhibit processing of viral proteins that become part of the HIV viral coat.

(2) Fusion inhibitors, drugs that block chemokine coreceptors (CCR5, CXCR4) on the cell membrane of the host cell. These coreceptors are necessary for HIV entry. One drug, Fuzeon, has been approved.

Possible future anti-HIV drugs

(7) Integrase inhibitors would prevent integration of the proviral DNA into the host DNA.

(11) Zinc finger inhibitors would disrupt the assembly of the inner core of the HIV virus.

(4) Antisense drugs, when inserted into immune cells, would act as mirror images of the HIV genetic code and prevent viral replication.

Some possible future HIV vaccines

- Vaccination against a combination of canary-pox virus and HIV genetic material. The canary-pox virus is used to carry HIV genes into the host cells, where they can code for production of HIV proteins. These, in turn, are attacked by the immune system of the host; antibodies are produced, and cytotoxic T cells are also mobilized to attack the HIV virus.
- Vaccination against the proviral DNA that codes for production of the viral coat. The proviral DNA is incorporated into the cells of the host, which then make HIV viral coat protein, to which the body raises antibodies.
- Vaccination against GP120 (called "Aidsvax").
- Vaccination against CD4 on the host cell membrane.

Even though there is, at present, no cure for HIV infection or AIDS, the opportunistic infections associated with HIV disease and caused by other microorganisms often can be treated and even cured. HIV-positive individuals who have access to antiviral medications can now live longer lives, and AIDS is beginning to be thought of as a long-term and manageable although ultimately fatal disease. Unfortunately, this complacency has led to a relaxation of safe-sex precautions against HIV infection in certain communities.

Precautions Against HIV Transmission

To lessen the chance that you could acquire the HIV virus, you should know as much as you can before you have sex with a person, and you should limit the number of your sexual partners. Do not have sex with prostitutes or with a person who has the HIV virus and/or AIDS itself. Avoid high-risk activities such as oral–genital contact or anal coitus. The use of male and/or female condoms (the latex rubber or polyurethane plastic variety but not the lambskin or membrane kind) can lessen (by up to 1000×) the chance of transmitting the virus. Do not use previously used needles and do not share anything that could be contaminated with blood, such as razor blades and toothbrushes. You should also avoid oral, vaginal, or anal contact with semen. Other sexually transmitted diseases that cause lesions in the genital area can increase your chances of being infected with the HIV virus if exposed. Thus, you should take precautions against being infected with an STD and abstain from sexual activity if any STD symptoms occur. Of course, abstaining from sexual contact altogether is the best preventative measure against contracting HIV.

Some General Aspects of Venereal Disease Infections

Inflammation and/or infection of the reproductive and urinary tract is a common component of many STDs. This section discusses these infections.

Vulvitis

Infection or inflammation of the vulva (*vulvitis*) can be caused by several sexually transmitted diseases. Infection with *Trichomonas* or *Candida* (see later) is a common cause of this malady, as can be the presence of parasites such as crabs or scabies. Infectious vulvitis can be accompanied by vaginitis or urethritis. Vulvitis can also be noninfectious, with the vulval membranes becoming itchy and inflamed. A common cause of noninfectious vulvitis is exposure to chemicals in spermicides, douches, soap, or toilet paper.

Vaginitis

Vaginitis can be a symptom of many of the aforementioned sexually transmitted diseases. Vaginitis, however, is also caused by other sexually transmitted organisms, such as *Candida albicans, Trichomonas vaginalis*, and *Haemophilus vaginalis*, all of which can occur normally in the vagina (see Chapter 2). Common physical symptoms of vaginitis include vaginal itching, burning, tenderness, and painful or difficult coitus. A vaginal discharge usually accompanies vaginitis. You should realize, however, that a vaginal discharge occurs normally. The normal cervical mucus mixes with vaginal cells and bacteria to produce a whitish discharge, which changes to pale yellow when exposed to air. The discharges associated with vaginitis have a color and odor different from this normal discharge.

Monilia vaginitis is a yeast infection of the vagina, also known as "candida," "moniliasis," or "vaginal thrush." It accounts for about one-half of the cases of vaginitis. About 75% of U.S. women will get a yeast infection sometime in their life. It is caused by the gram-positive fungus *C. albicans*, which normally is present in the vagina of 40% of women (see Chapter 3). When *C. albicans* multiplies rapidly, the signs of monilia vaginitis appear. *C. albicans* can also be present in males and can be transmitted during coitus.

Because this fungus (a yeast) thrives in a moist environment with a high pH, anything that makes the vagina more basic could lead to the spontaneous appearance of monilia vaginitis. Bacteria called *Doderlein's bacilli* (*Lactobacillus acidophilus*) normally are found in the vagina (see Chapter 2). These bacteria metabolize glycogen (a long-chain sugar) to lactic acid and thus help maintain an acidic vaginal condition. If antibiotics are given to a female, these bacteria may die, the vaginal pH increases, and a spontaneous monilia infection may occur. Such infections can also occur in the premenstrual period, during pregnancy, and in diabetics. About one-third of females with a venereal disease of another type also have monilia. Stress may make a woman more susceptible to yeast infection, and frequent coitus can aggravate the symptoms. There appears to be no increase in the incidence of yeast infection with the use of combination birth control pills.

In a female with monilia vaginitis, the vagina is itchy and raw, and the vagina, vulva, and anal regions can become reddened and swollen. White patches ("thrush") appear on the vulva, vaginal wall, and cervix, and a thick, white, cottage cheese-like vaginal discharge with a yeasty odor can occur. Coitus can be painful under these circumstances. Monilia can be transmitted by oral–genital contact and sometimes can occur in the mouth, accompanied by white patches, swelling, and pain. A fetus can contract *Candida* as it passes through the birth canal. This leads to the presence of the organisms in the newborn's digestive tract, causing thrush and digestive disorders. Males contracting *C. albicans* are usually asymptomatic.

The presence of *C. albicans* can be detected by taking a swab of the vaginal discharge and examining it under a microscope. Also, this yeast can be grown on a culture medium and the colonies examined. This affliction can be treated with fungicides such as nystatin (Mycostatin) in vaginal tablets, candicidin (Vanolud) in vaginal tablets or ointment, or with miconazole nitrate (Monistat cream). Periodic acidic douches can help, as can painting the area with a 1% gentian violet solution or using a tampon containing this dye. The resultant bright blue color wears off in a few days. Application of a yogurt douche containing the bacterium *L. acidophilus* also may be helpful.

Haemophilus vaginalis is a gram-negative bacterium that can occur spontaneously in the vagina (see Chapter 2) or can be transmitted sexually; it can also occur in males. This bacterium causes *H.vaginitis*, the third leading kind of infectious vaginitis. In females, the presence of large numbers of *H. vaginalis* produces a scanty vaginal discharge accompanied by mild burning and itching. Then, a profuse discharge occurs, which is dirty white or gray and foul-smelling. Urination and coitus are painful at this stage. About 80% of males are asymptomatic. *H. vaginitis* can be treated with sulfa creams or the antibiotics ampicillin or tetracycline.

Nonspecific vaginitis means just what it implies, i.e., vaginitis for which a specific cause is not identified. Most often, however, the culprit is *Escherichia coli*, a bacterium commonly present in the digestive tract. Vaginitis caused by *E. coli* is characterized by a gray, foul-smelling vaginal discharge. Poor hygiene often is the cause. Vaginitis can also be caused by estrogen withdrawal (as after menopause) or allergic responses to chemicals.

A woman can take several precautions against vaginitis. Proper hygiene (washing and drying the vulva thoroughly) is important. After defecation, a female should never wipe the anus from back to front as this could facilitate the passage of *E. coli* to the vulva. Also, a couple should never switch from anal to vaginal coitus without first washing the genitals, and a condom should be used for anal coitus. The use of feminine hygiene deodorant sprays should be avoided as they can irritate the vagina. Wearing tight nylon hose, such as panty hose, can decrease air circulation and may favor development of vaginitis. A low carbohydrate diet reduces sugar levels, which decreases the chance of vaginitis. Finally, a woman can douche periodically with a mildly acidic solution.

Urethritis and Cystitis

Urethritis is a common component of sexually transmitted diseases, especially those caused by bacteria such as *Chlamydia trachomatis, Ureaplasma urealyticum, Haemophilus vaginalis*, and *Mycoplasma genitalium*. These are often referred to as *nonspecific urethritis*. The disease organisms can pass up the urethra to the urinary bladder, causing its infection or inflammation (cystitis). Bladder infections can also be caused by *E. coli* or other bacteria normally resident in feces. This can be the result of poor hygiene as *E. coli* commonly are present in fecal material. Because of a "milking" action of the penis on the urethra during coitus, frequent sexual activity can facilitate passage of *E. coli* into the urethra and eventually into the urinary bladder. Manual or oral sexual stimulation, as well as the use of objects such as vibrators or dildos, can also increase the risk of urethritis and cystitis in females.

Symptoms of cystitis include a desire to urinate frequently, a burning pain when urinating, hazy urine often tinged with blood, and backache. Cystitis can be treated with an antibiotic or a sulfa drug. A medication for pain often is given with the antibiotic, and this medication contains a dye that turns the urine bright orange-red. Urination soon after coitus seems to lessen the chance of developing urethritis and cystitis.

Cervicitis

Many sexually transmitted diseases can infect or inflame the cervix, a condition called *cervicitis*. Chronic low-level cervicitis is common in women after childbirth, accompanied by a mild pus-filled cervical discharge, backache, and pelvic pain. Women taking combination pills with high levels of estrogen also tend to develop this affliction. Cervicitis usually can be treated with antibiotics, but it often disappears spontaneously. Of concern is the possibility that cervicitis may predispose a female to developing cervical cancer.

Pelvic Inflammatory Disease

Several sexually transmitted diseases can invade the uterus and eventually the oviducts. Inflammation and infection of the oviducts (salpingitis) are accompanied by low abdominal pain, vomiting, fever, and disturbed menstruation. Salpingitis can lead to the formation of scar tissue, which blocks the tubes and can lead to ectopic pregnancy and infertility (see Chapters 10 and 16).

The infection can then spread to many pelvic organs, a condition called *pelvic inflammatory disease*. This condition is very serious and can lead to death. Pelvic inflammatory disease can be caused by venereal disease organisms such as *N. gonorrhoeae, C. trachomatis*, and *M. genitalium*, but it also can be caused by nonvenereal infections due to bacteria such as *Streptococcus, Staphylococcus, Escherichia*, and *Mycobacterium*. Gonorrhea, however, is responsible for 65 to 75% of cases of pelvic inflammatory disease. The incidence of pelvic inflammatory disease increases after a nonsterile abortion, after coitus within a few days of an induced abortion, or during usage of an IUD (see Chapter 14). Pelvic inflammatory disease can be treated with antibiotics. This can kill the causative bacteria but does not eliminate scar tissue already formed.

Prostatitis

The prostate gland can become infected and inflamed, a condition called *prostatitis* (see Chapter 4). About 80% of the time, bacterial prostatitis is caused by *E. coli*, but it can also be due to *N. gonorrhoeae, T. vaginalis*, and other microorganisms. The prostate glands of many young men are chronically infected, often with no obvious symptoms. Other men develop symptoms such as urethral discharge, fever, chills, a tender groin, an increase in the frequency and desire to urinate, lower back pain, and perianal aching. Bacterial prostatitis is the most common cause of recurring urinary tract infection in males. Men with bacterial prostatitis can suffer spasmodic, painful contractions of the prostate during orgasm. Antibiotic treatment for infectious prostatitis must be specific for the particular bacteria present.

Psychological Aspects of Sexually Transmitted Disease

As we have seen, the effects of a sexually transmitted disease on a person's physical health will vary with the type of disease organism involved, ranging from little or no discomfort to lifetime impairment or death. Similarly, the adverse effects of having a sexually transmitted disease on a person's emotional health will vary to some extent with the disease type. A person having syphilis may feel shame or guilt about contracting such a "filthy" problem, but he or she may not feel that way about having, for example, nonspecific urethritis. A person with a sexually transmitted disease may feel that he or she is a "bad person" or that he or she is being punished for "improper" behavior. Often, this relates to the individual's basic feeling about sexual activity.

A person's life frequently will be disrupted after contracting a sexually transmitted disease. Usually, the individual must refrain from sexual activity for a while to avoid transmitting the disease to other people. Even if protective devices such as barrier contraceptives are used, the presence of the disease may mean a change in sexual habits and interactions. Also the infection or pain associated with the infection can cause distraction during coitus or even avoidance of sexual activity. The person may be faced with the difficult task of telling a potential sex partner or spouse about the problem before having intimate contact. Even more difficult, a physician or other health professional must question a person with a venereal disease and determine recent sexual contact(s) in order for the other person(s) to be treated. In all cases, the sexual partner(s) of an infected person should also be treated, even if they have no symptoms.

Contraction of a sexually transmitted disease often is a major disruption to a person's life. The pain of this process, however, can be alleviated by (1) having a clear understanding about the disease, (2) admitting one's responsibility in the matter, (3) seeking medical and psychological help, and (4) preventing transmission of the disease to others.

Preventing Sexually Transmitted Disease

The best way to avoid contracting a sexually transmitted disease is to abstain from sexual contact, an approach not acceptable to many people. What else can you do to lessen the odds that you will suffer from one of these diseases? One important way is to limit your sexual activity to people whom you know well and can trust. This is why faithful married couples, often but not always, are free of this problem. If you find it difficult to limit your sexual activity in this way, it is helpful to discuss the topic and even to inspect a sex partner's genital region before coitus; this can even be part of foreplay if approached in a tactful, caring manner. Good hygiene, including washing the genital region before coitus, can help, as can urination before or after coitus. As mentioned earlier, some contraceptive devices, such as the condom, diaphragm, and spermicides, can help prevent transmission of the disease organisms. None of these measures will guarantee freedom from these diseases, but they will lessen the incidence of them. All of these preventive measures require a thoughtful approach to one's sex life, although the passion of the moment sometimes wins the battle with rationality.

Sexually transmitted diseases can have serious physical and psychological effects. Everyone should be concerned with decreasing the incidence of these diseases in the general population as well as preventing any spread to himself or herself or to his or her sexual partner(s). If anyone has any questions or thinks that he or she has contracted such a disease, he or she should contact a physician or one of the many venereal disease clinics that exist in each state.

Chapter Summary

Sexually transmitted diseases are those that are transmitted during coitus or other genital contact. The incidence of several of these diseases is on the rise in the United States, and some have reached epidemic proportions and present a major public health hazard. Sexually transmitted diseases, with the organisms causing them in parentheses, are (1) gonorrhea (*Neisseria gonorrhoeae*), (2) syphilis (*Treponema pallidum*), (3) herpes genitalis (herpes simplex virus, type 2), (4) acquired immunodeficiency syndrome (human immunodeficiency virus), (5) chlamydia and other forms of nonspecific urethritis (*Chlamydia trachomatis, Ureaplasma urealyticum, Haemophilus vaginalis, Mycoplasma genitalium*), (6) genital warts (human papilloma virus), (7) chancroid (*Haemophilus ducreyi*), (8) lymphogranuloma venereum (*Chlamydia trachomatis*), (9) granuloma inguinale (*Calymmatobacterium granulomatis*), (10) monilia vaginitis (*Candida albicans*), (11) *haemophilus* vaginitis (*Haemophilus vaginalis*), (12) *Trichomonas* vaginitis (*Trichomonas vaginalis*), (13) molluscum contagiosum (a pox virus), (14) viral hepatitis, type B (hepatitis virus B), (15) pediculosis pubis (*Phthirus pubis*), and (16) scabies (*Sarcoptes scabiei*).

Symptoms of some of these afflictions in females include vulvitis, vaginitis, urethritis, cystitis, cervicitis, endometritis, and salpingitis. Infection can be widespread, resulting in pelvic inflammatory disease. In males, common symptoms include urethritis, prostatitis, and orchitis. Some diseases, such as syphilis and herpes genitalis, are characterized by the appearance of sores in the genital region. Advanced stages of some afflictions, such as syphilis, are dangerous and can kill, and one (AIDS) is almost always fatal.

Many of these diseases that involve bacterial infections can be treated with antibiotics. Parasitic infestations are treated with insecticides. For other afflictions, such as herpes genitalis, genital warts, and AIDS, no cure has been developed.

Contracting a sexually transmitted disease can disrupt one's emotional health and personal relationships. Gaining an understanding of the disease, as well as seeking treatment and dealing with guilt, is advisable. Precautions against sexually transmitted disease include proper hygiene, limiting one's sexual activity, and the use of barrier contraceptives.

Further Reading

Aral., S., *et al.* (1988). Recurrent genital herpes: What helps adjustment? *Sex. Trans. Dis.* **15**, 164–166.

Baltimore, D. (1995). Lessons from people with nonprogressive HIV infection. *N. Engl. J. Med.* **332**, 259–260.

Blattner, W. V., *et al.* (1988). HIV causes AIDS. *Science* **241**, 515–516.

Caldwell, J. C., and Caldwell, P. (1996). The African AIDS epidemic. *Sci. Am.* **274**(2), 62–68.

Centers for Disease Control and Prevention (1992). 1993 revised classification system for HIV infection and expanded surveillance case definition for AIDS among adolescents and adults. *Morbid. Mortal. Week. Rep.* 1992, no. RR-17.

Centofanti, M. (1995). New drug sweeps clean in HIV model. *Sci. News* **148**, 324.

Christensen, D. (1999). Why AIDS? The mystery of how HIV attacks the immune system. *Sci. News* **155**, 204–205.

Christensen, D. (1999). AIDS virus jumped from chimps. *Sci. News* **155**, 84.

Christensen, D. (2002). What activates AIDS? *Sci. News* **161**, 360–361.

Cohen, J. (1995). New clues to how some people live with HIV. *Science* **270**, 917–918.

Cohen, J. (1996). Receptor mutations help slow disease progression. *Science* **273**, 1797–1798.

Dan, B. B. (1990). Sex, lies, and chlamydia rates. *J. Am. Med. Assoc.* **264**, 3191–3192.

Ezzell, C. (2001). Hope in a vial. *Sci. Am.* **286**(6), 38–47.

Fackelmann, K. (1995). Staying alive: Scientists study people who outwit the AIDS virus. *Sci. News* **147**, 172–174.

Facklemann, K. (1998). Stamping out syphilis. *Sci. News* **154**, 202–204.

Harder, B. (2003). Testing times: The importance of identifying HIV infections before it's too late. *Sci. News* **164**, 347–348.

Kingman, S. (1995). Human papilloma virus vaccine tested in cervical cancer. *J. NIH Res.* **7**(7), 46–47.

Martens, M., and Faro, S. (1989). Update of trichomoniasis: Detection and management. *Med. Aspects Hum. Sex.* Jan., 73–79.

Marx, J. (1989). Drug-resistant strains of AIDS virus found. *Science* **243**, 1551–1552.

Melnick, N., *et al.* (1993). Changes in sexual behavior by young urban heterosexual adults in response to the AIDS epidemic. *Public Health Rep.* **108**, 582–588.

Mills, J., and Masur, H. (1990). AIDS-related infections. *Sci. Am.* **263**(2), 50–59.

Nowak, M. A., and McMichael, A. J. (1995). How HIV defeats the immune system. *Sci. Am.* **273**(2), 58–65.
Pinkerton, S. D., and Abramson, P. R. (1997). Condoms and the prevention of AIDS. *Am. Sci.* **85**, 364–372.
Schrager, L. K., and Fauci, A. S. (1995). Trapped but still dangerous. *Nature* **377**, 680–681.
Scientific American (Oct., 1988). Vol. 259, No. 4, entire issue on AIDS.
Stine, G. J. (2002). "AIDS Update, 2002." Prentice Hall, Englewood Cliffs, NJ.
Steinberg, J. (1992). AIDS redefined: What is the meaning of this? *J. NIH Res.* **4**(11), 45–46.
Washington, E. E., *et al.* (1987). *Chlamydia trachomatis* infections in the United States: What are they costing us? *J. Am. Med. Assoc.* **257**, 2070–2072.
Weiss, R. A., and Jaffee, H. W. (1990). Duesberg, HIV and AIDS. *Nature* **345**, 659–660.

Advanced Reading

Baker, D., *et al.* (1989). Clinical evaluation of a new herpes simplex virus ELISA: A rapid diagnostic test for herpes simplex virus. *Obstet. Gynecol.* **73**, 322–325.
Blower, S., *et al.* (2003). Forecasting the future of HIV epidemics: The impact of antiretroviral therapies and imperfect vaccines. *AIDS Rev.* **5**, 113–125.
Cates, W. Jr., *et al.* (1999). Estimates of the incidence and prevalence of sexually transmitted diseases. *Sex. Trans. Dis.* **26**(Suppl. 4), 52–57.
Centers for Disease Control and Prevention (CDC) (2003). Sexually transmitted disease surveillance, 2002. Atlanta, GA: U.S. Dept. of Health and Human Services Centers for Disease Control and Prevention.
Deacon, N. J., *et al.* (1995). Genomic structure of an attenuated quasi species of HIV-1 from a blood transfusion donor and recipients. *Science* **270**, 988–991.
Dicker, L.W., *et al.* (2003). Gonorrhea prevalence and coinfection with chlamydia in women in the United States, 2000. *Sex. Trans. Dis.* **30**, 472–476.
Dimitrov, D. S. (1996). Fusin: A place for HIV-1 and T4 cells to meet. *Nature Med.* **2**, 640–641.
Diseker, R. A., III, *et al.* (2000). Circumcision and STD in the United States: Cross-sectional and cohort analyses. *Sex. Transm. Infect.* **76**, 474–479.
Franco, E. L. (1995). Cancer causes revisited: Human papillomavirus and cervical neoplasia. *J. Natl. Cancer Inst.* **87**, 779–780.
French, P.P., *et al.* (2003). Use-effectiveness of the female versus male condom in preventing sexually transmitted disease in women. *Sex. Transm. Dis.* **30**, 433–439.
Gold, D., *et al.* (1988). Chronic-dose acyclovir to suppress frequently recurring genital herpes-simplex virus infection: Effect on antibody response to herpes simplex virus type 2 proteins. *J. Infect. Dis.* **158**, 1227–1234.
Golden, M.R., *et al.* (2003). Update on syphilis: Resurgence of an old problem. *JAMA* **290**, 1510–1514.
Hammerschlag, M. (1989). Chlamydial infections. *J. Pediat.* **114**, 727–734.
Heath, S. L., *et al.* (1995). Follicular dendritic cells and human immunodeficiency virus infectivity. *Nature* **377**, 740–744.
Huet, T., *et al.* (1990). Genetic organization of a chimpanzee lentivirus related to HIV-1. *Nature* **345**, 356–359.

Kaplan, A.H. (2002). Assembly of the HIV-1 core particle. *AIDS Rev.* **4**, 104–111.

Khabbaz, R. F., *et al.* (1992). Simian immunodeficiency virus needlestick accident in a laboratory worker. *Lancet* **340**, 271–273.

Kinloch-de Loes, S., *et al.* (1995). A controlled trial of zidovudine in primary human immunodeficiency virus infection. *N. Engl. J. Med.* **333**, 408–413.

Kovacs, J. A., *et al.* (1995). Increases in CD4 T lymphocytes with intermittent courses of interleukin-2 in patients with human immunodeficiency virus infection. *N. Engl. J. Med.* **332**, 567–575.

Kulkarni, P.S., *et al.* (2003). Resistance to HIV-1 infection: Lessons learned from studies of highly exposed persistently seronegative (HEPS) individuals. *AIDS Rev.* **5**, 87–103.

Kuviat, N., *et al.* (1989). Prevalence of genital papillomavirus infection among women attending a college student health clinic or a sexually transmitted disease clinic. *J. Infect. Dis.* **159**, 293–301.

Luby, E. C., and Klinge, V. (1985). Genital herpes: A pervasive psychosocial disorder. *Arch. Dermatol.* **121**, 494–497.

Luzuriaga, K., and Sullivan, J.L. (2002). Pediatric HIV-1 infection: Advances and remaining challenges. *AIDS Rev.* **4**, 21–26.

Margolis, S. (1984). Genital warts and molluscum contagiosum. *Urol. Clin. North Am.* **11**, 163–170.

Mujeeb, S.A., and Aetaf, A. (2003). The AIDS pandemic. *N. Engl. J. Med.* **349**, 814–815.

Nerurkar, L., *et al.* (1983). Survival of herpes simplex virus in water specimens collected from hot tubs in spa facilities and on plastic surfaces. *J. Am. Med. Assoc.* **250**, 3081–3083.

Rietmeyer, C., *et al.* (1988). Condoms as physical and chemical barriers against human immunodeficiency virus. *J. Am. Med. Assoc.* **259**, 1851–1853.

Saksena, N.K., and Potter, S.J. (2003). Reservoirs of HIV-1 *in vivo*: Implications for antiretroviral therapy. *AIDS Rev.* **5**, 3–18.

Saultz, J., and Toffler, W. (1989). *Trichomonas* infections in men. *Am. Family Physician* **39**, 177–181.

Schneider, A. (1988). HPV infection in women and their male partners. *Contemporary OB/GYN* (Nov.), 131–144.

Simmone, G., *et al.* (1997). Potent inhibition of HIV-1 infectivity in macrophages and lymphocytes by a novel CCR5 antagonist. *Science* **276**, 276–279.

Swinker, M. L., *et al.* (1988). Prevalence of *Chlamydia trachomatis* cervical infection in a college gynecology clinic: Relationship to other infections and clinical features. *Sex. Transmitted Dis.* **15**, 123–126.

Van der Vleet, V. (2003). The AIDS threats facing Southern Africa. *S. Afr. Med. J.* **93**, 490.

Voelker, R. (2003). As AIDS cases increase, health experts warn against missed prevention opportunities. *JAMA* **290**, 1304–1306.

Weinstock, H., *et al.* (2004). Sexually transmitted diseases among American youth: Incidence and prevalence estimates, 2000. *Persp. Sex. Reprod. Health* **36**, 6–16.

Glossary

Abdominal pregnancy An ectopic pregnancy wherein implantation has occurred in the abdominal cavity.

ABO incompatibility When the mother's immune system rejects her fetus's tissues because of a difference in ABO blood type.

Abortifacient A substance that induces abortion.

Abortion See **Induced abortion** and **Spontaneous abortion**.

Abruptio placenta Premature detachment of the placenta from the uterus.

Acetaminophen A nonaspirin pain reliever.

Acidophils Cells in the adenohypophysis that stain with acidic dyes. There are acidophils that secrete growth hormone and that secrete prolactin.

Acne Condition in which sebaceous glands become clogged and infected, producing pimples and blackheads; common in pubertal teenagers.

Acquired immunodeficiency syndrome (AIDS) A usually fatal, sexually transmitted disease caused by human immunodeficiency viruses.

Acrosin A protease enzyme present in the sperm acrosome that, when released during the acrosome reaction, breaks down protein of the zona pellucida.

Acrosomal space Region in the sperm head, between the outer and inner acrosomal membranes, that contains enzymes important to fertilization.

Acrosome Thick cap-like vesicle in the head of a sperm; contains enzymes important in sperm penetration and fertilization.

Acrosome reaction When the outer acrosomal membrane of the sperm head fuses with the sperm plasma membrane, forming a composite membrane with openings through which pass acrosomal enzymes necessary for sperm penetration. See also **Composite membrane; Outer acrosomal membrane**.

Activational effects Those effects of sex hormones that cause the appearance of sex differences in physiology and behavior; are reversible (i.e., disappear seasonally or at certain stages of reproductive cycles).

Activin An inhibin-like polypeptide produced by ovaries and testes; stimulates FSH secretion.

Acupuncture The stimulation of sensation by inserting needles into specified skin regions.

Adenohypophysis Epithelial lobe of the hypophysis; consists of the pars distalis, pars intermedia, and pars tuberalis.

Adolescence The socially defined period of youth between puberty and social adulthood.
Adrenalectomy Surgical removal or destruction of the adrenal glands.
Adrenarche Maturation and increased activity of the adrenal glands prior to puberty.
Adrenogenital syndrome An inherited condition characterized by excessive secretion of androgens from the adrenal glands, producing precocious development in both sexes and masculinization of the female; congenital adrenal hyperplasia.
Adultery Having voluntary coitus with someone other than one's spouse.
Affinity The strength of binding of a ligand to its receptor.
A-frame orgasm See **Uterine orgasm**.
Afterbirth Maternal and fetal tissue (amnionic membrane; placenta) expelled from the mother in stage 3 of labor.
After-pains Painful uterine contractions occurring for a few days after delivery of a child.
Agammaglobulinemia A sex-linked, recessive, inherited disorder characterized by the inability to form antibodies.
Agonist A molecule that binds and activates a receptor.
AIDS-related complex (ARC) A syndrome caused by the HIV virus; not immediately life-threatening but usually develops into AIDS itself.
Albumin Protein found in the human body; when in oviductal fluid, it may facilitate sperm activation.
Aldosterone Steroid hormone, secreted by the adrenal glands, that causes salt and water retention by the kidneys.
Allantois Extraembryonic membrane that contributes to the formation of the placenta.
α-fetoprotein An estrogen-binding protein, present in the blood of newborn rats, that keeps estrogen from masculinizing the female brain.
5α-reductase The enzyme that converts testosterone to dihydrotestosterone.
Alprazolam A prescription antianxiety drug used to treat premenstrual syndrome.
Amenorrhea Absence of menstruation; see also **Primary amenorrhea, Secondary amenorrhea**, and **Oligomenorrhea**.
Amines Group of nitrogen-containing compounds formed from ammonia.
Amniocentesis Method of sampling amnionic fluid, and the fetal cells within this fluid, to test for chromosomal and genetic fetal disorders.
Amnion Extraembryonic membrane that forms a closed, fluid-filled sac about the embryo and fetus.
Amnionic fluid Fluid, secreted by the amnion, that bathes the embryo and fetus; also called amniotic fluid.
Amphetamines A group of stimulant drugs.
Amygdala A structure in the brain that is a component of the limbic system and involved in the generation of emotion.
Amyl nitrite A fruity-smelling volatile liquid, the fumes of which can be inhaled to intensify an orgasm.
Anabolic steroids Synthetic androgens utilized to promote muscle growth.
Anal coitus Insertion of the penis through the anus into the rectum.
Analgesic Substance that diminishes the sense of pain.

Analog A chemical compound with similar structure, but not necessarily similar function, to another compound.
Anaphrodisiac An external substance that lowers sexual desire.
Androgen-binding protein (ABP) A protein, secreted by the Sertoli cells, that binds to androgens present in the testicular fluid.
Androgenesis When a sperm penetrates an ovum and does not fuse with the oocyte pronucleus but instead develops into an embryo with haploid cells derived from the father.
Androgen insensitivity syndrome See **Testicular feminization syndrome**.
Androgen replacement therapy (ART) Administration of an androgen to treat androgen insufficiency.
Androgens Substances that promote development and maturation of male sex structures; secreted by the testes and ovaries as well as the adrenal glands.
Androgyny When a person has both feminine and masculine characteristics.
Andropause Decline of testosterone secretion and related symptoms in older men.
Androstenedione A weak androgen that serves as a precursor for estrogen synthesis in the ovaries; present in male and female blood.
Androstenes Possible male sex pheromones.
Anencephaly Congenital defect resulting in a stillborn fetus with no brain.
Anesthetic Substance that diminishes all sensations.
Anestrus Condition in which estrogen levels are low and a female is not sexually receptive.
Aneuploidy Embryos with either one too few (45; monosomy), or one too many (47; trisomy) chromosomes.
Anorexia nervosa A psychophysiological eating disorder characterized by distorted body image, self-induced starvation, emaciation, and often infertility.
Anosmia Absence of sense of smell.
Anovulatory cycle Infertile menstrual cycle with no ovulation; often occurs in the first months after menarche.
Anoxia Absence of oxygen.
Antagonist A chemical that, when it binds to a receptor, blocks the activation of the receptor, preventing a biological response.
Anteflexion Referring to the uterus, it is the normal forward bend to this organ.
Anterior commissure Nerve fiber tract connecting left and right brain.
Anterior pituitary gland See **Pars distalis**.
Antibodies Proteins in blood that are made in reaction to foreign chemicals, cells, or viruses and neutralizes them, thus producing immunity.
Antiestrogen A substance that blocks the action of estrogens.
Antigen A foreign substance in the body that causes production of an antibody to it.
Antidiuretic hormone (ADH) See **Vasopressin**.
Antihistamines Drugs that block the production or action of histamine, a potent vasodilator.
Antihypertensive drugs Drugs that lower blood pressure.
Antral cavity Cavity in tertiary ovarian follicles that is filled with antral fluid.
Antral fluid Fluid in antral cavity; secreted by granulosa cells, but mainly is a filtrate of blood plasma; follicular fluid.

Antral follicle A mature, preovulatory tertiary follicle containing a large antrum.
Apgar score Rating given to the respiration, muscle tone, reflex irritability, heart rate, and skin color of a newborn baby.
Aphrodisiac Any substance that stimulates sexual desire.
Apnea Temporary cessation of respiration.
Apoptosis An orderly, genetically programmed series of events leading to the death of a cell; programmed cell death.
Arcuate nucleus Brain region in the hypothalamus; contains GnRH cells and the **Surge center**.
Arcuate uterus Uterus with a slight dent on the fundus.
Areola The pigmented region of the skin surrounding the nipple.
Arginine vasotocin (AVT) Polypeptide hormone, secreted by the fetal neurohypophysis, that may regulate secretion of amnionic fluid; also found in adult pineal gland.
Aromatase An enzyme that converts testosterone to estradiol.
Artificial insemination The artificial introduction of semen into the vagina or uterus; sperm can be provided by the husband or another male (donor).
Assisted hatching See **Zona drilling**.
Asymptomatic phase of HIV disease Relatively long period of absence of external HIV symptoms, even though immune system is slowly being destroyed.
Atherosclerosis Deposition of fatty or crystalline deposits in the lining of blood vessels.
Atresia Process by which ovarian follicles die and degenerate.
Autoallergic orchitis Inflammation of the testes due to a man's allergic reaction to his own sperm that leak into his tissues after a vasectomy.
Autosomes Any chromosomes that are not sex chromosomes.
Axon Extension of a neuron that conducts nerve impulses and secretory products away from the cell body.
Azoospermia Absence of sperm in semen.
B_1-Glycoprotein A placental protein that can be used to detect the future possibility of spontaneous abortion.
Back labor Back pain during stage 1 of labor, caused by the back of the fetal head pressing against the mother's sacrum.
Bacterial prostatitis See **Prostatitis**.
Bacterial vaginosis A vaginal bacterial infection; linked to premature delivery in pregnant women.
Barbiturates Drugs, such as methaqualone, that are used as hypnotics and sleeping pills.
Barr body A condensed mass of chromatin in female cells, resulting from inactivation of all X chromosomes but one. Thus, a Barr body is not present in male cells, but there is one in normal female cells and more than one in female cells with greater than two X chromosomes; sex chromatin.
Bartholin's glands See **Greater vestibular glands**.
Basal body temperature method Determining the safe and fertile periods in the menstrual cycle utilizing the rise in body temperature during the luteal phase; see also **Rhythm methods**.
Basement membrane An acellular membrane underlying an epidermis; in the testis, the membrane encasing each seminiferous tubule.
Basophils Cells present in the adenohypophysis; stain with basic dyes.

Bed nucleus of the stria terminalis (BNST) Hypothalamic region that is feminized in male transsexuals.
Bendictin Drug used to treat morning sickness in pregnant women. Evidence shows that it has teratogenic effects.
Benign prostatic hyperplasia (BPH) Noncancerous form of prostate enlargement in older men.
Benign tumor Tumor in which the cells are not cancerous.
Beta-3 A uterine protein that could be used in diagnoses of endometriosis.
Bicornuate uterus A forked uterus that is the result of a developmental error.
Bifid penis A forked penis produced by a developmental error.
Bilirubin Yellowish breakdown product of destroyed red blood cells.
Bioassay Technique that measures levels of a specific hormone in blood or tissue using a biological response to that hormone.
Biotic potential Ability of a population to increase under optimal environmental conditions.
Bipedalism Walking erect on the hind limbs only.
Birth canal The cavities of the cervix and vagina through which the fetus passes during parturition.
Birth defects See **Congenital defects**.
Birth rate The rate at which young are born into a population.
Birthing room Hospital room that provides a home-like setting for labor and delivery.
Bisexual A person who has had sexual contact with both sexes as an adult.
Blastocoel Fluid-filled cavity within the blastocyst; blastocyst cavity.
Blastocyst Stage in embryonic development after the morula stage; consists of an inner cell mass, an outer trophoblast layer, and a blastocoel.
Blastocyst cavity See **Blastocoel**.
Blastomere separation A process of separating cells of the early preembryo so that they each form an entire embryo.
Blastomeres The cells produced by mitotic division of a zygote.
Blood–testis barrier A specialized barrier formed by Sertoli cells that separates interstitial blood vessels from the interior of the seminiferous tubules, allowing a unique chemical microenvironment within the seminiferous tubules.
Bloody show Small amount of pinkish fluid discharged from the vagina during early labor; consists of mucus and blood.
B-lymphocytes White blood cells of the immune system; involved in antibody production.
Body stalk Embryonic structure that connects the embryo to the chorion. It contains the allantois and blood vessels that become part of the umbilical cord.
Bradley method A form of natural childbirth.
Braxton–Hicks contractions Mild and irregular uterine contractions during pregnancy.
BRCA1 Gene on chromosome 17; involved in breast and ovarian cancer.
BRCA2 Gene on chromosome 13; involved in breast cancer in women and men.
Breaking of the bag of waters Natural rupture of the amnionic sac, usually in the initial stages of labor; can be done artificially by the physician or midwife.
Breakthrough bleeding See **Spotting**.
Breast self-examination A method by which a woman can check herself for breast lumps.
Breast-feeding When an infant suckles the breast to obtain milk.

Breasts See **Mammary glands**.

Breech presentation When the feet or buttocks of the fetus are presented at the cervix.

Broad ligaments Paired ligaments that suspend the body of the uterus from the pelvic wall.

Bromocriptine Drug that inhibits prolactin secretion; used to treat male and female infertility and to inhibit lactation.

Brown fat Fat present in the newborn that, when broken down, gives off heat to help the baby maintain body temperature.

Buccal smear test When the cells lining the newborn's mouth are examined for the presence of X and Y chromosomes.

Bulbocavernosus muscle Muscle at the base of the penis that contracts rhythmically during ejaculation.

Bulbourethral glands Paired male sex accessory glands that produce a small amount of mucus; Cowper's glands.

Calendar method Determining the safe and fertile periods in the menstrual cycle utilizing a woman's variation in cycle length and the predicted time of ovulation; see also **Rhythm methods**.

Calmodulin Protein in seminal plasma that may render sperm in the epididymis capable of being capacitated.

Calymmatobacterium granulomatis A Gram-negative bacterium that causes granuloma inguinale; *Donovania granulomatis*.

Cancer Condition in which certain cells lose the ability to control their growth and multiplication and are abnormal in appearance. The cells of a cancerous tumor can remain in place or can spread (metastasize) to other regions of the body and become malignant.

Candida albicans Yeast-like fungus, normally present in the mouth, intestines, and vagina, which causes monilia vaginitis.

Cantharides Drugs such as "Spanish fly" that irritate the urinary tract and thus may have an aphrodisiac-like effect.

Caput Thick, soft swelling on the part of a newborn's head that rested against the cervix.

Carcinogen A cancer-causing agent.

Cardiovascular system disease Disease that can result in death due to myocardial infarction, pulmonary embolism, or cerebral hemorrhage; more common in combination birth control pill users, especially those over 35 or who smoke.

Carnitine Constituent of seminal plasma from the epididymis; breaks down fatty acids, the products being used as nutrients for sperm.

Carrying capacity The number of individuals who can be supported by a given environment.

Castration Surgical removal of the gonad; see also **Orchidectomy** and **Ovariectomy**.

Caudal Conduction anesthetic injected into the outside membrane of the spinal cord low in the back.

Cavernous urethra See **Spongy urethra**.

Celibacy Abstention from sexual activity.

Center for Disease Control and Prevention (CDCP) Institute of the United States government charged with the prevention of infectious diseases.

Central effect of a hormone The action of a hormone on the central nervous system.
Central nervous system The brain and spinal cord.
Cephalohematoma Blood clot between the skull bone and bone membrane in a newborn.
Cerebral palsy Spastic muscle paralysis due to brain damage.
Cerebrospinal fluid Viscous fluid in the ventricles of the brain and the central canal of the spinal cord.
Cervical biopsy Surgical sampling of a small piece of cervical tissue for further examination.
Cervical canal A narrow passage in the uterine cervix that connects the uterine cavity with the vagina.
Cervical cap A mechanical barrier to sperm transport through the cervix; similar to but smaller than a diaphragm.
Cervical crypts Deep recesses in the wall of the cervix; serve as reservoirs for sperm.
Cervical cysts Small, noncancerous, pimple-like growths in the cervical lining.
Cervical dilation Widening of the external cervical os from 0.3 to 10 cm in diameter during stage 1 of labor. Also, artificial dilation of the cervical canal before an induced abortion or other medical procedures involving the uterus.
Cervical dysplasia Nonmalignant cellular changes in the cervix.
Cervical effacement Thinning of the wall of the cervix during stage 1 of labor.
Cervical effacement and dilation Stage 1 of labor.
Cervical mucus method Determining the safe and fertile periods in the menstrual cycle using changes in cervical mucous volume and consistency; see also **Rhythm methods**.
Cervical polyps Tear-shaped, noncancerous growths from the cervical lining.
Cervicitis Inflammation or infection of the cervix.
Cervix See **Uterine cervix**.
Cesarean delivery Surgical incision into the uterus through the abdominal wall to remove the fetus.
Chancre An ulcer-like sore seen, for example, in the primary stage of syphilis and chancroid.
Chancroid A sexually transmitted disease caused by the bacterium *Haemophilus ducreyi.*
Chemoradiation Chemotherapy used in combination with radiation for the treatment of cancers.
Chlamydia trachomatis A sexually transmitted microorganism that causes some cases of nonspecific urethritis as well as lymphogranuloma venereum.
Chloasma Brown facial spotting seen in some users of the combination pill and in pregnant women.
Cholecystokinin A chemical neurotransmitter in the brain.
Chlordiazepoxide A minor tranquilizer (Librium).
Cholesterol Precursor steroid for the synthesis of all steroid hormones.
Chordee When the penis is bowed or bent.
Chorion Extraembryonic membrane that, along with the allantois, forms the fetal portion of the placenta.
Chorionic villi Finger-like projections of the chorion that come in contact with maternal blood in uterine sinusoids of the placenta.

Chorionic villus sampling (CVS) Procedure used to sample and evaluate fetal cells.

Chromosomal aberrations Abnormal chromosome number or structure.

Chromosomal deletion Absence of a piece of chromosome; see also **Chromosomal translocation**.

Chromosomal translocation The breaking off of a piece of a chromosome and its attachment to a nonhomologous chromosome.

Circumcision Surgical removal of the prepuce of the penis in the male or of the clitoris; sometimes along with the labia, in the female.

Classical hemophilia A sex-linked, recessive, hereditary disease characterized by failure of blood to clot.

Clitoral glans Tip of the clitoris; homologous to the glans of the penis in the male.

Clitoral orgasm An orgasm resulting from stimulation of the clitoris.

Clitoral prepuce Flap of skin partially covering the clitoris; homologous to the prepuce (foreskin) of the penis.

Clitoral shaft Main body of the clitoris: partially homologous to the shaft of the penis in the male.

Clitoris A small, erectile structure embedded in tissue at the upper junction of the labia minora; homologous to the penis.

Clomiphene An antiestrogenic drug that commonly is used to treat female infertility.

Cloning To duplicate an organism asexually.

Cocaine Recreational drug used as a stimulant.

Coitus Strictly speaking, sexual intercourse between a male and female during which the penis is inserted into the vagina; **Vaginal coitus** or **Sexual intercourse**. See also **Anal coitus; Femoral coitus; Mammary coitus; Oral coitus**.

Coitus interruptus Withdrawal of the penis so that ejaculation occurs outside the vagina and vulva.

Coitus obstructus Avoidance of ejaculation by pressing on the penile urethra, thus causing retrograde ejaculation into the male's bladder.

Coitus reservatus When a male stops short of ejaculation and allows penile detumescence to occur within the vagina.

Collagenase Enzyme that destroys collagen fibers in connective tissue.

Colostrum The secretion of the mammary glands during late pregnancy and the first 2 days after delivery. Contains important nutrients and maternal antibodies for the newborn.

Colpotomy Incision through the vaginal wall; used for tubal sterilization.

Combination pill An oral contraceptive containing a synthetic estrogen and progestogen.

Combined menstrual cycle Cyclic changes in the ovaries, uterus, pituitary and hypothalamus during the menstrual cycle.

Composite membrane See **Acrosome reaction**.

Concealed ovulation Ovulation not accompanied by an associated courtship signal.

Conception See **Fertilization**.

Conceptus The products of conception, including the zygote, preembryo, embryo, fetus, and extraembryonic membranes.

Condom Barrier method of contraception using a rubber or membranous sheath fitted over the penis or in the vagina.

Conduction anesthetic Anesthetic injected into tissue.
Condylomata acuminata See **Genital warts**.
Conformational change A change in the 3-dimensional shape of a protein or region of a protein.
Congenital adrenal hyperplasia See **Adrenogenital syndrome**.
Congenital defects Physiological or anatomical defects present at birth; birth defects.
Congenital hypothyroidism Thyroid hormone deficiency present at birth; untreated, it can lead to growth restriction and mental retardation.
Congenital syphilis Condition in which a child is born with syphilis.
Conjoined twins Identical twins in which, during their development, parts of their bodies remain fused; Siamese twins.
Contraception Prevention of pregnancy by abstinence or the use of substances, devices, or surgical procedures that prevent ovulation, fertilization, or implantation.
Contraceptive patch An adhesive patch that delivers hormones through the skin.
Copper IUD An intrauterine device made of flexible plastic with a wrapping of copper wire.
Corona glandis Rounded edge on the back of the glans penis.
Corona-penetrating enzyme Enzyme from acrosome that dissolves material between cells of the corona radiata.
Corona radiata Thin layer of granulosa cells; surrounds the oocyte in tertiary follicles and the ovum after ovulation.
Corpora atretica Structures formed from atretic follicles.
Corpora cavernosa Paired columns of spongy, erectile tissue present in the shaft of both the clitoris and penis.
Corpus albicans Connective tissue-filled structure that is the remnant of the regressed corpus luteum.
Corpus luteum A yellowish endocrine gland formed from the wall of an ovulated follicle.
Corpus hemorrhagicum Empty ovarian follicle existing after ovulation and before its transformation into a corpus luteum.
Corpus spongiosum Spongy erectile tissue in the shaft of the male penis; it is homologous to the labia minora in the female.
Cortical cords Sex cords in the cortex of developing ovaries.
Cortical granules Granules in the cortical vesicles lying at the periphery of the ovum cytoplasm; sperm penetration causes them to release acrosin inhibitor and a protease, both of which prevent more sperm from penetrating the egg.
Cortical reaction Release of cortical granules.
Cortical vesicles Membrane-bound bodies in the ovum; release cortical granules after sperm penetration.
Corticosteroids Steroid hormones, such as cortisone and cortisol, that are secreted by the adrenal glands.
Corticotropin (ACTH) Polypeptide hormone secreted by basophils of the adenohypophysis; stimulates secretion of steroid hormones (corticosteroids) from the adrenal glands.
Corticotropin-releasing hormone (CRH) Polypeptide from the hypothalamus; increases secretion of corticotropin.

Cortisol Steroid hormone, secreted by the cortex of the adrenal glands, that raises blood sugar levels.

Courtship behaviors Behaviors that serve to bring males and females together for mating.

Cowper's glands See **Bulbourethral glands**.

Cremaster A muscle layer in the scrotum that contracts upon sexual stimulation.

Cremasteric reflex Contraction of the cremaster muscle of the scrotum when the inner thighs are stroked.

Crib death See **Sudden infant death syndrome**.

Cri-du-chat syndrome Chromosomal deletion on chromosome 5, resulting in a syndrome of physical abnormalities; affected newborns cry like a hungry kitten.

Critical body fat hypothesis Theory that a minimal ratio of fat weight to whole body weight is necessary for menarche or menstrual cycles to occur.

Critical period Window of time when an effector can produce a result; e.g., when a hormone can irreversibly influence development, or when a teratogen can disrupt normal development.

Crowning When the top of the fetal head appears in the birth canal and does not recede between contractions.

Crown–rump length (CR) Distance between the crown of the head and the rump of the embryo.

Crude birth rate Number of individuals born in a population in a given amount of time.

Crude death rate Number of individuals dying in a population in a given amount of time.

Cryptorchid testes Testes that remain in the body cavity during childhood; undescended testes.

Culdoscopy Incision through the vaginal wall; used in tubal sterilization.

Cumulus oophorus Column of granulosa cells that attaches the oocyte to the wall of tertiary follicles; leaves with the ovulated ovum.

Cunnilingus When the tongue or mouth is used to stimulate the female external genitalia.

Curettage Scraping the lining of the uterus using a metal instrument (curette).

Curette Metal instrument used to scrape the lining of the uterus.

Cyanosis When the blood of a newborn does not contain enough oxygen, leading to a bluish tint to the skin; can be caused by patent ductus arteriosus or open foramen ovale ("blue babies"), or may be present briefly in normal newborns.

Cyproterone acetate An antiandrogen that blocks binding of androgens to receptors in target cells.

Cystic follicle A large fluid-filled or solid sac derived from an unovulated graafian follicle.

Cystic mastitis Development of one or more fluid-filled sacs (cysts) in the breasts.

Cystitis Inflammation or infection of the urinary bladder.

Cytomegalovirus Virus that causes cytomegalic inclusion disease.

Cytoplasmic transfer Microinjection of cytoplasm from a healthy donor egg into the deficient egg of a woman undergoing assisted reproduction.

Cytosine A nitrogenous base present in DNA and RNA.
Cytotoxic ("killer") T lymphocytes White blood cells that directly attack and destroy foreign cells.
Cytotrophoblast Cellular component of the trophoblast; gives rise to the chorion.
Death rate Rate at which individuals die in a population.
Decidua basalis Maternal part of the placenta.
Decidua capsularis Part of the endometrium that grows over and covers the developing fetus.
Decidua parietalis Pregnant endometrium not contained in the placenta.
Deciduoma response Tumor-like growth of uterine tissue over the implanted embryo.
Dehydroepiandrosterone (DHEA) Weak androgen serving as a precursor for estrogen and testosterone synthesis; synthesized by ovaries, testes, and "fetal zone" of fetal adrenal glands.
Delayed puberty When puberty has not begun by age 14 (testicular growth) or 18 (skeletal growth) in males, or by age 14 (breast development) or age 15 (skeletal growth) in females.
Delta sleep-inducing peptide Neurotransmitter made by GnRH cell in the hypothalamus.
Dendrite Extension of a neuron that conducts a nerve impulse toward the cell body.
Dermoid cyst Lump of tissue within an ovarian follicle, containing embryonic organs; possibly derived from fertilization of an oocyte while it still is in the ovary.
Diabetes mellitus Condition in which lack of insulin produces high blood sugar and copious urination.
Diaphragm A mechanical barrier to passage of sperm into the cervix; fits into the vagina to cover the cervix.
Diazepam A tranquilizer (Valium).
Diencephalon Portion of the brain containing the hypothalamus, thalamus, epithalamus, and third ventricle.
Diesterase An enzyme in the uterus that breaks down glycerylphosphocholine, thus supplying nutrients for sperm.
Diethylstilbestrol (DES) A potent synthetic estrogen.
Digital rectal exam (DRE) When a physician palpates the prostate gland through the wall of the rectum.
Dihydrotestosterone (DHT) A potent androgen derived from testosterone.
Dilation and curettage (D&C) Surgical procedure of first dilating the cervix and then scraping out the contents of the uterus; used for induced abortion during the late first trimester.
Dilation and evacuation (D&E) Method of late second or third trimester abortion.
Dilation contractions Contractions in stage 1 of labor that dilate (as well as efface) the cervix.
Dilator Metal instrument used to dilate the cervical canal before uterine curettage.
Diploid Containing a double complement of chromosomes; in human cells, there are 23 pairs (46 chromosomes); 2N.
Diuretics Drugs used to increase urine flow.

Dizygotic twins Twins resulting from fertilization of two separate ova by two separate sperm; fraternal twins or nonidentical twins.

DNA binding domain The portion of a steroid receptor molecule that binds to cellular DNA.

Döderlein's bacillus A gram-positive bacterium normally found in the vagina; *Lactobacillus acidophilus*.

Domain Within a protein, a region with a distinct structure and, often, a specific biochemical function.

Donovania granulomatis See ***Calymmatobacterium granulomatis***.

Dopamine Neurotransmitter in the brain; stimulates GnRH secretion.

Dosage-sensitive sex reversal gene Gene on the X chromosome; involved in sex determination.

Dose–response relationship When a response to a given hormone is predictably related to the amount of the hormone present.

Double penis Two penises produced by a developmental error.

Double uterus Two uteri, with a single vagina or two vaginas being present; produced by a developmental error.

Doubling time Time in years it will take to double population size at a given population growth rate.

Douching Rinsing the vagina with a solution; not an effective contraceptive measure.

Down's syndrome An affliction caused by trisomy of chromosome 21. These individuals have physical disorders and mental retardation; **Mongolism**.

Downregulation In endocrinology, the process by which a cell decreases its number of receptors for a hormone, thus decreasing its sensitivity to the hormone.

Duchenne's muscular dystrophy Sex-linked, recessive, inherited disorder leading to degeneration of skeletal muscle and early death.

Ductal carcinoma *in situ* Noninvasive form of breast cancer limited to the lining of breast ducts.

Ductus arteriosus Fetal blood vessel that shunts blood from the pulmonary trunk to the descending aorta, bypassing the fetal lung; closes after birth.

Ductus deferens See **Vas deferens**.

Ductus epididymis Main duct of the body of the epididymis.

Ductus venosus Fetal blood vessel that directs oxygenated blood away from the fetal liver.

Due date Expected date of birth as calculated from first day of last menstrual period or day of conception; term.

Dysmenorrhea Painful menstruation caused by uterine cramps. See **Primary dysmenorrhea, Secondary dysmenorrhea**.

Dyspareunia Difficult or painful coitus.

Eclampsia See **Toxemia**.

Ectoderm Outermost of the three primary germ layers of the embryo; gives rise to the nervous system, the external sense organs (ear, eye, etc.), the skin, and the mucous membranes of the mouth and anus.

Ectopic pregnancy Implantation of an embryo outside the uterus.

Ectromelia Fetal condition characterized by total absence of limbs.

Edema Excessive accumulation of extracellular water in tissues.

Effacement contractions Contractions in stage 1 of labor that efface (thin out) the wall of the cervix.

Egg activation Metabolic activation and cytoplasmic changes in an egg in preparation for embryonic development, normally triggered by fertilization.

Egg donor A woman who provides eggs for use in assisted reproduction.

Ejaculation Expulsion of semen from the male sex accessory ducts; see also **Female ejaculation**.

Ejaculation reflex Reaction when stimuli associated with sexual arousal activate the ejaculatory center in the spinal cord, which in turn causes ejaculation.

Ejaculatory center A group of neurons in the spinal cord that controls ejaculation; **Spinal nucleus of the bulbocavernosus**.

Ejaculatory ducts Small paired ducts in the male that receive the ampulla of each vas deferens and the duct from each seminal vesicle. The ejaculatory ducts empty into the prostatic urethra.

Ejaculatory incompetence When a man is unable to ejaculate, even though he may have an orgasm.

Electra complex In psychoanalytic theory, stage in a female child's life when she is in love with her father and jealous of her mother.

Embryo Stage of prenatal development between implantation (end of 2nd week) through the 8th week after fertilization.

Embryo biopsy Removal of one or two cells of an early embryo for culture and genetic screening.

Embryo transfer Assisted reproductive technology method used to place a pre-embryo in the mother's uterus.

Embryonic disc Early embryo, derived from the inner cell mass of the blastocyst, that contains the primary germ layers.

Embryonic period Stage of pregnancy from the 2nd through the 8th week of pregnancy.

Embryonic stem cell A totipotent cell cultured from the inner cell mass of a blastocyst.

Emergency contraception Post-coital hormonal contraception used as an emergency measure to prevent pregnancy after unprotected sex; **Morning-after pill**.

Emigration Leaving a place of residence in a defined population for elsewhere.

Emission stage of ejaculation First stage of the ejaculatory response involving contractions of the male sex accessory ducts and glands.

Encephalization Evolutionary enlargement of the brain in the hominid line of primates.

Endemic syphilis See **Nonvenereal syphilis**.

Endocrine glands Ductless glands that secrete hormones into the blood.

Endocrine system System consisting of the endocrine glands; includes the hypophysis, pineal gland, gonads, placenta, thyroid and parathyroid glands, adrenal glands, digestive tract, kidneys, pancreas, and thymus.

Endocrine-disrupting contaminants (EDCs) Environmental pollutants that disrupt normal development or functioning of the endocrine cells or their target cells.

Endocrinology The scientific discipline involving the study of endocrine glands and their secretions.

Endoderm The innermost of the three primary germ layers; gives rise to the digestive and respiratory systems.

Endometrial ablation When the endometrium is removed by heat cauterization.

Endometrial biopsy Surgical sampling of a small piece of endometrium for further examination.

Endometrial hyperplasia Unusually thick uterine endometrium due to prolonged estrogenic stimulation and consequent cell multiplication.

Endometrial polyps Mushroom-like growths of the endometrium.

Endometriosis Abnormal growth of uterine cells outside the uterus; can cause infertility.

Endometritis Inflammation or infection of the endometrium.

Endometrium The uterine lining, which consists of the stratum functionalis and stratum basalis.

Endorphins Natural, opiate-like substances that decrease pain and also inhibit GnRH secretion.

Engagement of the presenting part Dropping of the fetus between the pelvic bones so that its head rests against the neck of the cervix; occurs near term.

Enkephalins Natural, opiate-like substances that decrease pain.

Ensoulment Supposed event wherein a developing human acquires a soul and thus becomes a person.

Environmental resistance Limitation of population growth by factors in the environment.

Epiblast The upper layer of the bilaminar disk of the preembryo; this founder cell population gives rise to all cell types of the embryo.

Epidermal growth factor Substance present in human milk and other tissues that promotes tissue growth.

Epididymis (pl., Epididymides) Comma-shaped organ on the posterior surface of each testis; its duct transports sperm from the testis to the vas deferens.

Epidural A conduction anesthetic injected into the outside membranes of the spinal cord.

Epinephrine A stress hormone secreted by the adrenal glands.

Episiotomy Surgical incision of the perineum during parturition to reduce the possibility of tearing the perineal tissues.

Epstein–Barr virus Virus that causes infectious mononucleosis.

Erectile dysfunction Failure to gain or maintain an erection; impotence.

Erection The stiffening and enlargement of the penis (or clitoris), usually as a result of sexual arousal.

Erection center A group of neurons in the lower end of the spinal cord that controls erection of the penis.

Erection reflex When an erotic stimulus causes nerve activation of the erection center, which in turn activates the erection mechanisms.

Erogenous zones Areas of the body that are especially sexually sensitive to tactile stimuli.

Erotic stimuli Stimuli that cause sexual arousal.

Erythroblastosis fetalis Fetal condition in which immature red blood cells are predominant in the blood; caused by Rhesus disease.

Escherichia coli A bacterium normally present in the human intestinal tract.

Estradiol-17β The major natural estrogen, secreted by the ovaries, testes, and placenta.

Estrenes Possible human pheromones.

Estriol An estrogen secreted in large amounts by the placenta.
Estrogen-replacement therapy (ERT) Estrogens administered to women with low estrogen secretion, such as after menopause; often given with a progestogen.
Estrogens Substances that promote maturation of female reproductive organs and secondary sexual characteristics; secreted by the ovaries, testes, adrenal glands, and placenta.
Estrone An estrogen secreted in small amounts by the ovaries, placenta, and fat.
Estrous behavior A cyclic period of sexual receptivity ("heat") in females occurring near the time of ovulation and related to high levels of estrogen in the blood.
Estrous cycle A reproductive cycle in which the female exhibits sexual receptivity (estrus) only near the time of ovulation.
Estrus The female sexually receptive period.
Eugenics The artificial manipulation of gene frequencies in a population by selective breeding.
Eunuch An orchidectomized (castrated) male.
Excitement phase The initial stage in the human sexual response cycle that follows effective sexual stimulation.
Exocrine glands Glands that secrete substances into ducts that empty into body cavities and onto body surfaces.
Exogenous Originating outside the body; supplied to the body by an external source.
Exponential growth Growth at an increasing rate based on a "compound-interest" principle.
Expulsion of the fetus Stage 2 of labor.
Expulsion of the placenta Stage 3 of labor.
Expulsion stage of ejaculation The second stage of the ejaculation process, during which semen is expelled.
External cervical os Opening of the cervical canal into the vagina.
External fertilization When fertilization occurs *in vitro* (in a dish or test tube).
External genitalia In the female, the mons pubis, labia majora and minora, vaginal introitus, hymen, and clitoris (also called the **Vulva**); in the male, the penis and scrotum.
Extirpation Removal of tissue or organs.
Extracellular domain The region of a protein that extends into the extracellular space.
Extraembryonic membranes Four membranes (yolk sac, amnion, allantois, chorion) derived from, but not part of, the developing embryo.
Fallopian tubes See **Oviducts**.
False labor Contractions that occur near term; can be rhythmic but do not last or cause much cervical dilation or effacement.
False pregnancy See **Pseudocyesis**.
Family planning programs Programs with the goals of eliminating the birth of unwanted children and limiting family size.
Feedback loop When production of a product either increases or decreases further production of that product.
Fellatio Taking the penis into the mouth and stimulating it for erotic purposes; **Oral coitus**.

Female climacteric The sum of changes (psychological, emotional, hormonal and physical) occurring in a menopausal woman.

Female condom A polyurethane sheath that is placed in the vaginal canal and thus blocks sperm transport.

Female ejaculation Expulsion of fluid from the lesser vestibular glands into the vestibule during orgasm.

Femaleness Biological characteristics of the female sex.

Feminine Social definition of what are female characteristics; often used to mean "womanly."

Femoral coitus Insertion of the penis between the thighs of another person.

Fern test Laboratory method of determining the phase of the menstrual cycle using degree of crystallization in dried cervical mucus.

Fertility rate The number of children a woman has in her lifetime.

Fertilization When the pronuclei of a haploid ootid and sperm fuse to form a diploid zygote; conception.

Fetal alcohol syndrome Condition of newborn caused by the mother's chronic ingestion of alcohol during pregnancy, characterized by small size and retardation.

Fetal blood sampling Diagnostic method in which a small sample of fetal blood is obtained to test for genetic or biochemical disorders.

Fetal distress Abnormal cardiovascular signs of the fetus during labor and delivery.

Fetal ejection reflex Stimulation of oxytocin release by mechanical stimulation of the uterus, cervix, or vagina, resulting in further uterine contractions.

Fetal fibronectin A protein that helps attach the placenta to the uterine wall.

Fetal microchimerism Persistence of tiny amounts of fetal cells in the mother, sometimes for years or decades after pregnancy.

Fetal monitoring Continuous evaluation of vital signs of the fetus during labor delivery.

Fetal zone Region in fetal adrenal glands that secretes large amounts of a weak androgen (dehydroepiandrosterone).

Feto–placental unit Integrated system of fetus and mother that combines to produce secretion of estrogens from the placenta.

Fetoscopy Method of fetal scanning using direct observation with an optical instrument inserted through an abdominal incision.

Fetus Offspring at stage in prenatal development from the 9th week until term.

Fibrin See **Fibrinogenase**.

Fibrinogen See **Fibrinogenase**.

Fibrinogenase Enzyme in seminal plasma; converts fibrinogen to fibrin, and thus causes semen coagulation.

Fibrinolytic enzyme Enzyme in seminal plasma; breaks down fibrin and thus causes semen liquefaction.

Fibroadenomas Benign breast lumps.

Fibroids Benign tumors of the uterine smooth muscle.

Fimbriae Finger-like projections on the edge of the infundibulum of each oviduct.

First meiotic arrest When the first meiotic division in primary oocytes is stopped before completion; condition present in all primary oocytes.

First meiotic division First phase of meiosis, during which reduction division occurs; produces a secondary oocyte in females and a secondary spermatocyte in males.

First polar body Small haploid cell produced by unequal first meiotic division of oocyte; degenerates before or after dividing again.

Flagellum A thin, filamentous organelle used by cells, such as sperm, for locomotion.

Fluorescent in situ hybridization (FISH) A method to analyze results after amniocentesis.

Fluoxetine Used to treat premenstrual syndrome; Prozac.

Follicle-stimulating hormone (FSH) Glycoprotein hormone secreted by basophils of the adenohypophysis that, together with LH, maintains gonadal function in both sexes.

Follicle-stimulating hormone-releasing hormone (FSHRH) See **Gonadotropin-releasing hormone**.

Follicular fluid See **Antral fluid**.

Follicular phase Period in menstrual cycle beginning at the end of menstruation and ending with ovulation.

Follistatin An ovarian polypeptide that inhibits FSH secretion.

Fontanels Soft spots in the skull of newborns.

Foramen ovale Opening between the right and left atria of the fetal heart that allows blood flow to equalize between the right and left sides of the heart; closes soon after birth.

Forceps delivery Assisting expulsion of fetus by using a tong-like instrument called a forceps.

Foreplay The preliminary stages of coitus, involving kissing, touching, and caressing between partners.

Foreskin See **Penile prepuce**.

Fornix A circular recess at the upper end of the vagina that encircles the uterine cervix.

Fragile X syndrome An inherited disorder in which an X chromosome has an abnormally long piece; second leading cause of mental retardation.

Fraternal twins See **Dizygotic twins**.

Frigidity Term previously used to describe a woman who is disinterested in sex or is unable to achieve sexual gratification.

Functional magnetic resonance imaging (FMRI) A method used to visualize regions of brain activity.

Fusin A type of chemokine cell surface receptor, used by the HIV virus to bind to and enter T cells; CXCR4.

Fusion inhibitor One of a class of anti-HIV drugs that inhibits the ability of the HIV virus to bind to a host cell.

Galactorrhea Condition where prolactin levels are high in a nonpregnant female, producing milk secretion and amenorrhea.

Galactosemia An inherited metabolic disorder resulting in the inability to digest galactose, a sugar found in milk.

Galanin A neurotransmitter secreted by GnRH neurons in the hypothalamus.

Gamete intrafallopian transfer (GIFT) Transfer of a woman's oocytes from her ovary to her oviduct below the point where the oviduct is blocked.

Gametes Spermatozoa in the male, and an ovulated ovum in the female; they are haploid.

Gamma-aminobutyric acid (GABA) A brain neurotransmitter that inhibits GnRH secretion.

Gap junction A connection between two cells that allows the passage of ions and small molecules between the cells.

Gastroschisis When there is a hole in a newborn's abdominal wall, through which the intestines protrude.

Gay A male homosexual.

Gender identity The awareness of or belief about one's maleness or femaleness.

Gender role See **Sex role**.

General adaptation syndrome A set of physiological responses to any long-term stressor, mainly involving increased adrenal gland activity; proposed by Hans Selye.

Genetic mosaic When populations of cells of an individual differ genotypically.

Genistein A soy product that is being studied for its ability to inhibit the growth of cancer cells.

Genital ridges Embryonic ridges on the dorsal wall of the abdominal cavity; will form the gonads.

Genital tubercle Structure of the indifferent external genitalia of the early embryo that will form the penis or clitoris; phallus.

Genital warts Virus-induced warts in the genital region that can be sexually transmitted; *Condylomata acuminata*.

Germinal stage When a preembryo is present; the first 2 weeks of development.

Germinal vesicle The oocyte nucleus.

Germinal vesicle breakdown Disintegration of the nuclear membrane of an oocyte at the start of final oocyte maturation; LH-induced.

Gestation See **Pregnancy**.

Glans penis The soft bulbous end of the penis.

Glycerylphosphocholine Constituent of seminal plasma, secreted by the epididymis, that is a nutrient source for sperm.

GnRH agonists See **GnRH analogs**.

GnRH analogs Synthetic molecules similar in structure to gonadotropin-releasing hormone (GnRH). Some inhibit, and others stimulate, gonadotropin secretion.

GnRH antagonists Artificial chemicals that block GnRH receptors.

GnRH-associated peptide (GAP) A molecule produced in the hypothalamus, which stimulates FSH and LH and inhibits prolactin secretion.

GnRH inhibitory agonists Chemicals that lower the number of GnRH receptors in the pituitary gland.

GnRH pulse generator Neurons in hypothalamus that synapse with GnRH cells and cause pulsatile GnRH secretion.

GnRH stimulatory agonists Chemicals that mimic GnRH action.

Gonadostat Area in hypothalamus that regulates feedback control of GnRH secretion.

Gonadotropic hormones Pituitary hormones that affect gonadal function, including FSH and LH: **Gonadotropins**.

Gonadotropins See **Gonadotropic hormones**.

Gonadotropin-releasing hormone (GnRH) Neurohormone from the hypothalamus that increases synthesis and release of both LH and FSH; same as luteinizing hormone-releasing hormone and follicle-stimulating hormone-releasing hormone.

Gonads Structures that produce the male gametes (testes) and female gametes (ovaries).

Gonococcal ophthalmia neonatorum A condition in which a newborn's eyes are infected with *Neisseria gonorrhoeae.*
Gonococcus A common name for the bacterium *Neisseria gonorrhoeae*, the cause of gonorrhea.
Gonorrhea A sexually transmitted disease caused by the bacterium *Neisseria gonorrhoeae.*
Grafenberg spot A small spot in the anterior vaginal wall that, when stimulated, causes sexual arousal.
Graafian follicle A mature preovulatory follicle; see **Antral follicle**.
Grandmother hypothesis Theory of the evolutionary advantage of human menopause.
Granuloma inguinale A rare sexually transmitted disease caused by *Calymmatobacterium granulomatis.*
Granulosa cells Cells of the membrana granulosa of ovarian follicles.
Grasp reflex Flexion of a newborn's fingers when its palm is touched.
Greater vestibular glands Mucus-secreting glands in the female vestibule, homologous to the Cowper's glands of the male; Bartholin's glands.
Growth hormone (GH) Protein hormone, secreted by acidophils of the adenohypophysis, that stimulates incorporation of amino acids into protein (tissue growth) and raises blood sugar.
Guanethidine An antihypertensive drug that blocks the sympathetic nervous system and can cause erectile dysfunction.
Gubernaculum Ligament connecting the testes with the scrotum; plays a role in testicular descent in the late fetus.
Guevedoces Inherited form of male pseudohermaphroditism resulting from an absence of the enzyme 5α-reductase in external genitalia tissues so that they are unable to change testosterone to 5α-DHT and grow.
Gumma A soft, gummy tumor, as seen in cases of tertiary stage syphilis.
Gynandromorph A true hermaphrodite in which one side of the reproductive tract is male and the other side is female.
Gynecomastia Excessive development of the mammary glands in the male.
Gynogenesis When a sperm penetrates an egg, the male pronucleus does not form, and a haploid (N) embryo develops from the unfertilized ovum.
Haploid Containing only a single complement of chromosomes; in man, 23 chromosomes, or N; present in gametes.
Hegar's sign Softening of the lower part of the uterus occurring about 4 weeks after a missed menses; a probable sign of pregnancy.
Helper T (CD4) lymphocytes White blood cells that support the remainder of the immune system; HIV binds to the CD4 protein on these cells, enters them, and kills them.
Hematopoietic stem cell A stem cell from which all red and white blood cells develop.
Hemochorial placenta Type of placenta, present in the human, in which maternal blood in sinusoids directly bathes the chorionic villi.
Haemophilus ducreyi A gram-negative bacterium that causes chancroid.
Haemophilus vaginalis A gram-negative bacterium that is one cause of vaginitis and nonspecific urethritis.
***Haemophilis* vaginitis** Inflammation of the vagina caused by *Haemophilus vaginalis*; can be sexually transmitted.
Hepatitis Inflammation or infection of the liver.

Hermaphrodite A person with both ovarian and testicular tissue.

Herpes genitalis A sexually transmitted disease caused by herpes simplex virus, type 2.

Herpes simplex virus, type 1 Virus that causes cold sores.

Herpes simplex virus, type 2 Virus that causes herpes genitalis.

Heterosexual A person who is sexually attracted to and engages in sexual activity primarily with members of the opposite sex.

Heterosexual behavior Sexual activity with a member of one's opposite sex.

Heterozygous Condition in which a cell contains two different alleles on homologous chromosomes.

Heterozygous dominant genotype When cells have a dominant and recessive allele for a given trait.

High-definition imaging (HDI) A very sensitive ultrasound method used to detect breast lumps.

Hilus cells Sparse, testosterone-secreting cells present in the hilus of the ovary; homologous to the Leydig cells of the testes.

Hirsutism Abnormal hairiness, especially in women.

Histones Non-DNA proteins in chromosomes.

HIV disease The usually lethal disease caused by HIV, of which AIDS is the final stage.

Home birth Labor and birth at home, with or without medical supervision.

Homeostasis Regulation of body functions at a steady state.

Homophobia Fear of, or repulsion to, homosexuality in oneself or others.

Homosexual A person who is sexually attracted to and engages in sexual activity primarily or exclusively with members of the same sex.

Homosexual behavior Sexual activity with a member of one's own sex.

Hormone receptors Proteins on the cell membrane or in the cytoplasm or nucleus of cells in target tissues for hormones. Each kind of receptor binds to a specific hormone or group of hormones.

Hormone replacement therapy (HRT) The therapeutic use of estrogens and progestogens to replace hormones no longer produced by the ovaries after menopause.

Hormones Substances secreted by endocrine glands into the blood, which travel to specific target tissues and affect the growth and function of those tissues.

Hot flashes Intense feeling of warmth and profuse sweating that occur periodically in most menopausal and postmenopausal women and in andropausal men.

Huhner's test See **Postcoital test**.

Human chorionic gonadotropin (hCG) Protein hormone secreted by the blastocyst and placenta; similar in biological activity to luteinizing hormone (LH).

Human immunodeficiency virus (HIV) Virus that causes HIV disease and AIDS; consists of two types, HIV-1 and HIV-2.

Human menopausal gonadotropin (hMG) The combination of FSH and LH found at high concentrations in the blood and urine of postmenopausal women.

Human papilloma virus (HPV) Virus that causes genital warts as well as penile and cervical cancer.

Human placental lactogen (hPL) Protein secreted by the placenta; has biological actions similar to prolactin and growth hormone.

Hyaluronidase Enzyme present in the sperm acrosome; breaks down acid that cements cells of the corona radiata together.

H-Y antigen Protein, coded by a gene or genes on the Y chromosome, that is present on the cell membrane of male cells.

Hydatidiform mole Tumor-like uterine growth resulting from implantation of a trophoblast without an embryo. The cells of these moles are diploid, but the chromosomes are derived entirely from the father.

Hydrocele Accumulation of body fluid in a sac-like cavity, especially in the scrotum when the inguinal canal does not close after testicular descent.

Hydrocephaly A congenital defect characterized by excessive fluid accumulation in the brain.

2-Hydroxyestrone Nonestrogenic estrogen metabolite.

16-Hydroxyestrone Estrogenic estrogen metabolite.

Hymen A membrane that partially closes the vaginal introitus.

Hypoblast The lower layer of the bilaminar embryonic disk of a human preembryo; gives rise to extraembryonic structures such as the lining of the yolk sac.

Hypophysectomy Removal or destruction of the hypophysis.

Hypophyseal portal veins Small veins in the hypophysis that carry hypothalamic neurohormones to the adenohypophysis.

Hypophysis Gland lying in the sphenoid bone and connected to the hypothalamus; consists of the adenohypophysis and neurohypophysis; pituitary gland.

Hypophysiotropic area (HTA) Region in the hypothalamus; contains cell bodies of the neurosecretory neurons that control the function of the adenohypophysis.

Hypospadias Failure of the embryonic urethra to close so that the urethral meatus opens on the lower surface of the penis or into the vagina.

Hypothalamo-hypophysial portal system Vascular system connecting the median eminence of the hypothalamus with the adenohypophysis; consists of the primary and secondary capillary plexi connected by the hypophysial portal veins; see also **Primary capillary plexus** and **Secondary capillary plexus**.

Hypothalamus Floor of the diencephalon; contains cells that regulate function of the hypophysis and some aspects of sexual behavior.

Hysterectomy Surgical removal of the uterus.

Hysterosalpingogram An X-ray procedure using a dye to evaluate the shape of the uterus and the fallopian tubes.

Hysteroscopic tubal sterilization A method of female sterilization using small coils implanted into the proximal fallopian tube ends; scar tissue develops and permanently closes the tubes.

Hysterotomy Surgical method of abortion used after the 20th week of pregnancy, during which the fetus is removed through an incision in the abdomen and uterus; minicesarean.

Ibuprofen An antiprostaglandin painkiller.

Identical twins See **Monozygotic twins**.

Idiopathic CD4 lymphopenia (ICL) A disease of the immune system with symptoms similar to HIV disease.

Immature newborn When the newborn birth weight is between 0.50 kg (1 lb, 1.5 oz) and 0.99 kg (2 lb, 3 oz).

Immigration Moving to a new place of residence from another defined population.

Immunoassay pregnancy test Pregnancy test combining anti-hCG with a woman's urine.

Imperforate hymen Condition in which tissue completely blocks the vaginal introitus; must be removed surgically.

Implantation Apposition and attachment to, and finally invasion of, the endometrium by a blastocyst; nidation.

Impotence See **Erectile dysfunction**.

***In vitro* fertilization** See **External fertilization**.

Incompetent cervix Condition in which the cervix is unable to support a growing fetus to term; more common after induced abortions that used a metal dilator.

Incubation period Time between when an infectious microorganism enters the body and the appearance of disease symptoms.

Indifferent gonads Embryonic gonads, consisting of a cortex and medulla, that will give rise to testes or ovaries.

Induced abortion Termination of pregnancy by artificial means.

Induced labor Labor artificially induced by breaking the bag of water (amnion) and/or by administering Pitocin or prostaglandins.

Induced ovulation Ovulation that only occurs as a result of copulation or stimuli associated with the presence of a male.

Infanticide Killing a newborn or infant; used in the past to control population and to eliminate undesired children.

Infertility Failure to produce and maintain a pregnancy after about a year of unprotected coitus.

Injectable progestogen Female contraceptive method using an injected progestogen.

Inguinal canals Openings in the lower pelvic wall through which pass the testes during their descent; vasa deferentia pass through these canals in the adult male, as do the round ligaments in females.

Inguinal hernia Extension of abdominal contents through an open inguinal canal into the scrotal sac.

Inhibin Substance found in fluid of testes and associated ducts as well as in antral fluid in the ovaries; inhibits FSH secretion.

Inner acrosomal membrane Inner membrane of acrosome.

Inner cell mass Portion of the blastocyst that gives rise to the embryo as well as the amnion, yolk sac, and allantois.

Intact dilation and extraction (D&X) A method of late-term abortion developed as an alternative to dilation and evacuation.

Integrase Enzyme that incorporates HIV DNA into host cell DNA.

Internal cervical os Opening of the uterine cervix into the uterine cavity.

Intersexuality When a person has ambiguous reproductive structures.

Interstitial cells In the male, see **Leydig cells**. In the female, see **Ovarian interstitial cells**.

Interstitial nuclei 1, 2, 3, and 4 of the hypothalamus (INAH) Hypothalamic nuclei perhaps involved in human sexual behavior and orientation.

Interstitial cell-stimulating hormone (ICSH) See **Luteinizing hormone**.

Intraamnionic saline Abortion method used to terminate second trimester pregnancies, also called "salting out." Besides (or in addition to) saline, the substances urea, glucose, or prostaglandin can be injected into the amnionic fluid to induce abortion.

Intracervical ring A contraceptive device, containing progesterone, that is placed within the cervix.
Intracytoplasmic sperm injection (ICSI) See **Sperm microinjection**.
Intradermal progestogen implant Female contraceptive method using implantation of progestogen-containing tubes under the skin.
Intramural oviduct Region of oviduct within the uterine wall.
Intrauterine device (IUD) Foreign object, usually flexible plastic, placed into the uterus for contraception; some can include copper or progestogen.
Intromission Penetration of the penis into the vagina during vaginal coitus.
Ischemia Inadequate flow of blood and oxygen to a region of the body.
Jaundice Yellowish skin color due to the deposition of bilirubin.
Juveniles Individuals who have not reached puberty.
Kallmann's syndrome A human genetic disorder of GnRH cell migration in the embryo, resulting in infertility and the inability to smell.
Kegel exercise Voluntary contraction of the pubococcygeus muscle used to strengthen its tone.
Klinefelter's syndrome One of the more common developmental disorders of the reproductive tract in which a male has a chromosomal abnormality (47; XXY), undeveloped testes, and gynecomastia.
Labia majora Outer, major lips of female vulva; homologous to the male scrotum.
Labia minora Inner, minor lips of the female vulva; homologous to the corpus spongiosum of the male penis.
Labioscrotal swelling Part of the indifferent external genitalia of the early embryo that gives rise to the scrotum in males or the labia majora and mons pubis in females.
Labor Process leading to delivery that involves uterine contractions and fetal movement; divided into three stages: (1) cervical effacement and dilation, (2) expulsion of fetus, and (3) expulsion of placenta.
Lactation Secretion of milk by the mammary glands.
Lactiferous duct Duct of mammary tissue lying between the mammary ampulla and nipple.
Lactobacillus acidophilus See **Döderlein's bacillus**.
Lamaze method Prepared childbirth method involving knowledge of biological processes, controlled breathing and relaxation, and assistance by a partner.
Laminaria tent Cylinder of dried, sterilized seaweed used to dilate the cervix slowly by absorbing moisture.
Lanugo Fine downy hair covering the fetal skin.
Laparoscopy Examination of abdominal cavity with a tiny viewing tube and illuminator inserted through its wall.
Laparotomy Incision through the abdominal wall; used for tubal sterilization.
Larynx Voice box.
Latent stage of syphilis Third stage of syphilis, characterized by a relatively long-lasting absence of symptoms.
Lateral cervical ligaments Paired ligaments that attach the cervix to the pelvic wall.
Leboyer method Method of childbirth and newborn care in which the delivery environment is made as nonstressful as possible.
Lectins Proteins, derived from certain plants, that bind to sugars in the zona pellucida and prevent sperm penetration.

Leptin A peptide hormone produced by fat cells that acts on the hypothalamus to suppress appetite and also influences fertility in males and females.

Lesbian A female homosexual.

Lesch–Nyhan syndrome A sex-linked, recessive, inherited disorder characterized by physical abnormalities and self-mutilation.

Lesser vestibular glands Fluid-secreting glands opening into the female vestibule; are homologous to the prostate gland of the male; Skene's glands.

Leukocytes White blood cells.

Leukorrhea A whitish, sticky vaginal or uterine discharge containing white blood cells.

Leydig cells Cells in the interstitial spaces between seminiferous tubules; secrete androgens: interstitial cells.

Libido See **Sex drive**.

Ligand A molecule that binds to a specific site on a receptor molecule.

Ligand-binding domain The region of a protein receptor to which the ligand attaches.

Lightening Movement of the fetus down into the pelvic cavity 2 or 3 weeks before labor begins.

Limbic system Part of the brain that influences sex drive.

Linea nigra A dark line that appears on the lower abdominal wall from the pubic area upward toward the navel of some pubertal females.

Lipotropin (LPH) A polypeptide hormone, secreted by basophils of the adenohypophysis, that causes fat to be metabolized to fatty acids and glycerol.

Lithopedion Calcified remnants of an abdominal ectopic pregnancy; "stone baby."

Lochia Vaginal discharge characteristic of women after they have delivered a child.

Lordosis Female mating posture in rodents.

Lumen The interior cavity or space within an organ.

Lumpectomy See **Partial mastectomy**.

Luteal cells Endocrine cells of the corpus luteum.

Luteal phase Period in menstrual cycle between ovulation and the onset of menstruation.

Luteal phase deficiency Insufficient hormone (especially progesterone) secretion by the ovaries during the luteal phase; a cause of early miscarriage and infertility.

Luteinization Process by which granulosa cells of the graafian follicle switch from being primarily estrogen-secreting to primarily progesterone-secreting.

Luteinized cysts Abnormal lumps of luteal tissue in the ovary.

Luteinizing hormone (LH) Glycoprotein hormone secreted by basophils of the adenohypophysis that, together with FSH, maintains gonadal function in both sexes; interstitial cell-stimulating hormone.

Luteinizing hormone-releasing hormone (LHRH) See **Gonadotropin-releasing hormone**.

Luteolytic factor Substance that causes the corpus luteum to regress.

Lymphogranuloma venereum A sexually transmitted disease caused by a strain of *Chlamydia trachomatis*; see also **Reiter's syndrome**.

Lymphocytes White blood cells involved in immunity.

Lysergic acid diethylamide (LSD) A hallucinogen that arouses brain centers and mimics sympathetic nervous system arousal.

Macrophages Large, mobile, phagocytosing cells of the immune system.
Major histocompatibility complex genes (MHCs) Genes that code for proteins that "display" antigens to the immune system.
Maleness Biological characteristics of the male sex.
Malignant Adjective applying to metastasizing cancerous growth; dangerous, likely to cause death if not treated.
Mammae See **Mammary glands**.
Mammary alveoli Glandular structures of the mammary glands; secrete milk.
Mammary ampulla Wide storage tube for milk existing between a mammary duct and a lactiferous duct in the mammary gland.
Mammary coitus When a man inserts his penis between a woman's breasts.
Mammary duct Duct in mammary gland between the secondary mammary tubule and the mammary ampulla.
Mammary glands Paired structures on the thoracic wall of both sexes; in the female, they secrete milk; breasts or mammae.
Mammogram Use of an X-ray to detect breast abnormalities.
Masculine Having the qualities and characteristics of a male; often used to mean "manly."
Mastectomy Surgical removal of a breast; can be a **partial mastectomy** ("lumpectomy"), or removal of entire breast alone (**simple mastectomy**) or with underlying muscle and axillary lymph nodes (**radical mastectomy**). Only the breast and lymph nodes are removed in a **modified radical mastectomy**.
Masturbation Self-stimulation of the genitals to produce sexual arousal.
Maternal microchimerism Persistence of small numbers of maternal cells in her progeny after birth.
Maternal recognition of pregnancy When death of the corpus luteum is prevented by hCG during early pregnancy.
Mating behaviors Behaviors associated with coitus.
Maturation promoting factor A protein complex that includes cyclin B and cyclin-dependent kinase; triggers cells to progress through meiotic and mitotic cell cycles.
Meconium Greenish fecal material present in the fetal large intestine.
Median eminence Region in floor of the hypothalamus, near the infundibulum, where neurohormones are released from neurosecretory neurons of the hypophysiotropic area of the hypothalamus.
Medical abortion See **Medication abortion**.
Medication abortion Early termination of pregnancy using pharmaceuticals only.
Medullary cords Cords in the medulla of the indifferent gonad of the embryo that give rise to the seminiferous tubules in males and the rete ovarii in females.
Meiotic maturation The process by which immature oocytes become fertilizable eggs, including the reinitiation of meiosis 1, transition from meiosis I to meiosis II, and arrest at metaphase of meiosis II.
Meiosis A type of cell division found in the gonads, in which the chromosome number is reduced from 46 (2N) to 23 (N).
Melanin Brown pigment that is synthesized from the amino acid tyrosine.
Melanophores Pigment cells that contain melanin.
Melanophore-stimulating hormone (MSH) Protein hormone, secreted by basophils of the adenohypophysis, that stimulates melanin synthesis.

Melatonin Hormone synthesized and secreted by the pineal gland during darkness; inhibits reproductive function.

Membrana granulosa Single or multiple layer of granulosa cells between the theca and the zona pellucida of the ovarian follicle.

Membranous urethra The portion of the urethra passing through the urogenital diaphragm.

Menarche The onset of menstruation at puberty.

Menopause The cessation of menstruation in the human female, usually occurring between the ages of 45 and 55.

Menses See **Menstrual phase**.

Menstrual cramps Painful, spasmodic contractions of the uterus, possibly caused by unusually high uterine prostaglandin levels during the menstrual phase.

Menstrual cycle A monthly reproductive cycle in which periodic uterine bleeding occurs, as in humans and some other primates.

Menstrual extraction See **Menstrual regulation**.

Menstrual phase Period of menstrual cycle when menstruation occurs; menses.

Menstrual regulation Removal of the endometrial surface layer with or without an embryo by aspirating the uterus with a syringe attached to a plastic tube; used to induce abortion in the early first trimester; menstrual extraction.

Menstruation Discharge of blood along with sloughed cells of the uterine lining from the uterus through the vagina; normally occurs at monthly intervals.

Mesoderm Middle of the three primary germ layers of the embryo; gives rise to the muscular, skeletal, circulatory, excretory, and reproductive systems.

Mesonephric ducts See **Wolffian ducts**.

Mesonephric tubules Structures of the embryonic kidney (mesonephros) that form the vasa efferentia of the testes.

Mesonephros Embryonic kidney, which regresses before birth except that its tubules and ducts (wolffian ducts) develop into portions of the adult male reproductive tract.

Metabolic rate Rate at which tissues use oxygen.

Metastasis Spread of malignant cells from a tumor to other regions of the body.

Methotrexate Drug used to induce early abortion.

Microcephaly A congenital defect characterized by a relatively small brain.

Micropenis An extremely small penis, often due to an underdeveloped fetal anterior pituitary gland.

Mifepristone A drug that blocks the action of progesterone; used in combination with misoprostol in early medical abortions; **RU486**.

Milk-ejection reflex Stimulation of oxytocin secretion by suckling, oxytocin then causing ejection of milk from the nipple.

Milk line Paired lines on the ventrum of the embryo containing potential pairs of mammary glands.

Mini-cesarean See **Hysterotomy**.

Minilaparotomy “Band-aid” method of tubal sterilization using a small abdominal incision just above the pelvic bone.

Minimata’s disease Cerebral palsy caused by mercury poisoning.

Minipill Female oral contraceptive containing only a low amount of progestogen.

Miscarriage See **Spontaneous abortion**.

Misoprostol Drug used to induce early abortion.

Mitosis Process of cell division in which a diploid cell divides to produce two diploid progeny.

Mittelschmerz Mild abdominal pain associated with ovulation.

MMP-9 An enzyme that helps dissolve the amnionic sac during late pregnancy.

Modified radical mastectomy Surgical removal of a breast and surrounding lymph nodes.

Molluscum contagiosum A virus-induced disease characterized by pinkish skin warts.

Monestrus Having only one estrous cycle a year.

Mongolism See **Down's syndrome**.

Monilia vaginitis Inflammation of the vagina caused by *Candida albicans*; can be sexually transmitted.

Monogamy Marriage of two people for a theoretically indefinite period; in biology, when an animal has a single mate for one breeding cycle.

Monosomy A condition of aneuploidy in which an individual has one too few (45) chromosomes.

Monozygotic twins Twins usually resulting from separation of an inner cell mass of a blastocyst into two, after which two identical individuals develop; identical twins.

Mons pubis An elevated cushion of fat in the female, covered with skin and pubic hair; mons veneris.

Mons veneris See **Mons pubis**.

Morning-after pill See **Emergency contraception**.

Morning sickness Nausea experienced by some women in the first few weeks of pregnancy.

Moro reflex Reflexive movement of a newborn's arms and legs in response to being startled; startle reflex.

Morula Stage in prenatal development when there is a solid ball of eight or more cells.

Mucosa A mucous membrane lining a body tube, such as the fallopian tube; tunica mucosa.

Mucous plug Barrier of mucus in the cervix that is lost during early labor.

Mucus Viscous substance on surface of membranes that moistens, lubricates, and protects; made up of mucin (a mucopolysaccharide) and water; serves as a barrier to foreign agents and a partial barrier to sperm transport in the cervix.

Müllerian-inhibiting substance (MIS) Glycoprotein, secreted by the Sertoli cells of the embryonic testes, that causes regression of the müllerian ducts in males.

Müllerian ducts Embryonic ducts that form the sex accessory ducts of the adult female; paramesonephric ducts.

Multifetal pregnancy reduction A procedure to reduce the number of fetuses in the uterus, to increase the chances that the remaining fetuses will survive and develop normally.

Multiparous Adjective applied to a woman who has borne more than one child.

Mumps Viral infection of the salivary glands that can also infect the testes of an adult male and cause infertility.

Muscularis The smooth muscle layer of a body organ; tunica muscularis.
Mutagen A chemical or other factor that changes the genetic components of cells.
Mycoplasma genitalium A sexually transmitted bacterium that can cause nonspecific urethritis.
Myoepithelial cells Cells surrounding each alveolus of the mammary gland; their contraction leads to milk ejection.
Myoid cells Smooth muscle-like cells surrounding seminiferous tubules in the testis.
Myometrium The smooth muscle layer of the uterine wall.
Myotonia Increased muscular tension.
Naloxone An inhibitor of endorphin secretion.
Narcotics Addictive drugs that induce stupor or arrest activity, such as heroin, morphine, and methadone.
Natural family planning Use of one or more methods to determine the "safe" period in a menstrual cycle to avoid conception.
Natural selection The differential reproduction of certain individuals in a population because they are better (or not as well) adapted to the environment.
Negative feedback When production of a product decreases further production of that product.
Neisseria gonorrhoeae A Gram-negative bacterium, also called gonococus, that causes gonorrhea.
Neonatal death Infant death within the first day of life.
Neonate A newborn child.
Neoplasm See **Tumor**.
Neurohormones Hormones synthesized and secreted by neurosecretory neurons.
Neurohypophysis Neural portion of the hypophysis; posterior pituitary gland or pars nervosa.
Neurons Nerve cells.
Neuropeptide Y A brain neurotransmitter that stimulates GnRH secretion and reproduction.
Neurosecretory neuron Nerve cell that is specialized for synthesis and release of neurohormones.
Neurosecretory nuclei Paired clusters of neurosecretory neuron cell bodies in the hypothalamus.
Neurotransmitter Substance secreted by a regular neuron that crosses the synapse between two neurons and stimulates or inhibits nerve impulses in the next neuron.
Nidation See **Implantation**.
Nipple A conical structure on the breast; bears the opening of the lactiferous ducts through which milk is ejected.
Nocturnal emissions Involuntary ejaculation of semen, usually at night during sleep.
Nonbacterial prostatitis See **Prostatitis**.
Nondisjunction The failure of homologous chromosomes to separate during meiosis.
Nongonococcal urethritis See **Nonspecific urethritis**.
Nonidentical twins See **Dizygotic twins**.

Nonnucleoside reverse transcription inhibitors (NNRTIs) Antiretroviral drugs used against HIV; they interfere directly with reverse transcriptase.

Nonoxynol-9 A spermicide.

Nonspecific urethritis A sexually transmitted inflammation or infection of the urethra that is not due to the presence of gonorrhea.

Nonspecific vaginitis Inflammation or infection of the vagina for which the cause can vary.

Nonvenereal syphilis A disease of children in arid regions of the world caused by a bacterium similar or identical to *Treponema pallidum*. It is transmitted by direct (nonsexual) body contact or when drinking or eating; endemic syphilis.

Norepinephrine A neurotransmitter secreted by neurons of the sympathetic nervous system and central nervous system.

Nucleoside reverse transcription inhibitors (NRTIs) Antiretroviral drugs used against HIV; modified nucleosides, when incorporated into DNA, interfere with DNA synthesis and viral replication.

Nulliparous Adjective applied to a woman who has never borne a child.

Nurse midwife Trained nurse who assists in home deliveries.

Octoxynol 9 A spermicide.

Oedipus complex In Freudian theory, the sexual attraction of a little boy to his mother.

Oligomenorrhea When a woman periodically misses a menstrual period.

Oligospermia When sperm count is below the fertile level in a male.

Onuf's nucleus See **Spinal nucleus of the bulbocavernosus**.

Oocyte Female germ cell after beginning meiosis.

Oocyte maturation inhibitor Proposed substance in antral fluid that inhibits oocyte maturation until the LH surge.

Oogenesis Process of meiosis that results in a female gamete.

Oogonia (sing., Oogonium) Immature, mitotically dividing female germ cells; present only in fetal ovaries.

Oophorectomy See **Ovariectomy**.

Ootid Haploid female germ cell after completion of the second meiotic division.

Ootid pronucleus Nucleus of female germ cell after completion of meiosis and before fusion with sperm pronucleus.

Opportunistic infections of HIV disease Infections resulting from HIV immunosuppression but not caused by HIV itself.

Oral coitus Contact of the mouth or oral cavity with the genitals; see also **Cunnilingus** and **Fellatio**.

Orchidectomy Surgical removal of the testes.

Orchitis Inflammation or infection of the testes.

Organizational effects Those effects of hormones that occur early in development and cause irreversible sex differences in the brain or other tissues.

Orgasm The peak of arousal in sexual activity.

Orgasmic dysfunction Failure to achieve orgasm.

Orgasmic phase Third stage in the human sexual response cycle, during which orgasm occurs.

Orgasmic platform The area, including the outer third of the vaginal barrel and the labia minora, which displays vasocongestion during the female plateau phase.

Osteoporosis Loss of calcium and phosphorus from bones; common in older people, especially postmenopausal women and andropausal men.
Ostium Internal opening of the oviductal infundibulum.
Outer acrosomal membrane Outer membrane of acrosomal cap on sperm head; see also **Acrosome reaction**.
Ovarian cortex Outer, more dense layer of the ovarian stroma.
Ovarian cycle Cyclic changes in the ovaries during the menstrual cycle.
Ovarian cyst An abnormal growth of ovarian follicular tissue.
Ovarian follicle A bag or sac of tissue in the ovary that contains the oocyte.
Ovarian hyperstimulation syndrome A serious medical complication of superovulation, including ovarian swelling and abdominal fluid accumulation.
Ovarian interstitial cells Clumps of endocrine cells in the ovarian stroma that are derived from the theca of atretic follicles.
Ovarian medulla Central, less dense region of the ovarian stroma.
Ovarian stroma Connective tissue framework of the ovary.
Ovariectomy Surgical removal of ovaries; oophorectomy.
Ovaries The gonads of the female.
Oviductal ampulla Widened portion of the oviduct between the infundibulum and the isthmus.
Oviductal infundibulum Widened end of the oviduct that captures the ovulated egg.
Oviductal isthmus Narrow portion of the oviduct between the ampulla and the intramural region.
Oviductal muscularis Layer of smooth muscle in the oviductal wall.
Oviductal serosa Thin external covering of the oviduct.
Oviducts Paired tubes that lead from the ovaries to the upper portion of the uterus; same as fallopian tubes or uterine tubes.
Ovotestis Gonad containing both testicular and ovarian tissue.
Ovulation Process by which an ovum is extruded from a mature ovarian follicle.
Ovum An ovulated, haploid secondary oocyte, in second meiotic arrest, before it has been penetrated by a sperm.
Ovum activation Reinitiation of meiosis in the ovum as a result of sperm penetration.
Oxytocin Polypeptide hormone secreted by the neurohypophysis; causes contraction of smooth muscle in reproductive tissues of both sexes.
Papule A small elevation on the skin.
Papanicolaou test Test by which cells of the uterine cervix are removed with a swab and examined for the presence of cervical cancer; Pap smear.
Pap smear See **Papanicolaou test**.
Paracervical A local anesthetic injected into both sides of the cervix.
Paracrine A chemical messenger that acts locally and is not transported in the blood.
Paramesonephric ducts See **Müllerian ducts**.
Paraventricular nucleus (PVN) Region of the hypothalamus that synthesizes oxytocin and vasopressin.
Pars distalis Major portion of the adenohypophysis; anterior pituitary gland.
Pars intermedia Portion of adenohypophysis between the pars distalis and the neurohypophysis; often absent in adult humans.
Pars nervosa See **Neurohypophysis**.

Pars tuberalis Portion of adenohypophysis existing as a collar of cells around the pituitary stalk.

Parthenogenesis Development of an embryo from an unfertilized egg; virgin birth.

Partial hydatidiform mole Implantation of a trophoblast and a dead embryo; cells of these moles are triploid.

Partial mastectomy Survical removal of a lump from within the breast; lumpectomy

Parturition Birth.

Patent ductus arteriosus When the ductus arteriosus fails to close after birth.

Paternal behavior Care-giving behavior of the father to his child.

Paternal leucocyte immunization Treatment for female infertility that is caused by immunological rejection of the fetus.

Pearl's formula Method of expressing failure rate of contraceptive measures per 100 woman-years.

Pediculosis pubis A condition in which one is parasitized by the louse *Phthirus pubis*; also called "crabs"; can be sexually transmitted.

Pelvic inflammatory disease Widespread infection of the female pelvic organs.

Penicillinase An enzyme, produced by certain bacteria, that inactivates penicillin.

Penile agenesis Congenital absence of a penis due to the lack of a genital tubercle in the embryo.

Penile body See **Penile shaft**.

Penile implant A tube inserted within the penis; can be made erect by manipulating a pump in the scrotum.

Penile prepuce The loose flap of skin that partially covers the penile glans; foreskin.

Penile shaft The cylindrical, erectile portion of the penis; penile body.

Penis Portion of the male external genitalia that serves as an intromittent and urinary organ.

Peptide A short chain of amino acids.

Perganol Human menopausal gonadotropin.

Perimenopause Initial appearance of menopausal symptoms.

Perimetrium Thin membrane covering the outside of the uterus.

Perineum Pelvic floor; the space between the anus and scrotum in the male, and the anus and vulva in the female.

Peripheral effect of a hormone Action of a hormone on tissues outside the central nervous systern.

Perivitelline space Region between the zona pellucida and the vitelline membrane of an oocyte.

Persistent pulmonary hypertension Condition in the newborn in which blood flow through the lungs is resisted by the presence of thickened pulmonary arteries; can be caused by fetal exposure to aspirin or indomethacin.

Peyronie's disease A condition in which the penis develops a fibrous ridge along its top or side, causing curvature.

Phallus The male organ of copulation; the penis.

Phenylalanine An amino acid, levels of which are high in the blood and urine of an infant with phenylketonuria.

Phenylketonuria (PKU) An inherited disorder in protein metabolism that can cause mental retardation.

Pheromone A chemical produced by one individual of a species that changes the behavior or physiology of another member of the same species.

Phimosis Constriction of the penile prepuce.

Phocomelia Condition in which a child is born with hands and feet but no arms or legs.

Phthirus pubis A blood-sucking crab louse that parasitizes humans and causes pediculosis pubis (crabs).

Physiological jaundice Yellowing of the skin and eyes of a newborn because of excessive deposition of the pigment bilirubin, a breakdown product of red blood cells.

Phytoestrogens Plant estrogens.

Pineal gland Single glandular outpocketing of the epithalamus that synthesizes and secretes melatonin.

Pitocin Trade name of an artificial oxytocin used to induce or hasten labor.

Pituitary gland See **Hypophysis**.

Pituitary stalk The tissue connecting the hypophysis with the brain.

Placenta A structure, composed of fetal and maternal tissue, attached to the inner surface of the uterus and connected to the fetus by the umbilical cord; serves as a respiratory, excretory, and nutritive organ for the fetus, and also secretes hormones important in pregnancy.

Placenta previa Uncommon condition in which the placenta is formed low in the uterus and partially or completely covers the opening of the cervix; cesarean delivery often is required.

Placentophagia Eating of the placenta (afterbirth).

Plantar reflex When a newborn flexes its toes after someone touches the sole of its foot.

Plasma membrane Outer cellular membrane.

Plateau phase The second stage in the human sexual response cycle; immediately precedes orgasm.

Polyandry A mating strategy in which a female has two or more male mates; in humans, when a woman is married to more than one male.

Polychlorinated biphenyls (PCBs) Environmental pollutants that have been associated with lowering sperm production in American males.

Polycystic ovarian syndrome Condition in which several large abnormal follicles exist in each ovary.

Polyestrus Having more than one estrous cycle a year.

Polygamy When the spouse of either sex is married to more than one person at one time.

Polygyny A mating strategy in which a male has two or more female mates; in humans, when a man is married to more than one woman.

Polypeptide Small chain of amino acids.

Polyploidy Having more than two full sets of homologous chromosomes.

Polyspermy Fertilization of an ovum by more than one sperm.

Polythelia When more than two mammary glands persist in an adult.

Population growth rate Change in number of individuals in a population in a given amount of time.

Portal system A vascular system that directly connects two capillary beds.

Positive feedback When production of a product increases further production of that product.

Positive signs of pregnancy Signs that definitely indicate pregnancy, including detection of a fetal heartbeat, visualization of the fetus with ultrasound or fetoscopy, and feeling the fetus move.

Positron emission tomography (PET) Used to observe brain activity.

Postcoital estrogen High dose of estrogen (oral or injected) given soon after coitus to prevent pregnancy; morning-after pill.

Postcoital test (Huhner's test) When a woman's cervical mucus is examined for her husband's sperm number and viability within 2 hr of coitus.

Posterior pituitary gland See **Neurohypophysis**.

Postmature newborn A newborn with clinical signs of a pathologically prolonged pregnancy.

Postpartum After birth.

Postpartum depression Depression (fits of crying, anxiety) in a woman appearing soon after she has had a child.

Postpartum dysphoric disorder A transitory syndrome of psychological difficulties commonly occurring within a few days of delivery, including low mood, anxiety, mood changes, and confusion.

Postpartum psychiatric syndrome Severe psychosis developing in a woman soon after she has had a child.

Postreproductive adults Those too old to have offspring.

Postterm infant An infant born more than 2 weeks after its due date.

Potassium nitrate See **Saltpeter**.

Precocious puberty When puberty occurs before age 6–7 in females or age 9 in males.

Preembryo The developing organism between fertilization and the beginning of gastrulation.

Preeclampsia See **Toxemia**.

Preemptive mastectomy Having a bilateral mastectomy to avoid future risk of breast cancer.

Pregnancy The condition of carrying a developing preembryo, embryo, or fetus in the uterus; gestation.

Pregnancy tests Tests that utilize detection of human chorionic gonadotropin (hCG) in blood or urine to determine if a woman is pregnant.

Pregnancy wastage Infertility due to chronic miscarriage.

Pregnanediol A metabolite of progesterone found in urine.

Pregnenolone Precursor steroid for synthesis of steroid hormones; derived from cholesterol.

Preimplantation genetic diagnosis Genetic testing on a cell taken from a preembryo prior to transfer of the embryo to the uterus during IVF; used to screen preembryos for serious inherited disorders.

Premature ejaculation Ejaculation before, just at, or soon after intromission.

Premature newborn A newborn with a birth weight between 0.99 kg (2 lb, 3 oz) and 2.49 kg (5.5 lb).

Premature ovarian failure Cessation of menstruation and onset of menopause before age 40 caused by early ovarian failure.

Premenstrual syndrome A group of physical and emotional symptoms associated with the late luteal phase of the menstrual cycle.

Premenstrual tension Emotional components of premenstrual syndrome.

Prepared childbirth Term used for several kinds of "natural" childbirth procedures.
Prepro-GnRH The large protein that gives rise to GnRH and GAP.
Presumptive signs of pregnancy Possible signs of pregnancy, including secondary amenorrhea, morning sickness, and breast changes.
Preterm infant An infant born before the end of full gestation; often defined as born before 37 weeks of the mother's last menstrual period.
Priapism Persistent, abnormal erection of the penis, usually without sexual desire.
Primary amenorrhea When a female has not reached menarche by age 16.
Primary capillary plexus Cluster of capillaries in the median eminence that receive neurohormones secreted in that region; see also **Hypothalamo–hypophysial portal system**.
Primary dysmenorrhea Dysmenorrhea caused by prostaglandin-induced uterine cramping.
Primary erectile dysfunction When a man is unable to gain an erection now or in the past.
Primary follicles Ovarian follicles consisting of a primary oocyte surrounded by a single layer of cuboidal granulosa cells and a thin theca.
Primary germ layers The layers of cells (ectoderm, mesoderm, endoderm) in the early embryo that give rise to all adult tissues.
Primary oocyte An oocyte in first meiotic arrest.
Primary orgasmic dysfunction When a woman is unable to have an orgasm now or in the past.
Primary sex ratio Ratio of male to female embryos determined at, or shortly after, fertilization.
Primary sexual characteristics Those reproductive structures that distinguish male from female and are involved in gamete exchange (mating), including the gonads, sex accessory ducts and glands, and external genitalia.
Primary spermatocyte Male germ cell before the first reduction division of meiosis.
Primary stage of syphilis Initial stage of syphilis, during which a sore (chancre) develops.
Primiparous Giving birth for the first time.
Primordial follicles Ovarian follicles consisting of a primary oocyte surrounded by a single layer of flattened granulosa cells.
Primordial germ cells Cells that migrate from the yolk sac to the indifferent gonads of the embryo, where they give rise to the oogonia of the ovaries or the spermatogonia of the testes.
Probable signs of pregnancy Indications that a woman, in all probability, is pregnant; include Hegar's sign and a positive pregnancy test.
Proctitis Inflammation or infection of the rectum.
Progestasert An IUD that contains a progestogen.
Progesterone The major natural progestogen.
Progestogens Substances that promote secretory function of the uterus; secreted by the ovaries and placenta; progestins.
Progestins See **Progestogens**.
Programmed cell death See **Apoptosis**.
Prolactin (PRL) Protein hormone, secreted by acidophils of the adenohypophysis, that stimulates milk synthesis in the mammary glands.

Prolactin release-inhibiting hormone (PRIH) Hormone from the hypothalamus that decreases synthesis and secretion of prolactin.
Prolapsed uterus When the uterus falls down from the normal position.
Proliferative phase Stage of build-up of the endometrial lining prior to ovulation.
Pronucleus The haploid nucleus of egg or sperm prior to fertilization.
Prophylactic measure A measure taken to prevent occurrence of a disease or disorder.
Prostaglandins (PG) Family of small 20-carbon atom molecules derived from fatty acids that have various effects on the reproductive system of both sexes.
Prostate gland An unpaired male sex accessory gland that produces 13 to 33% of seminal plasma.
Prostate-specific antigen (PSA) A protein produced by the prostate and secreted into the blood; used to screen for abnormal prostate growth.
Prostatic urethra Portion of the urethra enclosed by the prostate gland.
Prostatic utricle Small blind pouch, present in the adult male prostate gland, that is a remnant of the müllerian duct system.
Prostatitis Inflammation of the prostate gland; can be the result of a bacterial infection (bacterial prostatitis) or due to unknown, nonbacterial causes (nonbacterial prostatitis).
Protease An enzyme that hydrolyzes proteins to peptides and/or amino acids.
Protease inhibitor One of a class of anti-HIV drugs designed to inhibit the enzyme protease and interfere with viral replication.
Proteins Long chains of amino acids linked by peptide bonds.
Proviral DNA HIV DNA inserted into a host cell's DNA.
Pseudocyesis When some presumptive signs of pregnancy appear in a nonpregnant woman; has a psychosomatic origin; false pregnancy.
Pseudohermaphrodite An individual who has gonads that agree with chromosomal sex but has ambiguous external genitalia.
Puberty The biological transformation that takes a person from being a sexually immature child to a sexually mature adult; sexual maturation.
Pubescence The state of a child between the onset of pubertal changes and the completion of sexual maturation.
Pubococcygeus muscle Muscle in the pelvic floor that forms support for the pelvic organs.
Pudendal block When an anesthetic is injected into the pudendal nerves in the vaginal wall to numb the vagina.
Puerperium The period of confinement of a woman after she has given birth.
Purulent Pus-filled.
Radical mastectomy See **Mastectomy**.
Radioimmunoassay Procedure for hormone assay utilizing competitive binding of nonradioactive and radioactive hormone to an antibody to that hormone.
Raphe Ridge in the midline of the scrotum.
Recombination The formation of new gene combinations as a result of crossing-over between homologous chromosomes.
Reduction division Process of the first meiotic division during which the diploid (2N) condition is changed to a haploid (N) condition.
Refractory period A temporary state after orgasm when a male is not responsive to sexual stimuli.

Reiter's syndrome Serious, advanced symptoms of lymphogranuloma venereum.

Relaxin Polypeptide hormone secreted by the corpus luteum and placenta during pregnancy; relaxes the pubic symphysis, inhibits uterine contractions, and helps dilate the cervix.

Release-inhibiting hormone (RIH) A hormone, secreted by neurosecretory neurons of the hypothalamus, that decreases synthesis and secretion of a hormone of the adenohypophysis.

Releasing hormone (RH) A hormone, secreted by neurosecretory neurons of the hypothalamus, that increases synthesis and secretion of a hormone of the adenohypophysis.

Replacement therapy Administration of a gland extract or pure hormone that reverses the effects of extirpation of an endocrine gland.

Reproductive adults Sexually mature individuals.

Reproductive potential Maximal number of offspring an individual is capable of producing in a given amount of time.

Reserpine A drug that lowers blood pressure; can reduce fertility and cause sexual dysfunction.

Resolution phase The last stage in the human sexual response cycle, during which the sexual system returns to its unexcited state.

Respiratory distress syndrome When a newborn's lungs do not have surfactant and it is unable to breathe properly.

Rete ovarii Vestigial remnants of the medullary cords present in the medulla of the adult ovary.

Rete testis Tiny tubules within the testis that transport sperm from the seminiferous tubules to the vasa efferentia.

Retroflexion Referring to a uterus that is tilted backwards.

Retrograde ejaculation The movement of semen into the bladder instead of out of the body during ejaculation.

Retrolental fibroplasia Disorder of the newborn leading to partial or complete blindness; caused by exposure to too much oxygen.

Reverse transcriptase Enzyme that converts HIV RNA into HIV DNA.

Reverse transcriptase inhibitor An antiretroviral viral drug that acts to inhibit the process of reverse transcription in viruses such as HIV.

Rhesus disease Condition of fetus in which the mother's cells are producing antibodies to the fetal red blood cells.

Rhythm methods Contraceptive methods based on prediction of safe and fertile periods in the menstrual cycle; see also **Calendar, Basal body temperature, Cervical mucus**, and **Sympto-thermal methods**.

Ritotrene Drug used to prevent premature labor.

Rooting reflex When the hungry newborn keeps its head in contact with an object that touches its mouth or cheek.

Round ligaments Paired, cord-like ligaments that connect the top of the uterus with the pelvic wall.

RU 486 See **Mifepristone**.

Rubella German measles virus.

Saddle block A conduction anesthetic injected into the subdural space of the spinal cord low in the back.

Salpingectomy Surgical removal of the oviducts.

Salpingitis Inflammation or infection of the oviducts.

Saltpeter A chemical falsely accused of being an anaphrodisiac; potassium nitrate.

Sarcoptes scabiei A mite that parasitizes human skin and causes scabies.

Scabies Skin condition caused by the mite *Sarcoptes scabiei*; can be sexually transmitted.

Scrotum The pouch, suspended from the groin, that contains the testes and its accessory ducts.

Sebaceous glands Skin glands at the base of hair follicles that secrete an oily substance called sebum; their activity increases during puberty.

Sebum Oily secretion of sebaceous glands.

Secondary amenorrhea When a woman, who had menstruated previously, fails to menstruate for at least 6 consecutive months; see also **Amenorrhea**.

Secondary capillary plexus Capillary bed in pars distalis; delivers releasing and release-inhibiting hormones.

Secondary dysmenorrhea Dysmenorrhea not caused by elevated prostaglandin levels during menstruation.

Secondary erectile dysfunction When a man is unable to gain an erection at least 25% of the time that the opportunity is present.

Secondary follicles Ovarian follicles consisting of a primary oocyte surrounded by several layers of granulosa cells and a single theca.

Secondary mammary tubules Small ducts in the mammary gland that carry milk from the alveoli to the mammary ducts.

Secondary oocyte A haploid oocyte produced by completion of the first meiotic division.

Secondary orgasmic dysfunction When a woman fails to have an orgasm in selective situations.

Secondary sex ratio Ratio of males to females at birth.

Secondary sexual characteristics External, sex-typical male or female structures (excluding the external genitalia) that are not directly involved in gamete production and transport; serve as courtship signals.

Secondary spermatocyte Haploid male germ cell that is the product of reduction division of primary spermatocytes.

Second meiotic arrest Occurs after the secondary oocyte has begun the second meiotic division; ends only after sperm penetration of the ovum.

Second meiotic division The phase of oocyte meiosis in which a haploid secondary oocyte divides to produce a haploid ootid and second polar body.

Second messenger A short-lived cytosolic molecule that triggers a biochemical response within the cell; released in response to a "first messenger" such as a hormone at the cell surface.

Second polar body Small haploid cell that is a product of the second meiotic division of an oocyte; degenerates before or after dividing.

Secondary stage of syphilis The second stage of syphilis, characterized by a widely distributed rash.

Secretory phase Stage of endometrial development following the proliferative phase, in which the endometrium produces secretions in preparation for implantation of a fertilized egg.

Sella turcica Cup-shaped depression in the sphenoid bone at the base of the skull in which lies the hypophysis.

Semen Mixture of sperm and secretions of male sex accessory structures (seminal plasma) that leaves the male urethra during ejaculation; seminal fluid.

Semen coagulation Increase in viscosity of semen within 1 min of ejaculation.

Semen liquefaction Increase in semen fluidity several minutes after ejaculation.

Seminal fluid See **Semen**.

Seminal plasma Fluid portion of semen secreted by male sex accessory glands.

Seminal vesicles Paired male sex accessory glands that produce seminal plasma rich in fructose and prostaglandins.

Seminiferous epithelium Layer of cells lining each seminiferous tubule.

Seminiferous tubules The small tubes, within the testes, that contain male germ cells; sites of sperm production.

Septic abortion Spontaneous abortion (miscarriage) caused by uterine infection.

Septic pregnancy Condition in which infection occurs within the uterus of a pregnant woman.

Sequential pill Type of oral contraceptive in which an estrogen alone is taken, followed by a combination of an estrogen and progestogen.

Serial monogamy Having one spouse at a time, but more than one in a lifetime.

Seroconversion The appearance of detectable antibodies in the blood after infection; after this stage in HIV disease, one is considered HIV-positive.

Serosa The outer connective tissue covering of an organ; tunica serosa.

Sertoli cells Cells in seminiferous tubule that produce androgen-binding protein and support male germ cells; sustenacular cells.

Sex-determining region gene Y (SRY) Gene on Y chromosome that codes for a protein (testis-determining factor) which causes testis formation.

Sex-steroid binding globulins Blood proteins that bind to gonadal steroid hormones.

Sex accessory ducts Those internal structures in the male or female that transport or support germ cells or embryos.

Sex accessory glands Those glands that secrete into sex accessory ducts in either sex.

Sex-change operation Operation that constructs artificial external genitalia into those characteristic of the opposite sex.

Sex chromatin See **Barr body**.

Sex chromosomes Pair of homologous chromosomes that are XX in female somatic cells and XY in male somatic cells.

Sex drive The motivation or desire to behave sexually; libido.

Sex flush The vasocongestion of the skin that begins during the excitement phase of the sexual response cycle.

Sex-influenced trait An inherited trait, the expressivity of which is influenced by sex hormones.

Sex-linked trait An inherited trait controlled by an allele located on a sex chromosome (usually the X chromosome).

Sex role The public manifestation of one's gender identity; gender role.

Sex steroid binding globulin A plasma protein that binds sex steroids with high affinity.

Sexual dimorphism When the male and female of a species differ in appearance.
Sexual dysfunction Any psychological or physical condition that inhibits sexual gratification.
Sexual intercourse See **Coitus**.
Sexual maturation See **Puberty**.
Sexual monomorphism When both sexes are similar in appearance or structure.
Sexual reproduction Reproduction involving meiosis and fertilization.
Sexual response cycle The sequence of four phases of the human sexual response after effective erotic stimuli are present.
Sexuality One's sexual attitudes, feelings, and behavior.
Sexual selection A form of natural selection in which behavioral or morphological traits become magnified and are used as courtship signals or in competition among males.
Sexually dimorphic nucleus (SDN) A hypothalamic region that is larger in males than in females after age 4; **INAH 1**
Sexually indifferent stage Stage in the early embryo when the gonads, sex accessory glands, and external genitalia are similar in both sexes.
Sexually transmitted diseases (STDs) Infectious diseases that can be transmitted from one person to another during coitus or other genital contact; venereal diseases.
Sexually transmitted infections (STIs) See **Sexually transmitted diseases**.
Short luteal phase syndrome When the corpus luteum dies prematurely during the menstrual cycle.
Siamese twins See **Conjoined twins**.
Sickle cell disease An inherited disorder that causes a severe form of anemia.
Signal transduction The biochemical events that convert a signal from outside the cell and result in a functional change within the cell.
Simian immunodeficiency viruses (SIVs) African primate viruses related to, and perhaps evolving into, HIVs.
Simple mastectomy See **Mastectomy**.
Skene's glands See **Lesser vestibular glands**.
Sleep-inducing peptide A sedative-like substance present in human breast milk.
Smegma Cheesy secretion from glands present under the male prepuce.
Sodomy A legal term, defined differently in various states, but usually referring to "acts against nature" such as anal or oral coitus or coitus with animals.
Somatostatin Another name for growth hormone release-inhibiting hormone.
Sperm See **Spermatozoon**.
Sperm hyperactivation Increase in sperm motility, probably caused by contact with oviductal fluid.
Spermatic cords Paired structures in males that suspend the testes and scrotum from the pelvic wall and contain the vas deferens, nerves, and blood vessels.
Spermatid Haploid male germ cell that matures into a spermatozoon, a process called spermiogenesis.
Spermatogenesis The process of sperm formation occurring in the testis, whereby spermatogonia are transformed by meiosis to spermatids.
Spermatogonium (pl., Spermatogonia) Diploid male germ cell that divides mitotically before becoming a primary spermatocyte.

Spermatozoa See **Spermatozoon.**
Spermatozoon (pl., Spermatozoa) A mature male germ cell; sperm.
Sperm capacitation Process by which sperm acquire the ability to penetrate the zona pellucida of a recently ovulated ovum; probably occurs in the uterus or oviduct.
Sperm granuloma Lump formed after a vasectomy because of an autoallergic response to sperm leaking into surrounding tissues.
Spermiation Process by which mature sperm are released into the lumen of the seminiferous tubule from the Sertoli cells in which they were embedded.
Spermicide Agent placed in vagina that acts as a barrier to sperm transport and kills sperm.
Spermiogenesis Maturation of a spermatid to a spermatozoon.
Sperm microinjection When a single sperm is injected with a needle into an egg; also called intracytoplasmic sperm injection (ICSI).
Sperm head Portion of sperm that contains the nucleus and acrosome.
Sperm midpiece Portion of sperm that contains the mitochondria; between the sperm neck and tail.
Sperm neck Portion of the sperm between the head and midpiece.
Sperm plasma membrane Outer cell membrane of the sperm.
Sperm pronucleus Sperm nucleus immediately before fusion with the haploid ovum pronucleus.
Spinal anesthetic Injection of an anesthetic into the subdural space around the spinal cord; used during labor and delivery.
Spinal nucleus of the bulbocavernosus See **Ejaculatory center.**
Spinnbarheit Threadability of cervical mucus.
Splenium Posterior end of corpus callosum of the brain.
Sponge barrier contraceptive A form of barrier contraceptive that is placed in the external cervical os.
Spongy urethra Portion of the urethra in the penis; cavernous urethra.
Spontaneous abortion Spontaneous loss of a preembryo, embryo or fetus; miscarriage.
Spontaneous erections Penile erections that occur with increasing frequency during puberty without the presence of sexual stimuli.
Spontaneous ovulation Ovulation that occurs cyclically, not requiring copulation or presence of a male.
Spotting Slight blood loss from the uterus around the time of ovulation; breakthrough bleeding.
Stable population A population in which input (births plus immigration) equals output (deaths plus emigration).
Standardized birth rate Number of individuals born to a specific age group of a population in a given amount of time.
Standardized death rate Number of individuals dying within a specific age group of a population in a given amount of time.
Staphylococcus aureus Bacterium present in the vagina of some women; associated with toxic shock syndrome.
Startle reflex See **Moro reflex.**
Status orgasmus A sustained orgasm in a woman, lasting 20 sec or longer.
Stein–Leventhal syndrome Female condition caused by inability of the ovaries to convert androgens to estrogens; characterized by polycystic ovaries, infertility, obesity, and excessive body hair.

Steroid A class of molecules derived from cholesterol.
Steroidogenesis Biosynthesis of steroid hormones in the gonads, adrenal glands, and placenta.
Steroidogenesis stimulating protein (STP) Protein secreted by Sertoli cells of the testes that enhances testosterone secretion.
Stigma Small avascular region on the surface of a graafian follicle; future site of ovulation.
Stillbirth When a fetus is dead at birth.
Stratum basalis Outer layer of the uterine endometrium that is not shed during menstruation.
Stratum functionalis Inner layer of uterine endometrium that is shed during menstruation.
Strawberry marks Small reddish blotches on the skin of a newborn that go away in time.
Stress The physiological response to a stressor.
Stressor Any set of circumstances that disturbs the normal homeostasis of the body.
Sucking reflex Sucking movements of a newborn's mouth that are elicited by touching its lips.
Sudden infant death syndrome (SIDS) Sudden unexpected death of an infant, probably due to failure of its respiratory mechanisms; crib death.
Superfetation Development of a second fetus after one has already started development in the uterus.
Superovulation The use of gonadotropic drugs to promote the development of multiple mature follicles in the ovary.
Suprachiasmatic nucleus (SCN) Hypothalamic region involved in control of daily and seasonal rhythms.
Supraoptic nucleus (SON) Nucleus in hypothalamus that secretes oxytocin and vasopressin.
Surface epithelium Thin epithelial covering, such as that covering the ovary.
Surfactant Lipoprotein present in the newborn's lungs; allows the air sacs to fill by affecting surface tension.
Surge center Region in hypothalamus that controls the LH surge in females.
Surgical abortion Termination of a pregnancy with the use of surgical instruments including vacuum aspiration.
Surrogate pregnancy When a woman contracts to carry the child of another couple to term, after which she relinquishes the infant to the couple.
Suspensory ligaments of Cooper Ligaments in breasts that provide support.
Sustentacular cells See **Sertoli cells**.
Sweat glands Skin glands; these can be apocrine, causing a typical odor, or eccrine, secreting a salty fluid involved in temperature regulation.
Symptomatic phase of HIV disease Stage of HIV infection when some opportunistic infections appear but a person does not as yet have AIDS.
Sympto-thermal method Method of determining safe and fertile periods in the menstrual cycle using cervical mucus, basal body temperature, and other signs of ovulation; see also **Rhythm methods**.
Synapse Site of communication between a nerve cell and some other cell.
Syncytiotrophoblast Syncytial component of the trophoblast.
Syncytium Tissue with cells that have communicating cytoplasm.

Syndrome A group of symptoms relating to a single cause or event.

Syphilis A sexually transmitted disease caused by the bacterium *Treponema pallidum*.

Tamoxifen An anti-estrogen drug used to treat breast cancer.

Tanner stages Stages of physical development in children and adolescents, used to determine stage of puberty.

Target tissues Tissues containing cells that have specific receptors for a certain hormone and exhibit a growth or physiological response to that hormone.

Taxol An anticancer drug originally extracted from the yew tree.

Tay–Sach's disease An autosomal recessive, inherited disorder of the nervous system, most common in Jews of central European ancestry.

Tenting effect Expansion of the vagina during the female sexual response cycle; caused by elevation of the uterus.

Teratogen A chemical or other substance that causes abnormalities in embryonic or fetal growth and development.

Term See **Due date**.

Tertiary follicles Large ovarian follicles in which there is an antral cavity; the theca of these follicles is divided into a theca interna and externa.

Tertiary stage of syphilis The life-threatening, final stage of syphilis.

Testes (sing., Testis) The male gonads; site of spermatogenesis and androgen secretion; testicles.

Testicles See **Testes**.

Testicular artery Artery in each spermatic cord that supplies blood to the testis and epididymides.

Testicular biopsy Surgical sampling of a small piece of testicular tissue.

Testicular feminization syndrome (TFM) Form of male pseudohermaphroditism that is caued by an inherited absence of androgen receptors in target cells; androgen insensitivity syndrome.

Testicular lobules Compartments within the testis; contain the seminiferous tubules.

Testicular veins Two veins in the spermatic cord that drain blood from the testis and epididymis.

Testis-determining factor (TDF) Substance coded for by the SRY gene on the Y chromosome that causes testis formation in the embryo.

Testosterone The major natural androgen.

Testosterone enanthate A synthetic androgen that may be used as a contraceptive injection for men.

Test-tube baby Baby conceived outside the mother, the resultant preembryo then being implanted into the mother's uterus for development until parturition.

Tetrahydrocannabinol (THC) The active, hallucinogenic chemical in marijuana and hashish.

Thalidomide Drug formerly prescribed for pregnant women as a sedative or to treat morning sickness until it was learned that it had severe teratogenic effects.

Theca Connective tissue covering of the ovarian follicle and corpus luteum.

Theca externa External, dense, vascular, connective tissue layer of the theca of tertiary ovarian follicles.

Theca interna Internal glandular layer of the theca of tertiary ovarian follicles.

Therapeutic abortion When an abortion is induced in consideration of the woman's physical health.

Thermography Method of detecting tumors using the heat they produce.

Thyrotropin (TSH) Glycoprotein hormone, secreted by basophils of the adenohypophysis, that stimulates the thyroid glands to synthesize and secrete thyroid hormones.

Thyrotropin-releasing hormone (TRH) Hormone of the hypothalamus that increases synthesis and secretion of thyrotropin.

Thyroxine A major thyroid hormone.

Toxemia Sometimes life-threatening complication of pregnancy. Early stage (preeclampsia) is characterized by hypertension, edema, and the presence of protein in the urine; late stage (eclampsia) is accompanied by coma, convulsion, and sometimes death.

Toxic shock syndrome A disease caused by the presence of certain strains of the bacterium *Staphylococcus aureus* in the vagina; has been associated with use of highly absorbent tampons.

Transcervical balloon tuboplasty Unblocking an obstructed oviduct by inflating a small balloon inserted into the oviductal lumen.

Transition contractions Strong contractions during stage 1 of labor that dilate the cervix from 7 to 10 cm.

Transition dilation The final phase of stage 1 labor, during which the cervix dilates from 7 to 10 cm.

Transmembrane domain The region of a protein that resides within the cell membrane.

Transsexual One who feels that he or she is trapped in the body of the wrong gender; sex-role inversion.

Transverse presentation When the shoulders and arms of the fetus are engaged in the cervix.

Treponema pallidum Spirochete bacterium that causes venereal syphilis and also may cause nonvenereal syphilis.

Treponema pertenue Spirochete bacterium that causes yaws.

Trichomonas vaginalis A pear-shaped flagellate protozoan, normally found in the vagina, which causes trichomonas vaginitis.

Trichomonas vaginitis Inflammation of the vagina due to *Trichomonas vaginalis*; can be sexually transmitted; **Trichomoniasis**.

Trichomoniasis Infection with *Trichomonas vaginalis*.

Trimester Pregnancy is divided into 3-month intervals, and each interval is a trimester.

Triploidy An embryo with an extra set of chromosomes in each cell (3N).

Trisomy A condition of aneuploidy in which an individual has one too many (47) chromosomes.

Trophoblast Outer layer of blastocyst that plays a role in implantation and secretes hCG.

True hermaphrodite A person who has an ovary on one side and a testis on the other, or an ovotestis on one or both sides.

Tubal pregnancy Ectopic pregnancy within an oviduct.

Tubal sterilization Surgical blockage of the oviducts; used as a contraceptive measure.

Tubuli recti Straight ducts within the testis; transport sperm from the seminiferous tubules to the rete testis.

Tumescent The condition of being swollen.
Tumor A lump or mass containing cells that have lost their control of growth and multiplication; **Neoplasm**.
Tunica albuginea Thin connective tissue covering of ovary between surface epithelium and the ovarian cortex. Also, the thin internal covering of each testis.
Tunica dartos Muscle in the scrotum that contracts in response to a lowering of scrotal temperature.
Tunica vaginalis The thin membrane lying directly over the tunica albuginea of each testis.
Turner's syndrome Females with cells having only one sex chromosome (XO condition); individuals are short and sterile, with underdeveloped genitalia and sex accessory structures.
Ultrasound High-frequency sound used for fetal scanning and other purposes.
Umbilical arteries Two fetal blood vessels in the umbilical cord; carry deoxygenated blood to the placenta.
Umbilical cord Cord connecting the fetal circulation to the placenta.
Umbilical herniation Temporary extrusion (herniation) of the intestine into the umbilical cord during early fetal development.
Umbilical vein Fetal blood vessel in the umbilical cord; carries oxygenated blood from the placenta to the fetus.
Upregulation In endocrinology, an increase in the number of receptors for a hormone, thus increasing the cellular response to the hormone.
Ureaplasma urealyticum An organism, with viral and bacterial properties, that may cause some cases of nonspecific urethritis.
Urethra The duct that carries urine from the urinary bladder to the outside.
Urethral bulb The enlarged end of the spongy urethra in the male.
Urethral meatus The external opening of the urethra in the penis; urethral orifice.
Urethral orifice Opening of the urethra into the vestibule in the female or the terminal end of the penis in the male, through which urine passes.
Urethritis Inflammation or infection of the urethra.
Urinary incontinence Uncontrollable leakage of urine from the urethra.
Urogenital folds Structure of the embryonic external genitalia that gives rise to the labia minora in the female and the ventral aspect of the penile shaft in the male.
Urogenital sinus Embryonic structure that gives rise to the urinary bladder and urethra in both sexes, as well as the lower two-thirds of the vagina in the female.
Uterine cervix Lower region of the uterus between the uterine corpus and the vagina.
Uterine corpus Body of the uterus, between the uterine fundus and cervix.
Uterine cycle Cyclic changes in the uterine endometrium during the menstrual cycle.
Uterine fundus Dome-shaped top of the uterus.
Uterine glands Tubular glands in the uterine endometrium that secrete fluid and nutrients during the luteal phase of the menstrual cycle.
Uterine isthmus External constriction of the uterus that demarks the junction between the uterine corpus and cervix.

Uterine ligaments Bands or cords of tissue that support the uterus.

Uterine orgasm Orgasm that results from stimulation of the Grafenberg spot in the vaginal wall.

Uterine stroma The connective tissue framework of the endometrium.

Uterine tubes See **Oviducts**.

Uterosacral ligaments Paired ligaments attaching the lower uterus to the sacrum.

Uterotubal junction Muscular constriction at the entrance of the intramural oviduct into the uterine cavity.

Uterus Pear-shaped female organ located in the pelvic cavity; receives the oviducts and connects with the vagina; source of menstrual flow and site of pregnancy; also known as the "womb."

Vaccine Foreign substance that, when administered to a person, renders that person immune to that substance in the future.

Vacurette Tube used to suck out the uterine contents during a vacuum aspiration abortion.

Vacuum aspiration Method of inducing abortion during the late first trimester by suction applied through a tube (vacurette) inserted into the cervix; vacuum curettage.

Vacuum curettage See **Vacuum aspiration**.

Vacuum extraction Assistance in expulsion of a fetus by use of a suction device attached to the top of the fetal head.

Vagina Tube extending from the uterine cervix to the vestibule.

Vaginal coitus See **Coitus**.

Vaginal introitus Opening of the vagina into the vestibule.

Vaginal lubrication When a clear fluid, leaking from blood vessels in the vaginal wall, appears in the vagina a few seconds after sexual arousal.

Vaginal orgasm Orgasm resulting from stimulation of the vagina; see **Uterine orgasm**.

Vaginal pessary A device placed in the vagina to support a sagging vagina and uterus.

Vaginal ring A doughnut-shaped, progestogen-containing contraceptive device that is placed within the vaginal cavity.

Vaginal smear A small sampling of vaginal tissue used for detection of changes in vaginal lining during the menstrual cycle.

Vaginismus Strong, spasmodic, often painful contractions of the vagina or its surrounding muscles.

Vaginitis Inflammation or infection of the vagina.

Varicella-zoster virus Virus that causes chickenpox in children and shingles in adults.

Varicocele Enlarged testicular veins in the spermatic cord, which can press on the vasa deferentia and reduce fertility.

Varicose veins When veins enlarge and swell on the skin surface.

Vasa efferentia Series of ducts in the testis and head of the epididymis that convey sperm from the rete testis to the ductus epididymis.

Vas deferens (pl., Vasa deferentia) Duct carrying sperm from the epididymis to the ejaculatory duct; ductus deferens.

Vasectomy Surgical sterilization by interruption of the vasa deferentia.

Vasoactive-intestinal peptide (VIP) A common neurotransmitter in the nervous system.

Vasocongestion Pooling or engorgement of blood in tissue blood vessels.

Vasopressin Polypeptide hormone secreted by the neurohypophysis; causes the kidneys to retain water; antidiuretic hormone.

Venereal diseases See **Sexually transmitted diseases**.

Ventromedial hypothalamic nucleus (VHN) Hypothalamic brain area perhaps involved in female sexual behavior.

Vernix caseosa Protective coating on skin of a fetus, consisting of oil and sloughed skin cells.

Vestibule The space in the female between the labia minora, into which open the vaginal introitus and urethral orifice.

Viral hepatitis, type B Infection of the liver by a virus that can be sexually transmitted.

Vitamin B_6 Member of the B complex of vitamins that influences the function of the nervous system.

Vitelline envelope Protective glycoprotein layer surrounding an egg; **Zona pellucida** in mammals.

Vomeronasal organ (VNO) Paired structure in nasal cavities; senses pheromones.

Vulva Female external genitalia.

Vulvitis Inflammation or infection of the vulva.

Wet nursing Breast-feeding a child not your own, often for hire.

Wharton's jelly Jelly-like supporting substance within the umbilical cord.

Whitten effect Induction and synchronization of rat estrous cycles by the presence of a male rate.

Window period In HIV disease, time between HIV infection and appearance of HIV antibodies in the blood.

Witch's milk Fluid secreted by the mammary glands of male and female infants.

Wolffian ducts Embryonic precursors of the male sex accessory ducts and seminal vesicles; mesonephric ducts.

Woman-year Use of a contraceptive by one woman for 1 year; used to express failure rates of contraceptive measures; see also **Pearl's formula**.

X-chromosome inactivation When, in female cells, all but one X chromosome condenses into a clump (Barr body) and is genetically inactive.

Xenoestrogens Estrogens that are foreign to the human body.

***Xq28* gene** A region of 200 genes on the X chromosome that relate somehow to male sexual orientation.

Yaws A skin disease common in children in tropical regions; caused by the bacterium *Treponema pertenue*; usually not sexually transmitted.

Yohimbine A reputed aphrodisiac that stimulates the parasympathetic nervous system.

Yolk sac Extraembryonic membrane in embryo that contains a slight amount of yolk and gives rise to the primordial germ cells.

Yolk sac cavity Cavity, within the yolk sac, that is derived from the blastocoel.

Zona drilling Removing a tiny portion of the zona pellucida from an ovum to facilitate external fertilization.

Zona pellucida Acellular membrane between the oocyte and membrana granulosa of an ovarian follicle; surrounds the ovulated egg prior to implantation.

Zona reaction Change in chemical nature and structure of the zona pellucida caused by ovum cortical granule secretion; prevents polyspermy.

ZP3 Receptor on surface of the ovum zona pellucida that binds to the sperm head and initiates the acrosome reaction.

Zygote Diploid, fertilized cell produced at conception.

Zygote intrafallopian transfer (ZIFT) When a woman's egg is fertilized in a dish with the husband's sperm, and the zygote is transferred to the wife's oviduct below the point of obstruction.

Illustration and Table Credits

Chapter 1

Fig. 1-2: Adapted from *Principles of Human Anatomy*, 3rd ed., by Gerard J. Tortora (New York: Harper & Row, Publishers, 1983). **Fig. 1-6**: Adapted from *Principles of Human Anatomy*, 3rd ed., by Gerard J. Tortora (New York: Harper & Row, Publishers, 1983). **Fig. 1-7** and **1-11**: Adapted from E.G. Rennels and D.C. Herbert, "The Anterior Pituitary Gland – Its Cells and Hormones," *Bioscience* 29 (7) (1979): 409. © American Institute of Biological Sciences. **Fig. 1-12**: Adapted from A.V. Schally, "Aspects of Hypothalamic Regulation of the Pituitary Gland," *Science* 202 (1978), 18–28. © Nobel Foundation, Stockholm, Sweden. **Box 1-2**: Adapted from Schwanzel-Fukuda, M., *et al.* (1992). Biology of luteinizing hormone-releasing hormone neurons during and after their migration from the olfactory placode. *Endocr. Rev.* **13**, 623–634.

Chapter 2

Fig. 2-1: Adapted from *From Woman to Woman*, by Lucienne Lanson and illustrated by Anita Karl (New York: Alfred A. Knopf, Inc., 1975). **Fig. 2-2**: Adapted from K. L. Moore, *The Developing Human: Clinically Oriented Embryology*, 3rd ed. (Philadelphia: W. B. Saunders Company, 1982). **Fig. 2-3**: Adapted from *General Endocrinology*, 6th ed., by C. Donald Turner and Joseph T. Bagnara (Philadelphia: W. B. Saunders Company, 1976). **Fig. 2-4** and **2-5**: Reproduced from W. Bloom and D. W. Fawcett, *A Textbook of Histology*, 10th ed. (Philadelphia: W. B. Saunders Company, 1975). **Fig. 2-7**: Adapted from K. L. Moore, *The Developing Human: Clinically Oriented Embryology*, 3rd ed. (Philadelphia: W. B. Saunders Company, 1982). **Fig. 2-8** and **2-9**: Adapted from K. L. Moore, *The Developing Human: Clinically Oriented Embryology*, 3rd ed. (Philadelphia: W. B. Saunders Company, 1982). **Fig. 2-10**: Adapted from *From Woman to Woman*, by Lucienne Lanson and illustrated by Anita Karl (New York: Alfred A. Knopf, Inc., 1975). **Fig. 2-11** and **2-12**: Adapted from *Principles of Human Anatomy and Physiology*, 3rd ed., by Gerard J. Tortora (New York: Harper & Row, Publishers, 1983). **Fig. 2-13**: Adapted from A.L.R. Findlay, "Lactation," *Research in Reproduction* vol. 6, no. 6, copyright 1974; used with permission of the International Planned Parenthood Assoc. and Dr. A.L.R. Findlay. **Fig. 2-14**: Adapted from MacMahon, B., *et al.* Etiology of breast cancer. A review. *J. Natl. Cancer Inst.* 50: 21–42, by permission of Oxford University Press. **Fig. 2-15**: Modified with permission of the American Cancer Society. **Table 2-2**: Adapted from R. C. Kolodny, W. H. Masters, V. E. Johnson, and M. A. Biggs. *Textbook of Human Sexuality for Nurses* (Boston: Little, Brown & Company, 1979), p. 154. **Box 2-1**: Adapted from Eaton, S. B., and Eaton, S. B., III (1999). Breast cancer in an evolutionary context. *In* "Evolutionary Medicine" (W. R. Trevathan *et al.*, eds.), pp. 429–442. Oxford Univ. Press, New York.

Chapter 3

Fig. 3-2: Adapted with permission of the publisher from C. Grobstein, "External Human Fertilization," *Scientific American* 240(6), 57–67; copyright © 1979 by Scientific American, Inc.; all rights reserved. **Fig. 3-3**: Adapted from *The Cycling Female: Her Menstrual Rhythm*, by Allen Lein (San Francisco: W.H. Freeman and Co., Pub. 1979). **Fig. 3-4**: Adapted from R. A. Steiner and J. L. Cameron, "Endocrine Control of Reproduction," in *Textbook of Physiology, Vol. 2*, edited by H. D. Patton *et al.* (Philadelphia: W. B. Saunders Co., 1989). **Box 3-1**: From R.G. Kessel and R.H. Kardon, *Tissues and Organs: A Text-Atlas of Scanning Electron Microscopy*, WH Freeman and Co, San Francisco, p. 288 (1979). **Box 3-2**: Adapted from Frisch, R. E., *et al.* (1993). Magnetic resonance imaging of overall and regional body fat, estrogen metabolism, and ovulation of athletes compared to controls. *J. Clin. Endocrinol. Metabol.* **77**, 471–477.

Chapter 4

Fig. 4-2: Reprinted from Alexander, P. Spence, and Elliott B. Mason, *Human Anatomy and Physiology*, 1st ed. (Menlo Park, Calif.: Benjamin Cummings Publishing Company, Inc., 1979). **Fig. 4-3 a**: Adapted from W. Bloom and D. W. Fawcett, *A Textbook of Histology*, 10th ed. (Philadelphia: W. B. Saunders Company, 1975). **b**: Adapted with permission of the publisher from L. B. Arey, *Developmental Anatomy* (Philadelphia: W. B. Saunders Company, 1965). **Fig. 4-4**: Adapted from K. L. Moore, *The Developing Human: Clinically Oriented Embryology*, 3rd ed. (Philadelphia: W. B. Saunders Company, 1982). **Fig. 4-5**: Reproduced from *Tissues and Organs: A Text-Atlas of Scanning Electron Microscopy* by Richard G. Kessel and Randy H. Kardon (San Francisco: W. H. Freeman & Company, Publishers, 1979); photograph courtesy of Dr. Richard G. Kessel. **Fig. 4-8**: Adapted from L. Weiss and R. O. Greep, *Histology*, 4th ed. (New York: McGraw-Hill Book Company, 1977). **Fig. 4-9**: From National Cancer Institute website. **Fig. 4-10**: Adapted from *Principles of Human Anatomy* 3rd ed., by Gerard J. Tortora (New York: Harper & Row, Publishers, Inc., 1983). **Table 4-1**: Modified from Garnick, M. B. (1994). The dilemmas of prostate cancer. *Sci. Am.* **270**(4), 72–81. **Box 4-2**: Adapted from

Freeman, S. (1990). The evolution of the scrotum: A new hypothesis. *J. Theoret. Biol.* **145**, 429–445.

Chapter 5

Fig. 5-1: Courtesy of Dr. George Henry and the Reproductive Genetics Center, Denver, Colorado. **Fig. 5-3**: Adapted with permission of the publisher from J. D. Wilson, F. W. George, and J. E. Griffith, "The Hormonal Control of Sexual Development," *Science* 211, 1278–1284. Copyright © 1981 AAAS. **Fig. 5-5**: Adapted from K. L. Moore, *The Developing Human: Clinically Oriented Embryology*, 3rd ed. (Philadelphia: W. B. Saunders Company, 1982). **Fig. 5-6** and **5-7**: Adapted with permission of the publisher from J. D. Wilson, F. W. George, and J. E. Griffith, "The Hormonal Control of Sexual Development," *Science* 211, 1278–1284. Copyright © 1981 AAAS. **Fig. 5-8**: Adapted from P. J. Hogarth, *Biology of Reproduction: Tertiary Level Biology* (New York: John Wiley & Sons, Inc., 1978). **Fig. 5-9**: Photograph courtesy of Dr. Howard W. Jones. **Fig. 5-10** and **5-12**: Adapted from K. L. Moore, *The Developing Human: Clinically Oriented Embryology*, 3rd ed. (Philadelphia: W. B. Saunders Company, 1982). **Box 5-1**: Adapted from Roldan, E. R. S., and Gomendio, M. (1999). The Y chromosome as a battle ground for sexual selection. *Trends Ecol. Evol.* **14**, 58–62.

Chapter 6

Fig. 6-1: Reproduced from J. M. Tanner, "Growing Up," *Scientific American* 229(3), 34–43; Copyright © 1973 by Scientific American, Inc.; all rights reserved. **Fig. 6-2** and **6-3**: Adapted from B. T. Donovan and J. J. Van der Werff Ten Bosch, *Physiology of Puberty*, in *Monographs of the Physiological Society*, no. 15, edited by H. Barcroft, H. Davson, and W. D. M. Paton (Baltimore: The Williams & Wilkins Company, 1965). Editorial Board for Monographs of the Physiological Society. **Fig. 6-4**, **6-5**, and **6-6**: Reproduced from J. M. Tanner, *Growth at Adolescence*, 2nd ed. (Oxford: Blackwell Scientific Publications, 1973); reprinted with permission of Blackwell Scientific Publications. **Fig. 6-7** and **6-8**: Adapted from M. M. Grumbach, G. D. Grave, and F. E. Mayer, "Hypothalamic-Pituitary Regulation of Puberty: Evidence and Concepts Derived from Clinical Research," in *Control of the Onset of Puberty*, edited by M. M. Grumbach, G. D. Grave, and F. E. Mayer (New York: John Wiley & Sons, Inc., 1974). **Fig. 6-9**: Adapted from E. D. Weitzman, R. M. Boyar, S. Kapen, and L. Hellman, "The Relationship of Sleep and Sleep Stages to Neuroendocrine Secretion and Biological Rhythms in Man," in *Recent Progress in Hormone Research*, 31, 399–446, edited by R. O. Greep (New York: Academic Press, Inc., 1975). **Fig. 6-10**: From Wyshak, G., and Frisch, R. E. (1982). Evidence for a secular trend in age of menarche. *N. Engl J. Med.* **303**, 1033–1035. **Table 6-2**: From Frisch, R. E. (1991). Body weight, body fat, and ovulation. *Trends. Endocrinol. Metab.* **2**, 191–197. **Table 6-3**: Reproduced from J. M. Tanner, *Growth at Adolescence*, 2nd ed. (Oxford: Blackwell Scientific Publications, 1973); reprinted with permission of Blackwell Scientific Publications. **Box 6-2**: From Belsky, J., *et al.* (1991). Childhood experience, interpersonal development, and reproductive strategy: An evolutionary theory of socialization. *Child Dev.* **62**, 647–670.

Chapter 7

Fig. 7-1: From Sievert, L. L. (2001). Aging and reproductive senescence. *In* "Reproductive Ecology and Human Evolution" (Ellison, P. T., ed.), pp. 267–292. Aldine de Gruyter, New York. **Box 7-1**: Used with permission of Forbes, L. S. (1997). The evolutionary biology of spontaneous abortion in humans. *Trends Ecol. Evol.* **12**, 446–450.

Chapter 8

Fig. 8-1: Reproduced from W. H. Masters, V. E. Johnson, and R. C. Kolodny, *Biological Foundations of Human Sexuality* (HarperCollins, New York, 1993). **Fig. 8-2**: Reproduced from W. H. Masters and V. E. Johnson, *Human Sexual Response* (Boston: Little, Brown & Company, 1966). **Fig. 8-3**: Reproduced from W. H. Masters, V. E. Johnson, and R. C. Kolodny, *Biological Foundations of Human Sexuality* (HarperCollins, New York, 1993). **Box 8-1**: Reproduced from R. Taylor, "Brave New Nose: Sniffing Out Human Sexual Chemistry," in *Journal of NIH Research*, 6(Jan.), 47–51 (1994). **Box 8-2**: Reproduced from cover of *Discovery*, Nov., 1993.

Chapter 9

Fig. 9-1: Courtesy of Dr. R. Yanagimachi. **Fig. 9-3**: Adapted from C.R. Austin, "Fertilization," in *Reproduction in Mammals, Book 1, Germ Cells and Fertilization*, Eds. C.R. Austin and R.V. Short (Cambridge, England: Cambridge University Press, 1972). **Fig. 9-4**: Adapted from *Reproduction and Human Welfare: A Challenge to Research*, by R.O. Greep, M.A. Koblinsky, and F.S. Jaffe: Cambridge, Mass.: MIT Press, 1976. Reprinted by permission from Ford Foundation. **Table 9-1**: Adapted from B. Goldstein, *Human Sexuality* (New York: McGraw-Hill Book Company, 1976). **Box 9-1**: Adapted from Ralt, D., *et al.* (1991). Sperm attraction to a follicular factor(s) correlates with human egg fertilizability. *Proc. Natl. Acad. Sci. USA* **88**, 2840–2844.

Chapter 10

Fig. 10-1: Adapted from K. L. Moore, *The Developing Human: Clinically Oriented Embryology*, 3rd ed. (Philadelphia: W. B. Saunders Company, 1982). **Fig. 10-2**: Reproduced with permission of the publisher from C. Grobstein, "External Human Fertilization," *Scientific American* 240(6), 57–67; copyright © 1979 by Scientific American, Inc.; all rights reserved. **Fig. 10-3**: Adapted from A. McLaren, "The Embryo," in *Reproduction in Mammals, Book 2, Embryonic and Fetal Development*, edited by C. R. Austin and R. V. Short (Cambridge, England: Cambridge University Press, 1972). **Fig. 10-4**: Adapted from K. L. Moore, *The Developing Human: Clinically Oriented Embryology*, 3rd ed. (Philadelphia: W. B. Saunders Company, 1982). **Fig. 10-5**: Adapted from P. Beaconsfield, G. Birdwell, and R. Beaconsfield, "The Placenta," *Scientific American* 243(2),

(1980), 94–102; copyright © 1980 by Scientific American, Inc; all rights reserved. **Fig. 10-7** and **10-8**: Adapted from K. L. Moore, *The Developing Human: Clinically Oriented Embryology*, 3rd ed. (Philadelphia: W. B. Saunders Company, 1982). **Fig. 10-9**: Reproduced from H. B. Taussig, "The Thalidomide Syndrome," *Scientific American* 207(2), 29–35; copyright © 1962 by Scientific American, Inc.; all rights reserved. **Fig. 10-10** and **10-11**: Reproduced from K. L. Moore, *The Developing Human: Clinically Oriented Embryology*, 3rd ed. (Philadelphia: W. B. Saunders Company, 1982). **Fig. 10-12**: Adapted from Alexander Spence and Eliott B. Mason, *Human Anatomy and Physiology*, 1st ed, fig. 29.18, p. 809 (Menlo Park, CA: Benjamin Cummings Publ Co, 1979). **Fig. 10-13**: Reproduced from K. L. Moore, *The Developing Human: Clinically Oriented Embryology*, 3rd ed. (Philadelphia: W. B. Saunders Company, 1982). **Fig. 10-14**: Adapted from *Principles of Human Anatomy*, 3rd ed, by Gerard J. Tortora (New York: Harper & Row, Publishers, 1983). **Fig. 10-15**: Adapted from R. B. Heap, "Role of Hormones in Pregnancy," in *Reproduction in Mammals, Book 3, Hormones in Reproduction*, edited by C. R. Austin and R. V. Short (Cambridge, England: Cambridge University Press, 1972). **Table 10-4**: Adapted from *Principles of Human Anatomy*, 3rd ed., by Gerard J. Tortora (New York: Harper & Row, Publishers, 1983); Exhibit 29-2, p. 753. **Box 10-3**: Reproduced from J. M. Riddle, J. W. Estes, and J. C. Russell, "Ever Since Eve . . . Birth Control in the Ancient World," in *Archeology*, March/April, pp. 29–35 (1994).

Chapter 11

Fig. 11-1 and **11-2**: From Lam, D. A., and Miron, J. A. (1994). Factors contributing to the seasonality of human reproduction. *In* "Human Reproductive Ecology: Interactions of Environment, Fertility, and Behavior" (K. L. Campbell and J. W. Wood, eds.), Annals of the New York Academy of Sciences, New York. **Fig. 11-5**: Adapted from D. Ewy and R. Ewy, *Preparation for Childbirth. A Lamaze Guide* (Boulder, CO: Pruett Publ Co, 1972). **Fig. 11-7**: Adapted from K. L. Moore, *The Developing Human: Clinically Oriented Embryology*, 3rd ed. (Philadelphia: W. B. Saunders Company, 1982). **Fig. 11-8**, **11-10**, and **11-11**: Adapted from W. G. Birch, *A Doctor Discusses Pregnancy*, Chicago: Budlong Press, 1963. **Table 11-2**: Adapted from W. H. Masters, V. E. Johnson, and R. C. Kolodnoy, in *Biological Foundations of Human Sexuality*, p. 92. (New York: Harper Collins College Publishing, 1993). **Box 11-1**: From Smith, R. (1999). The timing of birth. *Sci. Am.* **280**(3), 68–75; copyright © 1999 by Scientific American, Inc., all rights reserved. **Box 11-2**: From Trevathan, W. R. (1987). "Human Birth: An Evolutionary Perspective." Aldine de Gruyter, Hawthorne, NY.

Chapter 12

Fig. 12-1: Adapted from *Principles of Anatomy*, 3rd ed., by Gerard J. Tortora (New York: Harper & Row, Publishers, Inc., 1983). **Fig. 12-2**: Adapted from *General Endocrinology*, 6th ed., by C. Donald Turner and Joseph T. Bagnara (Philadelphia: W. B. Saunders Company, 1976). **Table 12-1**: Adapted from *Maternal Nutrition and the Course of Pregnancy. Summary Report*, NIH, Publication (HSM) 72-5600,

p. 8 (Washington, DC: US Gov Printing Office, 1971). **Table 12-2**: From A. L. R. Findlay (1974), Lactation, *Research in Reproduction*, Vol. 6, No. 6, map. Reprinted with permission of the International Planned Parenthood Federation and Dr. A. L. R. Findlay; 1974. **Box 12-2**: Adapted from Howie, P.W., and McNeilly, A.S. (1982). Effect of breastfeeding patterns on human birth intervals. *J. Reprod. Fert.* **65**, 545–557.

Chapter 13

Fig. 13-5: Reproduced from P. Corfman and E. Huyck, *Population Research at the U.S. National Institutes of Health: A Response to a National and International Problem*, DHEW Publ. No. (NIH) 75–781, 1978. **Fig. 13-9**: Adapted from C. J. Avers, *The Biology of Sex* (New York: John Wiley & Sons, Inc., 1974). **Table 13-2**: Data from the Population Reference Bureau. **Table 13-3** and **13-4**: From D. J. Bogue, *Mass Communication and Motivation for Birth Control* (Chicago: University of Chicago Community and Family Study Center, 1967). **Box 13-1**: Redrawn from Glasier, A. (2002). Contraception – past and future, Historical perspective. *Nature Cell Biology & Nature Medicine*, Fertility Suppl, s3–s6. **Box 13-2**: Reproduced from G. Feeney, "Fertility Decline in East Asia," in *Science*, Vol. 266, pp. 1518–1523. Copyright © 1994 AAAS.

Chapter 14

Fig. 14-1: Adapted from Campbell, NA (1993). *Biology*, 3rd Ed. Benjamin Cummings Publ Co. **Fig. 14-4**: Reproduced from R. G. Edwards, "Control of Human Development," in *Reproduction in Mammals, Book 5, Artificial Control of Reproduction*, edited by C. R. Austin and R. V. Short (Cambridge: Cambridge University Press, 1972). **Fig. 14-6**: Reproduced from J. S. Hyde, *Understanding Human Sexuality* (New York: McGraw-Hill Book Company, 1979).

Chapter 15

Fig. 15-1, **15-2**, and **15-3**: Adapted from Elam-Evans, L.D., *et al.* (2003). Abortion surveillance – United States, 2000. *MMWR Surveill. Summ.* **52**, 1–32. **Fig. 15-5**: Reproduced from J. S. Hyde, *Understanding Human Sexuality* (New York: McGraw-Hill Book Company, 1979).

Chapter 16

Fig. 16-1: Adapted from R.G. Edwards, Control of Human Development, in *Reproduction in Mammals, Book 5, Artificial Control of Reproduction*, eds. CR Austin and RV Short. Cambridge, England: Cambridge Univ. Press, 1972. **Box 16-1**: Adapted from Stupp, PW, and Warren, CW, Seasonal differences in pregnancy outcomes: United States, 1971–1989. *Ann N Y Acad Sci.* 1994 Feb 18;709:46–54. **Box 16-2**: Redrawn from Katz, P., Nachtigall, R., and Showstack, J. (2002). The economic impact of the assisted reproductive technologies. *Nature Cell Biology & Nature Medicine*, Fertility Suppl, s29–s32.

Chapter 17

Fig. 17-2: Reprinted from Richardson, S. (1995). S/He-brains. *Discover* June, 36. **Fig. 17-3**: Adapted from (I and II) Swaab, D. F., and Hofman, M. A. (1995). Sexual differentiation of the human hypothalamus in relation to gender and sexual orientation. *Trends Neurosci.* **18**, 264–270, with permission from Elsevier, and (a, b, and c) Pansky (1988) Review of Neuroscience, 2nd ed., Macmillan Publ. Co.: NY. **Fig. 17-4**: Reproduced with permission of the publisher from Gladue *et al.* (1984), "Neuroendocrine Response to Estrogen and Sexual Orientation," *Science* 225, 1496–1499; © 1984 AAAS. **Fig. 17-5**: From D. F. Swaab and M. A. Hoffmann, "Sexual Differentiation of the Human Hypothalamus: Ontogeny of the Sexually Dimorphous Nucleus of the Preoptic Area," in *Developmental Brain Research* 44, 314–318 (1988). **Fig. 17-6**: Adapted from Swaab, D. F., and Hofman, M. A. (1995). Sexual differentiation of the human hypothalamus in relation to gender and sexual orientation. *Trends Neurosci.* **18**, 264–270, with permission from Elsevier. **Box 17-1**: Reproduced from D. Kimura, "Sex Differences in the Brain," *Scientific American* (Sep.), 120–121 (1992); copyright © 1992 by Scientific American, Inc., all rights reserved.

Chapter 18

Fig. 18-1: Data from the CDC. **Fig. 18-2**: Adapted from Stine, GJ (2002). *AIDS Update 2002*. Prentice Hall: New Jersey. **Fig. 18-3** and **18-4**: Reproduced with permission of the publisher from J. S. Hyde, *Understanding Human Sexuality* (New York, McGraw-Hill Book Company, 1979). **Fig. 18-5**: Data from UNAIDS. **Fig. 18-6** and **18-7**: From Stine, GJ (2002). *AIDS Update 2002*. Prentice Hall: New Jersey. **Fig. 18-8** and **18-9**: Data from the CDC. **Fig. 18-10**: Redrawn from Stine, GJ (2002). AIDS Update 2002. Prentice Hall: New Jersey. **Fig. 18-11**: Adapted from Nowak, M. A., and McMichael, A. J. (1995). How HIV defeats the immune system. *Sci. Am.* **273**(2), 58–65; copyright © 1995 by Scientific American, Inc., all rights reserved. **Fig. 18-12** and **18-13**: Adapted from Stine, GJ (2002). *AIDS Update 2002*. Prentice Hall: New Jersey. **Table 18-1**, **18-2**, **18-3**, and **18-5**: Adapted from Stine, GJ (2002). *AIDS Update 2002*. Prentice Hall: New Jersey. **Box 18-2**: Redrawn from Stine, GH (2002). *AIDS Update 2002*. Prentice Hall: New Jersey.

Index

T